Pacific Salmon Life Histories

2

PLATE 2. Developing pink salmon eggs and hatched alevins. *Photograph by Marj Trim*

PLATE 3. Mature male and female pink salmon, British Columbia. *Photograph by J-G. Godin*

PLATE 1 (*previous page*). Pre-spawning sockeye salmon holding along the banks of the Adams River, British Columbia. *Photograph by Marj Trim*

3

Pacific Salmon Life Histories

Edited by C. Groot and L. Margolis
Department of Fisheries and Oceans
Biological Sciences Branch
Pacific Biological Station, Nanaimo
British Columbia, Canada

UBCPress
Vancouver

Published in co-operation with the Government of Canada, Department of Fisheries and Oceans

Printed in Canada on acid-free paper ∞

ISBN 0-7748-0359-2

Canadian Cataloguing in Publication Data
Main entry under title:
Pacific salmon life histories

Includes bibliographical references and index.
ISBN 0-7748-0359-2

1. Pacific salmon. I. Groot, C. (Cornelis), 1928- II. Margolis, L., 1927-

QL638.S2P33 1991 597'.55 C91-091310-2

UBC Press
University of British Columbia
6344 Memorial Rd
Vancouver, BC V6T 1Z2
(604) 822-3259
Fax: (604) 822-6083

CONTENTS

COLOUR PLATES

PREFACE

*C. Groot and L. Margolis**

PACIFIC SALMON (genus *Oncorhynchus*) are an important biological and economic resource of countries of the North Pacific rim. The geographic distribution of these salmon extends from San Francisco Bay, in California, northward along the Canadian and Alaskan coasts to rivers draining into the Arctic Ocean, and southward down the Asian coastal areas of the USSR, Japan, and Korea.

There are seven species of Pacific salmon. Their scientific and common names are:

Oncorhynchus nerka (Walbaum): sockeye
Oncorhynchus gorbuscha (Walbaum): pink
Oncorhynchus keta (Walbaum): chum
Oncorhynchus tshawytscha (Walbaum): chinook
Oncorhynchus kisutch (Walbaum): coho
Oncorhynchus masou (Brevoort): masu, sakura-masu
Oncorhynchus rhodurus (Gunther): amago, biwamasu

The first five species reproduce on both the Asian and North American continents and the last two occur only in Asia. Because of their attractiveness as a food and sport fish, Pacific salmon have been introduced far beyond their native geographic range in the northern hemisphere as well as to the southern hemisphere.

Pacific salmon spawn in gravel beds in rivers, streams, or along lake shores. They generally migrate to sea after an early freshwater life and are widely distributed over the North Pacific Ocean and Bering Sea during the marine years of their lives. Most perform extensive migrations while at sea. Some populations or subpopulations of certain species stay in coastal waters or remain in fresh water.

Upon maturation, after one to seven years (depending on the species and stock), Pacific salmon usually return to their home rivers and natal breeding grounds. Spawning takes place between late summer and early winter, after which all of them die, except some amago and non-anadromous masu males, which are known to spawn several times. Anadromy and the strong tendency for homing to the river of origin have resulted in the development of many reproductively isolated subpopulations in Pacific salmon, which are referred to as stocks.

Pacific salmon are harvested commercially by Canada, Japan, North and South Korea, the United States, and the USSR in coastal waters. Japan also harvests salmon on the high seas of the western North Pacific Ocean and the central Bering Sea under controlled measures established by two treaties. These are: (1) the International Convention for the High Seas Fisheries of the North Pacific Ocean, involving Canada, Japan, and the United States; and (2) the USSR-Japan Fisheries Agreement.

Some Pacific salmon are harvested by Native peoples as a food fish during upstream migration in rivers in Canada, the United States, and the USSR, and many, especially coho and chinook salmon, are captured by sport fishermen in coastal waters of North America. Recently, Pacific salmon have also become important in fish farming.

The annual all-nation commercial catch of Pacific salmon has been approximately 800,000 tonnes in recent years, with a landed value of about CDN $2.5 billion. In North America, the recreational value of the Pacific salmon fishery is also significant. Production of farmed Pacific salmon is continually increasing in Canada, the United States, and Japan, and also in southern hemisphere nations, such as Chile and New Zealand, to which Pacific salmon have been introduced.

*Department of Fisheries and Oceans, Biological Sciences Branch, Pacific Biological Station, Nanaimo, British Columbia, Canada V9R 5K6

The need for responsible management in international and national waters, the desire to enhance depleted stocks as a result of overfishing and habitat deterioration, and a general focus on understanding basic principles of fish population biology have led to an explosive increase in biological knowledge of Pacific salmon over the last fifty years. An abundance of information on ocean distribution, migrations, habitat preference, life cycles, breeding behaviour, and other characteristics has been gathered and published in a great variety of journals and report series.

This volume brings together and summarizes much of the available biological information on the life histories of the seven Pacific salmon species. The level of knowledge for the different species is not equal because the intensity of investigation has been dependent on the economic and recreational importance of the species.

We were fortunate to find authors who had longstanding experience with the species of Pacific salmon. Some were close to the end of their official working careers when plans for this book were initiated, and we are grateful that they were able to find time in their busy lives to bring together the information presented here. All the contributors have a deep interest in this unique group of fish and in their expansive and complex life histories. Ernest Salo and Fumihiko Kato, authors of the chapters on chum salmon and masu and amago salmon, respectively, died before the completion of this volume. It is sad that they could not see the final products of their labours.

In dealing with the information on the life histories of the different species of Pacific salmon, we allowed the authors a certain degree of latitude on how to approach their specific subject matter. However, it was agreed that, in general, information on the following topics should be covered in the texts: distribution, relative abundance, transplants, ascent to spawn, incubation, emergence, fry migration, freshwater residence, early sea life, offshore migration, high-seas residence, maturation and return migration, upstream migration, homing and straying, and dominance cycles. Information on age, size, growth, food, diet, survival, mortality, abundance, predation, and competition has been included in the above aspects of the life histories. In general, the literature has been covered up to 1985–86, and, in some cases, more recent literature has been reviewed.

The scientific names for steelhead/rainbow trout (*Salmo gairdneri*) and cutthroat trout (*Salmo clarki clarki*) have recently been changed. They are now included in the genus *Oncorhynchus* and have been renamed *O. mykiss* and *O. clarki*, respectively. We have followed these name changes in the texts, but have not included the life histories of these species in this volume because they are still known as trout rather than salmon.

Different systems have been developed to designate the age of Pacific salmon. We have followed the European system in this volume in which the winters in fresh water after hatching and the winters in salt water are identified and separated by a period. Thus, an age 1.2 fish has spent one winter in fresh water and two in salt water, and an age 0.3 fish migrated to sea soon after emergence from the stream gravel and has spent three winters in the ocean. In some cases the authors in this volume have also referred to the age of salmon as two-year-olds, three-year-olds, and so on, which refers to the year of life since the eggs were deposited in the gravel.

Two other age designation systems are in use in the literature. In the Gilbert and Rich system, the first number indicates the age of the fish in years starting from the egg stage and the second one the number of winters in fresh water; this is designated as A_b, where A represents the former and the subscript b, the latter. In the Soviet system the first number indicates how old the fish is in number of winters since emergence and the second one identifies the numbers of winters spent in fresh water; this is designated as x_y, where x represents the former and the subscript y, the latter. The following table shows how the different age notations compare.

Method	Age								
Year-olds	3	4	5	3	4	5	3	4	5
European	0.2	0.3	0.4	1.1	1.2	1.3	2.0	2.1	2.2
Gilbert-Rich	3_1	4_1	5_1	3_2	4_2	5_2	3_3	4_3	5_3
Soviet	2_0	3_0	4_0	2_1	3_1	4_1	2_2	3_2	4_2

Because the locations of many small streams, rivers, lakes, and other geographic localities mentioned in the various chapters may not be readily known to the reader, we have included a geographical index at the end of the volume.

A book of this magnitude can only be completed with the support of many colleagues. We gratefully acknowledge the efforts of the following people for their assistance in reviewing the manuscripts: E.L. Brannon of the University of Idaho; T.G. Northcote of the University of British Columbia; J.H. Helle of the U.S. National Marine Fisheries Service at Auke Bay, Alaska; H.H. Wagner of the Oregon Department of Fish and Wildlife; Fumihiro Koga of the Seikai National Fisheries Research Institute; Seizo Hasegawa of the Japan Sea National Fisheries Research Institute; G.J. Duker of Northwest and Alaska Fisheries Science Centers; and T.D. Beacham, K.D. Hyatt, G.F. Hartman, R.J. LeBrasseur, J. McDonald, and the late F.C. Withler of the Pacific Biological Station, Canada Department of Fisheries and Oceans (DFO). We also acknowledge the help of Gordon Miller, librarian of the DFO Biological Sciences Branch, Pacific Region, for checking the reference lists; Kazuya Nagasawa, National Research Institute of Far Seas Fisheries, Shimizu, and Victor Tumanov and O.A. Bulatov, Pacific Research Institute of Fisheries and Oceanography, Vladivostok, for providing information for the geographical index; and R.J. LeBrasseur for preparing the general index.

Harry Heine magnificently captured the different life history stages of the Pacific salmon in water-colour, and we appreciate his permission to use his paintings in this volume. A number of typists, illustrators, and photographers contributed to the production of the finished manuscripts and we owe them our gratitude. Specifically, we note the contributions of Pat Baglo, Suzanne Benoit, Sherry Greenham, Laurie Mackie, Jean McLeod, Jennifer Nielson, George Pasek, Madeleine Sherry, Michelle Steele, Ann Thompson, Barbara Wanke, and Evan Warneboldt.

Special thanks are due to the staff of UBC Press, especially Jean Wilson and Holly Keller-Brohman, for the assistance received in seeing this volume through to publication.

We sincerely thank Drs. J.C. Davis (Science Director, Pacific Region, DFO) and R.J. Beamish (Director, Biological Sciences Branch, Pacific Region, DFO) for their support for this project.

CONTRIBUTORS

Robert L. Burgner obtained his B.S. and Ph.D. from the University of Washington in 1942 and 1958, respectively. Since then, he has held many positions at the Fisheries Research Institute at this university, culminating in his appointment in 1968 as professor and director of the institute. During his distinguished career he has been the principal investigator in various projects, including studies of sockeye salmon in Alaska and Washington lakes, of marine ecology at Amchitka, and of salmon and steelhead migrations on the high seas. He has been a member of many study groups and committees, such as the IBP Coniferous Biome Study and the Scientific and Statistical Committee of the North Pacific Fishery Management Council, and, most notably, he has served on various committees of the International North Pacific Fisheries Commission. Upon retirement in 1984, the University of Washington bestowed upon Dr. Burgner the title of professor emeritus.

Cornelis Groot studied biology at the universities of Amsterdam and Leyden in Holland. He emigrated to Canada in 1956 and joined the Fisheries Research Board of Canada (now the Department of Fisheries and Oceans) at the Pacific Biological Station, Nanaimo, BC, as an ethologist. There, he worked on problems related to the major threat posed to salmon by the proposed construction of multiple hydroelectric dams on the Fraser River. He then turned his attention to long-distance migration, orientation, and navigation of Pacific salmon, and obtained his Ph.D. from the University of Leyden in 1965. In 1966 Dr. Groot was appointed biological director of the Netherlands Institute of Sea Research but decided to return to Canada and rejoin the staff of the Pacific Biological Station. He has worked at this institute until the present time. From 1973 to 1980, he was research co-ordinator for the Federal-Provincial Salmonid Enhancement Program, and, more recently, he has been appointed Climate Change Co-ordinator, examining the possible impacts that global climate changes can have on the fisheries resources of western Canada. He has also been a visiting professor at Simon Fraser University, Burnaby, BC, and at the Bamfield Marine Station.

Michael C. Healey began studying biology at the University of British Columbia in Vancouver, where he obtained his B.Sc. and M.Sc. degrees. After receiving a Ph.D. from the University of Aberdeen, Scotland, in 1969, he held a postdoctoral fellowship at the Pacific Biological Station in Nanaimo, BC. His professional career began at the Freshwater Institute in Winnipeg in 1970 where he was program leader of the Fish Populations Program. In 1974 he returned to the Pacific Biological Station where he became leader of the Strait of Georgia Program and later the Salmon Ecology Program. His studies on juvenile salmon in streams, estuaries, and coastal areas brought him much recognition in the scientific community. In 1986 he turned his attention to ocean stages of Pacific salmon, specifically of chinook and coho salmon, and became co-ordinator of the Marine Survival of Salmon (MASS) Program. In 1990 Dr. Healey moved to the University of British Columbia to become director of the Westwater Research Centre and professor in the Department of Oceanography.

William R. Heard has directed and conducted research on Alaska salmon for thirty years and has also served on many research planning and policy groups concerned with salmon resources. He received a B.S. in Zoology from Oklahoma A. & M. College and an M.S. in Fisheries from Oklahoma State University. Since 1967 he has been employed at the Auke Bay Fisheries Laboratory of the

National Marine Fisheries Service at Auke Bay, Alaska, and has worked in areas of salmon enhancement and ocean ranching, and studied life histories of pink, chum, and coho salmon in intertidal experimental spawning channels. At present, he is program manager for the Salmon Enhancement and Marine Recruitment Studies Program at the Auke Bay Fisheries Laboratory, and his current research activities include early marine salmon ecology, and chinook and sockeye salmon enhancement. He has published widely in the areas of ecology, life history, behaviour, zoogeography, management, and enhancement of a variety of fishes, especially Pacific salmon.

Fumihiko Kato was born in China and returned with his family to Japan after the Second World War. He studied in the faculty of agriculture at Tokyo University where he obtained his M.Sc. in Fisheries in 1969. He subsequently joined the Fishery Agency of Japan as a research scientist at the Japan Sea Regional Fisheries Laboratory in Niigata Prefecture. There, he spent seventeen years on various assignments related to fisheries research in the Sea of Japan. Although he studied the marine ecological problems of a variety of fish species, his main interest was in the population dynamics of masu salmon. In 1985 he was appointed chief of the Coastal Fisheries Development Section of the Seikai Regional Fisheries Laboratory in Nagasaki Prefecture. As leader of the Japanese Flounder Resource Enhancement Program, he contributed to the nationwide enhancement of fisheries resources. Mr. Kato died prematurely in 1990 at the age of forty-five, three months after he was appointed research co-ordinator of resource management at the National Institute of Fisheries Science.

Leo Margolis received his Ph.D. from McGill University in 1952. In that year he joined the Pacific Biological Station in Nanaimo, BC, where he has occupied positions of research scientist and head of various research sections and divisions. He is currently senior scientist at the Pacific Biological Station and an adjunct professor at Simon Fraser University, Burnaby, BC. Actively engaged for more than thirty-five years in research on Pacific salmon biology, Dr. Margolis has been closely associated with the high-seas salmon work of the International North Pacific Fisheries Commission almost since its inception in 1953 and has served on many of the Commission's committees and sub-committees. He has had a long history of involvement, on behalf of Canada, with various Pacific rim countries of the northern and southern hemispheres, in a wide range of fisheries-related issues. His salmon research has focussed on stock identification, ocean migrations and distribution, and other aspects of the sea life of these fish. Additionally, he is recognized globally for his research in aquatic parasitology. He has authored more than 140 original research and review articles and has edited five special volumes or series in fish parasitology and fisheries science. In 1975 he was elected a Fellow of the Royal Society of Canada and in 1990 he was appointed an Officer of the Order of Canada. He was President of the Canadian Society of Zoologists in 1990–91.

Ernest O. Salo obtained a B.S. in Zoology and a Ph.D. in Fisheries at the University of Washington. He completed his doctoral thesis while serving as a biologist for the Washington Department of Fisheries and as assistant supervisor of the Washington State Salmon Hatcheries. In 1955 he accepted a faculty position at California State University at Humboldt and continued there as chairman of the Division of Natural Resources from 1961 to 1965. He returned to the University of Washington in 1965 as a faculty member of the Fisheries Research Institute and continued as professor emeritus, School of Fisheries, after semi-retirement in 1985. Dr. Salo was especially dedicated to student education and development, working with and guiding graduate students as well as improving and teaching undergraduate courses in fisheries. He served as director of the Big Beef Creek Research Station in Washington from 1965 to 1985 and was the first director of the Center for Streamside Studies at the University of Washington, from 1987 to 1988. He assisted the Chilean government in the initial phases of establishing salmon culture in South American waters and was internationally renowned as one of the foremost authorities on salmonid biology and forestry-fisheries interactions. He passed away in July 1989 in Seattle.

Federick Keith Sandercock completed his undergraduate studies at the University of Toronto, and his graduate studies at the University of British Columbia in Vancouver. Soon after obtaining his Ph.D. in 1969 he was employed with the Fisheries Service of the Department of Fisheries and Forestry of Canada (now Department of Fisheries and Oceans). Dr. Sandercock was actively involved in the initial development of the Federal-Provincial Salmonid Enhancement Program (SEP), which has the objective of substantially increasing the Pacific salmonid resources of British Columbia. In SEP, he became chief of Enhancement Operations and played a key role in organizing and supervising the hatchery and spawning channel programs, and was responsible for several new initiatives in salmon culture methods. Although deeply involved in the field of artificial salmon propagation, Dr. Sandercock continued to keep in touch with the biological issues of salmon in their natural environment. Specifically, he became a specialist in the field of coho salmon biology and life history.

Life History of Sockeye Salmon

CONTENTS

HEINE
76

5

PLATE 5. Ocean phase of sockeye salmon. *Photograph by K. Cooke*

6

PLATE 6. Spawning sockeye salmon, Iliamna Lake, Alaska: female (*top*), male (*bottom*). *Photograph by Thomas C. Kline*

PLATE 4 (*previous page*). Sockeye salmon life history stages. *Painting by H. Heine*

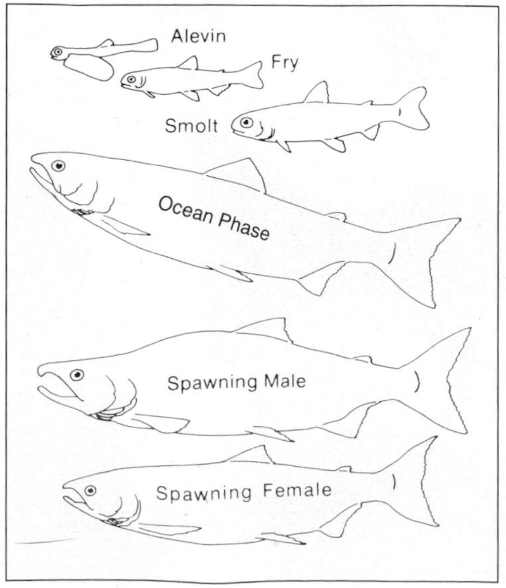

LIFE HISTORY OF SOCKEYE SALMON

(*Oncorhynchus nerka*)

Robert L. Burgner*

INTRODUCTION

THE SOCKEYE SALMON, *Oncorhynchus nerka* (Walbaum), is the third most abundant of the seven species of Pacific salmon, after pink salmon (*O. gorbuscha*) and chum salmon (*O. keta*). Sockeye contributed about 17% by weight and 14% in numbers to the total salmon catch in the North Pacific Ocean and adjacent waters during the period 1952–76 (INPFC 1979). According to Hart (1973), citing W.E. Ricker, the common name *sockeye* (adopted by the American Fisheries Society) is a corruption of the name used by various Indian tribes of southern British Columbia and originally printed as *sukkai*. Other common names include red salmon (Alaska); blueback salmon (Columbia River); nerka and krasnaya ryba (USSR); benizake and benimasu (Japan); and kokanee, little redfish, silver trout, and himemasu (Japan) for the non-anadromous form.

Sockeye salmon exhibit a greater variety of life history patterns than other members of the genus *Oncorhynchus*, and characteristically make more use of lake rearing habitat in juvenile stages. Although sockeye are primarily anadromous, there are distinct populations called kokanee which mature, spawn, and die in fresh water without a period of sea life. Typically, but not universally, juvenile anadromous sockeye utilize lake rearing areas for one to three years after emergence from the gravel; however, some populations utilize stream areas for rearing and may migrate to sea soon after emergence. Anadromous sockeye may spend from one to four years in the ocean before returning to fresh water to spawn and die in late summer and autumn. The sockeye also shows a wide variety of racial adaptations to rather specialized spawning and rearing habitat combinations.

It had been generally accepted that the sockeye, along with the other members of the genus, evolved from the more primitive trout genus *Salmo* (Neave 1958; Foerster 1968; Hoar 1976).[1] This was believed to have occurred during a series of geographical isolations around the North Pacific rim as ocean elevations fell and rose during glaciation changes early in the Pleistocene period. There is disagreement among ichthyologists (e.g., Thorpe 1982) concerning the probable sequence of derivation of the *Oncorhynchus* species, their relative evolutionary status, and their degree of specialization. Whatever the evolutionary chain of events, the sockeye salmon has characteristic anatomical, physiological, and behavioural attributes that clearly demark it from its sibling species.

The adaptations of sockeye to lake environments appear to require more precise homing to spawning areas, both as to time and location, than is found in the other species of Pacific salmon. Although available spawning localities are more restricted because of the usual requirement of a lake rearing environment for the juveniles, the overall success of this adaptation is indicated by the fact that the sockeye is much more abundant than

*Fisheries Research Institute, University of Washington, Seattle, Washington 98195

1 In view of evidence of a close genetic link between trout of Pacific lineage and the Pacific salmon, they have recently been combined under the single genus *Oncorhynchus* (Smith and Stearly 1989).

chinook (*O. tshawytscha*) and coho salmon (*O. kisutch*), which utilize stream rearing environments as juveniles. Behaviourally, juvenile sockeye in fresh water do not need the territorial stream behaviour as displayed by juvenile coho and chinook salmon, but do exhibit schooling tendencies more characteristic of pelagic feeding fishes.

Other distinctions of sockeye include growth rate and size at maturity. Sockeye do not exhibit the rapid marine growth of coho or pink salmon, which mature and return to fresh water after a single winter in the ocean, or of chinook salmon, which attain a much larger average size at maturity.

The flesh of sockeye is a darker red than that of the other salmon species, a colour long considered to be a marketing attribute of the canned product. Perhaps the most noticeable characteristic of the sockeye among Pacific salmon is the remarkable transformation in external coloration of both sexes, and in body shape of the male, as they reach maturity in the spawning environment (Plate 1). The green heads and bright red bodies – the nuptual coloration of spawning sockeye – are indeed distinctive (Plates 4 and 6).

The primary spawning grounds of sockeye salmon in North America extend from tributaries of the Columbia River to the Kuskokwim River in western Alaska, and, on the Asian side, the spawning areas are found mainly on the Kamchatka Peninsula, USSR (Figure 1). During their feeding and maturation phase in the ocean, sockeye range throughout the North Pacific Ocean, Bering Sea, and eastern Sea of Okhotsk north of 40°N. There is considerable intermingling of Asian and North American populations, and of North American populations from Bering Sea and Gulf of Alaska streams. There is a general southward shift in ocean distribution with the approach of winter and a northward movement with the warming of ocean surface waters in the spring. Maturing sockeye return to their respective spawning rivers at different times, from late spring to midsummer. Spawning times range from late July through January, but are primarily from midsummer until late autumn. As with all species of Pacific salmon, sockeye die shortly after completion of spawning.

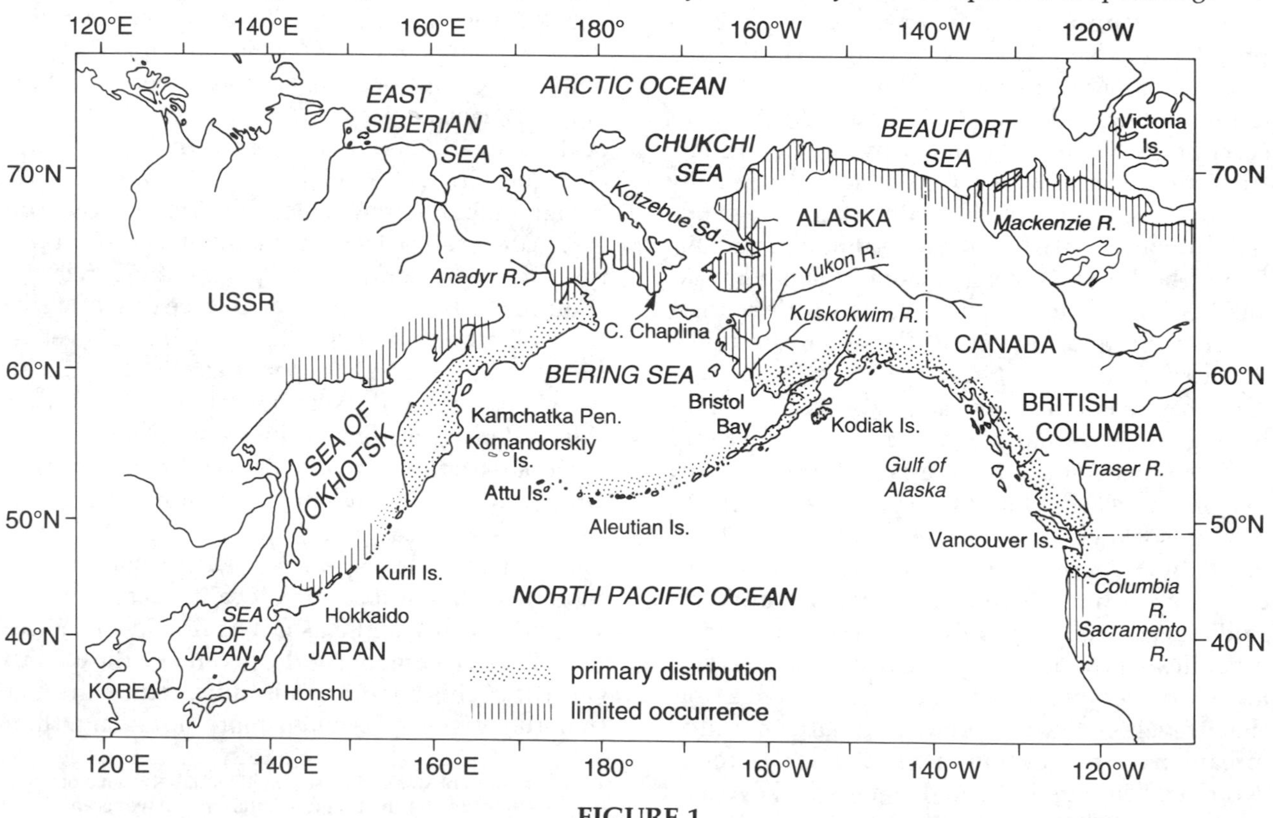

FIGURE 1
Coastal and spawning distributions of sockeye salmon

Distribution of Spawning Stocks

Geographically, populations of sockeye range widely in the temperate and sub-arctic waters of the North Pacific Ocean and northern adjoining Bering Sea and Sea of Okhotsk (Figure 1). In North America, spawning populations are found in streams from the Sacramento River in California (Hallock and Fry 1967) north to Kotzebue Sound (Atkinson et al. 1967; McPhail and Lindsey 1970), but occur in commercially important numbers only from the Columbia River to the Kuskokwim River in the Bering Sea. Although sockeye salmon are indigenous to the Yukon River drainage, they are found there in very low numbers (McBride et al. 1983). A few sockeye were reported in marine waters of the Canadian Arctic at Bathurst Inlet and Holman Island (west coast of Victoria Island), but they were considered to be strays (Hunter 1974). To the west, spawning occurs in streams on Attu Island, the westernmost of the Aleutian Islands.

In the western North Pacific Ocean and Bering Sea, the spawning distribution of sockeye salmon extends northeastward along the Far Eastern coast to the Anadyr River, southward to the southern tip of the Kamchatka Peninsula, and westward to the Okhota and Kukhtuy rivers of the northwest coast of the Sea of Okhotsk (Figure 2); also, some sockeye occur in the streams of the northern and southern Kuril Islands (Hanamura 1966) and the Komandorskiy Islands (Shmidt 1950). Sockeye are found in very small numbers in the Anadyr River but reportedly are more abundant in the Tumansk River, about 35 km south of the Anadyr estuary (Kaganovsky 1960). Sockeye spawn in low numbers in the Taui, Ola, and Gizhiga rivers on the north coast of the Sea of Okhotsk (Klokov 1970). The coastal marine range is reported to be from Cape Chaplina in the northern Bering Sea southward along the Kamchatka Peninsula and Kuril Islands to the north coast of Hokkaido, then north along western Kamchatka to the north coast of the Sea of Okhotsk (Figure 1) (Hart 1973).

The distribution of spawning stocks and areas of greatest abundance of sockeye salmon are a consequence of the unique specialization of the species. Unlike the other *Oncorhynchus* species, sockeye

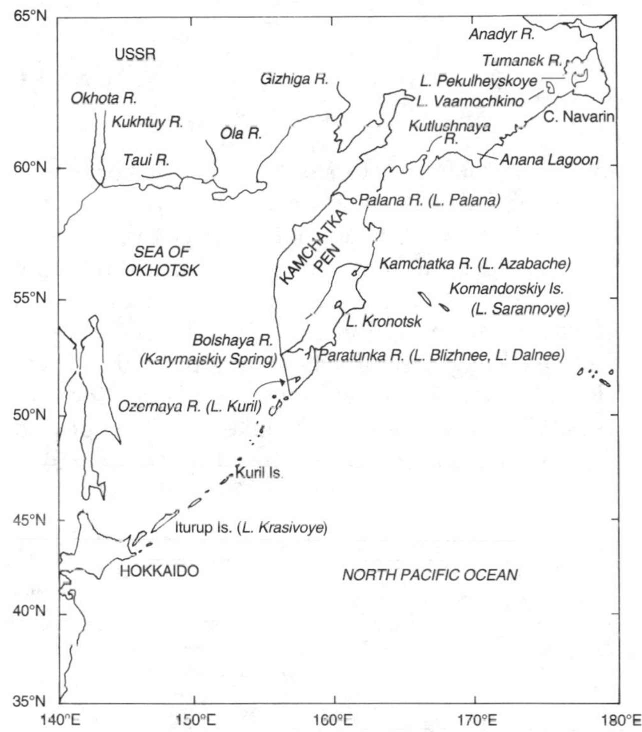

FIGURE 2

Primary sockeye spawning areas in the USSR

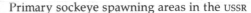

salmon usually spawn in areas associated with lakes, where the juveniles rear in the limnetic zones before they smoltify and migrate to sea. Because of this specialization, sockeye distribution and abundance are in most instances related to the availability of north temperate rivers with accessible lakes in their watersheds. It is not surprising, then, that the two largest spawning complexes of sockeye salmon in the North Pacific rim are found in the Bristol Bay watershed of southwestern Alaska and the Fraser River drainage of British Columbia, both possessing extensive lake rearing areas accessible to salmon. The Bristol Bay stocks provided over 50% of North American sockeye runs for the period 1950-84 (Rogers 1986). Other important production areas in North America are the Skeena River, Nass River, Somass River, and the Rivers and Smith inlet areas in British Columbia, and the Chignik, Karluk, and Copper rivers and tributaries of Cook Inlet in Alaska. In the USSR, the two predominant sockeye-producing river systems are the Ozernaya River in southwestern Kamchatka and the Kamchatka River along the central east coast of the peninsula.

COMPARISON OF SOCKEYE SALMON SPAWNING SYSTEMS

In the Bristol Bay watershed a series of short rivers drain single lakes or lake chains lying at low elevation at the base of mountain ranges (Figure 3). Iliamna Lake, the largest of the sockeye-producing lakes of the world, is 2,622 km^2 in area and lies at an elevation of only 15 m above sea level (Table 1). Lake Aleknagik, the lower lake of the Wood River chain, is 10 m in elevation, and extreme tides reverse the river flow into the lake for a brief period during the day. Except for the Alagnak River system, the lakes lie at less than 100 m altitude. Lake Chauekuktuli of the Tikchik River system is farthest from the sea, some 257 km from tidewater. Most of the Bristol Bay sockeye lakes are deep, extending well below sea level as so-called cryptodepressions formed by past glacial action (Table 1).

Whereas the Bristol Bay lakes lie at low elevations and short distances from tidewater, the Fraser River lakes are mostly at higher elevations and several hundred kilometres above tidewater (Figure 4). The Fraser River, nearly 1,600 km in length, originates in the central plateau region of

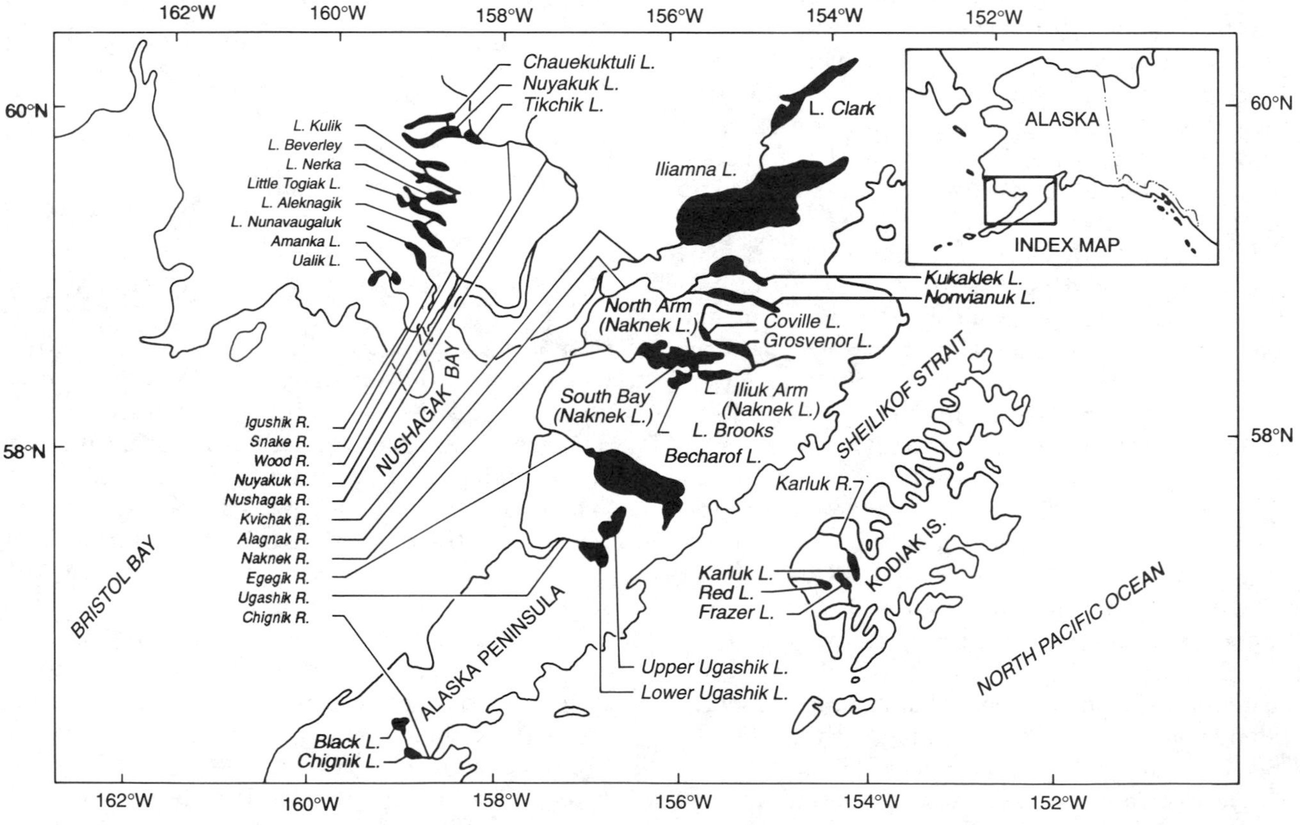

FIGURE 3
Sockeye lake systems of southwestern Alaska

TABLE 1
Physical data on important or well-studied sockeye nursery lakes of the world

Location	River system	Nursery lake	Distance from sea (km)	Elevation (m)	Lake area (km²)	Depth: Maximum (m)	Depth: Mean (m)
Washington	Columbia	Wenatchee	842	569	10	73	46
		Osoyoos	986	278	23	64	14
	Quinault	Quinault	54	56	15	300	
	Cedar	Washington	11	6	88	65	33
British Columbia	Fraser	Cultus	88	41	6	42	32
		Harrison	129	10	218		151
		Shuswap	483	347	310	162	62
		Chilko	644	1172	200	366	108
		Quesnel	723	725	270		158
		Bowron	1064	945	10		16
		Fraser	965	640	55	31	13
		Stuart	977	678	360		20
	Somass	Sproat	11	29	41		59
		Great Central	27	82	51	273	212
	Rivers Inlet	Owikeno	2	14	96	369	172
	Smith Inlet	Long	2	15	10		73
	Skeena	Babine	380	708	490	186	55
		Lakelse	113		13	32	8
	Nass	Meziaden	209	243	40		
Central Alaska	Copper	Tazlina	290	545	ca. 150		
Cook Inlet	Kenai	Kenai	109	133			
	Kasilof	Tustumena	24	33	291	270	
Kodiak Is.	Karluk	Karluk	39	106	40	126	49
	Dog Salmon	Frazer	11	108	17	59	33
	Ayakulik	Red	26	62	8	45	29
Alaska Pen.	Chignik	Chignik	19	5	22	64	26
		Black	32	15	39	6	3
Bristol Bay	Ugashik	Lower Ugashik	51	3	208		
		Upper Ugashik	69	3	177		
	Egegik	Becharof	45	15	1132		
	Naknek	Naknek	56	10	610	173	41
		Brooks	103	19	75	79	45
		Grosvenor	150	31	73	107	50
		Coville	161	33	33	53	19
	Alagnak	Kukaklek	113	246	176		
		Nonvianuk	101	192	121		
	Kvichak	Iliamna	80	15	2622	301	44
		Clark	203	67	267	262	103
	Wood	Aleknagik	32	10	83	110	43
		Nerka	63	21	201	164	39
		Little Togiak	95	23	6	77	30
		Beverley	122	30	90	188	55
		Kulik	153	43	45	160	77
	Nuyakuk	Tikchik	220	95	53	45	15
		Nuyakuk	237	95	144	283	113
		Chauekuktuli	257	98	82	268	111

(continued on next page)

TABLE 1 (continued)

Location	River system	Nursery lake	Distance from sea (km)	Elevation (m)	Lake area (km²)	Depth Maximum (m)	Depth Mean (m)
	Igushik	Amanka	80	9	35	65	23
		Ualik	100	15	39	72	28
	Snake	Nunavaugaluk	72	14	89	162	57
	Togiak	Togiak	77	67	ca. 40		
		Upper Togiak	114	92	ca. 9		
Kamchatka	Ozernaya	Kuril	40	106	76	309	176
	Kamchatka	Azabache	48	90		33	18
	Kronotsk	Kronotsk	32	371	ca. 230	ca. 600	ca. 60
	Paratunka	Dalnee			1.4	60	32
		Blizhnee			3.5	37	16

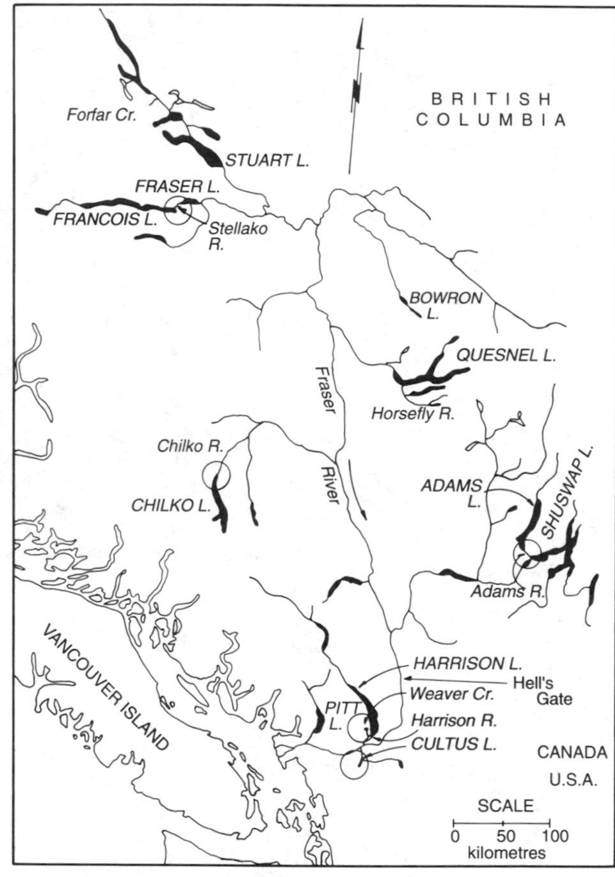

FIGURE 4
Sockeye lakes of the Fraser River watershed, British Columbia

British Columbia and drains a watershed of over 230,000 km² (Killick and Clemens 1963). It is divided by the narrow Hell's Gate canyon through the coastal mountain range into the Lower Fraser area and the extensive area of the Upper Fraser and its tributaries, which contain most of the sockeye lakes. The approximate migration distances from tidewater and elevations of the spawning areas for major populations are indicated in Table 1.

Because of the extensive and varied terrain in which the Fraser River system lakes lie, they present more diverse habitat for the spawning runs of sockeye than do the lakes of the Bristol Bay area. This may in part explain the greater diversity in run timing, where each lake population tends to have its individual charactistics. Whereas 80% of the Bristol Bay sockeye run passes through the bay in 12 to 14 days (Burgner 1980), the sockeye run to the Fraser River enters the river over nearly a three-month period (Killick 1955). The matter of run timing and its cause is complex, as will be discussed later.

The Skeena River system is the second largest sockeye system in British Columbia and presents a complex of tributary streams and lakes with sockeye runs of varying size (Brett 1952). Of these, Babine Lake and its tributaries (Figure 5) are most important, and the sockeye spawning areas there have now been supplemented by artificial spawning channels at Fulton River and Pinkut Creek to better utilize the extensive rearing potential of this lake (West and Mason 1987).

In the early 1900s the Columbia River basin still supported sockeye runs well in excess of one million fish. However, runs declined dramatically to relative insignificance as the result of the combined effect of the commercial fisheries in the lower river, fishing on the spawning grounds, and complete blocking of most spawning areas. Of the eight lake systems originally producing sockeye in the mid and upper Columbia River drainage, only three are producing today (Fulton 1970).

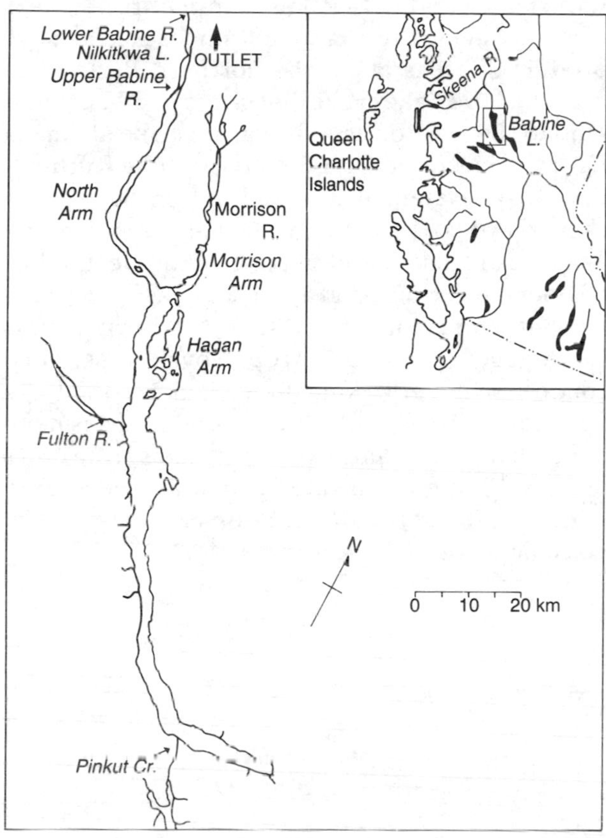

FIGURE 5
Babine Lake, principal sockeye lake of the Skeena River system, British Columbia

Of the other important sockeye lake systems in North America, several are noteworthy for their uniqueness. Chignik and Black lakes of the Chignik River system in the southern Alaska Peninsula (Figure 3) contrast greatly. Black Lake is shallow (mean depth, 3 m) (Burgner et al. 1969), all spawning occurs in tributary streams, and the sockeye enter the lake in June and early July. Chignik Lake is deeper, colder in summer, murky from volcanic terrain runoff, spawning occurs in stream and lake beach areas, and the sockeye do not begin entry until early July. The two lake populations exhibit distinctly different characteristics of age and growth. In Cook Inlet, the two major sockeye lakes – Kenai and Tustumena – receive heavy glacial runoff, but spawning occurs primarily in the clear-water tributaries. This situation is similar to that at Owikeno Lake in Rivers Inlet, British Columbia. On Kodiak Island, sockeye enter Karluk Lake over an extended time period, utilize a complex of stream and beach spawning areas, and show different run timings to separate spawning areas. The Karluk Lake sockeye population was formerly much more productive; however, the stocks in the middle of the run were heavily harvested in the early part of this century and have never fully recovered (Thompson 1950; Gard et al. 1987). Of the major sockeye runs, the Copper River run in the northern Gulf of Alaska is the earliest in timing, beginning to move upriver in early June en route to distant lake tributaries. The much smaller sockeye run into Lake Quinault on the Olympic Peninsula, Washington, is unique in that the run extends from late January through July, peaking in late May or early June, yet spawning does not begin until November and continues through January (Johnson 1977).

Of the North American lakes now producing anadromous sockeye, two have significant artificially introduced runs.[2] A run was established in Frazer Lake on Kodiak Island by successfully overcoming a barrier to upstream migration through construction of fish ladders and by transplanting eggs, fry, and adult sockeye into the lake. In 1978, twenty-one years after the first introduction, runs entering this lake numbered 142,000 sockeye (Blackett 1979). Lake Washington also harbours an introduced run, apparently established from Baker Lake stock in a series of transplants beginning around 1940. Although kokanee were in the watershed historically, no anadromous sockeye were observed until returns from planted juveniles were seen. The run increased dramatically in the mid-1960s and has ranged from 118,000 to 626,000 adults over the years 1970–82 (Starr et al. 1984).

The primary sockeye spawning systems of Asia (Figure 2) are in distinct contrast to North American systems. Lake Kuril, situated near the southern tip of the Kamchatka Peninsula, is by far the

2 Kokanee, the non-anadromous sockeye, have been successfully introduced into a large number of lakes in North America and elsewhere (Seeley and McCammon 1966).

most important of the production areas. It is a deep calderal lake of volcanic origin lying in the mountains at an elevation of about 100 m at the head of 60-km-long Ozernaya River (Krokhin and Krogius 1937). At 75 km^2 in surface area, the lake is only 1/35 the size of Iliamna Lake, Alaska, yet has produced runs of sockeye as high as 10 million fish (Egorova et al. 1961). Spawning occurs primarily in upwelling groundwater areas along the lake shore, but it also occurs in the upper outlet river, in tributary springs, and in the lower reaches of tributary streams. The Ozernaya River sockeye enter the river over an unusually prolonged period, from the end of May until the beginning of October, and the period of spawning is even more prolonged (Egorova 1970a).

The Kamchatka River, also situated in a volcanic area, is the largest river on the Kamchatka Peninsula and the second most important Asian sockeye spawning area. It differs from other major sockeye spawning-rearing areas in that juvenile sockeye rear in springs and side channels of the river. However, some also rear in lakes, particularly in Lake Azabache near the river mouth (Bugaev 1983). The timing of the spawning run to the Kamchatka River (early June-late July, mainly 10-30 June; Bugaev 1987) is much earlier than that to the Ozernaya River and is of shorter duration.

Asian sockeye production areas of lesser importance include the relatively small Dalnee and Blizhnee lakes on the Paratunka River, southeastern Kamchatka, and Lake Palana and the spring areas of the Bolshaya River in west Kamchatka. Sockeye producing areas along the northern Bering Sea coast include Lakes Pekulheyskoye and Vaamochkino to the west of Cape Navarin, a lake at Anana Lagoon near Cape Olyutorskiy, and the Kutlushnaya River system at the head of Korf Bay (V. Bugaev, Ko TINRO, Petropavlovsk, USSR, pers. comm.). Sockeye are also found in Lake Sarannoe, the Komandorskiy Islands, and as far south as Lake Krasivoye, Iturup Island, in the southern Kuril Islands (V. Bugaev, Ko TINRO, Petropavlovsk, USSR, pers. comm.). In general, Kamchatka is devoid of large lake rearing areas, and sockeye production is more dependent on stream, lake beach, and spring areas of upwelling groundwater which are associated with the porous nature of the extensive terrain of volcanic origin.

AGE COMPOSITION

The age composition of sockeye salmon returning from the ocean to spawn in their respective spawning grounds varies considerably among populations and from year to year within a population. Healey (1987) noted that sockeye females display more variation in age and size at maturity, both between and within populations, than do the other Pacific salmon.

The length of time juveniles spend in fresh water after emergence may vary from a few weeks, in the case of some unusual river populations, to three years for portions of some lake-rearing juvenile populations, the most common ages being one or two years. Smoltification and migration seaward normally occur in spring. Some populations, such as most Fraser River stocks, smoltify almost exclusively as yearlings, whereas other populations, such as in the Egegik River system of Bristol Bay, smoltify primarily as two-year-olds or in their third year if measured from time of egg deposition. In some populations the length of stay in fresh water varies considerably as in the Kvichak River system in Bristol Bay, where the proportion of yearling and two-year-olds shifts from brood year to brood year. Reasons for this will be discussed in a later section.

The length of stay of sockeye in the ocean varies from one to four years. Sockeye returning after one year in the ocean are much smaller, are predominantly males, and as such are termed "jacks." "Jills" are less common, comprising only 4% of the combined "jack-jill" samples in the Fraser River from 1915 to 1960 (Killick and Clemens 1963); but in the unusual sockeye population utilizing the

Okanagan River, tributary to the Columbia River, jills have been nearly as abundant as jacks in some years (Major and Craddock 1962). Jacks can comprise a significant proportion of the return from a brood year in some populations, such as in the Columbia and Fraser river systems, but are rare or non-existent in other populations, such as in the Chignik River run. In the more southern populations, such as in the Columbia and the Fraser rivers, the predominant ocean age at return is two years, whereas in Bristol Bay populations the proportion of returns after three years is also important. In some populations, such as in the Kamchatka River and Chignik River systems, 3-ocean age fish (i.e., fish that have spent three winters at sea) greatly predominate.

The designation of age in Pacific salmon has resulted in a great deal of confusion among researchers, but has been clarified considerably by Koo (1962a). There are three age designation systems or formulae in common use for sockeye and other Pacific salmon (Koo 1962a; Foerster 1968). The earliest adopted is the Gilbert-Rich designation (Gilbert and Rich 1927), in which the total age is given with the freshwater age as a subscript. An example of this would be "5_3." The subscript "3" indicates that the fish had migrated to sea in the spring of its third year of life from the time it was deposited in the gravel as an egg. It would have spent one winter in the gravel developing from egg to alevin prior to emergence and two winters in the nursery lake as a free-swimming juvenile. In the ocean it would have spent two summers and two winters before returning to spawn in the third summer. This life history would normally be reflected on the scales or otoliths of the fish by growth patterns indicating two winter bands, or annuli, formed in fresh water and two formed in the ocean. The numeral "5" indicates that the fish has completed its fourth winter of life after emergence from the gravel and will have reached five years of life from time of deposition as an egg, if it lives to time of spawning.

The age formula still in use by Soviet scientists for Pacific salmon is an early European method used for Atlantic salmon (*Salmo salar*), and reflects only the number of winter annuli on the scales. The 5_3Gilbert-Rich formula thus becomes 4_2 by the alternative formula.

The third age designation system, discussed in detail by Koo (1962a), is the European formulation used for Atlantic salmon and which is now commonly used for Pacific salmon in International North Pacific Fisheries Commission publications as well as by the Fisheries Research Institute and some other agencies. The formula corresponding to a 5_3 Gilbert-Rich formula is 2.2. The decimal point is used to separate the number of winters spent by the juvenile fish in fresh water from the number of winters spent in the ocean, and reflects the number of corresponding annular marks found on the scale or otolith of the fish. The age of the adult in reference only to its freshwater life can be expressed as 2., or in reference only to its marine life as .2. If desirable, a plus sign can be used to designate summer growth beyond the annular mark, e.g., 2+. or .2+. A designation of 0.3 or 0+.3 would indicate a salmon migrating to sea as a fry, or underyearling, and returning from the ocean after three winters at sea. This last age designation system will be used in this chapter.

Foerster (1968) listed the growth patterns observed among anadromous sockeye and their designation in the three notation systems previously discussed. He also listed the most common age groups found in a number of sockeye river systems. Table 2 contains a similar list, with age designated by the European formula preferred by Koo (1962a). The age group that greatly predominates in most years is underlined. Listing of a second or third age group indicates that it commonly may exceed 20% of the return run.

The age group combinations present in sockeye runs vary among river systems, and often among races within a river system. For example, the substantial differences in age composition that occur among lake system populations of sockeye within Bristol Bay and among spawning populations in an individual lake are illustrated in Figures 6 and 7. The adaptive significance of different age combinations in sockeye populations is a subject receiving increased attention (Healey 1987; Rogers 1987).

TABLE 2
Most common age groups of sockeye salmon, by river system

River system or stock	Age groups	Remarks	Source
Washington			
Columbia R.	1.2, 1.1		Anas & Gauley (1956)
Quinault R.	1.2, 1.3		Johnson (1977)
L. Washington	1.2, 1.3	Several years	Jim Ames, Wash. Dept. Fish. (pers. comm.)
British Columbia			
Fraser R.	1.2	Catch samples 1915-60	Killick & Clemens (1963)
Harrison Rapids (Fraser)	0.2	Several years	Gilbert (1918-20); Schaefer (1951)
Skeena R. (Babine L.)	1.3, 1.2	Variable, on the average, about equal; brood 1961-77 returns	McDonald & Hume (1984)
Nass R.	2.2, 1.2		Foerster (1968)
Rivers Inlet (Owikeno L.)	1.3, 1.2	5 years	Healey (1987)
Smith Inlet	1.3, 1.2	5 years	Healey (1987)
S.E. Alaska			
Chilkoot L.	1.3	1976-82	Bergander (1985)
Chilkat L.	1.3, 2.3, 2.2	1976-82	Bergander (1985)
East Alsek R.	0.3, 0.2	1981-85	McBride & Brogle (1983); McBride (1984, 1986); Riffe et al. (1987)
Alsek R.	1.3, 1.2	1981-85	Riffe et al. (1987)
Situk R.	1.3, 1.2, 2.2, 2.3	1966, 1968, 1971, 1974-79, 1981, 1983-85 escapements	"
Copper R.	1.3, 1.2	1982, 1983	Sharr (1983); Sharr et al. (1985)
Central Alaska			
Susitna R.	1.3, 1.2	1972-84	Cross et al. (1987)
Kenai R.	1.3, 2.3, 1.2, 2.2	1972-84	"
Kasilof L.	1.3, 1.2, 2.2	1972-84	"
Karluk L.	2.2, 2.3, 3.2	1922, 1924-49	Rounsefell (1985b)
Frazer L.	2.2, 2.3, 1.2, 1.3	1963-86	Kyle et al. (1988)
Chignik R. (Chignik L.)	2.3, 2.2	Several years	Various FRI & ADF&G reports
Chignik R. (Black L.)	1.3, 2.3, 1.2	Several years	"
Western Alaska			
Kvichak R.	2.2, 1.2, 2.3, 1.3	14 years	Various ADF&G Tech. Data Reports & Bristol Bay annual Management Reports
Naknek R.	2.2, 1.3, 2.3, 1.2	14 years	"
Wood R.	1.2, 1.3, 2.2	14 years	"
Nuyakuk R.	1.3, 1.2, 2.2	14 years	"
Igushik R.	1.3, 1.2, 2.2, 2.3	14 years	"
Egegik R.	2.2, 2.3, 1.3, 1.2	18 years	"
Ugashik R.	2.2, 1.2, 1.3, 2.3	18 years	"
Togiak R.	1.3, 1.2	18 years	"
Kuskokwim R.	1.3, 2.3, 1.2	1982-85	Huttunen (1984a, 1985-87)
Kanektok R.	1.3, 1.2	1982-85	"
Goodnews R.	1.3, 1.2	1982-85	"
Kamchatka			
Ozernaya R.	2.3, 2.2, 3.2, 3.3	1940-59	Egorova et al. (1961)
Bolshaya R.	1.3, 1.2, 1.4	1935-47	Semko (1954a)
Kamchatka R.	1.3, 2.3	1957-60	Hanamura (1966)
Lake Dalnee	1.2, 2.2, 1.3	1934-43	Krogius & Krokhin (1948)

Source: Adapted from Burgner (1987), Table 2
Notes: Underlined numbers indicate that the age group is always predominant. Inclusion of an age group indicates that it has exceeded 20% in one or more years examined.

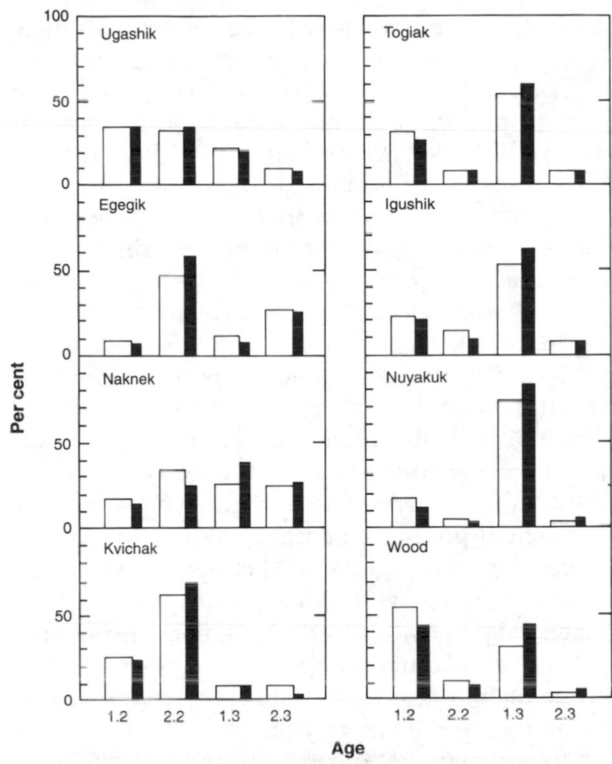

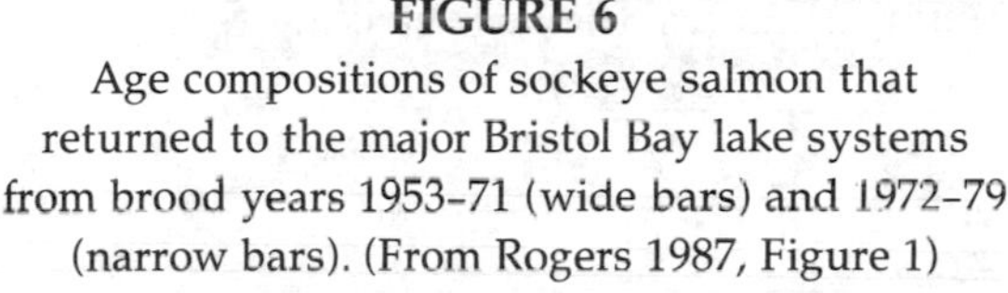

FIGURE 6

Age compositions of sockeye salmon that returned to the major Bristol Bay lake systems from brood years 1953-71 (wide bars) and 1972-79 (narrow bars). (From Rogers 1987, Figure 1)

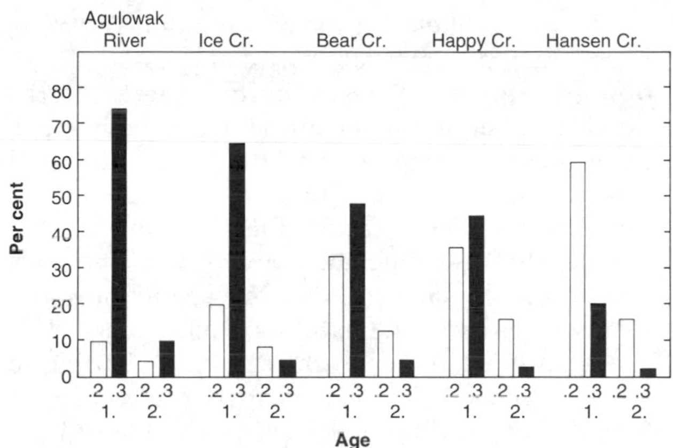

FIGURE 7

Age compositions of sockeye salmon returns to five spawning grounds in Lake Aleknagik from brood years 1953-79. (From Rogers 1987, Figure 8)

Sexual Dimorphism

As it begins its home journey from the sea, the sockeye salmon is sleek and silvery with a blue-black top of the head, silvery white jaws and opercular area, a steel-blue back, silvery sides extending above the lateral line, and shading to white on the belly (Plates 4 and 5). Unlike the pink, coho, and chinook salmon, its dorsal and caudal fins are without spots. The bases of the caudal fin rays show some silver sheen but not as heavily as the coho or chum salmon. The general body shape, including the caudal peduncle, is more slender than that of the coho salmon. The caudal peduncle is thicker and deeper, however, than that of the chum salmon. It is easily distinguished from the pink salmon by its larger scales, and from the chinook salmon by its generally smaller size, lack of dark pigmentation on the gums, and lack of spots on back and tail. The two sexes are coloured alike and are difficult to distinguish externally except for the slightly more elongate snout and jaw of the male.

As sockeye salmon near their home rivers, they begin a marked sexual dimorphism (Plates 4 and 6). The female retains the fusiform shape, but with swelling of the abdominal area as the egg skeins enlarge, and with receding of the gums around the teeth and slight elongation of the snout. The male, in contrast, becomes laterally compressed, developing a prominent fleshy hump anterior to the dorsal fin. The snout becomes longer and hooked, the upper jaw elongates, and the canine-like teeth grow out from the receding gums. The most striking change is in coloration (Plates 4 and 6). In the female, the head becomes olive green, the back bright red, the sides greyish-black, and the abdomen white. The male exhibits a more distinctive greenish head and a bright red back shading to blackish-red on the sides, with white ventrally. Fine dark speckling develops on the tail of both sexes. Also, the skin thickens and the scales begin to be resorbed. These changes are more pronounced in the male than in the female. The rapidity of transformation in shape and coloration varies among different races and appears to be dependent on the length of time between entry into fresh water and actual spawning.

The more conspicuous coloration and shape of the male sockeye salmon appear to favour survival and reproduction by the more fusiform females where bear predation is prevalent. It has been noted that sex ratios of bear-killed sockeye in the Wood River lakes are heavily skewed towards male sockeye (R.L. Burgner, unpublished data).

Variation in spawning coloration occurs among sockeye populations. For instance, the abdomen may be blackish rather than white. Perhaps the most unusual variation is an apparent mimicry of chum salmon coloration that has developed in a population of sockeye that spawns in Weaver Creek, British Columbia (G. Duker, National Marine Fisheries Service, Seattle, Washington, pers. comm.). Sockeye and chum salmon spawn in the creek during the same time of the season but apparently do not interbreed.

Foerster (1968) called attention to the fact that Smirnov (1958, 1959b) related the red colour manifestation to the presence of high levels of carotenoid and lipoid pigments in skin and muscle tissues. Smirnov argued that because sockeye tend to spawn in areas of slower water flow and reduced oxygen content, these pigments make possible the more efficient uptake of oxygen from the environment because they are active oxygen carriers and catalyze the oxygen-reduction process. Whatever may be the significance of the prespawning changes, they are striking to behold.

Upstream Migration

The timing and rate of upstream migration of maturing sockeye varies considerably among sockeye stocks. The ascent of rivers to spawn is the final part of the homing migration. Gilhousen (1960) stated that "not only spawning time but also migration time is a consequence of the process of maturation. Maturation, then, is a physiological process whereby sockeye become sensitive to certain facets of their environment, interacting with them to bring about the sequence of migration and spawning necessary for survival." He assumed that sockeye maturation and the resulting migration is largely a response to the pattern of the changing length of days. He reiterated Royal's (1953) hypothesis that for a given race, the occurrence of a rather sharp peak or mode in timing, both in migration and spawning, indicated that the fish in the peak usually encounter an advantageous set of conditions for survival extending over a period of relatively short duration. Royal noted that approximately two-thirds of the migration of a typical Fraser River population will appear within a period of a week to 12 days, and he argued that racial differences in migration timing and strong homing tendencies capitalize on different conditions in freshwater habitats. For Fraser River sockeye, run timing appears to be synchronized with the specific temperature regime of the home stream so that spawning will occur at an appropriate time for development and emergence of alevins in spring (Miller and Brannon 1982). It is inferred that salmon migrations of longer duration with a broader peak have a more extended period of optimal survival conditions.

For Fraser River sockeye, the peaks of passage of individual spawning races into the river occur in succession from the first few days in July to the end of August or the beginning of September (Gilhousen 1960). The chronological order of their migration, spawning, and death shows remarkable consistency (Killick 1955). June and July runs (to Bowron, Stuart, and Quesnel lakes) travel at a rapid rate of 48–51 km/d from the lower river fishery to the spawning grounds, 630–970 km upriver. August runs (to Chilko and Stellako rivers) travel more slowly at 35 km/d and delay more at the entrance to their spawning grounds, 550–885 km upriver. The latest run (Adams River) enters the Fraser River in September after some milling at the river mouth and travels the 386 km to the spawning ground at a rate of 37 km/d. Chilko fish have the highest climb against the river currents; their spawning grounds are at an elevation of approximately 1,200 m.

Not only is there a consistent difference in timing and migration rate among spawning runs of the Fraser River, but the initial chronological order of migration (early, middle, and late) within races is maintained, in general, from the time of escapement from the commercial fishery at the river mouth to the time of spawning (Killick 1955). As an illustration of this consistency, Killick estimated that 82% of a group of Bowron Lake sockeye escaping the Fraser River fishery over a three-day period to migrate an additional 970 km would eventually spawn within a nine-day period. Although males tend to migrate earlier, no difference was found between the sexes in migration rate.

The chronological consistencies of migration and spawning of individual Fraser River sockeye runs have permitted an unusual degree of precision in management of the fisheries according to the strength of annual runs to individual spawning grounds. Such chronological separation is not observed in Bristol Bay sockeye runs, which pass together through the fishery over a short time period even though spawning occurs over a much broader time interval.

Salmon usually cease feeding before entering their natal streams and depend on their energy reserves for migration, maturation of gonads, spawning, and redd (nest) defense until death. They undergo remarkable physical, chemical, biochemical, and physiological changes during migration, which extensively deplete body reserves of fat and protein. Idler and Clemens (1959) measured such changes during migration for Stuart and Chilko lake races of Fraser River sockeye. Stuart Lake fish reached the spawning grounds after approximately 24 days of travel from Albion in the lower river, a distance of 1,023 km and an elevation gain of 709 m, and averaged 12 days on the spawning grounds. The Chilko run travelled for 18 days over the 596 km, gaining an elevation of 1,200 m, and averaging 25 days on the spawning grounds before death.

The two populations exhibited similar changes during their migration and spawning. Body fat reserves (excluding viscera) averaged 14%–15% at initiation of migration and were slightly higher in the female. Most of this reserve was used in the upriver migration, and at death the females had utilized well over 90% of their reserves. The males consumed less than 90%. Initial body protein reserves (about 19% of body weight) were reduced by an average of 55%–61% at time of death for females and 34%–42% for males. The females thus expended more energy during sexual development, spawning, and redd defense, and ended up with less reserves. The males showed a large net increase in body water content up to completion of spawning, whereas the females showed very little change. Both sexes of both races lost body water from the time of completion of spawning until death. The overall weight loss of the female body is over 30% from river entry until death.

In their ascent to spawn, sockeye take advantage of slower water and eddies along the stream banks to conserve energy. They travel in schools, and during heavy migrations (such as those seen in the clear trunk rivers of Bristol Bay) they form a continuous moving band on one or both sides of the river, migrating steadily and uniformly close to the bottom where the current is slower. Taking advantage of this behaviour, fishery biologists introduced a method of counting from towers along the stream banks in Bristol Bay rivers to supplant construction of troublesome weirs across the streams (Thompson 1962; Becker 1962). Migration rates past the tower sites located above tidewater are similar to those reported above for Fraser River sockeye. During the peak of the run in years of heavy migration, daily migrations past the Kvichak River counting towers frequently exceed a million sockeye for a number of days (e.g., 13 consecutive days in 1980; Yuen et al. 1984).

Spawning

Spawning Habitat

Among the species of Pacific salmon, the sockeye salmon exhibits the greatest diversity in adaptation to a wide variety of spawning habitats. The selection of habitat and timing of spawning by a sockeye colony are linked to success of survival, not only during spawning and incubation of the eggs and alevins, but also in the chain of freshwater and marine environments to which the progeny are subsequently exposed. In most instances, but not all, the subsequent environment for the juveniles is a lake or lake chain, and the behaviour of the juveniles after emergence depends on the location of the spawning area in relation to the lake rearing area to be utilized. The circumstances surrounding the initial establishment of a spawning colony and the subsequent adaptive behaviour of the progeny can only be surmised. However, the continued use of a specific spawning environment by a sockeye race or colony depends on the precise homing ability of the species, in which straying to other potential spawning locations is minimal.

Characteristically, sockeye spawning areas are adjacent to lake rearing areas. In this aspect, the sockeye salmon differs from the other five anadromous Pacific salmon species, which normally do not depend on lake rearing during the juvenile stage. The character and suitability of the areas utilized for spawning depend greatly on the geology, topography, climate, water chemistry, and runoff pattern of the watershed. Headwater rivers, tributary creeks, rivers between lakes, outlet rivers, spring areas, and submerged areas of lake beaches are utilized to varying degrees. The sockeye salmon is the only species of the genus *Oncorhynchus* to spawn extensively in shoal beach areas along lake shores, typically in areas of upwelling groundwater that provide circulation through the redd, or nest. Spring-fed ponds and side channels also tend to be more heavily utilized by sockeye salmon than by the other species, particularly in Kamchatka and Bristol Bay. In some Bristol Bay areas the annual spawning activity and nest digging of sockeye tend to maintain and enlarge the spring pond area by churning up the gravel and by cutting into the gravel banks. Semko (1954a) observed similar effects of spawning activity in Kamchatkan spring areas. In situations in which spawning occurs in a river at the outlet of a lake, the progeny must be able to migrate upstream into the lake if they are to utilize it as a rearing area. In these instances, the progeny are dependent for survival not only on the characteristics of the intragravel environment but also on river flow conditions for upstream migration after emergence and on behavioural adaptations.

The relative importance of each type of spawning area varies greatly among lake systems and from year to year within systems. This was examined by Burgner et al. (1969) for sockeye lakes in southwestern Alaska. For instance, the distribution of spawning sockeye salmon was examined among three types of spawning areas in the four main lakes of the Wood River system during the years 1955–62. The average percentages of sockeye utilizing these areas are shown in Table 3.

TABLE 3

Distribution of spawning sockeye salmon among spawning areas in lakes of the Wood River system

Lake	Tributary creeks (%)	Lake beaches (%)	Rivers between lakes (%)
Aleknagik	40	3	57
Nerka	26	47	27
Beverley	8	87	5
Kulik	0	59	41

The differences in spawning distribution reflected the relative availability of the spawning area types in the four lakes. In this instance, the juveniles emerging in the rivers between lakes migrated downstream to the lake below. In spite of these great differences in type of area utilized, and variation from year to year, the number of spawners per hectare of lake nursery area was quite similar among the four lakes.

Differences in the proportion of different age groups occurred among the three spawning area types. The two large river spawning areas between lakes consistently had a higher proportion of the older, larger, ocean age .3 spawners than did the beach and creek areas, presumably because of the greater ability of larger fish to spawn in the coarse river gravels (Rogers 1987). In Iliamna Lake, in which sockeye run strengths are strongly cyclic, escapements to large streams have varied by a factor of 10, whereas escapements to the lake beaches varied by a factor of about 50 (Burgner et al. 1969).

Lake-beach spawning is important in most sockeye lakes of Bristol Bay, Kodiak Island, and the Alaska Peninsula. On the Kamchatka Peninsula, lake beaches accommodate the major portion of the spawning population in Lake Kuril, the most productive sockeye lake in the USSR (Krokhin and Krogius 1937), and in the Paratunka lakes (Krogius and Krokhin 1948). The occurrence of extensive beach spawning around islands is perhaps unique to Iliamna Lake, Alaska, where spawning occurs in coarse rubble material, and circulation around the eggs and alevins is dependent on wind-driven lake currents (Kerns and Donaldson 1968; Olsen 1968). Beach spawning is less important in the major lakes of the Fraser and Babine river systems in British Columbia (Foerster 1968; McDonald and Hume 1984) but is of primary importance in Cultus Lake (Ricker 1966), in Great Central Lake on Vancouver Island (Barraclough and Robinson 1972), and in lakes of the upper Skeena River drainage (Brett 1952).

Rivers between lakes are utilized heavily for spawning in the Bristol Bay area because they provide more stable flow than headwater drainage streams unbuffered by lake reservoirs. Small spring-fed creeks and spring pond areas also tend to be heavily utilized, again apparently because of stable flow and temperature conditions. In the larger abraded rivers, spawning tends to occur in the finer gravel of side channels. Spawning in glacial-fed rivers is generally avoided, probably because survival of eggs and alevins would be low in the silt-clogged gravel interstices of the stream bed.

For two major spawning systems in British Columbia, the Fraser and Skeena rivers, most spawning takes place in tributary streams, trunk streams between lakes, and at lake outlets. For the progeny of sockeye that spawn in outlet rivers, various behaviour patterns ensue. The main spawning in Chilko Lake occurs in the outlet river, and after emergence the fry migrate upstream along the river banks to enter the lake rearing area. In the Adams River, below Adams Lake, the fry move downstream into Shuswap Lake to rear.

Characteristics of Kamchatkan spawning areas reported by Soviet investigators have been summarized by Foerster (1968). Although shore spawning is extensive in many of the lakes, important spawning populations occur in river areas where there is no lake access. The adaptation to non-lake rearing environments is probably most extensive in the Kamchatka River where the juveniles overwinter in non-freezing springs, spring ponds, creeks, and side channels in instances where no lake is available. Similar situations are found in the Mulchatna River, Bristol Bay (M.L. Nelson, Alaska Department of Fish and Game, Anchorage, Alaska, pers. comm.) and the transboundary Stikine River of southeastern Alaska and British Columbia (McCart et al. 1980; Wood et al. 1987).

Examples of a more extreme adaptation to the lack of a lake rearing area are found in the lower Harrison River, British Columbia, and the East Alsek River along the Yakutat coast of Alaska. In both instances the juveniles begin their migration to sea soon after emergence. In the Harrison River, the juveniles are prevented by rapids from migrating upstream into Harrison Lake (Gilbert 1920, 1922, 1923; Schaefer 1951). It has now been determined that age 0. migrants tarry to feed and grow in tidewater sloughs of the Fraser River estuary (Dunford 1975; Levy and Northcote 1982; Macdonald 1984; Birtwell et al. 1987). In the East Alsek River, which has runs as high as 180,000 sockeye (McBride and Brogle 1983), spawning takes place in a short, primarily spring-fed river with limited rearing habitat. Predominant ages are 0.3 and 0.2, whereas in the main Alsek River, predominant ages of the lake rearing populations are 1.3 and 1.2. The East Alsek River originates on the moraine deposit area of Malaspina Glacier, as do three other small rivers - the Akive, Italio, and Lost rivers -

where smaller sockeye populations have developed that also migrate seaward in their first summer.

River spawning of sockeye in the Stikine River and its principal tributary, the Iskut River, is remarkable because of the highly turbid character of the rivers due to glaciation. The majority of the progeny remain over winter in the river, migrating at age 1., but some migrate at age 0. (Wood et al. 1987). These latter so-called "sea type" sockeye apparently feed in the stream for several months before migration.

Timing of Spawning

Sockeye salmon generally spawn in late summer and autumn. Within this period, time of spawning for different stocks can vary greatly, apparently because of adaptations to the most favourable survival conditions for spawning, egg and alevin incubation, emergence, and subsequent juvenile feeding. In Bristol Bay, the duration of mass migration of mature adults into the lake systems lasts only four weeks in late June to July (Royce 1965; Mundy 1979; Burgner 1980). Once in the lake system, the adult salmon segregate to their individual spawning areas. Although timing of spawning varies little from year to year within a specific spawning area, there are great differences in timing among spawning areas (Marriott 1964; Demory et al. 1964; Gilbert 1968). Within the Wood River lake chain, for instance, spawning begins in late July in the smaller streams of Lake Aleknagik, in early to mid-August in the tributaries of Nerka and Beverley lakes, in late August in the trunk rivers between the lakes, and in late August to mid-September in most lake beach areas. The timing of spawning varies among the beaches, and some spawning continues into October (Gilbert 1968). In Iliamna Lake, some spawning continues into December (Demory et al. 1964), and sockeye have been observed under the ice after freeze-up begins in late fall. The timing of spawning appears to be dependent to some degree on the temperature regimen in the gravel where the eggs are incubated. This varies distinctly among spawning area types.

The most prolonged period of spawning for a single sockeye population occurs in the Ozernaya River basin (Egorova 1970a, 1970b). Although in most lakes there are distinct sub-populations, or stocks, identified by times of spawning and location and type of spawning grounds, there is an absence of such seasonal groupings in Lake Kuril and its tributaries. The main population is considered to be a single homogeneous stock with an affinity for lake dwelling. Spawning continues from the end of June or July until early February, with the main spawning occurring from September to November. Perhaps the absence of distinct stocks is due to the uniformity of incubation temperature regimes among the spawning areas.

In the Fraser River, the differences among stocks in timing of spawning is related directly to the temperature regime of the spawning site (Figure 8) (Brannon 1987). Spawning is later in the warmer incubation environments, with a range of four months in time of peak spawning between the earliest and latest spawning stocks. Since rate of development of the embryos is faster at the higher incubation temperatures, this flexibility among stocks in time of spawning is an adaptation that helps to synchronize emergence timing of the progeny the following spring (Miller and Brannon 1982; Brannon 1984, 1987).

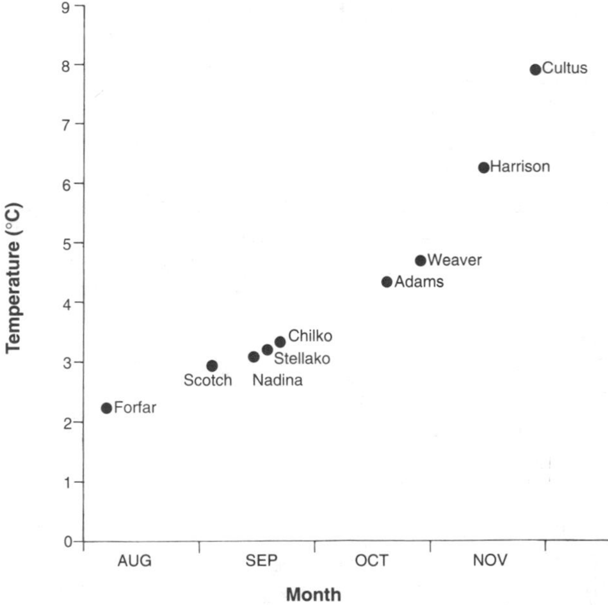

FIGURE 8

Spawning times and mean incubation temperatures of nine Fraser River sockeye stocks. (From Brannon 1987, Figure 1)

Spawning Substrate

The composition of spawning substrate utilized by sockeye salmon varies widely. In Knutson Bay (Iliamna Lake, Alaska), shore spawning occurs to a depth of nearly 30 m in coarse granitic sand in an area of strong upwelling groundwater (Olsen 1968; R.L. Burgner, unpublished data). In contrast, in the same lake around Triangle Island, mass spawning takes place over large angular rubble, too large to be moved by salmon in the normal digging process. The spawned eggs settle in the crevices between the rocks (Kerns and Donaldson 1968; Olsen 1968). Foerster (1968) reported somewhat similar spawning over coarse bottom too large to dislodge in the Kispiox River (Skeena River drainage, British Columbia). Generally, however, spawning along lake beaches and in streams takes place in gravel small enough to be readily dislodged by digging, and the digging process tends to remove the silt and clean the gravel where the eggs are deposited. This gravel-cleaning process is more efficient in streams where the current carries the dislodged fine materials downstream than in still waters of lake shores and spring areas. In spring ponds, where algal growth is often very thick, the digging activities are effective in removing the algae.

Water depth does not seem to be a critical factor to the female sockeye salmon in selecting a spawning site. In the small streams and spring ponds of Alaska, it is common to observe pairs of salmon in the spawning process with their dorsal surfaces protruding from the water. Sometimes the water is so shallow that the hump-backed males, in particular, have difficulty swimming about without stranding. Even in the larger rivers, spawning depths are generally not great because riffle areas are preferred. However, spawning on lake beaches can extend to considerable depths, as was indicated for Knutson Bay. More commonly, spawning along lake beaches and fluvial fans in the Bristol Bay area is at depths of less than 3–4 m. Similar depths are reported for spawning in Lake Kuril (Krokhin and Krogius 1937).

It is clear that sockeye can detect upwelling groundwater areas along lake beaches and in spring pond areas in which to spawn (Krogius and Krokhin, 1948, 1956b). This was also demonstrated in studies of spawning and winter survival of eggs and alevins in lake beaches at Lake Nerka and Little Togiak Lake in the Wood River system of Bristol Bay (R.L. Burgner, unpublished data). Beaches without upwelling groundwater were not utilized, and in those beaches where upwelling occurred, spawning was heaviest in the areas of strongest upwelling. Similar conclusions were drawn from observations of spawning distribution in the extensive spring pond area at the head of Pick Creek, Lake Nerka (Mathisen 1962).

Spawning and Egg Deposition

The spawning behaviour exhibited by sockeye salmon is characteristic of the genus. The female selects a redd site, digs a depression ("nest") in the gravel substrate and deposits a batch of eggs, which are simultaneously fertilized by an accompanying male or males. She then covers the eggs by further digging and repeats the digging and spawning process one to several times. Finally, she covers the completed redd containing several nests of eggs and then guards the redd site until near death. By this process the eggs are assured almost 100% fertilization and are provided an incubation environment for development until hatching and emergence. Details of the spawning behaviour of sockeye salmon have been studied by a number of scientists, notably Kuznetsov (1928), Mathisen (1962), Hanson and Smith (1967), and McCart (1969, 1970) for anadromous sockeye, and Schultz (1935) for kokanee.

To dig the nest the female turns on her side and rapidly flexes her body and tail in a series of violent rhythmic movements interspersed by resting periods and testing of the nest with her anal and pectoral fins. In the digging process, sand, debris, and small gravel are lifted from the nest by the suction and current created as the female alternately presses her tail against the bottom and lifts it off during her flexing movements. In running water, the currents aid in the displacement of the finer substrate materials.

The excavated nest is oblong with its longest dimension parallel to the stream current. Mean dimensions given for sockeye nests in a Babine Lake stream were 102 x 86 cm (McCart 1969) and in a spring pond in Bristol Bay they were 76 x 51 cm (Mathisen 1962). During nest preparation, activities of the female attract and lead to acceptance of a male; in addition, intruding females, fish predators, and undesirable males are driven away.

Males tend to arrive on the spawning grounds earlier, remain alive longer, and are capable of spawning for a longer period than the length of time required for a female to complete egg deposition (Mathisen 1962). The dominant male accepted by the female fights off other males by direct or ritualistic fighting. One or more males may take a "satellite" position (Mathisen 1962) behind or to the side of the primary pair and attempt to participate in the spawning act. The dominant male responds to the female's digging and nest-testing activities by courting her. Male courting involves repeated acts of gentle touching of the female's side with the snout, positioning of the body against the female, and quivering vigorously for a second or two while in that position. This in turn stimulates the female to continue preparing the nest and finally to release her eggs (Mathisen 1962).

At spawning, the pair lie side by side with vents positioned close together near the bottom of the nest. During the spawning act, jaws are agape and backs are arched slightly, the body is tilted outward, dorsal fins undulate, and tails vibrate rapidly. Eggs and sperm are released simultaneously. The spawning act is commonly preceded by a few trial spawnings without release of eggs or sperm (Mathisen 1962; McCart 1969).

After the spawning act, normally lasting 10–12 seconds (Mathisen 1962), the female makes rapid digging movements on the upstream edge of the nest which bury the eggs quickly. The eggs, upon extrusion into the bottom of the nest, are temporarily adhesive, which keeps them attached to each other and the gravel during burial (Sheridan 1960). McCart (1969) noted that digging to cover the eggs differs from nest digging in the following ways: (1) greater frequency of digging, (2) fewer body flexures per dig, (3) increased body orientation at an angle to the current, and (4) less tendency to alternate sides while digging.

Mathisen (1962) found that the number of eggs deposited in each nest, or egg pocket, to be usually between 500 and 1,100, the depth of the egg pockets to be mainly 15–23 cm below the completed redd surface, and the number of nests per female to be between three and seven, with four and five most common. After covering a nest, the female goes through a resting period of generally several hours over the redd location before excavating the next nest. The male may remain with the female until completion of egg deposition or may move to another female after the spawning act. The successive pockets of eggs deposited by a female may thus be fertilized by the same male, by different dominant males, or partly by satellite males.

Once egg deposition is completed, the female remains over the redd until near death, driving away females and often males (Hanson and Smith 1967). Before dying, the males sometimes gather in quiet stream areas or pools downstream (Hanson and Smith 1967; R.L. Burgner, unpublished data).

Some variations in the spawning behaviour of sockeye salmon have been observed in lake-beach spawning grounds. Most beach areas are characterized by upwelling groundwater and absence of current, and the nests are typically round rather than oval in shape. The females tend to orient in various directions, and excavated gravel is distributed around the periphery of the redd (McCart 1969). Satellite male positions are less defined. The most extreme deviation in spawning pattern, however, has been observed in Iliamna Lake on island beaches (Kerns and Donaldson 1968). Because upwelling groundwater is lacking (Olsen 1968), spawning occurs in rather exposed rocky beaches where intragravel circulation is apparently provided by wind-generated lake currents and seiches. Substrate varies from coarse gravel or small, broken rock to large, broken rock or round boulders. In 1965, with heavy spawning, it was observed that the salmon generally spawned in dense groups of 10 to 50 fish over the immovable rock substrate and with little display of territorial defense by either sex. In a group of milling fish, one or more of the females would be seen with their vents down in rock interstices, emitting eggs, while the males in close proximity were observed releasing milt. There was little evidence of attempted digging by the females or of redd defense following spawning (Kerns and Donaldson 1968).

Size is a factor in mate selection by females, and in the ability of the male to assert dominance over rival males. Hanson and Smith (1967) found that for age 1.2 and 1.3 sockeye in a Babine Lake stream, fish of all lengths tended to mate with similar sized fish, that smaller males were less successful in holding mates against larger intruders, and that small males were more often alone than were large males and were more frequently in a satellite position. Small females were found to mate with large males but spent more time alone than did large females. Jacks (age 1.1) and the still smaller kokanee occupied satellite positions in attempting to spawn with the larger fish. Mathisen (1962) also found that the size and vigour of the male decided the outcome. The genetic implications of such selection depend on the degree to which size and age of parents are reflected in size and age of surviving offspring within a spawning colony.

The average length of time that female sockeye spend on the spawning grounds was 15 days in two Kamchatkan areas (Kuznetsov 1928; Krogius and Krokhin 1948). In Hanson Creek, a small shallow tributary of Lake Aleknagik, Alaska, the sockeye matured off the creek mouth, and most females began spawning within a day of entering the stream. The average life span was determined to be 5.6 days for females and 6.6 days for males once they entered the stream (Mathisen 1962). Heavy predation by glaucous-winged gulls (*Larus glaucescens*) and brown bears (*Ursus arctos*) on live fish shortened the average life span of Hanson Creek sockeye in the stream. In the short lateral spawning streams of Kodiak Island where brown bear predation is heavy, stream life of the sockeye is shortened by behaviour adapted to avoid the bears (Gard 1971). Salmon remain in the lake, off the creek mouths, until they are ready to spawn. The ripe fish then make one or more daytime journeys into the stream and return to the lake in late afternoon where they spend the night, which is the period when the bears are most active. The females are spawned out within two days after they establish their redds (Clark 1959). The average length of stream life was not reported. In Forfar Creek (Stuart Lake, a Fraser River tributary), the life span from creek entry to death was 12–19 days (Killick 1955). In the Kvichak River system, Hartman (1959) reported that female spawners remained on or over their redds for 7.75 days. Tsunoda (1967) found for Hidden Creek spawners (Lake Brooks, Alaska) that males lived for 9–12 days and females 7–14 days.

The success of spawning, in terms of egg deposition in the gravel, depends upon the degree of egg retention, the efficiency of fertilization, and the success of egg burial by the female.

Some egg retention in the female ovary occurs during spawning, but with healthy females it is generally less than 5% (see Foerster 1968, for a summary of several investigations). Thus, under normal circumstances, egg loss due to retention is minor. However, disease, parasitism, injury, delay in migration due to stream blockage, high stream temperatures, high spawning density, or untimely death such as by bear or gull predation can significantly increase average egg retention.

Heavy pre-spawning mortalities from disease or parasitism have occurred in numerous instances in certain Fraser River spawning areas. Pre-spawning mortalities reached 90% in the Chilko River population in 1963 and 62% in the Horsefly River run in 1961 (Williams 1973). Timing of the runs and temperatures in the Fraser River during the upriver migration appear to be factors which are related to pre-spawning loss (IPSFC 1976). Infectious hematopoietic necrosis (IHN), the bacterial disease caused by *Flexibacter columnaris*, the protozoan parasite *Ichthyophthirius multifiliis*, and a myxosporean have been variously associated with occurrence of heavy pre-spawning mortalities.

A classic example of pre-spawning mortalities caused by delays in migration due to stream blockage is provided by the effects of the obstruction in the Fraser River canyon at Hell's Gate from 1911 to 1913. This obstruction resulted from a rock slide during railroad construction (Thompson 1945). In this disastrous situation, later alleviated by construction of the Hell's Gate fishways, sockeye salmon were blocked or delayed in migration at certain flows and vast numbers died without reaching their spawning destinations.

Regarding effect of spawning density, West and Mason (1987) found that for sockeye spawning in the Fulton River spawning channels (Babine Lake), partial egg retention and spawning failure were positively and significantly correlated with each other and with density of females ($p<.001$). However, both values combined averaged less than 3% of total eggs available for deposition.

Concerning salmon losses to brown bears, Shuman (1950) estimated a minimum kill of 94,000 unspawned salmon on Karluk Lake spawning areas in 1947. This amounted to 19.4% of the Karluk River spawning run in that year. Since brown bear predation is generally much heavier on male sockeye than on females (Gard 1971; R.L. Burgner, unpublished data), the effect on egg deposition by the females was presumably much lower. Merrell (1964) concluded that "even under conditions of intense bear activity on a small stream where salmon are abundant and vulnerable, bears catch mostly spawned out fish and have little effect on ultimate production." Gard (1971) corroborated this conclusion.

Egorova (1970b) reported that in the Ozernaya River, sockeye spawning is restricted basically to nighttime, which she believed is an adaptation to reduce charr predation on eggs during deposition.

Efficiency of ova fertilization during spawning is very high among sockeye, as with other salmonids, approaching 100% (Mathisen 1962; Egorova 1970b). The millions of sperm contained in the milt ejaculated by the male salmon during the spawning act apparently provide an excess to fertilize the pockets of eggs deposited by the female. Only one spermatozoon is required to fertilize each egg. After the spermatozoon enters the micropyle canal in the egg capsule, other spermatozoa are prevented from entering (Velsen 1980). Fertilization must occur quickly after the female releases the eggs, before they swell and "harden" upon immersion in water.

Losses occurring during egg deposition or following egg deposition due to dislodgement are more difficult to assess. Crowding on the spawning grounds and "wave" spawning can result in repeated excavations of the same gravel by subsequent females (superimposition) and can cause heavy egg loss (Krogius and Krokhin 1948). Selifonov (1987) found a high correlation between spawner abundance and percentage egg mortality of Lake Kuril sockeye over a 40-year period. Where spawning occurs over an extended period of time, e.g., weeks, successive waves of spawners can displace deposited pockets of eggs. The displaced eggs may die due to physical shock, exposure, displacement into less favourable incubation conditions, or predation. As with eggs of the other salmon species, sockeye eggs are vulnerable to mechanical shock until the "eyed" stage is reached.

Superimposition is minimized by the territorial defense of the redd by the female following egg deposition, which protects the redd for a few days. Female territory is partly a function of spawner density (Mathisen 1962). D.E. Rogers (Fisheries Research Institute, University of Washington, Seattle, Washington, pers. comm.) and I have both observed that territorial defense by spawning pairs decreases under situations of high spawner density. This also appeared true of the island beach spawners observed by Kerns and Donaldson (1968). In Bristol Bay stream spawning areas, estimates of the capacity of streams to support spawning sockeye were based on a density of one female/2 m^2 (Burgner et al. 1969). In flow-controlled Fulton River and Pinkut Creek spawning channels (Babine Lake), maximum fry production per unit area was achieved at a spawner density of about one female/m^2 (West and Mason 1987).

Fecundity and Egg Size

Fecundity

Pacific salmon are distinctive because of the comparatively small number and large size of eggs relative to fish size and the fact that they spawn only once before death. Fecundity varies among species, among stocks of a given species, and also with size and ocean age of the female in an individual stock. The determination of fecundity of females in a population is, therefore, a basic measurement to determine the survival rates of any given life stage. For sockeye, average fecundities range from about 2,000 to 2,400 eggs per female (Lake Blizhnee, Kamchatka Peninsula) to 5,000 (Karymaiskiy Spring, Kamchatka Peninsula), varying with average age and size of the fish (Foerster 1968). Fecundity in kokanee is much lower and may range from about 300 to less than 2,000 eggs depending on fish size (Seeley and McCammon 1966).

Semko (1954a) was the first to suggest that fecundity is higher for those species of salmon that, as juveniles, spend a longer time in fresh water. Rounsefell (1957) remarked that fluvial anadromous *Oncorhynchus* show a lower number of eggs for their weight than lacustrine anadromous sockeye. He suggested that, since size of eggs can be obtained only at the sacrifice of number, "in each ecological situation there is some point at which, on the average, the forces favoring size are exactly balanced by those favoring number."

It is not known at what point the final fecundity of a female is set. At the time of seaward migration the female smolt contains many more ova than finally develop in the mature fish. The ultimate size, and probably the ocean age, in many populations, will undoubtedly be controlled by growing conditions encountered in the ocean. Since gonadal development accelerates in the final month or so in the ocean, fecundity may well be set at a very late stage in life. Data on Babine Lake sockeye suggest that fecundity is more closely associated with the terminal year of ocean life than with the penultimate year (West and Mason 1987).

The primary criteria determining fecundity within populations are size of females and the number of years spent in the ocean. Females of a given ocean age show no consistent or significant difference in fecundity in relation to duration of juvenile freshwater life (Rounsefell 1957).

In those instances in which sockeye migration into a lake extends over a considerable period of time, later migrants are larger and have a higher fecundity than earlier migrating fish of the same ocean age. A striking example of this is found in the Ozernaya River stock which migrates to Lake Kuril on the Kamchatka Peninsula. Egorova (1970a) stated that the Ozernaya stock is generally homogeneous without distinct seasonal races. The fish begin entering the river in late May and continue into October, with the major portion of the run occurring in midsummer. Samples to determine length, weight, and fecundity for early, middle, and late season fish were collected over a six-year period (Egorova, in Hanamura 1966). Of the two major age groups, 2.2 and 2.3, the females showed consistent increases in size and fecundity within an age group as the season progressed (Table 4). The age 2.2 fish were consistently smaller and had fewer eggs than the 2.3 fish. However, the .2 age females consistently had a higher fecundity per kg than the .3 age fish. Data from Babine Lake, British Columbia, indicated a similar seasonal trend in size and fecundity within ocean-age groups (H.D. Smith, Canadian Department of Fisheries and Oceans, Vancouver, British Columbia, pers. comm.). In both these instances, the size-fecundity relationship may not change within season for a given ocean-age group, i.e., the difference in fecundity with time within season appears to be largely a function of increased fish size.

TABLE 4

Seasonal changes in size and fecundity of Ozernaya River sockeye salmon

	Av. fork length (cm)		Av. weight (kg)		Av. fecundity (no. of eggs)		Eggs/kg	
Migration season	Age 2.2	2.3	Age 2.2	2.3	Age 2.2	2.3	Age 2.2	2.3
Early	51.7	56.6	1.71	2.20	2807	3115	1642	1416
Middle	52.2	57.9	1.82	2.49	3247	3691	1784	1482
Late	53.3	59.0	2.00	2.61	3314	3983	1657	1526

Source: Data of T.V. Egorova for years 1950–55, weighted by annual sample size. From Hanamura (1966), Table 21

The sockeye run to the Okanagan River, a Columbia River tributary, provides an unusual case in that age .1 females comprise a substantial portion of the spawning escapement in some years, the remainder being age .2. The average fecundity of age .2 females in 1957 was about 50% more than that of age .1 females (2,887 versus 1,928) (Major and Craddock 1962). Since age .2 sockeye in the Columbia River weigh about 70% more than age .1 sockeye (1.7 versus 1.0 kg, sexes combined) (Fulton 1970), fecundity per kg is apparently much higher in the age .1 females.

Among populations there are differences in fecundity due to size and age of the population members as well as differences in the number of eggs for a given size of fish (Figure 9 and Table 5). The average number of eggs per cm of fish length ranges from 49.6 for age .3 females in samples from Karluk Lake to 83.1 eggs for age .3 females from Pedro Bay (Iliamna Lake). Fecundity per unit length of a female sockeye tends to be highest in the Bristol Bay district of Alaska, particularly in the Iliamna (Kvichak), Togiak, and Tikchik lake areas, followed by the Wood River and Naknek River systems. Fecundity per unit length is intermediate in the age .3 females from the Chignik Lake area (Alaska Peninsula). Unit fecundity in the age .2 fish from the Ozernaya River (eastern Kamchatka) overlaps with that of the Naknek River system but is lower for the predominant age .3 sockeye. Female sockeye from Bear Lake (Kodiak Island) and Babine Lake (British Columbia) have similar unit fecundities, followed by Karluk Lake, and finally by age .2 fish from Chilko Lake (Fraser River), which have the lowest unit fecundity.

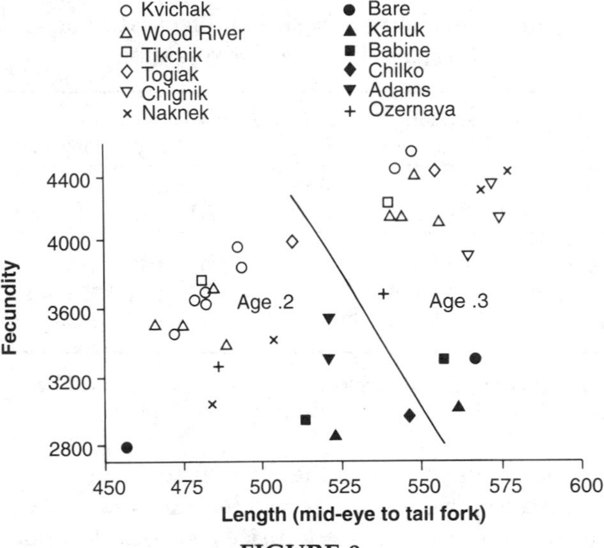

FIGURE 9

Length-fecundity comparison among sockeye populations for ocean age .2 and .3 females (mean values)

Although these data suggest considerable differences in fecundity among lake system populations, data are needed on between-year variability in size-fecundity relationships within populations so that between-population differences can be properly tested. Such information is provided by length-fecundity regressions for Fulton River and Pinkut Creek sockeye stocks (Babine Lake) (Figure 10). Combining the age 1.2 and 1.3 data for several years, West and Mason (1987) found that (1) significant annual variation in fecundity occurred within the Fulton and Pinkut stocks; (2) variation in length accounted for an average of 48% of variation in fecundity; and (3) annual variation in fecundity amounted to ±230 eggs, or 7.4% of the average

TABLE 5

Data on fecundity per unit length for ocean age .2 and .3 female sockeye salmon, ranked from highest to lowest (lengths converted to mid-eye to tail fork)

Lake systems	Spawning locality (M = mixed)	Sample size	Year	Fecundity per unit length (eggs/cm)	Source
		Ocean age .2			
Iliamna	Iliamna R.	40	1965	80.6	1
Togiak	M	15	1965	78.5	1
Nuyakuk	M	12	1965	78.2	1
Wood R.	M	18	1965	77.1	1
Iliamna	Finn Bay	63	1964	76.9	1
Iliamna	Knutson Bay	42	1965	76.6	1
Iliamna	M	44	1965	76.3	1
Iliamna	Cabin Bay	33	1964	76.2	1
Wood R.	Agulukpak R.	6	1966	74.9	1
Iliamna	Moose Bay	42	1965	74.7	1
Wood R.	Pick Cr.	38	1966	73.5	1
Iliamna	Copper R.	38	1965	73.1	1
Wood R.	M	64	1964	69.6	1
Shuswap	Adams R.	39	1951	68.0	2
Ozernaya	(Mid-season)	65	1950-55	66.9	3
Shuswap	Adams R.	79	1950	63.6	2
Naknek	M	45	1964	63.1	4
Bare	M	17	1952-55	60.8	5
Naknek	M	38	1963	57.9	4
Babine	M	97	1965	57.1	6
Karluk	M	70	1938-39	54.6	7
Chilko	Chilko R.	468	1956-65	54.2	8
		Ocean age .3			
Iliamna	Pedro Bay	61	1964	83.1	1
Iliamna	Moose Bay	14	1964	82.1	1
Wood R.	Agulukpak R.	22	1966	80.2	1
Togiak	M	33	1965	79.9	1
Nuyakuk	M	23	1965	78.2	1
Wood R.	M	38	1965	77.2	1
Naknek	M	52	1963	77.0	4
Wood R.	Pick Cr.	12	1966	76.4	1
Chignik	M	29	1965	76.1	1
Naknek	M	34	1964	75.9	4
Wood R.	M	29	1964	74.0	1
Chignik	Chignik L.	17	1961	72.3	1
Chignik	M	29	1961	69.1	1
Ozernaya	(Mid-season)	196	1950-55	68.6	3
Babine	M	80	1965	59.6	6
Bare	M	27	1952-55	58.9	5
Karluk	M	80	1938-39	49.6	7

Sources:
1 Fisheries Research Institute, unpublished
2 Ward and Larkin (1964)
3 Hanamura (1966), Table 21
4 C. DiCostanzo, Auke Bay Laboratory, Auke Bay, AK
5 Nelson (1959)
6 H.D. Smith, Pacific Biological Station, Nanaimo, BC
7 Rounsefell (1957:457)
8 J. Roos, Internat. Pacific Salmon Fish. Comm., New Westminster, BC

fecundity. Although significant annual variations within stocks were found, they were minor compared to differences in fecundity among stocks from different river systems reported on by others (West and Mason 1987).

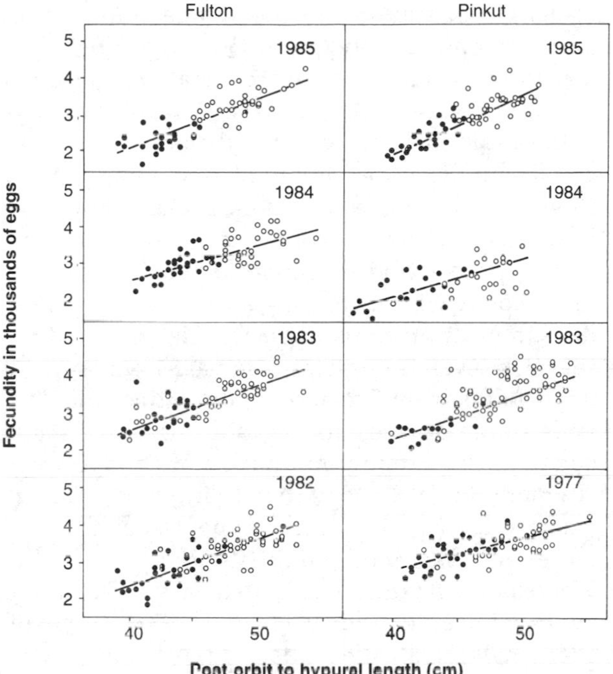

FIGURE 10

Fecundity on length regressions for Fulton River and Pinkut Creek sockeye stocks; • = age 1.2 females, o = age 1.3 females. (From West and Mason 1987, Figure 3)

In fishing districts such as Bristol Bay, where the primary gear consists of gillnets, the fishery is selective for size and hence affects the relative fecundity of fish taken in the catch and in the escapement to the spawning grounds. The Bristol Bay fishery is particularly selective towards the larger of the age .2 females (Koo 1962b; Mathisen et al. 1963). Therefore, the average fecundity of age .2 females taken in the catch is expected to be greater than the average fecundity of those that escape to the spawning grounds.

Egg Size

Sockeye salmon eggs in redds can usually be distinguished from those of other salmon by their darker orange-red colour and smaller size (Hanamura 1966). Pink and chum salmon eggs are lighter orange, and both chum and chinook eggs are much larger. Coloration of coho eggs is similar but they are less transparent than sockeye eggs because of a "milky opaqueness" covering a much greater part of the egg membrane. Egorova (1970b) commented that sockeye eggs are more brightly coloured by carotenoid pigment than the eggs of the other Far Eastern salmon but are close to those of masu (*O. masou*) in brightness of colour and quantity of pigments.

Among the Pacific salmon, female sockeye have the highest fecundity and the smallest egg size per given size of fish. Within populations, the mean size of eggs has been found to increase with length of female. In studies of sockeye at Pick Creek (Lake Nerka, Alaska), Mathisen (1962) found that for females returning to spawn after two or three winters in the ocean, both fecundity and egg size were functions of fish length, and fecundity was inversely related to the size of the eggs. For Skeena River sockeye, Bilton (1971) determined that there was no difference between mean weights of water-hardened fertilized and unfertilized eggs. He also found that mean weight of eggs increased with length of female for age 1.2 and 1.3 fish, combined, at Scully Creek (Lakelse Lake), and for age 1.2 and 1.3 fish, separately, in the Lower Babine River. Of the females sampled at both Scully Creek and the Lower Babine River, the age 1.2 females averaged smaller in length and in individual egg weight than the age 1.3 females. Bilton (1971) also found that mean length and weight of young sockeye three months after hatching were both positively and significantly correlated with egg weight and length of the female parent.

In three Fraser River populations studied by Meade and Woodall (1968), the proportional differences noted in egg weight were reflected in final body weight upon emergence. The mean dry weight of eggs of Cultus Lake sockeye was only 57% that of Upper Pitt River sockeye, and the mean dry weight of the newly emerged fry from the respective spawning channels was only 61%. Brannon (1972) stated that among six Fraser River sockeye populations, the large variations in size at emergence were population characteristics determined by egg size.

Brannon (1987) presented mean egg weight data for age 1.2 sockeye in nine Fraser River stocks spawning in 1969 (Table 6). He concluded that the time required for yolk absorption among fry from eggs of the different populations is correlated with egg size (r = .88). However, there were notable inconsistencies which he attributed to population-specific relationships between egg size and incubation time. Within a population year class, he found that egg weight could vary in excess of 30% among females, but that the length of the incubation period to yolk absorption was very similar among female egg lots and not significantly correlated with egg weight. Thus, he concluded that egg size differences among sockeye stocks relate primarily to reproductive strategies other than development rate.

Differences in size of sockeye eggs were used by Robertson (1922) to demonstrate racial differences among three nearby spawning populations in Fraser River tributaries, which he presented as further proof of the home stream theory. For seven years, measurements of number of eggs per metre trough were taken during spawn-taking at the three localities. Counts at each locality were similar from year to year and there was no overlap in counts among localities (Table 6). At Morris Creek, tributary to Harrison River, the run of large, early-spawning sockeye had medium sized eggs; the very small, late-spawning sockeye of Harrison Rapids had very large eggs; whereas the small, late-spawning fish at nearby Cultus Lake had exceptionally small eggs (Robertson 1922).

Very large differences in egg size exist between anadromous sockeye and dwarf, lake-resident sockeye. For Lake Dalnee populations on the Kamchatka Peninsula, Smirnov (1959a) reported that loose eggs of anadromous sockeye weighed 84–130 mg (average 103 mg), whereas eggs of the dwarf sockeye weighed 51–69 mg (average 58.5 mg), only about half as much (Table 6). Embryos of the dwarf sockeye were smaller at corresponding developmental stages, and hatching occurred at an earlier morphological stage and over a shorter period of time.

Sizes of ripe, water-hardened eggs are reported in the literature in various ways, e.g., in number per metre trough and per 6-inch (about 15 cm) trough, in diameter, in weight, or in dry weight. Some egg size measurements are listed in Table 6.

Female Size, Fecundity, and Egg Size

It is of interest to consider the adaptive significance of the relationships among age and size at maturity, fecundity, and egg size in female sockeye (Healey 1987). As noted earlier, age and size at maturity is highly variable, both within and between populations. Size is primarily determined by length of stay in the ocean; age .3 fish are on average considerably larger than age .2 fish. This larger size of age .3 fish is presumably achieved at the risk of higher average ocean mortality because of the additional year spent in the ocean, although, as will be discussed later, little is known about this. Healey (1987) noted that sockeye appear to have allocated more surplus energy to growth than to egg production, because fecundity increases more slowly with increasing size in sockeye than it does in most other fish species. He observed further that, even though large female size may be an important survival strategy, populations display a wide range of sizes and ages at maturity, which suggests that some combination of sizes and ages is a more stable strategy than a single size and age.

Large female size tends to be associated with large egg size and slightly higher fecundity, whereas small females tend to have small eggs and slightly lower fecundity. Healey (1987) suggested that small females, because of their smaller eggs (requiring less oxygen), may be able to exploit poorer quality gravels if unable to compete with large females for good quality gravel. Large females may exploit the best quality gravels and spawn in faster flowing water, but their eggs may be at a disadvantage during years of unusually high or low flows because of scouring or siltation of the redds. Thus, a mix of sizes (and ages) of fish may be the best survival strategy over time. Among spawning areas in the Wood River system, Alaska, Rogers (1987) observed that small streams tend to have lower percentages of age .3 spawners than medium-sized streams, and that the largest streams had the highest percentage of age .3 spawners. Large fish are apparently at a disadvantage with respect to predation by gulls and bears in the small streams, but are better able to cope with the faster flow and larger substrate of the large streams. Thus, environment has had an influence on the genetic adaptations of individual populations.

TABLE 6
Measurements of sockeye ova

Locality	Year	Sample size	Female age	Mean fork length (cm)	Egg diam. (mm)	Egg wt. (g)	Dry wt. (mg)	Dry wt. S.D.	Source
Fraser R. system									
Upper Pitt R.	–			(64–69)			64.68		1
	1969	20	1.2	58.9	6.11		61.37	6.48	2
Cultus L	–			(56–60)			36.38		1
	1969	20	1.2	55.9	5.30		33.07	2.85	2
	1914–20			56.9	av. 5.45				3
	1932				5.77				4
	1933				5.29				4
Harrison R. Rapids	1969	20	0.2	54.1	5.98		65.08	7.40	2
	1914–20			56.6	6.60				3
Weaver Cr.	–			(58–65)			57.13		1
	1969	20	1.2	60.3	5.83		46.73	3.90	2
Morris Cr.	1914–20			58.7	av. 6.09				3
Chilko R.	1969	20	1.2	56.0	5.73		40.43	3.86	2
Horsefly R.	1969	20	1.2	58.4	5.27		35.06	2.27	2
Raft R.	1969	20	1.2	59.8	5.62		41.67	3.55	2
Stellako R.	1969	20	1.2	58.3	5.34		33.59	2.94	2
Nadina R.	1969	20	1.2	55.1	5.53		33.55	3.29	2
Skeena R. system									
Scully Cr.	1964	10	1.2	49.6		0.1235			5
	1964	9	1.3	59.1		0.1508			5
Lower Babine R.	1965–66	90	1.2	54.0		0.1448			5
	1965–66	42	1.3	62.0		0.1643			5
Bristol Bay									
L. Iliamna									
mainland beach	1965		.2		5.3				6
Triangle Is.	1965		.2		5.6				6
L. Nerka, Pick Cr.	1951–52	42		57.4	5.7				7
	1951–52			50.5	5.3				7
	1951–52			64.0	6.0				7
Kamchatka									
Karymaiskiy Spr.,			1.2			0.129			8
Bolshaya R.			1.3			0.122			8
			1.3			0.120			8
L. Dalnee									
anadromous						0.103			9
residual						0.058			9
Ozernaya R.					5.6–5.8				10

Sources:

1 Mead and Woodall (1968). Fork lengths are ranges in sample. Sample sizes and age not given.
2 Brannon (1987), and International Pacific Salmon Fisheries Commission (unpublished)
3 Robertson (1922). Means for years 1914–20. Lengths adjusted per Killick and Clemens (1963).
4 Foerster (1968). Sample size, age, length not given.
5 Bilton (1971). Lengths converted from hypural length per Bilton et al. (1967) formula. Average egg wt. values calculated from Bilton (1971), Table 2.
6 Olsen (1968). Sample size, fish length not given.
7 Mathisen (1962). Lengths converted from MEFT, egg diameter converted from trough number. Egg diameters based on length-egg size regression. Ages not given.
8 Semko (1954a).
9 Smirnov (1959a).
10 Egorova (1970b).

INCUBATION

Bams (1969) pointed out that the use of gravel beds for larval incubation is successful only because both adult and larval stages have developed many adaptations which optimize survival in the gravel. He commented further that while some mode of anchoring the eggs is required, the selected mode, by burial deep in gravel, is extreme, particularly in relation to oxygen availability. Clearly, there are selective benefits to this intergravel existence, such as protection from predation. Other potential benefits include protection from freezing, fluctuating flows and desiccation, or shifts in gravel that result in displacement of eggs or alevins.

Losses during incubation are generally influenced by the degree of crowding during spawning and by environmental conditions. Crowding may also result in egg deposition in marginally suitable spawning areas or increased egg density in more suitable spawning areas. Marginal areas may have decreased intragravel flow because of decreased permeability, or weaker intragravel flow, because of weaker upwelling in instances of spring and lake beach spawning areas. In instances of upwelling groundwater, weaker flow may not adequately protect the eggs and/or alevins from temperature extremes and freezing.

The rate of development of eggs and alevins in the redd is a function of the temperature regimen. Ievleva (1951) and Velsen (1980) described the stages of embryonic development of sockeye salmon in relation to thermal age, and Olsen (1968) and Egorova (1970b) presented the developmental stages of sockeye alevins. (Velsen and Olsen both include photographs of developmental stages, and Egorova includes drawings.) Rucker (1937) measured the effect of temperature on growth in length and weight of kokanee embryos. In general, the stages of development for sockeye salmon are similar in appearance to those of other species of Pacific salmon, but the rates of development differ. At constant temperatures (not normally found in natural spawning areas), the number of temperature units (TU) required for hatching and emergence varies with the temperature level. Velsen (1980) reported that, at a constant temperature of 10°C, the incubation period to 50% hatch among the five salmon species found in British Columbia ranged from about 47 to 65 days (470 to 650 TU's), with that for sockeye being the longest. Some selected development rate data are presented in Table 7.

A fluctuating temperature regimen is the normal condition in the natural environment, and this regimen varies greatly within and between environments. In groundwater-fed spring, pond, and lake-beach environments, temperatures may fluctuate little during development. For instance, in Pick Creek ponds (tributary to Lake Nerka, Alaska), the intragravel temperature in the actively upwelling areas remains near 4°C throughout the year, and it tends towards constancy in actively upwelling beach spawning areas (Figure 11). In contrast, the intragravel temperature in nearby Elva Creek, a small trunk stream between Elva Lake and Lake Nerka, is about 10°C at time of sockeye spawning in August-September, drops to near 0°C for a long period during winter, and then rises gradually to about 4°C at emergence time in early spring (Figure 11). Many variations of temperature regimens are found in sockeye spawning redds but, in general, spawning occurs during periods of declining temperatures in late summer or autumn, development occurs during winter at lowered temperature, and emergence ensues during rising water temperatures.

Within the temperature ranges normally encountered during incubation in natural spawning grounds, fewer TU's are required at lower temperatures. A curvilinear relationship between incubation temperature and rate of embryonic development results in an increasing extent of compensation as temperature drops (Figure 12). When temperatures drop below 6°C, compensation becomes increasingly curvilinear, which substantially tempers the severity of very low temperatures (Dong 1981; Brannon 1987). The overall effect of this adaptation is to reduce the influence of year-to-year variability in temperature

TABLE 7
Selected data on development rate of sockeye eggs and alevins

Stock	Brood year	Mean temp. (°C)	Temp. regimen	Development stage	Period of incubation (days)	Degree-days (°C)	Source
Kennedy L., Vancouver Is. (hatchery)	1928	3.6	Variable	First hatch	110	393	Foerster (1968), Table 31
	1928	4.3	Variable	First hatch	106	457	"
	1929	5.0	Variable	First hatch	116	578	"
	1929	9.7	Variable	First hatch	74	720	"
Cultus L., Fraser R. (hatchery)	1928	2.2	Variable	First hatch	156	347	"
	1929	3.9	Variable	First hatch	171	667	"
Fulton R., Babine L. (laboratory)	–	5.0	Constant	50% hatch	119	595	Velsen (1980), Table 1
	–	8.0	Constant	50% hatch	80	640	"
	–	11.0	Constant	50% hatch	57	627	"
Upper Pitt R., Fraser R. (spawning channel)	1966	4.9	Variable	Peak emergence	223	1094	Mead & Woodall (1968)
Weaver Cr., Fraser R. (spawning channel)	1966	5.8	Variable	Peak emergence	173	1006	"
Cultus L., Fraser R. (spawning channel)	1966	8.1	Constant	Peak emergence	142	1150	"
Averages of 9 Fraser R. (stocks incubated at Cultus L. research hatchery)		0.25	Constant	Yolk absorption	444	111	Brannon (1987), Table 1
		1.6	Constant	Yolk absorption	337	539	"
		2.3	Constant	Yolk absorption	290	667	"
		3.4	Constant	Yolk absorption	241	821	"
		5.6	Constant	Yolk absorption	174	974	"
		7.6	Constant	Yolk absorption	135	1024	"
		9.7	Constant	Yolk absorption	110	1065	"
		12.1	Constant	Yolk absorption	97	1175	"
		13.1	Constant	Yolk absorption	92	1199	"
		14.1	Constant	Yolk absorption	88	1250	"
Gulkana R., trib. to Copper R., Alaska (incubation boxes)		3.2	Variable	Peak emergence	225	723	Roberson & Holder (1987)

cycles, which otherwise would result in a wide range in time of emergence.

For sockeye spawning in areas bathed with upwelling groundwater of constant temperature, variation in emergence timing would be expected to occur primarily from differences in time of egg deposition. However, at Cultus Lake, where beach spawning extends for well over a month, eggs of the early-spawning females required several more days incubation time at constant temperature (8°C) than those of late-spawning fish (Brannon 1987). This difference between development rates of embryos of early and late spawners at Cultus Lake is a unique adaptation that has the effect of reducing the range in emergence timing in a situation where incubation temperature is essentially constant.

Because there is undoubtedly an optimum time for fry emergence in order to coincide with favourable feeding and survival conditions, the timing of spawning in relation to environmental temperature is important for optimum timing of emergence (Bams 1969; Godin 1982; Brannon 1984, 1987). There are also other survival factors at play.

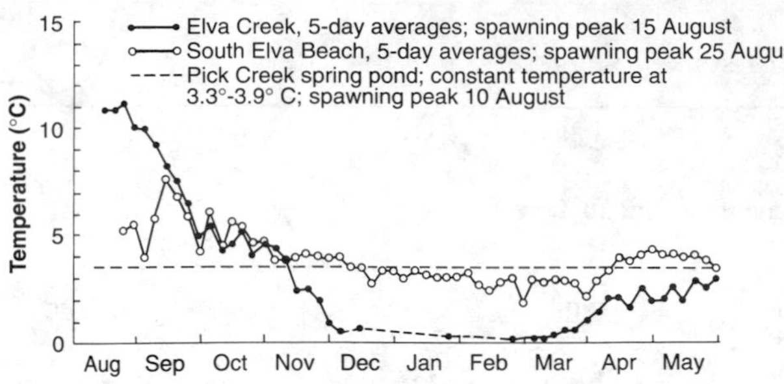

FIGURE 11

Intragravel temperatures during incubation in three sockeye spawning areas, Lake Nerka, 1956-57

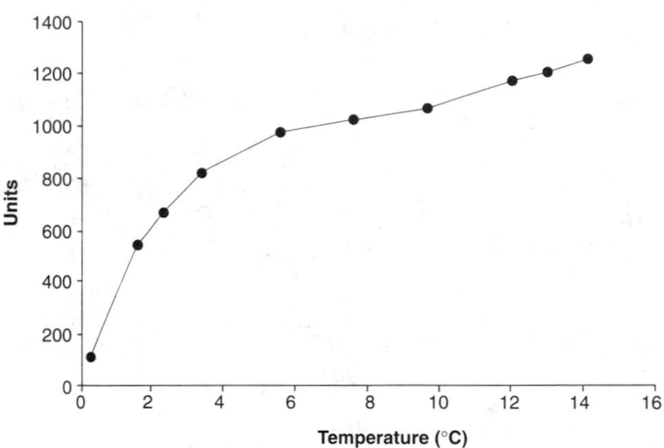

FIGURE 12

Relationship between incubation temperature (°C) and cumulative temperature units required for sockeye embryonic development to yolk absorption stage at constant temperatures. (From Brannon 1987, Figure 2)

For instance, in Iliamna Lake, island beach spawning occurs earlier in the summer than mainland beach spawning. Upwelling groundwater provides over-winter circulation in the sockeye redds along mainland beaches. However, island spawning areas are dependent on wind-generated lake currents to bathe the developing embryos, and it is important that hatching occurs before the lake freezes over in early winter and wind-driven circulation ceases (Olsen 1968). Further, certain populations spawn over an extended period of time, with distinct bimodality in spawning timing; fry emergence in spring is also distinctly bimodal and protracted. Meadow and Canyon Creek populations at Karluk Lake are examples of runs that exhibit this behaviour (Hartman et al. 1967). The significance of this pattern is not clear.

Many sockeye populations spawn in lakeshore, stream, and spring areas where flow is provided by upwelling groundwater. Such water often has a lower oxygen content and lower pH than usual stream spawning habitat (Foerster 1968, citing Krokhin and Krogius 1937; Krogius and Krokhin 1948; Semko 1954a; Smirnov 1958). Adaptations of the sockeye embryo for survival under reduced oxygen levels are believed to include small egg size, a higher carotenoid pigmentation which facilitates rapid transfer of oxygen to the embryo, and an elaborate and dense network of capillaries covering the yolk sac to assist in oxygen transfer (Smirnov 1950; Soin 1956, 1964).

Sockeye embryos have a longer incubation period than the other species of Pacific salmon and use up a higher percentage of their yolk sac material before hatching (Smirnov 1958, 1964). These developmental features may be further adaptations to the sockeye's choice of spawning environment. Oxygen consumption increases following hatching (Smirnov 1958), and any subsequent reduction in water flow may have serious effects on survival unless the alevin can move about to locate improved flow conditions. More TU's are apparently required for embryonic development if oxygen levels in the redd drop towards marginal conditions for survival (Brannon 1965).

Bams (1969) noted that the energy supply available to the egg and subsequent alevin while in the gravel is a fixed quantity, determined by the original amount of yolk in the egg. He explained that the rate of yolk absorption is independent of variations in metabolic demand, and hence the predetermined time of emergence under the temperature regime is unaffected. With an unchanging supply rate, a higher than normal metabolic demand will consume part of the yolk material which otherwise would have been used for growth. Since tissue differentiation takes precedence over growth, the end result is a smaller, but fully formed fry at time of emergence.

By observing sockeye and pink alevins in special glass-sided gravel tanks, Bams (1969) detected several adaptive responses apparently developed to cope with development in and escape from the intragravel environment. He classified them as "behavioural" and "physiological" responses. Behavioural responses included "ventilation swimming" (slow swimming in place for varying lengths of time, apparently stimulated by increased levels of CO_2); "coughing" to expel matter through the mouth (a gill-cleaning mechanism which reverses the flow of water across the gills); "emergency movement" through the gravel, which could be induced by greatly reducing the intragravel water flow in the tank; and negative phototaxis, which keeps the alevins below the gravel surface until time for emergence. Physiological responses included: (1) hatching at an earlier than normal stage of development under conditions of reduced O_2 availability, apparently because the hatched larvae can obtain much more O_2 than the unhatched larvae; (2) secretion of mucus on the gill surfaces to rid them of fine silt; and (3) "spontaneous" initiation of emergence when the internal migratory drive builds up. Thus, alevins in the gravel redds have ways of compensating for sub-optimal environmental factors such as lowered O_2 levels, buildup of CO_2, decreased water flow, increased siltation, and light.

EMERGENCE

Until the onset of emergence behaviour, salmonid alevin behaviour can be described as "markedly photonegative, positively geotactic and thigmokinetic" (Godin 1982, citing Noakes 1978). These responses keep them below the gravel surface, and emergence is dependent upon a weakening or reversal of these responses. Initially, upon emergence, the sockeye alevin, termed "fry" if its yolk supply has been totally used (Hurley and Brannon 1969)[3], retains a predominantly photonegative response, and emergence usually occurs after nightfall (McDonald 1960; Hartman et al. 1962; Heard 1964; Bams 1969; Hamalainen 1978). The fry accumulate just below the gravel surface and emerge when the inhibitory effect of daylight is removed (Bams 1969). McDonald (1960) demonstrated that initiation of evening emergence could be delayed by artificial lighting of the gravel surface. Under natural conditions, peak outmigrations occurred during the darkest hours of the night, and the period of nightly outmigration narrowed as dusk became later and dawn became earlier with the advance of spring (McDonald 1960; Hartman et al. 1962; McCart 1967). If the fry did not migrate immediately, they remained buried in the gravel or hid in shady areas when daylight came (Hartman et al. 1962). The strong tendency of certain salmonid species to emerge mainly at night is considered by several investigators (Bams 1969; Godin 1982) to be an antipredator adaptation, because the visual acuity and capture efficiency of the fish predators decrease as the light diminishes.

With the onset of darkness, the sockeye fry emerging from the gravel behave as individuals. In tributary streams, emerging fry exhibit a marked negative rheotaxis, actively swimming downstream as individuals (McDonald 1960; Hartman et al. 1962). In some lake outlet spawning areas, the emerging fry swim laterally in an attempt to reach the river banks and avoid being swept downstream (McCart 1967; Brannon 1972; Clarke and Smith 1972). The emergence behaviour of fry in lakeshore spawning areas has not been reported. In Little Togiak Lake, Bristol Bay, the sockeye fry move into shallow water shortly after emergence (R.L. Burgner, unpublished data), but in Great Central Lake, Vancouver Island, they apparently move directly to deeper water (Barraclough and Robinson 1972). In most stream situations the fry proceed downstream without delay (McCart 1967), although exceptions have been observed in certain streams of the Wood River system of Bristol Bay, where some fry remain in the stream for extended periods and show substantial growth (Burgner

3 In field observations, the term "fry" is used once the fish has emerged from the gravel even though some yolk supply may still remain (Brannon 1972).

1962a; Burgner and Green 1963; Rogers 1968). At least a portion of the fry in spring spawning areas in tributaries of Lake Azabache and in Karymaiskiy Spring (Bolshaya River, Kamchatka Peninsula) emerge from the spawning gravel and make the transition to external feeding within the springs (Synkova 1951; Simonova 1972). Emerging fry in some stream populations without lake access may also delay their migration to sea following emergence.

Several investigators discuss seasonal timing of sockeye fry emergence in terms of optimizing the timing of dispersal into their feeding habitat, particularly to take advantage of the seasonal peak abundance of zooplankton of appropriate size. It is postulated that fry emerging earlier or later than the optimum may suffer greater mortality, and thus that timing is a response to this selective pressure (Bams 1969; Godin 1982; Miller and Brannon 1982; Brannon 1984). Bams (1969) proposed that the survival value in entering the lake early is to take advantage of feeding in the lake as long as possible during the summer, thus achieving larger size in preparation for spring smoltification. However, too early an entry into the lake, before the increase in plankton abundance, probably decreases survival chances because of slower growth and hence greater susceptibility to predators. The annual timing of fry migration and its seasonal pattern is a function of the seasonal timing of the adult spawning period, ecological factors (primarily temperature) within the incubation habitat that affect development rate and alevin behaviour, and transit time needed by the fry to reach their feeding habitat (Godin 1982). Fry reaching their feeding habitat by downstream migration need little lag time, but fry from outlet river spawning areas may require a first stage period of growth in the stream before migrating upstream into their nursery lake.

The interplay of timing of spawning, incubation temperatures, and emergence timing is well illustrated for Fraser River populations. Brannon (1984, 1987) provided evidence that sockeye returning to cold streams spawn earliest, and those in warmer streams spawn progressively later to compensate for the differences in embryonic development rate. Fry emergence apparently begins in early to mid-April in most instances, peaks in early to mid-May, and ends in late May to early June (data from Brannon 1972). Emergence from Cultus Lake beach areas is more prolonged, reflecting the prolonged spawning period. The approximately forty-five separate Fraser River stocks monitored annually enter the trunk stream at separate times over a six-month period. However, peak emergence of fry from all stocks is condensed to within a short time frame in the spring, some stocks having incubated sixteen or more weeks longer than others.

Generally, as one moves from the lower Fraser River northward to spawning areas at higher elevation, spawning time is progressively earlier, reflecting the cooler mean incubation temperatures of the spawning streams (Brannon 1984). This same timing trend in spawning is evident within lakes between inlet and outlet spawning streams. Stocks were found to spawn three to seven weeks earlier in the inlet streams, which are generally cooler, in order to achieve similar emergence timing synchronized with the productivity cycle of the nursery lake. Miller and Brannon (1982) believe that the whole system of life cycle time windows hinges on the timing of emergence of fry from the spawning beds.

As is often the case in nature, there are a number of as yet unexplained exceptions to the rule. For instance, in Lake Brooks (Bristol Bay watershed), spawning occurs at the same time in two small tributary streams, Hidden and One-Shot creeks, yet peak emergence is three weeks later in the cooler One-Shot Creek (Hartman et al. 1967). In Karluk Lake, Kodiak Island, distinct summer and fall runs of sockeye spawn in two terminal streams, Canyon and Meadow creeks, and the fry in both streams emerge in separate early and late waves, peaking in early May and early July. About sixty days separate the midpoints of the two spawning periods and the two seasonal migrations of fry. The timing of the two separate seasonal waves of fry migration is nearly identical in the two streams (Hartman et al. 1967). Small lateral streams on Karluk Lake receive only summer spawners, and the emergence timing in early May matches that of the early wave in the two terminal streams. Beach spawning occurs in the fall, concurrent with that in the terminal streams (Hartman et al. 1967).

A third interesting exception to the rule of concise spring emergence is found in Lake Kuril. Spawning occurs in this lake along its beaches, in small inlet rivers, and in the upper section of the outlet river. The Lake Kuril sockeye population is characterized by very prolonged periods of spawning migration, spawning, and emergence of fry from the redds (Egorova 1970a, 1970b). As noted earlier, spawning lasts from the end of June to early February, with the main spawning extending from September to November. The period of emergence extends from mid-March to the end of September and peaks in July–August. Emerging fry are seen regularly as early as late February. The prolonged emergence is suggested as a means of reducing competition for available food during initial feeding. Juvenile sockeye in Lake Kuril are slow-growing and smoltify generally at age 2+ (Selifonov 1970).

It has been proposed that fry emergence and migration is later in more northern latitudes to correspond to the seasonal period of maximal food production in the rearing habitat (Godin 1982). However, information available for sockeye lakes lying between 47°N and 58°N does not clearly support the proposal of a latitudinal cline in emergence and migration timing. Information on the seasonal availability of zooplankton in the respective lake nursery areas might help to clarify the timing sequences of fry activity. It appears that peak emergence does not occur as early as April, except in the introduced Cedar River run in Lake Washington (Hamalainen 1978), but primarily takes place in the month of May.

Salmonid alevins have a higher specific gravity than water and must attain a state of neutral buoyancy following emergence in order to change to a feeding, free-swimming existence. For sockeye alevins this is accomplished by a series of rapid swimming bursts to the water surface to gulp air to fill the swim bladder with atmospheric air (Bams 1969). Repeated cruises to the water surface, alternating with resting periods, are necessary to achieve neutral buoyancy. This activity is restricted to the dark hours of the day, apparently as an adaptive response to predation (Bams 1969), and precedes normal migration. After attaining neutral buoyancy and before downstream migration, fry remaining in streamside holding areas position themselves near the water surface and face into the current (Hartman et al. 1962).

There are considerable differences among populations of sockeye fry in the amount of yolk material still unutilized at the time of emergence from natural spawning beds (Brannon 1967; McCart 1967; Mead and Woodall 1968; Hurley and Brannon 1969; Brannon 1972). Godin (1982) suggested that the ecological significance of this remains unclear. However, Brannon (1972) noted that, in Fraser River fry populations, a reserve supply is more common in those populations that must spend more time and energy to reach nursery areas following emergence. He considered this understandable because the fish tended to be concentrated in narrow zones with relatively limited food supplies during migration. The presence of large yolk reserves is also a racial characteristic of fry populations emerging from Babine River, outlet to Babine Lake in the Skeena River system (McCart 1967). During two years of sampling in the Upper Babine River, nearly 20% of the downstream migrants sampled had yolk sacs classified as "large." These fry spend a period of days or weeks in the river before migrating upstream into Babine Lake and then continue to migrate uplake for considerable distances. Although external feeding and growth takes place in the river, the initial yolk reserve provides an additional safeguard to assure that the fry reach their nursery lake.

Fry Movement and Dispersal

Fry Migration

During the period of migration following emergence, sockeye exhibit rapid transformations in response to external stimuli. During emergence and the initial downstream migration, the fry in stream areas operate as individuals, in most cases under the cover of darkness. In most outlet river spawning areas, an upstream migration must follow the initial downstream movement. In both situations there is rapid change from solitary to schooling behaviour and from negative to positive phototactic behaviour. There is also a change from negative to positive rheotaxis. These changes are exemplified in the behaviour of fry in the Upper Babine River, outlet to Babine Lake (McCart 1967). Downstream movement of newly emerged fry, moving as individuals and dispersed across the river, occurred during the hours of darkness. The return movement of fry, moving upstream in schools along the banks to gain access to their Babine Lake rearing area, occurred almost exclusively during daylight hours. As will be discussed, the speed and strength of the change in rheotaxis varies among sockeye races according to the migratory route the fry must follow from their incubation area to their nursery area.

The route and complexity of migration pattern depends on the location of the ultimate nursery area with respect to the spawning area from which the fry emerge. Because sockeye primarily use lakes as nursery areas, they display considerably more varied and often more complex fry migration patterns to reach their destination than do other Pacific salmon. For fry emerging from lakeshore or island spawning grounds, the initial movement is simply one of dispersal along the shore, as is typical in Bristol Bay lakes (Burgner 1962a), or offshore into deep water, as at Cultus Lake (Brannon 1972). For fry emerging from terminal or lateral tributary spawning areas, the initial movement downstream into the nursery area is only a matter of moving with the current of the incubation stream and can usually be accomplished overnight (McDonald 1960; Hartman et al. 1962; Stober and Hamalainen 1980). Although rapid evacuation of the tributary stream is the most common behaviour, in some localities a portion of the fry may remain for a time in the stream before moving down to the nursery lake (Burgner 1962a; Rogers 1968; Simonova 1972).

Fry emerging from lake outlet spawning areas show a variety of responses (Brannon 1972). In many systems, upstream migration into the lake occurs. Initially, upon emergence and surfacing to achieve neutral buoyancy, the fry are swept downstream some distance by the current before reaching the river bank. Unlike the tributary stream situation, the fry must actively swim towards the stream banks. After a period of holding and sometimes extended feeding, the fry eventually return upstream along the stream bank. This pattern is widespread and occurs in the Chilko River (Andrew and Geen 1960; Brannon 1967, 1972); the Lower Babine River (McCart 1967; Clarke and Smith 1972); the Karluk River on Kodiak Island (Raleigh 1967; VanCleve and Bevan 1973a); the Whannock River draining Owikeno Lake on the central British Columbia coast (Foskett 1958); the Ozernaya River of Lake Kuril (Egorova 1970a, 1970b); and the Wood River, outlet of the Wood River lake chain, Bristol Bay (R.L. Burgner, unpublished data). At Wood River, the slackening or reversal of river flow during high tides may ease migration upstream into Lake Aleknagik. A modified migration pattern is found in fry emerging from the Little River, outlet to Shuswap Lake (Brannon 1972), and from the Upper Babine River, outlet to Babine Lake (McCart 1967). In these systems a portion of the fry descend downstream into a lake where they accumulate before returning upstream to their final rearing lake, i.e., from Little Shuswap Lake into Shuswap Lake and from Nilkitkwa Lake into Babine Lake, respectively. Fry in the Lower Babine River may migrate upstream into Nilkitkwa Lake and continue up the Upper Babine River into Babine Lake (McCart 1967).

An apparently more complex migration is accomplished by fry emerging in Weaver Creek, which flows into Morris Slough. The latter empties into the Harrison River, outlet to Harrison Lake, British Columbia. The newly emerged fry are carried downstream at night to the slower velocities of Morris Slough. They then actively swim downstream about 1 km to Harrison River, reverse their orientation to current, and migrate 5 km upstream in Harrison River to the lake (Brannon 1972). Since concentrations of fry arrived in Harrison Lake only four or five days after emergence, Brannon concluded that the upstream migration occurs immediately without holding or displacement downstream in the river. As a result of further tests, Brannon et al. (1981) concluded that lake odours, elapsed time, and river current were stimuli for the fry to reverse their orientation and migrate upstream to the lake. A similar migration path, downstream from a tributary into the Ozernaya River and a subsequent upstream migration into Lake Kuril, was reported by Egorova (1970b).

An alternate behaviour of fry often occurs in outlet spawning streams in situations where the stream is a trunk stream between two lakes. In such situations the downstream lake is used as the nursery area, and no upstream migration is necessary. Examples are trunk streams in the Wood River lakes (Burgner 1962b; Burgner and Green 1963) and Stellako River, draining from Francois Lake into Fraser Lake in the Fraser River system (Brannon 1967, 1972). In these instances water flow is swift enough along the stream banks to limit access of the fry upstream from most of the spawning area, and there is no evidence that the fry attempt an upstream migration.

Other behaviour alternatives occur in situations where there is no ready lake access. A well known example is the sockeye race that spawns in Harrison River rapids (Schaefer 1951), discussed earlier. Because the fry are unable to ascend to the lake, they proceed to the Fraser River estuary for feeding soon after hatching and enter the marine environment during their first summer. Consequently, they show no freshwater annulus on their scales upon return as adults. In the Yakutat area of Alaska, the East Alsek River, without lake access, is also unique in that most of the returning adult sockeye are 0. age fish. In other river systems without lake access, a portion of the fry over-winter in spring areas, side channels, and sloughs, but their specific migratory behaviour has not been determined. Examples are the Kamchatka River (Bugaev 1978), the Mulchatna River in Bristol Bay (M.L. Nelson, Alaska Department of Fish and Game, Dillingham, Alaska, pers. comm.), and the Stikine River (Wood et al. 1987). In the Kamchatka River a complex array of fry behaviour patterns apparently ensues. Some fry proceed to sea without over-wintering, some remain in the area of the spawning grounds during the winter, and some proceed downstream to Lake Azabache to rear in the lake (Bugaev 1983).

This wide variety in utilization of incubation and nursery locations requires that sockeye fry possess complex mechanisms for control of migration behaviour (Brannon 1972). Initially, attempts were made to explain migration patterns of sockeye fry on the basis of universal responses to such environmental factors as light, temperature, and stream velocity in relation to size, age, and swimming ability. More recently, differences among races in migratory behaviour of fry in the Fraser River system (Brannon 1967) and Karluk Lake (Raleigh 1967) were determined to have a genetic basis. Brannon (1972) concluded from observations on six fry populations from Fraser River tributaries "that mechanisms controlling sockeye fry migration are genetically based, and involve racially specific velocity response patterns, with olfactory stimulation having a directing influence on rheotactic behavior. In the presence of current, light and temperature influence only timing and intensity of responsiveness. Once sockeye fry enter the lake environment, in the absence of current, light appears to become the major phenomenon directing orientation."

Current was found to be an important environmental factor determining differences in direction of stream migration within and between populations. In some cases, a change in stream velocity could reverse the rheotactic response of the fry. Upstream migration in rivers with a strong current requires visual orientation, and in such cases migration occurred almost entirely during daylight (Brannon 1972). Neither age nor size were factors affecting response to current, although increased age and/or size might be required before movement upstream was possible. When given the choice, fry demonstrated a preference for lake source over tributary source water (Bodznick 1978a), exhibiting an odour discrimination ability in which calcium ions were one component of the odorant (Bodznick 1978b).

The findings of the Fraser River studies conducted by the International Pacific Salmon Fisheries Commission had strong implications for rehabilitation programs, because innate behaviour patterns influence the suitability of donor stock for initiating new runs (Brannon 1972).

Fry Dispersal in the Rearing Area

During movement towards or into nursery lakes, fry exhibit directional preferences which are innate and population-specific with respect to the geography of the particular river-lake system (Brannon 1972; Brannon et al. 1981). This has the effect of orienting the fry in the direction that will disperse them quickly into their lake feeding areas. Direction finding during daytime is achieved by celestial orientation using the sun. During periods of heavy overcast and at night the fry can use the earth's magnetic field to determine compass direction (Quinn 1980, 1982a; Quinn and Brannon 1982). The fry thus have a sun compass and a magnetic compass to direct their migration in open water. The importance and the operation of these two orientation systems differs among stocks (Quinn and Brannon 1982).

Movement of the fry into the nursery area may be direct and immediate, or sequential, the latter involving occupation of intermediate feeding areas for a period of time. The plasticity of response suggests definite racial adaptations to a variety of different environmental conditions. Direct movement into the limnetic zone is common. This is exemplified by fry emerging from Cultus Lake beaches (Brannon 1972) and from the Cedar River into Lake Washington (Woodey 1972). Intermediate feeding and growth can occur along outlet river banks before upstream migration into the nursery lake, as in Chilko River, British Columbia, or in a downstream lake such as Little Shuswap Lake before upstream migration into Shuswap Lake (Brannon 1972).

In other populations, pronounced sequential feeding in littoral areas within the lake occurs, as in the Wood River lakes, Alaska (Burgner 1962a; Rogers 1968; Pella 1968) and in Lake Dalnee in Kamchatka (Krogius and Krokhin 1956b). In early spring the emerging sockeye fry enter the littoral areas of the lakes, where they feed for a month or so before moving offshore to take up a pelagic existence. During July the bathymetric distribution of the sockeye fry changes as they shift offshore to the limnetic zone, where they exhibit distinct diel vertical migrations (Pella 1968). While in the littoral zone the fry may be found on sunny days close in along the shallow beaches in a few centimetres of water (R.L. Burgner, unpublished data). This behaviour was also noted by McDonald (1969) in Babine Lake, where those fry onshore were found in very shallow water during the day and did not appear to avoid bright sunlight. This was in contrast to fry in pelagic areas, which sought deeper water during the day than at night.

Midsummer interlake migrations of a portion of the fry population may occur. In the Naknek Lake system, Bristol Bay, a summer redistribution of fry occurs from Lake Coville at the upper end of the system. In some years, this lake receives as much as 65% of the escapement into the Naknek River system, but it has only about 3% of available rearing area (Burgner et al. 1969; Ellis 1974). Midsummer interlake migration of fry is an annual occurrence between Black and Chignik lakes in the Chignik Lake system and is, to some extent, proportional to the annual abundance of fry in Black Lake (Narver 1966; Burgner et al. 1969; Parr 1972; Marshall and Burgner 1977). Midsummer migrations of fry also occur from two small lakes, Lynx and Hidden, in the Lake Nerka watershed in Alaska (Burgner 1962a). These migrations are quite clearly population adjusting mechanisms that have developed in situations where the rearing area of the initial nursery lake is limiting. This type of midsummer population adjustment is not apparent in the chain of four Wood River lakes (Bristol Bay), which are interconnected by short rivers. In these lakes the rearing areas tend to be proportional to spawning population size. There could be survival disadvantage in interlake migrations that would result in higher fry densities in the lower lakes in the drainage. Such migrations also bear the risk of heavier predation than when they occur at the smolt stage.

In-lake dispersions of fry to utilize the zooplankton resources of the limnetic nursery area show various patterns among sockeye lakes. In lakes of the Fraser River, fry may migrate near shore (Chilko and Shuswap lakes), or through the limnetic area without littoral residence (Cultus Lake), or by a combination of both (Fraser Lake) (Goodlad et al. 1974). On first entering Shuswap and Fraser lakes, the fry remain near the entrance point for some time, but migrate at a rate of approximately 1 km/d once movement is initiated. In the Wood River lakes, within-lake distribution is not uniform by late summer because average length of young of the year increases in each of these lakes from west to east (Burgner et al. 1969). Since the major spawning grounds are in the western portions of these lakes, the apparent difference in growth rate is more likely the result of the widespread dispersal through the lakes of the older and larger fish that were among the first to emerge. In Iliamna Lake, the largest lake in Alaska (2,622 km^2), the major spawning areas are in or near the eastern end of the lake, and most age 0 sockeye rear in the eastern two-thirds of the lake at least through September (Mathisen 1966; Burgner et al. 1969). Those fish that do not smoltify the following spring are most abundant at age 1 in the central part of the lake in early summer and are also abundant in the western part by late summer. This progressive movement from east to west during lake residence would have the effect of reducing competitive stress for food and space between age 0 and age 1 sockeye juveniles during lake residence, a mechanism that may be particularly important during adjacent years of high abundance in the Kvichak River population cycle.

The most complete record of dispersion and lake distribution of sockeye fry was made in Babine Lake (McDonald 1969; McDonald and Hume 1984). This lake is 150 km long and up to 7 km wide. A portion of the fry migrating into the main lake basin from Fulton River and Pinkut Creek were marked by clipping of ventral fins. Recoveries were made during extensive sampling in the lake by purse seine during three fishing periods each year: late June-early July, late August-early September, and early October. Additional recoveries of marked fish were made during smolt sampling at the lake outlet.

Although the pattern of fry movement and distribution varied between years, certain characteristics were typical. Upon leaving the Fulton River, the fry may remain onshore for a time or proceed directly into open waters, some reaching the opposite shore. Initially the fry dispersed rapidly away from the Fulton River mouth, located on the west shore about mid-lake, and they spread mostly in a southward direction, which is away from the lake outlet. By July, some marked fry were recovered 72 km away from their home river. Estimated average rate of travel of the release group away from the river was about 1 km/d. After this initial southward shift, the fry dispersed more uniformly over the lake during the remainder of the summer and autumn.

It is of interest that no marked Fulton River fry were recovered in October sampling of the 45-km-long outlet arm of the lake, even though its entrance is only 25 km north of Fulton River. The North Arm is the rearing area of Babine River spawners, roughly 44% of the Babine Lake escapement. The surface area of this arm is only 11% of the total lake area and lake growth is significantly less than in the main lake. Hence, there would be an apparent growth disadvantage to the main-lake fry if they were to disperse into the North Arm. Thus, the patterns of rapid initial dispersion to alleviate congestion around the river mouth, the spreading over the lake to utilize the nursery grounds, and the apparent avoidance of the more crowded North Arm serve to enhance growth and survival of Fulton River fry.

Rapid initial dispersion is also characteristic of fry of the introduced Lake Washington sockeye population. Over 90% of the run spawns in the Cedar River at the south end of the lake. The fry enter the lake and quickly move offshore and northward, so that by early summer densities are highest towards the north end of the lake (Woodey 1972; Dawson 1972). From the northward shift of peak densities of fry abundance, determined by acoustic surveys, Dawson calculated a rate of movement of 0.4 km/d during the period 10 April to 22 June, 1970. Quinn (1980) demonstrated experimentally that Cedar River fry have an initial northward directional preference, appropriate for movement of fry into the lake.

Freshwater Life

Food and Feeding

Sockeye salmon juveniles typically spend one or more growing seasons in the limnetic zone of a nursery lake before smoltification. The transition in feeding behaviour and diet from the time of emergence of the fry from stream or lakeshore to the time of smoltification takes many forms. In general, it is a shift from dependence on dipteran insects to pelagic entomostracan zooplankton. Both the items of diet and the feeding behaviour of juvenile sockeye are modified in a variety of ways by environmental conditions and the presence of competitor and predator species.

Although alevins may ingest some food before emergence, feeding is generally initiated after the fry have passed the swim-up stage. As stated earlier, some populations of sockeye fry emerge from the natural environment with considerable yolk reserves while others have little or no yolk left at emergence. In the Babine River, where the former situation is typical, McCart (1969) examined the stomachs of fry accumulating along the banks to determine the proportions containing food among non-yolk and yolk fry. He found that, although the presence of yolk material did not preclude feeding, fry containing yolk material did not feed to the same extent as fry with yolk absorbed. Under experimental feeding conditions at Cultus Lake, sockeye alevins consumed food shortly after hatching. Growth was not significantly influenced unless food was withheld after yolk reserves were completely utilized (Hurley and Brannon 1969). Sockeye fry from the Fulton River spawning channel, Babine Lake, suffered significant mortality if held for more than two weeks without external food (Bilton and Robins 1973). Thus, commencement of feeding is critical when yolk reserves have been depleted. The presence of yolk reserves after emergence helps sustain the fry in situations where energy needs for migration may exceed that available in food supplies.

The diet of recently emerged fry is a function of habitat and available food. In the Upper and Lower Babine rivers, the diet reflects the availability of lake plankton; the copepods *Diaptomus* and *Cyclops*, in that order, were the mainstay of the diet (McCart 1969). The relative abundance of individual plankters in plankton samples was reflected in their relative abundance in fry stomachs. Chironomids, particularly pupae, were the most important insect food. Gut content analyses clearly showed that the major peaks of feeding occurred during the day and the lows during the night.

The diet of Babine River fry contrasts with that reported for initial feeding situations where lake plankton species are not readily available. For example, in the stream spawning areas tributary to Lake Azabache, Kamchatka, the fry feed actively in springs and spring pond areas from the time they begin swim-up in April until they descend to the lake by the end of June. The predominant food components in stomachs of fry during this period were chironomid larvae (Simonova 1972). Less frequently, chironomid pupae and larvae of other dipterans, stoneflies, and caddis flies were encountered. The fry disproportionately utilized the smaller forms of chironomid larvae.

Fry in the littoral area of Lake Aleknagik, Bristol Bay, fed primarily on dipteran larvae, pupae and adults, cyclopoid copepods, and cladocerans during their spring and early summer period of inshore feeding (Rogers 1968). This initial feeding in the warmer shallows of the littoral zone occurs at a time when yearling sockeye are still abundant in the limnetic zone and when the spring zooplankton bloom is just beginning (Rogers 1968). It is undoubtedly an adaptation to take advantage of the best initial conditions for feeding, growth, and survival. Similar feeding behaviour is reported for Lakes Kuril and Dalnee (Krogius and Krokhin 1956b) and Lake Azabache (Simonova 1972).

The spring to summer transition of underyearling sockeye from the littoral habitat to a pelagic existence was studied in detail in Lake Aleknagik by use of beach seine, bottom trawl, limnetic tow net, and hydroacoustics (Pella 1968). In spring and early summer, from the time of lake entry to the beginning of July, fry inhabit, almost exclusively, the shallow waters of the littoral zone, preferring depths less than 10 m. Midsummer (early July to end of July) is a period of transition. Beach seine catches of fry decline in daytime, whereas daytime bottom trawl catches at 3- and 6-m depths first increase and then decrease during this period. This coincides with the appearance of fry in numbers in the limnetic zone at night. Their presence in the surface waters of the limnetic zone at midsummer is a diel phenomenon. The echograms from Pella's (1968) hydroacoustic transects clearly demonstrated that the fry appear in the surface waters at dusk and leave at dawn, descending to the bottom or to waters beyond the echosounder's range at 25 m. It was not determined whether the sockeye fry at mid-lake simply sounded vertically into depths beyond detection by the echosounder or returned to the shelf and sides of the lake. Pella's schematic representation of the change in bathymetric distribution of sockeye fry during the summer is presented in Figure 13. This diel migration, characteristic in other Bristol Bay lakes as well, is accompanied by daytime schooling in deep water and by dispersal of the schools during their nighttime sojourn in the epilimnion. The transition from littoral to limnetic zone may also involve a shift from primarily daytime feeding to crepuscular feeding, although this has not been clearly established. The period of daylight is prolonged, and hours of dusk and darkness are few in the late spring and early summer at the latitude of the Bristol Bay lakes (57°N - 61°N).

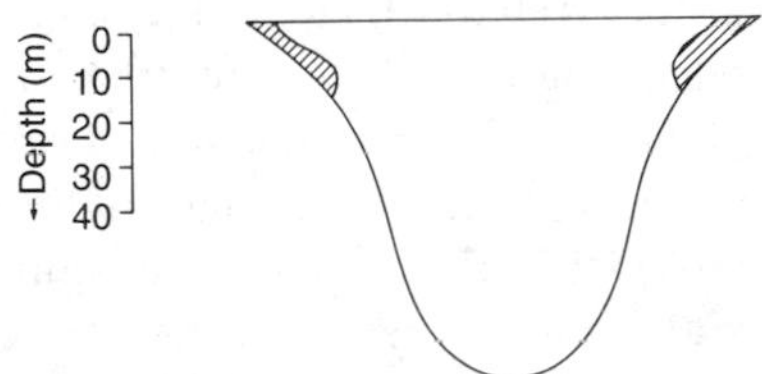

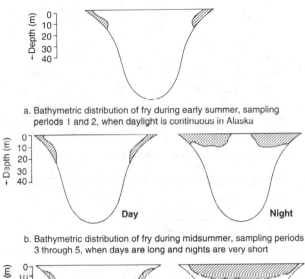

a. Bathymetric distribution of fry during early summer, sampling periods 1 and 2, when daylight is continuous in Alaska

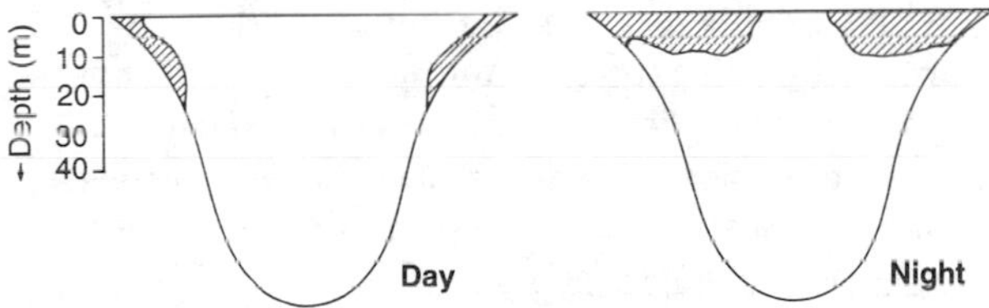

b. Bathymetric distribution of fry during midsummer, sampling periods 3 through 5, when days are long and nights are very short

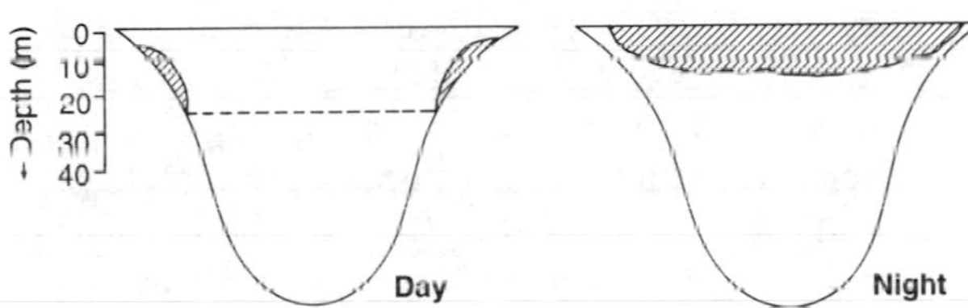

c. Bathymetric distribution of fry during late summer, sampling periods 6 through 8, as nights become increasingly long (distribution below 25m is unknown in daytime)

FIGURE 13

Schematic representation of the bathymetric distribution of underyearling sockeye during summer in Lake Aleknagik (cross-section of the lake; hatched areas represent regions occupied by sockeye; vertical scale is greatly exaggerated). (From Pella 1968, Figure 21)

Juvenile sockeye feeding in the limnetic zone are visual predators and thus require light to see their prey (Brett and Groot 1963; Doble and Eggers 1978; Eggers 1977, 1978, 1982; Eggers et al. 1978). Because they must seek their prey under sufficient conditions of light intensity, they are themselves vulnerable to piscivore predators. In order to survive and grow during their period of freshwater residence (usually one or two years), the limnetic-feeding juvenile sockeye may change the timing, duration, and location of feeding; the prey organisms to pursue upon encounter; and their behaviour when not feeding (Eggers 1978). Within the lake environment, they encounter seasonal and vertical gradients of prey abundance, prey species composition, light intensity, and temperature. In addition, they must interact with competitor species for food and space and must minimize vulnerability to potential predators. As a result, juvenile sockeye salmon exhibit complicated seasonal and diel patterns of feeding behaviour that vary greatly among populations in different lake environments. Light intensity and temperature apparently serve as orienting stimuli during vertical migrations of juvenile sockeye in lakes, controlling their day and night vertical distributions, respectively. In order for these diel vertical migrations to have evolved, ecological benefits must have been realized that compensated for the energetic costs of vertical migration and the foregone feeding opportunities in the lake euphotic zone where food is more visible and abundant (Levy 1987). As will be discussed, metabolic advantage and predator avoidance are selective forces that appear to account in part for the observed vertical movement phenomena of juvenile sockeye during their stay in the limnetic zones of lakes (Brett 1971a, Eggers 1978, Biette and Geen 1980; Levy 1987).

Upon transition to pelagic feeding, the juvenile sockeye target entomostracan zooplankton which, in the oligotrophic Bristol Bay lakes, also undergo distinct rhythmic diel vertical migrations (Burgner 1958; Hoag 1972; Carlson 1974). *Bosmina*, *Daphnia*, and *Cyclops* were the most frequent genera in the summer diet of Wood River lakes sockeye, followed by the calanoid copepods *Diaptomus* and *Eurytemora* (Rogers 1968). Copepodid and nauplius stages of copepods occurred infrequently. Insects, predominantly adult Tendipedidae, were an important component of the age 0 sockeye diet in the Wood River lakes but were of lesser importance in the diet of Iliamna Lake juveniles. In Iliamna Lake, *Cyclops scutifer* and *Bosmina coregoni* usually dominated the summer diet and in vertical zooplankton hauls (no. 6 mesh) (Hoag 1972). Calanoid copepods

(*Diaptomus gracilis* and *Eurytemora yukonensis*) and the cladocerans *Daphnia longiremis* and *Holopedium gibberum* were of lesser importance. In Hoag's (1972) study, sockeye fry contained proportionally more *Bosmina*, sockeye yearlings contained more *Cyclops*, and fry and yearlings contained less calanoid copepods than did the zooplankton hauls. However, the degree of discrepancy was relatively small. In Brooks Lake, another oligotrophic lake in the Bristol Bay watershed, four species of zooplankton (*D. longiremis, B. longirostris, B. coregoni,* and *Cyclops sp.*) made up over 96% of the summer plankton, and the three genera showed similar predominance in the stomachs of juvenile sockeye salmon (Merrell 1964). Numerically, insects - primarily midges - comprised only 4% of the gut contents.

Because insects included in the diet are of larger average size than zooplankton, the former generally play a greater role in juvenile sockeye diet than their numerical abundance indicates. This was evident in a three-year study by Parr (1972) of summer food habits of juvenile sockeye and competitors in the Chignik lakes (Alaska Peninsula). Parr converted counts of food organisms to relative caloric values to allow direct comparison of consumption of different food organisms of unequal size. In shallow Black Lake (mean depth 3 m), insect pupae and adults (mainly tendipedid midges) were the most important dietary component, followed by *Bosmina* and *Cyclops*. In the contrasting environment of Chignik Lake (mean depth 26 m), *Cyclops* was most important, followed by insect pupae-adults, *Bosmina*, and the calanoid, *Eurytemora*. The importance of insects in the diet may also change seasonally. Barraclough and Robinson (1972) found that, in Great Central Lake, Vancouver Island, chironomids were a frequent dietary component during the late winter-early spring, forming up to 100% of the diet by weight during some winter periods. However, from late summer to late fall insects were a minor part of the diet.

The complexity of summer and autumn diel vertical migrations and feeding behaviour of underyearling sockeye is well illustrated by studies at Babine Lake. The diel migrations and feeding habits were summarized by Narver (1970) as follows:

> From early July to September, from about 1.5 hr after sunrise to 1.5 hr before sunset, the young sockeye were in two distinct layers at about 20 and 35 m, each layer about 6 m thick. About 1.5 hr before sunset the two layers began to ascend. About 0.5 hr after sunset all fish were within 3 m of the surface, and during darkness they were dispersed throughout the top 5–15 m with most fish being below the thermocline. Soon after the first light of dawn the fish usually tended to move toward the lake surface and then descended rapidly to the daytime depths.
>
> By early October the pattern had changed markedly. During daylight the fish were still found in roughly two layers at about 24 and 40 m. However, the evening ascent did not commence until about 0.5 hr before sunset, the ascent was much slower, and the fish did not come to the surface but were dispersed between 9 and 27 m. This change in behaviour was temporally associated with a cooling of the epilimnion and a decrease in intensity of feeding . . . Young sockeye fed most intensively in the evening as they approached the surface and again at dawn just as they commenced the descent. At those times the most common food item was *Daphnia longispina*, followed by *B. coregoni* and *H.* [*Heterocope*] *septentrionalis*. These three species were strongly selected by young sockeye, since they were numerically much less abundant than other limnetic zooplankton species. In August, terrestrial insects were occasionally of major importance. At midday the upper layer of sockeye (about 20 m) was feeding on *H. septentrionalis* whereas the lower layer (about 35 m) was not feeding or was feeding at a low intensity.

Several aspects of the diel behaviour of juvenile sockeye in Babine Lake are of interest. Narver (1970) concluded that the diel change in intensity of underwater illumination played an important role in timing of diel vertical movement of sockeye juveniles. Their dispersion after dark was apparently due to a reduction in the intensity of illumination below that required for schooling. As for their food, the major part of the zooplankton standing stock was found above 9 m depth, and only *Heterocope* and *B. coregoni* undertook major diel vertical movements. (Interestingly, the movement of *Bosmina* was reversed, i.e., near surface during the day and at depth at night, in contrast to findings of Carlson (1974) in Iliamna Lake.) The primarily crepuscular feeding behaviour of the juvenile sockeye in warm surface waters (17°C) during midsummer and daily descent through the thermocline to temperatures near 5°C discounts their preference for physiologically optimum temperatures (about 15°C), as experimentally determined by Brett (1971a). Brett, who reviewed data on the influence of thermal regimes in Babine Lake on observed and possible growth rates of young sockeye, and on their feeding and digestion rates, concluded "that sockeye have evolved a pattern of thermoregulation (effected by vertical migratory behaviour) peculiarly adapted to maximizing growth, through the selective pressure of bioenergetic efficiency."

Seasonal changes in feeding behaviour of underyearling sockeye have been studied in detail in two lakes that remain essentially ice-free in winter, i.e., ultraoligotrophic Great Central Lake on Vancouver Island, which may freeze over during winter for periods of one to two months, and mesotrophic Lake Washington in western Washington, which remains ice-free. Growth of juvenile sockeye contrasts greatly between the two lakes; yearling smolts from Lake Washington are several times heavier (av. wt. 18.5 g versus under 4 g for Great Central Lake smolts (Eggers 1978)).

In Great Central Lake the diet of underyearling sockeye changes seasonally from the time of first feeding in April through the course of the year (Barraclough and Robinson 1972). Figures 14 and 15 show these changes in relative numbers and relative weights of the different organisms consumed from April 1970 to March 1971, and Figure 16 shows the overall change in number and weight of organisms per fish. Similar changes occurred in the diet of yearling sockeye, although the yearlings were more selective in cropping the larger forms of zooplankton (Barraclough and Robinson 1972). They also consumed larval fish (*Cottus sp.*) during summer months.

Such seasonal diversity reflects the seasonal successions of food organisms, their visibility and depth distribution in the water column during the diel feeding migrations of the juvenile sockeye, the preferential selection exercised by the fish, and the ability of the prey organisms to avoid capture. Feeding in Great Central Lake occurred primarily during the evening twilight period and in the epilimnion, with secondary feeding at dawn, prior to

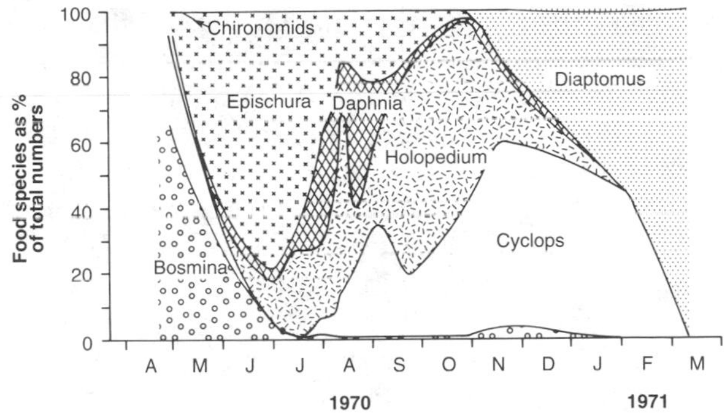

FIGURE 14

Seasonal changes in food consumed by underyearling sockeye salmon, Great Central Lake, expressed as a percentage by numbers of organisms. (From Barraclough and Robinson 1972, Figure 6)

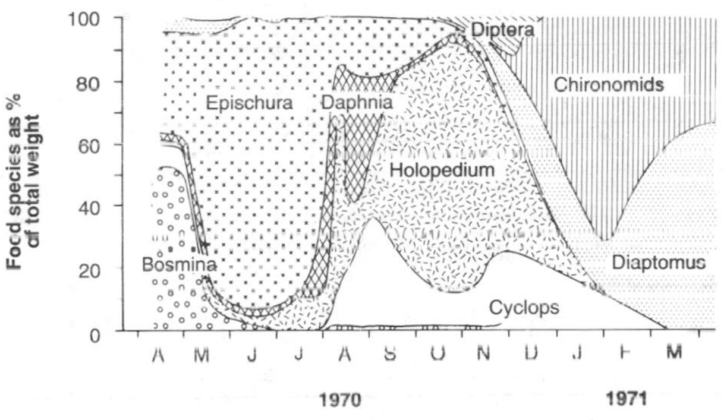

FIGURE 15

Seasonal changes in food consumed by underyearling sockeye salmon, Great Central Lake, expressed as a percentage by weight of organisms. (From Barraclough and Robinson 1972, Figure 7)

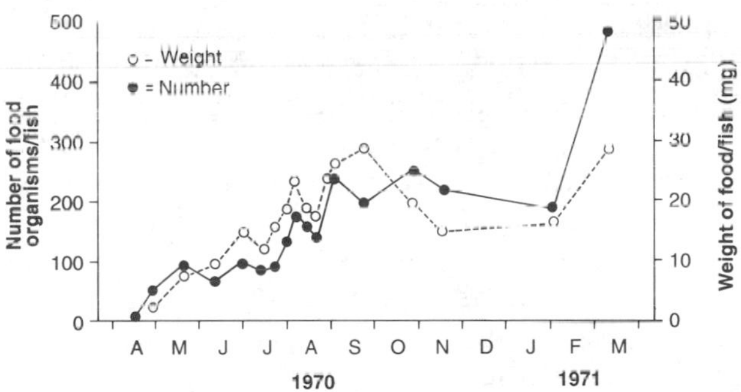

FIGURE 16

Seasonal changes in number and weight of food consumed per fish for underyearling sockeye, Great Central Lake. (From Barraclough and Robinson 1972, Figure 5)

return of the fish to daytime depths. The fish did not feed during darkness at any time of the year. Their observed vertical distribution indicates that the fish tended to avoid temperatures below 4°C and above 18°C (LeBrasseur et al. 1978). As in Babine Lake, the juvenile sockeye reside at depths below the thermocline during the daytime. In contrast, however, the fish in Great Central Lake were rarely at the surface during feeding except during early winter and during the spring migration. LeBrasseur et al. (1978) stated: "From July through September, when epilimnial temperatures exceeded 17°C, the sockeye apparently made little if any direct use of the zooplankton in the epilimnion where approximately half the zooplankton biomass was concentrated (LeBrasseur and Kennedy 1972)."

In Lake Washington, the juvenile sockeye remain in the hypolimnion during summer temperature stratification. With the progression of summer, epilimnial temperatures rise above 20°C and the thermocline deepens progressively. Consequently, the depths occupied by the sockeye increase as the thermocline depth increases (Woodey 1972). This depth trend is shown in relation to ambient light in Figure 17. However, for most of the year, sockeye are below 20 m even when epilimnial temperatures are within their tolerance ranges, and the extent of the diel feeding migration is less than 13 m (Eggers 1978). Eggers concluded that, because Lake Washington is a zooplankton-rich environment, the sockeye can meet their energy requirements by rising to feed relatively deep in the water column at the lower edge of the region of high zooplankton abundance, and they do so during a single feeding period near dusk. There was no dawn feeding period as in Babine and Great Central lakes.

Schooling of juvenile sockeye in Lake Washington occurs during the day throughout the summer and autumn growing season. The schools disperse as dusk approaches and re-form with increasing light intensity after dawn. Sockeye do not school during the winter growth period but resume schooling during the pre-smolt growth period before smolt migration (Eggers 1978).

Distinct prey size preferences were exhibited by juvenile sockeye salmon feeding on zooplankton in Lake Washington (Eggers 1978). "Preference" was considered to be "that component of prey selection that cannot be attributed to differential rates of prey encounter and evasion" that result from differential visibility due to prey size, shape, colour, and motion, or from differential evasive ability among prey. The extent of prey selection by juvenile sockeye in Lake Washington was much greater than would be expected if sockeye consumed prey as encountered in the visual field. Eggers (1982) demonstrated that there is a distinct minimum size threshold for prey occurring in sockeye stomachs, and that the threshold varied seasonally with the relative availability of large and small prey types (Figures 18 and 19). This was considered consistent with foraging theory, which

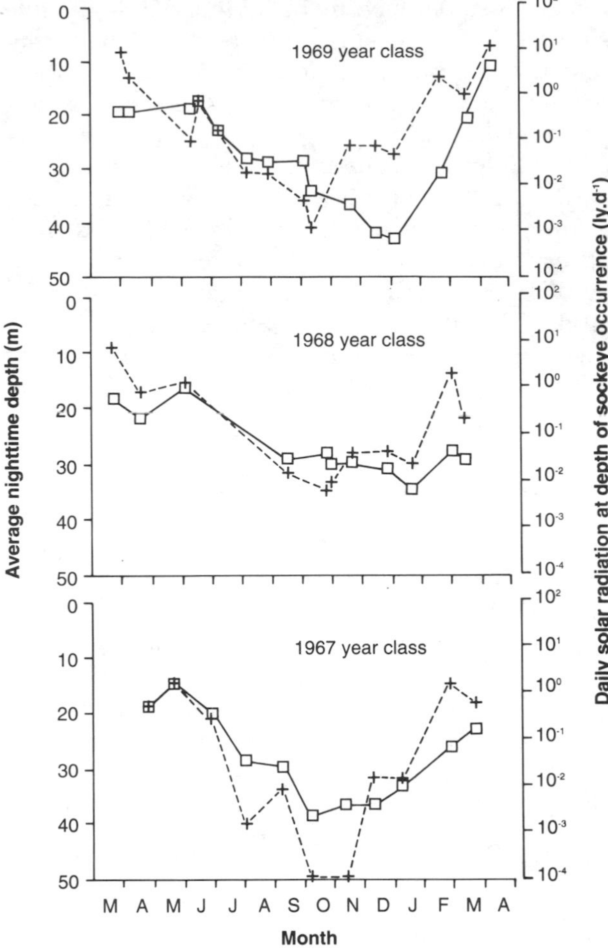

FIGURE 17

Nighttime mean depth of occurrence (solid line) of juvenile sockeye in Lake Washington, together with index of ambient light intensity (dashed line) at depth of occurrence. (From Eggers 1978, Figure 4)

predicts that the predator will maximize energy consumed per unit foraging time if prey below threshold size are ignored (Werner 1972; Charnov 1976; Eggers 1977, 1982). The threshold increases with abundance of large prey types and decreases with prey density. The increase in diet frequency of smaller forms during certain times was considered to reflect the decreasing marginal benefit to the fish of ignoring small prey.

Of the three large zooplankters, the more motile cladoceran *Diaphanosoma* and the copepod *Epischura* were under-represented in the sockeye diet relative to the cladoceran *Daphnia*. The under-representation of evasive prey in the diet may reflect either the preference of the sockeye against the evasive prey or their inability to capture these large forms (Eggers 1982). The selection against small prey was consistent with size preference of the fish to increase foraging efficiency, and not to possible escape of small prey through the gill rakers. Eggers (1982) summarized the feeding behaviour of juvenile sockeye in Lake Washington as follows: "Sockeye are showing a clear preference for large nonevasive prey. However, this preference is not fixed but is extremely dynamic, and is a consequence of seasonal variability in availability of large prey types. Small as well as evasive prey are pursued and eaten at times of the year when the large nonevasive forms are rare or absent from the water column. This size preference seems to represent an active decision on whether to pursue at the encounter point in the predation cycle (Holling 1959). This decision is presumably dependent on the perceived foraging success under the current prey regime." He theorized further that. "There may be extreme selection for efficient foraging on the part of planktivorous fish. The act of feeding incurs risk of predation. Planktivores that visually encounter and pursue prey individually of necessity forage in lighted areas of the water column and thus are vulnerable to sight-feeding piscivores while foraging (Eggers 1978). Any strategy that minimizes the time needed to meet energy requirements will reduce risk of predation."

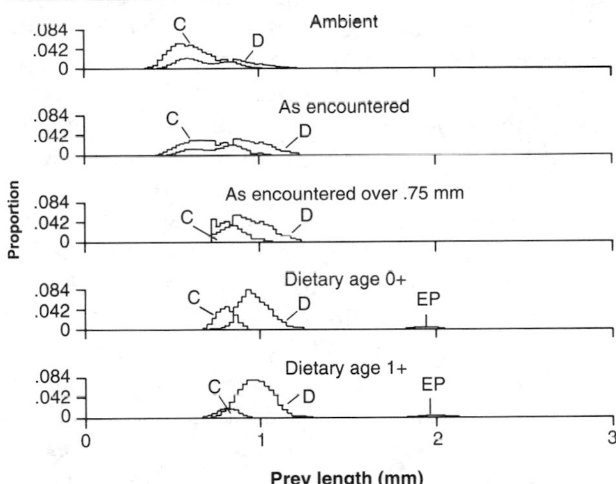

FIGURE 18

Zooplankton size distributions determined in Lake Washington on February 1975 sampling trip. Ambient: as occurring in plankton samples; as encountered: corrected for relative visibility to juvenile sockeye; and dietary: as found in stomachs. Abbreviations: *C* = *Cyclops*, *D* = *Diaptomus*, *EP* = *Epischura*. (From Eggers 1982, Figure 3)

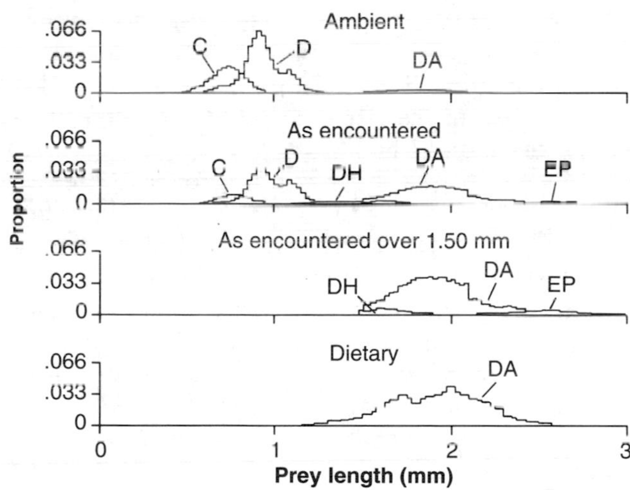

FIGURE 19

Zooplankton size distributions determined in Lake Washington on August 1975 sampling trip. *DA* = *Daphnia*, *DH* = *Diaphanosoma*; other abbreviations as in Figure 18. (From Eggers 1982, Figure 5)

Residence Time and Growth in Fresh Water

The annual growth attained by juvenile sockeye and length of residence in fresh water varies greatly among populations in different lake systems as well as between years within individual lakes. Factors affecting growth are highly complex and include (1) size and species composition, visibility, motility, seasonal abundance, and distribution of the available food supply in the water column; (2) water temperature and thermal stratification of the lake; (3) photoperiod and length of growing season; (4) relative turbidity of the lake waters and available light intensity in the water column; (5) intra- and interspecific competition; (6) parasitism and disease; (7) feeding behaviour of the juvenile sockeye to minimize predation; and (8) migratory movements to seek favourable feeding environments. As will be discussed later, growth influences duration of stay in fresh water before smoltification, and within many lake populations the larger members of a year class tend to migrate to sea earlier in the spring or migrate a year earlier than smaller members. In the more southern and interior lake systems, such as the Fraser River watershed, smoltification after one year of lake residence is nearly universal, with few carry-overs for another year (Table 2) (Killick and Clemens 1963). In Bristol Bay there is much variability in age at smoltification both among lake system populations (Table 2) and among year classes within lake systems, e.g., Kvichak River system (Rogers and Poe 1984; Eggers and Rogers 1987). Size is not strictly the determinant for duration of stay in fresh water, because some populations with very poor freshwater growth in their first year migrate as yearlings (e.g., Lake Owikeno, British Columbia), whereas other populations exhibiting good first-year growth migrate predominantly after a second year of growth (e.g., Lake Becharof, Bristol Bay) (see Tables 2 and 8).

Sockeye fry at the beginning of lake life are between 25 and 31 mm in length and weigh between 0.1 and 0.2 g (LeBrasseur et al. 1978; Rogers 1979; McDonald and Hume 1984), but their subsequent growth rate varies widely among populations. Comparative size data for age 1. sockeye smolts demonstrate the wide range among lake populations in freshwater growth (Table 8). Until recently, yearling smolts produced in Frazer Lake (Kodiak Island) and in Lake Washington were among the largest found; both populations originated from introduced sockeye runs and both lakes had high standing crops of entomostracan crustacea (Eggers et al. 1978; Blackett 1979). The smallest yearlings (av. 59-61 mm) are found in the glacial waters of Owikeno Lake, British Columbia (Ruggles 1966). Mean weights of yearling smolts in Frazer Lake and Lake Washington were about eleven and nine times greater, respectively, than mean weight of yearling Owikeno Lake smolts (2 g). In spite of their great difference in growth, the Lake Washington and Owikeno Lake populations migrate seaward almost entirely as yearlings, whereas the Frazer Lake smolts largely remain an additional year in fresh water before smoltification (Table 2). The latter tendency may be a genetic carry-over from the source stocks (Karluk and Red lakes), which smoltify primarily at age 2. (In recent years, a sharp drop in size of Frazer Lake smolts has occurred with buildup of the sockeye population spawning in the lake (Kyle et al. 1988).)

For those populations producing substantial proportions of age 2. smolts, the additional year of growth in fresh water results in longer fish; however, average weight is not doubled as a result of the second year's growth (Table 8). This occurs partly because the yearlings remaining in the lake are smaller in mean size than the yearling smolts (Barnaby 1944; Nelson 1959; Burgner 1962b). Other factors, such as lowered feeding and/or food conversion efficiency with increase in size or a change in feeding strategy, may reduce specific growth rate. Specific growth rate (% wt increase/d) is a complex function of temperature, ration size, and fish size (Brett et al. 1969; Brett 1971a, 1971b; Shelbourn et al. 1973; Brett and Shelbourn 1975; Brett 1976; Biette and Geen 1980). Brett and Shelbourn (1975) concluded that specific growth rate remained steady when juvenile sockeye were on a fixed ration, if the rations were fully consumed.

Considerable attention has been directed in physiological and field studies towards understanding why juvenile sockeye exhibit their various diel and seasonal feeding behaviour and depth preferences in different lake environments. Feeding experiments over a range of constant temperatures revealed that young, small sockeye show a sharp growth optimum at 15°C and that growth conversion efficiency (% flesh from food) peaks at a lower temperature: 11.5°C (Brett 1971a). Lower temperatures suppress conversion efficiency, growth rate, maximum meal size, and digestion rate. There exists a general "zone of efficiency" of food conversion greater than 20% at constant temperatures ranging from 5° to 17°C. Under condi-

TABLE 8
Comparative size data for sockeye smolts from various lake systems

Lake or lake system	Age	Mean length (mm)			Mean weight (g)			Source
		No. yr.	Avg.	Range	No. yr.	Avg.	Range	
Wash. State								
Washington	1	4	125	120–129	4	18.5	16.7–26.5	Bryant (1976)
Fraser River								
Cultus	1	21	82	68–94	21	6.2	3.0–8.6	Eggers (1978)
Chilko	1	20	82	73–101	20	4.6	3.1–8.4	Eggers (1978)
Fraser	1	1	90	–	1	7.8	–	Goodlad et al. (1974)
Shuswap	1	7	74	–	7	4.0	–	Goodlad et al. (1974)
BC Coastal								
Babine (main lake)	1	22	79	75–83	22	4.9	3.9–5.8	H.D. Smith, Canada Dept. Fisheries and Oceans (pers. comm.)
Owikeno	1	4	60	59–61	1	2.0	–	Ruggles (1966)
Long	1	–	–	–	5	2.7	1.2–4.9	Hyatt & Stockner (1985)
Vancouver Is.								
Great Central	1	8	69	64–78	9	3.8	2.7–5.3	LeBrasseur et al. (1978)
	2	–	–	–	6	8.1	4.0–14.3	LeBrasseur et al. (1978)
Kennedy	1	–	–	–	1	2.6	–	Hyatt & Stockner (1985)
Henderson	1	–	–	–	5	3.2	2.3–4.3	Hyatt & Stockner (1985)
Hobiton	1	–	–	–	5	4.0	3.3–4.8	Hyatt & Stockner (1985)
S.E. Alaska								
Falls	1	1	67	–	1	2.4	–	Koenings et al. (1984)
	2	1	70	–	1	2.8	–	
	3	1	76		1	3.5	–	
Copper River								
Summit	1	1	92	–	–	–	–	Roberson & Holder (1987)
Tokun	1	2	65	63–67	2	2.3	2.1–2.4	McDaniel et al. (1985)
Prince Wm. Sd.								
Eshamy	1	2	75	73–77	2	3.5	3.2–3.7	McDaniel et al. (1985)
	2	2	99	97–100	2	7.8	7.1–8.4	
Cook Inlet								
Larson	1	1	77	–	1	3.7	–	Lebida (1984)
Hidden	1		143	–	–	27.3	–	Koenings et al. (1984)
	2	–	200	–	–	83.9	–	
Big	1	–	132	–	–	25.5	–	Koenings et al. (1984)
	2	–	166	–	–	48.1	–	
Russian	1	–	84	–	–	5.1	–	Koenings et al. (1984)
	2	–	93	–	–	6.5		
Kenai	1	–	62	–	–	2.1	–	Koenings et al. (1984)
	2	–	72	–	–	3.1	–	
Crescent	1	2	69	68–69	2	2.8	2.7–2.8	Koenings et al. (1985)
	2	2	76	76–76	2	3.7	3.6–3.8	
Kasilof	1	5	70	68–73	5	2.9	2.7–3.3	Flagg et al. (1985)
	2	5	85	82–90	5	5.1	4.8–5.3	Flagg et al. (1985), Figs. 7, 8
Packers	1	3	83	74–96	3	5.2	3.4–7.9	Koenings & Burkett (1987b)
	2	3	97	88–104	3	7.5	5.4–9.4	
	3	1	113	–	1	11.6	–	

(continued on next page)

TABLE 8 (continued)

Lake or lake system	Age	Mean length (mm) No. yr.	Avg.	Range	Mean weight (g) No. yr.	Avg.	Range	Source
Tustumena	1	5	70	68-73	5	2.9	2.8-3.3	Koenings & Burkett (1987b)
	2	5	85	82-88	5	5.1	4.8-5.2	
Kodiak-Afognak								
Karluk	1	11	111	100-122	-	-	-	Barnaby (1944), years 1925-36
	2	12	134	127-142	-	-	-	
	3	12	144	137-152	-	-	-	
	1	5	107	102-110	5	12.7	10.7-14.5	Drucker (1970), years 1961-68
	2	8	118	113-128	8	15.0	12.4-21.0	
	3	8	130	123-142	8	19.9	15.3-26.7	
	1	4	102	96-112	4	10.8	8.3-14.8	ADF&G (1979-82)
	2	4	113	103-120	4	14.1	9.4-18.5	
	3	4	128	113-147	4	19.3	11.7-29.1	
Red	1	-	85	-	-	5.8	-	Koenings et al. (1984)
	2	-	111	-	-	12.8	-	
Frazer	1	14	135	113-163	14	22.7	12-35	Blackett (1984)
	2	14	162	142-185	14	39.1	23-62	
	3	9	185	154-201	5	51.4	30-76	
Little Kitoi	1	6	64	43-75	-	-	-	Meehan (1966)
	2	6	72	69-83	-	-	-	
Ruth	1	5	104	88-119	-	-	-	Meehan (1966)
	2	5	125	114-137	-	-	-	
Bristol Bay								
Wood River	1	24	83	71-91	5	5.2	3.5-7.6	Bucher (1984)
	2	22	100	90-114	5	8.6	6.8-10.1	
Iliamna-Clark	1	28	88	80-98	25	5.9	4.2-8.4	Bill (1984)
	2	27	109	97-122	25	10.9	7.5-16.4	
	3	1	124	-	1	17.5	-	
Naknek	1	23	100	91-113	23	9.1	6.9-13.1	Huttunen (1984b)
	2	23	112	100-120	23	12.6	10.1-14.7	
	3	3	121	109-131	2	21.3	19.1-23.5	
Becharof	1	7	103	99-107	2	9.2	9.1-9.2	Bue (1984)
	2	7	120	115-130	2	15.8	14.6-17.1	
	3	7	129	115-145	2	21.3	19.1-23.5	
Ugashik	1	16	91	81-97	16	6.5	5.0-7.7	Eggers (1984)
	2	16	114	104-125	16	12.4	9.6-15.9	
	3	4	131	125-138	4	18.4	14.3-22.5	
Kamchatka								
Dalnee	1	9	116	106-133	9	16.4	12.5-23.0	Krogius & Krokhin (1948)
	2	9	154	138-161	9	36.2	28.3-43.7	
	3	1	203	-	1	79.6	-	
Kuril	2	12	93	85-102	-	-	-	Selifonov (1970)
	3	12	108	98-116	-	-	-	

Source: Adapted from Burgner (1987), Table 1

tions of reduced daily ration, maximum food conversion efficiency is attained at progressively lower temperatures as food ration becomes more restricted (Brett et al. 1969; Brett 1971a; Biette and Geen 1980). Extending this to the field situation in Babine Lake, Brett (1971a) compared the observed seasonal increase in average weight of underyearling sockeye with the expected growth at the temperatures occupied by the juveniles (Figures 20 and 21). He concluded that food intake was below the point of daily satiation (full ration), and that the ration in the field condition, which dictates much of the energetic physiology of young sockeye, was restricted.

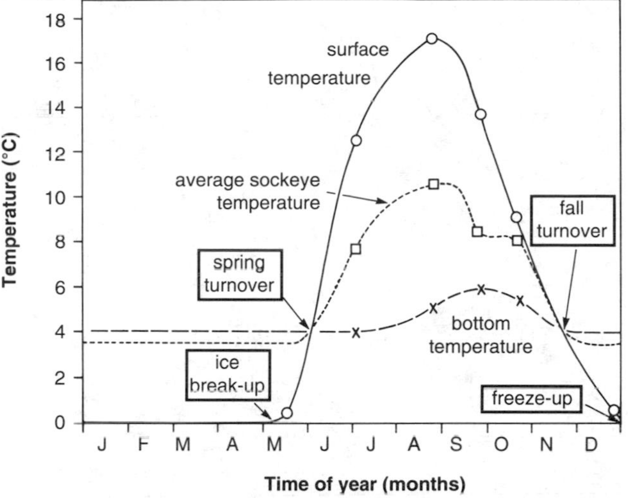

FIGURE 20

Computed average temperatures experienced by juvenile sockeye during vertical migration in Babine Lake, compared with surface and bottom temperature, 1967. (From Brett 1971a, Figure 7)

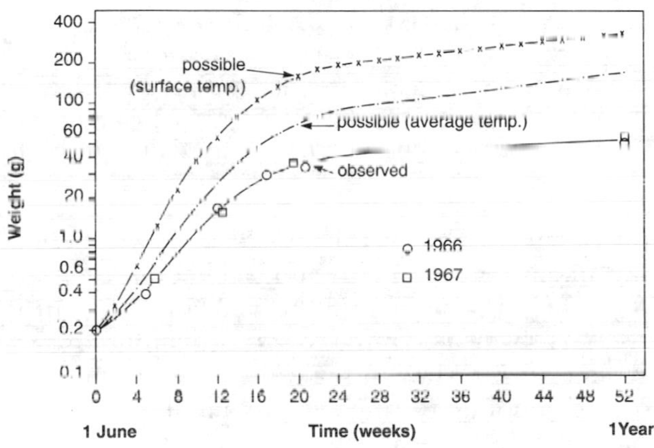

FIGURE 21

Observed growth rate of juvenile sockeye in Babine Lake compared with computed possible growth rates for "satiation ration" at average temperatures experienced by the fish and at surface temperatures. (From Brett 1971a, Figure 8)

For those sockeye populations migrating almost exclusively as yearlings, the seasonal changes in average size during lake residence and average size at smoltification in the spring can be examined as an approximation of average individual growth rate. This assumes that sampling is non-selective, that in-lake mortality is not size-selective, and that the hold-over of smaller juveniles is insignificant. Since these assumptions may be violated to some degree, the size increases shown by sampling indicate apparent average growth rate rather than true growth rate between sampling periods. Apparent growth rate patterns in different populations are discussed below for lake populations with different growth rates.

In Lake Owikeno, the growth rate of sockeye juveniles is extremely low. The lake has a high turbidity from glacial silt, relatively low temperature, and a low standing crop of zooplankton (Ruggles 1966). By autumn the underyearling sockeye have attained a weight of only 1 g, and achieve their smolt weight of about 2 g only by an acceleration of growth during early spring, prior to seaward migration. This results in a distinctive scale growth pattern that can be used to distinguish this stock from many others in British Columbia (Ruggles 1966).

Growth is considerably better in Babine Lake, where mean smolt weight over a 16-year period varied from 4.6 to 5.8 g (McDonald and Hume 1984). Growth is rapid in spring and summer, declines in autumn, and declines still further over winter (Figure 22). Relative increases in weight were greatest from May to July, and relative increases in length were greatest from July through August. Broods that produced small underyearlings in autumn tended to have better growth during the winter, but there was a suggestion of an inverse relation between winter growth and biomass that may have been an influencing factor (McDonald and Hume 1984). Because there was no late winter sampling of the population, a possible upsurge in growth in early spring was not detected.

Juvenile sockeye in Lake Washington show three distinct growth stanzas during lacustrine residence (Figure 23). These are: (1) a period of near-exponential growth occurring from the time of entry into the lake until mid-autumn; (2) a period of low or negative growth in late autumn and winter; and (3) a period of high growth rate in spring prior to smoltification and seaward migration (Eggers 1978). The feeding behaviour differs greatly among these growth stages. During the

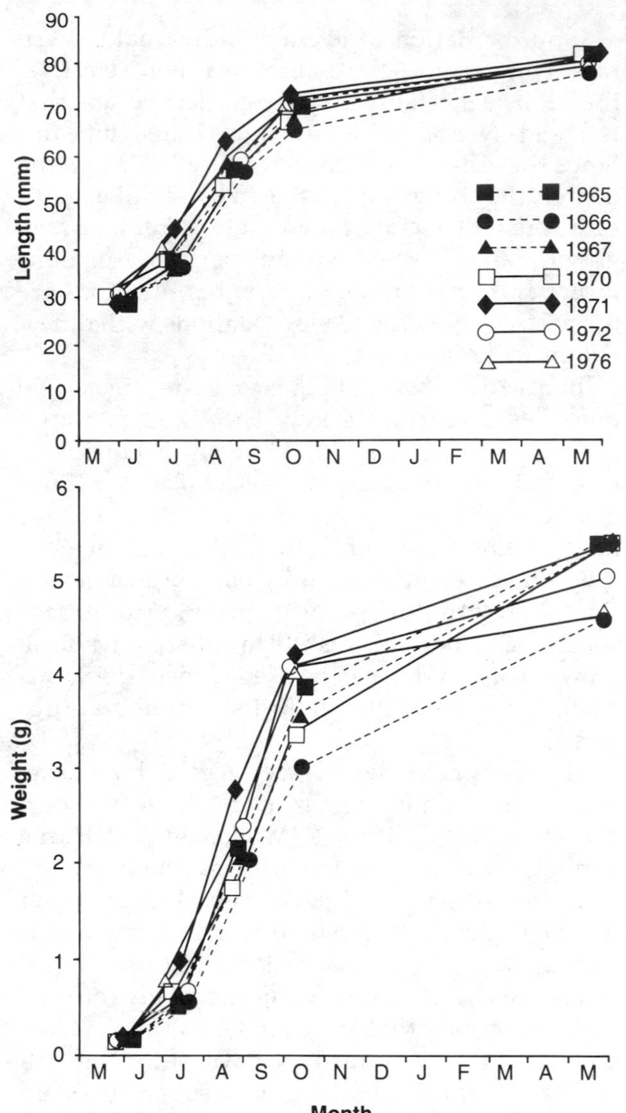

FIGURE 22

Increase in mean length and weight of juvenile sockeye from fry to smolt stage in the main basin of Babine Lake. (From McDonald and Hume 1984, Figure 6)

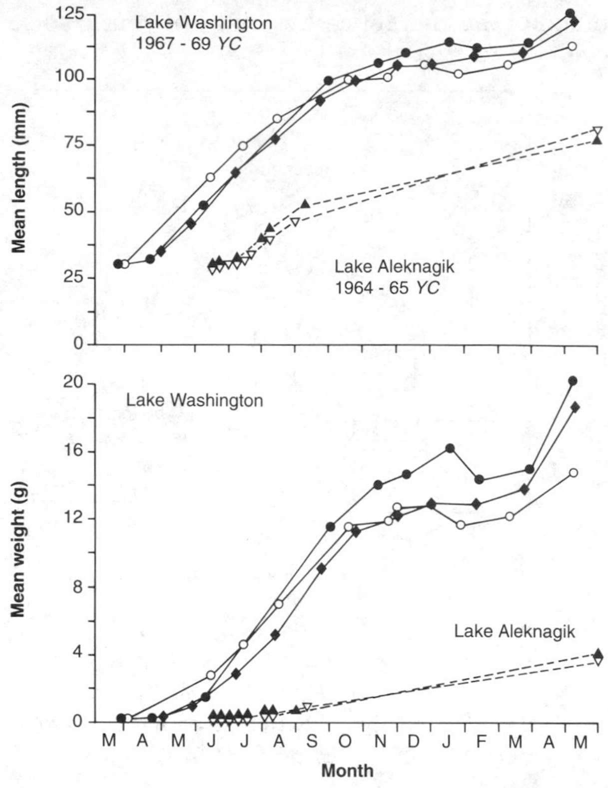

FIGURE 23

Comparison of seasonal growth trajectories of length and weight of juvenile sockeye for Lake Washington and Lake Aleknagik. *YC* = year class. (From Eggers 1978, Figure 6)

first period the fry feed intensely, with a low incidence of empty stomachs in the population towards the end of the daily feeding period (Doble and Eggers 1978). In the winter growth stage, the rate of prey ingestion for actively feeding fish is much lower, and a high incidence of empty stomachs indicates that they may undergo a fasting-feeding cycle within a period of several days (Eggers 1975). During the pre-smolt growth period the amount of food in the stomach increases by an order of magnitude, and the incidence of empty stomachs is nearly zero. Sockeye juveniles do not school during the winter growth period but resume schooling during the pre-smolt growth period before migration (Eggers 1978). Eggers hypothesized that, because of their high energy reserves, it may be advantageous for the juvenile sockeye to minimize feeding during winter and remain deep in the water column at low light intensity to minimize piscivore predation. That the juveniles resume a more active exploitation of zooplankton in late winter, at a time when water temperature and zooplankton abundance are lower than in December, suggests that they are responding to declining energy reserves (Eggers 1978). Lake Washington has a high biomass of zooplankton per unit area of lake, and planktivores (primarily juvenile sockeye) crop less than 2% of the annual zooplankton production (Eggers et al. 1978). Thus, Lake Washington represents an extreme case, compared to most other sockeye systems, in that the growing season is long, growth is very good, and zooplankton standing crop is very high.

The seasonal growth pattern of underyearling sockeye in Great Central Lake (Figure 24) indicates a different feeding strategy, which may be associated with less favourable growing conditions than exist in Babine Lake and Lake Washington. In both years shown (one influenced by artificial enrichment of the lake), there is no indication of cessation of feeding and growth during winter months. The effect of predators on the survival of juvenile sockeye in Great Central Lake is thought to be small (Robinson and Barraclough 1978), and the fish may thus direct more attention to effective feeding.

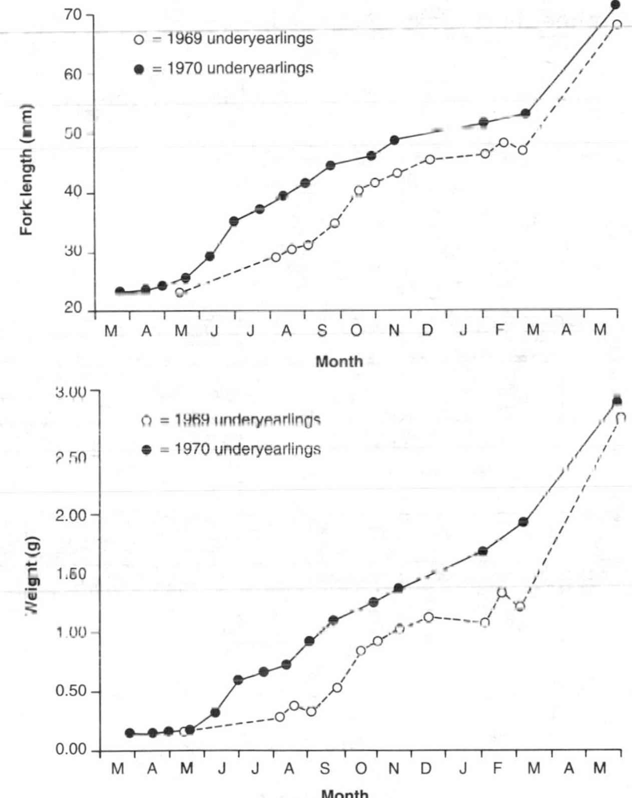

FIGURE 24

Increase in mean length and weight of underyearling sockeye from fry to smolt stage in Great Central Lake, 1969 and 1970 year classes. (Adapted from Barraclough and Robinson 1972, Figures 11 and 12)

The effect of length of growing season can be illustrated in a comparison of seasonal growth trajectories between populations in Lake Washington and Lake Aleknagik (Wood River system, Bristol Bay) (Figure 23). The period of rapid size increase begins about 2 1/2 months later in Lake Aleknagik, and ends earlier in the autumn because surface water temperatures usually drop below 4°C by the end of October (Rogers 1979). Most of the growth has occurred by autumn (Rogers 1979), and rapid increase in growth does not begin again until early summer, after the main smolt migration is nearly completed (Burgner 1962b).

Within a sockeye-producing lake, growth of juvenile sockeye may be directly related to temperature and inversely related to fish population density. It is often difficult to distinguish between these effects, particularly in lakes with high population density. Reliable estimates of differences between years in juvenile size and growth rate are usually easier to obtain than reliable estimates of juvenile population density. Therefore, the number of parent spawners per unit rearing area (e.g., spawners/ha, or females/ha if sex ratio information is available) is often used as an index of population density of progeny the following summer (Burgner 1964; Burgner et al. 1969; Hartman and Burgner 1972; Goodlad et al. 1974; Rogers et al. 1980). This assumes, of course, that over-winter survival of eggs and alevins is relatively constant from year to year and that older year classes or competitor species do not have overriding influences on growth of the underyearling sockeye. The best evidence of the influence of annual differences in seasonal temperature on growth comes from populations in lakes with short growing seasons, as noted below.

In Lake Aleknagik a significant part of the annual variation in lengths of sockeye salmon fry in the littoral zone at the beginning of summer is explained by surface water temperatures (Rogers 1973a). Since water temperature in late June is largely a function of the date of ice breakup and amount of solar radiation, the growth attained in the spring by sockeye fry is primarily a function of climatological conditions. A month later, in the latter part of July, annual mean length of sockeye fry prior to movement into the limnetic area is more highly correlated with annual mean beach seine catch in the littoral area, indicating that intraspecific competition is now of more importance (Rogers 1973a). The early season influence of climatic conditions is also apparent in the amount of new summer growth appearing on scales of yearling smolts (Burgner 1962b). Differences between years in amount of new growth attained by a given calendar date during early summer was related to time of breakup of lake ice and lake temperatures subsequent to breakup.

In Iliamna Lake, which has the world's largest sockeye runs, the growth of sockeye fry, as estimated by their mean length on 1 September annually, is more closely related to temperature than to the size of the parent escapement (Rogers and Poe 1984).

In Chilko Lake, British Columbia, a deep, cold lake at high elevation with a short growing season, temperature during fry emergence and early lake residence accounted for much of the recorded fluctuation in freshwater growth, as indicated by circulus counts on the scales of returning adult salmon (Goodlad et al. 1974). However, inclusion of data for years of high sockeye density (over 9 females/ha) substantially reduced the correlation coefficient, indicating an influence of higher population densities on growth as well (reduced from 0.716 to 0.56). In two other sockeye rearing lakes of the Fraser River system, growth was inversely related to estimated population density, and in only one (Fraser Lake, also with a short growing season), was there evidence of an early season temperature effect on growth (Goodlad et al. 1974). At Chilko Lake, the growing season was the shortest among the three lakes. Variation in lake heating during early spring, as indicated by May temperatures, provided an index of subsequent lake temperatures during the short growing season (Goodlad et al. 1974). Presumably, the reduced growth associated with cooler temperatures resulted from a combined influence of temperature on buildup of the food supply and on the feeding energetics of the juvenile sockeye.

If the nursery areas exert a limiting effect on the population size of juvenile sockeye, then progeny of large spawning populations should show evidence of food shortage. This effect has been observed in some of the Bristol Bay and Fraser River lakes, and, more recently, in Frazer Lake, Kodiak Island (Kyle et al. 1988). Summer growth exhibited by underyearling sockeye in the Wood River lakes provides the first example. The five main lakes of the Wood River system are similar in nursery area quality (water chemistry, primary production, zooplankton standing crop per unit lake area) and over a period of years have nearly the same average abundance of spawning salmon/km^2 of lake area. However, in individual years the spawning population balance among lakes may be quite uneven, and differences in the growth of progeny then become apparent (Burgner 1962b, 1964). From 1958 on, samples of young salmon for growth studies have been collected annually within each lake by surface tow nets. The results show a distinct inverse relationship between numbers of parent spawners per unit nursery lake area and mean weight of progeny towards the end of their first summer in the lake (Figure 25).

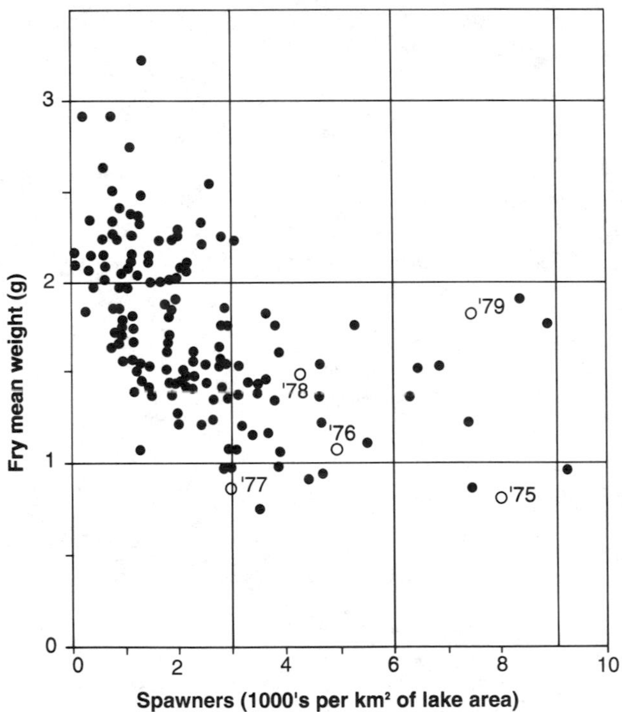

FIGURE 25

Mean weights of sockeye salmon fry on 1 September (1958–79) in each of the Wood River lakes versus the relative density of parent spawners (prior year). Observations from Little Togiak Lake following fertilization are indicated (o).
(From Rogers et al. 1980, Figure 11)

This effect of density on growth carries over to time of smolt migration. Mean weights of age 1. and 2. smolts from both the Wood and Kvichak rivers are inversely related to the density of parent spawners (Figure 26). Wood River smolts are mostly age 1. fish and are typically the smallest smolts in Bristol Bay. Population densities of sockeye range higher in the Wood and Kvichak river systems than in lakes of the Naknek and Ugashik river systems, where density dependent growth is not as obvious (Figure 27). The yearling Naknek River smolts show a greater range in size, probably a result of variable utilization of the more complex rearing environment of this multi-lake system.

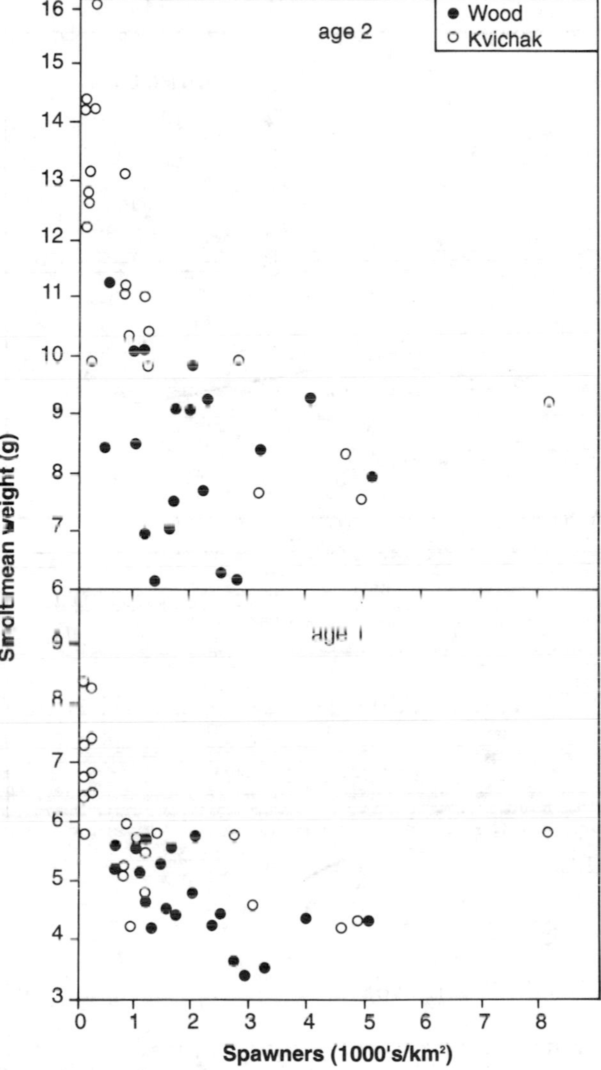

FIGURE 26

Plots of the mean weight of smolts on the relative density of parent spawners (number per km^2 of lake area) in the Wood River and Kvichak lake systems. (From Rogers 1980, Figure 2)

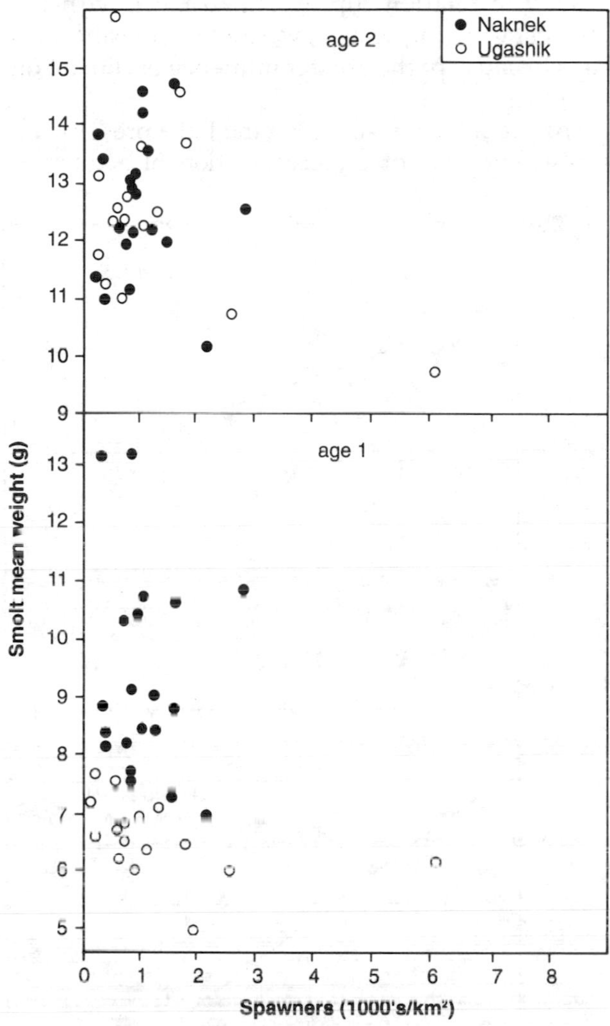

FIGURE 27

Plots of the mean weight of smolts on the relative density of parent spawners (number per km^2 of lake area) in the Naknek and Ugashik lake systems. (From Rogers 1980, Figure 3)

Increased density is also associated with reduced sockeye growth in Fraser River lakes. Goodlad et al. (1974) used parent spawners (in this case, females/ha) as an index of population density of fry produced, and used the freshwater scale circulus count of returning adult salmon as a growth index to examine density-growth relations for 20 annual spawning runs to each of four lakes (Figure 28). The inverse relationship between density and growth was significant at the 1% level in Shuswap and Cultus lakes, and at the 5% level in Fraser Lake. The relationship in Chilko Lake, with the shortest growing season, was not significant, perhaps because of the greater influence of climate on growth.

The main lake basin of Babine Lake presents an interesting case of underutilization of a nursery area because of the limited capacity of adjacent spawning grounds (McDonald and Hume 1984). Acting on Johnson's (1956, 1958, 1961) conclusion that the basin was underutilized, the Canadian government built artificial spawning channels that increased fry output nearly threefold. This increase was followed by roughly equivalent increases in the abundance of underyearlings and smolts, and there was no evidence of density-dependent mortality during the lacustrine stage (McDonald and

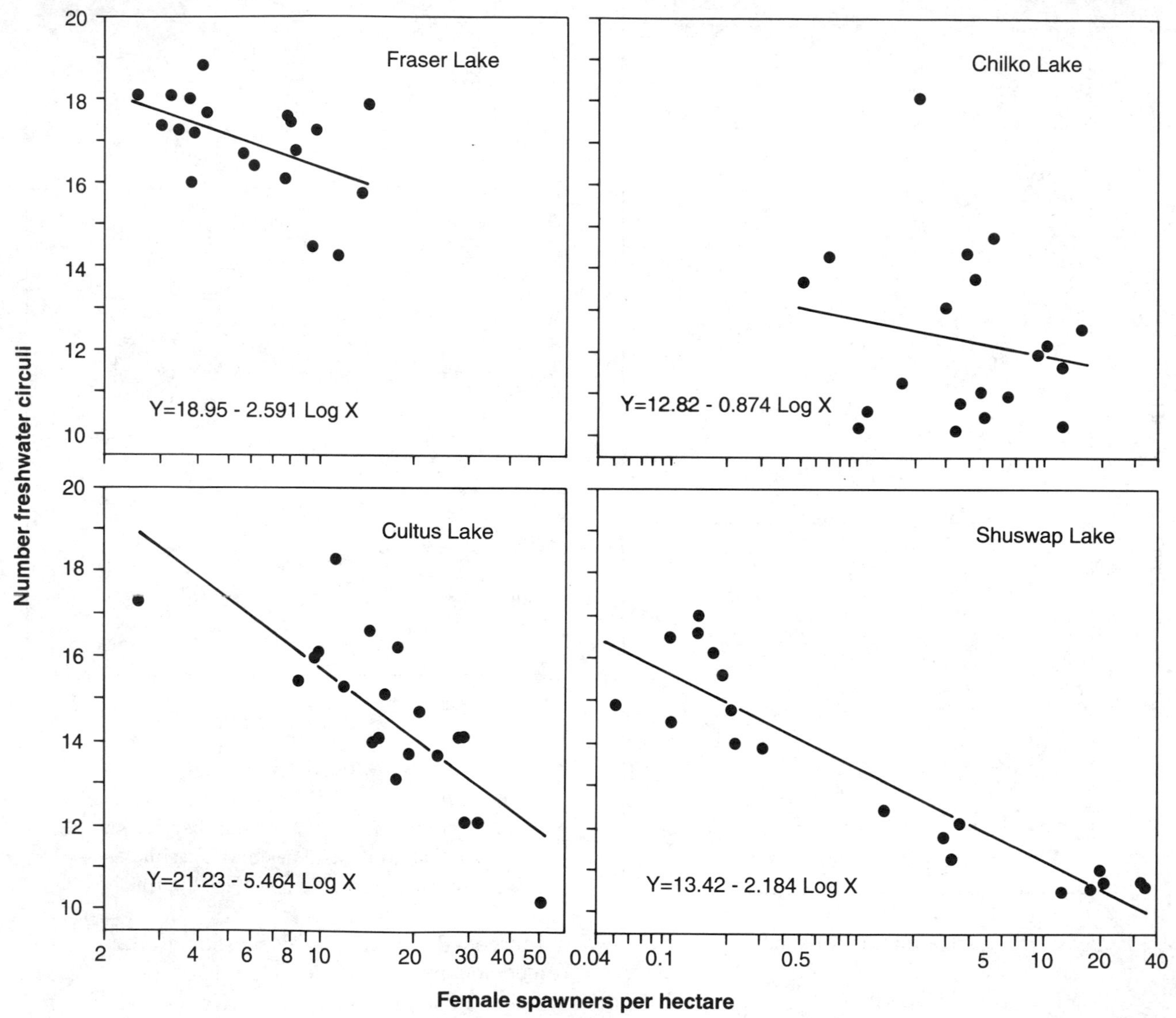

FIGURE 28

Female spawners/ha of lake area and freshwater scale growth of offspring for 20 years of spawning runs (1949–68) at Cultus, Chilko, Fraser, and Shuswap lakes. (From Goodlad et al. 1974, Figure 5)

Hume 1984; West and Mason 1987; McDonald et al. 1987). However, with this nearly threefold increase in average abundance, a decrease in average smolt size has occurred, suggesting that there has been an impact on the nursery area food supply (McDonald et al. 1987). This is perhaps to be expected because, although spawner densities per hectare of nursery area are still well below those seen in peak spawning years in major lakes of Bristol Bay and Fraser River, the fry production per spawner is much enhanced by the Babine Lake spawning channels (West and Mason 1987).

Although low spawner densities may be due to limitations in spawning area, overfishing, or low survival (progeny return per spawner), there are quite clearly great differences among lakes in rearing area capacity. In the Bristol Bay and Fraser River lake systems, average spawner densities per hectare of rearing area are generally below thirty adults (Table 9), and growth limitations are apparent above densities of 20/ha (Figures 24 and 26). However, average spawner densities several times higher are seen in Chignik and Karluk lakes (Table 9). During the period 1921-36, spawner densities in Karluk Lake averaged a phenomenal 282/ha and reached 634/ha (escapement data from Barnaby 1944). In spite of these high densities and the fact that most juveniles held over to age 2. or 3. before

TABLE 9

Comparisons of numbers of sockeye spawners per hectare of available lake rearing area in Pacific rim sockeye lake systems

Lake or lake system	Lake area (ha)	Spawning years	Spawning escapement (mean and range) (1000's)	Spawners/ha (mean and range)
Washington State				
Washington	8 760	1970-82	229(107-435)	26(12-50)
Fraser River				
Cultus	627	1949-72	19(3-56)	30(5-89)
Chilko	20 030	1949-72	199(19-596)	10(1-30)
Fraser	5 460	1949-72	62(25-141)	11(5-26)
Shuswap	30 960	1949-72	451(3-2120)	15(0.1-68)
Coastal BC				
Sproat	4 100	1970-82	69(25-215)	17(6-52)
Great Central	5 085	1970-82	136(15-253)	27(3-50)
Long	2 300	1970-82	99(21-214)	43(9-93)
Owikeno	9 600	1970-82	433(102-985)	45(11-103)
Babine	49 600	1946-82	688(102-888)	13(2-18)
Alaska				
Karluk	4 000	1921-53	902(400-2533)	266(100-633)
		1954-83	342(138-590)	86(35-148)
Chignik	6 100	1965-84	690(470-905)	113(77-148)
Iliamna-Clark	288 900	1964-83	6300(227-24326)	22(1-84)
Naknek	79 000	1964-83	1136(357-2665)	14(5-34)
Becharof	113 200	1964-83	834(329-1445)	7(3-13)
Ugashik	38 500	1964-83	683(39-3321)	18(1-86)
Wood R.	42 500	1964-83	1119(330-2969)	26(8-70)
Igushik	7 400	1964-83	383(60-1988)	52(8-269)
Nuyakuk	27 900	1964-83	426(20-3027)	15(1-108)
Kamchatka				
Kuril	7 600	1940-65	1512(300-4200)	199(39-553)
Dalnee	136	1937-68	23(1-100)	169(7-735)
Blizhnee	350	1938-60	36(4-83)	103(11-237)

Source: Adapted from Burgner (1987), Table 3

smoltifying, growth achieved by yearling and age 2. smolts was considerably better than that in Bristol Bay populations. Rounsefell (1958b) found no apparent density effect and concluded that the runs during this period were not sufficiently large for the smolts to have competed for the available food supply. A different situation exists in the Chignik Lake system (av. 113 spawners/ha). Here, juvenile growth is slow, particularly in Chignik Lake, and both inter- and intraspecific effects on growth have been found (Narver 1966; Parr 1972).

Because sockeye salmon spend a long time in fresh water as juveniles, they are infected by a great number of freshwater species of parasites (Margolis 1982a). Although the impact of parasitic infections on the health of wild Pacific salmon is generally poorly known (Margolis 1982a), two species of cestodes were observed to have an effect on growth of juvenile sockeye salmon. The better documented case is that of *Eubothrium salvelini*, a parasite acquired by the sockeye fry at an early stage and which develops to the ovigerous adult stage in the pyloric caeca of the juvenile sockeye during their freshwater period. In Babine Lake, approximately 30% of the sockeye yearlings are infected; the infected yearlings are typically smaller, less vigorous, have less stamina, and less capacity to adapt to sea water than uninfected smolts (Smith 1973; Boyce 1979; Boyce and Clarke 1983). In the Wood River lakes, plerocercoids of the cestode *Triaenophorus crassus* were found to parasitize a high percentage of the sockeye smolts (Burgner 1962b). This parasite is only found in sockeye in the Bering Sea drainage of western Alaska due to the co-presence of the northern pike, *Esox lucius*, its definitive host, in the regional lakes (Margolis 1982b). Because the plerocercoids destroy considerable tissue during development and encapsulation in the musculature of fingerling sockeye, they were suspected of severely weakening the fish if infection occurred while the fish were still small. There is evidence that charr predation is heavier on infected smolts during smolt migration (Rogers et al. 1972; Burke 1978).

Except for the initial period following emergence, described earlier, underyearling sockeye appear to be able to survive long periods of low food ration. Bilton and Robins (1971) determined that juvenile sockeye from Babine River stock could withstand many weeks of starvation under laboratory conditions and successfully resume growth when fed. Thus, mortality resulting directly from starvation is not likely once the juvenile sockeye have established a feeding regime in their lake environment.

Potential Competitor Species

Competition for common food or space resources during lake residence may exist in years when numbers of young sockeye in a year class are large, when there are two or more year classes utilizing the same resources, or when other species are also utilizing the resources. In the strict sense, competition exists only if the resources are in short supply, or if the fish seeking the resource harm one another in the process (Birch 1957). Interspecific competition implies an interaction between species that affects natality, survival, and/or individual growth rate in one or both species. If food is in short supply the juvenile sockeye or their competitors, or both, may grow more slowly. Interactive segregation may also occur, in which the interaction between two species forces them to magnify their differences in habitat or food selection (Nilsson 1965). Habitat choice, rate of growth, and the resultant length of stay in fresh water before smoltification influence survival of juvenile sockeye in fresh water primarily because these factors modify their vulnerability to predation. Predation, rather than starvation, is probably the primary source of mortality of juvenile sockeye in lakes.

Several non-salmonid species overlap with the juvenile sockeye in their lake distribution and food habits, and the fish communities and relative abundance of species vary from lake to lake. Species with overlapping food habits include the threespine and ninespine stickleback (*Gasterosteus aculeatus* and *Pungitius pungitius*), pond smelt (*Hypomesus olidus*), and pygmy whitefish (*Prosopium coulteri*). Lake whitefish (*Coregonus clupeaformis*), abundant in Morrison Lake, British Columbia, were reported to feed pelagically on the same plankters as young sockeye and to be a serious food competitor (McMahon 1948). Overall, the threespine stickleback is the most abundant species associated with juvenile sockeye salmon in the Wood, Kvichak, Naknek, Chignik, and Karluk river systems (Burgner et al. 1969), Lake Nunavaugaluk (Jaenicke et al. 1987), the Igushik River system (Rogers and Newcome 1975), the Nuyakuk River system (R.L. Burgner, unpublished data), as well as in several other Alaskan lakes studied. Threespine sticklebacks were slightly less abundant than ninespine sticklebacks in Chignik Lake but were much more abundant in nearby Black Lake. Pond smelt is sometimes locally abundant, e.g., in Coville Lake of the Naknek River system and in the Chignik lakes. It is the most abundant potential competitor in Lake Azabache in the lower Kamchatka River basin (Krogius and Krokhin 1956b). Pygmy whitefish, which are basically benthic, were highly abundant in trawl catches from Brooks Lake of the Naknek River system (Burgner et al. 1969). In Lake Dalnee, eastern Kamchatka, the ninespine stickleback occurs in large numbers in the littoral areas, and threespine sticklebacks are found mainly in the limnetic area (Krogius and Krokhin 1948). Sticklebacks are not present in Babine Lake, British Columbia, nor are any other potentially serious competitor of young sockeye (Withler et al. 1949). In Great Central Lake on Vancouver Island, threespine sticklebacks were found along shore but not in the limnetic zone (Manzer 1976). For 15 other sockeye lakes on Vancouver Island and the mainland coast of British Columbia, sampling indicated that only Alastair Lake had a dense population of threespine sticklebacks in the pelagic zone whereas Owikeno and Nimpkish lakes had lower population densities (Simpson et al. 1981). The threespine stickleback was observed to be the second most abundant fish in 17 lakes of coastal British Columbia, but it consistently occupied the limnetic zone during the summer growing season in only about half of these lakes (Hyatt and Stockner 1985). In the Fraser River system, no potential competitors of significance were reported by Goodlad et al. (1974) in the pelagic zones of Chilko, Shuswap, Fraser, and Cultus lakes.

Evidence of competitive interactions between juvenile sockeye (fry, underyearlings, and smolts) and limnetic threespine sticklebacks (juveniles and adults) was obtained in a series of experiments in which the two species were held separately and together in limnetic zone enclosures in oligotrophic Kennedy Lake, British Columbia (O'Neill and Hyatt 1987). The two species were stocked in the enclosures at densities within the range known to occur commonly in British Columbia coastal lakes. These two planktivore fish first reduced the abundance of the large entomostracan zooplankters and then maintained zooplankton communities which contained primarily nauplii and rotifers, prey seldom consumed. Sticklebacks larger than fry affected both quality and quantity of food available to sockeye, but stickleback fry affected only the quantity of food available. The authors concluded that, at these levels of predator and zooplankton prey, the limnetic-feeding sticklebacks would affect the growth and fitness of the juvenile sockeye.

In the Wood River lakes, the size of the combined populations of juvenile sockeye and threespine sticklebacks has an effect on the feeding and thus the growth of both populations (Rogers 1968, 1973a). An increased abundance of sockeye salmon fry resulted in a decreased growth of both species in early summer. In the littoral zone of lower Lake Aleknagik the diets of both species consisted predominantly of midges, but the sockeye fed more heavily on adult forms. In the limnetic zone the diets of both species consisted largely of entomostracans; again, winged insects were important in the diet of sockeye but were nearly absent from the diet of sticklebacks. Although annual variation in the relative abundance of the non-salmonid species did not appear to affect the growth of juvenile sockeye, there presumably was competition for food.

In a study of early summer food habits of the two species in three shoal area habitats of nearby Lake Nunavaugaluk (Bristol Bay), Jaenicke et al. (1987) found differences among the diets of sockeye fry and yearlings, and age 2 threespine sticklebacks. Adult insects (mainly chironomids) were an important food of the sockeye but not of the sticklebacks, and the sockeye fed less on benthos (mainly chironomid larvae), corroborating Rogers' (1968) finding that the sockeye fed more at the surface. Age 2 sticklebacks fed heavily on their own eggs. The major food of both sockeye and yearling sticklebacks was pelagic zooplankton, although there were differences in composition of species consumed. Potential growth effects of competition were not studied.

In the Naknek Lake system in Bristol Bay and at Karluk Lake (Kodiak Island), there was no evidence of density-dependent growth of sockeye. The impact of the several competitor species was not apparent, even though the Karluk Lake threespine sticklebacks were more abundant than sockeye in limnetic samples (Hartman and Burgner 1972).

In Black Lake of the Chignik Lake system, abundance of resident fishes apparently did not affect growth of sockeye underyearlings. The summer diets of sockeye, threespine and ninespine sticklebacks, and pond smelt overlap but differ in substantial ways, i.e., the sticklebacks depend heavily on aquatic insect larvae; the pond smelt on zooplankton; and the sockeye primarily on adult insects, to a lesser extent on zooplankton, and to a minor extent on insect larvae (Parr 1972). However, there was a strong inverse relation between the initial year-class strength of young sockeye in early summer and that of both threespine and ninespine sticklebacks. Apparently the stickleback fry had difficulty in establishing themselves in the presence of large numbers of sockeye fry, which enter the lake before the young sticklebacks hatch (Parr 1972). There was no evidence that the young sockeye were feeding to any extent on stickleback eggs or larvae.

In Great Central Lake considerable dietary overlap was observed between threespine sticklebacks and juvenile sockeye sampled together in littoral areas in May-July (Manzer 1976). It was noted, however, that during the year the two species occur together only in the littoral zone, and that sockeye salmon are found almost entirely in the limnetic zone. Therefore, it was inferred that because of the different distribution patterns of the two species, they are not serious competitors for food in Great Central Lake despite their similarity in diet.

The complexity of interactions of non-salmonids as competitor and buffer species to juvenile sockeye is apparent in Lake Dalnee (Krogius and Krokhin 1948, 1956b). Despite a relatively high plankton content in the lake and generally high growth rates for young sockeye, extensive intra- and interspecific competition was reported to have reduced their growth rate in some years and sometimes caused them to remain in the lake for a second or even a third year. Unusually large numbers of threespine sticklebacks competing for the same planktonic food were considered the chief reason for the very low numbers of seaward-migrant sockeye in three years (Krogius and Krokhin 1956a), although the mechanism of mortality apparently was not determined. The similarity in food of the two species in the pelagic zone of the lake was confirmed by Markovtsev (1972). Dolly Varden charr (*Salvelinus malma*) were thought to benefit young sockeye by destroying competing sticklebacks (threespine and ninespine). Sticklebacks were the dominant food of Dolly Varden during early summer, whereas juvenile sockeye were a minor food item (Savvaitova and Reshetnikov 1961). Hanamura (1966) noted that a higher abundance of sticklebacks apparently reduces the Dolly Varden predation on sockeye but potentially increases the competition for food between sticklebacks and sockeye.

Although interspecific competition for food and space does appear to occur between juvenile sockeye and other fish in some cases, intraspecific competition is more likely to influence growth and survival of juvenile sockeye because of their more direct intraspecific dietary overlap, schooling tendencies, and spatial associations. There are suggestions that interactive segregation between species occurs in some circumstances, but more conclusive evidence is needed. The potential for interspecific competition is greatest in instances where sockeye fry inhabit the littoral zone for some time before moving offshore. In the limnetic zone of most sockeye lakes, the sockeye greatly outnumber potential competitor species, and serious competition, if it occurs, is intraspecific in form. Kokanee, if present in abundance, are potential intraspecific competitors, particularly because several year classes are present in the lake simultaneously.

Predation in Fresh Water

Predation mortality in fresh water can best be discussed in relation to life cycle stages of the sockeye, beginning with the adult salmon as they enter the spawning grounds and carrying through the various juvenile stages: (1) egg deposition period, (2) eggs and alevins in the redds, (3) fry emergence and migration to the lake, (4) inshore fry stage, (5) pelagic feeding period, and (6) seaward migration. Because predation on adult sockeye by sea gulls and bears has been referred to earlier, this discussion will focus on predation on juvenile stages.

Many observations have been made of the feeding activities of predatory fishes and birds on eggs during sockeye spawning. However, most observers have concluded that the bulk of eggs eaten were dug up by late-arriving spawners and would have had a low chance of survival (Foerster 1968). Predatory fishes include trout, charrs, and cottids, while predatory birds include gulls, mergansers, and water ouzels (dippers). Eggs dislodged from the spawning of Adams River sockeye were found to be an important component of the diet of rainbow trout (*O. mykiss*) in lower Shuswap Lake during and following dominant-year spawning (Ward and Larkin 1964). For example, a portion of the eggs deposited in Adams River during heavy spawnings in the autumns of 1954 and 1958 were subsequently swept down into Shuswap Lake and served as a major diet item of rainbow trout. At Karluk Lake on Kodiak Island, both Arctic charr (*Salvelinus alpinus*) and Dolly Varden also scavenged heavily on dislodged sockeye salmon eggs (Morton 1982).

Less is known about predation on eggs and alevins in the redds. Physical and chemical factors such as desiccation, freezing, lowered oxygen resulting from siltation, reduced flow, and dislodgement by later spawning fish or shifting gravel are probably more important as mortality factors.

The period of fry emergence and migration to the nursery lake is a period of high vulnerability to predation by other fish species and by birds. This predation is only partially reduced by the nocturnal movements of downstream migrating fry. Mortality from predation varies considerably according to the spawning area type and location (creek, river, spring area, or lakeshore), the kinds and numbers of predators at hand, and physical conditions such as stream length, water turbidity, stream flow, and light conditions. In Scully Creek, Lakelse Lake, the principal predators were coho yearlings, cutthroat trout (*O. clarki*), Dolly Varden charr, and sculpins (*Cottus asper*) (Foerster 1968). Calculated loss due to predation during fry migration to the lake was estimated at 63%–84% during four years of observation. In Six Mile Creek, Babine Lake, rainbow trout were the principal predators, and it was estimated for one year that about two-thirds of the emerging fry became the prey of predators during their migration to the lake (Foerster 1968). In Karymaiskiy Spring (Bolshaya River, Kamchatka Peninsula), charr and juvenile coho (yearling and older) were principal predators (Semko 1954a, 1954b). Consumption of sockeye fry was reported to vary according to number of predators as well as relative numbers of sockeye, pink, and chum fry migrating from Karymaiskiy Spring. The latter two species served as buffers. Estimated predation rates on sockeye ranged between 13% and 91% during the eight-year study. In the Cedar River, Lake Washington, predators on sockeye fry include rainbow trout, juvenile coho and chinook salmon, sculpins, mountain whitefish (*Prosopium williamsoni*), and some birds (Stober and Hamalainen 1980). In two tests, fry mortality over a 33-km length of river was estimated at 25% and 69%. Although these sets of data probably do not encompass the potential ranges of predation and mortality encountered during the fry migration stage, they are sufficient to indicate that losses can be extensive.

During the inshore feeding stage, sockeye fry schools are often in company with a number of other prey species of fish, which in western Alaska and the Kamchatka lakes usually include three-spine and ninespine sticklebacks, cottids, and fry of Arctic charr or Dolly Varden charr. Potential predators, at least in western Alaska, include Arctic charr, rainbow trout, Dolly Varden charr, lake trout (*Salvelinus namaycush*), juvenile coho salmon, northern pike, and bird predators, particularly Arctic terns (*Sterna paradisaea*) and Bonaparte's gulls (*Larus philadelphia*) (Hartman and Burgner 1972). The habit of sockeye fry to remain in very shallow water and the presence of numerous other prey species buffers them to some extent from predator fish. In Shuswap Lake, where fry dispersal from the Adams River into the lake coincides with smolt migration out of the lake, Ward and Larkin (1964) hypothesized that rainbow trout predation on the fry was buffered by concurrent predation on the sockeye smolts. In the Naknek Lake system, Bristol Bay, where extensive downstream interlake migration of sockeye fry occurs in midsummer from Coville Lake into Grosvenor Lake, lake trout fed actively on the migrating fry (Ellis 1974). Interestingly, the fry migration occurred in both daylight and darkness, with no apparent behaviour to seek protection during darkness; however, in Grosvenor River, the migrant fry moved primarily at night (Ellis 1974).

The longest period of freshwater residence is the pelagic feeding period, which is ten or eleven months for those juveniles that smoltify as yearlings, but may be one or two years longer for the juveniles migrating at age 2. or 3. Feeding, schooling, and resting strategies that minimize exposure to predation during this longer period have obvious survival value. Eggers' (1982) feeding strategy model for Lake Washington sockeye emphasized predator avoidance as a key feature. Generally, the important predators on sockeye, such as Arctic charr, rainbow trout, and squawfish (*Ptychocheilus oregonensis*), tend to be benthic and are inshore feeders much of the time, which decreases their contact with the limnetic sockeye. In Bristol Bay lakes, Arctic charr concentrate around lake outlets to feed heavily on smolts during their seaward migration, then disperse and turn to benthos after the smolt migration (Nelson 1966; Moriarity 1977). As noted by Hartman and Burgner (1972), some predation in the limnetic area of lakes by such species as Arctic charr and lake trout is known to occur, but buffer forage fish species such as sticklebacks, pond smelt, and pygmy whitefish are more frequent in their diet. Predation in Cultus Lake, where buffer species are lacking, appears to be an exception.

The most conspicuous predation on juvenile sockeye occurs during the smolt migrations from the lakes. In Bristol Bay lakes the lake trout and Arctic charr congregate to take advantage of the feeding opportunity provided by the concentrated migrations of the smolts during passage between lakes or at the trunk river outlet (Hartman and Burgner 1972). These predators are joined by Arctic terns and Bonaparte's gulls, which dive to capture surfacing smolts. Glaucous-winged gulls and short-billed gulls (*Larus canus*) also feed on the smolts (Meacham and Clark 1979). Hartman et al. (1967) noted that "the simultaneous predation on sockeye smolts by fish from beneath and birds from above is readily observed along migration pathways. Often the activity of fish predators will cause the water to boil during migration of smolts" around lake outlet areas.

It has been hypothesized that such predation is "depensatory" (Neave 1953; Ricker 1954), being proportionately lower at higher levels of prey population, either because the predators become satiated during an evening feeding or because the predator population is limited in the number of smolts that it can capture. Depensatory rainbow trout predation on fry and smolts was hypothesized as a controlling factor perpetuating the strongly cyclic sockeye run to the Adams River (Ricker 1950; Ward and Larkin 1964; Larkin 1971). At Little Togiak River, Bristol Bay, where evening abundance of migrating smolts was usually low, daily consumption rates by Arctic charr increased with smolt abundance. However, in years with relatively greater smolt abundance, consumption rates by charr were disproportionately lower (Ruggerone and Rogers 1984). Thus, mortality as a percentage apparently increases with increased numbers of migrants up to a threshold density, above which mortality decreases. In Little Togiak River, smolts experience less risk of predation at daily migration abundances of about 20,000 smolts or greater; but migration densities of that magnitude are rare (Ruggerone and Rogers 1984).

That smolt predation by charr can be a significant source of mortality in the Wood River lakes was shown in a 1971 study at Agulowak River, where the largest population of charr concentrate during the interlake migrations of the smolts. The charr population of some 13,000–14,000 fish was estimated to have consumed 3–4 million smolts in 1971, or between one-third and two-thirds of the migrants in a year of relatively low smolt population magnitude (Rogers et al. 1972). Charr predation on smolts occurs at other locations in the lake system, most notably at the other interconnecting rivers, and overall predation is great enough to be a probable regulating mechanism in sockeye salmon production in the Wood River lake system. The potential impact of fish and bird predators on juvenile sockeye in Alaskan lakes was recognized many years earlier, and various predator control programs were instituted over a twenty-year period ending in 1940 (Meacham and Clark 1979), when their value was finally questioned (Hubbs 1940; DeLacy and Morton 1943). Since no evaluations of results of predator removal were conducted, their impact remains unknown. A new charr predation control program was initiated by the Alaska Department of Fish and Game in the Wood River lakes in 1975. This program involved the live capture and confinement of Arctic charr concentrated at river mouths during the smolt outmigration and subsequent release of the charr when the annual smolt migration was completed (Meacham and Clark 1979).

The nocturnal and seasonally highly concentrated migrations of sockeye smolts in the lake outlet rivers clearly result in reduced predation. Although at the latitude of Bristol Bay complete darkness seldom occurs during the period of migration from late May to early July, decrease in predator activity is still readily apparent during the darkest periods of the evening. Arctic terns and Bonaparte's gulls cease feeding with the onset of darkness (R.L. Burgner, unpublished data) and, at Little Togiak River, feeding activity of charr was observed to decrease substantially during the darkest one to two hours of the night (Ruggerone and Rogers 1984; R.L. Burgner, unpublished data). A shortened period of heavy smolt migration appears to have survival advantage over more protracted migrations at lower daily magnitude.

Kokanee and Residual Sockeye

Although the sockeye salmon is typically anadromous, usually migrating to the ocean after one or more years in a lake, there are two forms that remain in fresh water to mature and reproduce. The more distinct of the two is the kokanee, which may be found in lakes with or without anadromous runs. Their populations generally exist quite independent of anadromous runs. The second form is the so-called "residual" sockeye (Ricker 1938), which are, at least in part, progeny of anadromous parents. Except for their small size, the kokanee resemble anadromous sockeye in general appearance and in bright spawning coloration. The residual sockeye, on the other hand, are mostly males and do not develop strong secondary sex characteristics at time of spawning.

In the two lakes where residualism has been most studied (Cultus and Dalnee lakes), it was found that the progeny remaining behind in the lake to mature tended either to be the larger of the yearlings at normal smoltification time or the smaller of the yearlings that gained above average growth during their second growing season (Ricker 1938; Smirnov 1959a; Krokhin 1967). The residual component tends to mature at an earlier age than the anadromous component and the males earlier than the females. Because of their small size, female fecundity is much reduced and egg size is also smaller, resulting in less robust fry (Smirnov 1959a). Indications are that time of spawning is not synchronized precisely with the anadromous component, and success of reproduction may be low. The relative production of anadromous and residual progeny from spawning of residuals remains a question. The residuals in a sense resemble the precocious anadromous jack and jill sockeye in that they have a distorted population sex ratio and a tendency to mature early. Residual sockeye are absent or scarce in most sockeye lakes, particularly in more northern stocks.

The kokanee, on the other hand, have fully adapted to a freshwater existence and presumably diverted from a common anadromous stock in recent geological times (Ricker 1940); however, the process of evolution is not clear. In the Fraser River drainage, native kokanee are present in a number of lakes with anadromous sockeye runs; in the Columbia River drainage in Washington, Oregon, Idaho, and British Columbia, many lakes without anadromous sockeye in recent history contain kokanee (Nelson 1968); and on Vancouver Island, kokanee are also present in lakes with and without anadromous runs (Dymond 1936; Ricker 1940). In Washington State, Lakes Whatcom, Sammamish, and Washington contain native kokanee populations and no native anadromous sockeye. Native landlocked populations of kokanee are also found in lakes in the Yukon Territory (Nelson 1968), in three Kamchatkan lakes (Krokhin and Krogius 1936; Ostroumov 1977), and, formerly, in Lake Akan, Hokkaido (Oshima 1934). They are also found in a number of lakes inaccessible to anadromous sockeye, at least in recent geological times. Kokanee are uncommon in Alaska (Rounsefell 1958a; Nelson 1968). They have been introduced successfully into a number of lakes in the western United States, Hokkaido, and elsewhere (Seeley and McCammon 1966).

Kokanee differ from residual sockeye by having a normal sex ratio, a bright coloration at maturity, a greater absorption of scale margins at maturity, a lower infestation by the parasitic freshwater copepod genus *Salmincola* (Cultus Lake), and they spawn earlier (Ricker 1940). As noted, their secondary sex characteristics at spawning mirror those of anadromous sockeye. Spawning may occur both in streams and in lakeshore areas. It is generally segregated in time and area from anadromous spawnings. Size at maturity varies considerably among lake populations and with age. Among populations, mean lengths vary from 18 to 30 cm. The principal food of kokanee is similar to that of young sockeye, i.e., pelagic zooplankton and insects. Thus, there is a potential for intraspecific competition in lakes where both forms are present. As with anadromous sockeye, distinct sub-populations may develop within a single lake (Vernon 1957; Chernenko and Kurenkov 1980).

Seaward Migration

As the daily photoperiod and air temperatures increase in the spring and, in the more northern or higher elevation sockeye lakes, the ice cover melts and spring overturn occurs, the juvenile sockeye destined to migrate to sea are going through a preparatory phase. During this phase they undergo morphological, physiological, and behavioural modifications that result in a state of migratory readiness and disposition (Groot 1982). Groot stated: "From a functional point of view, the purpose of the preparatory phase is to ready the fish to i) make the move to the new habitat and ii) survive in the new habitat. The move to the new habitat [the ocean, in this instance] involves modifications in swimming ability, energy storage and utilization, and initiation of direction finding processes. Survival in the new habitat requires changes in osmoregulation, feeding, predator avoidance, and social interactions. The ultimate biological goal is increased survival from egg to adult stage."

The parr-smolt transformation involves changes in activity, in coloration, in shape, and in tolerance to sea water. Evidence suggests that the overall controlling force in juvenile salmonids is an innate, endogenous circannual rhythm (Hoar 1965, 1976; Wedemeyer et al. 1980; Groot 1982). In evaluating the results of several investigators, Groot (1982) concluded that silvering of the body is under strong endogenous control, whereas growth patterns and activity level are influenced by environmental factors. Increasing photoperiod is believed to be a major environmental controlling factor, synchronizing the endogenous rhythms with the seasonal changes, and temperature plays its part by regulating the rate of physiological responses to photoperiod by causing effects to appear sooner or later at higher or lower temperatures (Clarke et al. 1981; Groot 1982). As Hoar (1965) expressed it, the endocrine system forms the chemical link between the organism and the environment, with hormones affecting physiology and behaviour by elevating rates of metabolism, increasing growth, altering excitability, or changing alertness to stimuli (Hoar 1976; Groot 1982). The following is a summary of the salient points for the preparatory phase of seaward migration of sockeye smolts, modified from Groot's general summary for salmonids.

1 There is a strong endogenous component involved in triggering the diel and seasonal morphological, physiological, and behavioural changes during transformation from parr to smolt.
2 This endogenous rhythm only results in transformation when the juvenile sockeye reach a certain size (physiological age).
3 Daily photoperiod increase entrains the endogenous rhythm and thus controls the onset of smolting.
4 Temperature influences the onset of smolting through growth and regulates the magnitude and duration of the smolting process.
5 Hormones act as chemical links between the organism and the environment.
6 Hormone activities are rhythmic (circadian) and their effect on the fish depends on the phase relationship of their peaks of activity.
7 Entrainment of the hormonal rhythms to the diel cycle is through photoperiod, whereas the temporal relationship of the different rhythms is set by temperature.
8 The co-ordination of smolting processes leading towards migration readiness is attained in differ-

ent ways for separate stocks as a result of adaptations to environmental factors operating in the habitat.

The change in external appearance of the juvenile sockeye at smoltification is pronounced. In the early fry stage, body markings and coloration are quite distinct. Short, black, elliptical or oval parr marks are spaced vertically along the sides of the fish, extending to, or slightly below, the lateral line, and a row of definite black spots shows on both sides of the bluish or greenish back. Upon transition of the juveniles to a pelagic feeding regime in the lake, parr marks become less distinct, the sides more silvery, and the back darker, as is more characteristic of pelagic-feeding fish. But further changes occur at the time of smoltification; the fish become slimmer, more streamlined, and silvery. The parr marks are obscured by a thickening of the purine layers beneath the scales and deep in the dermis adjacent to the muscle, with an accompanying increase in the guanine:hypoxanthine ratio (Hoar 1976). Hoar notes that "although the adaptive significance of the silvering in a pelagic salmon seems obvious, its phylogenetic basis is entirely speculative."

Most sockeye lakes are covered by ice in winter, and the smolts usually begin to migrate within days of ice breakup and with the accompanying slight warming of lake outlet waters. Consequently, there is a cline from south to north in timing of the smolt migrations at lake outlets, influenced also by the altitude of the lake and the multi-basin characteristics of the lakes involved (Figure 29). As noted by Hartman et al. (1967), smolt migration is the earliest (usually late April) in Cultus Lake, which seldom freezes over (Goodlad et al. 1974). It is latest (late June and early July) at Tazlina Lake in the interior Copper River watershed, the northernmost of the lakes included. For single lake basins such as Cultus, Chilko, Karluk, and Brooks lakes, the smolt exodus covers a shorter time interval than for outlet lakes of multi-lake basins or lake chains, such as the Wood River and Naknek lakes (Hartman et al. 1967). The more extended migrations in the latter lakes result from differences in timing of ice breakup among the different basins (Burgner 1962b) as well as in migration distance to the outlet (Johnson and Groot 1963; Hartman et al. 1967; Groot 1972).

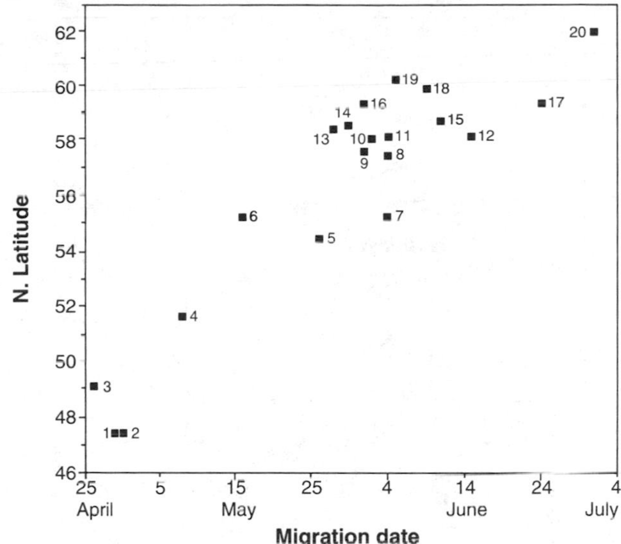

FIGURE 29

Mean date of sockeye smolt migration for North American lake systems located at different latitudes. *1* - Quinault, *2* - Washington, *3* - Cultus, *4* - Chilko, *5* - Lakelse, *6* - Nilkitkwa, *7* - Babine, *8* - Karluk, *9* - Ugashik, *10* - Becharof, *11* - Kitoi, *12* - Ruth, *13* - Auke, *14* - Brooks, *15* - Naknek, *16* - Kvichak, *17* - Wood River, *18* - Tikchik, *19* - Kasilof, *20* - Tazlina. (Data for 3, 5-8, 11-14, and 20 from Hartman et al. 1967, Figure 28)

The modes used by smolts to find their way to their lake outlet to commence seaward migration have been considered by a number of investigators (Burgner 1962b; Johnson and Groot 1963; Groot 1965, 1972; Hartman et al. 1967; Foerster 1968). The complexity of migration routes varies greatly from the single-basin lake to the multi-basin lake chain such as in the Wood River lakes of Bristol Bay (Figure 30). In this system the smolts originating in the uppermost lake (Kulik) must first find the lake outlet, then successively migrate southeasterly through Lake Beverley to the Agulukpak River, westerly, southeasterly, easterly, and westerly through Lake Nerka to the Agulowak River, and finally southeasterly through Lake Aleknagik to reach the trunk stream, the Wood River, a distance of about 150 km. Because only in Lake Aleknagik is the outlet at the very end of the lake, there are numerous opportunities for wrong turns. There are a number of opinions as to how smolts find the final lake outlet to descend to the sea (Foerster

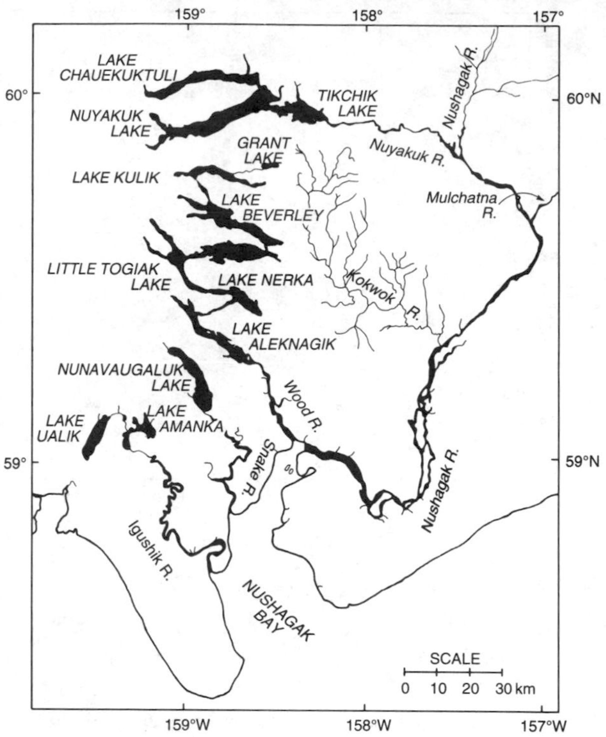

FIGURE 30
The Wood River lake system and adjoining river systems in the Nushagak district, Alaska

1968). These vary from the concept of a simple following of the shoreline until the current of the outflowing stream is detected to that of a migration directed by an innate compass orientation mechanism. Groot (1965, 1972) demonstrated that in multi-basin Babine Lake, seaward migration is a well-oriented movement from all parts of the lake to the outlet, and he established experimentally that sockeye smolts apparently have direction-finding systems for locating the outlet that include sun-compass and polarized light cues. A shift to a non-celestial mode of orientation was indicated with increasing cloud cover, apparently cued by the earth's magnetic field (Quinn 1982a; Quinn and Brannon 1982).

Concerning Groot's findings, Foerster (1968) comments: "The idea that young salmon possess a built-in compass and chronometer whereby, through some form of celestial navigation varying according to time of migration, they may respond to an inherited direction-preference orientation and make their way through and out of a lake to the sea is a novel and challenging one worthy of close attention and further investigation both experimentally in appropriate apparatus and in the field under natural conditions." From my own observations and those of D.E. Rogers (Fisheries Research Institute, University of Washington, Seattle, Washington, pers. comm.) in the Wood River lakes, it is probable that the migrants also make extensive use of shoreline orientation, since they are observed actively migrating close inshore and are not frequenting the deeper offshore areas of the lakes in the spring in the manner observed during summer-autumn feeding. Groot (1982) appropriately pointed out that "different mixes of direction finding mechanisms can exist within the migrations of one stock, as well as between similar migrations of separate stocks. The redundant nature of orientation systems and the wide range of interactions between components during evolution have resulted in complex spectra of direction finding in Pacific salmon." Concerning the navigational capabilities of juvenile salmonids, Hoar (1976) concluded that, among juvenile salmonids, sockeye seem to have the most highly developed navigational capabilities.

Additional observations by Groot (1972), utilizing sonar and time-lapse photography, established a diel periodicity in migratory activity of smolts in Babine Lake towards the lake outlet, with directed movements evident only during dusk, dawn, and early morning periods. During twilight observation periods the smolts swam close to their maximum continuous performance of about five to six body lengths/s (40–50 cm/s), determined experimentally by Brett (1965) for smolts of the size and in the water temperature range encountered in the lake. From tagging and recovery experiments in Babine Lake, Johnson and Groot (1963) calculated that the average migration rate was 5.1 and 7.8 km/d in two separate years. This suggested that the smolts would need to spend at least 6.5 to 8.5 hours each day migrating if they followed a direct course (Groot 1972). The daily rates of travel are similar to those reported by Burgner (1962b) for smolts in the Wood River lakes.

Juvenile sockeye travel in schools during their inlake and seaward migrations. Observations of smolt movement in the shallow outlet end of Lake Aleknagik indicated the presence of only a few schools during the day. In the early evening the number of schools increased but without decisively directional movement. From approximately twilight through the dark hours, movements became decisive with all schools swimming definitely and directly towards the lake outlet (Burgner 1962b). Because larger smolts tended to leave the lake earlier in the evening, it was hypothesized that the schools of larger fish either have a greater migration stimulus or traverse the final distance from the daytime milling area to the river more swiftly (Burgner 1962b).

In Iliamna Lake, the movement of juveniles towards the outlet at the western end of the lake is progressive (Mathisen 1966; Burgner et al. 1969; Hartman and Burgner 1972). The major spawning areas are in or near the eastern end of the lake. Most age 0 sockeye rear in the eastern two-thirds of the lake, at least through September. Those that do not leave the lake the following spring are most abundant as yearlings in the central part of the lake in early summer and are also abundant in the western part by August and September. By the following spring this westward movement must be complete because over-wintering yearling fish leave the lake as age 2. smolts in a matter of days during or immediately after lake ice breakup. The smolts migrating at age 1. lag behind, perhaps because many have not completed their migration to the lake outlet area by breakup time.

The annual exodus of smolts from the larger single basin lakes can be truly impressive since most of the migration occurs over a period of a very few days. For example, the estimated outmigration of smolts from Iliamna Lake totalled 270 million sockeye in 1978, with a weight of over 2000 t. Daily migrations can exceed 60 million smolts, with the bulk of the migration generally occurring during the four-hour period between 2200 and 0200 h. As Foerster (1968) noted for Cultus Lake, considerable variability occurs from year to year in regard to time when the migrations begin, when the majority leave the lake, and when the runs terminate. Factors causing variation in timing of sockeye smolt migrations include timing of lake ice breakup and subsequent water temperatures; extent and direction of wind action on the lake; and size, age, and physiological state of the smolts.

Data summarized by Foerster (1968) and Hartman et al. (1967) showed that smolt migrations in most lakes do not begin in strength until spring water temperatures rise above 4.4°C, although migrations may start at colder temperatures in some lakes. For example, migrations started at about 3.3°C at Chilko Lake in British Columbia, and more recent data for Iliamna Lake showed substantial migratory activity during years of late ice breakup at temperatures as low as 2°C (D.E. Rogers, Fisheries Research Institute, University of Washington, Seattle, Washington, pers. comm.). In these years, migrations in the outlet river (Kvichak River) often coincided with river flow of lake ice, driven to the outlet by downlake winds. Annual differences in migration timing appeared to result primarily from differences in lake temperature, although there was considerable year-to-year variation in threshold temperature for migration. In most sockeye lakes, migration is essentially over before water temperatures reach 10°C. This relation between seasonal temperatures and migration timing is shown in Figure 31 for the Wood River lakes.

Varying wind has been suggested as causing

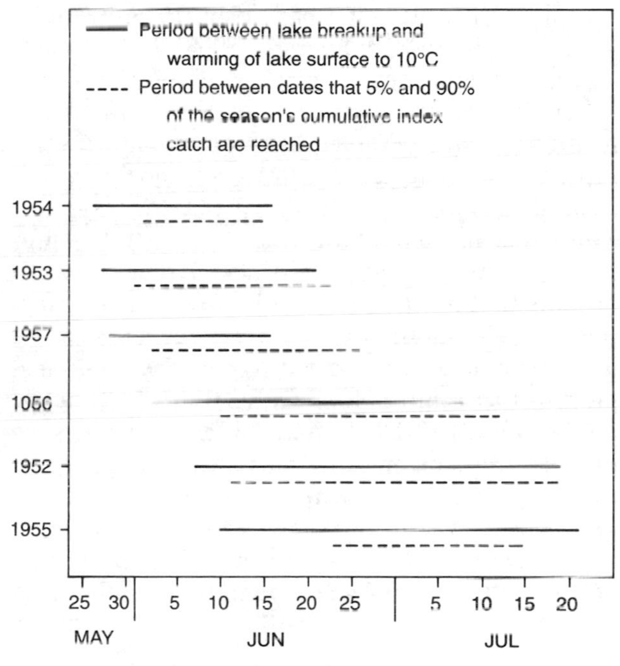

FIGURE 31

Relationship between spring climate and timing of the sockeye smolt migration, Wood River lakes, 1952 through 1957. Years arranged in sequence of ice breakup dates at Lake Aleknagik. (From Burgner 1962b)

fluctuations in the intensity of sockeye smolt migrations. Hartman et al. (1967) cited reports that smolt movements diminished when winds were offshore at the lake outlet but increased on days when winds were onshore, i.e., towards the lake outlet. These observations were made at Chignik, Ugashik, Becharof, Karluk, and Brooks lakes. Similar behaviour was reported from Lake Dalnee smolts (Krogius and Krokhin 1948). Hartman et al. (1967) suggested that when the wind is onshore, surface currents "pile up" warmer surface waters at the outlet, resulting in an increase in outlet stream temperatures and sometimes turbidity, which increases the intensity of migration. However, the opposite effect was reported at Cultus Lake, where winds blowing towards the outlet were reported to retard migration, presumably because of wave action and turbulence set up at the shallow lake outlet (Foerster 1968).

Within-season trends are often observed in size and age composition of migrating smolts. In many sockeye populations, smolts migrating earlier in the season tend to be larger than later migrants of the same year class (Gilbert 1916, 1918; Barnaby 1944; Burgner 1962b; Foerster 1968). In instances where exceptions have been noted, they appear to be either the result of the mixing of migrant groups in multi-basin watersheds or of the addition of rapid new spring growth in those smolts lagging behind (Dombroski 1954; Burgner 1962b; Foerster 1968). It is also commonly observed that the older sockeye smolts (which are also larger) tend to migrate first in the spring and that the smaller fish tend to remain an additional year or more in the lake before migrating seaward (Barnaby 1944; Krogius and Krokhin 1948; Burgner 1962b). As noted by Barnaby (1944), fish in the older age groups are usually the slower growing fish of the progeny from a particular spawning, and the urge to migrate seaward is related to the size and growth rate of fingerlings. Thus, the size and age composition pattern in the annual smolt migration is influenced heavily by the physiological readiness of the fish to undertake their transition from fresh to salt water. In most lake systems sockeye smolts migrate from the lake during the darkest hours of the day, between sunset and early morning (Kerns 1961; Burgner 1962b; Groot 1965; Hartman et al. 1967). Daily migrations usually peak around or before midnight, with well over 90% of the fish migrating in a period of four to five hours. The timing of daily peaks has been observed to shift with seasonal change in onset of darkness (Kerns 1961; Hartman et al. 1967), with degree of cloud cover (Foerster 1968), and with outlet turbidity caused by wind action (Hartman et al. 1967). Because daily migrations often tend to "tail off" after midnight, the pattern is probably also influenced by the daily accumulation of smolts waiting near the lake outlet to migrate (Burgner 1962b).

The migrating smolts leave the lake in schools, orient downstream, and swim faster than the current where the flow is uniform and quiet. They turn and pass downstream tail first in turbulent water or where weirs obstruct their movement (Barnaby 1944; Hartman et al 1967; Foerster 1968; R.L. Burgner, unpublished data). Often schools of smolts will hold above a weir for some time before passing through, tail first. Thus, the distance travelled downstream during the initial day is influenced by the time of day a smolt school initiates migration and the character of flow in the river. Rates of travel averaging up to 40 km/d over distances of about 575–650 km have been measured in the Columbia River (Anas and Gauley 1956). It is not known whether or not downstream migration to salt water, once begun, is continuous night and day. This may vary to some extent according to the degree of turbidity in the trunk river.

The schooling behaviour, the active rather than passive migration downstream, and the nocturnal movement are innate behavioural adaptations that decrease mortality from predation.

Early Sea Life

After smoltification and exodus from natal river systems in spring or early summer, young sockeye enter the marine environment where they reside for one to four years, usually two or three years, before returning to spawn. Depending on the stock, they may reside in the estuarine or near-shore environment before moving into oceanic waters, but in any event they are typically distributed in offshore waters by autumn following out-migration.

Sockeye smolts entering marine waters show a wide variety of adaptations to estuarine and near-shore conditions. They encounter differences in topography, salinity, sea temperature, turbidity, tides and currents, food abundance, predator populations, and river influence. These contrasting early marine environments have moulded different patterns of growth, survival and population behaviour of the juvenile sockeye. The annual variations in conditions encountered in individual environments at this early sea life stage are generally believed to be largely responsible for the variation seen in overall marine survival of cohort populations. Contrast in early marine environment and behaviour is exemplified by comparison of the Fraser River and Bristol Bay situations, described, respectively, by Straty (1974), Straty and Jaenicke (1980), and Groot and Cooke (1987).

Yearling Fraser River smolts enter the deep and relatively plankton-rich waters of the Strait of Georgia in April and May (Figure 32). Groot and Cooke (1987) investigated the seaward migration routes of Fraser River sockeye smolts. They found that the smolts first concentrate around the Fraser River mouth. By late May the smolts have dispersed north along the mainland coast and west towards the Gulf Islands. The smolts concentrate around the Gulf Islands, then move north. Most leave the Strait of Georgia in late June and July via the northern passages, to enter the open ocean. Their migration rate through the Strait of Georgia is estimated at 6-7 km/d (Groot and Cooke 1987). They then migrate northwestward along the coast of British Columbia and Alaska (Hartt and Dell 1986).

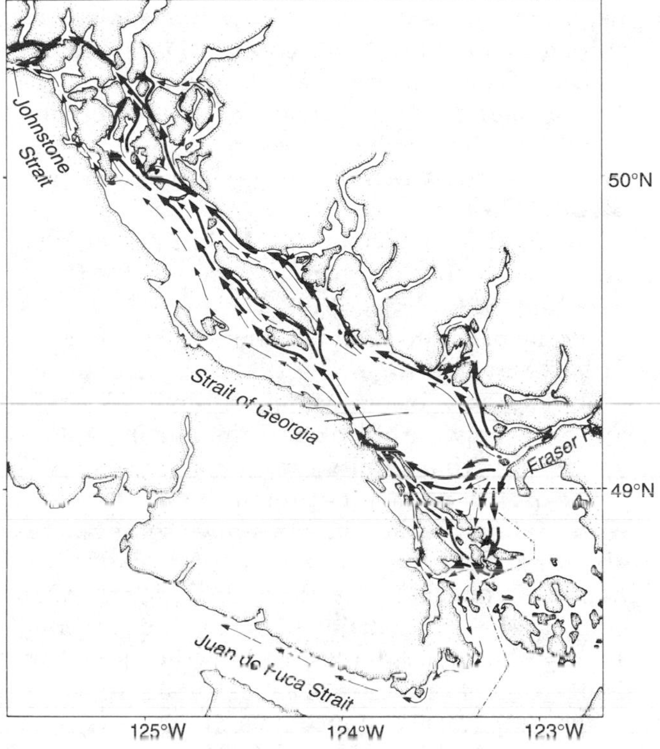

FIGURE 32

The assumed migratory routes of Fraser River sockeye salmon smolts in the Strait of Georgia on their way to the Pacific Ocean. (From Groot and Cooke 1987, Figure 5)

During the initial marine period in the Strait of Georgia, the yearling sockeye forage actively on a variety of organisms, apparently preferring copepods and insects, but also eating amphipods, euphausiids, and fish larvae when available (Healey 1980). Their growth rate is about 0.6 mm/d. In the Gulf Islands area, where they concentrate along with juvenile chum and pink salmon, interspecific differences in diet are seen. Chums concentrate on the larvacean *Oikopleura*, pinks on small copepods and invertebrate eggs, and sockeye on copepods, amphipods, and insects (Healey 1980).

A minor portion of the spring downstream migration is comprised of sockeye fry from spawning areas such as the Harrison River. The fry may tarry in marshes and sloughs of the Fraser River Delta for as long as five months, and those remaining until late summer-early fall reach an average size of 65-75 mm in fork length (Birtwell et al. 1987). These underyearlings enter the Fraser River plume beginning in late July and gradually disperse throughout the Strait of Georgia. They are found well dispersed in the strait in August and September and reach a size similar to the yearling smolts by September (Healey 1980). It is not known exactly when the age 0. sockeye leave the Strait of Georgia.

The conditions encountered by juvenile sockeye entering Bristol Bay are in sharp contrast to those experienced by Fraser River sockeye. The smolts enter the bay from several river systems primarily in late May and June. The shallow waters of the inner bay[4] and lower portions of the trunk rivers are extremely turbid due to the stirring action caused by winds and strong tidal currents. Because of river runoff, the inner bay exhibits a distinct salinity gradient. Juvenile sockeye are abundant in inner Bristol Bay for about two months, from late May through late July. After early August, the major population of juvenile sockeye has shifted seaward into the outer bay. The period of residence of a particular river stock varies with travel distance from the river of origin to the outer bay (roughly 145-280 km), and ranges from about two to six weeks (Straty 1974). Most of the fish are in the outer bay for an undetermined length of time after mid-August. The general seaward migration route in the inner bay is through the region of the most pronounced salinity gradients (Figure 33) and is generally near the coast rather than offshore in lower temperature areas (Straty and Jaenicke 1980). As a result of annual variations in river and sea temperatures, considerable year-to-year changes occur in the time that the smolts enter the bay and the length of residence in the inner and outer bays. Spring weather (water temperature) in the year of seaward migration directly influences size attained by the immature sockeye by the following summer (Rogers 1980).

4 Inner Bristol Bay includes that portion of the bay northeast of an approximately 220-km-long line between Port Heiden and Hagemeister Island.

The inner coastal waters contain a significantly lower abundance of zooplankton than offshore waters in the early summer, and a larger proportion of the sockeye sampled in the inner bay and inshore waters of the outer bay have empty stomachs than those sampled in offshore waters of the outer bay (Straty 1974). Scale studies revealed that juvenile Bristol Bay sockeye do not undergo true marine growth until they are seaward of the inner bay and have been at sea for at least four to six weeks. Once they enter the deeper, clearer, more saline, and plankton-rich waters of the outer bay, their growth is rapid. The later seaward migration timing and the significant delay in start of marine growth of Bristol Bay juvenile sockeye contribute to a lower total growth during the first growing season at sea as well as to a smaller size at comparable age of return than found for Fraser River sockeye salmon.

The distribution of juvenile sockeye salmon during their first year at sea in the eastern North Pacific Ocean and eastern Bering Sea has been determined primarily by purse seine sampling with fine-meshed nets; and off Kamchatka, it has been determined by surface gillnetting. The migrations of stocks entering the Gulf of Alaska will be described first (Figure 34). After entering the open sea during their first summer, the juveniles remain in a band relatively close to the coast. They first appear in catches near the outer coast of Vancouver Island and off southeastern Alaska in late June, indicating that migration into the open sea has just begun at that time. Limited sampling in the western Juan de Fuca Strait indicated that they migrate somewhat earlier than pink and chum salmon, which show the same general coastal movement (Hartt and Dell 1986). By July, they are found moving northwestward in the Gulf of Alaska within 40 km off the coast in a band that stretches for 1,800 km from Cape Flattery to Yakutat. Few are encountered along the coastal areas of central Alaska and along the south Alaska Peninsula, suggesting that stocks in these areas enter the sea somewhat later. No juveniles were taken far offshore. By August, they are still present from Cape Flattery to Yakutat but also appear along the coast of central and southwestern Alaska. They are still scarce or absent in areas further offshore. This band of migrating juvenile salmon includes a broad range of sizes. The size range was frequently

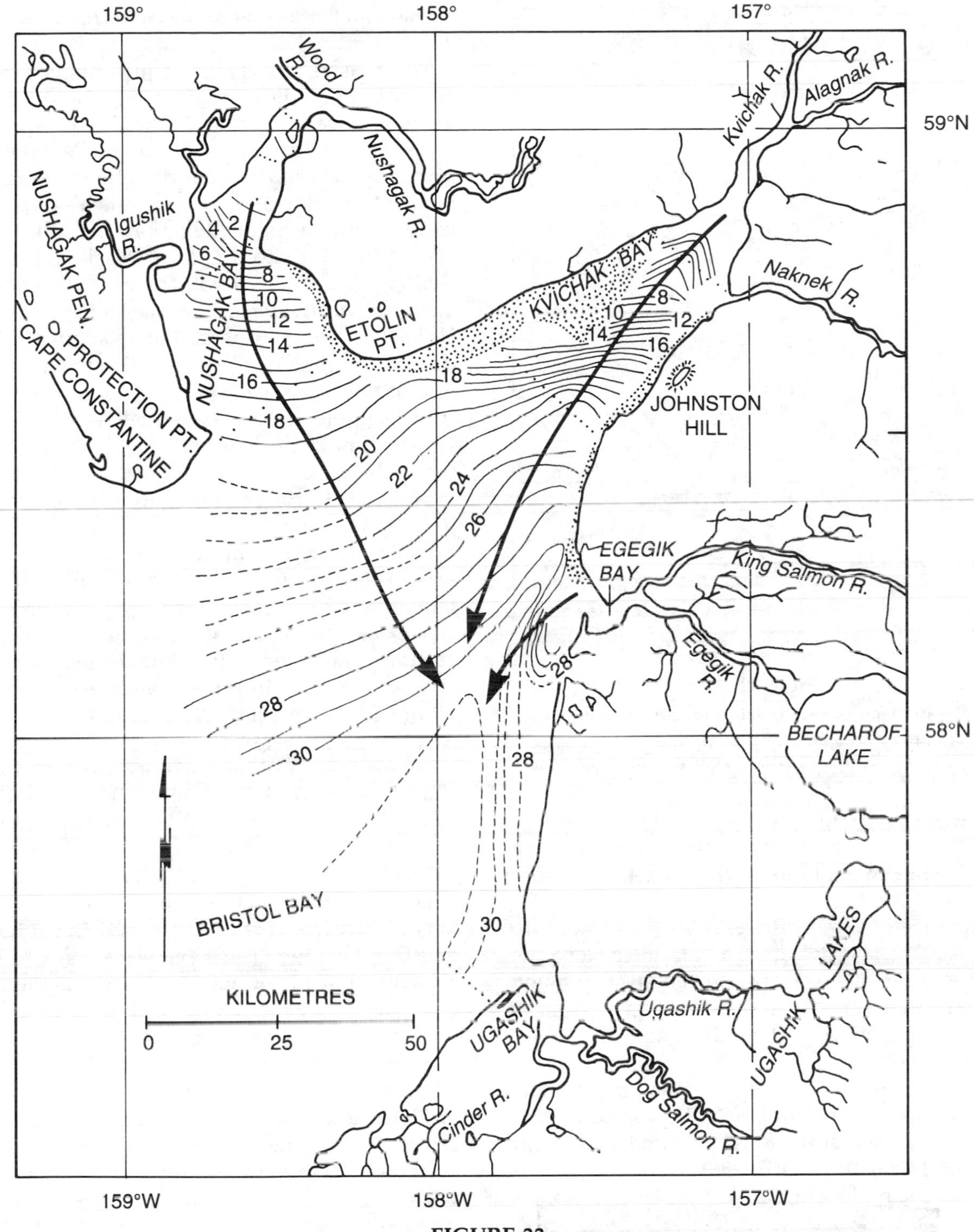

FIGURE 33

Sockeye salmon seaward migration route superimposed on horizontal surface salinity (‰) distribution (low tide), Bristol Bay, July and August 1966. (Adapted from Straty and Jaenicke 1980, Figure 8)

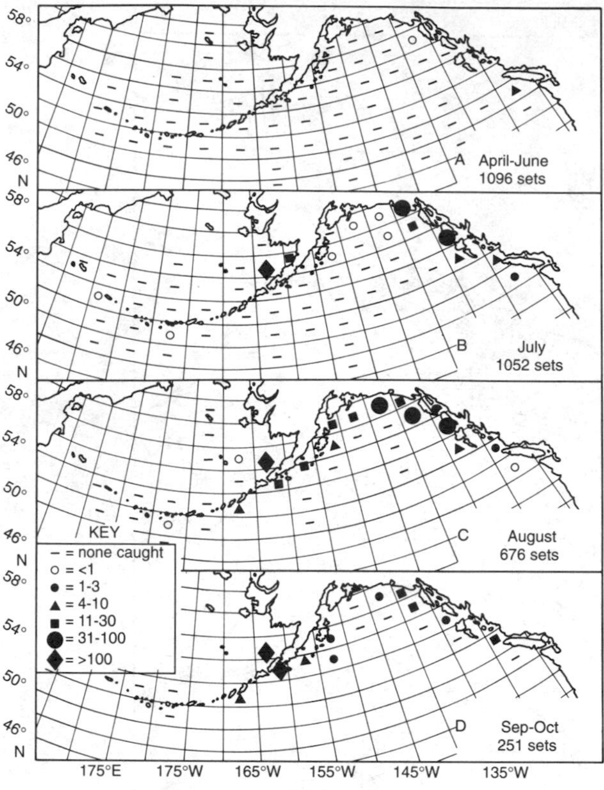

FIGURE 34

Mean catch per seine set of juvenile sockeye salmon by area and by time period; 3,075 sets, 1956–70. (From Hartt and Dell 1986, Figure 3)

bimodal, suggesting a mix of stocks that had recently entered the ocean together with larger fish that had begun migration earlier from more distant southern areas.

Evidence of the northwestward movement up the coast and southwestward movement along the Alaska Peninsula was supported by the much greater catch per haul in seine sets open towards the direction of migration. It is also shown by location of return of adult sockeye tagged as juveniles along this migration band (Figure 35). The average rate of travel for Fraser River sockeye during this coastal movement is estimated to be about 18.5 km/d (Hartt and Dell 1986).

Limited sampling in the North Pacific Ocean has shown that by September-October juvenile sockeye are still distributed primarily inshore. An offshore movement in late autumn or winter is conjectured from the location of age .1 sockeye in early spring. This general pattern of migration, mirrored by pink and chum salmon, is illustrated in Figure 36.

The coastal movement of juvenile sockeye in Bristol Bay and the eastern Bering Sea proceeds at a much slower pace than in the Gulf of Alaska. In outer Bristol Bay, catches of juvenile salmon in seines opened in different directions indicated a variable direction of migration, which sharply contrasted with the highly directional movements observed in the eastern Gulf of Alaska (Hartt and Dell 1986). The net rate of travel southwestward along the north side of the Alaska Peninsula and Unimak Island is estimated to be about 4–7 km/d during the four-month interval following entry into Bristol Bay in late spring. By September, substantial numbers of juvenile sockeye are still only 460–560 km from their estuaries of origin. They tend to remain within about 100 km of the shore during their feeding and migration movement. The full extent of migration in September is unknown because of lack of sampling farther west or northwest after August. Extended sampling in September and October is needed to determine the western extent of migration in late summer-early autumn (Hartt and Dell 1986). The presumed migrations during autumn and winter are shown in Figure 36. A southward movement through the Aleutian passes is indicated.

Less is known of the early marine life of Kamchatkan sockeye. According to Birman and Konovalov (1968), Ozernaya River sockeye feed for several months close to their natal stream, and by early October are joined by juveniles of Lake Palana which have migrated southward some 1,600 km distance from their river mouth. In September and October, age .0 sockeye were taken in Japanese research gillnets farther offshore in the Sea of Okhotsk, off the southern part of the coast of western Kamchatka (French et al. 1976). Juvenile sockeye were captured in September and October well offshore in the western Bering Sea and to the north of the Kamchatka River mouth towards Cape Olyutorskiy. If these were Kamchatka River fish, their movement in the Bering Sea was northeastward. Age .0 Kamchatkan juveniles are assumed to move south or southeast into the North Pacific Ocean by late autumn or early winter, but this has not been confirmed by sampling (French et al. 1976).

Little is known about the primary causes of early

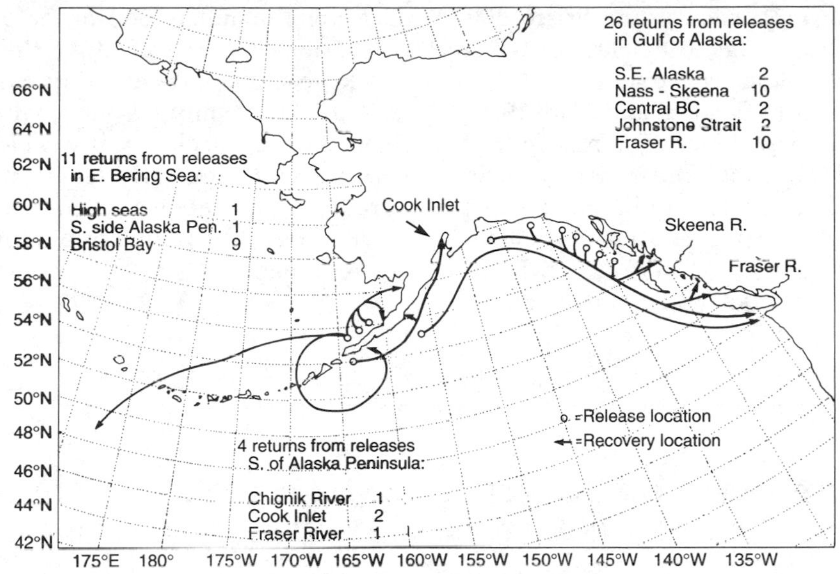

FIGURE 35

Release and recovery locations for 41 sockeye salmon tagged as juvenile fish and recovered two or three years later. (From Hartt and Dell 1986, Figure 22)

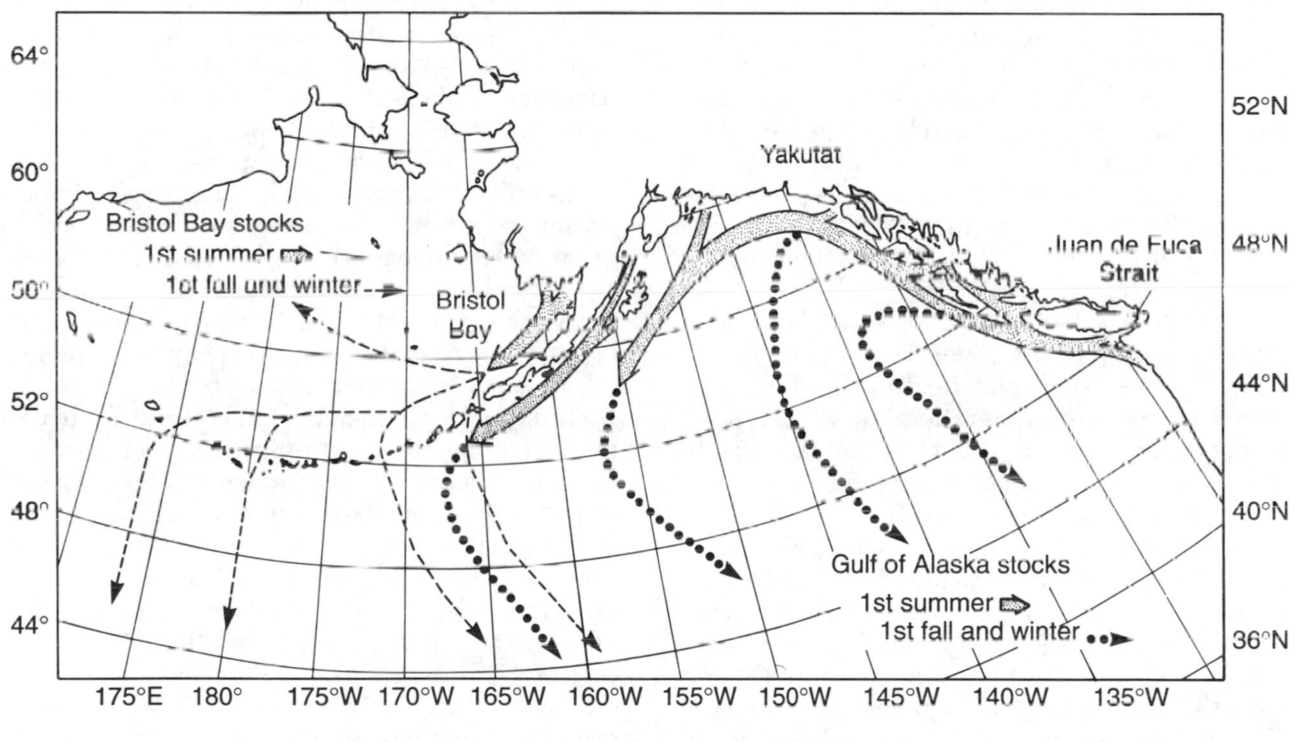

FIGURE 36

Diagram of oceanic migration patterns of some major stocks of North American sockeye, chum, and pink salmon during their first summer at sea, plus probable migrations during their first fall and winter. (Adapted from Hartt and Dell 1986, Figure 40)

marine mortality of Fraser River and Bristol Bay sockeye. In inner Bristol Bay, the beluga or white whale (*Delphinapterus leucas*) is one source of predation on sockeye smolts (Meacham and Clark 1979). Straty (1974) found no evidence of significant fish predation, but noted that some diving birds, which are extremely abundant in Bristol Bay and the eastern Bering Sea, are predators of sockeye. Other possible predators listed were seals, whales, porpoises, and adult chinook and coho salmon.

Much remains to be learned of the interactions of various factors affecting the early sea life of sockeye salmon. As noted by Straty and Jaenicke (1980), "The complexity, variability, and interaction of physical, chemical and biological processes operating during early sea life make it difficult to predict many cause-and-effect relations or to describe the niche of juvenile salmon in the marine environment."

Distribution, Migrations, and Origins of Sockeye in Offshore Waters

There is a large body of literature on the ocean life history of sockeye salmon. Most of what is known about the distributions, migrations, and origins of anadromous salmonids in the North Pacific Ocean has come from investigations by Canada, Japan, and the United States towards their respective research commitments to the International North Pacific Fisheries Commission (INPFC), established by the International Convention for the High Seas Fisheries of the North Pacific Ocean in 1953. Some information for the far western North Pacific Ocean is also available from research done by the USSR (e.g., Birman 1958, 1960).

The INPFC member nations began high-seas sampling and tagging programs in 1955, employing gillnet, purse seine, and floating longline gear. Although research operations have extended across the entire North Pacific Ocean, sampling in most years has been spotty in both time and space. Most of the sampling has been conducted from April to August. There has been some gillnet sampling in winter (January to March), but very little in autumn. Research was heaviest in the area around the central Aleutian Islands to determine the westward extent of North American salmon stocks and to judge the appropriateness of the 175°W abstention line as the eastern boundary of Japanese high-seas salmon fishing. Sockeye salmon quickly became the species of special concern in the INPFC investigations, as it was discovered early that the migratory routes of the important Bristol Bay stock extend far west of 175°W (INPFC 1958, 1959; Hartt 1960).

INPFC has twice called for the preparation of major reports updating and summarizing information on the ocean distribution and life history of each salmon species. Margolis et al. (1966) prepared the first report on sockeye salmon, and considerably more information was available for the second comprehensive report, prepared by French et al. (1976). The recent report summarized information from numerous investigations, including general studies of distribution and abundance, analyses of age, maturity and growth, food habits, and various stock identification analyses based on tagging experiments and parasitological, serological, scale pattern, and morphometric techniques.

The report by French et al. (1976) is still a fairly up-to-date summary of information on sockeye ocean life history, although much new information on migrations and distributions of various stocks in the western and central North Pacific Ocean south of about 48°N latitude has been acquired. In 1978, the North Pacific Fisheries Convention was renegotiated, and the region south of 56°N and east of 175°E was closed to Japanese high-seas salmon fishing. The new protocol placed heavy research emphasis on the Japanese landbased driftnet fishery area, south of 46°N, as there was little information on origins of salmonids in that region. Intensified tagging effort and analysis of scale growth patterns of fish collected in the landbased fishery area and vicinity have yielded new information on the southwestward extent of certain North American sockeye stocks and on the southeastward extent of Asian stocks.

General Aspects of Offshore Distribution

The movements of sockeye in offshore waters are complex and are affected by physical factors such as season, temperature, and salinity, and by biological factors such as maturity stage, age and size, availability and distribution of food organisms, and stock-of-origin (i.e., genetic disposition to specific migratory patterns). Figure 37 shows the 2°-latitude by 5°-longitude statistical areas in which sockeye have been found in 1956–83 INPFC research operations and miscellaneous nearshore studies. It is clear that sockeye salmon are widely distributed throughout the North Pacific Ocean and adjacent waters, including the Sea of Okhotsk, Bering Sea, and Gulf of Alaska. The distribution indicated in Figure 37 should be regarded as the minimum distribution because of paucity of sampling in certain areas and because of the small probability of capture in areas of low abundance near the extremes of the sockeye's distribution. In particular, sockeye must range farther north than indicated in Figure 37, as there are small spawning populations north of the Bering Strait (McPhail and Lindsey 1970). The distribution likely also extends slightly farther south in some longitudinal sectors than shown in Figure 37. In addition, sockeye might also occur in the western Sea of Okhotsk because reportedly there are (or were) small runs along the western coast of this sea (Birman 1960). However, Rukhlov (1982) provisionally indicated that the westward extent of sockeye distribution in the Sea of Okhotsk is about 145°E.

Distribution Relative to Major Oceanographic Features. The general high-seas distribution of sockeye with respect to the physical oceanographic regime (major circulation patterns, temperature, and salinity) was summarized by Foerster (1968), Bakkala (1971), Fujii (1975), and French et al. (1976). The gross oceanographic structure of the North Pacific Ocean and adjacent seas (Figure 38) features major water masses (or "domains," often

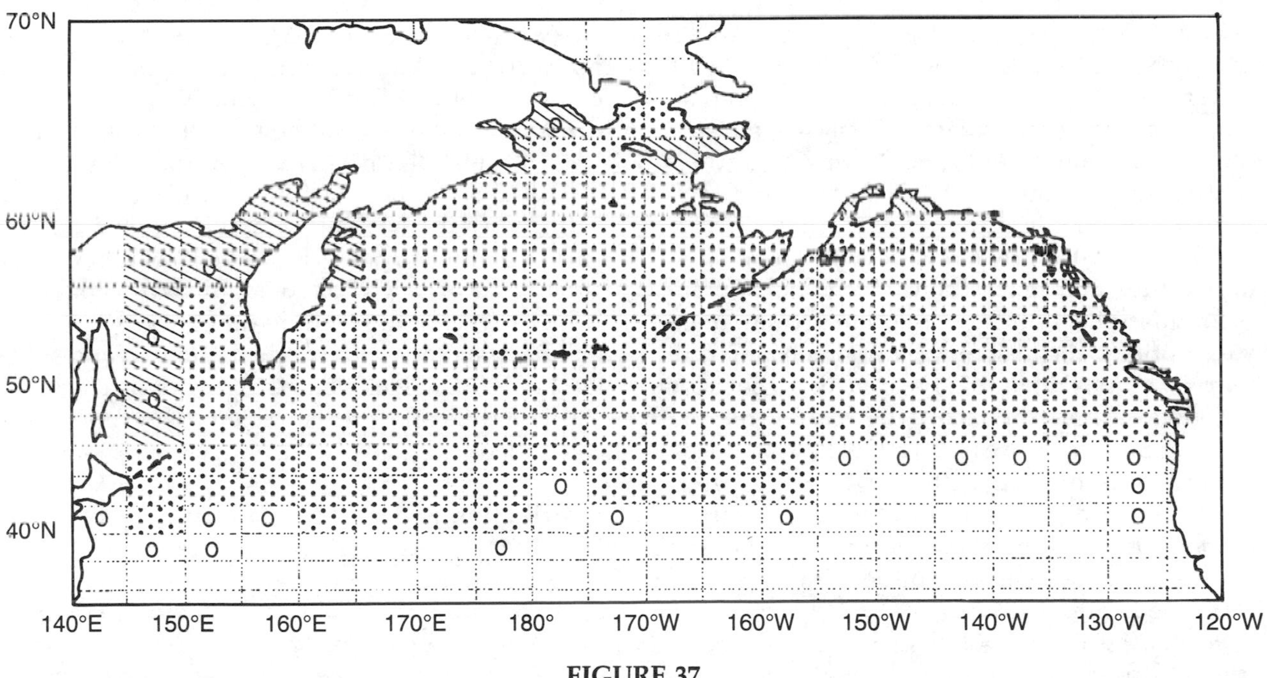

FIGURE 37

Overall ocean distributions of sockeye salmon. Stippled areas show where sockeye were caught in 1956–83 INPFC salmon research operations; zeros show areas where INPFC sampling was conducted but sockeye were not caught; hatched areas show where sockeye are known or strongly suspected to occur on the basis of distribution of spawning runs, or miscellaneous nearshore studies.

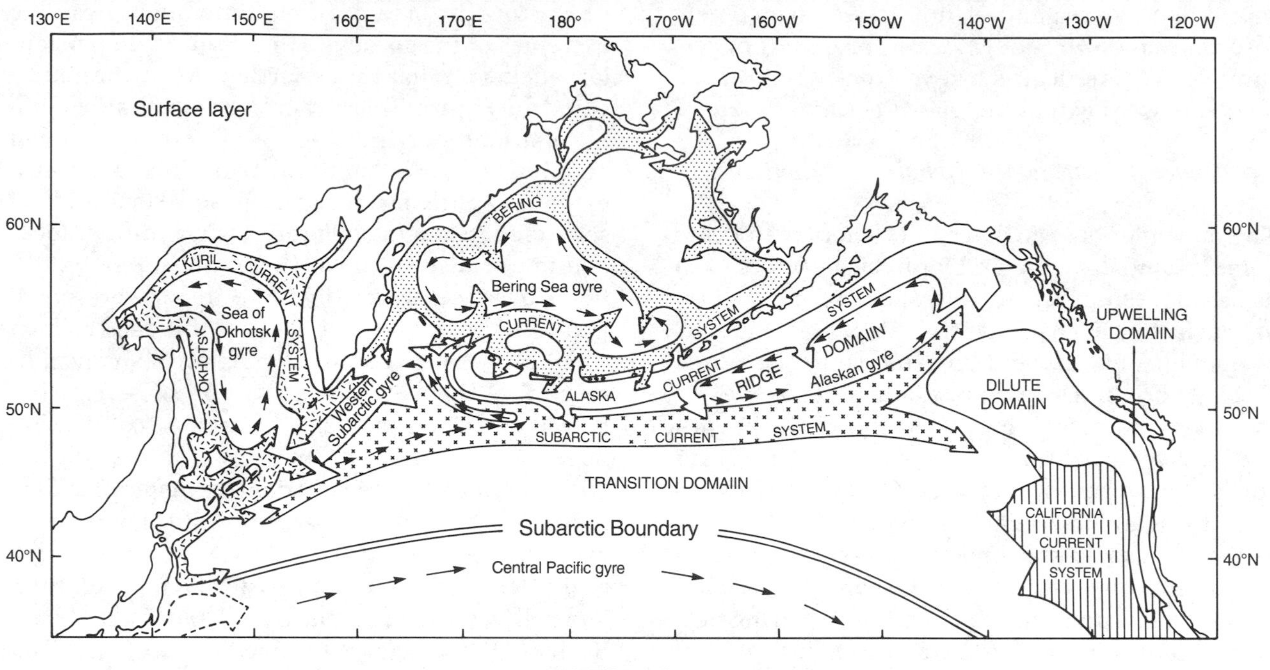

FIGURE 38

Schematic diagram indicating extent of surface layer domains and current systems in the Subarctic Pacific Region. (From Favorite et al. 1976, Figure 41)

bounded by temperature and/or salinity fronts) and major circulation systems (Favorite et al. 1976).

Over the course of a year, sockeye salmon are found throughout the Subarctic Pacific Region, and their seasonal movements and feeding migrations take them into a number of different oceanic environments. Their migrations and distribution have never been consistently linked to major oceanographic features, and general conclusions and inferences found in the literature are often accompanied by exceptions, inconsistencies, and anomalies. In this regard, Bakkala (1971) stated that "sockeye salmon have shown some relation to currents and other defined water areas, but they have not been exclusively associated with any single feature," and he concluded that the influence of defined oceanographic features on the distribution and migrations of sockeye has not been satisfactorily determined.

The southern limit of distribution of Pacific salmon appears to correspond roughly with the southern extent of the Transition Domain across much of the North Pacific Ocean, although there are differences between species as to how far south they may be found at any time. In their early work, Favorite and Hanavan (1963) suggested that the 34 ppt vertical isohaline front, marking the southern boundary of the Transition Domain, is, in effect, the southern limit of salmon distribution in early spring. Later, after northward seasonal migrations, the southern limit is farther north, roughly coinciding with the northern boundary of the Transition Domain (i.e., near the southern extent of cold subarctic water, 4°C or colder at 100 m). Bakkala (1971) suggested that immature sockeye are distributed farther south in spring than in mid-winter, but his data included very few southern gillnet sets and thus may be insufficient to permit a firm conclusion.

Distribution and Surface Salinity. French et al. (1976) concluded from several studies that there is little relation between high-seas distribution of sockeye and surface salinity. Sockeye are distributed across a wide variety of salinities (within the range of generally low salinities characterizing "salmon waters" of the Subarctic Pacific Region), and no consistent correspondence between distributional limits and salinities or salinity gradients has been observed. Favorite and Hanavan (1963) mentioned that the southern limit of salmon catches (all species) east of about 160°W in the summer of 1956 generally coincided with a band of minimum salinity (32.6 ppt). However, they did not conclude that this was a causal relationship, as west of 160°W the band approached the Aleutian Islands and salmon catches occurred well south of the band in that region.

Although surface salinity may not strongly influence the overall high-seas distribution of sockeye, it may have some direct or indirect influence in specific, local instances. For instance, French et al. (1976) mentioned that sockeye are generally in very low abundance over the continental shelf in the eastern Bering Sea (except during the times of seaward migration of juvenile and spawning migration of adult Bristol Bay sockeye), and they suggested that this might be related to the low salinity of the eastern Bering Sea caused by high runoff from western Alaska. Fujii (1975) concluded that the lower temperatures and higher salinities around the Aleutian passes in early spring essentially block the northward spawning migration of Bristol Bay sockeye, and that the schools migrate through the passes only after this cold and saline structure dissipates later in spring. He further concluded that the distribution of Bristol Bay sockeye during the spawning migration in the Bering Sea is influenced by salinity as well as by temperature. Burgner (1980), however, doubted that salinity or other factors serve to "block" the Bristol Bay migration, as the run timing varied remarkably little.

Distribution and Surface Temperature. The surface temperature structure of the North Pacific Ocean and Bering Sea is complex, and sockeye are found over a wide variety of temperature conditions even in the course of one season. Sea surface temperature, as salinity, has not been found to be a strong and consistent determinant of sockeye distribution, but it definitely influences distribution and migrations.

In general, sockeye tend to prefer cooler water than the other salmon species. Birman (1960) concluded that the temperature ranges of waters yielding catches of various salmon species in the northwestern Pacific in winter were for sockeye, 1.5°-6°C; chum, 1.5°-10°C; pink, 3.5°-8.5°C; and coho salmon, 5.5°-9°C. Manzer et al. (1965) compared "preferred" temperature ranges of the five principal salmon species and found that sockeye prefer the lowest temperatures, although there was considerable overlap with the other species, especially chum salmon. Burgner and Meyer (1983) analysed 1972-81 gillnet catch data from Japanese salmon research vessel operations and showed that the temperatures providing the greatest catches of each species were lowest for sockeye in all months from March to July.

French et al. (1976) summarized U.S. and Japanese research gillnet data and presented a lucid graphical depiction of sockeye distribution with respect to surface isotherms. They pointed out that the "preferred" temperature range for sockeye increases through the warming period, notwithstanding the fact that the fish migrate northward in spring and summer. In winter, catches of immature sockeye occurred in waters of 2°-7°C, and the largest average catches were at 3°-5°C (mode at 5°C). Catches of maturing fish were considerably larger at 2°-6°C (mode at 4°C). In spring, immatures were caught only in the 4°-10°C range (although sampling was done in waters of 1°-13°C), and the peak mean catch was at 7°C. Maturing fish in spring were caught in 1°-10°C waters, and the mode catch was at 6°C. The tendency for maturing fish to be in cooler waters was explained by the northerly migration of maturing fish into cooler Bering Sea waters in advance of the immature fish. By summer (July and August) the maturing component is in much lower abundance as the inshore runs progress, and the immature fish are widely distributed from about 46°N latitude in the central North Pacific Ocean to about 64°N in the Bering Sea. Greatest concentrations occurred in northeastern Pacific Ocean waters of about 9°-11°C and in Bering Sea waters of about 7°-10°C.

There are some correlations between overall sockeye distribution and sea surface temperature, although temperature influences distribution in concert with many other factors. Canadian researchers found that the southern and eastern limit of sockeye distribution in the North Pacific Ocean in winter was between the 6° and 7°C isotherms. Sockeye are found in much colder water in the far western North Pacific Ocean, where winter temperatures are much lower at the same latitudes. Birman (1958) concluded that the southern distribution of salmon in the western North Pacific

Ocean in summer is approximately along the 13.5°C isotherm. Harris et al. (1984) compared seine catches of sockeye south of 46°N in the central North Pacific Ocean in 1982 and 1983, and suggested that the apparently more southern extent of sockeye in 1983 was due to the colder local conditions in that year.

Research gillnet sampling in 1963 provided the only evidence that sockeye salmon are found in numbers in near-surface waters of the Bering Sea in winter months (French and Mason 1964). In fishing conducted in late January - early February, sockeye were taken at all stations fished extending north to 57°28'N in the central Bering Sea. Fish of ages .1, .2, and .3 were represented in these catches.

Surface temperature affects directly or indirectly the timing of sockeye migrations as well as actual distribution. Nishiyama (1984) determined that gonadal maturation of Bristol Bay sockeye salmon, which advances rapidly from late May through early July, was directly related to the mean water temperature in the central to southeastern Bering Sea in June. He demonstrated further that the peak return date of sockeye salmon to the Kvichak River of Bristol Bay was inversely related to the Bering Sea water temperature in June during the period 1965-72. Burgner (1980) similarly found an inverse relationship between mean air temperature in May in the Adak-Cold Bay area (an index of prevailing sea surface temperatures) and the median date of the Bristol Bay sockeye run in 1960 77. Blackbourn (1987) also found significant inverse relationships between late winter or spring sea surface temperatures in specific Marsden square areas in the Gulf of Alaska and the time of return of sockeye returning northward to Chignik, Upper Cook Inlet, Copper, and Skeena rivers. The run timings of several Fraser River sockeye stocks and the Quinault River (Washington) sockeye have also been closely correlated with winter to spring sea surface temperature in specific areas of the Gulf of Alaska (Blackbourn 1987), but in this case the correlation is positive. It is hypothesized that maturing Fraser River and Quinault River sockeye migrate north in spring before turning back south for their approach to the river. In a cold year, the fish are thought not to migrate as far north before reversing direction and, therefore, tend to approach their home rivers earlier than usual. Conversely, the more northern gulf stocks are believed to be displaced farther south in cold years and return northward later than usual. Blackbourn's (1987) analysis implies differences among stocks in their centres of winter distribution in the Gulf of Alaska.

Sea surface temperature also affects the route of the Fraser River spawning run returning from their ocean feeding grounds (Groot and Quinn 1987). When adult sockeye salmon return to the Fraser River they may either take a northern route via Queen Charlotte and Johnstone straits down the east side of Vancouver Island or a southern route on the Pacific side of Vancouver Island via Juan de Fuca Strait. The proportion of the total run using the northern route has varied from 2% to 80% over the years 1953–85 (IPSFC 1984; Groot and Quinn 1987). The annual variation is best predicted by ocean temperature conditions prior to or during the homeward journey. When conditions are warmer than usual, the returning sockeye tend to approach the coast farther north and use the route through Johnstone Strait. Thus, the adult sockeye do not necessarily return along the same route taken seaward as smolts (Groot and Cooke 1987).

Vertical Distribution. There has been little research directed at determining vertical distribution of salmon at sea. The available information from research and from the operation of commercial high-seas salmon fisheries, however, permit the conclusion that salmon generally occur in near-surface waters. The large and effective Japanese high-seas salmon fisheries at present use surface gillnet gear that fishes to a depth of about 8 m (Fukuhara 1971). Before 1972, Japan also had a longline fishery for salmon in the North Pacific Ocean, and that gear fished even closer to the surface (1–2 m).

Three main studies, summarized by French et al. (1976), have provided direct information on the vertical distribution of sockeye salmon. Manzer (1964) analysed catch data from gillnets set at depths down to 61 m in May to July 1959 and 1960 in the Gulf of Alaska. He found that sockeye were distributed throughout the depth range examined, but mainly in the upper levels. They tended to be caught closer to the surface at night, suggesting a diel vertical migration, but the differences between night and day were not great. Especially in the June and July sets, the catches tended to decrease abruptly near the bottom of the thermocline; catches at the deepest strata (37–49 m and 49–61 m) occurred only in May, when the thermocline was poorly developed. There were no obvious differences in vertical migration between age or maturity groups (at least for those fish taken by the gear). Machidori (1966) analysed data from operations in the northwestern Pacific Ocean and Bering Sea, made variously in June to August, 1962 and 1963. He found that both maturing and immature sockeye were distributed mainly in the upper 10 m, and also that there was a slight tendency for the fish to be shallower at night. In this experiment, based on 11 sets, no sockeye were caught deeper than 30 m. U.S. investigators set vertical gillnet panels in winter, spring, and summer 1968–69 in the central and northeastern Pacific Ocean and found that in each season at least 90% of the catch occurred in the top 15 m (French et al. 1976).

Distribution of Asian and North American Sockeye

Determination of Distribution. Three primary methods have been used to determine the ocean distribution and migration of sockeye by area of origin. These are tagging, parasite studies, and analyses of differences in growth patterns registered on the scales of the fish.

Tagging of sockeye salmon has contributed a very substantial body of information on the distribution of stocks, migration routes, rate of travel, growth, and survival (French et al. 1976). With respect to distribution of stocks, the tagging method relies on the principle that sockeye return to their stream of origin with a high degree of fidelity. Since 1956, well over 100,000 sockeye salmon have been captured, tagged, and released across the North Pacific Ocean and Bering Sea by Japan, Canada, the United States, and the USSR in studies of ocean distribution and migrations. Coastal recoveries of a portion of these fish, tagged either as immature fish destined to remain at sea for a year or more, or as maturing fish destined to return for spawning in the same year of tagging, provides much of our knowledge of high-seas and coastal migration by stock group.

Parasites have proven to be very useful as biological "tags" to extend our knowledge of ocean distribution of certain sockeye stocks of North American and Asian origin (Margolis 1963, 1982b; Konovalov 1971; French et al. 1976). To be useful as a tag in ocean migration studies, the parasite must have its origin in fresh water, have a restricted freshwater distribution, and survive in its salmon host for most or all of the ocean life of the salmon. Sockeye salmon, with the longest freshwater life as juveniles, harbour the largest number of freshwater parasite species (36) among the Pacific salmons (Margolis 1982b). After extensive sampling, identification, and analyses, it was determined that two parasite species are useful in providing evidence of the origin of their sockeye hosts sampled at sea. These are the plerocercoids of the cestode *Triaenophorus crassus*, found only in the somatic musculature of sockeye of Bristol Bay origin, and the intestinal nematode *Truttaedacnitis truttae*, found only in sockeye salmon of Kamchatkan origin. Both parasites infect juvenile sockeye in fresh water and survive within their sockeye hosts in the ocean environment. Several thousand sockeye sampled on the high seas were examined for presence of these parasites.

The third method of stock identification utilizes differences in the numbers and spacing of circuli of the sockeye scales and in the widths of the annual growth bands formed in fresh water and in the ocean. Results of these studies are reviewed by Margolis et al. (1966) and French et al. (1976). More recently, these studies have been extended, particularly in the southwestern portion of sockeye distribution in the North Pacific Ocean (Cook et al. 1981).

The known ocean distribution and overlap in ranges of Asian and North American sockeye based on these tagging, parasitological, and scale pattern studies are shown in Figure 39 by 2° latitude and 5°-longitude blocks. Asian sockeye remain generally west of 175°W, but parasite studies have revealed their presence slightly farther to the east in both the North Pacific Ocean and the Bering Sea (Margolis 1963). North American stocks, particularly of western Alaskan origin, are found southwest in the North Pacific Ocean to blocks bounded on the west by 160°E and in the Bering Sea to 170°E. Thus, Asian and North American sockeye overlap in distribution over fairly broad areas of the North Pacific Ocean and Bering Sea. In general, the centre of abundance of Asian fish is west of 175°E, and that of North American fish is east of this longitude.

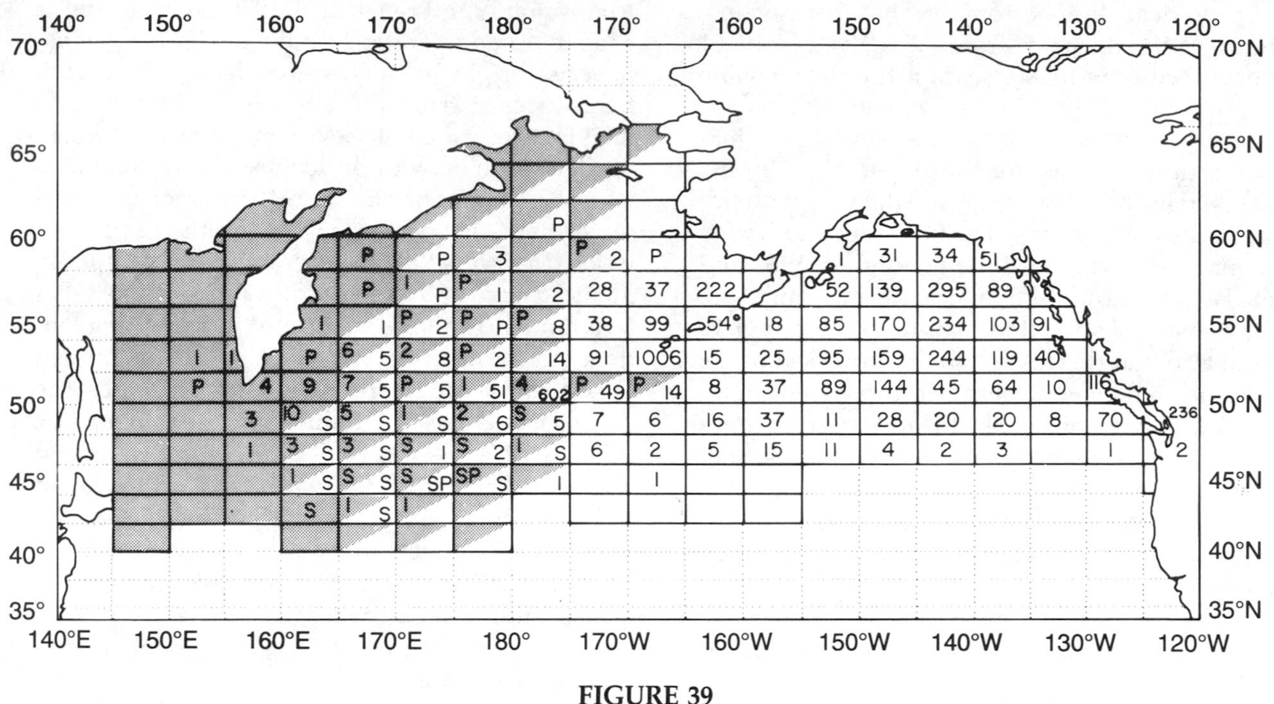

FIGURE 39

Known ocean distribution of Asian (shaded areas) and North American (open areas) sockeye salmon on the basis of tagging, parasitological analysis, and scale-pattern analysis. If occurrence of a continental stock group is proven by tagging experiments, the number of releases resulting in coastal recovery is shown. If there is no information from tagging, then a *p* means occurrence is known from detection of continent-specific parasite "tags," and an *s* means a statistically-significant estimate for the stock group was obtained in scale-pattern analyses (Cook et al. 1981). Lack of a symbol indicates the stock group's distribution was extrapolated to fill the overall distribution of sockeye as shown in Figure 37

Migrations of Major Stock Groups. Although information on ocean distribution of sockeye salmon is incomplete, especially for autumn and winter months, the vast amount of material available is adequate to provide tentative models of the oceanic migrations of major stocks (Royce et al. 1968; French and Bakkala 1974; French et al. 1976). The model diagrams presented here are modified from Figures 92–94 of French et al. (1976).

The first migration model shown is for Asian sockeye salmon (Figure 40). Most of the production comes from the Ozernaya River on the southwest coast of Kamchatka and from the Kamchatka River on the east coast, but the migrations depicted include the many additional small runs found along the Asian coast.

During the first eleven months at sea (July–May) juvenile sockeye move southward from the Sea of Okhotsk and the Bering Sea and eastward in the North Pacific Ocean to winter feeding areas in the western North Pacific Ocean (Figure 40A). (They are designated age .0 fish until 1 January, when they become age .1 fish.) The southward extent of this migration is not known precisely, but by spring they are found moving northward in the North Pacific Ocean and into the Bering Sea, coincident with seasonal warming of surface waters (Figure 40B). Sockeye apparently do not migrate back into the Sea of Okhotsk until their spawning migration.

In the fall of their second year at sea, they move southward again, this time from their summer feeding grounds to their winter habitat in the North Pacific Ocean (Figure 40C). By April to May the Asian sockeye (now age .2) are distributed approximately between 150°E and 177°W, and from about 43°N to 50°N. Maturing fish destined to spawn at age .2 now begin their homeward

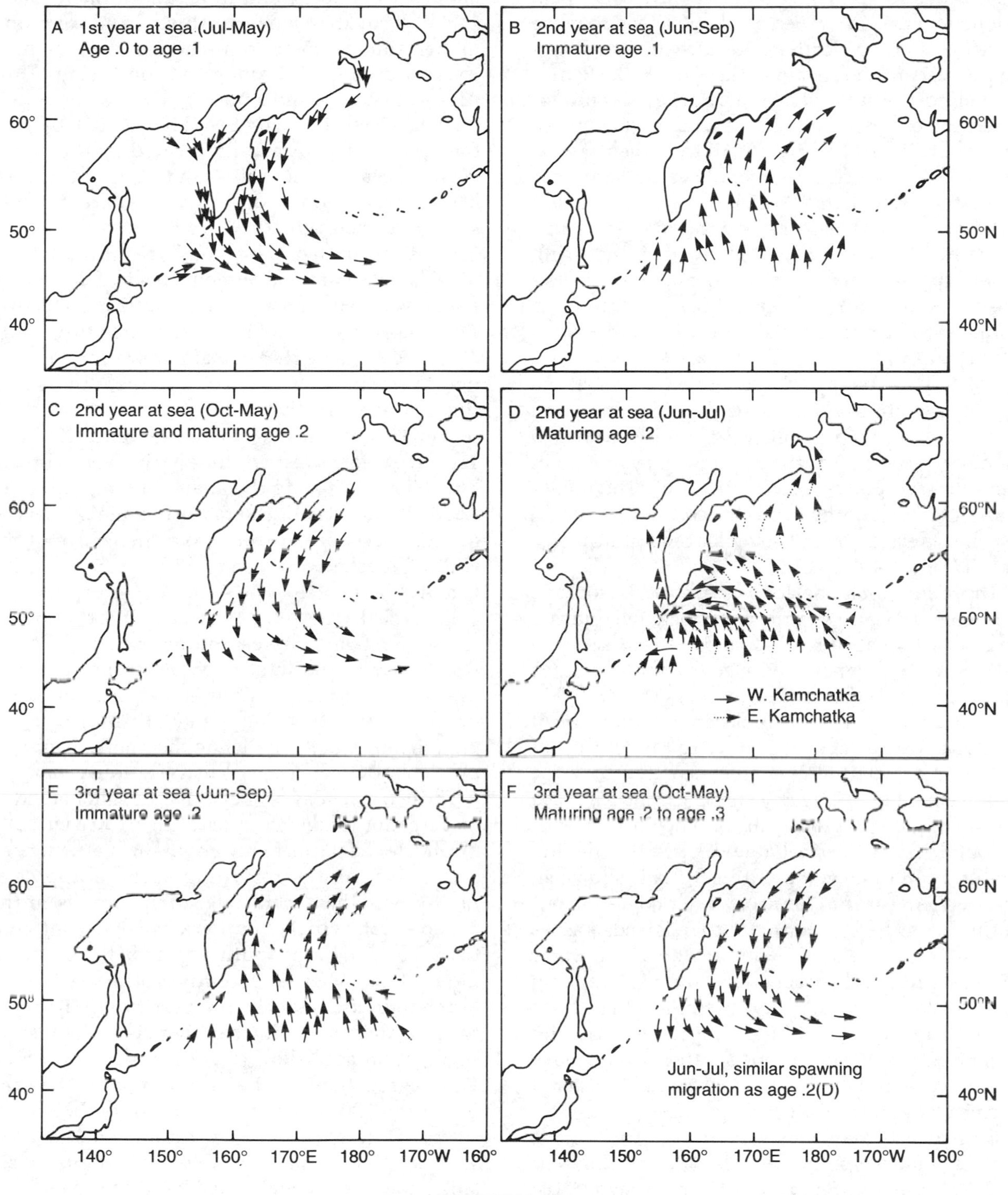

FIGURE 40

Model of migration of Asian sockeye salmon. (Adapted from French et al. 1976, Figure 92)

migrations to spawning areas (Figure 40D). Fish bound for areas in the Sea of Okhotsk tend to move northwest in the North Pacific Ocean, then swing southwest around the tip of the Kamchatka Peninsula or northern Kuril Islands. Sockeye bound for rivers of eastern Kamchatka and farther north move northward and westward from their winter areas. This migration is generally earlier than that of western Kamchatka stocks. The immature age .2 fish remaining at sea for one or more winters also move northward but lag behind their maturing counterparts. Their northward migration in the spring at the start of their third year at sea is similar to the migration they made the previous year (Figure 40E).

Most of the Asian sockeye salmon remain at sea for three winters. Their winter distribution is similar to that of the previous year, but is believed to be not quite as far south, and some may remain in the southern Bering Sea (Figure 40F). The homeward migration pathways for maturing age .3 sockeye are similar to those shown for the maturing age .2 fish (Figure 40D).

The migration model for western Alaska sockeye salmon (primarily Bristol Bay stocks) displays a wider east-west spread than for Asian sockeye (Figure 41). Information derived from scale pattern and tagging studies in recent years (Harris 1987) has extended the known ocean distribution of western Alaska sockeye farther to the southwest than was depicted by French et al. (1976).

Movement of Bristol Bay stocks during early sea life, discussed previously, is not known after mid-September but the probable distribution during the first ten to eleven months at sea is shown in Figure 41A. Autumn to early winter migration out of the Bering Sea through Aleutian Islands passes probably is motivated by lowering surface water temperatures and reduced food supplies. Tagging and age composition analyses suggest that those Bristol Bay stocks that begin migration early and enter from rivers more to the southwest may move farther west in the Bering Sea and enter the North Pacific Ocean farther west (Rogers 1986).

By April to May, western Alaska stocks have moved southward through the Alaskan Current and Ridge Area (Figure 38) to more favourable water temperatures and food sources in the Western Subarctic Intrusion or Transition areas waters. They are distributed primarily south of 50°N to about 43°N and extend in a band from 167°E to 140°W. They then move northward and westward in their summer migration. Those in the east move westward in a band along the south side of the Aleutian Islands, and a portion swings northward into the western Bering Sea (Figure 41B). By July, the age .1 fish are mainly located in the North Pacific Ocean north of 50°N in waters of the Alaskan Current and Ridge areas, where food in summer is more abundant.

In late autumn of the second year at sea, southward and eastward movements begin, followed by a segregation of immature and maturing sockeye. The majority of Bristol Bay sockeye mature at age .2. Those sockeye that remain at sea for an additional year move farther south, whereas those destined to mature the coming summer at age .2 remain in more northerly waters (Figures 41C and 41E). Their east-west winter distribution is broad. The following spring the spawning migration towards Bristol Bay of the maturing age .2 sockeye occurs over a broad front from the centre of the Gulf of Alaska to beyond the western Aleutian Islands (Figure 41F). The sockeye wintering in the Gulf of Alaska move westward then north through Aleutian Islands passes and northeast to Bristol Bay. Those to the west move north or northeast into the Bering Sea on their homeward route. The immature age .2 fish lag behind, displaying a migration movement paralleling their movements the previous summer as age .1 fish (Figure 41D).

Those Bristol Bay sockeye that remain behind to mature the following year at age .3 essentially duplicate the paths followed by the maturing age .2 sockeye. As the returning sockeye approach Bristol Bay, their main migration route is in the offshore waters of the southern half of the entrance to the bay and in the southern half of the bay itself (Straty 1975). There apparently is little segregation of the individual stocks in outer Bristol Bay, but towards the head of the bay there is a progressive segregation according to river system of origin. This segregation first begins to appear when the fish are still 200 km from the mouths of their home rivers. Differences in run timing among river stocks is almost imperceptible except for the Ugashik Lake stock, which consistently returns a few days later.

There is considerable overlap in migratory distributions of sockeye salmon originating in rivers of

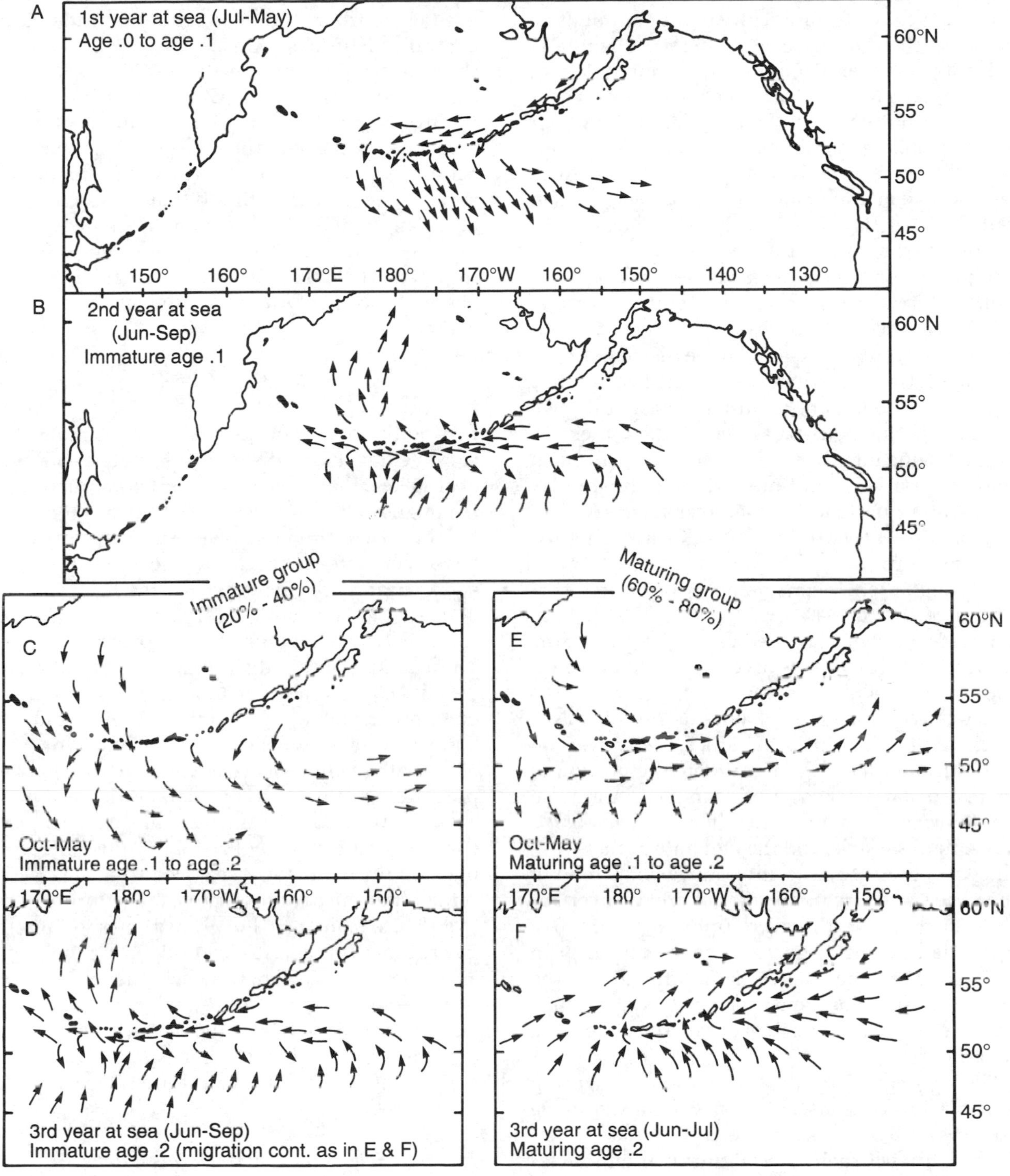

FIGURE 41

Model of migration of western Alaska sockeye salmon. (Adapted from French et al. 1976, Figure 93)

the northeastern Pacific Ocean from the Alaska Peninsula to the Columbia River, but as might be expected, there are also some differences. Figure 42 presents a generalized model of migration of northeastern Pacific Ocean sockeye stocks. These stocks do not enter the Bering Sea during their summer feeding migrations. It has now been determined by scale pattern analyses that central Alaskan sockeye migrate much farther to the west than southeastern Alaska and British Columbia-Washington stocks, and they may be intercepted in significant numbers in the Japanese landbased and mothership salmon fisheries operating west of 175°E. In the Gulf of Alaska, British Columbia-Washington stocks tend to be distributed farther south (to 46°N) than Alaskan stocks, but they utilize the general area south and east of Kodiak Island together with Alaskan stocks. Stocks in rivers flowing into the Gulf of Alaska display much more diversity in run timing than do Bristol Bay stocks, and runs to individual river sytems tend to be more protracted (e.g., Fraser, Karluk, Chignik, and Skeena rivers).

Although there is considerable overlap in distribution of Bristol Bay sockeye stocks with both Asian and Gulf of Alaska stocks, the timing of the migratory gyres of the three groups tends to reduce their simultaneous intermingling. This is particularly true for Bristol Bay and Gulf of Alaska stocks. As the Gulf of Alaska stocks swing west on their summer feeding migrations, the portion of the Bristol Bay stocks wintering in the Gulf are also moving west. They are generally well ahead of the Gulf-origin sockeye, and they migrate farther west. Likewise, the autumn-winter movements tend to be in the same directions but not greatly overlapping. In the west, the overlapping of Bristol Bay and Asian stocks appears to occur primarily in spring and summer stages of migration. Thus, the sockeye populations and their various age groups utilize a very broad swath of ocean in their feeding migrations.

There is considerable speculation concerning possible interactions between British Columbia and Bristol Bay stocks in the Gulf of Alaska that might influence their growth, survival, and cycles of abundance (Peterman 1984a, 1984b). The extensive intermingling that occurs among populations of widely different geographic origins has been amply illustrated by tagging. For example, Figure 43 shows a wide distribution of recoveries of sockeye tagged from a single longline set made in the central Gulf of Alaska on 27 April, 1962. However, the broader distribution of Bristol Bay sockeye across the North Pacific Ocean, their use of the Bering Sea, and the differences in timing of movements noted above reduce the possibility of interaction. Analyses of tag returns of Bristol Bay sockeye suggest that Bristol Bay stocks are less abundant than British Columbia stocks in the Gulf of Alaska (Rogers 1986). Thus, the long-term historic cycles of Fraser River and Kvichak River sockeye were not likely to have interacted to a great extent.

Rates of Travel

During the course of their ocean migrations, the sockeye travel long distances. As noted by Royce et al. (1968), they may well undertake annual feeding migrations in excess of 3,700 km and appear to be almost continuous travellers. Rates of travel have been determined from recoveries of tagged sockeye, and are calculated on the basis of days between tagging and recovery and the most direct route between tagging and recovery locations. These are minimum distances and rates of travel because considerable meandering is known to occur during feeding. In many instances, but not always, the migration rates are accelerated by travelling with the flow of ocean surface currents. This probably accounts to a considerable degree for the greater migration distances achieved by Fraser River juveniles than by Bristol Bay juveniles during their first summer at sea (Hartt and Dell 1986).

Migration speeds are greatest for maturing sockeye that are on their homeward journey. Bristol Bay sockeye average 46–56 km/d during their final 30–60 days at sea. There is a tendency towards more rapid movement of the migrants returning later in the season (Hartt 1966). The rates of travel of immature sockeye in their second and third summer at sea are less than those for mature fish (Hartt 1966).

Schedules of Return Migrations

The long ocean migrations of the different sockeye stocks terminate on remarkably consistent schedules. This is exemplified by the Bristol Bay stock

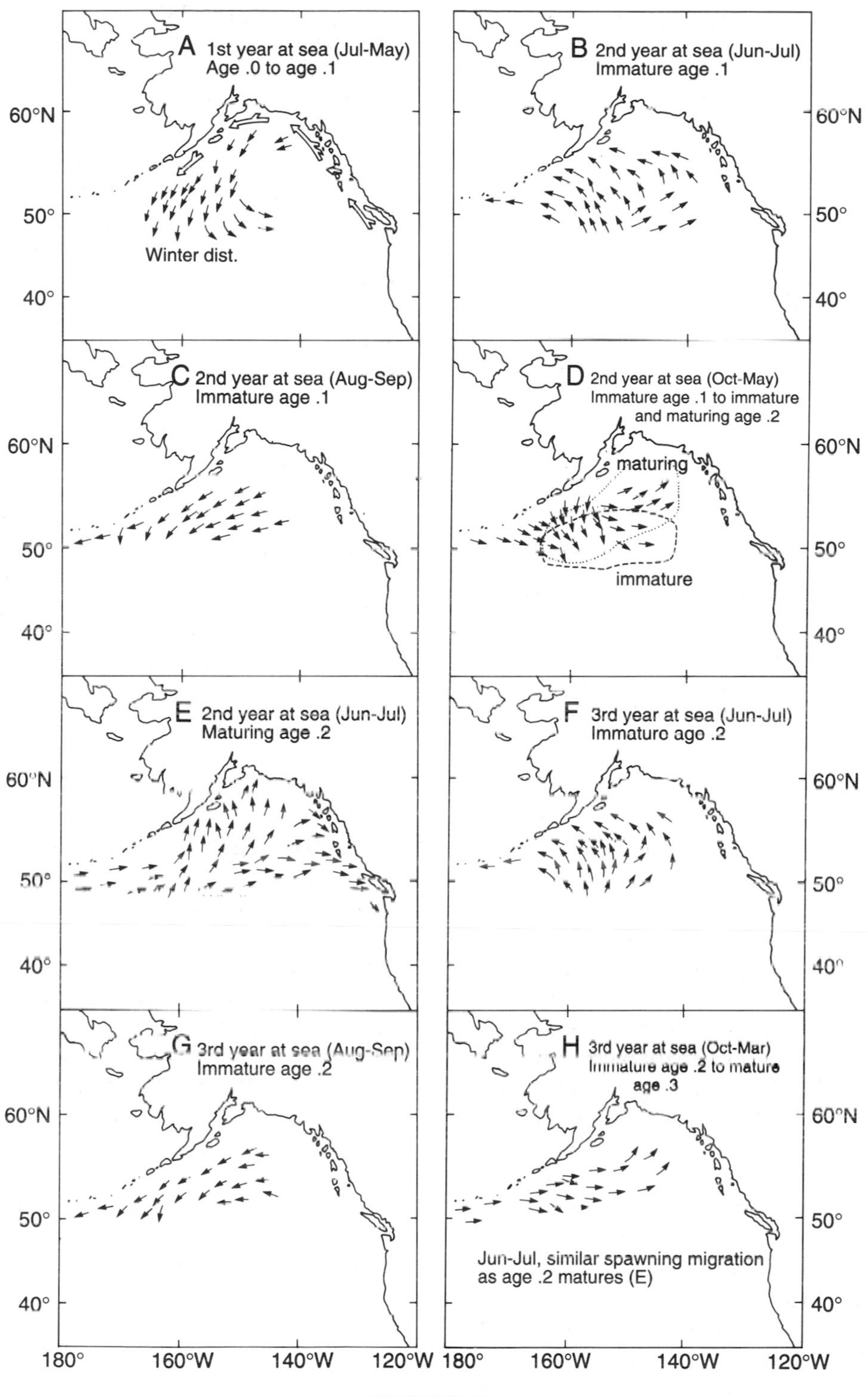

FIGURE 42

Model of migration of northeastern Pacific sockeye salmon.
(From French et al. 1976, Figure 94)

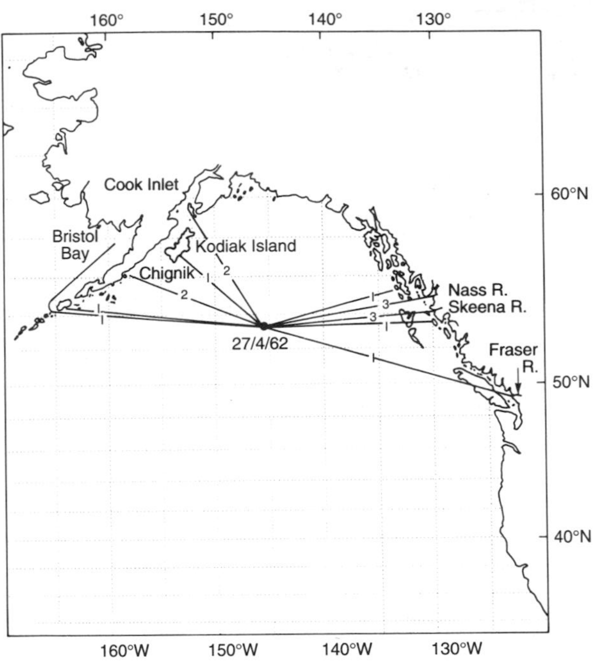

FIGURE 43

Distribution of recovered sockeye salmon tagged at the same place and time (27 April 1962). (From Neave 1964, Figure 2)

group which returns through the fishery annually during a very short time interval (Royce 1965; Royce et al. 1968; Burgner 1980). As noted earlier, each year the fish destined to mature tend to separate from those remaining at sea and begin an accelerated spring-early summer migration homeward to their rivers of origin. In May and early June, four to ten weeks before they arrive in Bristol Bay, they are still spread broadly over some 3,700 km of North Pacific Ocean in an east-west direction (Figure 41). Individuals or schools bound for a particular river system begin their homeward migrations from divergent points over vast ocean areas. Some must migrate to Bristol Bay from as far away as 2,200 km, either directly through Aleutian passes or by routes west, then north and northeast to circumnavigate the Alaska Peninsula (Royce et al. 1968). Despite this diverse distribution at sea, on average, 80% of the Bristol Bay run passes through the estuarine fishery within a two-week period each year (Burgner 1980). (The average for years 1956-76 was 12.9 days, s.d. 1.58 d.) The mean date of arrival also shows little variation. (The average for years 1956-76 was 4 July, range 29 June-10 July.) Interannual differences in early spring ocean surface temperatures account for over 50% of the deviation in Bristol Bay run timing, the runs being later when ocean temperatures are colder. Runs to individual river systems in Bristol Bay are slighty more compact in timing of return than are those for the rivers combined. This ability of sockeye to return homeward on such a precise schedule, one to four years after leaving the river as smolts, requires truly remarkable navigational and orientation skills that apparently are common to the genus *Oncorhynchus*.

Migration Mechanisms

The methods used by salmon to navigate on the high seas and return to their home streams in an orderly schedule remain one of the most intriguing mysteries of animal migration. Quinn (1982b) summarized the information and hypotheses presented by various investigators, and added his own model. His evidence in support of active orientation by the salmon include (1) the predictable movements by size, age, and season at sea, described earlier; (2) the ability of fish of different populations found together at a particular time and location (and thus experiencing common environmental cues) to disperse to a wide range of home streams; (3) the rapid homeward movements of maturing fish even though spread out over a broad geographical area and experiencing different environmental conditions; (4) the remarkable temporal precision of convergence on the mouth of their home stream with little straying; (5) the high rate of travel, approaching maximum sustainable speeds, suggesting strong orientation; and (6) the evidence of celestial and magnetic compass orientation abilities demonstrated for sockeye fry and smolts. Quinn noted that strict retracement of the outward path by the homeward bound salmon from ocean feeding areas does not occur, and that the broad distribution at sea of a single river population does not suggest a rigid set of pre-programmed compass preferences. To quote Larkin (1975), "It thus becomes necessary to postulate that salmon have a bi-coordinate system of navigation that enables them to know where they are and where they are to go (and when to leave in order to get there on time)."

The model proposed by Quinn (1982b) is "that salmon navigate using a map, based on the inclination and declination of the earth's magnetic field, a celestial compass with a backup magnetic compass, and an endogenous circannual rhythm adjusted by daylength." (Physiological conditions presumably determine at which age an individual sockeye salmon will choose to return to its home river.) As for the map, it is noted that east-west movements of a fish at northern latitudes result in it experiencing significant changes in the magnetic field's declination. To use this information it must be able to detect the magnetic field's horizontal component and to determine geographic north, presumably through a form of celestial compass orientation. The magnetic field probably provides the primary information on compass direction, because of the prevailing overcast conditions over the North Pacific Ocean and Bering Sea. However, the sun and polarized light may be a primary source of celestial orientation. In addition, "it is hypothesized that salmon possess a calendar, based on an endogenous circannual rhythm, synchronized by daylength or rate of daylight change, tied to a sense of latitude." (Quinn 1982b). Thus, the proposed calendar sense has endogenous and exogenous components. Along with this orientation mechanism the salmon must have a sense of their location in the ocean at any time with respect to the location of their natal stream. Finally, it appears that homing has learned components, because juveniles imprint on the odours of their home stream and are able to recognize these odours as they near their streams on their homeward journey as adults (Hasler et al. 1978). Quinn (1982b) suggested that they may also imprint on the local magnetic field at some critical time before migrating to sea. Whether or not Quinn's model represents the salmon's navigation system, these fish, including sockeye, are masterful navigators.

FEEDING AND GROWTH

Food

Foerster (1968) and French et al. (1976) summarized the stomach analysis data for sockeye reported by various investigators for many different areas of the ocean. Euphausids, hyperiid amphipods, small fish, and squid were the groups most frequently listed as main food items, with copepods, pteropods, and crustacean larvae listed as of lesser importance. The fish included lantern fish (Myctophidae) and juvenile cod (Gadidae) in the central North Pacific Ocean. In the eastern Bering Sea, juvenile sockeye (age .0) fed on larval capelin (*Mallotus villosus*), sand lance (*Ammodytes hexapterus*), and herring (*Clupea harengus pallasi*). Wing (1977) lists sand lance, herring, pollock (*Theragra chalcogramma*), and capelin as important food items for sockeye in southeastern Alaskan waters. Favorite (1970) reported on food dominance at 82 locations in the central North Pacific Ocean and Bering Sea from May to August 1960. Dominance at different stations varied for six taxonomic groups: amphipods, 43%; fish, 18%; squid, 16%; euphausiids, 12%; copepods, 7%; pteropods, 2%; and other, 2%. Two investigators reported markedly contrasting breakdowns of food items for sockeye in the northeast Pacific Ocean according to their ecological roles. LeBrasseur (1972) reported percentages of herbivores and primary and secondary carnivores as 35, 30, and 67, respectively, whereas Sanger (1972) derived percentages of 15, 80, and 5 for the same three groups.

The composition of food of sockeye depends quite clearly on the availability and relative abundance of the food items, which vary with season and location. McAllister et al. (1969) reported that sockeye tend to seek areas of high standing crops of macrozooplankton. They found consistently high abundance of large zooplankton within the Ridge zone south of the central Aleutian Islands, the area of known concentration of immature salmon in summer and fall. A geographic difference in food composition and weight consumed was also reported for maturing sockeye returning through the Bering Sea in June en route to western Alaska (Nishiyama 1974, 1984). In the basin area of central Bering Sea, food items were found to be more varied and included squid, fish larvae, amphipods, and euphausiids, whereas in the shelf area to the east they were almost exclusively euphausiids, with a small proportion of fish larvae, including walleye pollock. The average stomach content volume was greater in sockeye sampled from the shelf area, which appeared to coincide with the general trend of zooplankton biomass distribution between the two areas. The caloric value per unit weight of food consumed was also greater in the shelf area.

Annual differences in food consumed were reported by Ito (1964) for the period 1958-63 in samples from the Japanese mothership fishery. He found that the proportions of larger sized food items (squid and fish) dominated in even-numbered years and small food items (euphausiids, amphipods, and copepods) dominated in odd-numbered years. Such a relationship suggests a possible feeding interaction with maturing pink salmon, which are cyclically more abundant in odd-numbered years in that area. In comparing the food items in the stomachs of salmon to the distribution of food animals, Takeuchi (1972) found no strong indication of selective feeding among sockeye, pink, and chum salmon.

Ito (1964) noted no difference in food preference of sockeye by maturity stage in samples taken by gillnet in the Japanese salmon mothership fishery. However, euphausiids appeared to be more favoured by immature than maturing sockeye in samples taken by purse seine south of the eastern Aleutian Islands (Dell 1963) and in samples from various oceanographic domains in the eastern North Pacific Ocean in spring (LeBrasseur 1966).

Size and Growth at Sea

Sockeye from different regions differ in growth rate, age at maturity, and size at maturity. Sampling of identified individual stocks at sea and of combined stocks over a broad area of the ocean have brought out certain general features of the ocean growth pattern of sockeye salmon. This general pattern, for sexes combined, is shown in Figure 44. Maturing fish are larger than immature fish at a given age, growth in length is greatest during the first year at sea, and increase in weight is greatest during the second year.

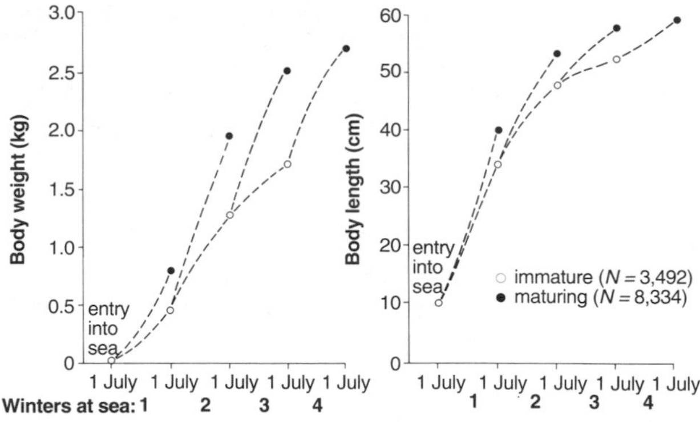

FIGURE 44

Estimated mean body weights and lengths of sockeye salmon on 1 July (Lander et al. 1966). Connecting lines indicate related stages, not actual growth. (From French et al. 1976, Figure 36)

Smolts of western Alaskan sockeye of ages 1. and 2. begin sea life with mean weights ranging from about 4 to 15 g and mean fork lengths from 7.7 to 12.5 cm. Repeated sampling over many years south of Adak Island in the central Aleutian Islands area has shown that the immature age 2.1 fish will a year later average slightly greater in mean length (35.4 cm) by 21 July than age 1.1 fish (32.9 cm). A slight difference in length (47.4 versus 46.6 cm) is still indicated between immature age 2.2 and 1.2 fish the following July (Rogers 1973b). These differences result primarily from the larger initial size and earlier seaward migration of the age 2. smolts. Male sockeye are larger in average size than females by the spring of their second winter at sea and maintain a size difference until maturity at age .2 or .3 (Figure 45).

The pattern of seasonal growth in sockeye has not been defined precisely because of the very limited ocean sampling which has been accomplished during the period late autumn to early spring. However, there is evidence from studies of scales and of seasonal trends in size that growth continues for most of the year. Sockeye scales show patterns of wider circuli in summer and of narrowed circuli, designated as "annual rings," at the end of winter (Figure 46). This could mean that the duration of the period of annual ring formation is somewhat proportional to the number of circuli

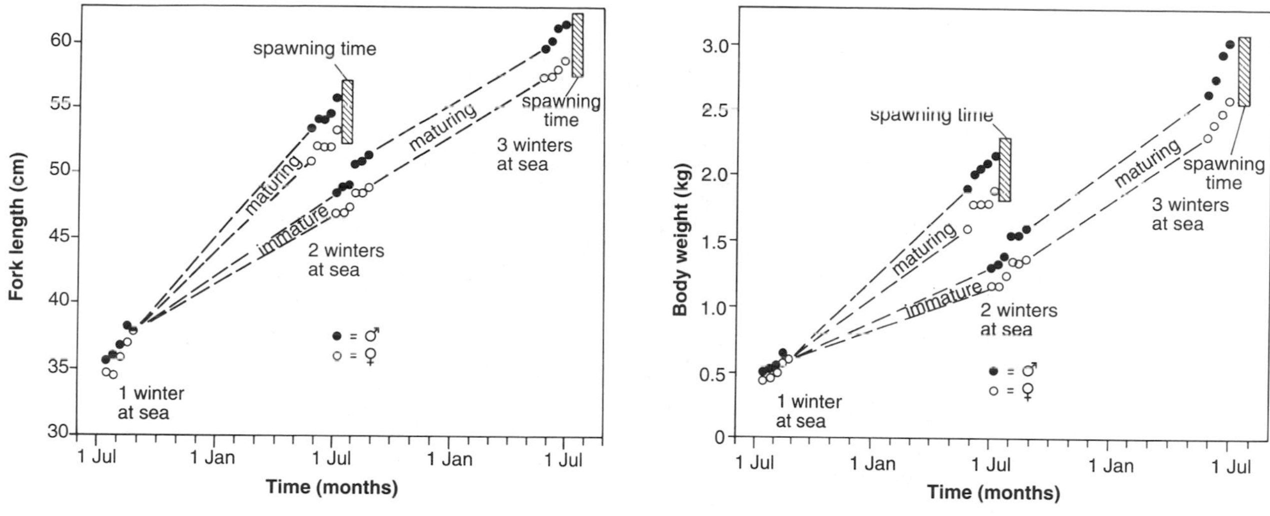

FIGURE 45

Mean fork length and body weight by 10-day period for each sampled life history group of western Alaska sockeye salmon. Dashed lines indicate related groups, not linear growth, for periods not sampled. (From Lander and Tanonaka 1964, Figures 16 and 17)

laid down to form this zone on the scale. However, the changes in width of circuli cannot be interpreted directly as reflecting changes in growth rate because it cannot be assumed that the circuli are laid down at a constant rate over the entire year. Bilton and Ludwig (1966), who examined scales of sockeye taken in the central Gulf of Alaska in January-early February 1964 and April 1965, concluded that the annual ring probably began to form some time between early November and January and that, on average, it is completed in January. All scales in samples taken in April showed new growth beyond the annulus. Scales of the age .2 fish tended to show more circuli after the annulus than did those of the age .3 fish in both sampling periods. These data suggest that if body growth slows down during annual ring formation, it is for a relatively short period of time. Data presented by French et al. (1976) on seasonal size of sockeye sampled in the North Pacific Ocean and Bering Sea also serve as evidence that growth continues most of the year (Figure 47). Fish of an ocean-age group sampled in winter were substantially larger than those sampled in September, and those sampled in April or May showed a substantial length increase over the winter samples. Thus, it is evident that growth is by no means confined to the late spring and summer months, and that, by their migrations, the sockeye are able to find suitable food at water temperatures conducive to growth over most of the year.

Differences in growth rates and maturity schedules occur in sockeye from different regions as a result of both genetic variability and differences in habitat occupied during ocean life. Maturing sockeye salmon from western Alaska taken on the high seas are slightly, but significantly, larger than Asian sockeye captured at the same time (Lander and Tanonaka 1964). Fraser River sockeye of age 1.2 attain more growth than western Alaska sockeye of the same age. Ocean growth within groups of stocks varies among brood years. For example, average size of western Alaskan fish of a given life history stage changes considerably in different years of sampling (Lander and Tanonaka 1964; Rogers 1973b).

There are certain characteristic stock growth conditions that are reflected in differences among populations in scale growth patterns. Mosher (1972), who developed a catalogue of sockeye scales from different Asian and North American coastal regions, noted a number of these differences. For instance, scales of Bristol Bay sockeye have relatively few circuli in the first ocean growth zone, and the summer growth circuli are broad, widely spaced and regular, particularly near the fresh-

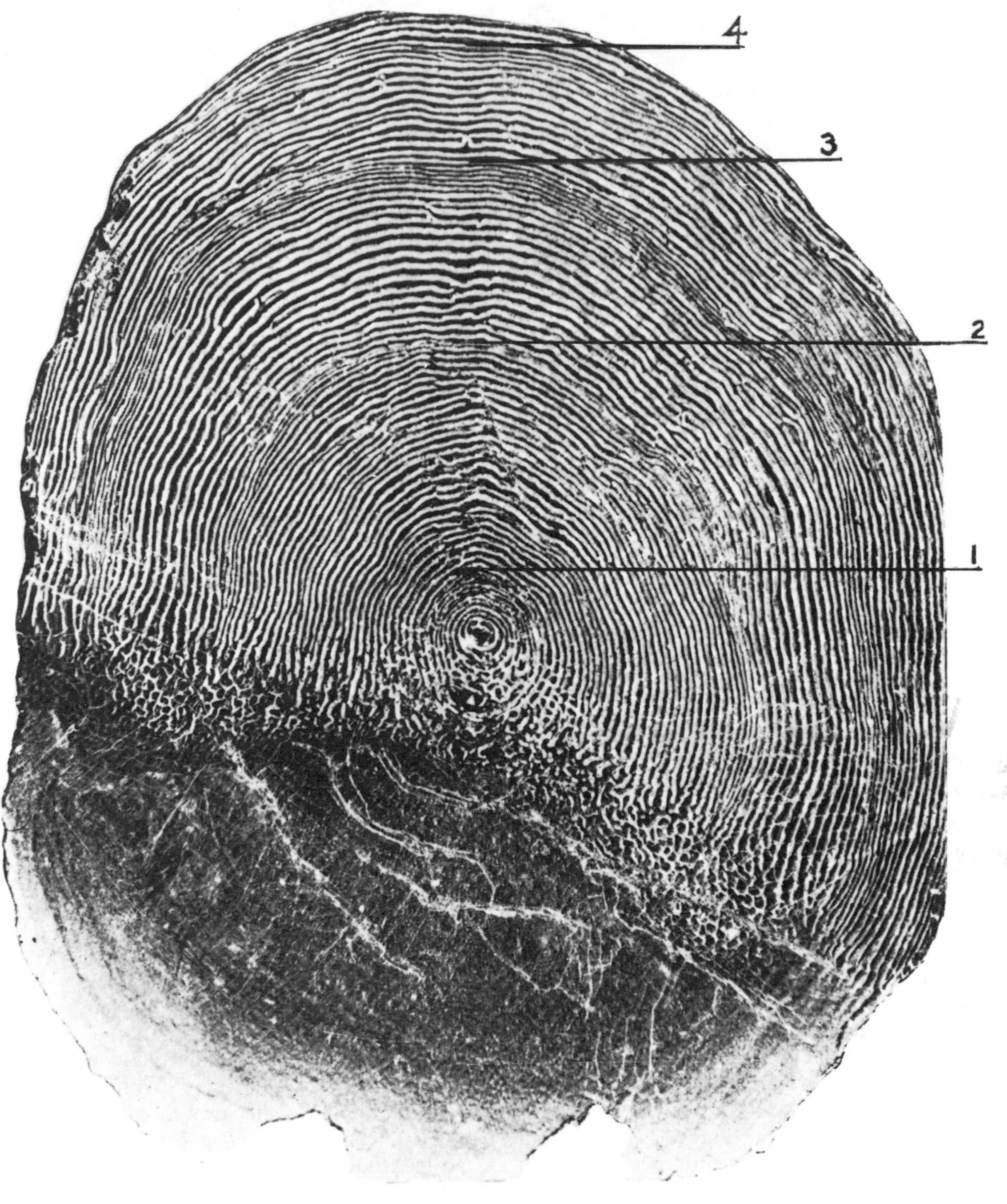

FIGURE 46
Photograph of a scale of a Fraser River sockeye female in its fifth year, age 1.3, captured on 18 June 1917. (From Gilbert 1918, Figure 12)

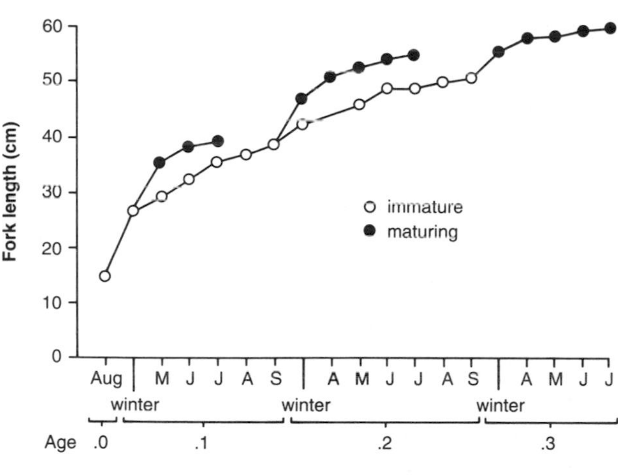

FIGURE 47

Average fork lengths of sockeye salmon taken at sea by ocean age and time periods. Data from gillnet catches, combined sexes, and for all areas in the North Pacific Ocean and Bering Sea. (From French et al. 1976, Figure 40)

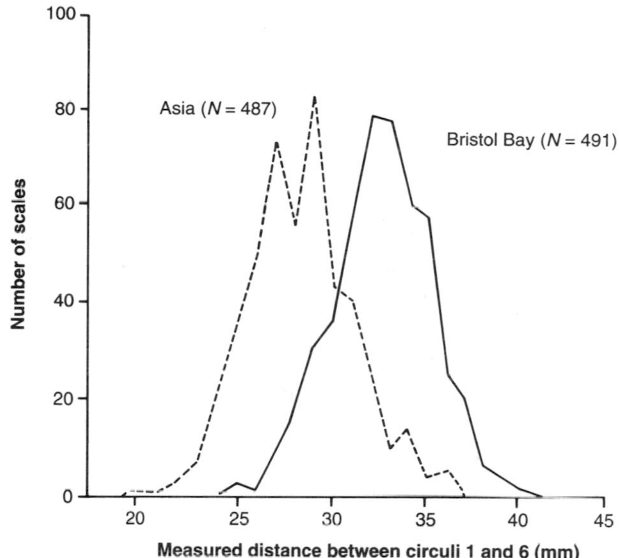

FIGURE 48

Frequency of measured distances between circuli 1 and 6 of the first ocean zones for samples collected from Bristol Bay and Asia, 1957 (scales were examined at 100x magnification). (From Anas and Murai 1969, Figure 13)

water zone. This latter difference was one of the characters used by Anas and Murai (1969) in discriminant function analyses to distinguish between Bristol Bay and Asian sockeye in samples taken in areas of intermingling of these stocks on the high seas (Figure 48). In another example, Mosher (1972) noted that sockeye from Rivers and Smith inlets and Nimpkish River have distinctive scales showing two large areas of contrasting texture: (1) freshwater and first ocean zones of fine-textured circuli, and (2) second and subsequent ocean zones with coarse-textured circuli of typical ocean growth. The greatest difference in ocean growth among stock groups of sockeye is in the first ocean zone, reflecting the period when they are environmentally the most segregated.

Size at Return

The size attained by sockeye salmon of a given race at sea tends to be quite uniform for a given length of stay in the ocean and varies, primarily, with the number of years spent in the ocean and, secondarily, with growth conditions encountered. Most sockeye salmon spend two or three years feeding in the ocean before their final summer of return. In Alaska, the Bristol Bay sockeye tend to be smallest in average size in the catch (2.56 kg) (INPFC 1979) and spend primarily two winters in the ocean. Because the gillnet fishery tends to be selective for larger sockeye, the average size of sockeye in the total run is less than that determined in catch samples. Chignik River sockeye, most of which spend three winters in the ocean, are the largest, averaging 3.16 kg. For Fraser River sockeye, which generally spend two winters at sea, the average size of the age .2 group in the commercial catch is 2.73 kg (Killick and Clemens 1963). The smallest sockeye on return are the Columbia River "blueback," which average only about 1.58 kg after two winters at sea.

The substantial variation among populations in size within an age class was noted by Healey (1987), who found that for 51 sockeye populations in Alaska, the mean length (mid-eye to fork of tail) of mature age 1.2 females ranged from 44.8 to 53.6 cm, and of age 1.3 females, from 50.8 to 59.5 cm. He observed that there was no correlation between mean size at age and mean age in the populations, as would be expected if size and maturity schedule were strongly linked.

SURVIVAL IN THE OCEAN

Estimates of total ocean mortality of sockeye salmon are based either on the percentage return as adults of fish marked as smolts, or, in instances where the total smolt migration has been estimated, on the total return of the brood-year adults. Thus, these estimates include river and estuarine mortality below the location of smolt marking or enumeration in the river or lake. In most, but not all, instances this location is at the lake outlet, and the length of river migration varies greatly among localities. A wide range of estuarine conditions is encountered among sockeye populations entering the marine environment around the North Pacific Ocean rim. The estimates also require the determination of numbers of adults in commercial catch and spawning escapement over the years of expected return from a brood year. This is complicated in most instances by the presence of other stocks and brood years in the fisheries and the need to apportion returns accordingly. Attention has also been given to estimated non-catch fishing mortality in the case of high-seas and some coastal fisheries. Where the estimates are based on return of marked individuals, corrections have usually been made for estimated differential lowered survival rate because of the effect of handling or of the mark itself. Marine mortality estimates for different stocks, smolt age groups, and brood years range from about 50% to over 95%.

Although inconsistencies are frequent, a gross examination of the data on adult return from smolt migrations indicates that, over the range of population data available, smolts of larger average size tend to have higher marine survival (Figure 49) (Ricker 1962). However, except for the Cultus Lake data presented by Foerster (1954), the effect of smolt size on survival is not at all clear within smolt age groups within populations, and the overall picture of increased survival of larger smolts is dependent on the inclusion of different smolt age groups (Karluk Lake) and differences among populations in freshwater growth. Lake Dalnee smolts, which exhibit the most freshwater growth, show the highest survival, although in this case the yearling smolts, which are much smaller than the age 2. smolts, had higher calculated average survival for the three comparable years of seaward migration (1938–40). For the Cultus Lake data, Foerster (1954) concluded that the negative correlation between magnitude of smolt migration and percentage return of adult spawners to the lake was indeed related principally to size of smolts at time of seaward migration.

The least variable and apparently most consistent data on marine survival are from the marking experiments of Barnaby (1944) at Karluk River. Smolts were marked about 6.4 km from the mouth of the river, which enters the open sea without a complex estuary. Smolts of ages 1., 2., 3., and 4. were released over a period of six years, the majority being ages 2. and 3. Total calculated returns from each year of smolt release varied between only 20.5% and 23.6% (uncorrected for possible differential mortality due to marking). The average return from the marking of age 2. smolts was 17.4% and for age 3. smolts, 25.7% (higher than age 2. for five of six release years). The consistency in these results may be due to the overall larger size of Karluk Lake smolts, the minimal opportunity for riverine or estuarine mortality, and the consistently larger size of age 3. smolts. However, the differential survival of the two age groups was probably influenced by three factors of unknown relative importance: (1) size and age at seaward migration, older smolts being larger in size; (2) timing of entry into the marine environment, older smolts tending to migrate earlier in the season; and (3) length of stay in the ocean, older smolts having a greater tendency to remain only two years at sea, but, at the same time, to return later in the season in the year of return (Burgner 1962b).

Hyatt and Stockner (1985) observed that populations of sockeye in some coastal British Columbia lakes produce smolts of the smallest mean sizes known for sockeye. For eight lakes over a 13-year period, mean weights varied between 1.1 and 3.4 g prior to lake fertilization. Limited data to date on smolt size and adult returns for four lakes after

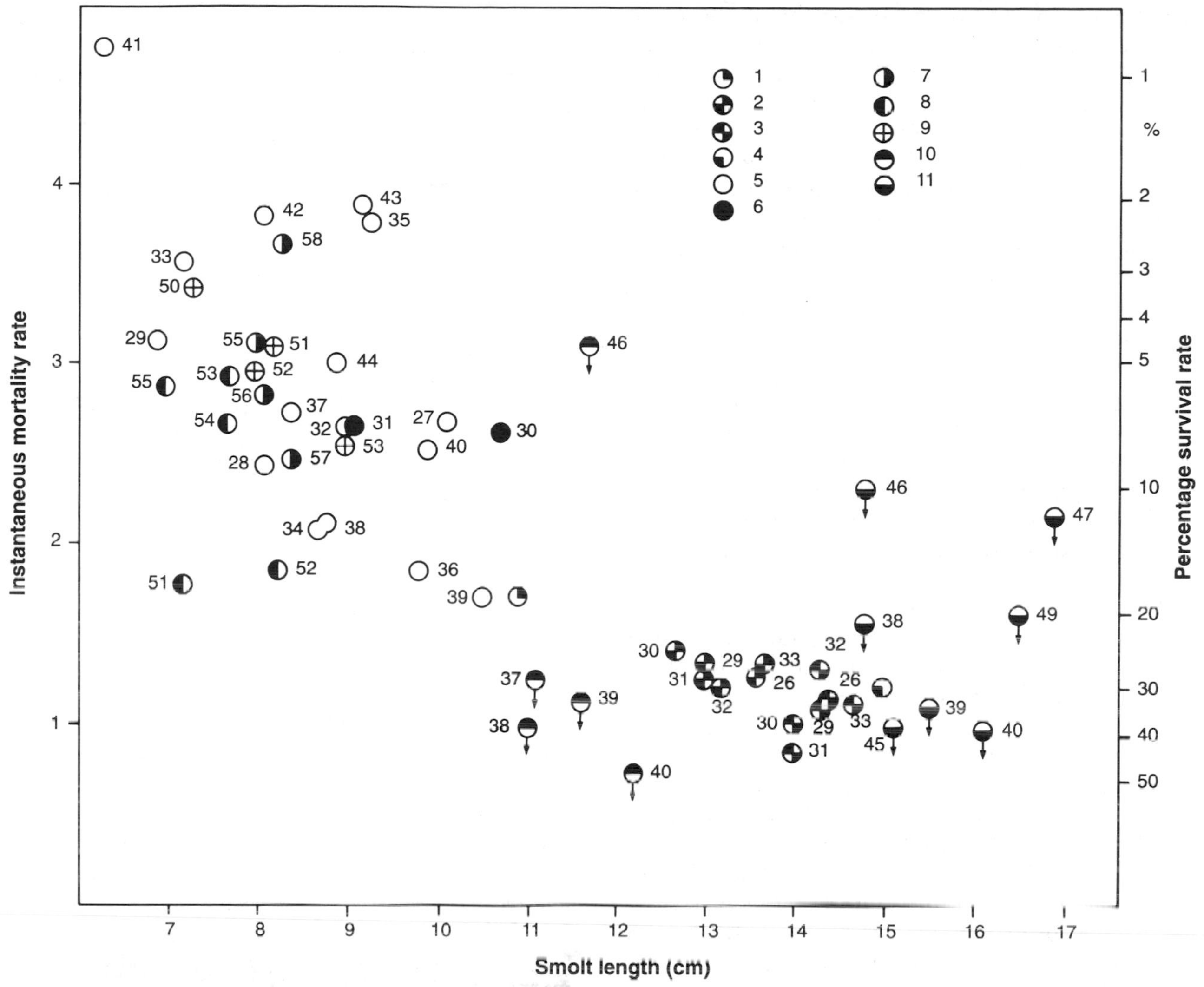

FIGURE 49

Estimates of instantaneous ocean mortality rates (left-hand scale) for sockeye of six stocks. Corresponding percentage survival rates are shown at right. The year beside each point is the year of the smolt migration. The points for Karluk sockeye include an adjustment for mark mortality. The clear points for Cultus Lake are based on percentage return to the lake as spawning escapement (Foerster, 1954, Table I) multiplied by the average factor 2.73 to adjust them to terms of total stock; the two black points are based on a complete enumeration of returning marked fish in the fishery and at the lake, adjusted for mark mortality. Lake Dalnee figures include some fishing mortality; the arrow pointing downward indicates that these points are all too high to be directly comparable with the rest. All other points represent natural mortality only (including mortality during downstream migration, and that during the upstream migration of fish which escaped the fishery), except that Bare Lake points may include a small amount of high-seas fishing mortality. *1* - Karluk (age 1. smolts, 1926–33 average); *2* - Karluk (age 2. smolts); *3* - Karluk (age 3. smolts); *4* - Karluk (age 4. smolts, 1926–33 average); *5* and *6* - Cultus; *7* - Babine; *8* - Chilko; *9* - Bare Lake; *10* - Dalnee (age 1. smolts); and *11* - Dalnee (age 2. smolts). (Adapted from Ricker 1962, Figure 1)

treatment indicated large variations between brood years in marine survival (4.9%–30.0%). They noted that the smallest smolts for which they measured marine survival (1977 brood year, Long Lake) had one of the highest marine survivals (30%). However, for the Wood River lakes (Alaska), Burgner (1962b) found an apparent lower marine survival of the smaller, stunted yearling smolts in a single brood-year migration in which two distinct size groups of yearlings were present.

In general, there are indications that increase in freshwater growth may decrease ocean age at return and hence decrease mortality incurred through an additional year of ocean life (Hyatt and Stockner 1985) (Figure 50). In contrast, for Bristol Bay sockeye, smolts having experienced better freshwater growth tend to migrate seaward at an earlier age (age 1.), and age 1. smolts tend to remain a year longer at sea than do age 2. smolts (Rogers 1984).

In spite of the inconsistencies among data sets, it seems that, in general, smolts of larger size will experience higher marine survival. Ricker (1976) provided an additional supporting argument, namely, that "within a year-class the average diameter of the freshwater zone of the scales, and the number of circuli in them, is greater among adult fish returning from the sea than it was among the smolts as they left the lake." However, Hyatt and Stockner (1985) concluded that "the effect of smolt size variations on adult abundance and age-at-return for populations of sockeye are only likely to emerge from studies or analysis of natural populations which provide large numbers of observations that span a wide range of smolt size or from studies of intensively manipulated populations where the potentially confounding role of interacting variables may be controlled."

Although many ingenious methods of estimation have been presented by various investigators, no satisfactory determination of the pattern of survival of sockeye populations during their marine life has been developed. The various methods were reviewed in detail by Ricker (1976), who presented a summary of estimates of mean monthly instantaneous rates of ocean mortality for sockeye salmon with comments on the nature of biases and degree of sampling error. Generally, it is hypothesized that natural mortality rates decrease with increase in size of fish as they become potential prey for fewer and fewer predators. Ricker expressed preference for the hypothesis (Mathews and Buckley 1976) that ocean mortality rate might be an inverse function of the weight of the fish as they grow. This seems to be supported by Henry's (1961) finding that differences among years in marine survival of age 1.2 Chilko Lake sockeye were directly correlated with first-year marine growth recorded on their scales.

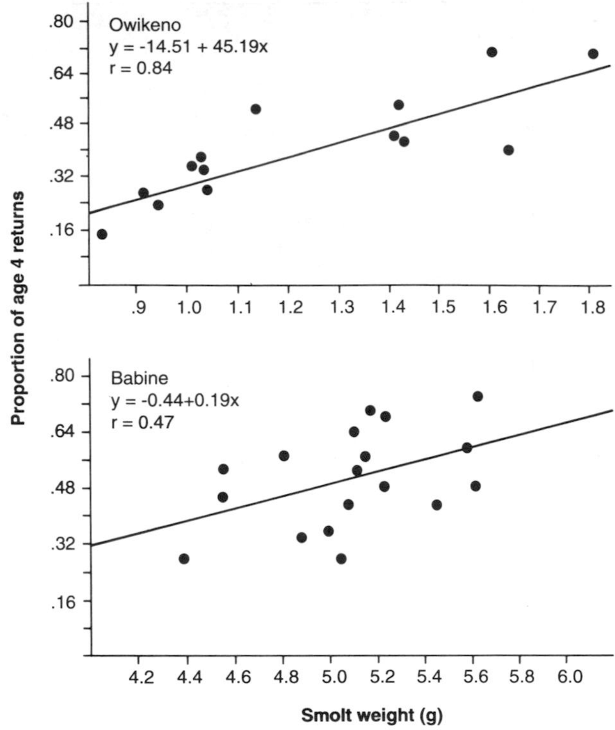

FIGURE 50

Proportion of brood-year adult returns migrating back at age 4 as a function of average smolt weight for that brood year for Owikeno Lake sockeye (D. Peacock, Department of Fisheries and Oceans, Prince Rupert, BC, pers. comm.) and Babine Lake sockeye (from Peterman 1982). (From Hyatt and Stockner 1985, Figure 9)

Variations in oceanographic conditions and in marine predator populations (fish, mammals, and birds) undoubtedly have affected the marine survival of sockeye populations in different ways around the North Pacific rim, but these effects are poorly understood. In studying trends in abundance of Bristol Bay sockeye since the 1920s, Rogers (1984) found that large changes in ocean surface temperature, particularly in the winter months when the fish were at sea, seemed to correspond to the major changes in abundance of the sockeye upon their return. He hypothesized that differences in winter ocean temperatures may have resulted in differences in winter distribution of sockeye and, hence, in their vulnerability to marine predators, particularly marine mammals.

Utilization and Abundance

Utilization

The adult sockeye salmon returning to the rivers to spawn have long been a preferred salmon species for commercial canning because of the attractive colour and firmness of their orange-red flesh, their good flavour, their uniform size, and concentrated run abundance. But long before the arrival of settlers on the North American Pacific coast and the beginning of canning, the aboriginal populations living along the coast and rivers relied heavily on sockeye and other salmon species as a staple food. The locations of villages and fish camps were often decided on the basis of access to returning sockeye salmon runs, on which the native villagers were heavily dependent as a component of their annual food supply. Many of the fish were preserved in smoked or dried forms as food for humans and their dogs.

Commercial utilization by white settlers initially consisted of salting the salmon, but heavy exploitation of sockeye began with the development of the canning industry. Sockeye was the principal species sought in early canning operations extending from the Fraser River northward. The first canneries on the Fraser River were established in 1870, on the Skeena River in 1877, southeastern Alaska in 1878, Nass River and Rivers Inlet in 1881, and Bristol Bay in 1884 (Foerster 1968). Additional canneries were built rapidly, and by 1900 most major North American sockeye runs were being heavily exploited. Sockeye salmon canning in the Kamchatka Peninsula developed in the early 1900s.

In North America, canned salmon remained almost the exclusive processed product until the 1960s, when salt-cured salmon roe, "sujiko," was developed as an export product to Japan, enhancing the value of the female salmon. In more recent years, improved refrigeration and freezing capabilities at shore-based and floating processing plants have made it possible to develop a fresh and frozen market for sockeye salmon. By the early 1980s, about half the sockeye salmon catch in Alaska was marketed fresh or frozen.

Sockeye salmon are exploited by intensive fisheries adjacent to rivers along the northwest coast of North America and the Kamchatka Peninsula. They are also exploited by high-seas gillnet fisheries conducted by Japanese fishermen. The catches by these fisheries reflect, in a general way, the relative abundance of the sockeye stocks from different areas, except that the catch of the high-seas fishery must be apportioned, insofar as it is possible, to the respective areas of origin. This has been the subject of extensive international research coordinated through INPFC.

Production Areas

Commercial catch data for the major fishing areas are presented in Figure 51 as average annual catches for the years 1970–83. The most important catch areas are Bristol Bay, the Japanese high-seas gillnet fisheries, and southern British Columbia and Washington State (essentially the fisheries for Fraser River sockeye). Studies of the origin of sockeye intercepted in the Japanese high-seas fisheries have revealed that they are primarily of Kamchatkan and Bristol Bay origin (Harris 1987). For the 1970–83 period, the catches were predominantly of Kamchatkan origin (principally Ozernaya and Kamchatka rivers).

Of the coastal areas shown in Figure 51, the largest production area in numbers of sockeye is western Alaska, the area north of the Alaska Peninsula bordering the eastern Bering Sea (INPFC

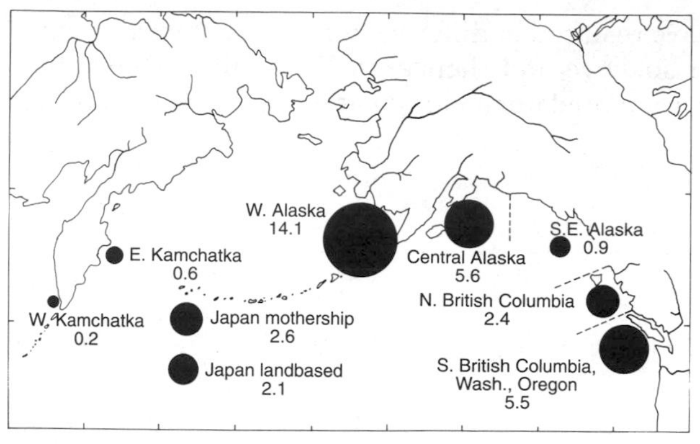

FIGURE 51

Average commercial catch of sockeye for the major fishing areas, 1970–83, in millions of fish

1972–86). Fishing is conducted in the bays with drift gillnets from small boats (less than 10 m in length) or by anchored "set" gillnets along the beaches. The river systems tributary to Bristol Bay, the Kvichak, Alagnak, Naknek, Egegik, and Ugashik rivers on the east side and the Nuyakuk, Wood, Igushik, and Togiak rivers to the west, are the producers of sockeye in this area. In central Alaska, the Chignik River on the south side of the Alaska Peninsula, Karluk and Red rivers on Kodiak Island, Kenai, Kasilof, and Susitna rivers in Cook Inlet, and Copper River on the eastern border of Prince William Sound are important sockeye producers. Fishing in these areas is done by purse seine and by drift and set gillnets. Southeastern Alaska has approximately 119 small production systems (McGregor 1983), the largest being Chilkoot and Chilkat lakes in Lynn Canal. In addition to the gear mentioned above, trap fishing is conducted by native associations in two localities.

In northern British Columbia, extending south to the north tip of Vancouver Island, the important production areas are the Skeena River, Nass River, Rivers Inlet, and Smith Inlet. The remaining area to the south in Figure 51 includes southern British Columbia, Washington, and Oregon, where production is dominated by the Fraser River system. Catches are made by troll gear off Vancouver Island; by purse seine, gillnet, and reef net (San Juan Islands) primarily in interior waters; and by gillnet off the mouths and in the tidal areas of the Fraser and Skeena rivers. Within the Fraser River watershed the major production areas are Shuswap, Chilko, Stuart, and Quesnel lakes. Together, the Fraser, Skeena, and Nass rivers, and Rivers Inlet systems account for about 87% of the total sockeye production in British Columbia (Starr et al. 1984).

On the Soviet Far East coast, about 90% of the sockeye production comes from two systems – the Ozernaya River (Lake Kuril) and the Kamchatka River on the west and east sides of the Kamchatka Peninsula, respectively (Krogius and Krokhin 1956b). Fishing is conducted by trap and beach seine, primarily at the river mouths.

Historical Catch Trends

Historical trends in North American coastal, Asian coastal, and high-seas catches of sockeye salmon are shown in Figure 52.

Catches in North America, primarily Alaska and British Columbia, have always been greater than Asian catches. North American catches peaked in the late 1930s, declined during the next two or more decades, then surged to a record high in the early 1980s. The recent record high catches in 1983–85 resulted primarily from an increase in run magnitudes of natural stocks in central and western Alaska.

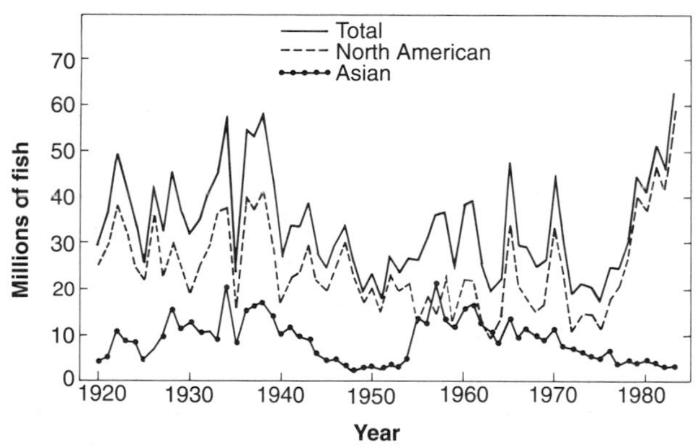

FIGURE 52

Total catch of sockeye salmon of Asian and North American origins by commercial fisheries of the USSR, Japan, the United States, and Canada, in millions of fish, 1920–83

Asian catches of sockeye during the period 1920–30 were made primarily by Japanese fishermen, fishing trap sites leased along the Kamchatka Peninsula, and secondarily by Soviet fishermen fishing traps and beach seines (Figure 53). The Japanese trap fishery continued until 1944. Statistics for the Soviet catch are not available prior to 1927 and are somewhat questionable for the period 1927–39. A Japanese mothership fishery operated off Kamchatka from 1929 through 1942 and a driftnet and trap fishery operated off the Japanese north Kuril Islands from 1933 through 1945. The combined catches of these Japanese fisheries are shown in Figure 53. The annual catches averaged over 13 million in the 1930s, and catches remained high until the Japanese trap, mothership, and north Kuril Islands fisheries phased out during World War II. Soviet catches averaged only 3.4 million sockeye during the period 1945–52 and dropped even lower after the Japanese fishery was resumed on the high seas.

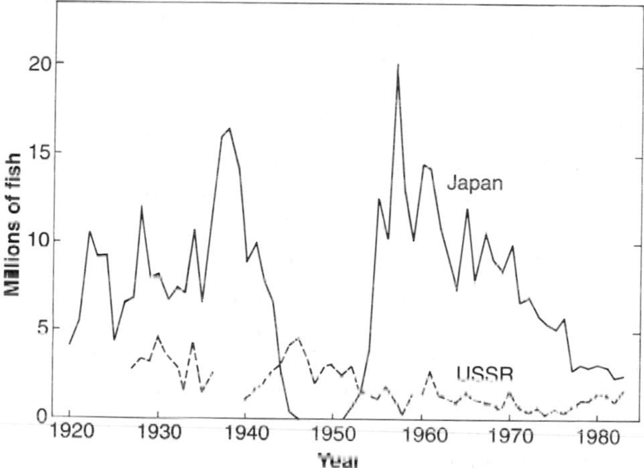

FIGURE 53

Catch of sockeye salmon by commercial fisheries of Japan and the USSR, in millions of fish, 1920–83

The Japanese mothership fishery started again in 1953 and a landbased driftnet fishery moved sufficiently offshore to begin substantial catches of sockeye in 1958. The combined catches of these fleets are also shown in Figure 53. Declines in sockeye catches since 1977 reflect, in part, the restrictions in effort and catch imposed by the USSR-Japan fisheries agreement, and restrictions in fishing areas resulting from renegotiation of the International Convention for the High Seas Fisheries of the North Pacific Ocean between Canada, Japan, and the United States (Harris 1987). The sockeye taken in these fisheries are primarily of Kamchatkan origin, although significant numbers of Alaskan sockeye are also caught.

Because of the much greater available freshwater rearing areas for sockeye in North America than in Asia, the North American sockeye stocks are much larger than those in Asia. The historical trends in commercial catch of North American sockeye salmon are shown in Figures 54 and 55.

In western Alaska, the catch pattern is heavily influenced by the cyclic production of the Kvichak River system. Prior to 1940 there were generally three adjoining years of high catches followed by one or two years of low catches. Since 1940 the Kvichak River sockeye pattern has indicated one or two years of relatively high catches followed by three years of low catches (Rogers and Poe 1984). An upsurge to a record western Alaskan catch (39.4 million sockeye) occurred in the last few years.

In central Alaska the early pattern is influenced by the decline of the Karluk River populations, and in later years catches ebbed in the late 1950s and early 1960s. The recent rise to a record catch of 11.4 million sockeye parallels the western Alaska trends and results primarily from the resurgence of Cook Inlet stocks.

Southeastern Alaska has lower annual catches (less than 3 million sockeye), with a drop in the catch level beginning in the 1940s. The recent resurgence seen in central and western Alaska is not apparent in this area.

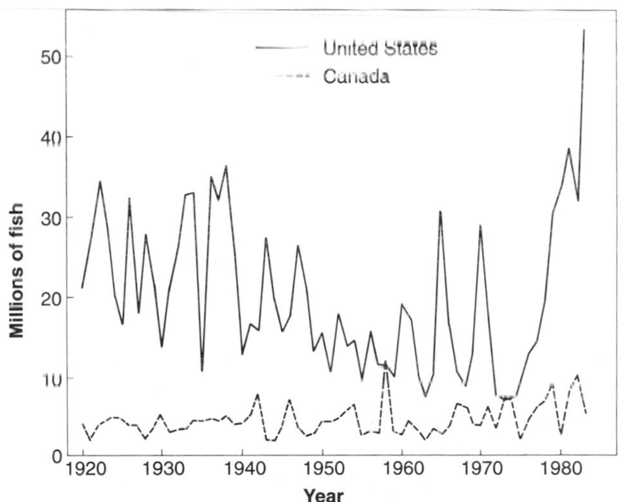

FIGURE 54

Catch of sockeye salmon by commercial fisheries of the United States and Canada, in millions of fish, 1920–83

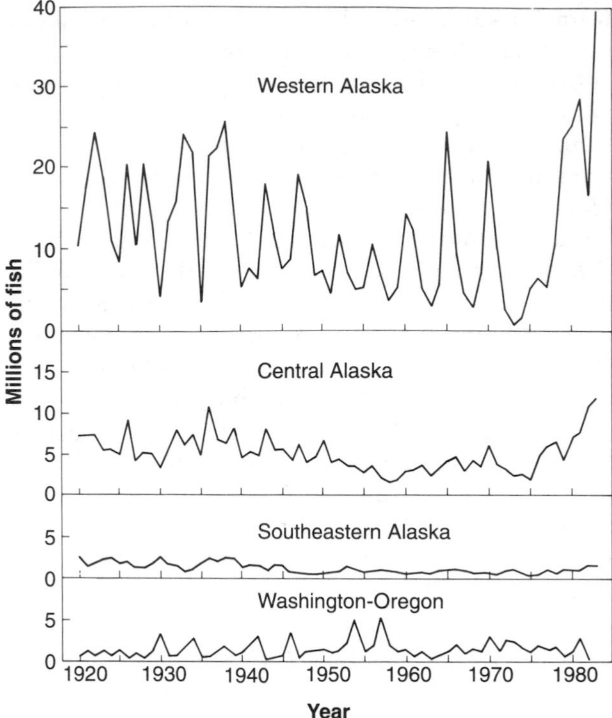

FIGURE 55

Catch of sockeye salmon by commercial fisheries in Alaska and Washington-Oregon, in millions of fish, 1920-83

The catch records for British Columbia (Figure 54) and Washington-Oregon (Figure 55) are incomplete prior to 1922 and hence do not reflect the early high catches in the Fraser River prior to the disastrous 1912-13 slides in the Fraser River canyon at Hell's Gate. The four-year cycle of the Fraser River run is reflected in the overall catch, and generally higher catches (to 10 million sockeye) occurred in British Columbia waters in recent years (Figure 54) as a result of restorative efforts on the Fraser and Skeena systems.

Sockeye salmon are not readily taken by sports gear in coastal areas. For example, Washington Department of Fisheries statistics for marine waters indicate that sport catches are less than 700 sockeye per year. Sport catches in British Columbia are also low. The sockeye are easier to catch once in fresh water, as is indicated by sport catches as high as 43,400 in Lake Washington by an intensive sport fishery, occurring during the period the sockeye are maturing in the lake prior to entry to spawning grounds (Hoines et al. 1985). Another intensive sport fishery occurs in the Kenai River, Cook Inlet, where some 40,000 sockeye are taken annually (1974-81 average) on their upriver migration to spawning areas (Cross et al. 1987).

Records of so-called subsistence catches of sockeye salmon are not universally maintained, but the catches are important, particularly to native peoples, in a number of localities. The largest catches are made by Fraser River Indian tribes during the upstream migrations of sockeye salmon. The average recorded annual subsistence catch for the years 1970-82 is 240,000, representing about 4.7% of the total annual catch of Fraser River sockeye (Starr et al. 1984).

Cycles and Regulating Mechanisms

Although populations of sockeye salmon generally tend to fluctuate around some general level, cyclic fluctuations in abundance are characteristic in some populations, and, in others, unexpected upward or downward trends in abundance occur. Processes that cause higher mortalities as sockeye densities increase and lower mortalities as densities decrease tend to stabilize a population around a fixed level, and are considered "compensatory." Other density-dependent processes that result in increased mortality as density decreases, and vice versa, are labelled "depensatory" (Neave 1953). Predation was discussed earlier as a likely depensatory process in some situations. Processes that appear to have impact on populations independent of their magnitude are termed "extrapensatory" (Neave 1953).

Ricker and Smith (1975) classified the mechanisms of interaction between year classes of sockeye as either direct or indirect. A direct mechanism is described as one in which a large year class reduces the survival rate of one or more subsequent year classes, for example, by exhausting their food supply, by direct cannibalism, or by the effect of infertile and dead eggs from a large spawning on the survival of eggs from subsequent spawnings. The example of an indirect mechanism is given as one where a small year class results in reduced abundance of a control agent (predator or parasite) and this permits a subsequent year class to have an excellent survival rate.

The optimum spawning escapement is usually considered to be the number of adult sockeye that will yield the greatest surplus of expected return over needed escapement. The concept of an optimum is built around the idea that over a considerable range of escapements the expected return is greater than the parent escapement, but that at higher levels of escapement the return of progeny per spawner is reduced as compensatory mortalities begin to increase. The form of this relationship can vary and usually is unclear because survival conditions encountered at the various stages are greatly influenced by environmental conditions that vary from year to year. However, compensatory processes are likely to operate as spawning and incubation grounds become crowded, or when competition for food in lake rearing areas results in reduced growth of juveniles and delay of seaward migration. A simplified curve is shown in Figure 56.

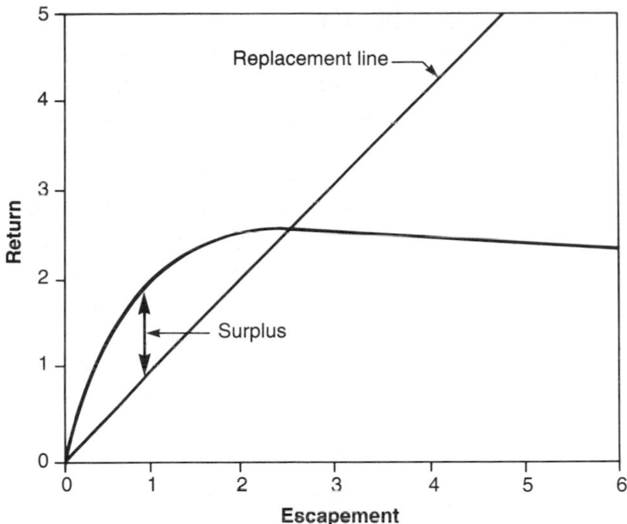

FIGURE 56

Theoretical return-escapement relation of sockeye salmon. (From Burgner et al. 1969, Figure 1)

Some populations of sockeye are noted for the extreme cyclic nature of the return runs. The most famous are the run to the Adams River, tributary to the Fraser River, and the run to the Kvichak River, Bristol Bay (Figure 57). These cycles have received particular attention because of the magnitude of their runs and because the highly cyclic returns create severe harvest and marketing problems. Can the cycles be "ironed out" to provide more uniform annual harvests, or is there danger in disrupting the processes that maintain high production every four or five years?

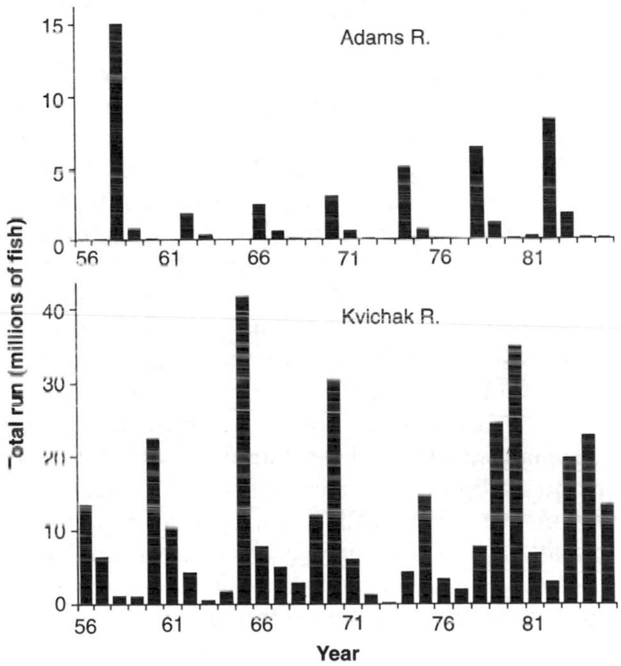

FIGURE 57

Sockeye salmon cycles, Adams River, BC, and Kvichak River, Alaska

The Adams River cycle is, at least superficially, the simpler and more consistent of the two. Following the wholesale destruction of upper Fraser River runs as a result of river blockage in the Fraser River Canyon, a new pattern of quadrennial cyclic dominance developed beginning in 1930 (Ricker 1950). Since 1938 the dominant runs have averaged about 250 times that of the smallest of the "off" years and about nine times that of the subdominant year. It was expected that mortalities affecting sockeye during the spawning-incubation stage were compensatory, i.e., mortality as a percentage was lower at the lower population levels of "off" years. Further, smolts resulting from "off" year spawnings were considerably larger in size than dominant year smolts, indicating that their lacustrine growth rate was higher and thus their survival potential should be at least as good. Mortality inflicted by rainbow trout on juvenile sockeye during lake residence and seaward migration was identified as the most probable depensatory mechanism balancing compensatory coefficients (Ward and Larkin 1964). Studies of the food habits and condition factors of rainbow trout in Shuswap Lake in the vicinity of the Adams River mouth supported this theory. The rainbow trout populations were believed to be held in check by heavier mortality during the two years of the cycle when sockeye are scarce. The pattern of events was modelled in simulation studies (Ward and Larkin 1964; Larkin 1971).

More recently, an alternate explanation of the cyclic pattern of Fraser sockeye runs has been advanced. It was noted by Ricker (1950) that prior to 1913, all upriver Fraser River stocks seem to have been dominant in the same year, implying a common depensatory mechanism(s) operating, perhaps in the lower Fraser River, the estuary, or the Strait of Georgia (Larkin 1971). One such depensatory mechanism, identified by Peterman (1980), is the native Indian food fishery that is conducted upriver from the Fraser River commercial fisheries, which takes a higher proportion of the small runs than of the dominant runs. It is now proposed by Walters and Staley (1987) that cyclic dominance in Fraser River salmon is maintained by overall high fishing rates, and that aboriginal fisheries may have been responsible for establishing the cyclic patterns inferred from historical records of the past century.

The cyclic run pattern for the Kvichak River sockeye is more complex. The Adams River cycle is consistently four years, and except for jacks returning in their third year, virtually all returning fish are four years of age. In the Kvichak River the cycle is normally five years, but occasionally shifts to four (Mathisen and Poe 1981). The sockeye age composition in the Kvichak River is highly variable from year to year because the progeny of a brood year may smoltify at either age 1. or age 2., and may spend from one to three winters in the ocean. Juveniles experiencing poor growth tend to smoltify at age 2., whereas Adams River juveniles consistently migrate at age 1. regardless of size. Most sockeye returning to the Kvichak River have spent two winters in the ocean. The most common age at return is 2.2; other significant ages are 1.2, 1.3, and 2.3 (Table 2). The shift in the cycle from five to four years has occurred from brood years producing a high proportion of age 1. smolts, returning at age 1.2 (e.g., 1952 and 1956 brood years) (Mathisen and Poe 1981).

Reconstruction of the history of total run patterns of sockeye to the Kvichak River is made difficult by the effects of a highly size- and sex-selective gillnet fishery, shifts in mesh size regulations from 5¾- to 5½- to 5⅜-inch, change in mesh materials from linen to synthetic, and lack of data on escapement numbers and age composition. Although reliable total run statistics are available only since 1956, it is evident from historical catch data that, unlike the Adams River situation, consecutive brood years often produced large returns, at least until the late 1930s. Then the run pattern changed to a single dominant year in the four or five years of the cycle. Annual inshore runs since 1956 have ranged from 250,000 to 42 million sockeye (Eggers and Rogers 1987).

It has been hypothesized that the Kvichak River cycle was maintained in some way by the supressing effect of the peak cycle broods on subsequent broods in Iliamna Lake, perhaps through exhaustion of lake food supply. Indirect evidence of this was that high escapements in the preceding brood year tended to depress production in the current brood year, and the growth of age 1. smolts was suppressed in post-cycle brood years. However, these effects have not been manifested in a statistically significant lower return per spawner for post-cycle brood years (Eggers and Rogers 1987). It was found that in Iliamna Lake the annual growth of underyearling sockeye, which determines their age at smoltification, was more closely related to lake temperature than to the size of the parent escapement (Rogers and Poe 1984). On the other hand, Eggers and Rogers (1987) found that the historical fishing pattern on the Kvichak sockeye was extremely depensatory. They concluded that depensatory fishing in conjunction with the suppression of production following large escapements has created and maintained the Kvichak cycle. Small escapements were often caused by excessive exploitation by domestic and foreign fisheries. Predation on juvenile sockeye in Iliamna Lake has not been studied sufficiently to determine whether or not predators have a significant depensatory effect.

The Skeena River sockeye run (primarily Babine Lake stock) presents another interesting example in cyclic run pattern. Although the juveniles smoltify almost entirely as yearlings, as do the Adams River sockeye, the length of stay in the ocean varies between one and three winters. Babine Lake sockeye returns at age 1.1 are almost entirely males (jacks) and averaged about 8% of the total returns for brood years 1961–77 (range 4%–16%) (McDonald and Hume 1984). The two primary age groups, 1.2 and 1.3, were on the average about equal in number in the Skeena River adult returns. However, for individual brood-year returns their ratio has varied greatly, from about 25:75 to 82:18 (Ricker and Smith 1975). In the early years of the fishery, until about the 1925 brood year, the returns were characterized by a distinct 5-year cycle of abundance. This was replaced by a 4-year cycle in returns from brood years in the 1930s and 1940s. Following disruption of escapements by a slide in the river in 1952, a 4 year cycle returned in the late 1950s (Figure 58). In the mid-1960s, spawning channels were built to capitalize on the underutilized rearing area of Babine Lake, and returns for the brood years 1961–77 have been characterized by a sharp difference in marine survival of smolts from odd- and even-numbered brood years (McDonald and Hume 1984). Rate of ocean survival of smolts from even-numbered brood years has been disappointing, averaging only 2.65% versus 6.24% for odd-numbered brood years.

An additional complicating factor appears to be a tendency for an alternation of age of return in successive generations of Skeena River sockeye salmon. It was determined by Bilton (1971) that the larger females (age 1.3) had larger eggs than the smaller females (age 1.2), that egg weight was positively correlated with initial size and subsequent growth of juveniles, and that juvenile growth in the lake and ocean was inversely related to age at maturity. He hypothesized that the 1.3 sockeye spawners tend to produce progeny that mature as age 1.2 fish, which in turn give rise to progeny that mature as age 1.3 fish, etc. A comparison of the age composition in brood-year escapements and in the returning progeny indeed support this hypothesis (Figure 59), although the relationship seems to have weakened following full spawning channel production in more recent years. Peterman (1982) demonstrated that larger smolts tended to return as adults at an earlier age, which also fits with Bilton's hypothesis.

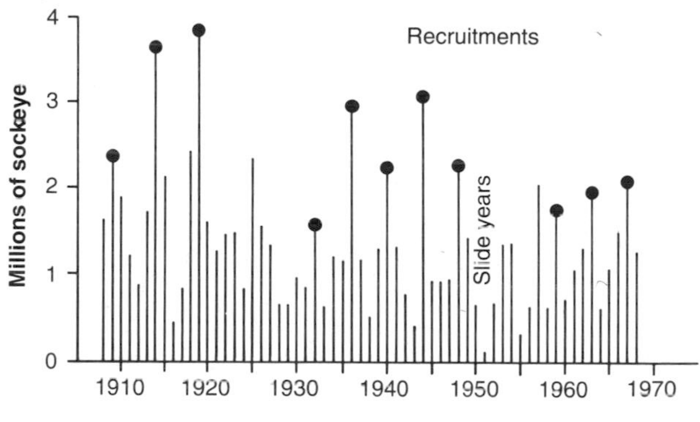

FIGURE 58

Estimated total recruitments of Skeena River sockeye salmon of the freshwater age 1. life history type. Recruitments exclude fish taken in marine statistical areas 3X and 3Y but include spawning stocks and subsistence catches. Recruitments are centred over the year of spawning. Solid dots represent regular sequences of "dominant" populations. (From Ricker and Smith 1975, Figure 1)

A number of other mechanisms have been examined to explain the historical cyclic sequences of events for Skeena River sockeye. The intestinal parasite *Eubothrium salvelini* was suspected as an indirect mechanism, because infected smolts weigh less and smaller smolts have a lower survival rate after they leave Babine Lake (Ricker and Smith 1975). However, they observed that the annual prevalence of parasitism in smolts was not related to population size of juveniles, and therefore the parasite did not seem to be acting as a depensatory mortality factor.

Available evidence does not suggest that the difference in even- and odd-year productivity originates in the lake, since there is no consistent difference in the mean size of smolts from even- and odd-year broods (McDonald and Hume 1984).

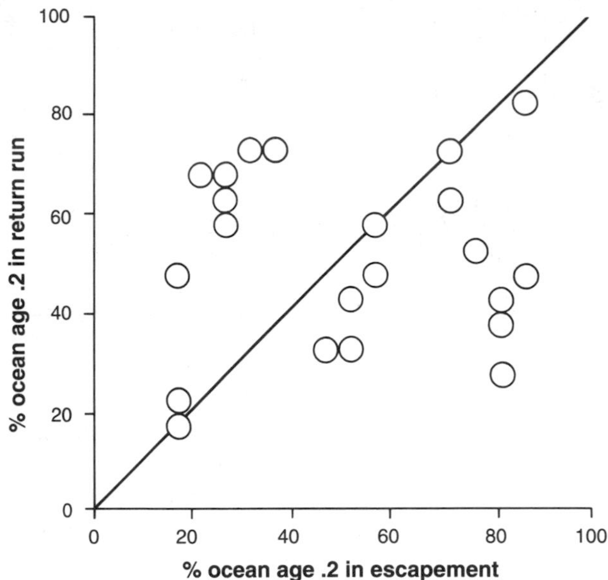

FIGURE 59

Percentage of ocean age .2 sockeye in the spawning escapement and in the returning progeny from brood year, Babine Lake. Brood years 1945–50 and 1953–67 (returns from 1951 and 1952 affected by river blockage from slide)

Peterman (1982) hypothesized that the difference in marine survival of even- and odd-year broods may be due to an interaction with pink salmon. He reasoned that pink salmon in northern British Columbia waters are generally more abundant in even-numbered years, so that their fry would be more abundant along the coast in odd-numbered years and may serve as a buffer to predation on odd-year broods of sockeye smolts. He also suggested that the returning adult pink salmon may actively prey on the outgoing even-year broods of sockeye smolts. But as McDonald and Hume (1984) pointed out, "It is difficult to relate either a 4- or 5-y cycle with consistent superiority of even- or odd-numbered brood years." Ricker (1982) suggested other factors that may be operating to cause fluctuations in age composition, including differential fishing mortality on the two ocean-age groups and changes in fishing seasons and techniques. However, a clear understanding of the cyclic fluctuations in Skeena River sockeye is still lacking.

The intensively studied sockeye of the Karluk River system present a fourth variation in complexity of reproduction. For its size, 40 km^2, Karluk Lake was once probably the most productive sockeye lake in North America, rivalling Lake Kuril (70 km^2) on the Ozernaya River in southwestern Kamchatka. Total runs to the Karluk River reached as high as 5.6 million sockeye (Rounsefell 1958b), compared with about 10 million to the Ozernaya River (Egorova et al. 1961). Sockeye in the Karluk River system have not fared well under exploitation, resulting in a drastic long-term decline in stock size (Figure 60). There has been considerable and sometimes heated controversy over the causes of this decline (e.g., Barnaby 1944; Thompson 1950; Rounsefell 1958b, 1973; Ricker 1972; VanCleve and Bevan 1973a, 1973b).

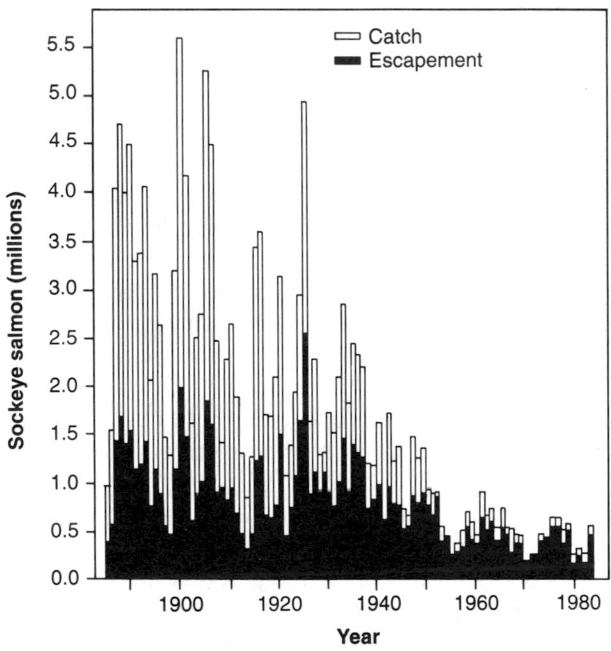

FIGURE 60

Karluk River sockeye salmon run, 1886–1984 (escapements estimated prior to 1922). (From Gard and Drucker 1985)

The life cycle of Karluk River sockeye shows some interesting variations from most other sockeye populations. Even though the juvenile sockeye experience good growth in Karluk Lake, most remain in the lake an additional year or two before smoltifying, instead of migrating as yearlings as do smolts in the Fraser and Babine river systems. Consequently, they are much larger at smoltification and, as discussed earlier, their marine survival, as estimated by fin marking experiments, is much higher and less variable (Barnaby 1944). Although there has been considerable fluctuation in magnitude of return runs, evidence of a consistent cyclic pattern of abundance is lacking. It is now clearly established that the Karluk River run is composed of sub-populations segregated, in part, by time and area of spawning. The spawning migration is protracted, extending from early June to mid-October. Electrophoretic studies have shown genetic differences between early and late spawning segments of the runs in the Karluk River; however, no such differences could be detected between groups of early-run fish or between groups of late-run fish (Wilmot and Burger 1985). It is now commonly accepted that overfishing of productive mid-season sub-populations was largely the cause of the initial decline in Karluk River sockeye and resulted in the bimodal early- and late-season run pattern still present today. This condition, first recognized by Thompson (1950), was apparently abetted during a later period by the operation of a weir at the lake outlet which interfered with movement of fry from the once important Karluk River spawning area upstream into Karluk Lake (Van Cleve and Bevan 1973a, 1973b).

Although other sockeye populations in central and western Alaska rebounded from low production levels in the 1960s and 1970s, the Karluk River run has failed to respond in like manner, in spite of a much reduced fishing effort. The mid-season runs have continued to be more heavily fished, reducing the opportunity for stock rebuilding (Gard et al. 1987). Possible causes for the lack of recovery are the loss of productivity from mid-season runs, reduction in lake fertility, or a change in the juvenile sockeye-predator-competitor balance. Shifts in sockeye smolt age composition have occurred (Barnaby 1944), and there is recent evidence that lake fertility has decreased and that the freshwater growth rate of the juvenile sockeye has declined (Koenings and Burkett 1987a). Although it has been hypothesized that the abundant three-spine stickleback population may be seriously competing with the sockeye salmon juveniles for food (McIntyre 1980), long-term changes in species balance have not been measured. Thus, the interactions of population-regulating mechanisms are still not clearly established for Karluk River sockeye (Gard et al. 1987).

Concluding Remarks

In the process of evolution, the sockeye salmon has diverged notably from the other *Oncorhynchus* species in its greater utilization of lacustrine environments both for spawning and rearing. Its greater adaptation to use of lake environments has resulted in reduced chances of interaction with other Pacific salmon species during spawning, incubation, and juvenile life in fresh water. The insertion of a lacustrine stage as a part of its usual life-history pattern has necessitated the evolution of complex timing for spawning, incubation, and rearing, which often involves intricate patterns of juvenile migration and orientation in fresh water not seen in the other *Oncorhynchus* species. In order to utilize the differing freshwater habitats, the sockeye has exhibited a wide range of adaptations to different sets of time- and space-linked environmental conditions. This range extends from those of the non-anadromous kokanee populations occurring naturally in many lakes to those of the so-called "ocean-type" sockeye stocks that migrate from their spawning stream environment to sea as underyearlings without any utilization of a lacustrine environment.

In adopting the limnetic environment for feeding and growth, planktivore juvenile sockeye take effective advantage of the available zooplankton resources. In most sockeye lakes sockeye are the predominant fish species utilizing this food resource and, in some lakes, they are virtually the sole occupant of the limnetic zone. In these instances, growth is limited essentially by environmental conditions affecting the fish or the food supply and, on occasion, by intraspecific competition. When a potential competitor species is present, it is most often the threespine stickleback, and its role as a competitor is under study in several lake systems. In utilizing the limnetic habitat, juvenile sockeye also tend to avoid many potential predators with more benthic and littoral feeding habits, and by means of their diel vertical migrations may obtain further protection from predatory fish and birds. Among lake system populations, variations are seen in the relative utilization of littoral areas prior to offshore migration, in the selective utilization of available entomostracan zooplankton and insect resources, in the diel migratory behaviour, in growth rate, and in length of stay in fresh water before smoltification. The success of sockeye in adapting to the lacustrine environment and to limnetic feeding is indicated by their abundance relative to other fish species present and to the magnitude of sockeye runs around the North Pacific rim.

The intra-stock migratory timing of sockeye as smolts heading to sea and as adults returning from the ocean shows remarkable consistency, and annual variation in timing is often explained or successfully predicted by variation in environmental conditions, primarily water temperature. Although the timing of return runs relates to timing of spawning in some river systems, such as the Fraser River, in other systems, such as Bristol Bay, where there is no such relationship, unknown run-timing strategies prevail. While timing of return of individual populations to their home stream is remarkably precise, the flexibility in their ocean migration is seen in their use of broad ocean areas and in the considerable brood-year variation in ocean age at return in many populations.

Although the bibliography on *O. nerka* is impressive, the sockeye salmon still retains many secrets of its life history. Much remains to be learned about the selective factors resulting in the varying life history and behaviour patterns within and among populations, about the factors limiting growth in the limnetic environment, about the interrelations with competitor and predator species, about the true nature of cyclic abundance phenomena, about patterns of estuarine and ocean migrations, and particularly about the causes of variation in marine mortality. Certainly, the sockeye remains a mysterious and fascinating species that will challenge the curiosity and ingenuity of biologists far into the future.

Acknowledgments

I gratefully acknowledge the encouragement, editorial reviews of the manuscript, and assistance provided by Drs. C. Groot and L. Margolis of the Pacific Biological Station, Nanaimo, who offered me the challenge of undertaking this summary of sockeye life history. It was much to my advantage to have the late Dr. R.E. Foerster's excellent 1968 treatise, *The Sockeye Salmon*, as a source reference and to have the manuscripts of the 1985 First International Sockeye Symposium made available to me by Chairman H.D. Smith for review. I also profited greatly from discussions with Dr. D.E. Rogers, Fisheries Research Institute, University of Washington, who provided me with resource data, and with C. Harris, of the same institute, who assisted me with writing the sections on marine migrations and distribution. Mrs. Carol Sisley prepared the many versions of the draft manuscript. Other personnel and resources of the institute were utilized as well. Critical reviews of my manuscript by Dr. K. Hyatt and J. McDonald, Pacific Biological Station, were also most helpful. I am particularly grateful for the assistance and patience of my wife, Marguerite, during the prolonged months of preparation and revision of this chapter.

References

Anas, R.E., and J.R. Gauley. 1956. Blueback salmon, *Oncorhynchus nerka*, age and length at seaward migration past Bonneville Dam. U.S. Fish Wildl. Serv. Spec. Sci. Rep. Fish. 185:46 p.

Anas, R.E., and S. Murai. 1969. Use of scale characters and a discriminant function for classifying sockeye salmon (*Oncorhynchus nerka*) by continent of origin. Int. North Pac. Fish. Comm. Bull. 26:157–192

Andrew, F.J., and G.H. Geen. 1960. Sockeye and pink salmon production in relation to proposed dams in the Fraser River system. Int. Pac. Salmon Fish. Comm. Bull. 11:259 p.

Atkinson, C.E., J.H. Rose, and T.O. Duncan. 1967. Pacific salmon in the United States, p. 43–223. *In*: Salmon of the North Pacific Ocean. Part IV. Spawning populations of North Pacific salmon. Int. North Pac. Fish. Comm. Bull. 23

Bakkala, R.G. 1971. Distribution and migration of immature sockeye salmon taken by U.S. research vessels with gillnets in offshore waters, 1956–67. Int. North Pac. Fish. Comm. Bull. 27:1–70

Bams, R.A. 1969. Adaptations of sockeye salmon associated with incubation in stream gravels, p. 71–87. *In*: T.G. Northcote (ed.). Symposium on Salmon and Trout in Streams. H.R. MacMillan Lectures in Fisheries. Institute of Fisheries, University of British Columbia, Vancouver, BC

Barnaby, J.T. 1944. Fluctuations in abundance of red salmon, *Oncorhynchus nerka* (Walbaum), of the Karluk River, Alaska. Fish. Bull. Fish Wildl. Serv. 50:237–296

Barraclough, W.E., and D. Robinson. 1972. The fertilization of Great Central Lake. III. Effect on juvenile sockeye salmon. Fish. Bull. (U.S.) 70:37–48

Becker, C.D. 1962. Estimating red salmon escapements by sample counts from observation towers. Fish. Bull. Fish Wildl. Serv. 61:355–369

Bergander, F. 1985. Sockeye salmon (*Oncorhynchus nerka*) stock assessment and evaluation in southeastern Alaska, 1982–1983. Alaska Dep. Fish Game ADF&G Tech. Data Rep. 142:27 p.

Biette, R.M., and G.H. Geen. 1980. Growth of underyearling sockeye salmon (*Oncorhynchus nerka*) under constant and cyclic temperatures in relation to live zooplankton ration size. Can. J. Fish. Aquat. Sci. 37:203–210

Bill, D. 1984. 1982 Kvichak River sockeye smolt studies, p. 2–13. *In*: D.M. Eggers and H.J. Yuen (eds.). 1982 Bristol Bay sockeye salmon smolt studies. Alaska Dep. Fish Game ADF&G Tech. Data Rep. 103

Bilton, H.T. 1971. A hypothesis of alternation of age of return in successive generations of Skeena River sockeye salmon (*Oncorhynchus nerka*). J. Fish. Res. Board Can. 28:513–516

Bilton, H.T., E.A.R. Ball, and D.W. Jenkinson. 1967. Age, size and sex composition of British Columbia sockeye salmon catches from 1912 to 1963. Fish. Res. Board Can. Nanaimo Biol. Stn. Circ. 25:4 p. and tables

Bilton, H.T., and S.A.M. Ludwig. 1966. Times of annulus formation on scales of sockeye, pink, and chum salmon in the Gulf of Alaska. J. Fish. Res. Board Can. 23:1403–1410

Bilton, H.T., and G.L. Robins. 1971. Response of young sockeye salmon (*Oncorhynchus nerka*) to prolonged periods of starvation. J. Fish. Res. Board Can. 28:1757–1761

——. 1973. The effect of starvation and subsequent feeding on survival and growth of Fulton Channel sockeye salmon fry (*Oncorhynchus nerka*). J. Fish. Res. Board Can. 30:1–5

Birch, L.C. 1957. The meaning of competition. Am. Nat. 91:5–18

Birman, I.B. 1958. On the occurrence and migration of Kamchatka salmon in the northwestern part of the Pacific Ocean, p. 31–51. *In*: Materialy po biologii morskovo perioda zhizni dalnevostochnykh lososei. Vsesoiuznyi Nauchno-issledovatel'skii Institut Morskovo Rybnovo Khoziaistvo i Okeanografii, Moscow. (Transl. from Russian; Fish. Res. Board Can. Transl. Ser. 180)

——. 1960. New information on the marine period of life and the marine fishery of Pacific salmon, p. 151–164. *In*: Trudy Soveshchaniia po biologicheskim osnovam okeanicheskovo rybolovostva, 1958. Tr. Soveshch. Ikhtiol. Komm. Akad. Nauk SSSR 10. (Transl. from Russian; Fish. Res. Board Can. Transl. Ser. 357)

Birman, I.B., and S.M. Konovalov. 1968. Distribution and migration in the ocean of a local stock of sockeye salmon, *Oncorhynchus nerka* (Walbaum), of Kurile Lake origin. Vopr. Ikhtiol. 8:728-736. (Transl. from Russian; Fish. Res. Board Can. Transl. Ser. 1219)

Birtwell, I.K., M.D. Nassichuck, and H. Beune. 1987. Underyearling sockeye salmon (*Oncorhynchus nerka*) in the estuary of the Fraser River, p. 25-35. *In*: H.D. Smith, L. Margolis, and C.W. Wood (eds.). Sockeye salmon (*Oncorhynchus nerka*) population biology and future management. Can. Spec. Publ. Fish. Aquat. Sci. 96

Blackbourn, D.J. 1987. Sea surface temperature and pre-season prediction of return timing in Fraser River sockeye salmon (*Oncorhynchus nerka*), p. 296-306. *In*: H.D. Smith, L. Margolis, and C.C. Wood (eds.). Sockeye salmon (*Oncorhynchus nerka*) population biology and future management. Can. Spec. Publ. Fish. Aquat. Sci. 96

Blackett, R.F. 1979. Establishment of sockeye (*Oncorhynchus nerka*) and chinook (*O. tshawytscha*) salmon runs at Frazer Lake, Kodiak Island, Alaska. J. Fish. Res. Board Can. 36:1265-1277

——. 1984. Progress report - Frazer Lake sockeye and chinook salmon enhancement, 1978 and 1979. Alaska Dep. Fish Game FRED Rep. 36:33 p.

Bodznick, D. 1978a. Water source preference and lakeward migration of sockeye salmon fry (*Oncorhynchus nerka*). J. Comp. Physiol. 127:139-146

——. 1978b. Calcium ions: an odorant for natural water discriminations and the migration behavior of sockeye salmon. J. Comp. Physiol. 127:157-166

Boyce, N.P. 1979. Effects of *Eubothrium salvelini* (Cestoda: Pseudophyllidea) on the growth and vitality of sockeye salmon, *Oncorhynchus nerka*. Can. J. Zool. 57:597-602

Boyce, N.P., and W.C. Clarke. 1983. *Eubothrium salvelini* (Cestoda: Pseudophyllidea) impairs seawater adaptation of migrant sockeye salmon yearlings (*Oncorhynchus nerka*) from Babine Lake, British Columbia. Can. J. Fish. Aquat. Sci. 40:821-824

Brannon, E.L. 1965. The influence of physical factors on the development and weight of sockeye salmon embryos and alevins. Int. Pac. Salmon Fish. Comm. Prog. Rep. 12:26 p.

——. 1967. Genetic control of migrating behavior of newly emerged sockeye salmon fry. Int. Pac. Salmon Fish. Comm. Prog. Rep. 16:31 p.

——. 1972. Mechanisms controlling migration of sockeye salmon fry. Int. Pac. Salmon Fish. Comm. Bull. 21:86 p.

——. 1984. Influence of stock origin on salmon migratory behavior, p. 103-111. *In*: J.D. McCleave, G.P. Arnold, J.J. Dodson, and W.H. Neill (eds.). Mechanisms of migration in fishes. NATO Conf. Ser. IV Mar. Sci. 14

——. 1987. Mechanisms stabilizing salmonid fry emergence timing, p. 120-124. *In*: H.D. Smith, L. Margolis, and C.C. Wood (eds.). Sockeye salmon (*Oncorhynchus nerka*) population biology and future management. Can. Spec. Publ. Fish. Aquat. Sci. 96

Brannon, E.L., T.P. Quinn, G.L. Lucchetti, and B.D. Ross. 1981. Compass orientation of sockeye salmon fry from a complex river system. Can. J. Zool. 59:1548-1553

Brett, J.R. 1952. Skeena River sockeye escapement and distribution. J. Fish. Res. Board Can. 8:453-468

——. 1965. The relation of size to rate of oxygen consumption and sustained swimming speed of sockeye salmon (*Oncorhynchus nerka*). J. Fish. Res. Board Can. 22:1491-1501

——. 1971a. Energetic responses of salmon to temperature: a study of some thermal relations in the physiology and freshwater ecology of sockeye salmon (*Oncorhynchus nerka*). Am. Zool. 11:99-113

——. 1971b. Satiation time, appetite, and maximum food intake of sockeye salmon, *Oncorhynchus nerka*. J. Fish. Res. Board Can. 28:409- 415

——. 1976. Scope for metabolism and growth of sockeye salmon, *Oncorhynchus nerka*, and some related energetics. J. Fish. Res. Board Can. 33:307-313

Brett, J.R., and C. Groot. 1963. Some aspects of olfactory and visual responses in Pacific salmon. J. Fish. Res. Board Can. 20:287-303

Brett, J.R., and J.E. Shelbourn. 1975. Growth rate of young sockeye salmon, *Oncorhynchus nerka*, in relation to fish size and ration level. J. Fish. Res. Board Can. 32:2103-2110

Brett, J.R., J.E. Shelbourn, and C.T. Shoop. 1969.

Growth rate and body composition of fingerling sockeye salmon, *Oncorhynchus nerka*, in relation to temperature and ration size. J. Fish. Res. Board Can. 26:2363–2394

Bryant, M.D. 1976. Lake Washington sockeye salmon: biological production and simulated harvest by three fisheries. Ph.D. thesis. University of Washington, Seattle, WA. 159 p.

Bucher, W.A. 1984. 1982 Wood River sockeye salmon smolt studies, p. 47–68. *In* D.M. Eggers and J.H. Yuen (eds.). 1982 Bristol Bay sockeye salmon smolt studies. Alaska Dep. Fish Game ADF&G Tech. Data Rep. 103

Bue, B.G. 1984. 1982 Egegik River sockeye salmon smolt studies, p. 28–40. *In* D.M. Eggers and J.H. Yuen (eds.). 1982 Bristol Bay sockeye salmon smolt studies. Alaska Dep. Fish Game ADF&G Tech. Data Rep. 103

Bugaev, V.F. 1978. Using the structure of zones of converged sclerites on scales as a criterion for differentiation of local groupings of the sockeye salmon, *Oncorhynchus nerka*, of the Kamchatka River basin. J. Ichthyol. 18:826–836

——. 1983. Spatial structure of the sockeye salmon *Oncorhynchus nerka* (Walbaum) populations in the basin of the Kamchatka River. Summary of dissertation. Biology Department, M.V. Lomonsov State University, Moscow, USSR. (Transl. from Russian; Can. Trans. Fish. Aquat. Sci. 5102)

——. 1987. Recommendations for rational exploitation of sockeye salmon (*Oncorhynchus nerka*) from the Kamchatka River, p. 396–402. *In*: H.D. Smith, L. Margolis, and C.C. Wood (eds.). Sockeye salmon (*Oncorhynchus nerka*) population biology and future management. Can. Spec. Publ. Fish. Aquat. Sci. 96

Burgner, R.L. 1958. A study of fluctuations in abundance, growth, and survival in the early life stages of the red salmon (*Oncorhynchus nerka* Walbaum) of the Wood River Lakes, Bristol Bay, Alaska. Ph.D. thesis. University of Washington, Seattle, WA. 200 p.

——. 1962a. Sampling red salmon fry by lake trap in the Wood River Lakes, Alaska, p. 315–348. *In*: T.S.Y. Koo (ed.). Studies of Alaska red salmon. Univ. Wash. Publ. Fish. New Ser. 1

——. 1962b. Studies of red salmon smolts from the Wood River Lakes, Alaska, p. 247–314. *In*: T.S.Y. Koo (ed.). Studies of Alaska red salmon. Univ. Wash. Publ. Fish. New Ser. 1

——. 1964. Factors influencing production of sockeye salmon (*Oncorhynchus nerka*) in lakes of southwestern Alaska. Int. Ver. Theor. Angew. Limnol. Verh. 15:504–513

——. 1980. Some features of ocean migration and timing of Pacific salmon, p. 153–164. *In*: W.J. McNeil and D.C. Himsworth (eds.). Salmonid ecosystems of the North Pacific. Oregon State University Press, Corvallis, OR

——. 1987. Factors influencing age and growth of juvenile sockeye salmon (*Oncorhynchus nerka*) in lakes, p. 129–142. *In*: H.D. Smith, L. Margolis, and C.C. Wood (eds.). Sockeye salmon (*Oncorhynchus nerka*) population biology and future management. Can. Spec. Publ. Fish. Aquat. Sci. 96

Burgner, R.L., C.J. DiCostanzo, R.J. Ellis, G.Y. Harry, Jr., W.L. Hartman, O.E. Kerns, Jr., O.A. Mathisen, and W.F. Royce. 1969. Biological studies and estimates of optimum escapements of sockeye salmon in the major river systems in southwestern Alaska. Fish. Bull. (U.S.) 67:405–459

Burgner, R.L., and J.M. Green. 1963. Study of interlake migration of red salmon fry, Agulowak River. Univ. Wash. Fish. Res. Inst. Circ. 182:13 p.

Burgner, R.L., and W.G. Meyer. 1983. Surface temperatures and salmon distribution relative to the boundaries of the Japanese drift gillnet fishery for flying squid (*Ommastrephes bartrami*). Univ. Wash. Fish. Res. Inst. FRI UW-8317:35 p.

Burke, J.A. 1978. Occurrence and effects of the parasite *Triaenophorus crassus* in sockeye salmon, *Oncorhynchus nerka*, of the Wood River lake system, Bristol Bay, Alaska. M.Sc. thesis. University of Washington, Seattle, WA. 73 p.

Carlson, T.J. 1974. The life history and population dynamics of *Bosmina coregoni* in Pedro Bay, Iliamna Lake, Alaska. M.Sc. thesis. University of Washington, Seattle, WA. 142 p.

Charnov, E.L. 1976. Optimal foraging; the marginal value theorem. Theor. Popul. Biol. 9:129–136

Chernenko, E.V., and G.D. Kurenkov. 1980. Differentiation of the stock of kokanee *Oncorhynchus nerka* (Walbaum) from Kronotsk Lake, p. 11–15. *In*: S.M. Konovalov (ed.). Populyatsionnaya biologiya i sistematika lososevykh. Sb. Rab. Akad. Nauk SSSR Dal'nev. Nauch. Tsentr. Inst. Biol. Morya 18. (In Russian)

Clark, W.K. 1959. Kodiak bear-red salmon relationships at Karluk Lake, Alaska. Trans. N. Am. Wildl. Conf. 24:337-345

Clarke, W.C., J.E. Shelbourn, and J.R. Brett. 1981. Effect of artificial photoperiod cycles, temperature, and salinity on growth and smolting in underyearling coho (*Oncorhynchus kisutch*), chinook (*O. tshawytscha*) and sockeye (*O. nerka*) salmon. Aquaculture 22:105-116

Clarke, W.C., and H.D. Smith. 1972. Observations on the migration of sockeye salmon fry (*Oncorhynchus nerka*) in the lower Babine River. J. Fish. Res. Board Can. 29:151-159

Cook, R.C., K.W. Myers, R.V. Walker, and C.K. Harris. 1981. The mixing proportion of Asian and Alaskan sockeye salmon in and around the land-based driftnet fishery area, 1972-1976. (Submitted to annual meeting. International North Pacific Fisheries Commission, Vancouver, British Columbia, November 1981). University of Washington, Fisheries Research Institute, Seattle, WA. 81 p.

Cross, B.A., W.E. Goshert, and D.L. Hicks. 1987. Origins of sockeye salmon in the fisheries of upper Cook Inlet in 1984 based on analysis of scale patterns. Alaska Dep. Fish Game ADF&G Tech. Fish. Rep. 87-01:120 p.

Dawson, J.J. 1972. Determination of seasonal distribution of juvenile sockeye salmon in Lake Washington by means of acoustics. M.Sc. thesis. University of Washington, Seattle, WA. 112 p.

Delacy, A.C., and W.M. Morton. 1943. Taxonomy and habits of the charrs, *Salvelinus malma* and *Salvelinus alpinus*, of the Karluk drainage system. Trans. Am. Fish. Soc. 72:79-91

Dell, M.B. 1963. Oceanic feeding habits of the sockeye salmon, *Oncorhynchus nerka* (Walbaum), in Aleutian waters. M.A. thesis. University of Michigan, Ann Arbor, MI. 40 p.

Demory, R.L., R.F. Orrell, and D.R. Heinle. 1964. Spawning ground catalog of the Kvichak River system, Bristol Bay, Alaska. U.S. Fish Wildl. Serv. Spec. Sci. Rep. Fish. 488:292 p.

Doble, B.D., and D.M. Eggers. 1978. Diel feeding chronology, rate of gastric evacuation, daily ration, and prey selectivity in Lake Washington juvenile sockeye salmon (*Oncorhynchus nerka*). Trans. Am. Fish. Soc. 107:36-45

Dombroski, E. 1954. The sizes of Babine Lake sockeye salmon smolt emigrants 1950-1953. Fish. Res. Board Can. Prog. Rep. Pac. Coast Stn. 99:30-44

Dong, J.N. 1981. Thermal tolerance and rate of development of coho salmon embryos. M.Sc. thesis. University of Washington, Seattle, WA. 51 p.

Drucker, B. 1970. Red salmon studies at Karluk Lake, 1968. U.S. Fish and Wildlife Service, Auke Bay, AK. 55 p.

Dunford, W.E. 1975. Space and food utilization by salmonids in marsh habitats of the Fraser River estuary. M.Sc. thesis. University of British Columbia, Vancouver, BC. 81 p.

Dymond, J.R. 1936. Some freshwater fishes of British Columbia. Rep. Br. Col. Comm. Fish. 1935:60-73

Eggers, D.M. 1975. A synthesis of the feeding behavior and growth of juvenile sockeye salmon in the limnetic environment. Ph.D. thesis. University of Washington, Seattle, WA. 217 p.

——. 1977. The nature of prey selection by planktivorous fish. Ecology 58:46-59

——. 1978. Limnetic feeding behavior of juvenile sockeye salmon in Lake Washington and predator avoidance. Limnol. Oceanogr. 23:1114-1125

——. 1982. Planktivore preference by prey size. Ecology 63:381-390

——. 1984. 1982 Ugashik River sockeye salmon smolts studies, p. 41-46. *In* D.M. Eggers and H.J. Yuen (eds.). 1982 Bristol Bay sockeye salmon smolt studies. Alaska Dep. Fish Game ADF&G Tech. Data Rep. 103

Eggers, D.M., N.W. Bartoo, N.A. Rickard, R.E. Nelson, R.C. Wissmar, R.L. Burgner, and A.H. Devol. 1978. The Lake Washington ecosystem; the perspective from the fish community production and forage base. J. Fish. Res. Board Can. 35:1553-1571

Eggers, D.M., and D.E. Rogers. 1987. The cycle of runs of sockeye salmon (*Oncorhynchus nerka*) to the Kvichak River, Bristol Bay, Alaska: cyclic dominance or depensatory fishing? p. 343-366. *In*: H.D. Smith, L. Margolis, and C.C. Wood (eds.). Sockeye salmon (*Oncorhynchus nerka*) population biology and future management. Can. Spec. Publ. Fish. Aquat. Sci. 96

Egorova, T.V. 1970a. The absence of seasonal groupings in sockeye salmon of the basin of the Ozernaya River. Izv. Tikhookean. Nauchno-Issled. Inst. Rybn. Khoz. Okeanogr. 78:43-47. (Transl. from Russian; Fish. Res. Board Can. Transl. Ser. 2351)

——. 1970b. Reproduction and development of sockeye in the Basin of Ozernaya River. Izv. Tikhookean. Nauchno-Issled. Inst. Rybn. Khoz. Okeanogr. 73:39–53. (Transl. from Russian; Fish. Res. Board Can. Transl. Ser. 2619)

Egorova, T.V., F.V. Krogius, I.I. Kurenkov, and R.S. Semko. 1961. Causes of fluctuations in the abundance of *Oncorhynchus nerka* (Walbaum) of the Ozernaya River (Kamchatka). Vopr. Ikhtiol. 1:439–447. (Transl. from Russian; Univ. Wash. Fish. Res. Inst. Circ. 159)

Ellis, R.J. 1974. Distribution, abundance, and growth of juvenile sockeye salmon, *Oncorhynchus nerka*, and associated species in the Naknek River system, 1961-1964. NOAA Tech. Rep. NMFS SSRF-678:53 p.

Favorite, F. 1970. Fishery oceanography. VI. Ocean food of sockeye salmon. Commer. Fish. Rev. 32:45–50

Favorite, F., A.J. Dodimead, and K. Nasu. 1976. Oceanography of the subarctic Pacific Region, 1960–71. Int. North Pac. Fish. Comm. Bull. 33:187 p.

Favorite, F., and M.G. Hanavan. 1963. Oceanographic conditions and salmon distribution south of the Alaska Peninsula and Aleutian Islands, 1956. Int. North Pac. Fish. Comm. Bull. 11:57–72

Flagg, L.C., P. Shields, and D.C. Waite. 1985. Sockeye salmon smolt studies, Kasilof River, Alaska 1984. Alaska Dep. Fish Game FRED Rep. 47:43 p.

Foerster, R.E. 1954. On the relation of adult sockeye salmon (*Oncorhynchus nerka*) returns to known smolt seaward migrations. J. Fish. Res. Board Can. 11:339–350

——. 1968. The sockeye salmon, *Oncorhynchus nerka*. Bull. Fish. Res. Board Can. 162:422 p.

Foskett, D.R. 1958. The Rivers Inlet sockeye salmon. J. Fish. Res. Board Can. 15:867–889

French, R., H. Bilton, M. Osako, and A. Hartt. 1976. Distribution and origin of sockeye salmon (*Oncorhynchus nerka*) in offshore waters of the North Pacific Ocean. Int. North Pac. Fish. Comm. Bull. 34: 113 p.

French, R.R., and R.G. Bakkala. 1974. A new model of ocean migrations of Bristol Bay sockeye salmon. Fish. Bull. (U.S.) 72:589–614

French, R.R., and J.E. Mason. 1964. Salmon distribution and abundance on the high seas–winter season 1962 and 1963. Int. North Pac. Fish. Comm. Annu. Rep. 1963:131–141

Fujii, T. 1975. On the relation between the homing migration of the western Alaska sockeye salmon *Oncorhynchus nerka* (Walbaum) and oceanic conditions in the eastern Bering Sea. Mem. Fac. Fish. Hokkaido Univ. 22:99–191

Fukuhara, F.M. 1971. An analysis of the biological and catch statistics of the Japanese mothership salmon fishery. Ph.D. thesis. University of Washington, Seattle, WA. 238 p.

Fulton, L.A. 1970. Spawning areas and abundance of steelhead trout and coho, sockeye, and chum salmon in the Columbia River basin - past and present. U.S. Nat. Mar. Fish. Serv. Spec. Sci. Rep. Fish. 618:37 p.

Gard, R. 1971. Brown bear predation on sockeye salmon at Karluk Lake, Alaska. J. Wildl. Manage. 35:193–204

Gard, R., and B. Drucker. 1985. Differentiation of subpopulations of sockeye salmon in the Karluk River system. Annu. Rep. School Fish. Sci. Univ. Alaska, Juneau 1983/84:7

Gard, R., B. Drucker, and R. Fagen. 1987. Differentiation of subpopulations of sockeye salmon (*Oncorhynchus nerka*), Karluk River system, Alaska, p. 408–418. *In*: H.D. Smith, L. Margolis, and C.C. Wood (eds.). Sockeye salmon (*Oncorhynchus nerka*) population biology and future management. Can. Spec. Publ. Fish. Aquat. Sci. 96

Gilbert, C.H. 1916, 1918, 1920, 1922, 1923. Contributions to the life history of sockeye salmon, nos. 3, 4, 6–8. Rep. Br. Col. Comm. Fish. 1915:S27-S64; 1917:Q33-Q80; 1919:U35-U68; 1921:W15-W64; 1922:T16-T49

Gilbert, C.H., and W.H. Rich. 1927. Second experiment in tagging salmon in the Alaska Peninsula fisheries reservation, summer of 1923. Bull. Bur. Fish. (U.S.) 42:27–75

Gilbert, J.R. 1968. Surveys of sockeye salmon spawning populations in the Nushagak District, Bristol Bay, Alaska, 1946–1958, p. 199–267. *In*: R.L. Burgner (ed.). Further studies of Alaska sockeye salmon. Univ. Wash. Publ. Fish. New Ser. 3

Gilhousen, P. 1960. Migratory behavior of adult Fraser River sockeye. Int. Pac. Salmon Fish. Comm. Prog. Rep. 7:78 p.

Godin, J.G.J. 1982. Migrations of salmonid fishes during early life history phases: daily and annual timing, p. 22–50. *In*: E.L. Brannon and E.O.

Salo (eds.). Proceedings of the Salmon and Trout Migratory Behavior Symposium. School of Fisheries, University of Washington, Seattle, WA

Goodlad, J.C., T.W. Gjernes, and E.L. Brannon. 1974. Factors affecting sockeye salmon (*Oncorhynchus nerka*) growth in four lakes of the Fraser River system. J. Fish. Res. Board Can. 31:871-892

Groot, C. 1965. On the orientation of young sockeye salmon (*Oncorhynchus nerka*) during their seaward migration out of lakes. Behaviour Suppl. 14:198 p.

——. 1972. Migration of yearling sockeye salmon (*Oncorhynchus nerka*) as determined by time-lapse photography of sonar observations. J. Fish. Res. Board Can. 29:1431-1444

——. 1982. Modifications on a theme - a perspective on migratory behavior of Pacific salmon, p. 1-21. *In*: E.L. Brannon and E.O. Salo (eds.). Proceedings of the Salmon and Trout Migratory Behavior Symposium. School of Fisheries. University of Washington, Seattle, WA

Groot, C., and K. Cooke. 1987. Are the migrations of juvenile and adult Fraser River sockeye salmon (*Oncorhynchus nerka*) in near-shore waters related? p. 53-60. *In*: H.D. Smith, L. Margolis, and C.C. Wood (eds.). Sockeye salmon (*Oncorhynchus nerka*) population biology and future management. Can. Spec. Publ. Fish. Aquat. Sci. 96

Groot, C., and T. Quinn. 1987. Homing migration of sockeye salmon, *Oncorhynchus nerka*, to the Fraser River. Fish. Bull. (U.S.) 85:455-469

Hallock, R.J., and D.H. Fry, Jr. 1967. Five species of salmon, *Oncorhynchus*, in the Sacramento River, California. Calif. Fish Game 53:5- 22

Hamalainen, A.H.E. 1978. Effects of instream flow levels on sockeye salmon fry production in the Cedar River, Washington. M.Sc. thesis. University of Washington, Seattle, WA. 90 p.

Hanamura, N. 1966. Sockeye salmon in the far east, p. 1-27. *In*: Salmon of the North Pacific Ocean. Part III. A review of the life history of North Pacific salmon. Int. North Pac. Fish. Comm. Bull. 18

Hanson, A.J., and H.D. Smith. 1967. Mate selection in a population of sockeye salmon (*Oncorhynchus nerka*) of mixed age groups. J. Fish. Res. Board Can. 24:1955-1977

Harris, C. 1987. Catches of North American sockeye salmon (*Oncorhynchus nerka*) by the Japanese high seas salmon fisheries, 1972-1984, p. 458-479. *In*: H.D. Smith, L. Margolis, and C.C. Wood (eds.). Sockeye salmon (*Oncorhynchus nerka*) population biology and future management. Can. Spec. Publ. Fish. Aquat. Sci. 96

Harris, C., K. Myers, C. Knudsen, R. Walker, N. Davis, W. Meyer, and R. Burgner. 1984. Monitoring migrations and origins of salmonids at sea-1983. Int. North Pac. Fish. Comm. Annu. Rep. 1983:88-108

Hart, J.L. 1973. Pacific fishes of Canada. Bull. Fish. Res. Board Can. 180:740 p.

Hartman, W.L. 1959. Red salmon spawning behavior. Sci. Alaska Proc. Alaska Sci. Conf. 9(1958):48-49

Hartman, W.L., and R.L. Burgner. 1972. Limnology and fish ecology of sockeye salmon nursery lakes of the world. J. Fish. Res. Board Can. 29:699-715

Hartman, W.L., W.R. Heard, and B. Drucker. 1967. Migratory behavior of sockeye salmon fry and smolts. J. Fish. Res. Board Can. 24:2069-2099

Hartman, W.L., C.W. Strickland, and D.T. Hoopes. 1962. Survival and behavior of sockeye salmon fry migrating into Brooks Lake, Alaska. Trans. Am. Fish. Soc. 91:133-139

Hartt, A.C. 1960. Pacific salmon in international waters. Trans. N. Am. Wildl. Conf. 25(1960): 339-346

——. 1966. Migrations of salmon in the North Pacific Ocean and Bering Sea as determined by seining and tagging, 1959-1960. Int. North Pac. Fish. Comm. Bull. 19:141 p.

Hartt, A.C., and M.B. Dell. 1986. Early oceanic migrations and growth of juvenile Pacific salmon and steelhead trout. Int. North Pac. Fish. Comm. Bull. 46:105 p.

Hasler, A.D., A.T. Scholz, and A.M. Horrall. 1978. Olfactory imprinting and homing in salmon. Am. Sci. 66:347-355

Healey, M.C. 1980. The ecology of juvenile salmon in Georgia Strait, British Columbia, p. 203-229. *In*: W.J. McNeil and D.C. Himsworth (eds.). Salmon ecosystems of the North Pacific. Oregon State University Press, Corvallis, OR

Healey, M.C. 1987. The adaptive significance of age and size at maturity in female sockeye salmon (*Oncorhynchus nerka*), p. 110-117. *In*: H.D. Smith, L. Margolis, and C.C. Wood (eds.). Sockeye salmon (*Oncorhynchus nerka*) population biology and future management. Can. Spec. Publ. Fish. Aquat. Sci. 96

Heard, W.R. 1964. Phototactic behaviour of emerging sockeye salmon fry. Anim. Behav. 12:382–389

Henry, K.A. 1961. Racial identification of Fraser River sockeye salmon by means of scales and its application to salmon management. Int. Pac. Salmon Fish. Comm. Bull. 12:92 p.

Hoag, S.H. 1972. The relationship between the summer food of juvenile sockeye salmon, *Oncorhynchus nerka*, and the standing stock of zooplankton in Iliamna Lake, Alaska. Fish. Bull. (U.S.) 70:355–362

Hoar, W.S. 1965. The endocrine system as a chemical link between the organism and its environment. Trans. R. Soc. Can. Ser. 4 3:175–200

——. 1976. Smolt transformation: evolution, behavior, and physiology. J. Fish. Res. Board Can. 33:1234–1252

Hoines, L.J., W.D. Ward, and C. Smitch. 1985. Washington State sport catch report 1984. Washington Department of Fisheries, Olympia, WA. 58 p.

Holling, C.S. 1959. Some characteristics of simple types of predation and parasitism. Can. Entomol. 41:385–398

Hubbs, C.L. 1940. Predator control in relation to fish management in Alaska. Trans. N. Am. Wildl. Conf. 5(1940):153–162

Hunter, J.G. 1974. Pacific salmon in Arctic Canada. Fish. Res. Board Can. MS Rep. Ser. 1319:17 p.

Hurley, D.A., and E.L. Brannon. 1969. Effect of feeding before and after yolk absorption on the growth of sockeye salmon. Int. Pac. Salmon Fish. Comm. Prog. Rep. 21:19 p.

Huttunen, D.C. 1984a. Abundance, age, sex, and size of salmon (*Oncorhynchus* spp) catches and escapements in the Kuskokwim area, 1982. Alaska Dep. Fish Game ADF&G Tech. Data Rep. 111:76 p.

——. 1984b. 1982 Naknek River sockeye salmon smolt studies, p. 14–27. *In*: D.M. Eggers and H.J. Yuen (eds.). 1982 Bristol Bay sockeye salmon smolt studies. Alaska Dep. Fish Game ADF&G Tech. Data Rep. 103

——. 1985. Abundance, age, sex, and size of salmon (*Oncorhynchus* spp) catches and escapements in the Kuskokwim area, 1983. Alaska Dep. Fish Game ADF&G Tech. Data Rep. 133:94 p.

——. 1986. Abundance, age, sex, and size of salmon (*Oncorhynchus* sp.) catches and escapements in the Kuskokwim area, 1984. Alaska Dep. Fish Game ADF&G Tech. Data Rep. 186:92 p.

——. 1987. Abundance, age, sex and size of salmon (*Oncorhynchus* sp.) catches and escapements in the Kuskokwim area, 1985. Alaska Dep. Fish Game ADF&G Tech. Data Rep. 212:83 p.

Hyatt, K.D., and J.G. Stockner. 1985. Responses of sockeye salmon (*Oncorhynchus nerka*) to fertilization of British Columbia coastal lakes. Can. J. Fish. Aquat. Sci. 42:320–331

Idler, D.R., and W.A. Clemens. 1959. The energy expenditures of Fraser River sockeye salmon during the spawning migration to Chilko and Stuart lakes. Int. Pac. Salmon Fish. Comm. Prog. Rep. 6:80 p.

Ievleva, M.Ya. 1951. Morphology and rate of embryonic development of Pacific salmon. Izv. Tikhookean. Nauchno- Issled. Inst. Rybn. Khoz. Okeanogr. 34:123–130. (Transl. from Russian; *In*: Pacific salmon: selected articles from Soviet periodicals, p. 236-244. Israel Program for Scientific Translations, Jerusalem, 1961)

International North Pacific Fisheries Commission (INPFC). 1958. Annual report for the year 1957. International North Pacific Fisheries Commission, Vancouver, BC. 86 p.

——. 1959. Annual report for the year 1958. International North Pacific Fisheries Commission, Vancouver, BC. 119 p.

——. 1972–1986. Statistical yearbooks 1970–1983. International North Pacific Fisheries Commission, Vancouver, BC

——. 1979. Historical catch statistics for the salmon of the North Pacific coast. Int. North Pac. Fish. Comm. Bull. 39:166 p.

International Pacific Salmon Fisheries Commission (IPSFC). 1976. Annual report for the year 1975. International Pacific Salmon Fisheries Commission, New Westminster, BC. 53 p.

——. 1984. Annual report for the year 1983. International Pacific Salmon Fisheries Commission, New Westminster, BC. 53 p.

Ito, J. 1964. Food and feeding habits of Pacific salmon (genus *Oncorhynchus*) in their oceanic life. Bull. Hokkaido Reg. Fish. Res. Lab. 29:85–97. (Transl. from Japanese; Fish. Res. Board Can. Transl. Ser. 1309)

Jaenicke, H.W., M.S. Hoffman, and M.L. Dahlberg.

1987. Food habits of sockeye salmon (*Oncorhynchus nerka*) fry and threespine stickleback (*Gasterosteus aculeatus*) in Lake Nunavaugaluk, Alaska, and a strategy to improve salmon survival and growth, p. 161–175. *In*: H.D. Smith, L. Margolis, and C.C. Wood (eds.). Sockeye salmon (*Oncorhynchus nerka*) population biology and future management. Can. Spec. Publ. Fish. Aquat. Sci. 96

Johnson, R.L. 1977. A management program for the Quinault sockeye salmon (*Oncorhynchus nerka*, Walbaum) fishery. M.Sc. thesis. University of Washington, Seattle, WA. 85 p.

Johnson, W.E. 1956. On the distribution of young sockeye salmon (*Oncorhynchus nerka*) in Babine and Nilkitkwa lakes, B.C. J. Fish. Res. Board Can. 13:695–708

——. 1958. Density and distribution of young sockeye salmon (*Oncorhynchus nerka*) throughout a multi-basin lake system. J. Fish. Res. Board Can. 15:961–987

——. 1961. Aspects of the ecology of a pelagic, zooplankton-eating fish. Int. Ver. Theor. Angew. Limnol. Verh. 14:727–731

Johnson, W.E., and C. Groot. 1963. Observations on the migration of young sockeye salmon (*Oncorhynchus nerka*) through a large, complex lake system. J. Fish. Res. Board Can. 20:919–938

Kaganovsky, A.G. 1960. Commercial fishes of the Anadyr River and the Anadyr estuary. Vestn. Dal'nevost. Fil. Akad. Nauk SSSR 1933(1/3):137-139. (Transl. from Russian; Fish. Res. Board Can. Transl. Ser. 282)

Kerns, O.E., Jr. 1961. Abundance and age of Kvichak River red salmon smolts. Fish. Bull. Fish Wildl. Serv. 61(189):301–320

Kerns, O.E. Jr., and J.R. Donaldson. 1968. Behavior and distribution of spawning sockeye salmon on island beaches in Iliamna Lake, Alaska, 1965. J. Fish. Res. Board Can. 25:485–494

Killick, S.R. 1955. The chronological order of Fraser River sockeye salmon during migration, spawning and death. Int. Pac. Salmon Fish. Comm. Bull. 7:95 p.

Killick, S.R., and W.A. Clemens. 1963. The age, sex ratio and size of Fraser River sockeye salmon, 1915–1960. Int. Pac. Salmon Fish. Comm. Bull. 14:113 p.

Klokov, V.K. 1970. The population dynamics of spawning schools of salmon on the north coast of the Sea of Okhotsk. Izv. Tikhookean. Nauchno- Issled. Inst. Rybn. Khoz. Okeanogr. 71:169–177. (Transl. from Russian; Fish. Res. Board Can. Transl. Ser. 2323)

Koenings, J.P., and R.D. Burkett. 1987a. An aquatic Rubic's cube: restoration of the Karluk Lake sockeye salmon (*Oncorhynchus nerka*), p. 419–434. *In*: H.D. Smith, L. Margolis, and C.C. Wood (eds.). Sockeye salmon (*Oncorhynchus nerka*) population biology and future management. Can. Spec. Publ. Fish. Aquat. Sci. 96

——. 1987b. Population characteristics of sockeye salmon (*Oncorhynchus nerka*) smolts relative to temperature regimes, euphotic volume, fry density and forage base within Alaskan lakes, p. 216–234. *In*: H.D. Smith, L. Margolis, and C.C. Wood (eds.). Sockeye salmon (*Oncorhynchus nerka*) population biology and future management. Can. Spec. Publ. Fish. Aquat. Sci. 96

Koenings, J.P., G.B. Kyle, and P. Marcuson. 1985. Limnological and fisheries evidence for rearing limitation of sockeye salmon, *Oncorhynchus nerka*, production from Packers Lake, Cook Inlet, Alaska (1973–1983). Alaska Dep. Fish Game FRED Rep. 56:122 p.

Koenings, J.P., J. McNair, and B. Sele. 1984. Limnology and fisheries evidence for area limitation of sockeye production in Falls Lake, northern Southeast Alaska (1981–1982). Alaska Dep. Fish Game FRED Rep. 23:59 p.

Konovalov, S.M. 1971. Differentiation of local populations of sockeye salmon *Oncorhynchus nerka* (Walbaum). Nauka Publishing House, Moscow, USSR. 229 p. (Transl. from Russian; Univ. Wash. Publ. Fish. New Ser. 5)

Koo, T.S.Y. 1962a. Age designation in salmon, p. 41–48. *In*: T.S.Y. Koo (ed.). Studies of Alaska red salmon. Univ. Wash. Publ. Fish. New Ser. 1

——. 1962b. Age and growth studies of red salmon scales by graphical means, p. 49–121. *In*: T.S.Y. Koo (ed.). Studies of Alaska red salmon. Univ. Wash. Publ. Fish. New Ser. 1

Krogius, F.V., and E.M. Krokhin. 1948. On the production of young sockeye salmon (*Oncorhynchus nerka* Walb.). Izv. Tikhookean. Nauchno- Issled. Inst. Rybn. Khoz. Okeanogr. 28:3–27. (Transl. from Russian; Fish. Res. Board Can. Transl. Ser. 109)

——. 1956a. Causes of the fluctuations in abundance of sockeye salmon in Kamchatka. Tr.

Probl. Temat. Soveshch. Zool. Inst. Akad. Nauk SSSR 6:144–149. (Transl. from Russian; Fish. Res. Board Can. Transl. Ser. 92)

——. 1956b. Results of a study of the biology of sockeye salmon, the conditions of the stocks and the fluctuations in numbers in Kamchatkan waters. Vopr. Ikhtiol. 7:3–20. (Transl. from Russian; Fish. Res. Board Can. Transl. Ser. 176)

Krokhin, E.M. 1967. A contribution to the study of dwarf sockeye *Oncorhynchus nerka* Walb., in Lake Dalnee (Kamchatka). Vopr. Ikhtiol. 7(33): 433–445. (Transl. from Russian; Fish. Res. Board Can. Transl. Ser. 986)

Krokhin, E.M., and F.V. Krogius. 1936. The lake form of sockeye salmon (*Oncorhynchus nerka*) from Lake Kronotsk in Kamchatka. C.R. (Dokl.) Acad. Sci. URSS 4(13) 2(106):89–92. (Also in Russian; Dokl. Akad. SSSR Nov. Ser. 4(13) 2(106): 87–90)

——. 1937. Lake Kuril and the biology of the sockeye salmon, *Oncorhynchus nerka* Walb., spawning in its basin. Tr. Tikhookean. Kom. Akad. Nauk SSSR 4:1–165. (In Russian; English summary)

Kuznetsov, I.I. 1928. Some observations on spawning of the Amur and Kamchatka salmons. Izv. Tikhookean. Nauchno-Prom. Stn. 2(3):1–195. (Transl. from Russian; Fish. Res. Board Can. Transl. Ser. 22)

Kyle, G.B., J.P. Koenings, and B.M. Barrett. 1988. Density-dependent, trophic level responses to an introduced run of sockeye salmon (*Oncorhynchus nerka*) at Frazer Lake, Kodiak Island, Alaska. Can. J. Fish. Aquat. Sci. 45:856–867

Lander, A.H., and G.K. Tanonaka. 1964. Marine growth of western Alaskan sockeye salmon (*Oncorhynchus nerka* Walbaum). Int. North Pac. Fish. Comm. Bull. 14:1–31

Lander, A.H., G.K. Tanonaka, K.N. Thorson, and T.A. Dark. 1966. Ocean mortality and growth. Int. North Pac. Fish. Comm. Annu. Rep. 1964:105–111

Larkin, P.A. 1971. Simulation studies of the Adams River sockeye salmon (*Oncorhynchus nerka*). J. Fish. Res. Board Can. 28:1492–1502

——. 1975. Some major problems for further study on Pacific salmon, p. 3–9. *In*: Symposium on Evaluation of Methods of Estimating the Abundance and Biological Attributes of Salmon on the High Seas. Int. North Pac. Fish. Comm. Bull. 32:3–9

Lebida, R.C. 1984. Larson Lake sockeye and coho salmon smolt enumeration and sampling, 1982. Alaska Dep. Fish Game FRED Rep. 35:31 p.

LeBrasseur, R.J. 1966. Stomach contents of salmon and steelhead trout in the northeastern Pacific Ocean. J. Fish. Res. Board Can. 23:85–100

——. 1972. Utilization of herbivore zooplankton by maturing salmon, p. 581–588. *In*: A.Y. Takenouti (ed.). Biological oceanography of the northern North Pacific Ocean. Idemitsu Shoten, Tokyo, Japan

LeBrasseur, R.J., and O.D. Kennedy. 1972. The fertilization of Great Central Lake. II: zooplankton standing stock. Fish. Bull. (U.S.) 70:25-36

LeBrasseur, R.J., C.D. McAllister, W.E. Barraclough, O.D. Kennedy, J. Manzer, D. Robinson, and K. Stephens. 1978. Enhancement of sockeye salmon (*Oncorhynchus nerka*) by lake fertilization in Great Central Lake: summary report. J. Fish. Res. Board Can. 35:1580–1596

Levy, D.A. 1987. Review of the ecological significance of diel vertical migration by juvenile sockeye salmon (*Oncorhynchus nerka*), p. 44-52. *In*: H.D. Smith, L. Margolis, and C.C. Wood (eds.). Sockeye salmon (*Oncorhynchus nerka*) population biology and future management. Can. Spec. Publ. Fish. Aquat. Sci. 96

Levy, D.A., and T.G. Northcote. 1982. Juvenile salmon residency in a marsh area of the Fraser River estuary. Can. J. Fish. Aquat. Sci. 39:270–276

Macdonald, A.L. 1984. Seasonal use of nearshore intertidal habitats by juvenile Pacific salmon on the delta-front of the Fraser River estuary, British Columbia. M.Sc. thesis. University of Victoria, Victoria, BC. 187 p.

Machidori, S. 1966. Vertical distribution of salmon (genus *Oncorhynchus*) in the northwestern Pacific. Bull. Hokkaido Reg. Fish. Res. Lab. 31:11–17. (Transl. from Japanese; U.S. Joint Publication Research Service; available at Northwest and Alaska Fisheries Center, Seattle, WA)

Major, R.L., and D.R. Craddock. 1962. Influence of early maturing females on reproductive potential of Columbia River blueback salmon (*Oncorhynchus nerka*). Fish. Bull. Fish Wildl. Serv. 61:429–437

Manzer, J.I. 1964. Preliminary observation on the vertical distribution of Pacific salmon (genus *Oncorhynchus*) in the Gulf of Alaska. J. Fish. Res.

Board Can. 21:891-903
——. 1976. Distribution, food, and feeding of the threespine stickleback, *Gasterosteus aculeatus*, in Great Central Lake, Vancouver Island, with comments on competition for food with juvenile sockeye salmon, *Oncorhynchus nerka*. Fish. Bull. (U.S.) 74:647-668

Manzer, J.I., T. Ishida, A.E. Peterson, and M.G. Hanavan. 1965. Salmon of the North Pacific Ocean. Part V. Offshore distribution of salmon. Int. North Pac. Fish. Comm. Bull. 15:452 p.

Margolis, L. 1963. Parasites as indicators of the geographical origin of sockeye salmon, *Oncorhynchus nerka* (Walbaum), occurring in the North Pacific Ocean and adjacent seas. Int. North Pac. Fish. Comm. Bull. 11:31- 56

——. 1982a. Pacific salmon and their parasites: a century of study. Bull. Can. Soc. Zool. 13(3):7-11

——. 1982b. Parasitology of Pacific salmon: an overview, p. 135-226. *In*: E. Meerovitch (ed.). Aspects of parasitology - a festschrift dedicated to the fiftieth anniversary of the Institute of Parasitology of McGill University, 1932-1982. McGill University, Montreal, Que.

Margolis, L., F.C. Cleaver, Y. Fukuda, and H. Godfrey. 1966. Salmon of the North Pacific Ocean. Part VI. Sockeye salmon in offshore waters. Int. North Pac. Fish. Comm. Bull. 20:70 p.

Markovtsev, V.G. 1972. Feeding and food relationships of young sockeye and three-spine stickleback in Lake Dalnee. Izv. Tikhookean. Nauchno-Issled. Inst. Rybn. Khoz. Okeanogr. 82:227-233. (Transl. from Russian; Fish. Res. Board Can. Transl. Ser. 2830)

Marriott, R.A. 1964. Spawning ground catalog of the Wood River system, Bristol Bay, Alaska. U.S. Fish Wildl. Serv. Spec. Sci. Rep. Fish. 494:210 p.

Marshall, S.L., and R.L. Burgner. 1977. Chignik sockeye studies. Univ. Wash. Fish. Res. Inst. FRI-UW-7733:156 p.

Mathews, S.B., and R. Buckley. 1976. Marine mortality of Puget Sound coho salmon (*Oncorhynchus kisutch*). J. Fish. Res. Board Can. 33:1677-1684

Mathisen, O.A. 1962. The effect of altered sex ratios on the spawning of red salmon, p. 137-248. *In*: T.S.Y. Koo (ed.). Studies of Alaska red salmon. Univ. Wash. Publ. Fish. New Ser. 1

——. 1966. Some adaptations of sockeye salmon races to limnological features of Iliamna Lake, Alaska. Int. Ver. Theor. Angew. Limnol. Verh. 16:1025-1035

Mathisen, O.A., R.L. Burgner, and T.S.Y. Koo. 1963. Statistical records and computations on red salmon (*Oncorhynchus nerka*) runs in the Nushagak District, Bristol Bay, Alaska, 1946-1959. U.S. Fish Wildl. Serv. Spec. Sci. Rep. Fish. 468: 32 p.

Mathisen, O.A., and P.H. Poe. 1981. Sockeye salmon cycles in the Kvichak River, Bristol Bay, Alaska. Int. Ver. Theor. Angew. Limnol. Verh. 21:1207-1213

McAllister, W.B., W.J. Ingraham, Jr., D. Day, and J. Larrance. 1969. Investigations by the United States for the International North Pacific Fisheries Commission-1967: oceanography. Int. North Pac. Fish. Comm. Annu. Rep. 1967:97-107

McBride, D.N. 1984. Compilation of catch, escapement, age, sex, and size data for salmon (*Oncorhynchus spp.*) returns to the Yakutat area, 1983. Alaska Dep. Fish Game ADF&G Tech. Data Rep. 126:98 p.

——. 1986. Compilation of catch, escapement, age, sex, and size data for salmon (*Oncorhynchus* sp.) returns to the Yakutat area, 1984. Alaska Dep. Fish Game ADF&G Tech. Data Rep. 164: 104 p.

McBride, D.N., and A. Brogle. 1983. Catch, escapement, age, sex and size of salmon (*Oncorhynchus spp.*) returns to the Yakutat area, 1982. Alaska Dep. Fish Game ADF&G Tech. Data Rep. 101:97 p.

McBride, D.N., H.H. Hammer, and L.S. Buklis. 1983. Age, sex and size of Yukon River salmon catch and escapement, 1982. Alaska Dep. Fish Game ADF&G Tech. Data Rep. 90:141 p.

McCart, P. 1967. Behavior and ecology of sockeye salmon fry in the Babine River. J. Fish. Res. Board Can. 24:375-428

——. 1969. Digging behavior of *Oncorhynchus nerka* spawning in streams at Babine Lake, British Columbia, p. 39-51. *In*: T.G. Northcote (ed.). Symposium on Salmon and Trout in Streams. H.R. McMillan Lectures in Fisheries. Institute of Fisheries, University of British Columbia, Vancouver, BC

——. 1970. A polymorphic population of *Oncorhynchus nerka* at Babine Lake, B.C., involving anadromous (sockeye) and non-anadromous (kokanee) forms. Ph.D. thesis. University of British Columbia, Vancouver, BC. 135 p.

McCart, P., D. Mayhood, M. Jones, and G. Glova. 1980. Stikine-Iskut fisheries studies, 1979. Prepared by P. McCart Biological Consultants Ltd. for B.C. Hydro and Power Authority

McDaniel, T.R., T. Kohler, and J. McDaniel. 1985. Lake Tokun fisheries program. Part I, p. 3–28. *In*: J.P. Koenings, T. McDaniel, and D. Barto (eds.). Limnology and fisheries evidence for rearing limitation of sockeye salmon, *Oncorhynchus nerka*, production from Lake Tokun, lower Copper River (1981–1984). Alaska Dep. Fish Game FRED Rep. 55

McDonald, J. 1960. The behaviour of Pacific salmon fry during their downstream migration to freshwater and saltwater nursery areas. J. Fish. Res. Board Can. 17:655–676

——. 1969. Distribution, growth, and survival of sockeye fry (*Oncorhynchus nerka*) produced in natural and artificial stream environments. J. Fish. Res. Board Can. 26:229–267

McDonald, J., and J.M. Hume. 1984. Babine Lake sockeye salmon (*Oncorhynchus nerka*) enhancement program: testing some major assumptions. Can. J. Fish. Aquat. Sci. 41:70–92

McDonald, P.D.M., H.D. Smith, and L. Jantz. 1987. The utility of Babine smolt enumerations in management of Babine and other Skeena River sockeye salmon (*Oncorhynchus nerka*) stocks, p. 280–295. *In*: H.D. Smith, L. Margolis, and C.C. Wood (eds.). Sockeye salmon (*Oncorhynchus nerka*) population biology and future management. Can. Spec. Publ. Fish. Aquat. Sci. 96

McGregor, A.J. 1983. Age, sex, and size of sockeye salmon (*Oncorhynchus nerka* Walbaum) catches and escapements in southeastern Alaska in 1982. Alaska Dep. Fish. Game ADF&G Tech. Data Rep. 100:124 p.

McIntyre, J.D. 1980. Further consideration of causes for decline of Karluk sockeye salmon (Processed report). U.S. Fish and Wildlife Service, National Fishery Research Center, Seattle, WA. 29 p.

McMahon, V.H. 1948. Lakes of the Skeena River drainage. VII. Morrison Lake. Fish. Res. Board Can. Prog. Rep. Pac. Coast Stn. 74:6–9

McPhail, J.D., and C.C. Lindsey. 1970. Freshwater fishes of northwestern Canada and Alaska. Bull. Fish. Res. Board Can. 173:381 p.

Meacham, C.P., and J. H. Clark. 1979. Management to increase anadromous salmon production, p. 377–386. *In*: H. Clepper (ed.). Predator-prey systems in fisheries management. Sport Fishing Institute, Washington, DC

Mead, R.W., and W.L. Woodall. 1968. Comparison of sockeye salmon fry produced by hatcheries, artificial channels and natural spawning areas. Int. Pac. Salmon Fish. Comm. Prog. Rep. 20:41 p.

Meehan, W.R. 1966. Growth and survival of sockeye salmon introduced into Ruth Lake after removal of resident fish populations. U.S. Fish Wildl. Serv. Spec. Sci. Rep. Fish. 532:18 p.

Merrell, T.R., Jr. 1964. Ecological studies of sockeye salmon and related limnological and climatological investigations, Brooks Lake, Alaska, 1957. U.S. Fish Wildl. Serv. Spec. Sci. Rep. Fish. 456:66 p.

Miller, R.J., and E.L. Brannon. 1982. The origin and development of life history patterns in Pacific salmonids, p. 296–309. *In*: E.L. Brannon and E.O. Salo (eds.). Proceedings of the Salmon and Trout Migratory Behavior Symposium. School of Fisheries, University of Washington, Seattle, WA

Moriarity, D.S. 1977. Arctic char in the Wood River Lakes. M.Sc. thesis. University of Washington, Seattle, WA. 68 p.

Morton, W.M. 1982. Comparative catches and food habits of Dolly Varden and Arctic chars, *Salvelinus malma* and *S. alpinus*, at Karluk, Alaska, in 1939–1941. Environ. Biol. Fishes 7:7–28

Mosher, K.H. 1972. Scale features of sockeye salmon from Asian and North American coastal regions. Fish. Bull. (U.S.) 70:141–183

Mundy, P.R. 1979. A quantitative measure of migratory timing illustrated by application to the management of commercial salmon fisheries. Ph.D. thesis. University of Washington, Seattle, WA. 85 p.

Narver, D.W. 1966. Pelagic ecology and carrying capacity of sockeye salmon in the Chignik lakes, Alaska. Ph.D. thesis. University of Washington, Seattle, WA. 348 p.

——. 1970. Diel vertical movements and feeding of underyearling sockeye salmon and the limnetic zooplankton in Babine Lake, British Columbia. J. Fish. Res. Board Can. 27:281–316

Neave, F. 1953. Principles affecting the size of pink and chum salmon populations in British Columbia. J. Fish. Res. Board Can. 9:450–491

——. 1958. The origin and speciation of *Oncorhynchus*. Proc. Trans. R. Soc. Can. Ser. 3, 52(5):25–39

——. 1964. Ocean migrations of Pacific salmon. J.

Fish. Res. Board Can. 21:1227–1244
Nelson, J.S. 1968. Distribution and nomenclature of North American kokanee, *Oncorhynchus nerka*. J. Fish. Res. Board Can. 25:409–414
Nelson, M.O. 1966. Food and distribution of Arctic char in Lake Aleknagik, Alaska, during the summer of 1962. M.Sc. thesis. University of Washington, Seattle, WA. 164 p.
Nelson, P.R. 1959. Effects of fertilizing Bare Lake, Alaska, on growth and production of red salmon (*O. nerka*). Fish. Bull. Fish Wildl. Serv. 60(159):86 p.
Nilsson, N.A. 1965. Food segregation between salmonid species in north Sweden. Inst. Freshwater Res. Drottningholm Rep. 46:58–78
Nishiyama, T. 1974. Energy requirements of Bristol Bay sockeye salmon in the central Bering Sea and Bristol Bay, p. 321–343. *In*: D.W. Hood and E.J. Kelley (eds.). Oceanography of the Bering Sea. Univ. Alaska Inst. Mar. Sci. Occ. Publ. 2
——. 1984. Two important factors in oceanic stages of salmon: food and temperature, p. 227–258. *In*: Proceedings of the Pacific Salmon Biology Conference (USSR, USA, Canada, Japan), Yuzhno-Sakhalinsk, USSR, 1978. Tikhookeanskii Nauchno-Issledovatel'skii Institut Rybnogo Khoziaistva i Okeanografii, Vladivostok, USSR
Noakes, D.L.G. 1978. Ontogeny of behavior in fishes: a survey and suggestions, p. 103–125. *In*: G.M. Burghart and M. Bekoff (eds.). The development of behavior: comparative and evolutionary aspects. Garland Publishing, New York, NY
Olsen, J.C. 1968. Physical environment and egg development in a mainland beach area and an island beach area of Iliamna Lake, p. 169–197. *In*: R.L. Burgner (ed.). Further studies of Alaska sockeye salmon. Univ. Wash. Publ. Fish. New Ser. 3
O'Neill, S.M., and K.D. Hyatt. 1987. An experimental study of competition for food between sockeye salmon (*Oncorhynchus nerka*) and threespine sticklebacks (*Gasterosteus aculeatus*) in a British Columbia coastal lake, p. 143–160. *In*: H.D. Smith, L. Margolis, and C.C. Wood (eds.). Sockeye salmon (*Oncorhynchus nerka*) population biology and future management. Can. Spec. Publ. Fish. Aquat. Sci. 96
Oshima, M. 1934. Life-history and distribution of the fresh-water salmons found in the waters of Japan. Proc. Fifth Pac. Sci. Congr. 5:3751-3773
Ostroumov, A.G. 1977. The kokanee, *Oncorhynchus nerka kennerlyi*, in the Vorovskaya River basin (Western Kamchatka). J. Ichthyol. 17:794–796
Parr, W.H., Jr. 1972. Interactions between sockeye salmon and lake resident fish in the Chignik lakes, Alaska. M.Sc. thesis. University of Washington, Seattle, WA. 103 p.
Pella, J.J. 1968. Distribution and growth of sockeye salmon fry in Lake Aleknagik, Alaska, during the summer of 1962, p. 45–111. *In*: R.L. Burgner (ed.). Further studies of Alaska sockeye salmon. Univ. Wash. Publ. Fish. New Ser. 3
Peterman, R.M. 1980. Dynamics of native Indian food fisheries on salmon in British Columbia. Can. J. Fish. Aquat. Sci. 37:561–566
——. 1982. Nonlinear relation between smolts and adults in Babine Lake sockeye salmon (*Oncorhynchus nerka*) and implications for other salmon populations. Can. J. Fish. Aquat. Sci. 39:904–913
——. 1984a. Cross-correlations between reconstructed ocean abundances of Bristol Bay and British Columbia sockeye salmon (*Oncorhynchus nerka*). Can. J. Fish. Aquat. Sci. 41:1814–1824
——. 1984b. Density dependent growth in early ocean life of sockeye salmon (*Oncorhynchus nerka*). Can. J. Fish. Aquat. Sci. 41:1825-1829
Quinn, T.P. 1980. Evidence for celestial and magnetic compass orientation in lake migrating sockeye salmon fry. J. Comp. Physiol. 137: 243–248
——. 1982a. Intra-specific differences in sockeye salmon fry compass orientation mechanisms, p. 79–111. *In*: E.L. Brannon and E.O. Salo (eds.). Proceedings of the Salmon and Trout Migratory Behavior Symposium. School of Fisheries, University of Washington, Seattle, WA
——. 1982b. A model for salmon navigation on the high seas, p. 229–237. *In*: E.L. Brannon and E.O. Salo (eds.). Proceedings of the Salmon and Trout Migratory Behavior Symposium. School of Fisheries, University of Washington, Seattle, WA
Quinn, T.P., and E.L. Brannon. 1982. The use of celestial and magnetic cues by orienting sockeye salmon smolts. J. Comp. Physiol. 147:547–552
Raleigh, R.F. 1967. Genetic control in the lakeward migrations of sockeye salmon (*Oncorhynchus nerka*) fry. J. Fish. Res. Board Can. 24:2613-2622
Ricker, W.E. 1938. "Residual" and kokanee salmon in Cultus Lake. J. Fish. Res. Board Can. 4:192–217

——. 1940. On the origin of kokanee, a freshwater type of sockeye salmon. Proc. Trans. R. Soc. Can. Ser. 3 34(5):121–135

——. 1950. Cycle dominance among the Fraser River sockeye. Ecology 31:6–26

——. 1954. Stock and recruitment. J. Fish. Res. Board Can. 11:559–623

——. 1962. Comparison of ocean growth and mortality of sockeye salmon during their last two years. J. Fish. Res. Board Can. 19:531–560

——. 1966. Sockeye salmon in British Columbia, p. 59–70. *In*: Salmon of the North Pacific Ocean. Part III. A review of the life history of North American salmon. Int. North Pac. Fish. Comm. Bull. 18

——. 1972. Hereditary and environmental factors affecting certain salmonid populations, p. 19–160. *In*: R.C. Simon and P.A. Larkin (eds.). The stock concept in Pacific salmon. H.R. MacMillan Lectures in Fisheries. Institute of Fisheries, University of British Columbia, Vancouver, BC

——. 1976. Review of the rate of growth and mortality of Pacific salmon in salt water, and noncatch mortality caused by fishing. J. Fish. Res. Board Can. 33:1483–1524

——. 1982. Size and age of British Columbia sockeye salmon (*Oncorhynchus nerka*) in relation to environmental factors and the fishery. Can. Tech. Rep. Fish. Aquat. Sci. 1115:32 p.

Ricker, W.E., and H.D. Smith. 1975. A revised interpretation of the history of the Skeena River sockeye salmon (*Oncorhynchus nerka*). J. Fish. Res. Board Can. 32:1369–1381

Riffe, R.R., S.A. McPherson, B.W. Van Alen, and D.N. McBride. 1987. Compilation of catch, escapement, age, sex and size data for salmon (*Oncorhynchus*) returns to the Yakutat area in 1985. Alaska Dep. Fish Game ADF&G Tech. Data Rep. 210:123 p.

Roberson, K., and R.R. Holder. 1987. Development and evaluation of a streamside sockeye salmon (*Oncorhynchus nerka*) incubation facility, Gulkana River, Alaska, p. 191–197. *In*: H.D. Smith, L. Margolis, and C.C. Wood (eds.). Sockeye salmon (*Oncorhynchus nerka*) population biology and future management. Can. Spec. Publ. Fish. Aquat. Sci. 96

Robertson, A. 1922. Further proof of the parent stream theory. Trans. Am. Fish. Soc. 51:87–90

Robinson, D.G., and W.E. Barraclough. 1978. Population estimates of sockeye salmon (*Oncorhynchus nerka*) in a fertilized oligotrophic lake. J. Fish. Res. Board Can. 35:851–860

Rogers, B.J. 1979. Responses of juvenile sockeye salmon and their food supply to inorganic fertilization of Little Togiak Lake, Alaska. M.Sc. thesis. University of Washington, Seattle, WA. 112 p.

Rogers, D.E. 1968. A comparison of the food of sockeye salmon fry and threespine sticklebacks in the Wood River Lakes, p. 1–43. *In*: R.L. Burgner (ed.). Further studies of Alaska sockeye salmon. Univ. Wash. Publ. Fish. New Ser. 3

——. 1973a. Abundance and size of juvenile sockeye salmon, *Oncorhynchus nerka*, and associated species in Lake Aleknagik, Alaska, in relation to their environment. Fish. Bull. (U.S.) 71:1061–1075

——. 1973b. Forecast of the sockeye salmon run to Bristol Bay in 1973. Univ. Wash. Fish. Res. Inst. Circ. 73-1:33 p.

——. 1980. Density-dependent growth of Bristol Bay sockeye salmon, p. 267–283. *In*: W.J. McNeil and D.C. Himsworth (eds.). Salmon ecosystems of the North Pacific. Oregon State University Press, Corvallis, OR

——. 1984. Trends in abundance of northeastern Pacific stocks of salmon, p. 100–127. *In*: W.G. Pearcy (ed.). The influence of ocean conditions on the production of salmonids in the North Pacific. Oreg. State Univ. Sea Grant Coll. Program ORESU-W-83-001

——. 1986. Pacific salmon, p. 461–476. *In*: D.W. Hood and S.T. Zimmerman (eds.). The Gulf of Alaska: physical environment and biological resources. U.S. National Ocean Service, Ocean Assessments Division, Alaska Office; U.S. Minerals Management Service, Alaska OCS Region, Washington, DC

——. 1987. The regulation of age at maturity in Wood River sockeye salmon (*Oncorhynchus nerka*), p. 78–89. *In*: H.D. Smith, L. Margolis, and C.C. Wood (eds.). Sockeye salmon (*Oncorhynchus nerka*) population biology and future management. Can. Spec. Publ. Fish. Aquat. Sci. 96

Rogers, D.E., L. Gilbertson, and D. Eggers. 1972. Predator-prey relationship between Arctic char and sockeye salmon smolts at the Agulowak River, Lake Aleknagik, in 1971. Univ. Wash. Fish. Res. Inst. Circ. 72-7:40 p.

Rogers, D.E., and N. Newcomb. 1975. Distribution,

abundance, and size of juvenile sockeye salmon and associated species in the Igushik lakes: final report. Univ. Wash. Fish. Res. Inst. FRI-UW-7508:46 p.

Rogers, D.E., and P.H. Poe. 1984. Escapement goals for the Kvichak River system. Univ. Wash. Fish. Res. Inst. FRI-UW-8407:66 p.

Rogers, D.E., B.J. Rogers, and J.F. Hardy. 1980. Effects of fertilization of Little Togiak Lake on the food supply and growth of sockeye salmon, p. 125–142. *In*: B.R. Melteff and R.A. Neve (eds.). Proceedings of the North Pacific Aquaculture Symposium. Alaska Sea Grant Rep. 82-2

Rounsefell, G.A. 1957. Fecundity of North American Salmonidae. Fish. Bull. Fish Wildl. Serv. 57:451–468

——. 1958a. Anadromy in North American Salmonidae. Fish. Bull. Fish Wildl. Serv. 58:171–185

——. 1958b. Factors causing decline in sockeye salmon of Karluk River, Alaska. Fish. Bull. Fish Wildl. Serv. 58:83–169

——. 1973. Comments on "Evaluation of causes for the decline of the Karluk sockeye salmon runs and recommendation for rehabilitation," by R. Van Cleve and D.E. Bevan. Fish. Bull. (U.S.) 71:561–659

Royal, L.A. 1953. The effects of regulatory selectivity on the productivity of Fraser River sockeye. Can. Fish Cult. 14:1–12

Royce, W.F. 1965. Almanac of Bristol Bay sockeye salmon. Univ. Wash. Fish. Res. Inst. Circ. 235:48 p.

Royce, W.F., L.S. Smith, and A.C. Hartt. 1968. Models of oceanic migrations of Pacific salmon and comments on guidance mechanisms. Fish. Bull. (U.S.) 66:441–462

Rucker, R.R. 1937. The effect of temperature on the growth of the embryos of *Oncorhynchus nerka* (Walbaum). M.Sc. thesis. University of Washington, Seattle, WA. 54 p.

Ruggerone, G.T., and D.E. Rogers. 1984. Arctic char predation on sockeye salmon smolts at Little Togiak River, Alaska. Fish. Bull. (U.S.) 82:401–410

Ruggles, C.P. 1966. Juvenile sockeye studies in Owikeno Lake, British Columbia. Can. Fish Cult. 36:3–21

Rukhlov, F.N. 1982. Life of Pacific salmons. Far Eastern Book Publishers, Sakhalin Branch, Yuzhno-Sakhalinsk, USSR. 110 p. (In Russian)

Sanger, G.A. 1972. Fishery potentials and estimated biological productivity of the Subarctic Pacific region, p. 561–574. *In*: A.Y. Takenouti (ed.). Biological oceanography of the northern North Pacific Ocean dedicated to Sigeru Motoda. Idemitsu Shoten, Tokyo, Japan

Savvaitova, K.A., and I.S. Reshetnikov. 1961. The food of different biological forms of the Dolly Varden char, *Salvelinus malma* (Walb.), in certain Kamchatka waters. Vopr. Ikhtiol. 1:127–135. (Transl. from Russian; Fish. Res. Board Can. Transl. Ser. 373)

Schaefer, M.B. 1951. A study of the spawning populations of sockeye salmon in the Harrison River system, with special reference to the problem of enumeration by means of marked members. Int. Pac. Salmon Fish. Comm. Bull. 4:207 p.

Schultz, L.P. 1935. The breeding activities of the little redfish, a landlocked form of the sockeye salmon, *Oncorhynchus nerka*. J. Pan-Pacific Res. Inst. 10; Mid-Pacific Mag. 48(1):67–77

Seeley, C.M., and G.W. McCammon. 1966. Kokanee, p. 274–294. *In*: A.C. Calhoun (ed.). Inland fisheries management. California Department of Fish and Game, Sacramento, CA

Selifonov, M.M. 1970. The growth of young sockeye salmon in Lake Kuril. Izv. Tikhookean. Nauchno-Issled. Inst. Rybn. Khoz. Okeanogr. 78:33–41. (Transl. from Russian; Fish. Res. Board Can. Transl. Ser. 2340)

——. 1987. Influence of environment on the abundance of sockeye salmon (*Oncorhynchus nerka*) from the Ozernaya and Kamchatka rivers, p. 125–128. *In*: H.D. Smith, L. Margolis, and C.C. Wood (eds.). Sockeye salmon (*Oncorhynchus nerka*) population biology and future management. Can. Spec. Publ. Fish Aquat. Sci. 96

Semko, R.S. 1954a. The stocks of West Kamchatka salmon and their commercial utilization. Izv. Tikhookean. Nauchno-Issled. Inst. Rybn. Khoz. Okeanogr. 41:3–109. (Transl. from Russian; Fish. Res. Board Can. Transl. Ser. 288)

——. 1954b. West Kamchatka salmon, their commercial exploitation and reproduction, p. 38–47. *In*: Trudy Soveshchaniia po voprosan lososevogo khoziaistvo dal'nego vostoka, 1953. Tr. Soveshch. Ikhtiol. Kom. Akad. Nauk SSSR 4. (Transl. from Russian; Israel Program for Scientific Translations, Jerusalem, 1960)

Sharr, S. 1983. Catch and escapement statistics for Copper and Bering River sockeye (*Oncorhynchus nerka*), chinook (*O. tshawytscha*), and coho salmon (*O. kisutch*), 1982. Alaska Dep. Fish Game ADF&G Tech. Data Rep. 98:45 p.

Sharr, S., D.R. Bernard, D.N. McBride, and W. Goshert. 1985. Catch and escapement statistics for Copper River, Bering River, and Prince William Sound sockeye, chinook, coho and chum salmon, 1983. Alaska Dep. Fish Game ADF&G Tech. Data Rep. 135:95 p.

Shelbourn, J.E., J.R. Brett, and S. Shirahata. 1973. Effect of temperature and feeding regime on the specific growth rate of sockeye salmon fry (*Oncorhynchus nerka*), with a consideration of size effect. J. Fish. Res. Board Can. 30:1191–1194

Sheridan, W.L. 1960. Frequency of digging movements of female pink salmon before and after egg deposition. Anim. Behav. 8:228–230

Shmidt, P.Y. 1950. Fishes of the Sea of Okhotsk. Tr. Tikhookean. Kom. Akad. Nauk SSSR 6:390 p. (Transl. from Russian; Israel Program for Scientific Translations, Jerusalem. 1965)

Shuman, R.F. 1950. Bear depredations on red salmon spawning populations in the Karluk River system, 1947. J. Wildl. Manage. 14:1–9

Simonova, N.A. 1972. The feeding of sockeye fry in spawning grounds and their food supply. Izv. Tikhookean. Nauchno-Issled. Inst. Rybn. Khoz. Okeanogr. 82:179–189. (Transl. from Russian; Fish. Res. Board Can. Transl. Ser. 2911)

Simpson, K., L. Hop Wo, and I. Miki. 1981. Fish surveys of 15 sockeye salmon (*Oncorhynchus nerka*) nursery lakes in British Columbia. Can. Tech. Rep. Fish. Aquat. Sci. 1022:87 p.

Smirnov, A.I. 1950. Importance of carotinoid pigmentation at the embryonic stages of cyprinids. Dokl. Akad. Nauk SSSR 73:609–612. (In Russian)

——. 1958. Certain features of the biology of propagation and development of the salmonid fish nerka, *Oncorhynchus nerka* (Walbaum). Dokl. Akad. Nauk SSSR 123:371–374. (Transl. from Russian; Fish. Res. Board Can. Transl. Ser. 229)

——. 1959a. Differences in the biology of reproduction and development of residual or dwarf sockeye and anadromous sockeye, *Oncorhynchus nerka* (Walbaum). Nauchn. Dokl. Vyssh. Shk. Biol. Nauk. 3:59–65. (Transl. from Russian; Fish. Res. Board Can. Transl. Ser. 266)

——. 1959b. The functional importance of prespawning changes in the skin of salmon (as exemplified by the genus *Oncorhynchus*). Zool. Zh. 38:734–744. (Transl. from Russian; Fish. Res. Board Can. Transl. Ser. 348)

——. 1964. Features of likeness and differences in the development of the Pacific salmon Salmonidae, *Oncorhynchus*. *In*: Problemy sovremmennoi embriologii. Izdatel'stvo Moskovogo Universiteta, Moscow, USSR. (In Russian)

Smith, G.R., and R.F. Stearley. 1989. The classification and scientific names of rainbow and cutthroat trouts. Fisheries 14:4-10

Smith, H.D. 1973. Observations on the cestode *Eubothrium salvelini* in juvenile sockeye salmon (*Oncorhynchus nerka*) at Babine Lake, British Columbia. J. Fish. Res. Board Can. 30:947–964

Soin, S.G. 1956. Respiratory significance of the carotinoid pigment in the eggs of salmonids and other representatives of the Clupeiformes. Zool. Zh. 35:1362–1369. (Transl. from Russian; Fish. Mar. Ser. (Can.) Transl. Ser. 4538)

——. 1964. Adaptational features in the development of fish in connection with different features of respiration. Vestn. Mosk. Univ. Ser. VI Biol. Pochvoved. 1964(6). (In Russian)

Starr, P.J., A.T. Charles, and M.A. Henderson. 1984. Reconstruction of British Columbia sockeye salmon (*Oncorhynchus nerka*) stocks: 1970–1982. Can. MS Rep. Fish. Aquat. Sci. 1780:123 p.

Stober, Q.J., and A.H. Hamalainen. 1980. Cedar River sockeye salmon production. Univ. Wash. Fish. Res. Inst. FRI-UW-8016:59 p.

Straty, R.R. 1974. Ecology and behavior of juvenile sockeye salmon (*Oncorhynchus nerka*) in Bristol Bay and the eastern Bering Sea, p. 285–320. *In*: D.W. Hood and E.J. Kelley (eds.). Oceanography of the Bering Sea with emphasis on renewable resources. Univ. Alaska Inst. Mar. Sci. Occ. Publ. 2

——. 1975. Migratory routes of adult sockeye salmon, *Oncorhynchus nerka*, in the eastern Bering Sea and Bristol Bay. NOAA Tech. Rep. NMFS SSRF-690:32 p.

Straty, R.R., and H.W. Jaenicke. 1980. Estuarine influence of salinity, temperature, and food on the behavior, growth, and dynamics of Bristol Bay sockeye salmon, p. 247–265. *In*: W.J. McNeil and D.C. Himsworth (eds.). Salmon ecosystems of the North Pacific. Oregon State University Press, Corvallis, OR

Synkova, A.I. 1951. On the food of Pacific salmon in Kamchatka waters. Izv. Tikhookean. Nauchno-

Issled. Inst. Rybn. Khoz. Okeanogr. 34:105-121. (Transl. from Russian; *In*: Pacific salmon: selected articles from Soviet periodicals, p. 216-235. Israel Program for Scientific Translations, Jerusalem, 1961; and Fish. Res. Board Can. Transl. Ser. 415)

Takeuchi, I. 1972. Food animals collected from the stomachs of three salmonid fishes (*Oncorhynchus*) and their distribution in the natural environments in the northern North Pacific. Bull. Hokkaido Reg. Fish. Res. Lab. 38:119 p. (In Japanese, English summary)

Thompson, W.F. 1945. Effect of the obstruction at Hells Gate on the sockeye salmon of the Fraser River. Int. Pac. Salmon Fish. Comm. Bull. 1:175 p.

——. 1950. Some salmon research problems in Alaska. Presented at Alaskan Science Conference of the National Academy of Sciences, National Research Council, Washington, 9-11 November, 1950. University of Washington, Fisheries Research Institute, Seattle, WA. 39 p.

——. 1962. The research program of the Fisheries Research Institute in Bristol Bay, 1945-1958, p. 1-36. *In*: T.S.Y. Koo (ed.). Studies of Alaska red salmon. Univ. Wash. Publ. Fish. New Ser. 1

Thorpe, J.E. 1982. Migration in salmonids, with special reference to juvenile movements in freshwater, p. 86-97. *In*: E.L. Brannon and E.O. Salo (eds.). Proceedings of the Salmon and Trout Migratory Behavior Symposium. School of Fisheries, University of Washington, Seattle, WA

Tsunoda, S. 1967. Movements of spawning sockeye salmon in Hidden Creek, Brooks Lake, Alaska. M.Sc. thesis. Oregon State University, Corvallis, OR. 52 p.

Van Cleve, R., and D.E. Bevan. 1973a. Evaluation of causes for the decline of the Karluk sockeye salmon runs and recommendations for rehabilitation. Fish. Bull. (U.S.) 71:627-649

——. 1973b. Reply to Rounsefell's "Comments on 'Evaluation of causes for the decline of the Karluk sockeye salmon runs and recommendations for rehabilitation,' by R. Van Cleve and D.E. Bevan." Fish. Bull. (U.S.) 71:661-663

Velsen, F.P.J. 1980. Embryonic development in eggs of sockeye salmon, *Oncorhynchus nerka*. Can. Spec. Publ. Fish. Aquat. Sci. 49:19 p.

Vernon, E.H. 1957. Morphometric comparisons of three races of kokanee (*Oncorhynchus nerka*) within a large British Columbia lake. J. Fish. Res. Board Can. 14:573-598

Walters, C.J., and M.J. Staley. 1987. Evidence against the existence of cyclic dominance in Fraser River sockeye salmon (*Oncorhynchus nerka*), p. 375-384. *In*: H.D. Smith, L. Margolis, and C.C. Wood (eds.). Sockeye salmon (*Oncorhynchus nerka*) population biology and future management. Can. Spec. Publ. Fish. Aquat. Sci. 96

Ward, F.J., and P.A. Larkin. 1964. Cyclic dominance in Adams River sockeye salmon. Int. Pac. Salmon Fish. Comm. Prog. Rep. 11:116 p.

Wedemeyer, G.A., R.L. Saunders, and W.C. Clarke. 1980. Environmental factors affecting smoltification and early marine survival of anadromous salmonids. Mar. Fish. Rev. 42(6):1-14

Werner, E.E. 1972. On the breadth of diet in fishes. Ph.D. thesis. Michigan State University, East Lansing, MI. 58 p.

West, C.J., and J.C. Mason. 1987. Evaluation of sockeye salmon (*Oncorhynchus nerka*) production from the Babine Lake development project, p. 176-190. *In*: H.D. Smith, L. Margolis, and C.C. Wood (eds.). Sockeye salmon (*Oncorhynchus nerka*) population biology and future management. Can. Spec. Publ. Fish. Aquat. Sci. 96

Williams, I.V. 1973. Investigation of the prespawning mortality of sockeye in Horsefly River and McKinley Creek in 1969. Int. Pac. Salmon Fish. Comm. Prog. Rep. 27(2):42 p.

Wilmot, R.L., and C.V. Burger. 1985. Genetic differences among populations of Alaskan sockeye salmon. Trans. Am. Fish. Soc. 114:236-243

Wing, B.L. 1977. Salmon food observations, p. 20-27. *In*: Southeast Alaska troll log book program: 1976 scientific report. Alaska Sea Grant Rep. 77-11

Withler, F.C., J.A. McConnell, and V.H. McMahon. 1949. Lakes of the Skeena River drainage. IX. Babine Lake. Fish. Res. Board Can. Prog. Rep. Pac. Coast Stn. 78:6-11

Wood, C.C., B.E. Riddell, and D.T. Rutherford. 1987. Alternative juvenile life histories of sockeye salmon (*Oncorhynchus nerka*) and their contribution to production in the Stikine River, northern British Columbia, p. 12-24. *In*: H.D. Smith, L. Margolis, and C.C. Wood (eds.). Sockeye salmon (*Oncorhynchus nerka*) population biology and future management. Can. Spec. Publ. Fish. Aquat. Sci. 96

Woodey, J.C. 1972. Distribution, feeding, and

growth of juvenile sockeye salmon in Lake Washington. Ph.D. thesis. University of Washington, Seattle, WA. 207 p.
Yuen, H.J., D.L. Bill, M.L. Nelson, R.B. Russell, and J. Skrade. 1984. Bristol Bay salmon (*Oncorhynchus* sp.)–1980: a compilation of catch, escapement, and biological data. Alaska Dep. Fish. Game ADF&G Tech. Data Rep. 128:113 p.

Life History of Pink Salmon

CONTENTS

HEINE '75

8

PLATE 8. Pink salmon fry shortly after emergence from the gravel. *Photograph by W.R. Heard*

PLATE 9. Albino pink salmon fry from Sashin Creek, southeastern Alaska. *Photograph by W.R. Heard*

PLATE 10. Adult pink-chum salmon hybrid. *Photograph by W.R. Heard*

PLATE 7 (*previous page*). Pink salmon life history stages. *Painting by H. Heine*

9

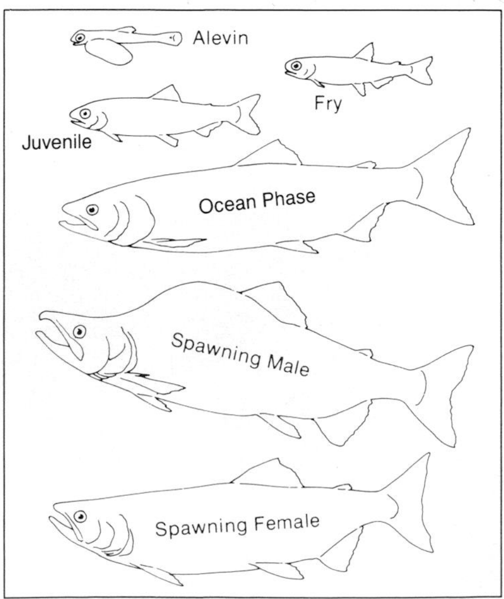

10

LIFE HISTORY OF PINK SALMON
(*Oncorhynchus gorbuscha*)

William R. Heard*

Introduction

THE PINK SALMON, *Oncorhynchus gorbuscha* (Walbaum), is the most abundant of the seven species of Pacific salmon, contributing about 40% by weight and 60% in numbers of all salmon caught commercially in the North Pacific Ocean and adjacent waters (Neave et al. 1967). Members of this species are anadromous and have the simplest or most specialized life cycle within the genus, depending on whether Pacific salmon originated from marine or freshwater ancestors. Upon emergence, pink salmon fry migrate quickly to sea and grow rapidly as they make extensive feeding migrations. After eighteen months in the ocean the maturing fish return to their river of origin to spawn and die.

Neave (1958) assumed that *Oncorhynchus* evolved from an ancestral freshwater form of Pacific *Salmo* during the Pleistocene, probably in the vicinity of the present-day Sea of Japan. Hoar (1958, 1976) and Chernenko (1969) further developed this argument and concluded that those *Oncorhynchus* species that rely least on the freshwater environment, namely, *O. gorbuscha* and *O. keta* (chum salmon), were the most specialized. Pink salmon have 52 chromosomes, fewer than other Pacific salmon (Simon 1963), which also may suggest specialization. Thorpe (1982), however, viewed the Salmonidae as relatively primitive teleosts, of probable marine pelagic origin, and about five million years old. This view is supported by Pliocene fossils from California and Oregon that are almost identical to present-day coho salmon, *Oncorhynchus kisutch* (Cavender and Miller 1972). Thorpe (1982) postulated that during evolution the salmonids tended towards greater dependence on fresh water and away from dependence on the sea. Under this scenario, the pink salmon, with the greatest dependence on the marine environment, is considered the least advanced extant *Oncorhynchus* species.

Pink salmon are distinguished from other Pacific salmon by (1) having a fixed two-year life span, (2) being the smallest of the Pacific salmon as adults (averaging 1.0–2.5 kg), (3) the fact that the young migrate to sea soon after emerging from the gravel (Neave 1955; Hoar 1956), and (4) developing a marked hump in large maturing males. This last characteristic is responsible for the vernacular name humpback salmon used in some areas.

Because of the fixed two-year life cycle, pink salmon spawning in a particular river system in odd and even years are reproductively isolated from each other and have developed into genetically different lines. In some river systems, like the Fraser River in British Columbia, only the odd-year line exists; returns in even years are negligible. In Bristol Bay, Alaska, the major runs occur in even years, whereas the coastal area between these two river systems is characterized by runs in both even and odd years.

There are rare exceptions to the two-year life cycle in pink salmon. Naturally occurring one-year-old (Ivankov et al. 1975; Foster et al. 1981;

*Auke Bay Laboratory, Alaska Fisheries Science Center, National Marine Fisheries Service, National Oceanic and Atmospheric Administration, Auke Bay, Alaska 99821

Hikita 1984) and three-year-old (Dvinin 1952; Anas 1959; Turner and Bilton 1968) pink salmon have been documented for indigenous populations. Three-year-old pink salmon occur in a transplanted stock in the Great Lakes of North America (Kwain and Chappel 1978; Wagner and Stauffer 1980), and laboratory populations have been raised to maturity at ages ranging from one to four years.

Pink salmon generally make less extensive spawning migrations in fresh water than other Pacific salmon. Thus, a large part of the freshwater production occurs within a few kilometres of the sea. Major runs also exist in large river systems, but even in such waters the spawning tributaries or mainstream spawning grounds are often in the lower reaches. In some areas a significant percentage of pink salmon spawning occurs in parts of the streams that are periodically covered by marine tidal waters.

The spawning grounds of pink salmon in North America range from central California to near the Mackenzie River in arctic Canada, and, on the Asian side, from North Korea to the Yana and Lena rivers in the USSR, which flow into the Arctic Ocean (Figure 1). During the ocean feeding and maturation phase, pink salmon are found throughout the North Pacific Ocean and Bering Sea north of about 40°N. Pink salmon populations originating from different coastal regions of the North Pacific rim countries occupy distinct ocean nursery areas (Takagi et al. 1981). During late autumn and winter pink salmon populations are found in the southern part of their ocean distributional range. In spring they generally move northward with the warming of ocean waters and return to their respective spawning rivers during the summer. Spawning generally takes place from August to November and the salmon die soon after.

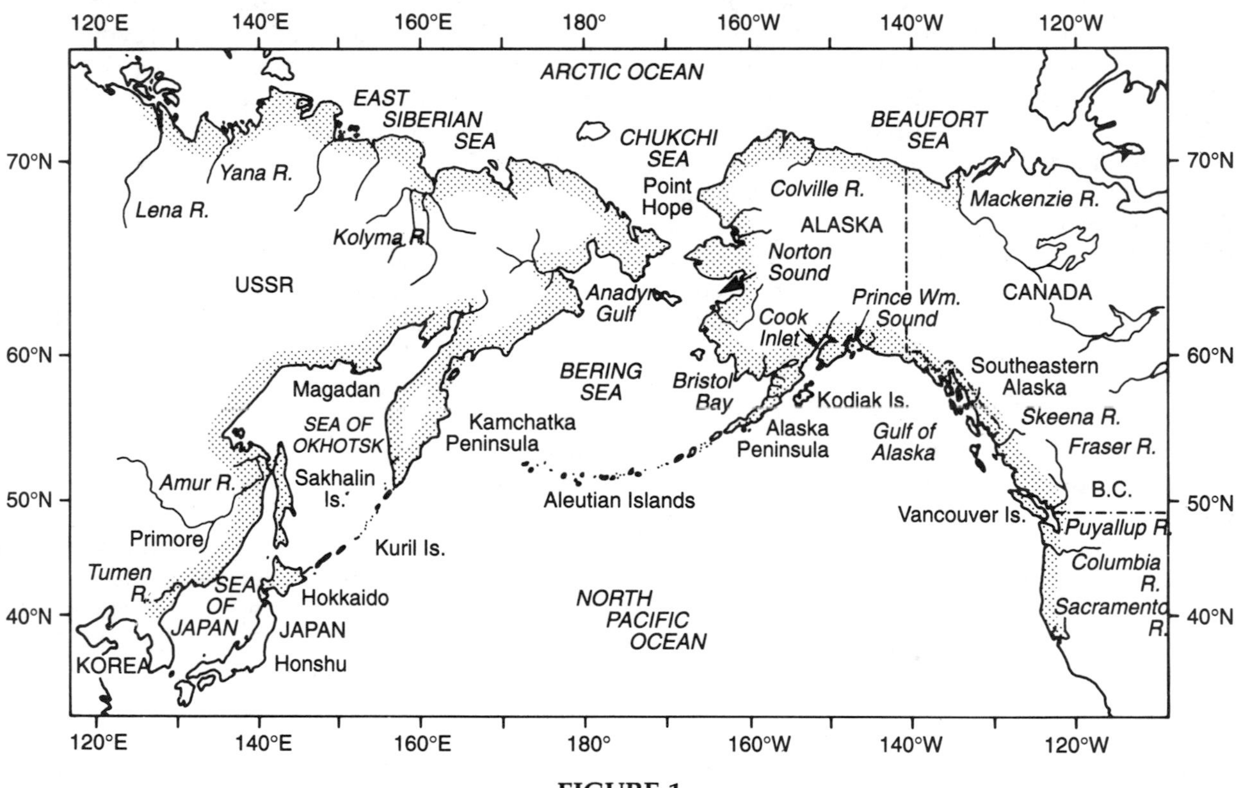

FIGURE 1

Indigenous spawning distribution of pink salmon throughout coastal waters of the North Pacific Ocean, Arctic Ocean, and adjacent seas

Distribution of Spawning Populations

The natural freshwater range of pink salmon includes the Pacific rim of Asia and North America north of about 40°N (Neave et al. 1967; Takagi et al. 1981). Within this vast area, spawning pink salmon are widely distributed in coastal streams of both continents up to the Bering Strait. North, east, and west of the Bering Strait, spawning populations become more irregular and occasional. Centres of large spawning populations occur at roughly parallel positions along the two continents from about latitudes 44°N to 65°N in Asia and about 48°N to 64°N in North America (Figure 1). Along both the Asian and North American coastlines pink salmon occupy ocean waters south of the limits of spawning streams.

Distribution in Asia

On the Sea of Japan coast of Hokkaido, Japan, the southernmost pink salmon spawning streams are the Teshio (Ishida 1967) and the Chitose rivers (Hikita and Terao 1967). On the Pacific Ocean coast of Japan, pink salmon spawn as far south as the Yurappu River in southern Hokkaido (Sano and Kobayashi 1953; Ishida 1967). No pink salmon spawning has been reported for the Pacific Ocean coast of Honshu (Ishida 1967); however, single specimens have been reported from the Akka River (Hoshiai and Sato 1973) and Ohkawa River (Kaeriyama and Hikita 1984) along the northeastern Honshu coast.

On the Asian continent the southern spawning limit of pink salmon is reported as the North Nandai River, North Korea, and the Tumen River at the boundary between Korea and the USSR (Shmidt 1950; Neave et al. 1967; Ishida 1967). Only small numbers are found in this region and to the north in the southern Primore District of the USSR (Milovidova-Dubrovskaya 1937). Larger runs of increasing commercial importance occur in northern Primore, north through the Amur District of the USSR mainland (Smirnov 1947; Kaganovsky 1949; Eniutina 1972), and across Tatar Strait along the Sea of Japan coast of Sakhalin Island.

The northernmost Asian area of significant pink salmon production is in the vicinity of the Anadyr Gulf along the Bering Sea coast. However, pink salmon have been reported as far west as the Lena River (Berg 1948; Andriyashev 1954). Major pink salmon runs occur in coastal streams throughout the greater Sea of Okhotsk basin, along the north coast through the Magadan District, and along the east and west coasts of the Kamchatka Peninsula.

Distribution in North America

The southernmost spawning of pink salmon in North America occurs in the Sacramento, San Lorenzo, and other streams of north-central California (Scofield 1916; Taft 1938; Hallock and Fry 1967). In most cases, the reports represent strays and not self-perpetuating runs, except for those from the Sacramento River, which may have very small runs (Hallock and Fry 1967). Pink salmon also occur irregularly in streams along the Oregon (Herrmann 1959) and Washington coasts, including the Columbia River, but none of these sustain significant runs.

Substantial spawning runs of pink salmon extend from the Puyallup River in Puget Sound, Washington, into adjacent waters of southern British Columbia and northward to Alaska (Vernon et al. 1964). In the Fraser River and in Puget Sound streams, runs occur only in odd-numbered years. Major spawning runs of pink salmon in both odd and even years occur throughout British Columbia north of the Fraser River (Aro and Shepard 1967; Ricker and Manzer 1974), although the even-year line is frequently dominant in different parts of the province (Neave 1952; Ricker 1962).

Large spawning populations of pink salmon occur each year throughout the coastal waters of Alaska (Atkinson et al. 1967; Fredin et al. 1974), including southeastern, central, and western Alaska. Often one line is more abundant than the other. The northernmost runs of commercial importance in North America are in the Norton Sound area at latitude 64°N (Regnart and Geiger 1974; Cunningham 1976; Regnart 1976; Kuhlmann 1977). North of the Bering Strait, pink salmon are reported in small numbers from streams of Kotzebue Sound (Smith et al. 1966; Regnart 1976) and further north and east along the Chukchi Sea coast (Craig 1984a), and along the Beaufort Sea coast east of Point Barrow (Corkum and McCart 1981; Craig 1984b). Although earlier reports from this region list pink salmon from the Mackenzie River or its delta (Dymond 1940; McPhail and Lindsey 1970), recent studies have not found pink salmon in the Mackenzie River area (Corkum and McCart 1981; Craig 1984b; Craig and Haldorson 1986).

Relative Abundance of Stocks

The best indices of pink salmon abundance throughout regions of the North Pacific are based on commercial harvest. Some factors to consider in relating catches of pink salmon to fluctuations in population abundance include the many different fisheries and fishing gears (INPFC 1979); peculiarities in odd- and even-year populations (Neave et al. 1967); the general lack of combined catch and escapement data for estimates of total run size of stocks (Royce 1962); changing harvest rates between years and between and within regions (Neave 1962); the harvest of different kinds of stocks, i.e., wild or hatchery stocks (McNeil 1980); and the presence of "returning" or "passing" stocks in the same fishery (Takagi et al. 1981). Despite these problems, catch statistics do reflect overall trends in abundance in different regions but may not relate to individual stocks.

During the period 1978–81, for which reasonably complete catch records are available, the total harvest of pink salmon in the North Pacific Ocean ranged from 126 to 179 million and averaged 160 million fish per year. By comparison, total harvest during the same period for the next two most abundant Pacific salmon, chum and sockeye (*O. nerka*), averaged 46 and 41 million fish per year, respectively (Table 1). The combined annual Asian and North American catches of pink salmon exceeded 200 million in the 1930s and early 1940s and were as low as 50 to 60 million during the 1950s (Figure 2). Overall pink salmon abundance as reflected by catch is greater in Asia than in North America (Figure 2).

North America

Pink salmon occur in greatest abundance along the North American coast in the central and southeastern Alaska region (Figure 3). Commercial harvests in these areas have fluctuated from 9 to 10 million to record highs of about 30 to 40 million fish. Abundance levels are lower in British Columbia (average harvest about 10 million) and drop off drastically in western Alaska and Washington (Figure 3). Pink salmon abundance was adversely affected for several years by severe tectonic disturbances to spawning grounds in 1964 in both the Prince William Sound and Kodiak Island districts (Roys 1971; Thorsteinson et al. 1971; Noerenberg 1971). Recent record catches in both districts suggest that stocks have adapted to the rearranged spawning environments.

TABLE 1

Annual and mean commercial catches of pink, chum, and sockeye salmon, 1978–81, by Canada, Japan, the United States, and the USSR (in thousands of fish)

	Canada	Japan	U.S.	USSR	Total and mean
Pink					
1978	10 488	16 864	53 856	45 270	126 478
1979	11 826	23 214	54 863	86 545	176 448
1980	8 162	21 470	63 284	65 473	158 389
1981	18 086	24 476	64 222	71 878	178 662
Mean					159 994
Chum					
1978	2 972	24 264	7 966	3 539	38 741
1979	866	32 203	5 960	4 297	43 326
1980	3 318	29 176	10 626	3 508	46 628
1981	1 123	36 599	12 079	4 702	54 503
Mean					45 799
Sockeye					
1978	7 211	3 175	19 536	1 257	31 179
1979	5 693	2 943	30 528	1 186	40 350
1980	3 117	3 202	33 865	1 512	41 696
1981	8 443	3 071	38 772	1 422	51 708
Mean					41 233

Source: International North Pacific Fisheries Commission (INPFC), Statistical Yearbooks for 1978–81; USSR data for 1980

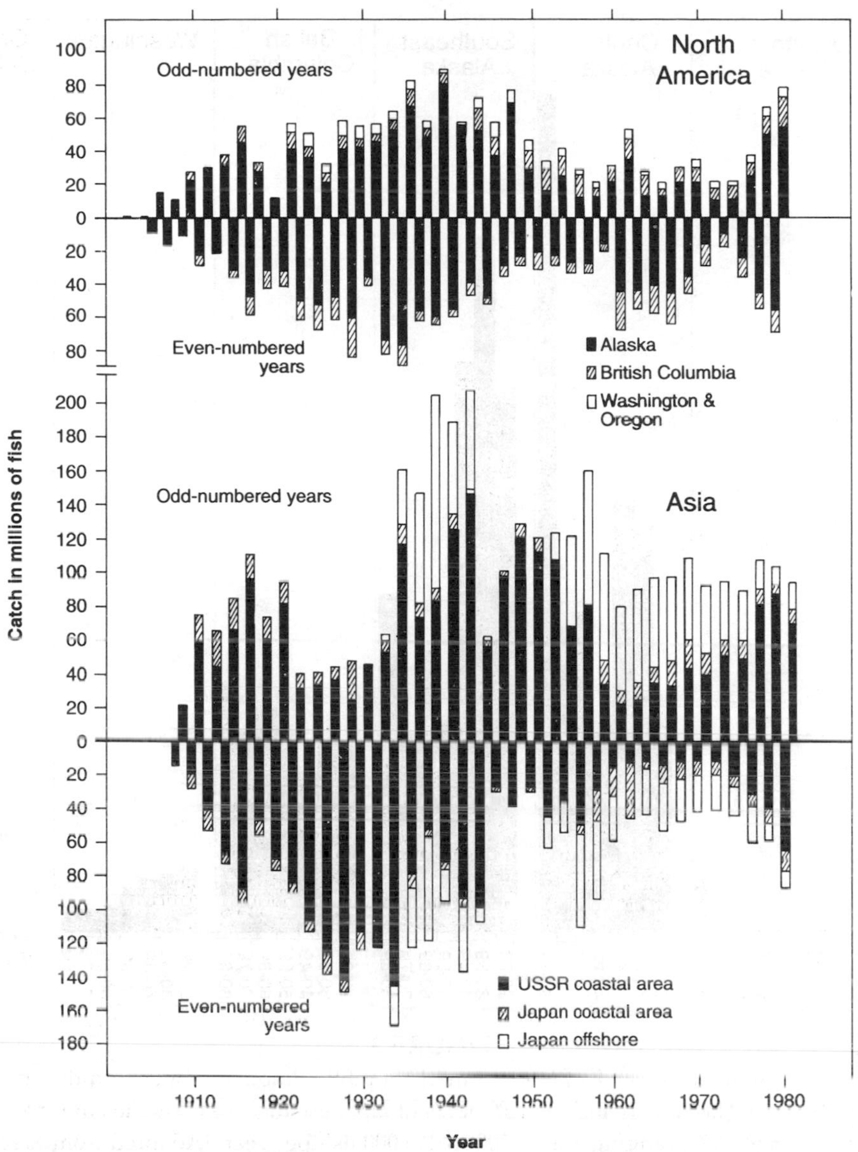

FIGURE 2
Historical commercial catch of North American and Asian pink salmon 1908–81. (Adapted from Takagi et al. 1981)

Rogers (1984) noted that recent increases of pink salmon abundance in the central and western Alaska regions are correlated with dramatic increases of all salmon in these regions. He found that winter temperatures were becoming warmer in the Gulf of Alaska and speculated that salmon are less concentrated in these waters during warm winters than in cold winters. This makes them less vulnerable to predation, the most likely cause of mortality during their oceanic life.

Asia

The history of the pink salmon harvest from Asian waters is more complex and has undergone a somewhat greater degree of change over time than that in North America. This is due in part to the effect of Japanese oceanic fisheries on mixtures of

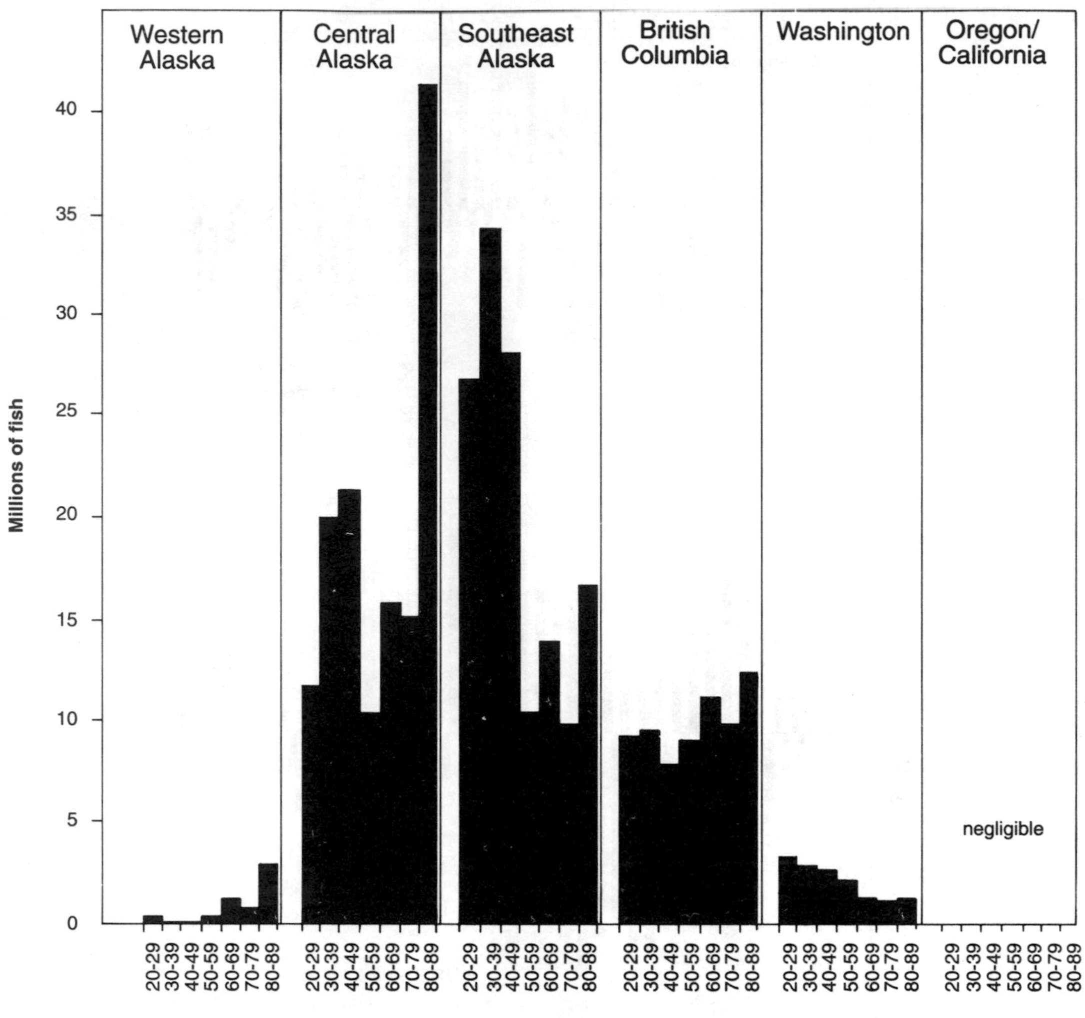

FIGURE 3

Average annual commercial catch of North American pink salmon by region and decade, 1920–29 to 1980–89. Western Alaska data include estimates of Japanese high-seas catches of pinks from that region from 1956 to 1977, ranging from <500 to 253,000 fish per year. (Adapted from Fredin 1980)

stocks originating from different regions of the Far East. Much of the Japanese coastal catch prior to the 1940s occurred along various USSR coastal regions. Since the 1950s coastal catches of pink salmon by Japanese fisheries, including those taken in the Sea of Japan, have remained relatively stable. However, the bulk of the harvest by Japan since the 1950s has been on the high seas. A summary of the months of operation of each of these Japanese fisheries through the mid-1970s, and some assumptions on the continental origin of fishes caught in them, are reviewed by Fredin et al. (1977) and Fredin (1980).

The commercial harvest of pink salmon in Asia reached peak levels in the 1930s and early 1940s with catches of over 200 million fish and then declined sharply (Figure 4). This was followed by a strong resurgence of catches in the 1950s which then stabilized throughout most of the 1960s and 1970s in a cyclic pattern. During this period, even-year catches averaged around 50 million fish and odd-year catches almost twice that level (Figure 2). In the late 1970s and early 1980s, there was an overall upswing in catches to an average of about

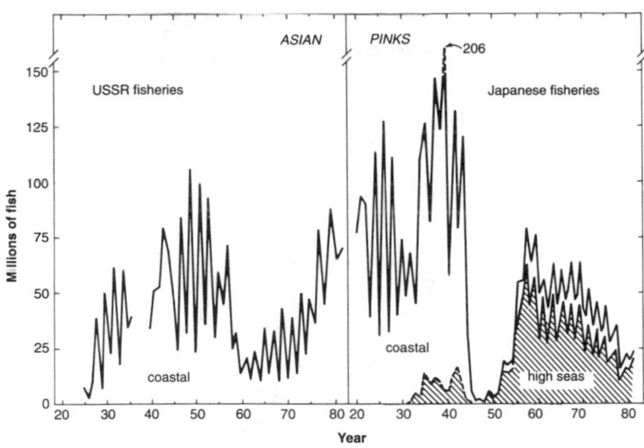

FIGURE 4
Commercial catch of Asian pink salmon by USSR and Japanese fisheries, 1920–81 (Japanese coastal catches include small numbers of masu salmon.) (Adapted from Fredin 1980)

100 million fish, perhaps as a result of the same factors that caused improvements in pink salmon harvests during this period in North America.

About 90% of the coastal harvest of pink salmon by the USSR comes from Kamchatka (including both east and west coasts), Sakhalin, and the Kuril Islands (Figure 5). Since the mid-1940s the USSR catch has shown a dominant odd-year pattern in catches from the Kamchatka region. During the smaller even-year cycles, a large percentage of the harvest comes from the Sakhalin and Kuril islands regions (Figure 5), where a shift from odd- to even-year dominance in catch occurred in 1976–77.

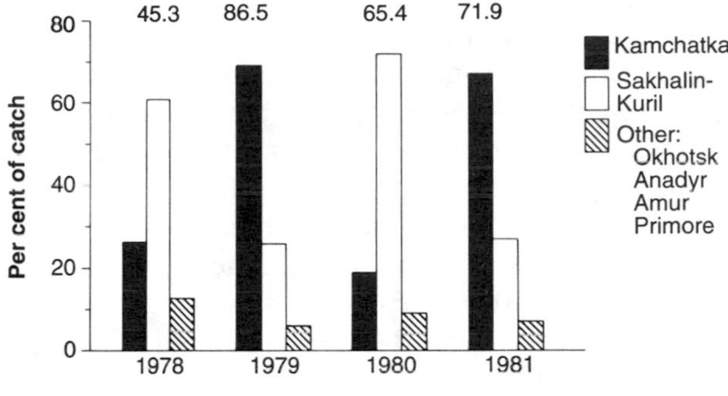

FIGURE 5
Percentages of USSR pink salmon catch from Kamchatka, Sakhalin-Kuril Islands, and all other regions, 1978–81. Total catch in millions of fish is shown above bars for each year. Data for 1978–79 are from INPFC (1979), for 1980–81 from unofficial USSR catch statistics

Transplants

Pink salmon have been widely transplanted both within and outside the endemic North Pacific Ocean area. Transplants within the natural range were frequently made to establish or strengthen off-year runs where the opposite line already existed. These transplants were based on the assumption that the recipient stream could sustain runs in the opposite-year line. Although many transplants were well planned and carefully conducted, the long-term success rate has been almost nil (Neave 1965; Withler 1982).

Transplants outside the natural range of pink salmon were made with the hope of establishing self-sustaining runs in other parts of the country or the world. The most successful transplant to date, that into the Great Lakes, was accidental. Most other transplants of pink salmon outside their natural range have been unsuccessful in establishing permanent runs, although, in some cases, adult returns occurred for a few years, often with hatchery-assisted reproduction.

Transplants within Natural Geographic Range

Significant runs of pink salmon in Puget Sound and southern British Columbia occur only in odd-numbered years. As a result, many attempts were made to establish even-year runs in these regions, particularly in Puget Sound. The development of hatchery techniques for artificially propagating Pacific salmon, especially the successful handling and moving of eyed eggs, made widespread transplantation possible (Withler 1982).

After development of early-day hatcheries in

Alaska (Roppel 1982), pink salmon eggs were shipped from Alaska to Washington in even years between 1914 and 1932 (excluding 1920). A total of about 85 million fry or fingerlings were released into numerous Puget Sound streams (Neave 1965). The results were all poor, with no reported adult returns.

Even-year transplants to Puget Sound resumed during the period 1948 through 1956 when approximately 1.6 million fingerlings, derived from eggs primarily from the Lakelse River (a tributary of the Skeena River in central British Columbia), were released from saltwater rearing facilities adjacent to Puget Sound streams (Ellis and Noble 1959). About 100–500 adult fish returned during each cycle year (Neave 1965). The program was discontinued after 1958-brood fingerlings released at Finch Creek in Hood Canal produced only a few returns in 1960.

In 1953 an odd-year transplant of pink salmon from the nearby Dungeness River stock was made to Finch Creek. Prior to this time, Finch Creek did not have a run in the "on" odd-year line like other streams in the vicinity. This transplant produced 1,958 returning adults in 1955 (Neave 1965) and formed the basis of an odd-year run to Finch Creek and the Hoodsport Hatchery that has continued until the present.

In British Columbia, about 36 different transplants of pink salmon eggs or fry between streams were attempted (Aro 1979). The largest transplant of eggs was to a newly developed spawning channel at Robertson Creek on the west coast of Vancouver Island. Transplants occurred in both odd and even years over a six-year period, 1959–64, with eggs originating from four central and southern British Colubmia streams, producing a total of 30 million fry. Initially, this project showed promise (MacKinnon 1960; Boyd 1964); however, it was abandoned when returns sufficient to replace eggs occurred in only one year (Withler 1982).

A transplant in the Queen Charlotte Islands, with eggs from Tlell River to nearby McClinton Creek in the Masset Inlet area of Graham Island produced 878,000 1931-brood fry and 506,000 1935-brood fry (Pritchard 1938). Only a few adults returned to the recipient stream, and some marked adults from the experiment were recovered in coastal fisheries.

Transplants of eggs to a spawning channel in the Big Qualicum River from the Cheakamus River in 1963, and from the Bear River in 1964, resulted in about 1.5 million and 3.0 million fry, respectively (Walker and Lister 1971). The 1965 adult returns were poor. In 1966, however, 11,940 pink salmon came back to the Big Qualicum River and were allowed to spawn naturally in the system. This run declined to 300 adults during the following cycle and then essentially disappeared (Withler 1982).

In 1954 and 1956, eggs from the Lakelse River were transplanted to a new spawning channel at Jones Creek in the lower Fraser River drainage (Neave and Wickett 1955; Wickett 1958a; Neave 1965). Although 1.1 and 0.3 million fry were produced in Jones Creek from the two even-year broods, and 2,800 adults returned from the 1954 brood to spawn naturally, this run quickly disappeared (Withler 1982). By contrast, the odd-year line in Jones Creek has maintained a fairly stable run since development of the spawning channel, usually averaging between two and five thousand adults annually (IPSFC 1982).

Another project to establish an off-year run of pink salmon was conducted at Bear River on Vancouver Island in 1975 (Withler 1982). The project grew in part from an earlier study at Tsolum River, also on Vancouver Island, which suggested that a paternal genetic component was involved in the homing ability of transplanted pink salmon (Bams 1976). Almost complete line dominance exists in Bear River and no off-year males were available in odd-numbered years. Cryogenically-stored sperm from on-year males (1974-brood) and sperm from artificially matured one-year-old males from the Bear River were used to fertilize transplanted eggs from the Glendale River on the mainland. In addition, a control group of "pure" Glendale River stock eggs fertilized by Glendale River males was hatched at the Bear River Hatchery to measure the influence of Bear River genes in the transplant study. Approximately 267,000 marked fry fertilized with Bear River milt, 128,000 marked "pure" Glendale River fry, and 1.2 million unmarked Glendale River fry (to serve as a predator buffer) were released in the Bear River. Although some marked adults from the study were recovered in commercial fisheries in 1977, none was recovered in the Bear River, in Glendale River, or in any adjacent streams (Withler and Morley 1982).

Several transplants of pink salmon were made in

Alaska, although many of them are not well documented. Three notable ones are addressed here. Sashin Creek on Baranof Island (southeastern Alaska) had abundant runs in both lines, but when the even-year line declined in the 1940s and early 1950s, all spawners in four cycle years were removed to study straying and possible repopulation of a barren stream (Merrell 1962; Harry and Olson 1963). In 1964, 1,906 adults seined from an estuary at Bear Harbor on Kuiu Island, 50 km distant, were placed in Sashin Creek upstream from a counting weir (McNeil et al. 1969). An additional 287 adult pink salmon of unknown origin migrated into the stream and spawned with the transplanted fish, and an estimated 310,000 fry were produced from natural spawning in Sashin Creek. In 1966, 5,761 adults returned to the stream, presumably progeny from the transplanted adults (Ellis 1969). Sashin Creek has since maintained reasonably strong runs in both odd-year and even-year lines, reaching escapements in excess of 50,000 adults in both 1979 and 1980 (Vallion et al. 1981).

Attempts were also made in Alaska to outplant pink salmon fry from hatcheries to streams where fishways were installed over barrier falls. The intent was to produce adult returns to newly accessible spawning grounds (McDaniel 1981; McDaniel et al. 1984). In 1975 and 1976, fry originating from Seal Bay Creek stock (Afognak Island), that spawn naturally downstream from a barrier falls, were transplanted above the falls in the same creek. Eggs were incubated at the Kitoi Bay Hatchery, approximately 60 km distant by coastline. Although 6,538 and 163 pink salmon returned to Seal Bay Creek in 1977 and 1978, respectively, most spawned in areas downstream from the point of release, rendering this transplant effort unsuccessful. Less than 1% of the marked returning adults were recovered at the Kitoi Bay Hatchery.

Transplants to Hobo Creek in Prince William Sound included 1.7 million 1979-brood fry from Jonah Creek stock incubated at the Port San Juan Hatchery and 7.0 million 1980-brood fry from the Port San Juan stock incubated at the Cannery Creek Hatchery (McDaniel et al. 1984). Hobo Creek and Jonah Creek are in northwestern Prince William Sound, about 40 km apart; the Port San Juan Hatchery is in southwestern Prince William Sound, approximately 100 km from Hobo Creek; and the Cannery Creek Hatchery is in the Port Wells fiord near Jonah Creek. Fry were transported by surface vessel to Hobo Creek from the incubation sites in recirculated fresh water, and 1.8% of the 1979-brood fry and 0.4% of the 1980-brood fry were marked before release. The estimated 1981 return of the adults from fry plants to Hobo Creek was 56,700 pink salmon, of which 7,000 (12.4%) returned to the transplant site at Hobo Creek and 49,700 (87.6%) returned to the incubation site at the Port San Juan Hatchery. The estimated 1982 return of adults from the fry plant was 23,868 pink salmon, of which 4,214 (17.7%) returned to Hobo Creek, 2,111 (8.8%) to the incubation site at the Cannery Creek Hatchery, and 17,543 (73.5%) were caught in fisheries near the incubation site. McDaniel et al. (1984) considered fry transplants to Hobo Creek successful in establishing a self-sustaining run, and 11,850 adults returned to the stream in 1983, mostly from progeny of the 1979-brood transplant.

Transplants of pink salmon within their natural range in the Far East either have not occurred often or have not been well documented. Sano and Kobayashi (1953) reported on a transplant of one million odd-year pink salmon eggs from the Shibetsu River to the Yurappu River in Hokkaido in 1951. A return rate of 0.5% was indicated in 1953. Most fish were caught along the coast near the Yurappu River, four marked adults were recovered along the sea of Okhotsk, one in a nearby river, and 22 were recorded in the Yurappu River.

In general, transplants of pink salmon to barren waters within their natural geographic range have failed. Only a few successes in Washington (Puget Sound) and southeastern and south-central Alaska have occurred. Reasons for the failure of other transplants are unknown but are likely related to the genetic characteristics of the odd-year or even-year stocks used.

Transplants outside Natural Geographic Range

The initial attempts to introduce pink salmon into the western North Atlantic Ocean were made during the period 1906–17 in many streams in Maine (Figure 6). Details are reviewed by Bigelow and Schroeder (1953), Neave (1965), and Ricker (1972). The first two odd-year cycle releases of about one million fry in several rivers (1907 and 1909) pro-

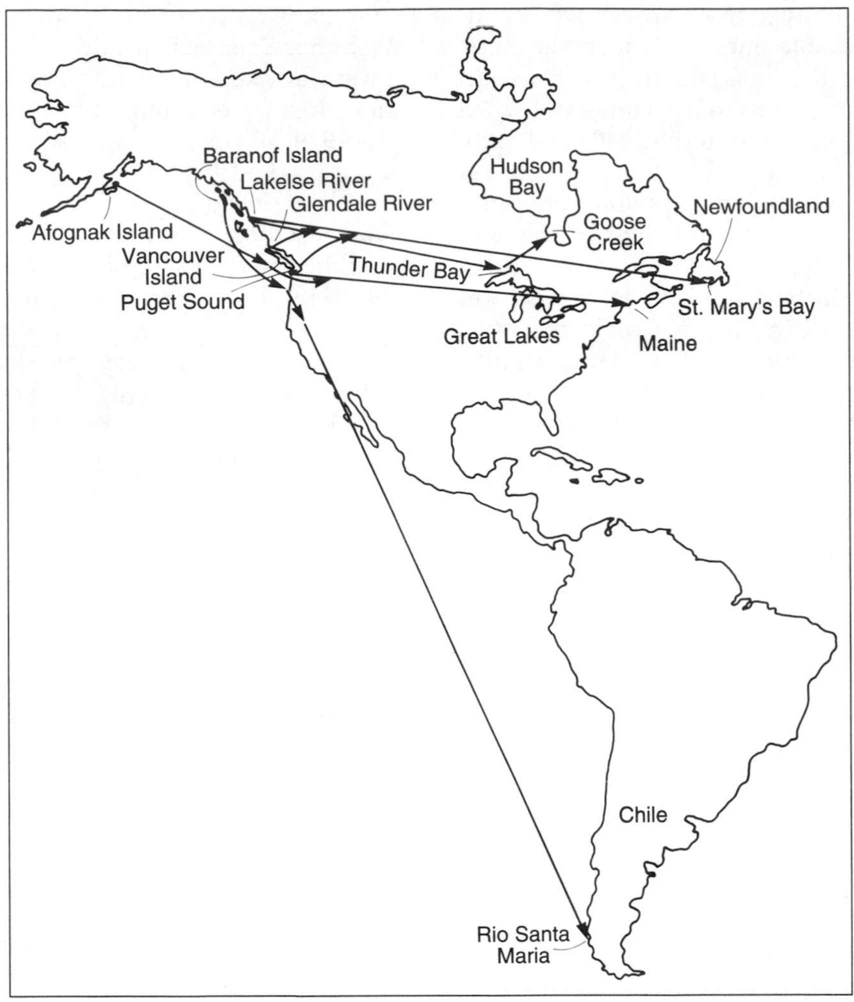

FIGURE 6

Some transplants of eastern Pacific pink salmon to areas outside their natural range

duced no adult returns. A second sequence of odd-year releases in 1913, 1915, and 1917 totalled about 13.5 million fry or fingerlings and resulted in substantial adult returns into several streams including the Dennys, Pembroke, and Penobscot rivers. Two even-year transplants in 1914 and 1916 produced 12.5 million fry; apparently no adults returned from this effort. Odd-year transplants to Maine originated from the Skagit River in Puget Sound, and even-year transplants originated from Afognak Island, Alaska (Figure 6). No Pacific transplants were made after 1917, although an odd-year run continued in Maine for at least four cycles, and some 2.6 million eggs were collected during this time from the Dennys and Pembroke rivers (Neave 1965). Undoubtedly, many odd-year pink salmon spawned naturally in Maine streams, and it was assumed that the species had established there. However, few pink salmon were seen in Maine after about 1927 (Bigelow and Schroeder 1953) and none occur there today. By comparison with other attempts to introduce this species into new areas, transplants of odd-year pink salmon in Maine did show a measure of success.

Another attempt to establish pink salmon runs on the Atlantic coast of North America involved three even-year and two odd-year transplants from British Columbia to St. Mary's Bay, Newfoundland from 1959 through 1966 (Lear 1975, 1980) (Figure 6). Controlled-flow spawning channels were used for planting eyed eggs in the North Harbour River. Even-year transplants from the Glendale and Lakelse rivers, of 2.5–5.9 million eggs per year, produced 10 million fry (Lear 1980). Odd-year transplants of 250,000 eggs from the Indian River in 1959 and 3.3 million eggs from the Lakelse River in 1965 resulted in 100,000 and 3.0 million fry, respectively. In 1967, 8,500 adults returned (Lear 1975) and spawned naturally, depositing 4.4 million eggs that produced 3.8 million fry. Since 1969, pink salmon in Newfoundland have been progeny from natural spawnings. However, runs have steadily declined (Lear and Day 1977), and it seems unlikely that any are still present today.

The Ontario Ministry of Natural Resources initiated a project to introduce both pink and chum salmon into Hudson Bay (Ricker and Loftus 1968) (Figure 6). In 1956, just over 0.5 million eyed eggs, or sac fry, and 224,000 fingerling pink salmon from the 1955-brood of the Lakelse River run were planted into Goose Creek, a tributary of Hudson Bay (Nunan 1967). Although downstream migrant fingerlings were collected in Goose Creek, no adult pink salmon were reported from Hudson Bay (Ricker and Loftus 1968).

A totally unexpected and most improbable result of the Hudson Bay project was the accidental establishment of a self-perpetuating odd-year pink salmon run in the Great Lakes (Schumacher and Eddy 1960; Schumacher and Hale 1962; Nunan 1967; Collins 1975; Emery 1981; Kwain and Lawrie 1981). Eyed eggs from the 1955-brood Lakelse River run were hatched at the Port Arthur Hatchery, Thunder Bay, Lake Superior, Ontario (Figure 7). In spring 1956, a small number of the resulting fry (100–350) escaped into Thunder Bay as a floatplane was being loaded for a flight to Hudson Bay (Figure 6). Later, about 21,000 fingerlings were discarded into the Port Arthur Hatchery drain that flows into the Current River and Lake Superior (Schumacher and Eddy 1960; Nunan 1967; Kwain and Lawrie 1981). In 1959, two mature male pink salmon were collected from two different Minnesota streams flowing into Lake Superior (Figure 7). These fish apparently were progeny from the undetected spawning in 1957 of survivors from the 1955-brood transplant releases (Schumacher and Eddy 1960). Since then, odd-year spawning pink salmon have dispersed throughout the Great Lakes, including a dramatic 1979 expansion into lower Lake Huron, Lake Erie, and Lake Ontario (Kwain and Lawrie 1981; Emery 1981) (Figure 7). A run of even-year spawners also began appearing in Lake Superior in the mid-1970s (Kwain 1978). It was initially postulated, and subsequently documented, that they were the progeny of three-year-old odd-year spawners (Kwain and Chappel 1978; Wagner and Stauffer 1980). Pink salmon are, at present, firmly established as part of the exotic Great Lakes fauna and are the first known populations of this species to complete and maintain their life cycle entirely in fresh water.

Many aspects of Great Lakes pink salmon biology, including spawning behaviour, egg and fry size, and emergence and downstream emigration behaviour (Kwain 1982), are similar to patterns found in the Pacific drainages. Differences such as lower fecundity, smaller size, variable ages at maturity, and different body shape seem related to less favourable growth than in native marine waters (Berg 1979). Berg (1979) also suggested that the small parental number and subsequent isolation, allowing for rapid genetic drift, may have permitted these unique features to develop. Charrett and Thomason (1987) found substantial biochemical genetic differences in Great Lakes pink salmon and the anadromous population in Lakelse River, British Columbia, from which they were derived.

The largest pink salmon transplant program on the western side of the Pacific Ocean occurred during the period 1956–78, when over 200 million eggs were shipped, primarily from Sakhalin Island, to the Kola Peninsula (Murmansk region) of the northeastern USSR (Bakshtansky 1980) (Figure 8). This project was undertaken to introduce both pink and chum salmon to the Barents and White seas areas of the Arctic Ocean (Kassov et al. 1960; Isaev 1961). The number of eggs transplanted annually averaged about 9.5 and 15.5 million in even- and odd-numbered years, respectively (Dyagilev and Markevich 1979).

Returns of adult pink salmon were highly variable, ranging from no fish in some years to one estimate of 0.5 million in 1973 (Dyagilev and Markevich 1979). Returns approached or exceeded 50,000 adults in 1960, 1965, 1971, 1973, 1975, and 1977 (Bakshtansky 1980). After comparing seasonal spawning and temperature patterns with early pink salmon embryology, Dyagilev and Markevich

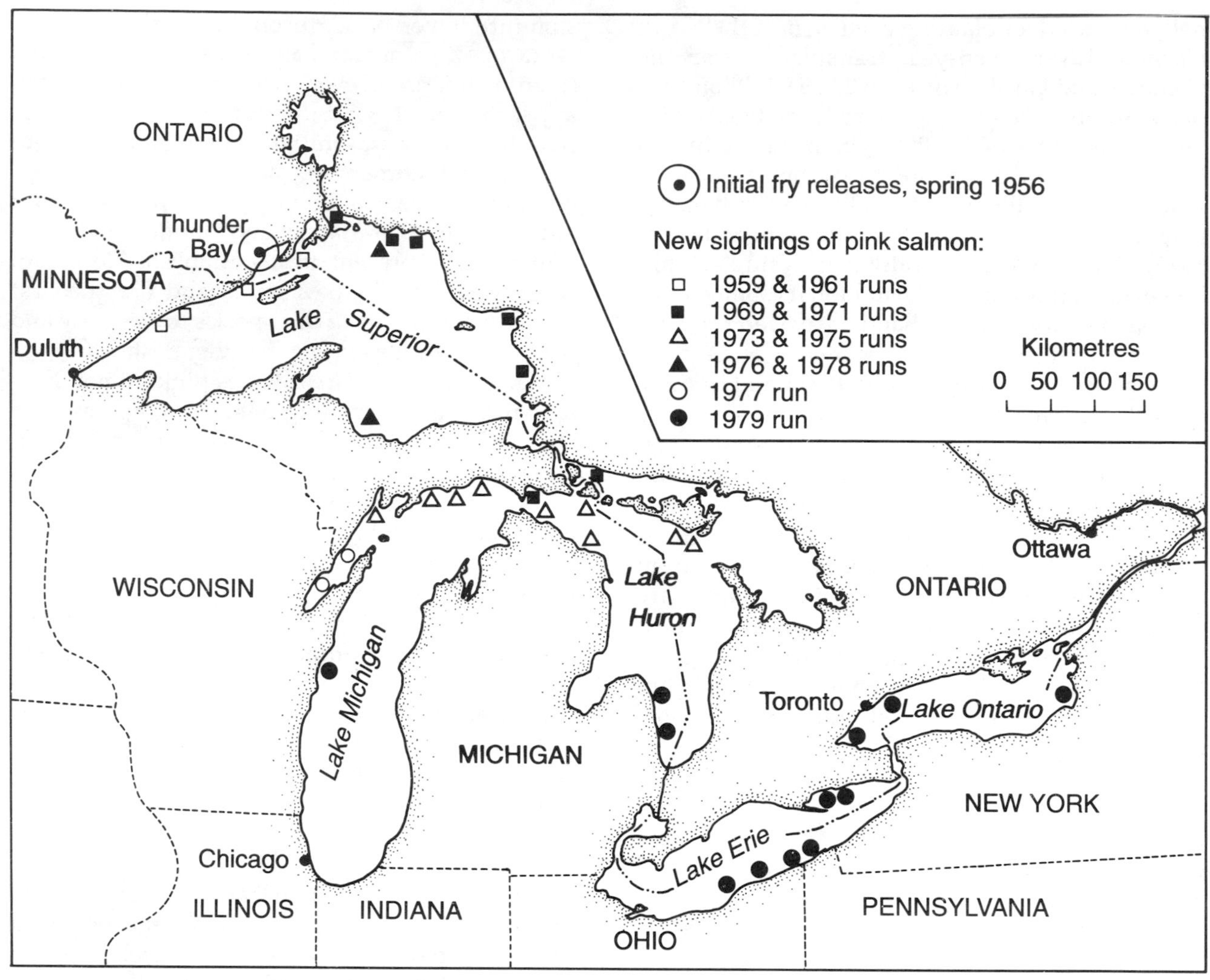

FIGURE 7

Dispersal of pink salmon in the Great Lakes of North America, 1956-79. (From Kwain and Lawrie 1981)

(1979) concluded that the even-year line consistently matured two to three weeks later than the odd-year line and that average Kola Peninsula temperatures caused constant high mortality in even-year eggs.

Adult returns from Kola Peninsula transplants were characterized by extreme straying, often over considerable distances from the streams where releases of fry or fingerlings were made. Isaev (1961) reported that large numbers of adult pink salmon entered many streams along the Barents and White sea coasts (Figure 8). More pink salmon strayed (at least in some years) into streams without fry releases than those that homed to their release site or stream. About 12,800 pink salmon were caught along the Norwegian coast in 1977; higher catches likely occurred in 1973-75, and many entered and spawned in streams in Finnmark, northern Norway, to the west and south of the Kola Peninsula (Berg 1961, 1977; Bjerknes and Vaag 1980a). Other strays were recovered further south and west along the coast of Scotland (Pyefinch 1962; Williamson 1974) and in waters of Iceland (Bjerknes and Vaag 1980b). Eastward, straying occurred to the Yenisey River and Kara Sea, where 1,119 adult pink salmon were caught in 1975 (Krupitskiy and Ustyugov 1977).

In general, transplants of pink salmon into

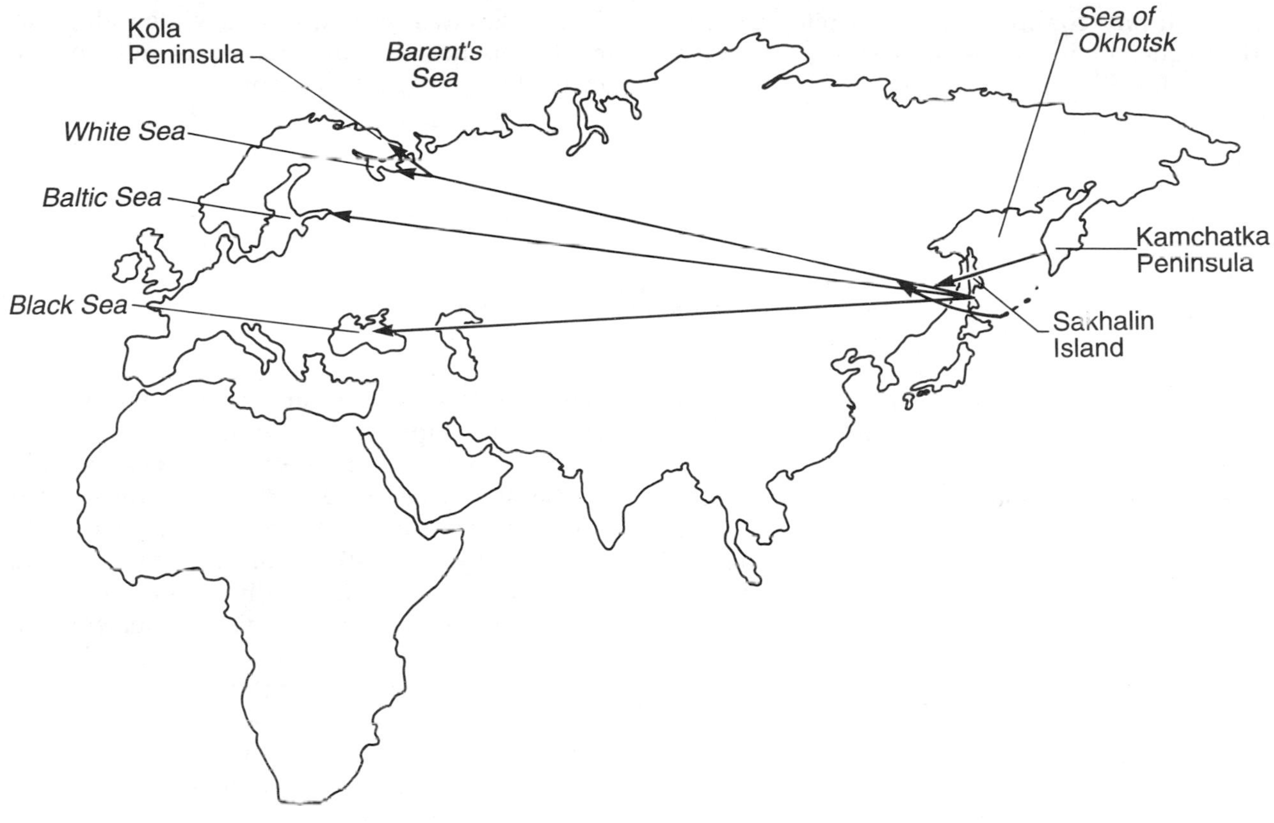

FIGURE 8

Some transplants of western Pacific pink salmon to areas outside their natural range

waters of northern Europe have shown a measure of success, especially in the odd-year line, but the final outcome remains to be determined. Dyagilev and Markevich (1979) pointed out that continued transplanting of unadapted odd-year pink salmon from the Far East, constantly mixed with "local" fish, only hinders selection and subsequent development of a well-adapted stock. Further, they indicated that an even-year run can only develop when eggs are imported from earlier spawning stocks and they suggested that Norway, with its milder climate, might be better suited for this line. Apparently no transplants from the Far East have been made to the Kola Peninsula since 1978.

Pink salmon were also transplanted from Sakhalin Island to the Black Sea (Neave 1965) and Baltic Sea (Solovjova 1976; Rimsh 1977) (Figure 8). In the Baltic Sea, fingerlings were released in several Latvian streams of the Bay of Riga from the 1972, 1973, and 1974 brood years; and releases totalled 1.8, 2.6, and 2.5 million for the three years, respectively (Solovjova 1976). Few data are available on the numbers of adults returning to streams, although many were reported in coastal Baltic waters (Rimsh 1977) and some along the Swedish coast (Berg 1977). No further information appears to be available in the English literature on the Black Sea transplant.

Lindberg and Brown (1982) reported on a pink salmon transplant to southern Chile in the Rio Santa Maria, Magallanes Province. About 0.6 million eyed eggs from Baranof Island were shipped to Chile in November 1981 (Figure 6). Low flows in the Rio Santa Maria caused heavy losses during the final incubation period and only 70,000 fingerlings survived. After release, schools of fingerlings were observed where the Rio Santa Maria enters the Strait of Magellan, but no returns have been recorded.

The scope of pink salmon transplants throughout the world has been extensive. However, most attempts were unsuccessful. It is unclear at present if any continuing runs of pink salmon, either from hatchery or natural propagation, have resulted from the Soviet transplants to the Kola Peninsula. The successful introduction into the Great Lakes was accidental and, while well documented, is not well understood in light of the many other transplant attempts.

Spawning Migration

Migrations of pink salmon into fresh water prior to spawning can vary in different stocks and regions according to seasonal timing, rates of migration, stage of maturity, distance of upstream movement, and response to environmental factors, including stream discharge, temperature, wind, and tide.

Seasonal Timing

As the spawning season approaches, schools of pink salmon begin to appear in bays and estuaries of small streams and nearshore waters influenced by large rivers. This inshore invasion of maturing fish throughout the North Pacific occurs mostly in the period June to September (Neave et al. 1967).

According to Takagi et al. (1981), coastal migrations of Asian pink salmon can be divided into three regional groups based on timing of fisheries. Migrations in the Sea of Japan group, including west Sakhalin, Primore, and Amur areas, occur in June and July. In the eastern Kamchatka group (mainly along the Bering Sea coast), coastal migrations are in July, and in the Sea of Okhotsk group (including the northern Sea of Okhotsk coast, western Kamchatka, eastern Sakhalin Island, southern Kuril Islands, and Hokkaido), migrations are mostly in July and August.

In North America no significant coastal migration of pink salmon occurs in June, although inshore runs may have started this early in parts of Alaska in the early 1900s (Royce 1962; Takagi et al. 1981). Inshore runs in western Alaska peak in late July, and the timing of spawning, which takes place in July and August, tends to be earlier in northern (Norton Sound) than in southern (Bristol Bay) areas (Atkinson et al. 1967). In central Alaska, inshore migration and commercial catches occur from mid-July until early August (Figure 9) with considerable variation between and within districts. Stream entry and spawning happens mostly in August and early September. Inshore runs in southeastern Alaska are primarily in July and August, and timing differs somewhat between northern and southern districts and between odd and even years (Figure 9). Generally, runs occur earlier in both years in the northern district. Spawning in southeastern Alaska is concentrated in late August and September. In addition to a north-south relationship, Royce (1962) and Sheridan (1962a) found that stream entry and spawning in southeastern Alaska varied with stream temperature within the region; early runs occurred in mainland streams with lower temperatures, late runs were mostly found on the outer islands with higher temperatures, and middle runs were present in streams mainly in the inner islands with intermediate temperature regimes (Figure 10).

In British Columbia, inshore migrations peak in late July, August, and early September. Spawning occurs mostly in September and October and tends to be progressively later from north to south. In Washington, the inshore migration (in odd years only) peaks in late August and early September, and spawning occurs mostly in late September and October.

Several authors have noted that in years when the population size of adult pink salmon in a region is large, the timing of inshore migration occurs later in the season than when the population size is average or small (Anonymous 1938; Davidson and Vaughan 1939, 1941; Eniutina 1972; Yefanov and Chupakhin 1982). This relationship is presumably determined by growth and onset of maturation in ocean nursery areas: in years of high abundance growth continues longer and maturity occurs later.

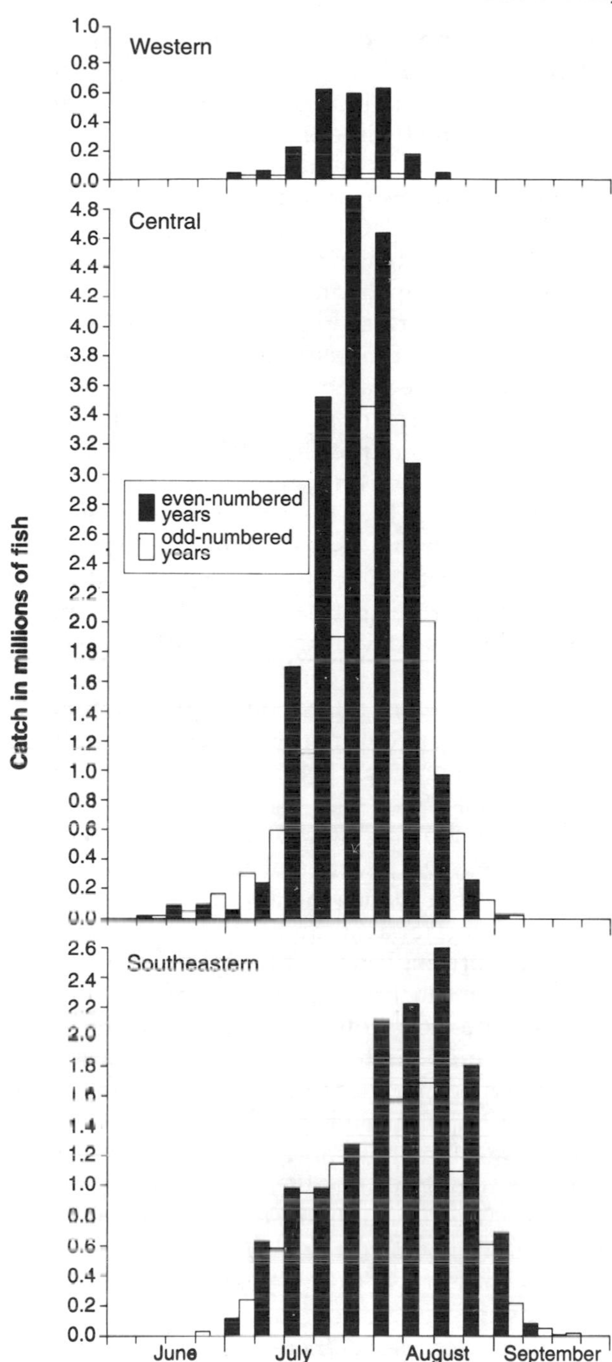

FIGURE 9
Average weekly catches of pink salmon in millions of fish, 1960-80, in western, central, and southeastern Alaska regions. (Data from INPFC Statistical Yearbooks, 1961-83)

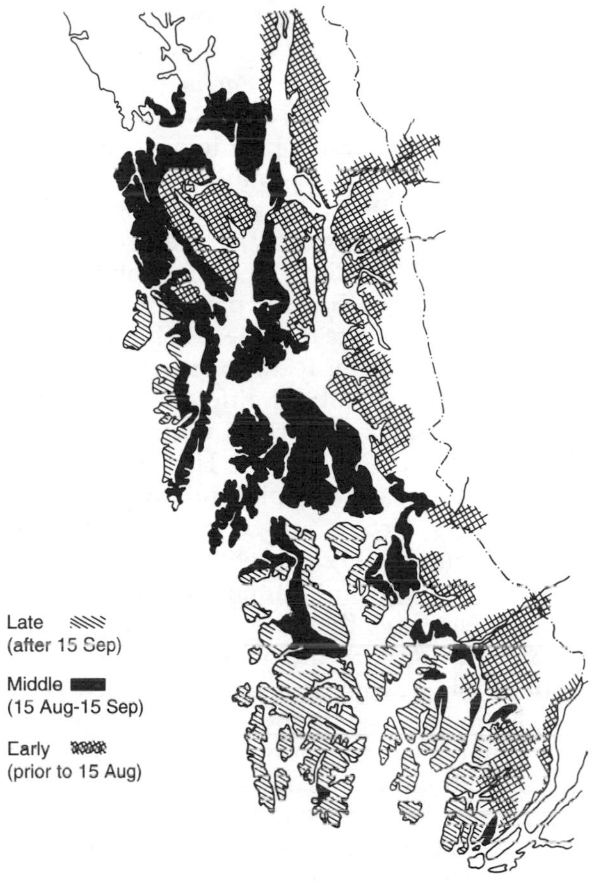

FIGURE 10
Southeastern Alaska showing areas where early, middle, and late run pink salmon spawning streams are located. (From Sheridan 1962a and Royce 1962)

In streams where timing of upstream migration extends over a period of several weeks, patterns of stream entry can often be divided into early and late segments. In the Fraser and Skeena rivers, a distinction is made between early and late segments (Neave 1966). In the Fraser River, mainstream and upriver spawners enter the river about two weeks earlier than lower tributary spawners (Ward 1959). Often the two segments use different spawning areas, and the early-run segment almost invariably spawns further upstream than the late-run segment (Ward 1959; Ivankov 1968a). Striking examples of this behaviour are found in even-year Prince William Sound streams, where early and late runs may spawn in closely adjacent but separate upstream and downstream areas. The later runs usually spawn in the intertidal portion of the stream (Helle et al. 1964; Helle 1970; Roys 1971).

Migration Behaviour

When pink salmon first appear in bays, estuaries, or near streams, the assemblage may contain schools or individuals still actively migrating to other more distant districts. Appearance of individuals or small schools in an estuary off a stream mouth, especially in the early part of the run, does not necessarily indicate that these fish will migrate into the nearby stream. Testing, proving, and milling about in the vicinity of different streams has become a recognized part of normal salmonid homing behaviour (Ricker 1972; Hasler and Scholz 1983). These first arriving pink salmon have been referred to as "scouts" (Dvinin 1952) or "messengers" (Eniutina 1972).

When returning pink salmon appear near the mouths of streams, they are usually seen swimming near the surface and often exhibit a characteristic leaping behaviour. Individuals leave the water, after a rapid swimming burst, in a forward leaping motion with their body initially oriented dorsoventrally, and then quickly rotated laterally, so the fish "falls" on its side or back (Berg 1948). A rapid series of jumps by the same fish often takes place. Reasons for this behaviour are unknown.

The rate that pink salmon migrate into streams can vary greatly. Steady numbers may move into a stream for several days, or the daily numbers of migrants may fluctuate widely. Often, in smaller streams, daily fluctuations in migrants are related to fluctuations in stream discharge (Hunter 1959). Large numbers of pink salmon may suddenly migrate into a stream at once. Berg (1948) cited an observation of a mass migration made by I.F. Pravdin in 1926: "Although the weather was calm and sunny, an extraordinary noise could be heard coming from the middle of the river . . . similar to the noise of boiling water splashing in a gigantic cauldron. The population of the fishing camps rushed to the river bank. Standing there, the fishermen feasted their eyes upon a tremendous school of fish, which went up the river, making a very loud noise, as if a new river had burst into the Bolshaya; the fish jumped out of the water continuously. The noisy stretch of fish was at least one verst long (1,067 m) and not less than 100 m wide, so that the size of the school could be estimated at several million specimens."

At Sashin Creek, pink salmon upon arrival do not migrate directly into the stream, but mill about in the bay for up to a month until sexually mature (Davidson et al. 1943). At Snake Creek (southeastern Alaska) and McClinton Creek (Queen Charlotte Islands, BC), however, under suitable streamflow conditions, upstream migration occurs within a few days of arrival, regardless of the state of sexual maturity (Davidson et al. 1943). At Snake Creek, lakes in the watershed provide stream flows suitable for upstream migration throughout the season. McClinton Creek has little watershed storage and pink salmon only migrate if stream flow is high; if the flow is low, they mill in the bay until it rains and the stream rises (Pritchard 1936a). After reaching maturity, pink salmon at all three streams attempt to migrate upstream regardless of streamflow conditions (Davidson et al. 1943). However, under extreme freshet conditions, with excessive currents upstream, migration of pink salmon is reduced or stopped until flows subside (Pritchard 1936a; Davidson et al. 1943; Dvinin 1952; Hunter 1959; Eniutina 1972).

Wind and tide may affect water levels and conditions in the lower reaches of rivers and adjacent estuaries and influence the upstream migration of pink salmon (Hanavan and Skud 1954; Eniutina 1972). Inshore movement of migrating adults in southeastern Alaska can speed up during increasing tides and slow down during decreasing tides (Davidson and Vaughan 1941), and wind direction may influence migration direction (Davidson and Christey 1938; Kaganovsky 1949).

Several authors have observed that fluctuations in stream temperatures during the migration period have little or no influence on upstream movement of pink salmon (Pritchard 1936a; Davidson et al. 1943; Hunter 1959). However, the intensity of the spawning migration may vary appreciably with sharp decreases or increases in water temperature. Sheridan (1962a) presented evidence to demonstrate that different seasonal temperature patterns in southeastern Alaska streams play an important role in the timing of upstream migration, spawning, subsequent downstream migration, and marine survival of fry from individual streams. He further noted that pink salmon in this region tend to spawn earlier or later in the season in specific streams, as water temperatures rise or fall to 10°C.

Pink salmon migrate into spawning streams primarily during daylight hours (Eniutina 1972; Semko 1954). Eniutina (1972) concluded that movement of pink salmon in spawning streams is determined by the physiological state (particularly the degree of maturity) and by a complex adaptive reaction to light, variations in water level, and, in some cases, temperature.

Distance of Upstream Migration

Pink salmon tend to spawn closer to the sea than other Pacific salmon, although many important spawning areas are located far inland. They are not particularly adept at leaping waterfalls and negotiating cascades or short high-velocity barriers. Often the upstream limit for spawning of pink salmon in coastal streams is a waterfall or rapids that other Pacific salmon can surmount.

In the Far East, spawning grounds in the Amur River are located upstream as far as 700 km in the Khungan River; the main Amur spawning area is in the Amgun River, a tributary about 200 km upstream from the mouth (Ishida 1966). In the Bolshaya River, pink salmon spawn 160 km upstream from the Sea of Okhotsk (Semko 1954); and in the Poronai River, the largest on Sakhalin Island, spawning apparently occurs, based on hatchery location, up to 175 km upstream. In North America, the upriver spawning grounds of pink salmon in the Fraser River at Seton Creek and Thompson River are about 250 km above the river mouth (Ward 1959), whereas in the Skeena River some spawners go above the Babine River counting fence, 480 km upstream (Godfrey et al. 1954). On the Snake River in southeastern Washington, Basham and Gilbreath (1978) reported spent female pink salmon at Little Goose Dam, 694 km from the Pacific Ocean, although no significant spawning runs of this species occur in the Columbia River drainage. In Alaska, the principal pink salmon spawning area of the Nushagak River system is the Nuyakuk River (Pennoyer 1970), about 200 km upstream from Bristol Bay. In the Kenai River of Cook Inlet, pink salmon spawn as far upstream as the Russian River (Atkinson et al. 1967), about 120 km from the stream mouth.

The extent and distance of upstream migration in some streams is influenced by size of the spawning population. Pink salmon in the Bolshaya River (Semko 1954), south Sakhalin streams (Dvinin 1952), Amur District streams (Eniutina 1972), Fraser River (Vernon 1962), and Sashin Creek (Merrell 1962) migrate more extensively and, in particular, go further upstream in years with large runs than in years with small runs.

Impediments to Migration

After pink salmon migrate past major fisheries and enter fresh water there may be important reductions in their numbers before they successfully spawn. Rock slides, poor water quality, drought, predators, parasites, and diseases can all take a toll.

In the Skeena River system of British Columbia, a large rock slide occurred in the Babine River in 1951, about 360 km upstream from the coast and 70 km below Babine Lake. The slide created a serious barrier to upstream migrant adult salmon in 1951 and 1952. In 1952, only 9 tagged pink salmon, 0.5% of the fish tagged below the slide, were recovered at the Babine Lake fence (Godfrey et al. 1954). After corrective action was taken at the slide during the winter of 1952–53, 23% of the pink salmon tagged in 1953 (13 of 56) were recovered at the fence (Godfrey et al. 1956). Although pink salmon suffered heavy losses at this slide in these two years, the effect on production was not as great as that in other species, especially sockeye salmon, because a relatively small portion of pink salmon in the Skeena system spawn in areas upstream from that affected by the slide.

Much more serious was the disastrous rock slide at Hell's Gate on the Fraser River in 1913. This blocked the migration of many millions of pink salmon that spawn in areas upriver from the slide, and for many years no pink salmon could reach the upper reaches of the river (IPSFC 1958; Withler 1982). Only after completion of fishways at Hell's Gate in 1945 was access to and re-invasion of upriver spawning areas possible (Vernon 1962). Beginning in 1947 and 1949, the first cycle years after completion of the fishways, one to two thousand pink salmon were recorded in upriver spawning areas (Withler 1982). Escapements above Hell's

Gate have increased steadily and reached levels of 1.4, 1.6, and 1.8 million in 1977, 1979, and 1981, respectively (IPSFC 1982). These upriver fish enter the Fraser River at the same time as downstream early-run fish that spawn in the lower main river, and there is little doubt that re-invasion of upper Fraser streams developed initially from straying of downstream early-run fish into upstream areas.

Withler (1982) described periodic significant increases in pink salmon production from the Kakweiken River in British Columbia, based on the operative or inoperative status of a fishway bypassing a waterfall. When operating properly, the fishway provided access to additional upstream spawning areas. Fishways have also been used to provide access for pink salmon over impassable falls and partial barriers and have allowed them to establish new upstream spawning areas in southeastern Alaska (Sullivan 1980), on Afognak Island (McDaniel 1981), and in Prince William Sound (McDaniel et al. 1984).

Roys (1971) described a different type of migration blockage for pink salmon spawning at Harrison Lagoon Creek in Prince William Sound. Here, during the great Alaska earthquake on 27 March 1964, the intertidal portion of the streambed subsided 2 m in elevation. An impassable falls, present before and after the earthquake, was located just upstream from productive pre-earthquake intertidal spawning gravels. During the first post-earthquake spawning period a significant number of pink salmon spawned at the new 0.6–1.2 m tide level (pre-earthquake 2.4–3.0 m tide level) where eggs cannot survive because of continuous high saline exposures.

Lack of rainfall can cause pre-spawning mortalities of pink salmon due to poor water quality and low stream flows. Poor water quality, especially low dissolved oxygen and high temperatures, can be exacerbated during drought conditions but can also develop under other circumstances.

Wickett (1958b) presented information indicating that in 1925, flows in many northern British Columbia streams were sufficiently low to prevent stream entry of pink salmon. This event likely set into motion a long-term decline in odd-year runs in the region, persisting over twenty years (Neave 1953). On Vancouver Island, low flows in the Tsolum River increased water temperatures to 17.6°C and reduced dissolved oxygen to less than 2.5 mg/1 in intertidal pools where unspawned pink salmon died after becoming trapped (Anonymous 1951). Murphy (1985) studied a 1981 die-off of pink and chum salmon in Porcupine Creek, Alaska, due to anoxia, and documented the occurrence of six other similar events in southeastern Alaska since 1949. In mid-August 1981, I observed the death of several hundred unspawned pink salmon trapped by low flows in the intertidal part of Lovers Cove Creek, southeastern Alaska (Figure 11).

Dvinin (1952) reported three occasions of heavy pre-spawning mortalities of upstream migrating pink salmon in the Far East. In 1947, a counting fence across the Magunkotan River in southern Sakhalin caused an accumulation during a large upstream migration of pink salmon that "led to a mass mortality." In 1949 a similar event occurred at the fence, in several tributaries of the river, and also in the Ochikho River "because of an exceptionally large and crowded accumulation of fish on the spawning grounds when the water level was extremely low and temperatures were high"; water temperatures to 25°C were recorded (Dvinin 1952). In 1934, on the Bolshaya River, Dvinin (1952) observed a mass pre-spawning mortality of pink salmon without obstructions, with normal stream flows, and, presumably, with nominal temperatures. Large numbers of pink salmon entered the estuary at 0300 h, and several hours later the surface of the river was almost completely covered with silvery bodies of pink salmon that were being carried away by the current. Dvinin assumed that lack of oxygen had been the cause of the mass mortality of this extremely numerous and dense accumulation of pink salmon.

Davidson (1933) described an unusual short-term event in Snake Creek, southeastern Alaska, on 6 August 1931, that killed about 5,000 of the 80,000 pink salmon in the stream. The event occurred just at sunset, was restricted to a 30-minute period in only one section of stream, and was attributed to asphyxiation from a temporary rise in carbon dioxide content of the water. The water temperature was 18.3°C and the evidence for carbon dioxide asphyxiation was based on a temporary drop in pH to 5.6 in the area where pink salmon and other fish died. The pH remained at 6.1 in other areas where fish were not affected.

FIGURE 11

Die-off of pink salmon trapped by low flow in the intertidal part of Lovers Cove Creek, August 1981

Parasitic lamprey attacks have been reported on adult pink salmon in both North American and Asian waters and this may result in pre-spawning deaths. In 1967 Williams and Gilhousen (1968) found evidence of attacks by Pacific lamprey, *Lampetra tridentatus*, on 20% of the adult pink salmon sampled in the lower Fraser River, but less than 2% had moderate or severe wounds. Pink salmon sampled in the Strait of Georgia had higher rates of lamprey attacks than those sampled in the Fraser River, suggesting a possible high mortality of wounded pink salmon just prior to entering the river. Birman (1950) found considerable evidence of attacks by Arctic lamprey, *Lampetra japonica*, on pink salmon in the Amur River estuary, where 21% of 275 and 44% of 234 fish sampled in 1948 and 1949 had lamprey scars. In the Great Lakes, Noltie (1987) found that sea lamprey (*Petromyzon marinus*) parasitism had a negative impact on breeding potential of pink salmon due to the poorer condition and smaller gonads of parasitized fish when compared with non-parasitized fish.

It is widely recognized that terrestrial vertebrates, especially bears, gulls, and eagles, feed on salmon, including unspawned fish, in fresh water. However, the effects of such predation, when it occurs, are difficult to measure. Frame (1974) found that predation by the black bear *Ursus americanus* at Olsen Creek amounted to 8% of the unspawned female pink and chum salmon in that stream in 1967. He also found that the bears selectively preyed on unspawned female salmon. Pink salmon apparently develop defensive behaviour to predator odours, and show fright responses to black bear odours (Forrester 1961).

Sex Ratio

Sex ratios in pink salmon populations tend to fluctuate considerably throughout the spawning season. A predominance of males is characteristic among the fish first entering the stream and those first occupying the spawning grounds. As the season progresses, the proportion of females entering the stream and on the spawning beds increases (Eniutina 1972). During the main part of the run the sex ratio tends to be close to 1:1, and towards the end of the run there are usually more females than males. Although the sex ratio for a particular stream or region may favour a predominance of males or females in any given year, the overall pattern tends to approach a 1:1 ratio. For example, Eniutina (1972) found that overall male to female ratios of pink salmon during the 1958 to 1961 spawning migration in Amur Sound were 1.1–1.2:1. In three of the four years, however, the sex ratio was considerably higher in favour of males at the beginning of the migration, starting as high as 2.9:1 in 1958.

Some spawning populations of pink salmon, however, have more females than males. At Hooknose Creek in British Columbia, Hunter (1959) found that females outnumbered males in nine of ten years. Golovanov (1982) reported similar findings for two streams in the Magadan region of the northern Sea of Okhotsk coast.

Ivankov (1968b) reported intraseasonal sex ratios of pink salmon in the southern Kuril Islands that ranged from 1.4 to 2.0 males per female for the period 1956 to 1964, and he attributed the preponderance of males to the selective removal of females by high-seas driftnet fishing. Females in this area tend to be larger than males, particularly at the beginning of the season.

Yefanov and Chupakhin (1982) questioned Ivankov's assumptions that the predominance of males of south Kuril pink salmon was due to the selective action of driftnets. Instead, they considered that it was due to a high rate of survival of males compared to females and suggested that the difference in survival is brought about by lower energy losses in the gonadal development of males.

Ricker et al. (1978) examined data reported by Todd and Larkin (1971) on the selective effect of gillnets in removing larger pink salmon from the Skeena River in 1968 and concluded that the effect was "not unduly serious because the rate of exploitation by gill net was unusually light that year - only 18%." Although no mention was made of overall effects of net selection on sex ratios, Todd and Larkin (1971) noted that net selection for larger Skeena River pink salmon in 1968 "was especially marked in females."

The annual percentages of males in pink salmon returns to Sashin Creek, Alaska, compiled for 37 years during the period 1941–80, ranged from 42% to 61% with a mean of 50.2% (S.D. 4.3%) (Olson and McNeil 1967; Vallion et al. 1981). Excluded from this analysis were escapements of less than 100 adults in 1952, 1954, and 1962, and in 1964 when pink salmon were transplanted to Sashin Creek from another area.

Nickerson (1979) examined sex ratios among spawning populations in 16 streams in Prince William Sound and partitioned them into four time-area categories, including early and late intertidal and early and late upstream spawners. He found a significant predominance of males in all groups except those categorized as late upstream spawners, but he could not separate effects of stream life and spawning behaviour on sex ratios among these partitioned run components.

The percentages of males in downstream migrant pink salmon fry were about 50% in the Kuril Islands (Yefanov and Chupakhin 1982) and Pritornaya River, Sakhalin Island (Shershnev and Zhul'kov 1979), and 52% in Hooknose Creek (Hunter 1959). Golovanov (1982) suggested that the sex ratio of migrant fry in streams along the north coast of the Sea of Okhotsk varies according to stock abundance, ranging from 50% males when the population was in a "satisfactory state" to 30% males in "periods of depression."

FECUNDITY

The fecundity of pink salmon has been studied in considerable detail throughout its range (Pritchard 1937; Foerster and Pritchard 1941; Neave 1948; Kaganovsky 1949; Dvinin 1952; Semko 1954; Rounsefell 1957; Hunter 1959; Grachev 1971; Ivankov and Andreyev 1969; Eniutina 1972). In general, egg content in mature females in a given year increases with length and weight of the fish. However, average numbers of eggs in a stock or region in different years, and between stocks or regions within the same year, varies considerably.

At maturity, pink salmon are the least fecund Pacific salmon, and presumably they also have the lowest overall mortality rate throughout their life cycle. The average number of eggs for a mature female ranges from about 1,200 to 1,900, depending on the region and year, throughout the natural range of the species (Table 2). Data sets for fecundities in Table 2 are based, in most cases, on from ten up to several hundred females per year for the populations indicated.

Foerster and Pritchard (1941) showed that for

TABLE 2
Summary of some studies examining fecundity of pink salmon at or near the time of spawning

Region	Stream	Data available: Year interval	Data available: *N**	Range of annual fecundities	Mean annual fecundity	Source
Central BC	Hooknose Cr.	1947–55	5 (O)	1316–1833	1636	Hunter (1959)
Central BC	Hooknose Cr.	1948–56	5 (E)	1341–1771	1549	Hunter (1959)
Northern BC	McClinton Cr.	1930–36	4 (E)	1535–1899	1748	Pritchard (1937)
Northern BC	McClinton Cr.	1930–40	6 (E)	1535–1899	1718	Foerster & Pritchard (1941)
S.E. Alaska	Sashin Cr.	1935–63	7 (O)	1908–2205	2038	Olson & McNeil (1967)
S.E. Alaska	Sashin Cr.	1936–60	5 (E)	1903–2227	2030	Olson & McNeil (1967)
S.E. Alaska	Sashin Cr.	1963–79	9 (O)	1653–2260	1923	Vallion et al. (1981)
S.E. Alaska	Sashin Cr.	1964–80	9 (E)	1510–2041	1857	Vallion et al. (1981)
Prince Wm. Sd.	Olsen Cr.	1963	1 (O)	–	1929	Helle (1970)
Prince Wm. Sd.	Olsen Cr.	1960–62	2 (E)	1815–1829	1822	Helle (1970); Helle et al. (1964)
West Kamchatka	Bolshaya R.	1937–43	4 (O)	1267–1512	1432	Kaganovsky (1949)
West Kamchatka	Bolshaya R.	1938–40	2 (E)	1310–1449	1380	Kaganovsky (1949)
West Kamchatka	Bolshaya R.	1941	1 (E)		1250	Semko (1954)
West Kamchatka	Karymaiskiy Sp.	1943–49	4 (O)	1445–1500	1477	Semko (1954)
West Kamchatka	Karymaiskiy Sp.	1942–50	4 (E)	1286–1600	1434	Semko (1954)
South Kuril	–	1955–67	7 (O)	1356–1702	1581	Ivankov & Andreyev (1969)
South Kuril	–	1956–66	6 (E)	1471–1827	1634	Ivankov & Andreyev (1969)
S.W. Sakhalin	–	1947–49	2(O)	1560–1576	1568	Dvinin (1952)
S.W. Sakhalin	–	1946–48	2 (E)	1197–1525	1361	Dvinin (1952)
S.W. Sakhalin	–	1947–49	2 (O)	1230–1306	1268	Dvinin (1952)
Amur	–	1933–35	2 (O)	1546–2020	1783	Kaganovsky (1949)
Amur	–	1934–36	2 (E)	1184–1262	1223	Kaganovsky (1949)
Amur	–	1951–63	6 (O)	1592–1832	1664	Eniutina (1972)
Amur	–	1950–62	6 (E)	1141–1696	1470	Eniutina (1972)
Amur	Iski R.	1951	1 (O)	–	1978	Eniutina (1954)
Amur	Amgun R.	1951	1 (O)	–	1528	Eniutina (1954)
Amur	Amgun R.	1952	1 (E)	–	1643	Eniutina (1954)
Amur	My R.	1952	1 (E)	–	1357	Eniutina (1954)

Note: *O = odd; E = even

each increase of one centimetre in length and one kilogram in weight of females, egg content increased by 57 and 472, respectively. At Hooknose Creek, from 1947 to 1957, Godfrey (1959) found that even-year females averaged 120 or 7.2% fewer eggs than odd-year females, a reflection of the larger average size of odd-year pinks in British Columbia. Eniutina (1972) reported an even greater disparity in average fecundity of odd-year and even-year pink salmon in the Amur region of the Far East. She indicated that Amur pink salmon in the ten-year period from 1925 to 1934 were characterized by sharp fluctuations in population abundance with strong even-year dominance; larger average size of fish in odd years; and by a fecundity, on average, of 500 more eggs per female in odd years than in even years. In subsequent years, disparity in line dominance and average size diminished, and in the period 1950–63 the difference in fecundity was only about 200 eggs greater in odd years than in even years (Table 2). The average fork length of female pink salmon in six even years from 1950 to 1962 and six odd years from 1951 to 1963 was 45.5 cm and 47.4 cm, respectively (Eniutina 1972).

Some of the variance in pink salmon fecundity apparently arises from different feeding conditions and growth rates throughout the marine period. Persov (1963) and Grachev (1971) determined that the number of oocytes in young salmon was significantly higher than in mature females. The distinction between "potential" and "final" or "ultimate" fecundity is based on the degeneration of some oocytes after migration of young to the sea. The rate of oocyte degeneration is associated with the rate of maturation and growth within different periods of marine life. Based on samples analysed from the Bering Sea, Sea of Okhotsk, Sea of Japan, and northwestern Pacific Ocean (Table 3), Grachev

TABLE 3

Average number of oocytes in pink salmon by gonad weight and month collected from ocean feeding areas in the Bering, Okhotsk, and Japan seas and the northwestern Pacific Ocean

Gonad weight (grams)	Aug (43)	Sep (307)	Oct (181)	Dec (49)	Feb (9)	Mar (67)	Apr (36)	May (118)	Jun (773)	Jul (764)	Aug (429)
< .05	2679	–	–	–	–	–	–	–	–	–	–
0.05–0.09	2525	2250	–	–	–	–	–	–	–	–	–
0.10–0.14	–	2332	2197	–	–	–	–	–	–	–	–
0.15–0.19	–	2357	2197	–	–	–	–	–	–	–	–
0.20–0.29	–	2309	2026	1889	–	–	–	–	–	–	–
0.30–0.39	–	2441	2308	–	–	–	–	–	–	–	–
0.40–0.49	–	2392	2556	2258	–	–	–	–	–	–	–
0.50–0.74	–	2489	2450	2354	–	–	–	–	–	–	–
0.75–0.99	–	–	2420	2354	–	–	–	–	–	–	–
1.00–1.49	–	–	2574	2536	–	–	–	–	–	–	–
1.50–01.99	–	–	–	2859	–	–	–	–	–	–	–
2.0–4.9	–	–	–	–	2161	2038	1863	–	–	–	–
5.0–9.9	–	–	–	–	–	2578	2086	2200	1521	–	–
10.0–19.9	–	–	–	–	–	2508	1820	1890	1813	1080	–
20.0–29.9	–	–	–	–	–	2253	1576	1382	1539	1814	–
30.0–39.9	–	–	–	–	–	–	–	1284	1408	2142	–
40.0–49.9	–	–	–	–	–	–	–	1612	1421	1911	1095
50.0–59.9	–	–	–	–	–	–	–	1314	1414	1721	1458
60.0–69.9	–	–	–	–	–	–	–	–	1476	1696	1626
70.0–79.9	–	–	–	–	–	–	–	–	1530	1777	1687
80.0–89.9	–	–	–	–	–	–	–	–	1565	1681	1629
90.0–99.9	–	–	–	–	–	–	–	–	1562	1748	1716
100–124	–	–	–	–	–	–	–	–	1700	1648	1634
125–149	–	–	–	–	–	–	–	–	–	1718	1702
150–174	–	–	–	–	–	–	–	–	–	1760	1734
>175	–	–	–	–	–	–	–	–	–	1765	1856

Source: From Grachev (1971)
Note: Numbers in parentheses are sample sizes.

(1971) concluded that much of the reduction in final fecundity of pink salmon occurs during the fall and early winter period. Grachev (1971) also found that pink salmon maturing early in their second summer are less fecund than those maturing later in the year. The weight of ovaries in sexually mature pink salmon ranges from about 50 to 300 g, with early maturing fish representing the lower end of this range. According to Grachev (1971), high growth rates in the second summer, without early maturation, probably permits accelerated development of "retarded" oocytes and reduces further oocyte degeneration. This allows later maturing fish, with heavier ovaries, to show apparent increases in final fecundity (Figure 12).

In conclusion, Neave (1948) assumed that the average number of eggs deposited by a female is closely related to the hazards of the life history. Fecundity, then, is a fundamental adaptive response to life history pattern and a basic mechanism helping to determine population abundance. As Neave (1948) pointed out, because sockeye salmon have twice as many eggs as pink salmon it would seem that "the egg of a pink salmon has

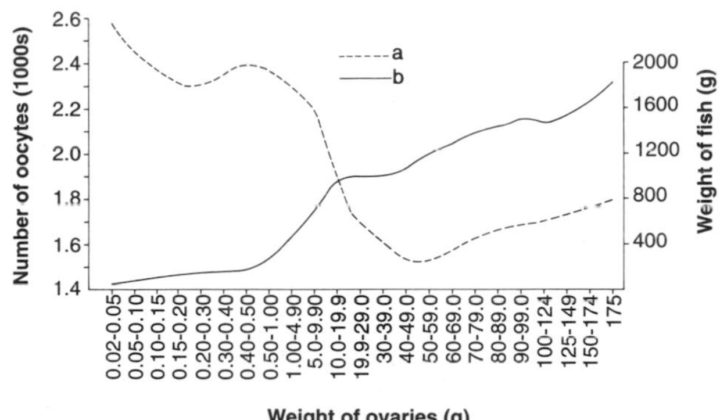

FIGURE 12

Changes in the number of oocytes in pink salmon in relation to weight of fish and gonads: *a*, number of oocytes; *b*, weight of fish. (From Grachev 1971)

twice as good a chance of completing the cycle as the egg of a sockeye." While this statement is a simplification of the differences between pink and sockeye salmon, it does emphasize differences in life history strategies, including fecundity, among species of Pacific salmon.

Spawning

Pre-spawning Changes in Pink Salmon

Pink salmon undergo a series of major morphological changes as they approach full sexual maturity and readiness to spawn. Sexually immature male and female pink salmon are alike in body form and colour. With the onset of maturity, however, the male develops striking secondary sexual characteristics that include an enormous hump on the back, a greatly enlarged head, and large teeth on both the upper and lower jaws that form a pronounced hooked kype. Some mature male pink salmon become compressed laterally in body form and develop a pronounced slab-sided appearance. Females during maturation undergo only minor morphological changes. Both males and females undergo a thickening of skin, absorption of scales, atrophy of alimentary canal and digestive organs, and striking changes in colour. During maturation, the colour of the pink salmon darkens in both sexes from a bright silvery appearance to a pale slate, brownish, or greenish-grey on the back and sides and a pale whitish colour ventrally. Often a trace of a reddish, lilac colour will be present on the flanks (Smirnov 1975). Small irregular black spots occur on the back, sides, dorsal, and caudal fins (Plates 3 and 7).

This remarkable transformation takes place in four to six weeks as schools mill about in estuaries or rest quietly in deeper pools of streams. Davidson (1935) likened morphological changes during maturity of pink salmon to a virtual metamorphosis. Foerster (1968) pointed out that changes during maturation of Pacific salmon "are so great that were one not cognizant of the transformation . . . one would not consider them (i.e., immature and mature fish) the same species."

In addition to the major morphological changes during maturation, complex changes take place in lipid composition (Kamyshnaya and Shatunovsky 1969; Olifirenko 1970); nucleic acid contents of organs and tissues (Berdyshev et al. 1969); blood constituents, including sodium, potassium, calcium, glucose, cholesterol, protein, and hematocrit levels (Hutton 1967; Triplett and Calaprice 1974); urea (Lysaya 1951); and hormone and enzyme functions.

Also, the carotinoid pigmentation in the skin of breeding salmon is intensified, which may lead to increased respiratory function of the skin while normal gill respiration is reduced (Smirnov 1959). Carotinoids originally deposited in the muscle tissue of Pacific salmon are transferred via blood serum during maturation and accumulate in the skin and, in females, in the gonads (Kitahara 1983). Soin (1956) correlated the intensity of carotinoid pigment in the eggs of three species with the respiratory needs of the eggs and the differences in environmental conditions of spawning grounds. He concluded that pink salmon with the least amount of egg carotinoid spawned in environments that deliver the most oxygen to the eggs.

The significance of the morphological changes in pink salmon are probably related to sex and species recognition (Soin 1954), and mating patterns that impose segregation from other *Oncorhynchus* species and prevent extensive interbreeding (Neave 1958). In addition, Soin (1954) suggested that "it seems that the behavior of the males - who almost motionlessly resist the current, staying behind the female - determines the shape of their body" and that "this shape makes for the highest degree of streamlining." He noted that the bodies of pink salmon males are more compressed than those of chum and masu (*O. masou*) salmon, because pink salmon spawn in the highest current velocity; he also pointed out that, as a rule, pink salmon will not spawn at flows less than 0.7 m/s, whereas others have found that pink salmon spawn at velocities up to 1.5 m/s (Semko 1954; Eniutina 1972).

The development of the hump in male pinks may provide species survival advantage. To some extent the hump must limit spawning in shallow water, especially at the streambed margins or "fringe areas" (Hunter 1959), which are the first to dry or freeze under low flows and severe winter conditions (Hunter 1949a). Semko (1954) noted a correlation between spawning depth and body size and found that larger pinks occupy deeper waters. The humped male also may serve as a predator buffer for females. On many riffles with large numbers of spawning pink salmon, the pronounced male hump, often extending above the water surface, is highly evident (Figure 13), and during a random predator chase, such as from a bear, attention most likely would be drawn first to a male.

In general, then, characteristics of the spawning environment and behaviour may have played the key role in evolution of the hump in male pink salmon. It is instructive to note similarities between pink and sockeye salmon spawning, for males in some sockeye populations develop humped backs almost as great as those of pink salmon. Similar to pink salmon, sockeye also tend to spawn in dense groups on riffles, and in both species details of redd[1] construction and spawning behaviour are similar. Hikita (1962) placed pink and sockeye salmon together on the same phylogenetic branch, partly because of the spawning male hump. Whatever selection factors led to development of the hump in males, they must have operated in both species.

Characteristics of Spawning Grounds

Pink salmon choose a fairly uniform spawning bed in small and large streams in both Asia and North America. Generally, these spawning beds are situated on riffles with clean gravel, or along the borders between pools and riffles in shallow water with moderate to fast currents. In large rivers they may spawn in discrete sections of main channels or in tributary channels. Pink salmon avoid spawning in quiet deep water, in pools, in areas with a slow current, or over heavily silted or mud-covered streambeds. As pointed out by Semko (1954), selection by pink salmon of places for egg deposition is determined by the optimal combination of two

1 A redd represents a trench in which a female has deposited several nests of eggs in succession. Nests are often referred to as "pockets" within a redd.

FIGURE 13

Pink salmon spawning riffle showing the humped back of males extending above the water surface. (*Top*: photograph by Jerrold Olson, National Marine Fisheries Service; *bottom*: photograph by Alaska Department of Fish and Game)

main interconnecting variables: depth of water and velocity of current.

On both the Asian and North American sides of the Pacific Ocean, pink salmon generally spawn at depths of 30–100 cm (Dvinin 1952; Hourston and MacKinnon 1956; Vasilenko-Lukina 1962; Eniutina 1972; Graybill 1979; Golovanov 1982). Well-populated spawning grounds of pink salmon are mainly at depths of 20–25 cm, less often reaching depths of 100–150 cm. In dry years, when spawning grounds are crowded, nests can be found at shallower depths of 10–15 cm.

Current velocities in pink salmon spawning grounds varied from 30 to 100 cm/s, sometimes reaching 140 cm/s (Smirnov 1975). Directly over the redds, about 5–7 cm from the surface, the velocity ranged from 30 to 140 cm/s with averages from 60 to 80 cm/s (Dvinin 1952; Soin 1954; Hourston and MacKinnon 1956; Vasilenko-Lukina 1962; Kobayashi 1968a; Smirnov 1975; Graybill 1979; Golovanov 1982).

Differences in characteristics of spawning sites of pink and chum salmon have been noted by several authors, including Semko (1954) for west Kamchatka, and Kobayashi (1968a) for Hokkaido. These authors observed that pink salmon selected sites in gravel where the gradient increased and the currents were relatively fast. In these areas, surface stream water must have permeated sufficiently to provide intragravel flow for dissolved oxygen delivery to eggs and alevins. Chum salmon, by contrast, tended to select spawning sites in areas with upwelling spring water and a relatively constant water temperature, without much regard to surface stream water. In general, intragravel oxygen values are higher for pink salmon spawning beds than for those of chum salmon (Kobayashi 1968a).

Pink salmon spawning beds consist primarily of coarse gravel with a few large cobbles, a large mixture of sand, and a small amount of silt (Dvinin 1952; Smirnov 1975). Hunter (1959), in describing the spawning grounds in Hooknose Creek, suggested that although a certain amount of fine sand and silt is present, the spawning grounds can be characterized as clean, coarse gravel.

In southeastern Alaska, McNeil and Ahnell (1964) systematically sampled streambed materials from pink salmon spawning grounds in eight different streams or sections within streams. Solid material was sorted through sieves and grouped according to particulate size by a volume displacement method. In general, the volume of streambed material consisting of solid particles that passed through a 0.8-mm sieve ranged from 5.7% to 20.0% and averaged 14.3%; the volume of material that passed through a 6.7-mm sieve and was retained by a 0.8-mm sieve ranged from 27.2% to 36.7% and averaged 31.4%; and the volume of material larger than 6.7 mm in diameter ranged from 43.4% to 63.4% and averaged 54.1%. Although there was considerable variation in these values from individual streams, McNeil and Ahnell (1964) found that the potential of a pink salmon spawning bed to produce fry successfully was directly related to streambed composition. This correlation was based on the permeability or ability of water to flow through the intragravel environment. They found an inverse relationship between the coefficient of permeability and the percentage of sand and silt in the streambed.

Rukhlov (1969) used a similar method to evaluate the composition of material taken from 61 pink salmon redds and of 126 samples taken from spawning grounds on five Sakhalin Island streams. He found that the percentage of material composed of sand was lower in the redds than in the streambed samples, and that the amount of large cobble material was greatest in the redds (Table 4).

TABLE 4

Comparative composition of streambed material from spawning grounds and redds of pink salmon from five Sakhalin Island streams

		Composition (%)			
	N	Sand	Gravel	Shingle	Cobble*
Spawning grounds	126	14.7 (1.5–40.3)	34.4 (0–55.1)	43.5 (0–74.8)	7.4 0–59.1)
Redds	61	11.5 (2.7–40.0)	36.2 (0–61.6)	39.6 (0–65.0)	12.7 (0–49.1)

Source: Adapted from Rukhlov (1969)
Notes: Numbers in parentheses are ranges.
*Defined by Rukhlov as larger than 100 mm in diameter

In rivers characterized by increased sand content, the survival rate for pink salmon eggs was less. In the Sakhalin samples, sand composed, on average, 11.5% of the material in the redds and 14.7% in the spawning grounds. When sand increased to 26% in the redds, the survival rate for eggs was less than 25% (Rukhlov 1969).

Direct comparisons for sand in redd composition between Rukhlov's (1969) and McNeil and Ahnell's (1964) reports are not possible because the exact sieve sizes that the Soviet scientist included in this category are not known. Both McNeil and Ahnell and Rukhlov also considered in some detail the dynamic effects on streambed composition of various factors such as precipitation patterns, floods, logging activity in the watershed, and redd building activity by spawning salmon.

Pink salmon spawn over a wide range of water temperatures. Although spawning by many individual stocks in different regions may occur within a restricted temperature range, perhaps in adaptive responses to environmental factors and thermal regimes of streams, no overall correlation exists between water temperature and egg deposition.

In the Far East, water temperatures during spawning of pink salmon ranged from 5° to 14°C in the Bolshaya River, and from 8.5° to 15.6°C in the Amur River (Kaganovsky 1949; Smirnov 1975). Most spawning apparently occurs at 11.5°–12°C and 8.5°–15.6°C, respectively, in these streams. In the Kurilka River, Kuril Islands, Ivankov (1967) reported that pink salmon begin spawning at 12.5°–13.5°C, continue spawning through the season as temperatures drop to between 6.6° and 7.8°C, and subsequently stop spawning at 5°C. Dvinin (1952) indicated that pink salmon spawn at temperatures from 7° to 19°C on south Sakhalin Island.

In southeastern Alaska, pink salmon spawn at temperatures ranging from 7° to 18°C, with much of the spawning occurring at temperatures around 10°C (Sheridan 1962a; Bailey and Evans 1971; Heard 1975; Vallion et al. 1981). In the Prince William Sound area (south-central Alaska), Helle et al. (1964) indicated that pink salmon spawned at water temperatures of about 9.0°–14.5°C, with fluctuations up to 5.5°C within a one-hour period due to tidal influence.

In British Columbia, Hunter (1959) reported that pink salmon spawning occurred at temperatures from 8° to 14°C in Hooknose Creek, with most spawning at 12°C. In the Fraser River, Andrew and Geen (1960) found that pink salmon spawning peaks at about 10°C. In general, when temperatures rise above 16°–17°C pink salmon spawning activity drops sharply (Vernon 1958; Smirnov 1975).

The proportion of pink salmon that spawn intertidally in short coastal streams can be as high as 74% in some regions. Hunter (1959) indicated that 15% of the spawning area in Hooknose Creek was intertidal, and Jones (1978) estimated that 13.7% of pink salmon spawning in southern southeastern Alaska was intertidal. In Prince William Sound, Noerenberg (1963) estimated that the proportion of pink salmon spawning in the intertidal zone over a ten-year period (1952–61), in even- and odd-numbered years, averaged about 74% and 46%, respectively. Vernon (1966) indicated that the lower extremity of spawning in the Fraser River was in the area of tidal influence. The extent of intertidal spawning of pink salmon in the Far East, if any, is poorly documented.

Details of intertidal spawning ecology for pink salmon have been reported by Hanavan and Skud (1954), Rockwell (1956), Helle et al. (1964), and Helle (1970). The general characteristics of the spawning grounds and spawning behaviour, apart from periodic tidal flooding accompanied by milling about of spawners during slack water periods, are similar to those upstream in freshwater areas. Intertidal spawners apparently return to the same redds when the tide recedes and freshwater flow over the streambed resumes.

Spawning Behaviour

The sequence of events during spawning involves selection of a redd site, preparation of nests, prespawning courtship behaviour, release and fertilization of eggs, burial of eggs, and defense of the redd. These activities are accompanied by a complex series of interactive behaviour patterns by both males and females.

Although males usually arrive first on the spawning riffle, it is the female that selects the redd site and proceeds with the business of digging and preparing the nests for egg deposition. The presence of the male, however, is apparently important in stimulating the female to begin nest construction. Studies by Mathisen (1962) on spawning behaviour of sockeye salmon showed that when ripe females were placed on spawning beds without males, the normal sequence of redd selection, construction, and defense was greatly disrupted.

Once a female has selected a site for spawning she proceeds with excavating the first nest of a redd by turning on her side and with vigorous lateral flexing movements, called digging (Fabricius 1963) or cutting (Jones and King 1949), forces streambed material up into the water column where it is carried downstream by the current. The vigorous downstroke of the posterior half of the body thrusts the water against the gravel with sufficient force to loosen it, and the upward flexion further assists the movement downstream of the displaced gravel by upward suction (Jones and King 1949). Soin (1954) described the digging behaviour of female pink salmon in the My River (USSR) as follows: "She takes up a position head against the current and makes an abrupt movement forward, rolling on either her left or right side, and raises gravel and sand from the bottom with powerful convulsive vertical strokes of the caudal fin. Gravel and sand are washed away downstream by the current, while the female moves on one side upstream for the entire length of the depression produced (1.5 to 2 m). Then the female returns to her initial position and repeats the entire procedure."

Male pink salmon occasionally make weak, ineffectual digging movements. Wickett (1959a) described digging by a dominant male that was followed by an attack on the male by the female. Digging behaviour by males shows the characteristics of a displacement activity (Tinbergen 1951).

In some situations pink salmon females perform preliminary or exploratory nest-digging movements (intentional digging) before actual nest construction. For this behaviour, females turn on their side and duplicate movements used later in digging the nest. This can be performed in schools at the mouth of a stream or in deeper pools before activity begins on spawning riffles. Usually it occurs without interactive behaviour between fish, often at mid-depth in the water column and some distance above the substrate. When intentional digging occurs near the gravel substrate it results in "false nests" (Figure 14). Construction of such false nests has been noted in the Fraser River, British Columbia, and the Nuyakuk River, Alaska (Vernon 1962). I have seen false nests in the Little Port Walter estuary (Alaska) along intertidal portions of the beach, when large runs of pink salmon returning to Sashin Creek are delayed from stream entry by freshets or weir operations.

Normally the nest and an area adjacent to it are defended by the female and the dominant male. Each usually attacks intruders of its own sex. One or more satellite males can be positioned just downstream from the spawning pair to form units which Chebanov (1980, 1982) called "nest groups."

Once females have selected a spawning site they make few movements away from it, except to defend against intruders. Males, by contrast, show a complex array of movements and activities on spawning grounds (Chebanov 1980). They are polygamous and can function both as leaders and satellites. Although small males make more movements on the spawning grounds (coefficient of intensity of movement), they only become nest group leaders with small females. The largest males, however, may be leaders in nest groups with both large and small females with equal frequency. Large males have the greatest spawning efficiency because they move about less on the spawning ground than small males, and a relatively large proportion of their movements result in leadership roles within nest groups (Chebanov 1980). The leading male pink salmon usually remains in the same nest group until one or more spawnings occur. Male pink salmon, however, can spawn with more than one female (Hunter 1959; Eniutina 1972). Thus, pink salmon show an assortative mating hierarchy determined by size of males and females (Chebanov 1980).

The defense of the immediate area around the nest by the dominant male involves aggressive rushes at intruders as well as chasing. Often the defender will butt or bite the intruder. During these defenses the large teeth and well-developed kype of the male pink salmon are used. On occasion, a leader male clamps his kype around the caudal peduncle of an intruder and the two fish become momentarily locked together; sometimes the current sweeps them downstream one or two metres before the encounter ends.

The number of males in a nest group depends partly on the sex ratio and partly on the stage of nest-building and spawning sequence of specific females. Under natural conditions, each female can be accompanied on average by three to five males (range two to ten males) (Soin 1954; Eniutina 1972;

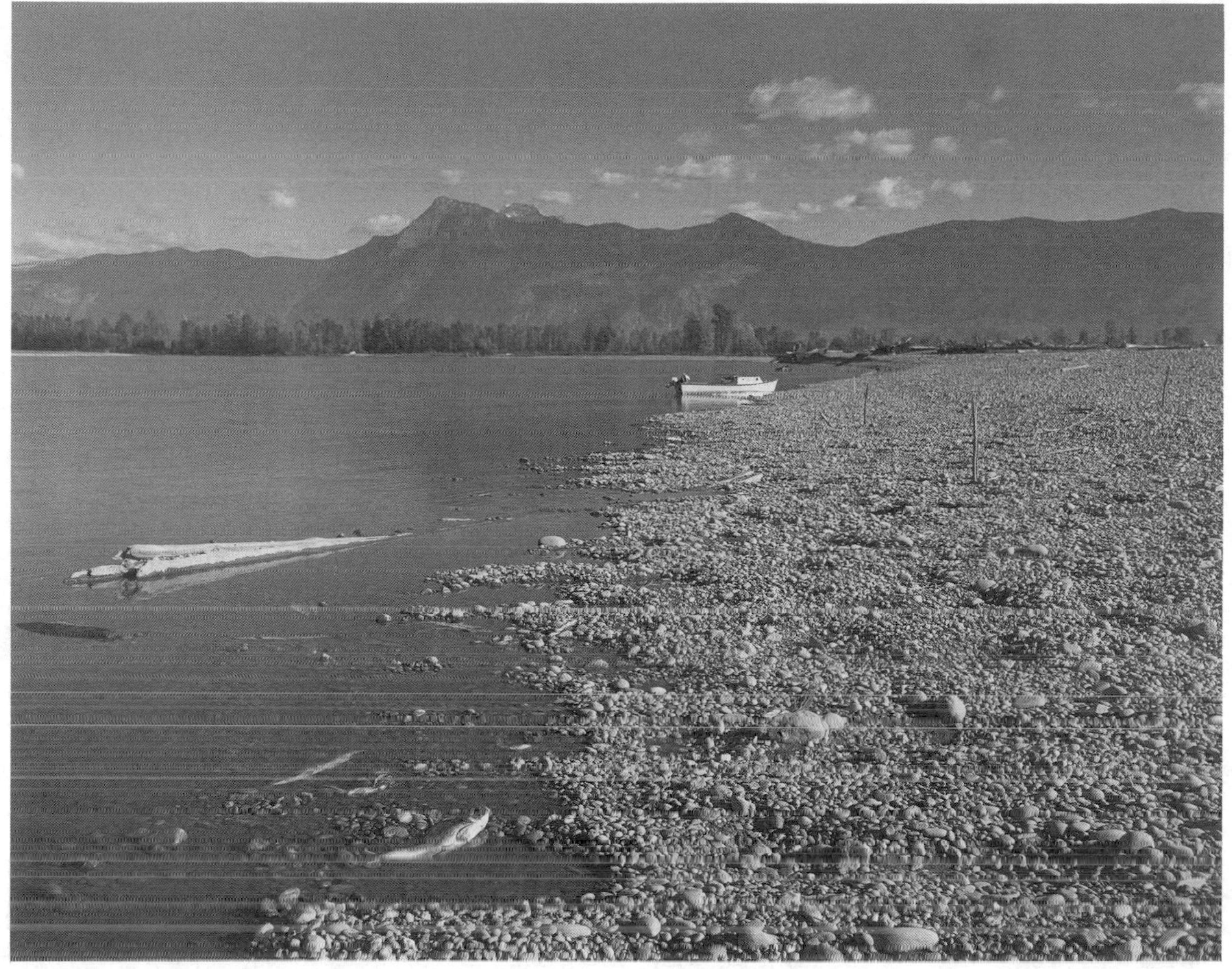

FIGURE 14
False nest dug by pink salmon on gravel bars along shallow margins of the Fraser River near Hope, British Columbia. Declining seasonal water levels expose the false nest after the spawning season is over. Studies by the International Pacific Salmon Fisheries Commission established that eggs were not deposited in these depressions. (Photograph courtesy of the International Pacific Salmon Fisheries Commission)

Smirnov 1975; Chebanov 1980). Often one large male attends the nest at the start of digging and from one to four males attend it when the nest is completed and the female near spawning (Kirkwood 1962).

Throughout the course of nest building and burial of eggs, a female pink salmon moves about 100 kg of gravel or 0.33 m^3 of streambed material, about one-third to one-half that of sockeye, chum, or coho salmon (Semko 1954; Eniutina 1972). In this regard, pink salmon, as the smallest member of *Oncorhynchus*, makes good use of the force of current in constructing a nest by selecting spawning sites in a relatively fast current.

The size of the pink salmon redd varies from 60 to 150 cm in width and 107 to 250 cm in length and depends on the size of the female, compaction of the gravel, and current velocity (Smirnov 1975). Eniutina (1972) indicated that the "relief" or gradient of the stream and the depth of water also determined redd size. The shape of the redd is usually elliptical (Vasilenko-Lukina 1962) or ovoid

(Kusnetzov 1928; Kaganovsky 1949), with a wide ridge or mound formed along the downstream margin.

McNeil (1967) estimated that the average area of pink salmon redds was 1.1 m^2, whereas Smirnov (1975) suggested that 1.5–2.0 m^2/female was necessary for effective use of spawning grounds. Eniutina (1972) noted that the density of redds varied from 1 per 1 m^2 to 1 per 10 m^2, depending on spawner density. Instantaneous densities greater than 0.8 females/m^2 were shown by Heard (1975) to influence behaviour patterns of pink salmon spawning, presumably including redd sizes. The term "instantaneous density" is defined as the density of spawners on the spawning grounds at any given time, in contrast to the overall annual density which may include repeated groups or waves of spawners throughout the spawning period.

Egg Deposition

As nest construction reaches completion, the female poises above the deepest part and then slowly drops vertically downward with her anal fin erect to test the condition of the nest with her fin and vent (Jones and King 1949). This behaviour, called "crouching" (Jones and Ball 1954), "anchoring" (Fabricius and Gustafson 1954), or "probing" (Tautz and Groot 1975) occurs periodically during nest construction and always immediately before eggs are released. When the female settles into the nest to probe, one or more males move alongside her and perform a series of rapid vibrating or quivering motions that presumably encourage and stimulate the female. Heard (1972) noted that before the female will release eggs, a suitable nest pocket and other appropriate stimuli must be present. To extrude eggs, the female takes up the crouching position, which looks like a probe, but with mouth agape, and with her vent deep in the nest. The male also quickly moves into the nest in a similar crouching position with mouth wide open and his body pressing close to the female. At this point the reproductive products are released with both partners vibrating their caudal penduncle and anal fin. The open mouths of both male and female provide resistance to the current and help the spawners to remain deep in the nest (Greeley 1933).

Immediately after spawning, the female increases the frequency of digging movements just upstream from the nest (Sheridan 1960; Smirnov 1975; Keenleyside and Dupuis 1988). The loosened gravel is carried downstream by the current and covers the eggs. Sheridan (1960) suggested that rapid covering of the eggs minimizes predation by other stream fishes and permits the eggs to be covered during the short time they retain their adhesive character.

Continued digging in front of the first nest gradually leads to excavation of a new pocket in which the next batch of eggs will be deposited. In total, a female pink salmon digs one to four nests which form a redd when all eggs are laid. She then changes her behaviour and starts to smooth out the redd with general digging motions to develop a shallow hump. The male leaves her and she continues to defend the redd area against other pink salmon, especially other females that are building nests nearby.

The spawning activity of pink salmon continues throughout the 24-hour diel period, but egg deposition occurs mainly at dusk and during darkness (Smirnov 1975; Chebanov 1980). This may be the reason that so few observations of actual egg deposition in pink salmon have been reported. McNeil (1967) speculated that pink salmon may spawn in darkness when territorial defense by females may be ineffective. Wickett (1959a) conjectured that low light intensity is normally required for egg deposition.

The individual groups of eggs deposited by one female pink salmon average 2.1 per redd (range 1–4) (Vasilenko-Lukina 1962). Eniutina (1972) reported that in the Iski River the eggs are mostly laid in two portions. Kaganovsky (1949), Soin (1954), and Smirnov (1975) suggested that the spawn is distributed in two or three and sometimes four nests, generally 20–40 cm apart. The numbers of eggs deposited in 48 pink salmon redds in the Ulika River of the Primore region ranged from 56 to 1,119 and averaged 513 eggs/nest (Vasilenko-Lukina 1962). The depth to which pink salmon eggs are buried in a nest by the female can range from 15 to 50 cm, with an average depth of 20–30 cm (Dvinin 1952; Vasilenko-Lukina 1962; Eniutina 1972).

Redd Life and Wave Spawning

After egg deposition is completed, the female pink

salmon remains at the redd for several days to defend the site, especially against other females. This behaviour prevents the immediate uncovering and loss of recently spawned eggs. During this post-spawning period the guarding female continues to make digging movements to finalize the shape of the redd.

Smirnov (1975) found that pink salmon spawning lasts from 1 to 8 days and that the female will remain near the redd from 10 to 13 days. Thus, the spawning phase, from initial selection and defense of the redd site until the female dies, can range from 11 to 21 days. McNeil (1962) studied 38 tagged female pink salmon on the Harris River in southeastern Alaska and found that they occupied the redd site for an average of 10.8 days. Ellis (1969) studied females in different sections of Sashin Creek and found that redd life varied considerably within stream sections and between years (Table 5). He suggested four possible reasons for the longer average time females spent on redds in 1966 than in 1963 and 1965: minimal pre-spawning activity, larger females, lack of floods, and lower spawner density.

TABLE 5

Average redd life in days for tagged female pink salmon in three study sections of Sashin Creek, southeastern Alaska, 1963–71

	Upper section		Middle section		Lower section	
Year	Females (*N*)	Redd life (days)	Females (*N*)	Redd life (days)	Females (*N*)	Redd life (days)
1963	13	8.3	36	6.6	37	8.1
1965	50	10.9	68	11.7	74	11.9
1966	4	14.2	41	17.8	71	20.6
1967	63	10.9	68	12.1	81	13.6
1968	95	10.3	116	9.8	–*	–
1969	66	12.7	85	12.3	33	13.9
1970	19	12.7	46	14.8	63	16.3
1971	42	11.1	111	11.1	111	12.5

Source: Data for 1963–66 from Ellis (1969); data for 1967–71 from National Marine Fisheries Service files
Note. *No spawning occurred in the lower section in 1968 when this part of the stream was artificially flooded to force spawners into upstream areas.

In most years there is a trend towards a shorter to longer redd life from the upper to lower sections of Sashin Creek (Table 5). The upper section of this stream has a steeper gradient with coarser streambed material than the middle and lower sections (Table 6). Also, the width of the wetted streambed tends to increase with gradient decrease from the upper to lower sections. Although average current velocities are not reported for these sections, it follows that, under various annual stream-flow regimes during spawning, velocities are faster in the upper, slower in the lower, and intermediate in the middle section. Although many factors likely account for variations in the redd life of females, the average current velocity encountered during spawning seems to play a role. The stronger the current, the shorter the redd life.

TABLE 6

Size composition of bottom materials* and average gradient in three study sections of Sashin Creek, southeastern Alaska

		Composition of bottom materials		
Section	Average gradient (%)	Cobbles† (%)	Pebbles and granules‡ (%)	Sands and silts§ (%)
Upper	0.7	81	16	3
Middle	0.3	61	26	13
Lower	0.1	47	36	17

Source: From McNeil (1966a)
Notes: *Materials >15.2 mm in diameter are excluded.
†Cobbles are >12.7 mm in diameter.
‡Pebbles and granules are 1.68–12.7 mm in diameter.
§Sands and silts are <1.68 mm in diameter.

A long redd life provides protection against repeated spawnings by subsequent females in the same area. A protracted spawning season, however, with repeated waves of spawners using the same spawning grounds, and a high annual density of spawners, can result in significant redd superimposition and loss of previously deposited eggs. In some river systems, such as the Magunkotan River and its tributaries on Sakhalin Island, three or more shifts of spawners can lay their eggs in the principal spawning grounds during a period of 1.5–2.0 months (Dvinin 1952). Such wave spawning was not observed when there was a moderate run of fish in the river.

FACTORS AFFECTING SPAWNING EFFICIENCY

Spawning success is dependent on the successful deposition of eggs in the intragravel environment of the spawning bed. Ellis (1969) defined spawning efficiency, or efficiency of egg deposition for pink salmon, as the percentage of the potential egg deposition for the entire escapement that is successfully buried in the gravel at the end of spawning. Neave (1953) listed two major mortality categories for pink salmon eggs prior to burial: (1) losses of unspawned adults due to predators, barriers, or insufficient water; and (2) losses of eggs due to retention or failure of fertilization. Ellis (1969) added to this list losses of improperly buried eggs that are washed from the stream or eaten by predators or scavengers.

Egg Retention

Normally, at low and moderate spawner densities, most eggs are extruded during spawning. However, high spawner densities with subsequent crowding sufficiently interrupt spawning and cause females to extrude only part of their eggs. The number of eggs retained in the female's body and not extruded during spawning, as a percentage of fecundity, has been estimated to range from less than 1% to over 40% (Table 7).

Semko (1954) and Helle et al. (1964) estimated egg retentions of 40% and 41.5% at Karymaiskiy Spring (Kamchatka) and Olsen Creek (Alaska), respectively, with high spawner densities. With low spawner densities, Helle et al. (1964) found egg retention to vary from 2.7% to 5.1% in three areas of the stream. At Hooknose Creek, Hunter (1959) showed that egg retention for six years, ranging from 0.1% to 2.7%, did not correlate with spawner density.

Although most spawning bed overcrowding for pink salmon is from other pink salmon, Semko (1954) also reported interspecific overcrowding from chum salmon in Karymaiskiy Spring of the Bolshaya River. Here, spawning grounds of the two species overlap to some extent. He noted that "the chum being larger and more vigorous fish, would interfere with the egg-laying conditions for the pinks, in consequence of which there would occur the loss of a large number of pink salmon females with almost a complete non-deposition of eggs." Rounsefell (1958) argued that interspecific crowding by large pink salmon escapements reduced sockeye salmon spawning success in the Karluk River in Alaska.

Although egg retention may on occasion cause the loss of some reproductive potential in pink

TABLE 7
Eggs retained in female pink salmon carcasses after spawning

Region	Stream	Year	Retained eggs % of fecundity	Source
N. Okhotsk coast	-	1971-79	0.2-5.0	Golovanov (1982)
Amur	Ulika R.	1957	1.5-2.0	Vasilenko-Lukina (1962)
	Samnya, Im, and Iski rivers	1960-62	1.0-18.0	Eniutina (1972)
	-	1963	23.0	Eniutina (1972)
Kamchatka	Karymaiskiy Sp.	1947	40.0	Semko (1954)
S. Sakhalin	-	-	1.0	Dvinin (1952)
Kamchatka	-	-	7.0-15.0	Dvinin (1952)
Prince Wm. Sd.	Olsen Cr.	1960-61	2.7-41.5	Helle et al. (1964)
Prince Wm. Sd.	Olsen Cr.	1962-63	10.0	Helle (1970)
S.E. Alaska	Lovers Cove Cr.	1966	4.1-18.4	Heard (1975)
S.E. Alaska	Sashin Cr.	1963-65	4.0-5.0	McNeil (1966a, 1968)
S.E. Alaska	Sashin Cr.	1966-67	0.5-1.5	Ellis (1969), Heard (1978)
Central BC	Hooknose Cr.	1951-56	0.1-2.7	Hunter (1959)

salmon, it is for the most part not a significant factor. I believe Semko (1954) put egg retention in the proper context: "the number of eggs lost during spawning is much greater than those wasted through non-deposition of all eggs by the female fish."

Superimposition, Spawner Density, and Egg Losses

The repeated spawning in the same location by subsequent waves of females leads to superimposition of redds and excavation of eggs (Smirnov 1947; Neave 1953; McNeil 1964a). Long bands of dead eggs stretching along the banks and being carried away by the current can often be seen in rivers with large numbers of spawners (Kaganovsky 1949; Dvinin 1952). Additional causes for egg losses during spawning are high stream flows and consumption of eggs by predators before the eggs are buried. Eniutina (1972) determined that egg losses of pink salmon in the Iski, My, and Im rivers (tributaries of the Amur River) ranged from 15.4% to 68.8% with a mean of 48.6%, and that these losses were due to fast-flowing currents (Table 8). Ellis (1969) attributed the inability of females to bury their eggs in Sashin Creek in 1966 to high water flows rather than to superimposition. Ivankov (1968a) reported egg losses during spawning in the Reidovaya River (Kuril Islands) of 43.5% and 34.0% in 1956 and 1957, years with relatively high and low pink salmon abundance, respectively.

TABLE 8

Estimated percentage of pink salmon eggs lost during spawning in Amur region streams

Year	Iski River	My River	Im River
1938	38.9	–	–
1939	15.4	–	–
1951	55.5	59.7	68.8
1952	–	43.5	–
1953	57.4	40.5	57.6

Source: Adapted from Eniutina (1972)

McNeil (1964a) derived a mathematical model for pink salmon egg mortality due to redd superimposition in the Harris River. He postulated an asymptotic relation between spawner density and mortality (as a percentage) based on an assumed random distribution of females on the spawning beds. If redd distribution were uniform, mortality from superimposition would be overestimated; whereas if redd distribution were contagious, mortality from superimposition would be underestimated. McNeil (1967) concluded that on the basis of release patterns of small buoyant balls buried in a streambed and subsequently excavated by digging females with an annual spawning density of 0.87 females/m^2, the distribution of redds was contagious. He also suggested that the degree of contagion might vary according to uniformity of habitat and spawner density.

Some authors have noted that high spawner densities and redd superimposition, although causing significant loss of eggs, may also have beneficial effects on the spawning bed environment by reducing streambed compaction and removing quantities of organic detritus and fine silt particulate material (Smirnov 1947, 1975; McNeil and Ahnell 1964). The capacity of spawning gravels for successfully incubating eggs and alevins to emergent fry is dependent on the delivery of an adequate oxygen supply to the intragravel environment (Wickett 1954). Oxygen delivery is primarily dependent on streambed permeability and intragravel flow (Terhune 1958; Wickett 1958b; Vaux 1968), which is in turn greatly influenced by the amount of fine sedimentation in the streambed (Cordone and Kelly 1961; McNeil and Ahnell 1964). Thus, the digging activity by female pink salmon, which is intensified at high spawner densities, may serve, in some cases, to clean the streambed and improve conditions for embryo development and survival.

Loose drift eggs on the spawning grounds, or unburied eggs within the redd, attract predators and scavengers. Reed (1967) found a somewhat ordered arrangement of four stream fishes, including juvenile coho salmon, rainbow trout (*O. mykiss*), Dolly Varden charr (*Salvelinus malma*), and prickly sculpin (*Cottus asper*), positioned at the downstream edge of pink and chum salmon redds. Although each of these species ate some salmon eggs, they also actively fed on aquatic insect larvae that were washed downstream after being uncovered by digging salmon. Reed (1967) concluded that Dolly Varden charr were not a major salmon egg predator, but speculated that the prickly sculpin might be.

Moyle (1966) found glaucous-winged gulls (*Larus*

glaucescens) to be the dominant gull feeding on loose drift salmon eggs in Olsen Creek, although other gulls, including mew gulls (*L. canus*) and Bonaparte's gulls (*L. philadelphia*), as well as Arctic terns (*Sterna paradisaea*), also fed on loose eggs. When feeding on drift eggs in a shallow riffle where the bird can stand, glaucous-winged gulls defend the feeding site against other gulls. Much of the time these gulls swam downstream with the current, bobbing for loose salmon eggs (Moyle 1966).

It is suspected that, in many cases, gulls account for a significant part of the "disappearance" of pink salmon eggs that are not successfully buried at the end of spawning. Gulls, of course, along with other scavengers, also actively feed on and account for the disappearance of spawned out salmon carcasses. The bulk of the eggs eaten under these conditions are generally "waste" eggs that could not otherwise survive (Foerster 1968).

Spawning Efficiency

Spawning efficiency can be determined when the number of females, average fecundity, and number or average density of eggs buried in the streambed at the end of spawning are known. The average spawning efficiency for pink salmon in Sashin Creek over thirteen years of records ranges from 17% to 54%, with a mean of 44% (Ellis 1969; Heard 1978; Vallion et al. 1981). Thus, as a result of fluctuating density, distribution, and stream-flow conditions, over half (56%) of the eggs brought into this stream by females are lost, on average, during spawning. This value is similar to the average of 48.6% for the nine yearly egg-loss estimates for Amur District streams (Eniutina 1972) (Table 8).

Neave (1948) suggested that "the average number of eggs deposited by a female is closely adjusted by the hazards of the life history." Thus, with respect to life history strategy, pink salmon, besides having the lowest fecundity of all Pacific salmon and the least dependence on fresh water, only need to bury successfully about half of their eggs during spawning.

Hybrids and Other Variants

Pink and chum salmon tend to choose separate redd sites and have different spawning requirements, but they frequently are found spawning at the same time and vicinity, often on the same riffle. Natural hybrids occur between these two species, but the level of hybridization is presumably low. Hunter (1949b) reported that 50 pink-chum hybrid adults were found at the Port John buying station in central British Columbia in 1949. One hybrid examined in detail had some characteristic features of each parent. Hybrid pink-chum adults have been observed in other parts of British Columbia (F. Jordan, Pacific Biological Station, Nanaimo, British Columbia, pers. comm.), in southeastern Alaska, and in Prince William Sound (J. Helle, National Marine Fisheries Service, Auke Bay, Alaska, pers. comm.). Man-made hybrids between these species have been studied in some detail both in Asia (Smirnov 1953; Pavlov 1959; Kamyshnaia 1961; Kobayashi 1964; Hikita and Yokohira 1964) and in North America (Foerster 1935, 1968; Simon and Nobel 1968).

Fertility and survival in anadromous F1 hybrids are substantial, as determined by Simon and Nobel (1968). These authors also studied survival of F2 hybrids and F1 hybrid backcrosses held in a hatchery. They found that some morphological features of F1 hybrid adults, such as hump and size of teeth in mature males and spots on the caudal fin, were intermediate to parental traits (Plate 10). Several meristic features, however, such as gill rakers, lateral line scales, and parr marks on fry were not intermediate but characteristic of one or the other parent. Hikita and Yokohira (1964) also found that parr marks in hybrid fry are either present or absent and suggested that this feature varied according to the male parent (Figure 15). This feature was also observed on Lakelse Lake pink x sockeye salmon crosses (R. Bams, Pacific Biological Station, Nanaimo, British Columbia, pers. comm.).

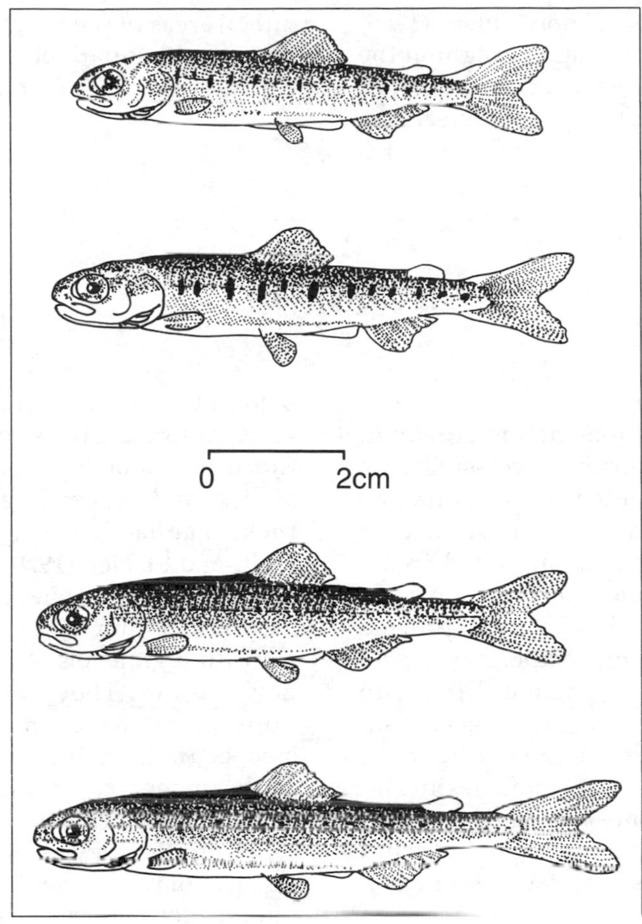

FIGURE 15
Pink-chum salmon hybrid fry: *top*, pink (female) x chum (male); *bottom*, chum (female) x pink (male). (From Hikita and Yokohira 1964)

Do pink and chum salmon mate with each other or do spermatozoa of one species "accidentally" fertilize eggs of the other? Accidental fertilization between species could occur with redds aligned in a current. Shuman (1950) found that pink salmon spermatozoa remained sufficiently viable to fertilize 17.9% and 4.8% of eggs of the same species placed at 8 and 16 m distances, respectively, downstream from the point of origin. Smirnov (1963) found that spermatozoa of pink salmon still fertilized 29%–50% of available eggs after 30 seconds in water and 15%–30% after 120 seconds in water. Even after 150 seconds some sperm viability was evident. Smirnov also found that some unfertilized eggs were still viable after 60 seconds to 8 minutes in water.

Although albinism is not unusual in salmonids, the absence of albinos in wild populations beyond juvenile stages indicates that they are ill-suited for survival under natural conditions. Dawson (1964, 1966), in an extensive review of colour variants and other aberrations in fishes, lists only one record of a wild adult albino salmon. During spring 1968, I observed approximately 150 albino pink salmon fry produced from natural spawning of the 1967 brood migrating downstream through a fry counting weir at Sashin Creek in southeastern Alaska (Plate 9). No albino adults or unusual colour variants have been reported in this stock.

Various colour variants have been observed from

time to time in adult pink salmon. Hikita (1965) reported several adult pink salmon caught in the North Pacific Ocean with aberrant colour blotches and mottled colour patterns on the dorsal and lateral areas of the fish. Ouchi and Kuroiwa (1963) described similar colour variants of adult pinks caught in the Sea of Japan.

INCUBATION

Fertilized eggs begin their five- to eight-month period of embryonic development and growth in intragravel interstices. To survive successfully, the eggs, alevins, and pre-emergent fry must first be protected from freezing, desiccation, streambed scouring or shifting, mechanical injury, and predators. Water surrounding them must be non-toxic and of sufficient quality and quantity to provide basic requirements of suitable temperatures, adequate supply of oxygen, and removal of waste materials. Collectively, these requirements are, on average, only partially met even under the most favourable natural conditions. Overall freshwater survival of pink salmon from egg to alevin, even in highly productive streams, commonly reaches only 10%–20%, and at times is as low as about 1%.

Factors Affecting Incubation

McNeil (1966b) studied the mortality of five broods of pink salmon eggs in spawning beds of three southeastern Alaska streams and found that the critical environmental factors were supply of dissolved oxygen, stability of spawning beds, and freezing. Rates of egg development, survival, size of hatched alevins, and percentage of deformed fry are related to temperature and oxygen levels during incubation. Temporary low stream temperatures or dissolved oxygen concentrations, however, may be relatively unimportant at some developmental stages, but lethal at others (Alderdice et al. 1958; Bailey and Evans 1971; Dyagilev and Markevich 1979). Generally, low oxygen levels are non-lethal early, but lethal late in development. Alderdice et al. (1958) observed that eggs subjected to low dissolved oxygen levels hatched prematurely at a rate dependent on the degree of hypoxia. Bailey and Evans (1971) found that spinal deformities occurred in eggs incubated at 3.0° and 4.5°C before gastrulation. Smirnov (1975) noted that over 50% of developing pink salmon eggs died at dissolved oxygen levels of 3–4 mg/l, and among those that hatched many alevins were deformed. Wells and McNeil (1970) reported that the largest and fastest developing pink salmon embryos and alevins in Sashin Creek came from spawning gravels with high levels of dissolved oxygen in intragravel water. They found larger alevins in the upper area, which had coarser substrate and steeper gradient, than in downstream areas (Table 6); the upper area was also characterized by higher levels of dissolved oxygen (Table 9).

Stream-flow levels during incubation and the severity of winter conditions influence successful pink salmon embryo development. Wickett (1958b) felt that high water levels during the early stages of incubation were necessary for good survival of embryos, especially at higher egg densities. This relation was based on changes in intragravel flow of oxygen-bearing water past the eggs. At Nile Creek on Vancouver Island, Wickett (1954) showed that the rate of flow through perforated standpipes set in the gravel was directly related to water level in the stream. He also found that survival of eggs placed in the standpipes was consistent with different oxygen supplies; in two columns of eggs with experimentally controlled flow through the standpipes, 60 layers of eggs survived with a velocity of 1,000 cm/h, whereas only two layers survived at a velocity of 8 cm/h (Wickett 1957). Not all oxygen delivery to eggs, however, is entirely dependent on the velocity of intragravel flow. O'Brien et al. (1978) recognized a second method of transport – natural convection – which develops from dissolved gas-induced density gradients across the egg surface. These authors suggested

TABLE 9
Dissolved oxygen content of intragravel water in the upper, middle, and lower areas of Sashin Creek (southeastern Alaska) in 1962, 1963, and 1965

		Dissolved Oxygen (mg/l)					
		Upper		Middle		Lower	
Date	Water tempera-ture (°C)	Mean	90% confidence limit of mean	Mean	90% confidence limit of mean	Mean	90% confidence limit of mean
1962							
23 August	13	6.9	±0.8	5.0	±1.0	5.2	±0.7
1963							
7 August	13	7.3	±0.6	5.1	±0.5	5.0	±0.6
13 September	12	8.8	±0.6	8.4	±0.6	8.3	±0.6
1965							
16 August	12	9.7	±0.7	8.9	±0.8	8.5	±1.1
13 September	11	6.2	±0.9	3.8	±0.8	2.3	±0.3
22 September	12	7.6	±0.8	4.3	±0.6	2.9	±0.6
Average	-	7.8	-	5.9	-	5.4	-

Source: From Wells and McNeil (1970)

that natural convection can serve as an emergency oxygen transport and metabolite disposal mechanism at low water velocities. With low egg densities and minimal overall biochemical oxygen demand, natural convection would allow some survival in areas with minimal streambed permeability and intragravel flows.

In streams of the Amur District, Eniutina (1972) found that "if the winter level of the rivers is high and they do not freeze completely to the bottom the oxygen concentration is approximately 7.5–10 mg/l, the carbon dioxide concentration 8–20 mg/l, and pH between 6.2 and 5.9. In winters of low water levels and little snow, when the rivers freeze completely to the bottom, their oxygen concentration may, however, fall almost to zero and their carbon dioxide concentration rise to 40 mg/l or more." The effects of severity of winter on successful pink salmon incubation in more northerly latitudes is thus influenced by temporal snowfall and temperature patterns. Freezing of spawning grounds is more extensive in the absence of snow cover than after an insulating snowfall (Davidson and Hutchinson 1943; Levanidov 1964).

Smirnov (1947) provided a dramatic example of the year-to-year variation in winter severity and of how eggs and fry sometimes die en masse because of low oxygen concentrations. He reported that severe freezing caused a decrease of oxygen content in the Beshenaya River (Amur region) from 94.5% of saturation in October to 7.0% of saturation in February during the winter of 1938–39. By contrast, during the following winter of 1939–40, the oxygen content of the river only fell to about 80% of saturation during the same period.

It is well known that large amounts of silt on and in spawning beds adversely influence survival of salmonid eggs and alevins (Cordone and Kelly 1961; McNeil and Ahnell 1964; Cooper 1965; Rukhlov 1969). The effect of silt on survival is generally indirect through reduced permeability of spawning gravels and inadequate intragravel flow to meet oxygen requirements. Wickett (1954) found good survival in the presence of much surface silt where there was spring upwelling through gravel; he concluded that "surface silt of itself did not seem to be lethal." In another paper, however, Wickett (1957) noted that incubation is only part of the cycle in the gravel, and reported that "fine material can hamper the emerging fry." Wickett (1959b) further mentioned that at Jones Creek "a heavy loss of alevins occurred in the spring of 1957 . . . due to nearly a foot of sand covering parts of the spawning bed. Redd sampling indicated 80% survival to hatching but only 10% of the eggs became migrants." Although not clearly indicated by Wickett, the decreased survival apparently was not due to the inability of fry to emerge through the surface deposition but to earlier losses during the alevin stage.

Embryo Development

As with all Pacific salmon eggs, the pink salmon egg is flaccid when extruded during spawning. Immediately upon leaving the body cavity the egg begins to absorb water in a process of imbibition and water-hardening (Hayes 1949; Zotin 1958), during which the egg swells into a turgid ellipsoidal shape, and the internal pressure rises (Alderdice et al. 1984). Fertilization can only occur within a short period after egg extrusion, generally less than 30 seconds (Shuman 1950; Nikolsky 1963; Smirnov 1963), and the outer membrane or egg capsule becomes a toughened, elastic, insoluble, but dissolved gas-permeable protective membrane. According to Hayes (1949), the uptake of water is virtually completed within one hour, but toughening of the outer membrane does not become marked until some two hours after transfer, and it continues for about thirty hours (see also Zotin 1958). Shortly after entering water, the egg capsule becomes highly adhesive for about twenty minutes, a feature that likely helps to retain eggs in the nest pocket until they are buried (Sheridan 1960). Smirnov (1975) showed that the weight of pink salmon eggs increased from 8.0% to 8.5% during hydration. After water uptake, pink salmon eggs vary in diameter from 4 to 7.9 mm and from 118 to 198 mg in weight (Yastrebkov 1966; Eniutina 1972; Smirnov 1975; Groot and Alderdice 1985). Wickett (1975) found that the average diameter of British Columbia pink salmon eggs was 6.9 mm.

Within about 3–8 hours after entering the water, the egg begins to form the plasmic disk or blastodisk (Smirnov 1954). Fertilization occurs (fusion of nuclei) within one hour after insemination (Jensen and Alderdice 1983). At roughly 12–18 hours, if the egg was fertilized, the blastodisk begins to undergo initial cell division. At usual spawning temperatures, which have a dominant influence on rate of development, the egg usually reaches the 4–16-cell stage within the first 24 hours (Velsen 1980).

Within a few minutes of water activation, and before cell division begins, eggs of salmonids become increasingly sensitive to movement or jarring (Jensen and Alderdice 1983) for a period of 15–30 days, depending on water temperatures. Smirnov (1954, 1975) has described in some detail the effects of mechanical agitation on developing pink salmon eggs. In general, this "critical tender stage" continues until gastrulation, myomere segmentation, and epiboly are completed, and faint eye pigmentation has developed (Jensen and Alderdice 1983). According to Ievleva (1951), eye pigmentation in pink salmon begins after about 330 C degree-days of development. Any significant movement or jarring of the egg before this stage is reached can kill the embryo. For example, if late-arriving females superimpose redds on egg pockets left by earlier spawners, eggs from the first spawning may die from mechanical shock even though they are not removed from the streambed.

The amount of shock required to kill an embryo during this time is slight. Smirnov (1954) found, in general, that from 30 minutes after fertilization through 16 days of incubation at temperatures ranging from 9.6° to 7.8°C, over 90% of the pink salmon eggs dropped into water from a height of 40 cm died from the shock. Following 18–19 days of incubation, mortality from the 40-cm drop fell to 63.1%, and after 22–24 days until hatching, loss from this form of shock fell to 5.5% or less. Jensen and Alderdice (1983) noted that sensitivity occurs in three stages in coho salmon eggs following water activation: a lower stage occurring between 10 and 45 minutes before any cleavage, an intermediate stage between 2 and 72 hours spanning early blastomere cleavage phases, and one of maximum sensitivity between 4 and 14 days spanning the gastrulation phase. Soin (1954) pointed out that "under no conditions may the eggs be transported during the period of gastrulation until the closing of the blastopore," which is the completion of epiboly.

Unfertilized eggs also undergo hydration, hardening of the egg membrane, and development of the blastodisk. If left undisturbed, the unfertilized egg can remain alive for up to 38 days at 12°–14°C (Soin 1953). Presumably, at lower temperatures they persist for more than four months under laboratory conditions (Soin 1954). Unfertilized eggs remain very fragile and sensitive to mechanical shock, similar to fertilized eggs before reaching the "eyed" stage of development. The prolonged life period of unfertilized eggs has a definite biological significance. Soin (1954) speculated that "if the unfertilized eggs perished earlier, their decomposition would bring about unfavorable respiratory conditions and the eggs which otherwise develop normally would perish too."

Temperature and Development

The relation between temperature and rate of development in salmonids has been examined in some detail (Hayes 1949; Seymour 1956; Alderdice and Velsen 1978; Crisp 1981; Murray and Beacham 1986; Beacham and Murray 1986). Development is faster at higher temperatures, and the relationship follows a general overall thermal sums pattern at intermediate temperatures.

For pink salmon, use of accumulated Celsius degree-days provide a rough approximation of thermal units from the time of spawning to hatching, and from hatching to emergence. Sheridan (1962a), using several data sources, estimated that the required temperature units for pink salmon hatching ranged from 100 to 162 degree-days, and for reaching migrant fry stage, ranged from 889 to 1,000 degree-days. Bailey et al. (1980) provided data on 1971-brood pink salmon incubation temperatures in hatchery gravel incubators at Auke Creek, Alaska (Figure 16) that permit the following estimates. Hatching occurred from 100 to 120 calendar days after fertilization, between roughly 530 and 610 degree-days. Emergence occurred from 204 to 220 calendar days after fertilization, roughly between 900 and 980 degree-days. For pink salmon incubated in the Tsolum River (British Columbia) and in hatchery gravel incubators using Tsolum River water, Bams (1972) estimated 970 and 950 degree-days from spawning to emergence, respectively.

Godin (1980a) incubated sibling pink salmon alevins in simulated redds from hatching to emergence at six temperatures ranging from 3.4° to 15.0°C. He found that cumulative thermal units from hatching to 50% emergence for these temperatures varied from 447 to 642 degree-days. Excluding the 15.0°C redd, which experienced a considerably higher temperature than most pink salmon stocks encounter during the alevin stage, the cumulative thermal units required to 50% emergence increased with each higher temperature, whereas the time to 50% emergence (in days) was negatively correlated with all temperatures (Table 10).

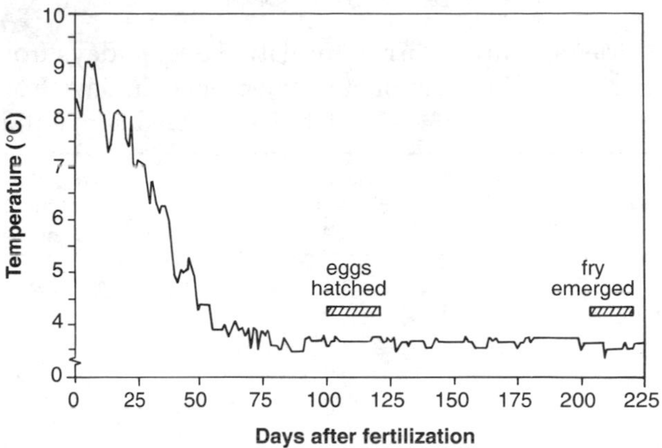

FIGURE 16

Incubation temperature of pink salmon eggs at Auke Creek Hatchery in southeastern Alaska from fertilization, 16 September 1971 (day 0) until termination of the experiment, 28 April 1972 (day 225) after fry emerged from incubators. Horizontal bars show when eggs hatched and fry emerged. (From Bailey et al. 1980)

There are likely many stock differences between and within pink salmon populations that permit adaptive adjustments in spawning, hatching, and emergence timing to thermal characteristics of individual streams. Beacham and Murray (1986) found that developmental differences in odd-year pink salmon from Vancouver Island and the Fraser River (southern British Columbia) were likely due to different physiological adaptations to water temperatures between these two groups of stocks. Sheridan (1962a) argued that different mean spawning times in southeastern Alaska streams with different thermal regimes have developed to coincide with optimum survival opportunities for fry emerging and migrating into estuaries the following spring.

At Auke Creek in southeastern Alaska, the pink salmon run comprises distinct early- and late-run spawners (Taylor 1980). Eggs deposited from these two groups experience different thermal histories due to differences in seasonal temperature regimes. The 1972-brood early-run eggs spawned on 11 August had a 50% emergence date of 30 March 1973, with 1,145 degree-days, whereas late-run eggs spawned on 22 September had a 50% emer-

TABLE 10

Mean temperature, cumulative degree-days from hatching to 50% emergence, duration of emergence period, and timing of 50% emergence (in days from hatching) for sibling pink salmon fry incubated in six simulated gravel redds maintained at different mean temperatures

Redd number	Mean temperature (°C)	Degree-days, hatching to 50% emergence	Duration of emergence period (days)	Timing of 50% emergence (days)
1	3.4	447	17	130.1
2	5.0	556	23	106.0
3	7.9	588	13	77.5
4	9.9	605	21	61.5
5	12.3	642	25	61.0
6	15.0	515	35	35.8

Source: Adapted from Godin (1980a)

gence date of 20 April 1973, with 935 degree-days (J. Taylor, National Marine Fisheries Service, Auke Bay Laboratory, Auke Bay, Alaska, pers. comm.). A difference of over 200 degree-days between fertilization and mean emergence in progeny of early- and late-run spawners may be an adaptation by the early-run fish to prevent fry emergence and migration into the estuary in January or February, a time when there would be little opportunity for growth and survival in this region.

Murray and Beacham (1986) examined the effects of varying incubation temperature regimes on six odd-year pink salmon stocks in southern British Columbia. By experimentally adjusting temperatures at the time of fertilization, hatching, and emergence, they found that the duration from fertilization to 50% emergence in the Chilliwack River stock varied from 98.4 to 287.0 days. These authors found that initial incubation temperatures have a greater effect on the rate of development to hatching and emergence than do later incubation temperatures.

Hatching

Salmonid embryos develop specialized unicellular glands for secretion of a proteolytic hatching enzyme to soften and weaken the egg membrane (Hayes 1942; Smith 1957; Popov and Zotin 1961). According to Popov and Zotin (1961) these cells initially appear in salmon embryos at the onset of eye pigmentation, shortly after the heart begins to beat. Action of the enzyme, secreted only at the time of hatching, is relatively rapid based on changes in mechanical and chemical properties of the egg membrane (Smith 1957). Release of the enzyme may be triggered by low oxygen levels and other stimuli (Hayes 1942). When the enzyme is released, a variety of movements of the embryo helps distribute the enzyme and aids in breaking through the capsule wall (Bams 1969; Poy 1970).

After hatching, the pink salmon alevin, with a well-developed circulatory system surrounding the yolk sac surface, which is capable of respiratory function (Soin 1954), now has a greatly increased capability for responding to continued developmental changes as well as changes in the intragravel environment. Freed from the constraints of gas transfer across the capsule, salmonid alevins have a much greater capacity for oxygen uptake after hatching than before (Hayes et al. 1951). They also diffuse micro-concentrations of carbon dioxide through "ventilation swimming" and disperse fine sediment by "coughing" a reverse flow of water across the gills, or by secreting mucus to entangle and remove silt particles on the gills (Bams 1969).

Intragravel movements of salmonid alevins relate to an array of phototactic, rheotactic, thigmotactic, and geotactic responses that vary with alevin age (Stuart 1953; Woodhead 1957; Marr 1963; Bams 1969; Dill 1969, 1982). Bams (1969) found that pink and sockeye salmon alevins develop both geotactically-directed emergence behaviour and positive rheotactic behaviour near the end of the intragravel period. Development of intragravel behavior in pink salmon, according to Dill (1982), follows three ontogenetic phases: righting phase, bottom-swimming phase, and surfacing phase. The bottom-swimming phase results in horizontal and downward dispersal of pink salmon alevins in the gravel.

The extent of vertical and lateral movement within the gravel for chum (Dill and Northcote 1970) and coho salmon (Dill 1969) was directly related to gravel size. Greater vertical movement occurred in large gravel with more interstitial space than in small gravel (Figure 17). Similar responses have been noted in pink salmon and other *Oncorhynchus* species.

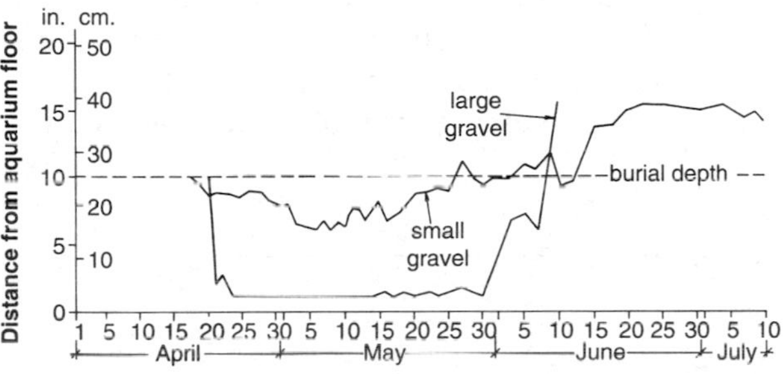

FIGURE 17

Mean vertical positions of coho salmon alevins in experimental aquaria, showing differences in alevin movement in large and small gravel. (From Dill 1969)

Bams (1969) found that alevins preferred an upright position beginning shortly after hatching and that they "never cease to work toward obtaining and retaining" this orientation. When pink salmon alevins were maintained at 10°C, Dill (1982) observed that this righting phase extended from hatching for about fifteen days. Although alevins can move fairly extensively in gravel according to Bams (1969), he also found that "under favourable conditions, which include darkness, even fairly large concentrations of alevins complete their development in the very crevice they hatched in, and their first movements through the gravel occurred at emergence."

Dill (1982) documented changes in length and weight (wet and dry) for pink salmon alevins from hatching through emergence at 10°C. Yolk weight, which represents 57% of alevin wet weight at hatching, declined to about 3% at the time of emergence from the gravel, and 49% of the available yolk was converted to tissue. Alevin length increased from 21.3 mm at hatching to 31.5 mm at emergence. Emergence of conspecifics ranged from 41 to 64 days post-hatch and averaged 52 days.

Intragravel Predators and Scavengers

Several studies have considered the effects of intragravel predators and scavengers on survival and abundance of salmonid eggs and alevins. McLarney (1964) found that coast-range sculpins (*Cottus aleuticus*) could readily move through coarse gravel in Sashin Creek. Larger sculpins, 7.0–8.4 cm long, fed indiscriminantly on live and dead pink salmon eggs; sculpins, 5.0–6.4 cm long, consumed only a few eggs, apparently due to difficulty in ingesting eggs too large for their gape. McLarney (1967) reported a seasonal movement of large sculpins into the coarse streambed area of upper Sashin Creek during the peak of pink salmon spawning. The ability of reticulate sculpins (*C. perplexus*) to move into gravel varied inversely with the size of the fish and directly with the size of the gravel (Phillips and Claire 1966).

Earp and Schwab (1954) described a heavy infestation of freshwater leeches (*Piscicola salmositica*) on eggs and fry of pink salmon at the Hood Canal Hatchery on Finch Creek in Washington. Apparently no damage occurred to unhatched eggs from attachment by one or more leeches. When leeches attached to alevins, however, they quickly became engorged with blood and the alevins invariably died. Losses from leech attacks in some hatchery baskets were estimated at 25%. The leeches attached to the head, eyes, and caudal peduncle of alevins.

Dead Embryos, Decomposition, and Scavengers

Several authors have found that dead pink salmon eggs persist, under certain conditions, in spawning gravels for many months (Hunter 1959; Ricker 1962; McNeil et al. 1964; Brickell 1971). In extreme cases, dead eggs can last from one spawning season to the next and may compete with the subsequent brood for oxygen needed for continued decomposition of the dead eggs. The persistence of dead eggs in gravel appears primarily related to the rate of aerobic bacterial decomposition, which in turn depends on oxygen delivery to the egg. If the egg shell is broken and oxygen is available, decomposition occurs at a faster rate than otherwise (Brickell 1971).

Several studies have considered the role of carnivorous invertebrates in the intragravel ecology of salmonid spawning beds, but most have difficulty in determining, with certainty, if the invertebrates are preying on live eggs or alevins, or scavenging dead ones (Briggs 1953; Stuart 1953; McDonald 1960; Astafeva 1964; Nicola 1968; Claire and Phillips 1968; Ellis 1970; Elliott and Bartoo 1981). The following observations relate to pink salmon. Helle et al. (1964) found that large numbers of oligochaetes were common in highly productive pink salmon spawning beds in Prince William Sound (Alaska), and Heard (1978) observed that the numbers of the planarian worm *Polycelis borealis* seasonally increased rapidly in pink salmon spawning beds where dead eggs and alevins were abundant. In experimental containers, planarians were not a hazard to live embryos. Elliott and Bartoo (1981) investigated the relationship of abundant chironomid larvae, *Polypedilum*, to pink salmon eggs and alevins and found that this dipteran was associated mostly with dead eggs heavily infested with fungus. The *Polypedilum* larvae apparently feed on fungus. Both Nicola (1968) and Ellis (1970) concluded that stonefly nymphs were scavengers and probably not predators on pink salmon embryos in the gravel.

Studies reported to date suggest that many invertebrates found in pink salmon spawning gravels have a beneficial effect on intragravel ecology through assisting in the removal of dead eggs and alevins, thus reducing overall oxygen demand.

Emergence

Emergence from gravel initiates the first migration in a highly complex series of major habitat changes of Pacific salmon (Godin 1982; Groot 1982). The movement of fry from gravel, like subsequent migrations, features innate responses and adaptive behaviour, differing somewhat according to species, which optimize survival opportunities.

Guidance Mechanisms, Geotaxis, and Rheotaxis

When pink salmon fry are ready to emerge they start to swim facing up, pushing discontinuously upward through gravel crevices (Dill 1982). Movements through gravel are slow and restrained, and alternate between periods of rest and activity. The course traversed through the streambed depends on endogenous factors as well as gravel composition, compaction, and perhaps micro-environmental differences in crevices. The area traversed in uniform gravel has the shape of an inverted cone. Bams (1969) observed intragravel progress by fry of 5 cm or more in length per minute under favourable conditions. He also documented a series of specialized locomotive movements that aided progress through the gravel, including "butting" behaviour, snake-like sliding movements, and different methods of reversing direction when fry entered a narrow, blocked crevice. The primary guidance mechanism for emergence of pink and sockeye salmon fry appears to be a positive geotactic response (Bams 1969).

Rheotaxis is a second guidance mechanism aiding emergence movements through gravel, although not all *Oncorhynchus* fry respond to intragravel currents in the same manner. Bams (1969) suggested that positive rheotaxis by pre-emergent pink salmon fry is probably used only when the geotactic mechanism is blocked by obstructions. Positive rheotaxis would tend to move pink salmon fry laterally upstream and eventually near the surface. Spawning riffles of this species are characterized by surface stream water flowing downward through the gravel (Semko 1954; Soin 1954; Sheridan 1962b; Kobayashi 1968a). Once free of gravel under natural conditions of darkness, pink salmon fry were found by Pritchard (1944a) and Neave (1955) to show strongly negative rheotaxis.

Other Factors Affecting Emergence

Pink salmon fry emerge predominantly at night (Figure 18), a behaviour considered to be an antipredator adaptation (Neave 1955; Godin 1980a, 1982). Shortly after hatching, salmonid alevins exhibit strong photonegative behaviour (Stuart 1953; Woodhead 1957; Bams 1969; Dill 1969; Carey and Noakes 1981), which may initiate and maintain the initial downward movement in pink salmon alevins (Dill 1982). Continuous light inhibits the emergence behaviour of both sockeye and pink salmon fry (Heard 1964; Bams 1969). Late in the season, however, pink salmon fry showed an increased tendency to emerge during daylight (Godin 1980a). The diel pattern of emergence is an important synchronizer of emergence of pink salmon fry from natural streambeds.

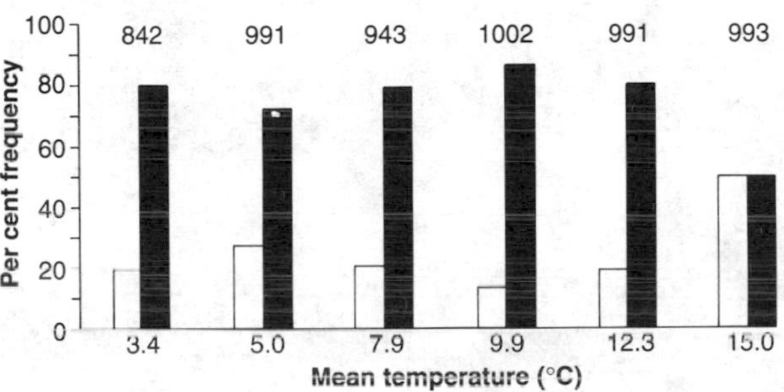

FIGURE 18

Frequency (per cent) of sibling pink salmon fry emerging from simulated redds during the day (open bars) and night (solid bars) of a 12-hour L:12-hour D cycle at different mean temperatures. The number of emerged sibling fry is indicated above the bars for each redd temperature. (From Godin 1980a)

Temperature also affects the emergence of pink salmon fry. In simulated redds maintained at temperatures from 3.4° to 12.3°C, Godin (1980a) found that about 80% of pink salmon fry emerged at night (Figure 18). However, in redds maintained at 15.0°C these fry emerged equally at night or day. Coburn and McCart (1967) transferred advanced pink salmon alevins to a 1.2-m-deep release tank in darkness and allowed them to emerge and migrate voluntarily. Water originated from a Skeena River tributary; the tank and alevins were covered and kept in darkness. They found that the number of fry leaving the tank without a normal diel light pattern was directly related to daily water temperature fluctuations. Few fry left when temperatures were low and large numbers left when temperatures were high. Bams (1969) suggested that "in nature the daily light and temperature cycles cause fry migrants to accumulate just below the surface of gravel beds towards dusk when the water temperature is high. Upon nightfall the inhibitory action of the light is removed, and the migrants enter the open water. In a matter of hours most of the accumulated fish have emerged, the daily run is past its peak, and numbers drop off to a low level. These latter fish are those that have only just reached the surface of the gravel. Towards dawn the water temperature is low, fish activity drops, and light will again inhibit any further emergence."

With the development of hatchery techniques by which many variables can be controlled, alevin density, flow patterns, velocity, and gravel characteristics have been found to influence pink salmon emergence (Bailey et al. 1980). For example, Bams (1979) reported that changes in emergence timing of pink salmon fry from round or crushed gravel were presumably related to greater void space in the latter. He also found cumulative emergence occurred earlier and that the rate of emergence increased with increased gravel size.

Involuntary "emergence" may occur in nature when floods scour streambeds and flush pink salmon alevins from the streambed (Pritchard 1938, Davidson and Hutchinson 1943; Khorevin et al. 1981). In other circumstances, reduced streamflows may drop water levels so that viable fry become trapped without access to free-standing water for emergence (Pritchard 1944a, 1948a).

At Kleanza Creek (British Columbia), Coburn and McCart (1967) found that increased turbidity from rising stream-flows in the absence of a normal diel photoperiod caused large numbers of pink salmon fry to leave, even when water temperatures were falling. These authors also observed that the experimental addition of silt to sockeye salmon release tanks at Scully Creek caused an immediate increase in fry output.

In simulation studies, early emergence of pink salmon fry has also been reported in response to experimentally elevated ammonia levels above 10 ppb (Rice and Bailey 1980), and to high densities of alevins in gravel incubators (Bailey et al. 1980). Fry emergence is therefore the result of behaviour processes which are influenced, both in a stimulating or inhibitory way, by environmental factors.

Initial Filling of Swim Bladder

A key moment of emergence behaviour for pink salmon occurs when the fry move into open water above the gravel and swallow air to achieve neutral buoyancy. Salmonids are physostomic, i.e., they have an open pneumatic duct connecting the esophagus with the swim bladder which must initially be filled by swallowing air (Tait 1960). Until this happens, fry are negatively buoyant. Initial filling of the swim bladder is synonymous with the "swim-up" fry phase, a term used in salmonid hatcheries. Dill (1982) has described this phenomenon for pink salmon.

When the fish first frees itself from the gravel, usually with a vigorous vertical swimming thrust or quick darting movement, it may rest on the gravel surface or make short swimming forays into the water column. Initial forays are followed by non-swimming as the fish drops back through the water column, presumably to rest. These characteristic swimming movements are often vigorous and mostly vertically oriented. Eventually, the fry swims upward to the surface of the water and performs rapid "hovering" (Dill 1982) swimming movements, often with the tip of the rostrum just out of the water. Here, small bubbles of air are taken into the mouth by gulping at the surface, at times with a lateral head thrust. Some of the air is transferred to the swim bladder as the fry "munches" on the bubble (Bams 1969). The process is usually repeated several times before individual fry have sufficient air in the swim bladder to achieve neutral buoyancy and to maintain a horizontal body position at rest in the water column.

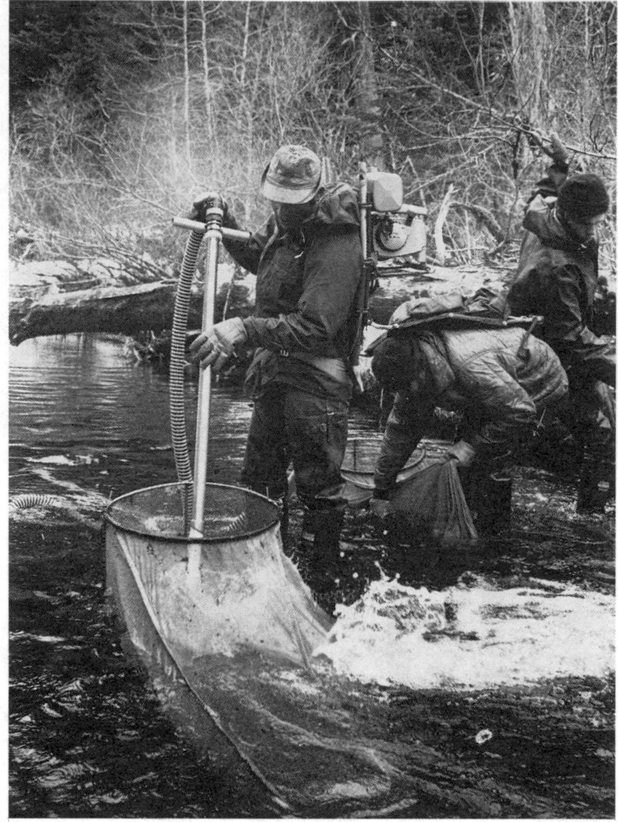

FIGURE 19

Pre-emergent pink salmon fry sampling using a backpack-operated hydraulic sampler to inject an air-water mixture into the streambed, forcing eggs, alevins, and fry to the surface where they are washed into collecting bags. (Photograph courtesy of the Alaska Department of Fish and Game)

Natural Densities of Pre-emergent Fry

The density of live pre-emergent pink salmon alevins and fry present in spawning gravels at the end of incubation provides a useful measure of incubation success. Pre-emergent fry densities from representative streams are used in some regions as an annual index of reproductive success to forecast future run strength. Kingsbury (1979) reviewed this process as used in several regions in Alaska. Areas were sampled randomly with a variant of the hydraulic streambed sampler (Figure 19) described by McNeil (1964b) to estimate egg and embryo densities in gravel. Timing of sampling varies by region and somewhat by seasonal weather conditions. In general, it is carried out from mid-February to mid-April when yolk reserves are about 80% absorbed. Roys (1968) summarized seven years of pre-emergent sampling in Prince William Sound and found alevin densities ranged from 102 to 315/m^2 in the intertidal zone, and from 136 to 474/m^2 in the freshwater zone; densities were lower in the intertidal zone in all years but one (Table 11). Jones (1978), however, found that intertidal areas in southeastern Alaska produced more fry per square metre than upstream areas. He also found that lake-influenced streams produced more fry per unit area than non-lake streams, presumably due to the stabilizing

TABLE 11

Proportions of pink salmon spawners using intertidal and freshwater areas of standard Prince William Sound index streams, mean pre-emergent fry density in these areas, weighted pre-emergent fry density by year, and subsequent adult returns to Prince William Sound for 1960–66 brood years

Brood year	Spawners use of streams (%)		Mean pre-emergent fry (/m²)		Weighted pre-emergent fry (/m²)	Adult return (millions)
	Intertidal	Fresh water	Intertidal	Fresh water		
1960	77	23	315.3	474.0	351.8	8.7
1961	35	65	180.4	317.2	269.3	6.6
1962	70	30	257.2	286.7	266.1	6.0
1963	46	54	118.5	194.9	159.7	3.4
1964	65	35	187.1	135.9	169.1	4.0
1965	37	63	110.0	177.2	152.3	3.8
1966	65	35	102.5	203.5	137.8	(3.1)

Source: From Roys (1968)

effect of the lake on flow and temperature.

Dangel and Jewell (1975) and Jones and Dangel (1982) summarized pre-emergent pink salmon fry densities in 12 southeastern Alaska districts from 1963 through 1981. Data were compiled annually from about 95 streams. Although numbers of pre-emergent fry varied widely between streams, districts, and years, they generally tended to reflect escapement levels from the previous season and effects of over-winter environmental conditions on progeny. During a four-year period, 1967–70 (Table 12), the mean number of pre-emergent fry by district ranged from 22/m² in District 112 in 1969 to 206/m² in District 107 in 1967 (Dangel and Jewell 1975). Maximum numbers recorded ranged from 1,375/m² in District 106 in 1968 to 4,950/m² in District 101 in 1970. In all years, streams, and districts, the minimum number of pre-emergent fry recorded for individual sample points was zero.

TABLE 12

Number of pre-emergent pink salmon fry samples taken, and mean and maximum annual numbers of fry collected per square metre of spawning ground from 12 districts in southeastern Alaska, 1967–70

District	1967			1968			1969			1970		
	N	Mean (/m²)	Maximum	*N*	Mean (/m²)	Maximum	*N*	Mean (/m²)	Maximum	*N*	Mean (/m²)	Maximum
101	515	88	2555	998	54	2225	1012	99	2650	934	111	4950
102	200	88	1635	329	61	2845	354	108	2710	345	195	2400
103	650	197	3360	779	92	2625	700	150	2870	1052	100	2525
105	137	64	1500	312	81	2740	314	73	2635	351	53	1440
106	245	158	3120	372	76	1375	459	127	2930	482	77	1575
107	350	206	2550	341	132	1535	286	173	1965	329	153	2740
109	199	135	1790	384	91	1895	357	134	1940	506	99	3150
110	223	176	4550	547	41	2005	318	177	2940	585	69	1885
111	160	151	2375	486	32	1400	475	115	3265	536	75	3075
112	424	172	2590	802	46	1610	825	22	2185	795	133	2400
113	550	114	3135	1003	76	2105	1026	65	2150	1069	89	2625
114	200	100	1590	514	58	2710	447	83	2280	410	201	3035

Source: Adapted from Dangel and Jewell (1975)

Fry Migration

Behaviour

After emergence and after achieving neutral buoyancy, pink salmon fry migrate quickly downstream at the surface. They spend, on average, less time in fresh water after leaving the gravel than other *Oncorhynchus* species. At emergence, pink salmon fry (Plates 7 and 8) are greenish along the back, silvery along the sides and belly, and without parr marks or pigmented spots (Bailey 1969; Eniutina 1972). Changes in behaviour occur rapidly during this period (Neave 1955; Hoar 1956, 1958, 1969; Hoar et al. 1957; J. McDonald 1960). These include marked changes from negative to positive or neutral responses to light and current and the development of schooling behaviour to replace individual responses. In a study on the extent of predation by crows on recently schooled groups of pink, chum, coho, and sockeye salmon fry in shallow experimental troughs, Hoar (1958) found much higher predation on pink salmon fry than on the other species. Pink salmon fry would not abandon the school when attacked, as did the others, to hide under stones in the simulated streambed. Hoar noted the obvious disadvantage of strong schooling behaviour in pink salmon fry in shallow streams and concluded that a rapid exodus from the rivers to the ocean is vital for their survival. The rapid behavioural changes during fry migration in response to light, current, and the development of schooling behaviour occur concomitantly as migrants reach tidewater at the stream mouth.

Marked schooling behaviour is evident when pink salmon fry reach the lower reaches of large rivers; (see J. McDonald (1960) for the Skeena River and Vernon (1966) for the Fraser River). Bakshtansky (1970), however, suggested that pink salmon fry migration in small rivers is performed by individual non-schooled fish that respond negatively to light, whereas schooling and exogenous feeding occur only in large rivers or estuaries.

Pritchard (1944a) recorded rates of travel of 4.6–6.1 m/s, and Wickett (1959c) observed speeds about 0.5 m/s in excess of the current. Under minimum illumination, a "bow-wave" (Neave 1955) or "wake" (Pritchard 1944a) could be seen as fry hurried downstream.

A broad area of bright illumination at night or an obstacle to migration such as a counting fence caused downstream migrant fry to swing about with head upstream and frequently bury themselves in the stream bottom (Pritchard 1944a; Neave 1955). Pritchard also noted that the "migration slowed abruptly" at McClinton Creek (Queen Charlotte Islands) during periods of bright moonlight.

Seasonal Timing

The seasonal timing of downstream migration of pink salmon fry varies widely by region and from year to year within regions and individual streams. Throughout the range of this species, downstream migrant fry can be found from late February in the Fraser River (Vernon 1966) to mid-August in some rivers of the Amur basin (Eniutina 1972). In some Amur streams, and in some years, the beginning of the fry migration starts when the rivers are still covered with ice (Eniutina 1972).

Figure 20 reviews the seasonal fry migration timing for 17 stocks of pink salmon, including the maximum seasonal and peak migration periods. The wide variation in seasonal timing is related to environmental features of the regions, especially mean annual stream and estuarine temperature regimes; stock characteristics, including timing and duration of spawning; and seasonal climatic aspects of individual brood years. In general, the reported peak periods for fry migration in British Columbia, southeastern Alaska, and Hokkaido occur from mid-April to mid-May, whereas those for Kamchatka, Sakhalin Island, the Amur basin, the Okhotsk coast, and the Kola Peninsula occur from late May to mid-June. In some Amur basin streams, the main fry migration extends into late June.

The duration of pink salmon fry migration for seven Sakhalin Island streams over a six-year period varied from 53 to 72 days (Tagmaz'yan 1971); the number of days showed a positive correlation with stream length.

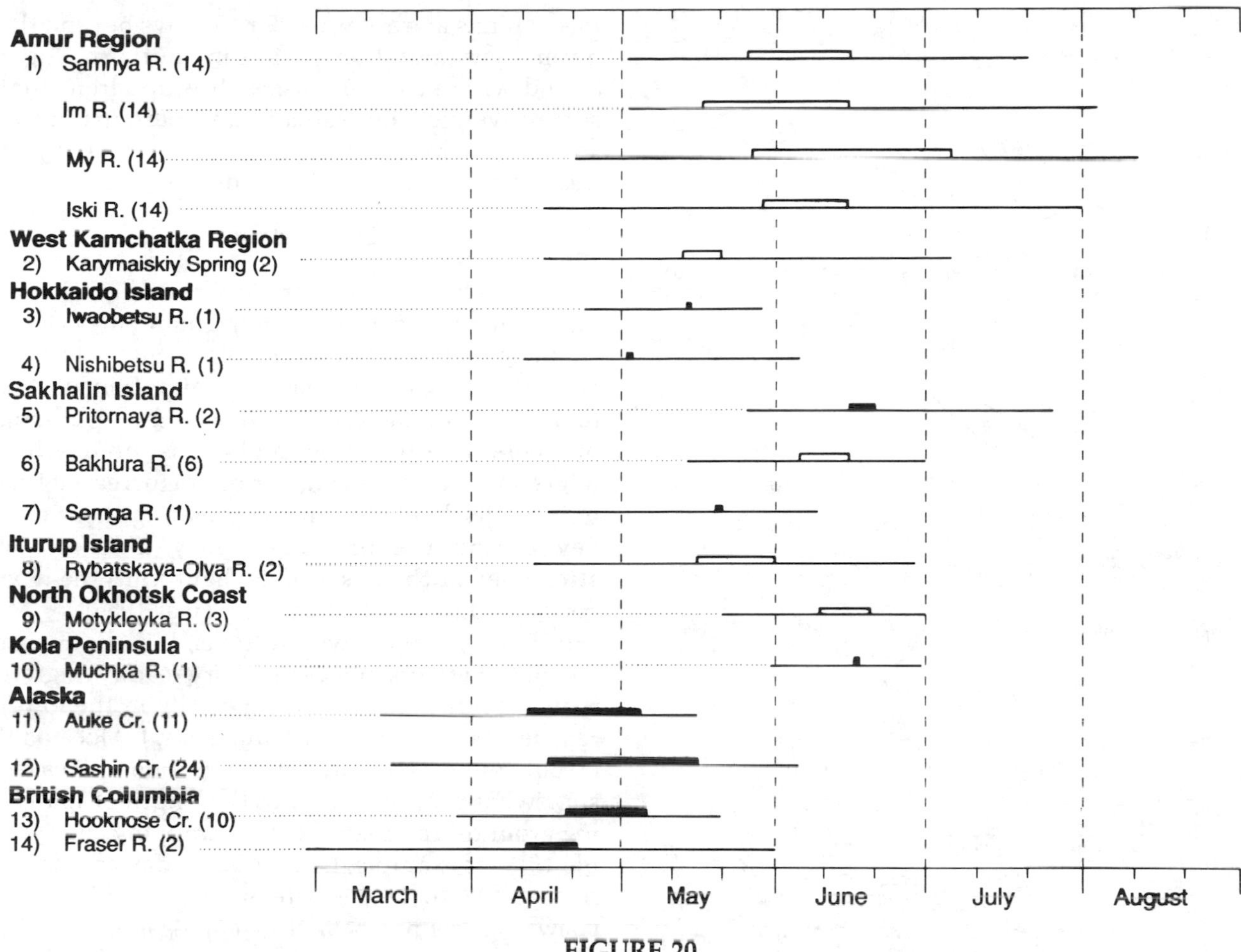

FIGURE 20

Seasonal timing of downstream pink salmon fry migrations reported for stocks in different regions. Horizontal line identifies the inclusive period of migration, closed bars above line include the range of dates when 50% of the migration occurred, open bars show range of dates when the "main" or "mass" migration occurred. Numbers in parentheses show number of annual observations for different streams. (Based on data from: (*1*) Eniutina 1972; (*2*) Semko 1954; (*3*) Kobayashi 1968a; (*4*) Kobayashi and Harada 1966; (*5*) Shershnev and Zhul'kov 1979; (*6*) Tagmaz'yan 1971; (*7*) Khorevin et al. 1981; (*8*) Ivankov 1968c; (*9*) Golovanov 1982; (*10*) Kamyshnaya 1967; (*11*) Taylor 1983; (*12*) Olson and McNeil 1967; (*13*) Hunter 1959; (*14*) Vernon 1966)

Cumulative annual fry migration curves based on daily numbers of migrants for Sashin Creek and Auke Creek (southeastern Alaska), the Fraser River (British Columbia), and the Pritornaya River (Sakhalin Island) (Figure 21) show that the slopes are somewhat steeper and more compressed seasonally for the smaller streams than for the Fraser River. This difference may be related to the length of spawning grounds upstream from stream mouth trapping areas, to greater numbers of stocks in the larger river, or both.

As noted in the previous section, once migrating, fry move quickly downstream out of fresh water to the estuary. However, not all fry from all stocks leave fresh water shortly after emergence. As indicated by the successful adaptation of the Lakelse River stock in the Great Lakes, some pink salmon can grow and reproduce successfully without leaving fresh water. Koo (1962) collected 108-mm-long pinks in August from Lake Aleknagik, Alaska, that had obviously grown well in fresh water. These "parr" had well-developed scale pat-

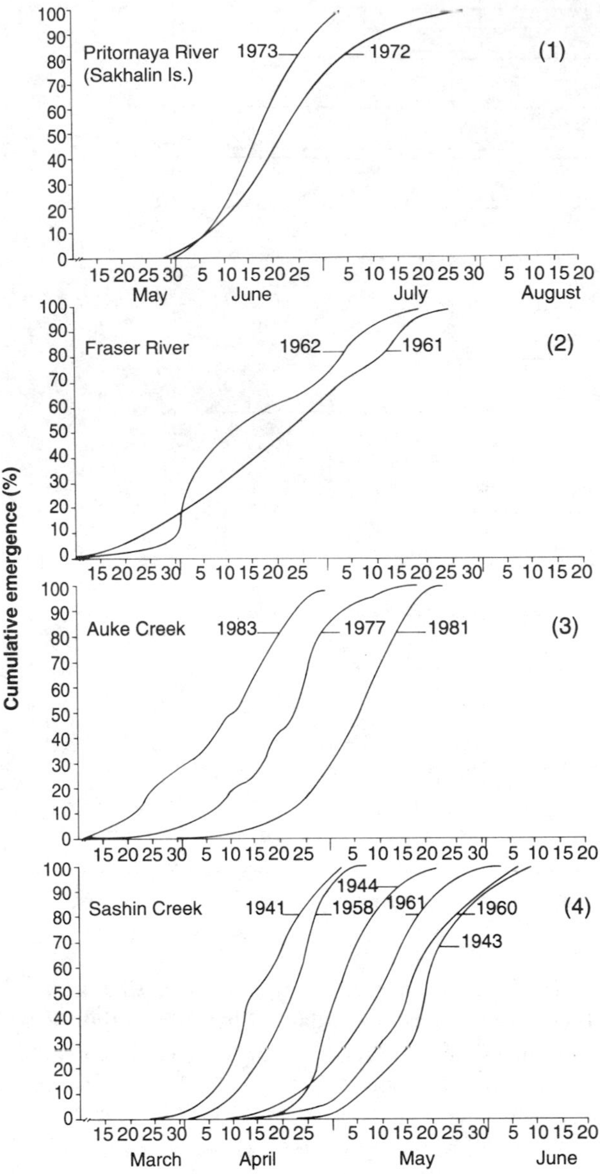

FIGURE 21

Cumulative annual downstream migration curves based on daily numbers of fry leaving some pink salmon streams. Years indicate broods; note monthly scale change for Pritornaya River. (Based on data from: (*1*) Shershnev and Zhul'kov 1979; (*2*) Vernon 1966; (*3*) Taylor 1983; (*4*) Olson and McNeil 1967)

terns similar to pinks caught in coastal waters. Ivankov (1968a) found some late-leaving pink fry actively feeding in the Kurilka River on Iturup Island in August, after the regular downstream migration ended in June. Senn and Buckley (1978) raised pinks in fresh water for 135 days beyond the normal fry migration period in southern Puget Sound without undue losses. It is apparent that some juvenile pink salmon, and perhaps entire specific stocks, are quite capable of remaining in fresh water for extended periods.

Diel Patterns

In small, short coastal streams, the migration of pink salmon fry from spawning ground to ocean is commonly completed in one night. Such streams are characterized by peak diel migrations in the first hours of darkness, followed by declining numbers the same night and almost no daytime migration, except during periods of increased turbidity. Unschooled fry not completing their journey during the first night may hide in the streambed at the onset of daylight until the next period of darkness (Neave 1955; Hoar 1956).

In larger, longer river systems, like the Fraser and Skeena rivers, pink salmon fry readily migrate seaward during daylight, especially in the lower reaches. In Skeena River tributaries, J. McDonald (1960) found that daytime migration increased somewhat proportionally to the length of spawning grounds upstream from the trapping sites (Table 13). For example, Lakelse River day catches of fry were zero at one site with only 5.6 km of spawning ground above the trapping area, but day

TABLE 13

The percentage of pink salmon fry caught during day and night at different locations in the Skeena River system and the length of spawning grounds located upstream from the trapping site

River site	Year	Fry caught: Night (%)	Fry caught: Day (%)	Length of spawning grounds upstream from site (km)
Lakelse upstream	1956	100	0	5.6
Lakelse downstream	1957	90	10	19.2
	1958	98	2	19.2
Kalum	1956	69	31	35.2
Kitwanga	1957	61	39	40.0
	1958	89	11	40.0
Kispiox	1956	35	65	88.0
	1957	53	47	48.0
	1958	69	31	80.0

Source: From J. McDonald (1960)

catches averaged 6% of total fry output over two years at a second site farther downstream, with a 19.2-km length of spawning ground. In the Kispiox River, with 88 km of spawning ground upstream of the trapping site, 65% of all migrant fry were caught during the day in 1956.

In the lower Fraser River, Vernon (1966) sampled migrant pink salmon fry during three 8-hour periods throughout the season to estimate total fry migration from this system. Although many vagaries are involved in a sampling effort of this magnitude, Vernon found that the portion of total numbers of fry caught during the three diel periods, 0500–1300 hours, 1300–2100 hours, and 2100–0500 hours, was approximately 30%, 50%, and 20%, respectively.

Under conditions of arctic daylight, Kamyshnaya (1967) found that some pink salmon fry migrated downstream in the Muchka River on the Kola Peninsula throughout the 24-hour period.

In laboratory studies, Godin (1980b) found important ontogenetic changes in daily rhythms of pink fry behaviour in the first few days following emergence, the time when wild fry begin their marine life. During the first week, fry exhibited irregular patterns of daily swimming activity with a tendency to swim near the surface more at night than at day. In the second week and thereafter, fry developed a distinct diurnal rhythm of swimming activity and a greater tendency to swim in the upper part of the water column during the day. These patterns were considered remnants of the negative photoresponse from the intragravel period, and new adaptations for diurnal feeding in nearshore habitats (Godin 1980b).

Vertical and Horizontal Distribution

Once the migration is under way, vertical and horizontal distribution of pink salmon fry in streams varies considerably, presumably according to behavioural responses and the physical characteristics of different streams. At Datlamen Creek on Graham Island, British Columbia, Neave (1955) reported that 62% of the nightly catch occurred in the upper one-third layer where the stream was about 0.5 m deep. Farther downstream in slack water under tidal influence, migrants were distributed uniformly from top to bottom. J. McDonald (1960) found pink salmon fry distributed throughout the total depth of water in Skeena River streams, with greatest catches made at intermediate depths. In shallower (generally 1.0–2.4 m deep) water, fry were found in the upper one-third of water depth, and a relatively greater proportion of fry were closer to the surface during the day than during the night. Eniutina (1972) found migrant pink salmon fry in the My River to be more or less uniformly distributed in the water column. In the lower Fraser, Vernon (1966) concluded that when the river was turbid, pink salmon fry were generally distributed more or less at random with respect to depth (up to 9.6 m). However, during periods of relatively clear water, fry were concentrated near the surface in daylight. In some of the larger, relatively clear Fraser River tributaries, Vernon (1966) also observed that the vertical distribution of migrant pink salmon fry was concentrated near the surface, except during periods of increased turbidity or in areas of turbulent flow. In these streams (Harrison, Vedder, and Thompson rivers) fry migration was restricted almost entirely to hours of darkness.

Horizontal distribution of migrant pink salmon fry is related primarily to current patterns in the stream. J.McDonald (1960) caught the greatest number of fry in Skeena River tributaries in the fastest water, usually near the centre of the stream. In the larger lower Fraser River, Vernon (1966) found fewer fry along the left bank, where the stream flow was lower, than in the mid- and right-bank sections of the river, which had faster flowing water.

Size of Fry, Feeding, and Diet

Throughout its Pacific range, the mean size of migrant pink fry varies from about 28 to 35 mm in fork length and 130 to 260 mg in weight (Table 14). Fry size varies among years within stocks and among stocks within years in the same region. In 1972, Shershnev and Zhul'kov (1979) found that the mean lengths of migrant pink salmon fry from the Pritornaya, Khvostovka, Bakhura, and Lesnaya rivers on Sakhalin Island were 31.0, 32.8, 34.0, and 33.1 mm, respectively. At the end of alevin growth, length increases decline to zero and weight starts to decline, representing negative growth. Bams (1970) developed a sensitive index for growth, length, and weight of salmonid larvae prior to exogenous feeding to express this relation.

TABLE 14
Mean annual fork length and weights of some migrant pink salmon fry populations

Region	Stream	Brood year	*N*	Mean fork length (mm)	Mean weight (mg)	Source
BC	Hooknose Cr.	1967	381	33.1	243	Bams (1970)
S.E. Alaska	Auke Cr.	1972	650	32.5	260	Bailey et al. (1976)
Amur	My R.	1950	200	32.8	190	Levanidov & Levanidova (1957)
		1952	300	33.2	245	"
		1953	500	33.3	197	"
	Iski R.	1952	300	32.7	236	"
		1953	257	30.3	167	"
	Im R.	1951	800	31.5	210	"
		1952	1260	32.5	230	"
		1953	800	31.8	220	"
	Samnya R.	1950	1000	33.5	250	"
		1951	700	34.1	220	"
		1952	397	34.2	-	"
N. Okhotsk	Turomcha R.	1962	-	30.4	160	Frolenko (1970)
		1963	-	29.7	135	"
		1965	-	30.0	132	"
		1967	-	28.0	143	"
	Taui R.	1965	-	31.3	157	"
Sakhalin	Pritornaya R.	1970	62	31.2	167	Shershnev & Zhul'kov (1979)
		1971	652	31.0	158	"
		1973	375	31.6	210	"
		1974	543	31.2	173	"
Hokkaido	Nishibetsu R.	1962	388*	30.3	174	Kobayashi & Harada (1966)
	Iwaobetsu R.	1965	-	34.9	220	Kobayashi (1968b)

Note: *Calculated from data in Table 6 (Kobayashi and Harada 1966) from two collection sites farthest downstream

Due to their rapid exodus from streams at emergence, pink salmon fry feed less in fresh water than other Pacific salmon. In short coastal streams, high percentages of stomachs are void of food (Table 15), and exogenous feeding for many pink fry begins in salt water (Kobayashi 1968b; Frolenko 1970; Bailey et al. 1975; Golovanov 1982). In larger rivers, where fry migration takes several days, the prevalence of food organisms in stomachs and feeding in fresh water increases (Levanidov and Levanidova 1957; J. McDonald 1960; Eniutina 1972).

Exogenous feeding for some pink salmon fry may occur in the gravel before emergence. Bailey et al. (1975) found food organisms (chironomid pupae and unidentifiable insects) in 4.3%, plant debris in 21%, and fine sand in 47% of pink salmon fry stomachs examined from redds. Exogenous feeding before emergence also occurs in fry of other *Oncorhynchus* species (Disler 1953; Dill 1967).

Sand is commonly found in stomachs of migrant pink fry (Table 15) although it is unclear if this is ingested before or after emergence or whether it is intentional or accidental. J. McDonald (1960) suggested that the presence of sand in stomachs may indicate accidental intake in turbulent water, but Levanidov and Levanidova (1957) implied that it "attests to an insufficient food supply." In Traitors Creek, Bailey et al. (1975) observed sand in considerably more stomachs taken from redds than from stomachs taken later from migrant fry. It seems likely that sand found in recently emerged migrant pink salmon fry stomachs could be ingested during emergence.

Many migrant fry have some residual yolk. The occurrence and amount of yolk generally decrease seasonally. Khorevin et al. (1981) found that 23% of

TABLE 15
Some food habit studies on migrant pink salmon fry in fresh water

Region	Stream and year of fry migration		Stomachs examined *N*	With food (%)	With sand (%)	Source
BC	Lakelse R.	1958	236	2	10	J. McDonald (1960)
	Kitwanga R.	1958	261	3	7	"
	Kalum R.	1956	54	11	21	"
	Kispiox R.	1958	1891	4	6	"
		1956	80	25	26	"
	Skeena R.	1958	51	43	2	"
		1956	40	68	0	"
S.E. Alaska	Traitors R.	1964	40	0	20	Bailey et al. (1975)
Amur	Samnya R.	1951	197	34	*	Levanidov & Levanidova (1957)
		1952	75	15	-	"
	Im R.	1952	284	23	-	"
	My R.	1951	108	8	-	"
N. Okhotsk	Turomcha R.	1963	-	6	-	Frolenko (1970)
		1964	-	20	-	"
		1966	-	13	-	"
		1968	-	10	-	"
	Taui R.	1966	-	20	-	"
	Ola R.	1962	-	40	-	"
Hokkaido	Iwaobetsu R.	1966	322	10	-	Kobayashi (1968b)
	Nishibetsu R.	1963	257†	37	‡	Kobayashi & Harada (1966)

Notes: *Authors report about 20% of empty stomachs contained sand at the end of migration.
†Some data excluded due to apparent error in translation of tables
‡Authors report some specimens with sand in stomachs during a flood.

pink salmon migrants in April from the Semga River on Sakhalin Island contained visible yolk. These fry, associated with the first spring floods, were considered by the authors to be physiologically not ready for migration. Later in May only 4% or less of migrant fry in the Semga River had yolk, even during freshets which might have dislodged them prematurely. Kobayashi and Harada (1966) found that the number of fry with yolk in the Nishibetsu River on Hokkaido ranged from 7.7% to 31.2% in April and May, with the percentage averaging less in May. In April, the amount of yolk averaged 2.2% - 7.3% of body weight, whereas in May, yolk averaged 1.1% - 2.3% of body weight. The proportion of migrants that started to feed ranged from 15% to 44% in April to 18% to 90% in May. On the northern Sea of Okhotsk coast, Golovanov (1982) found that residual yolk amounting to about 4% of body weight was present in 92% of the migrant pink fry at the start but fell to zero by the end of migration. He also found that only 5% - 6% of the migrants fed in fresh water.

Smirnov and Kamyshnaya (1965) reported that pink salmon fry on the Kola Peninsula remained and fed somewhat longer in streams after emergence than in other areas. This is likely brought about by the 24-hour polar day coinciding with emergence and downstream migration.

The principal food items eaten in fresh water by pink salmon fry are larval and pupal stages of dipteran insects, particularly chironomids (Levanidov and Levanidova 1957; J. McDonald 1960; Kobayashi and Harada 1966; Frolenko 1970). Other foods found in stomachs of migrants include larval mayflies (ephemeropterans), stoneflies (plecopterans), occasional water bugs (hemipterans), terrestrial insects, mites, and freshwater cladocerans and copepods. In spite of a generally low level of feeding activity by migrant pink salmon fry, individual fish are capable of intensive feeding, as indicated by Frolenko (1970) who found up to one hundred small chironomids in some stomachs.

Predation

A wide array of predators prey on migrant pink fry in streams and associated estuaries (Table 16). The general effect of predation on individual stocks is highly variable and poorly understood. In general, the most important predators on migrant fry are other fishes; these fish may be residents or migrants to or from the same stream or estuary. Often a given predator in one region or stream may prey heavily on fry, but elsewhere the same predator takes few fry.

Neave (1953) reviewed the effect of predation on pink fry at McClinton Creek and found an inverse relation between the size of fry populations and the percentage eaten ("depensatory mortality"). He concluded that the percentage of fry lost to predators increased with the distance fry travel as well as seasonally during the progress of fry migration.

Principal predators on pink fry at McClinton Creek were Dolly Varden charr and coho salmon

TABLE 16

Some predation studies and recorded predators of migrant pink salmon fry

Region	Stream	Predator		Source
BC	Hooknose Cr.	coastrange sculpin	*Cottus aleuticus*	Hunter (1959)
		prickly sculpin	*C. asper*	"
		coho salmon smolt	*O. kisutch*	"
		Dolly Varden charr	*Salvelinus malma*	"
		Cutthroat trout	*O. clarki*	"
		steelhead trout	*O. mykiss*	"
	Hooknose Cr.	crows	Corvidae	Hoar (1958)
	McClinton Cr.	coho salmon smolts	*O. kisutch*	Pritchard (1936b)
		Dolly Varden charr	*S. malma*	"
		cutthroat trout	*O. clarki*	"
	McClinton Cr.	caddis fly larvae	Trichoptera	Pritchard (1934a)
		water ouzel	*Cinclus mexicanus*	Pritchard (1934b)
		muskrat	*Ondatra zibethica*	Pritchard (1934c)
	on Graham Is.	water ouzel	*C. mexicanus*	Munro (1936)
		American merganser	*Mergus merganser*	Munro & Clemens (1937)
S.E. Alaska	Hood Bay Cr.	Dolly Varden charr	*S. malma*	Armstrong (1970)
		coho salmon smolts	*O. kisutch*	"
		Walleye pollock	*Theragra chalcogramma*	Armstrong & Winslow (1968)
	Sashin Cr. estuary	herring	*Clupea harengus*	Thorsteinson (1962)
Sakhalin Is.	Bakhura R.	Siberian charr	*S. leucomaenis*	Tagmaz'yan (1971)
		Dolly Varden charr	*S. malma*	"
		young masu salmon	*O. masou*	"
	Lyutoga R.	Siberian charr	*S. leucomaenis*	Khorevin et al. (1981)
		Arctic charr	*S. alpinus**	"
		young masu salmon	*O. masou*	"
Sakhalin Is.	several streams	Siberian charr	*S. leucomaenis*	Volovik & Gritsenko (1970)
		Sakhalin taimen	*Hucho perryi*	"
		young masu salmon	*O. masou*	"
		young coho salmon	*O. kisutch*	"
		pike	*Esox reicherti*	"
Kola Peninsula	several streams and estuaries	Sea trout	*Salmo trutta*	Bakshtansky (1980)
		young Atlantic salmon	*S. salar*	"
		pike	*Esox lucius*	"
		herring	*C. harengus*	"
		haddock	*Melanogrammus aeglefinus*	"
		saithe	*Pollachius virens*	"
		Arctic tern	*Sterna paradissaea*	"

Note: *According to translation editor, probably *S. malma*

smolts (Pritchard 1936b). The numbers of pink fry ranged from 1 to 23 per stomach and averaged from 6.1 to 6.6 per year per charr. At Hooknose Creek, Hunter (1959) estimated that 0.5 million pink and chum salmon fry were lost each year to predators. Dolly Varden, although present in the stream, were not as great a predator as were two species of cottids and coho salmon smolts. At Hood Bay Creek (Admiralty Island), Armstrong (1970) found 60% of 1,500 coho salmon stomachs with food. Of these, 421 averaged 1.9 pink or chum salmon fry per stomach. On the outlet stream from Eva Lake (Baranof Island, Alaska), he found that only three of 1,372 Dolly Varden examined contained young salmon. At Little Port Walter, the estuary at the mouth of Sashin Creek, Lagler and Wright (1962) examined 183 Dolly Varden stomachs during the pink fry migration period and found 143 contained food. Only four (2.8%) of them had remains of young salmon.

In the Far East, Dolly Varden, Siberian charr (*Salvelinu leucomaenis*), coho salmon smolts, and juveniles and smolts of masu salmon (*Oncorhynchus masou*) prey on pink fry in streams. However, as in North America, the effect of predators on migrant pink fry is varied. For example, predation on pink fry was not great (not more than one per predator) in the Semga River (Sakhalin Island) (Khorevin et al. 1981), whereas at Karymaiskiy Spring (West Kamchatka), Semko (1954) estimated that migrant pink fry lost to charr and coho in 1944, 1945, and 1946 amounted to 44.6%, 95.9%, and 64.5%, respectively, of the annual fry production.

Volovik and Gritsenko (1970) studied predation on pink salmon fry in Sakhalin Island streams and found that (1) the damage done by predators on the whole was insignificant; (2) Siberian charr were the most important predator, with coho and masu salmon less significant; (3) hatchery fry were preyed on to a greater extent than wild fry (at one hatchery it was estimated that 4 to 5 million pink fry might have been eaten by Siberian charr); and (4) the beneficial effect of charr scavenging large quantities of dead eggs in winter may offset damage done by preying on fry in spring. Because of differential predation on hatchery fry compared to that on wild fry, Volovik (1966) suggested that releasing fry close to the river mouth would increase hatchery effectiveness.

Fish predation on migrant pink fry in the rivers and estuaries played an important role in the relative success of pink salmon transplants to the Kola Peninsula (Bakshtansky 1964, 1965, 1980). Foremost among predators was herring, *Clupea harengus*, which moved into the mouths of rivers to feed on migrant pink fry. Bakshtansky (1965) found an inverse relation between fry survival and relative abundance of herring in the year of fry release. Herring were also found to prey on pink fry in Alaskan estuaries (Thorsteinson 1962) and may have affected the returns from transplants at Big Qualicum River in British Columbia (Walker and Lister 1971).

At Hooknose Creek, Hunter (1959) found that a relatively fixed number of pink salmon fry were eaten by a relatively fixed number of predators each year. In years of low fry abundance he estimated that up to 85% of the population was lost to predators. Peterman and Gatto (1978) examined Hunter's (1959) field data to determine the number of prey eaten per predator per unit time relative to prey density of pink and chum salmon fry. They found that the predators of fry at Hooknose Creek were operating, in seven of nine years, on the low end of their functional response curve, and were capable of causing high mortalities on even larger fry populations. In general, predators were not overly satiated at the fry densities encountered. Peterman and Gatto (1978) also found a form of competition among predators. For a given fry density, more fry are eaten per predator when predator numbers are small than when they are large. These authors stressed the need for experimental research on predation processes involving juvenile salmon, especially in conjunction with hatcheries and enhancement projects.

Freshwater Survival

Freshwater survival for individual stocks is determined from estimates of eggs brought to the spawning grounds by the parent brood (potential egg deposition) and from numbers of fry subsequently migrating seaward. Point estimates of survival during the freshwater phase of pink salmon life range from 0.1% to 43.4%, a 400-fold difference, whereas mean values for two or more years of data from the same stream range from 4.3% to 23.0%, roughly a five-fold difference (Table 17). Annual freshwater survival values of pink salmon spawn-

TABLE 17
Some reported freshwater survival rates from potential egg deposition to migrant fry for pink salmon

Region	Stream	Inclusive brood years	*N*	Survival (%) Mean	Range	Source
BC	Hooknose Cr.	1947–56	10	5.6	0.9–16.5	Hunter (1959)
	Nile Cr.	1951–53	3	13.4	0.4–32.3	Wickett (1962)
	Morrison Cr.	1943–45	2	5.2	4.7–6.7	Neave (1953)
	McClinton Cr.	1930–40	6	14.4	6.9–23.8	Pritchard (1948b)
	Fraser R.	1961–81	12	13.0	9.0–18.7	IPSFC (1986)
	Harrison R.	1957–65	5	13.9	3.7–22.2	IPSFC (1968)
	Vedder R.	1957–65	5	10.4	5.2–12.9	"
	Upper Seton Cr. channel	1961–81	11	51.8	21.7–85.5	Cooper (1977) IPSFC (1979, 1982, 1983)
	Lower Seton Cr. channel	1967–81	8	57.0	40.2–66.6	"
	Jones Cr. Channel	1955–61	4	42.1	30.0–63.0	MacKinnon (1963)
S.E. Alaska	Sashin Cr.	1940–59	19	5.1	0.1–22.8	Merrell (1962), Olson & McNeil (1967)
	Sashin Cr.	1961–79	19	9.0	0.2–20.2	Vallion et al. (1981)
	Auke Cr.	1971–81	11	4.3	0.2–12.3	Taylor (1983)
North Okhotsk	Motykleyka R.	1971–78	8	17.6	4.7–43.4	Golovanov (1982)
Sakhalin Is.	Lyutoga R.	1962–68	7	14.1	1.4–23.4	Kanid'yev et al. (1970)
	Khvostovka R.	1962–68	6	16.5	2.4–23.5	"
	Poronai R.	1962–68	7	21.9	3.3–29.3	"
	Inanusi R.	1962–68	7	7.6	2.9–13.7	"
	Lesnaya R.	1960–68	9	15.5	2.7–27.2	"
	Pritornaya R.	1972–73	2	23.0	6.0–40.0	Shershnev & Zhul'kov (1979)

ing in semi-natural channels in the Fraser River system ranged from 21.7% to 85.5% and averaged 42.1% to 57.0% over several years (Table 17).

No single environmental factor or specific event accounts for the fluctuations in freshwater survivals. There are, however, generalized patterns that apply throughout the range of the species. Under natural conditions, freshwater mortality is generally highest during spawning and tends to decrease during subsequent incubation, emergence, and fry migration periods. Neave (1953) recognized three types of causes of freshwater mortality in pink salmon: two density related and one independent of density. Density-related mortality included "compensatory" or density-dependent mortality that increases as the population level increases (operating during any freshwater life stage but most evident during spawning) and "depensatory" or inverse density-dependent mortality that increases as the population level decreases (the effect of predators eating a fixed number of fry when abundance is low). "Extrapensatory" or density-independent mortality was used by Neave to identify physical factors in the freshwater environment, especially those relating to stream flow and temperature conditions, that tend to operate at all levels independent of population density. Fluctuations in stream flows between the time of spawning and fry migration are the most important non-biological factor influencing pink salmon survival in fresh water (Neave and Wickett 1953; Wickett 1958b; Levanidov 1964)

Early Sea Life

Behaviour and Factors Affecting Initial Dispersal

Throughout their range, pathways for pink salmon fry to the open ocean are highly diverse and may involve complex migration routes. Newly formed schools may move quickly from the natal stream area or remain to feed along shorelines up to several weeks. The timing and pattern of seaward dispersal is influenced by many factors, including general size and location of the spawning stream, characteristics of adjacent shoreline and marine basin topography, extent of tidal fluctuations and associated current patterns, physiological and behavioural changes with growth, and, possibly, different genetic characteristics of individual stocks.

Newly emerged pink salmon fry show a preference for saline water over fresh water (Baggerman 1960; McInerney 1964) which may, in some situations, facilitate migration from the natal stream area. Time of migration from fresh water may affect dispersal relative to salinity responses. For example, Hurley and Woodall (1968) found progressively more rapid movement into sea water by late emerging fry. Girsa et al. (1980) noted that pink fry, after entering the White Sea from the Kola Peninsula, favoured low over high saline water. They intepreted this as a response to favourable feeding areas.

Neave (1966) indicated that initial marine dispersal of pink salmon fry up to 50–67 km from shore may be accomplished in a few days. Parker (1965) and Healey (1967) described the seaward pink fry migration in Burke Channel fiord (British Columbia) as an uneven saltatorial phenomenon. The fry covered 33 km or more in one or two days of active migration, followed by several days of non-migratory schooling and feeding in quiet bays and backwaters (Healey 1967). Shershnev et al. (1982) found that Sakhalin Island pink fry moved out quickly into open water up to 20 km offshore in Aniva and Terpenia bays, whereas in Iturup Island streams (Kuril Islands), fry stayed in the direct vicinity of the coast for 2.5–3.0 months. In Traitors Cove, Alaska, Bailey et al. (1975) found pink fry feeding up to two months in the estuary a few kilometres from the natal stream. Although some early season fry may have migrated seaward from this estuary, schools remained in the same area for several weeks. Limited dispersal during early sea life was also noted in Auke Bay by Landingham (1982) and near Evans Island in Prince William Sound by Cooney et al. (1978). The latter area is only about 10 km from waters of the Gulf of Alaska. In contrast, Skeena and Fraser River fry are transported in plumes of riverine water across open water for up to 50 km to other protected areas, where they then feed and grow for two to three months before migrating to the open ocean (Manzer and Shepard 1962; Phillips and Barraclough 1978; Healey 1978, 1980, 1982).

Early marine schools of pink fry, often in tens or hundreds of thousands of fish, tend to follow shorelines and, during the first weeks at sea, spend much of their time in shallow water of only a few centimetres deep (Figure 22) (LeBrasseur and Parker 1964; Healey 1967; Gerke and Kaczynski 1972; Bailey et al. 1975; Cooney et al. 1978; Simenstad et al. 1980). Kaczynski et al. (1973) suggested that this onshore period involved a distinct ecological life history stage in both pink and chum salmon. In many areas throughout their ranges, pink and chum fry of similar age and size comingle in both large and small schools during early sea life. Not all early stage schooling, however, is confined to shorelines. Healey (1967) noted that during active migration in Burke Channel "schools were seen moving well offshore." Healey also noted, based on phosphorescence in the water, that schools of fry remained intact at night.

According to Karpenko (1987), juvenile pink salmon in the Bering Sea off the northeastern Kamchatka coast are found in one of three hydrological zones during their three to four months of marine life: (1) the littoral zone, up to 150 m from shore; (2) open parts of inlets and bays, from 150 m to 3.2 km from shore; and (3) the open part of Karaginskiy Gulf, 3.2 to 96.5 km from shore. Distribution within these regions is seasonally related to the size of pinks, with an offshore movement of larger fish in August and September (Karpenko 1987).

Foods and Feeding

Pink salmon fry routinely obtain large quantities of food sufficient to sustain rapid growth from a broad range of habitats providing pelagic and epibenthic foods. Collectively, diet studies show that pink salmon are both opportunistic and generalized feeders and, on occasion, they specialize in specific prey items. Diel sampling of stomachs showed fewer and more digested food items at night than during the day, indicating that juvenile pinks are primarily diurnal feeders (Parker and LeBrasseur 1974; Bailey et al. 1975; Simenstad et al. 1980; Godin 1981a; Shershnev et al. 1982). Godin (1981a), Simenstad et al. (1980), and Parker and

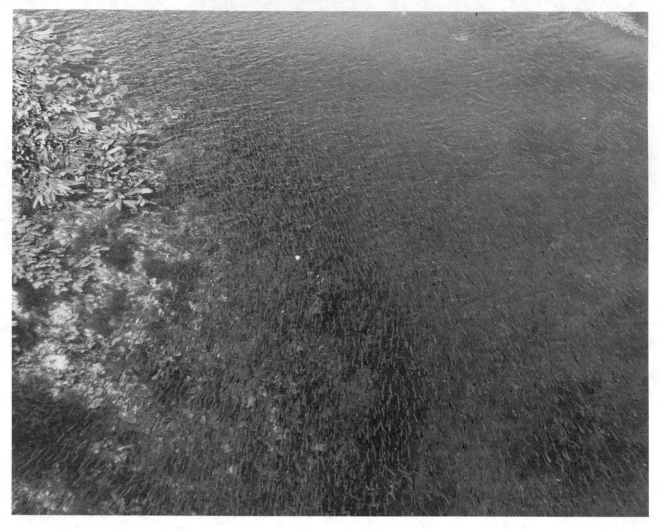

FIGURE 22

School of pink salmon fingerlings, during the early marine nearshore phase, swimming along the northeast shore of Galiano Island in the Strait of Georgia (early May, 1968). (Photograph courtesy of the International Pacific Salmon Fisheries Commission)

LeBrasseur (1974) found that peak feeding of pink salmon juveniles occurred at dusk. Stomachs contained the least food at 0500 hours (Parker and LeBrasseur 1974). Bailey et al. (1975) saw fry dimpling the surface on moonlit nights, presumably feeding, at Traitors Cove, Alaska. Actively feeding schools of pink fry may move in a circular pattern, remaining days or weeks in the same bay or cove (Healey 1967; Cooney et al. 1978). They may take advantage of natural food delivery and concentrating mechanisms, such as eddies and current shear lines, in such areas (Healey 1982).

Counts of pink and chum fry along nearshore transects in southeastern Alaska demonstrated that specific nursery areas where young fry held and fed could be differentiated from other transition areas where fry transit quickly but do not remain (J.E. Bailey, National Marine Fisheries Service, Auke Bay, Alaska, pers. comm.). These nursery areas comprised irregular shorelines with complex eddies, which in some areas continuously replenished the nursery with zooplankton and provided shelter from wind-generated waves and currents and strong tidal currents. Similar observations have been made with respect to aggregations of pink fry in other areas (Gerke and Kaczynski 1972; Parker and LeBrasseur 1974; Cooney et al. 1978; Healey 1982).

Takagi et al. (1981) reviewed in detail most food studies of pink salmon conducted to the mid-1970s throughout the North Pacific Ocean during their first summer at sea (age .0). These and several important new studies of juvenile pink salmon diets are summarized in Table 18.

In general, the diet of pink fry in shore-oriented nursery areas in North America is heavily dependent on calanoid, cyclopoid, and epibenthic harpacticoid copepods. The relative importance of the different forms depends on the particular area and year (Kaczynski et al. 1973; Bailey et al. 1975; Cooney et al. 1978; Kron and Yuen 1978; Simenstad et al. 1980; Godin 1981a; Landingham 1982). The importance of harpacticoids and calanoid copepods in the early marine diets of pink fry has also been noted in the Far East (Frolenko 1970; Takagi et al. 1981; Shershnev et al. 1982). Karpenko (1979) collected juvenile pink salmon from Bering Sea waters in two areas along the eastern Kamchatka coast in 1974–75. In Karagin Bay, fry, 32–55 mm in length, fed principally on calanoid copepods and mysids, whereas in Avacha Bay, juvenile pink salmon, 57–96 mm long, fed on calanoids and pteropods. Harpacticoid copepods were not mentioned as food of pink salmon from these collections. However, in July 1978 collections of pink salmon stomachs from near the mouth of the Anapka River (eastern Kamchatka), Karpenko (1982a) indicated that the principal foods eaten were insects and harpacticoids. In other Bering Sea studies along the Kamchatka coastline, the principal foods eaten by juvenile pink salmon were macrozooplankton, including a variety of crustaceans, along with larval and juvenile fishes (Karpenko and Piskunov 1984; Karpenko 1987; Karpenko and Maksimenkov 1988). Along the Sea of Okhotsk coast of the Kamchatka Peninsula near the Utka River, Karpenko and Safronov (1985) found that the early marine diet of pink salmon consisted primarily of small calanoid copepods and larval fishes.

In general, the relation between pelagic and epibenthic copepods in the nearshore diet of pink salmon fry may be largely one of availability and habitat type (Kron and Yuen 1978). Along shallow beach nurseries, with cobble-sand-mud substrates and low gradient shorelines, harpacticoids are often among the more important diet items. In other nearshore nurseries, with boulder-bedrock substrates and steep gradient shorelines, calanoids and other pelagic zooplankters are usually more important in the diet. Tidal and eddy currents likely play a significant role in food delivery to the latter habitats.

Besides copepods in various forms, other principal prey items eaten by juvenile pinks during their shore-oriented early marine phase from March to June include barnacle nauplii (cirripeds), mysids, amphipods, euphausiids, decapod larvae, insects, larvaceans, eggs of invertebrates and fishes, and fish larvae (Table 18).

Feeding Rates

Because pink salmon fry grow rapidly during early sea life, they must consume large amounts of food (LaBrasseur and Parker 1964; Healey 1980). LaBrasseur (1969) estimated that an average daily food ration of 10%–12% of body weight would be required for pink fry to average a 4.5% daily growth rate. Later, Parker and LeBrasseur (1974) suggested that an average daily food requirement of about 7%–8% of body weight would achieve a similar rate of growth.

Healey (1982) examined seasonal differences in average weights of stomach contents as a percentage of body weight for juvenile pink salmon in nearshore and offshore habitats of the lower Strait of Georgia in 1976. He found that stomach content weights from nearshore and offshore areas averaged 1.52% and 1.73% of body weight, respectively, before 15 May, and 2.80%–2.45%, respectively, between 15 May and 15 June. These values were not significantly different between habitats but were higher in both habitats in the later time period. Assuming gastric turnover rates of three to four times a day, a rough estimate of a daily ration from 4.6% to 11.2% of body weight was obtained. In Departure and Hammond bays near Nanaimo, British Columbia, daily rations of pink salmon fry were estimated by Godin (1981a) to be 13.1% and 6.6%, respectively, of dry body weight per day. Simenstad et al. (1980) estimated the daily food ration of 0.61 g pink fry in Puget Sound at 30.2% of body weight per day.

In laboratory tests, Godin (1981b) found that

TABLE 18

Information on the stomach contents of age .0 pink salmon sampled from various areas in the North Pacific and adjacent waters

Year	Month	Area	Number examined	Fishes	Eggs	Larva-ceans	Sagit-tas	Squids	Ptero-pods	In-sects	Deca-pods	Euph-ausiids	Amphi-pods	My-sids	Cirri-pedes	Cope-pods	Ostra-cods	Clado-cerans	Oth-ers	Source
			Relative importance of prey animals in stomachs																	
Percentage composition in volume																				
1955	Jun-Aug	Chatham Sound	537	5	+	40	-	-	-	1	4	3	1	+	6	31	-	-	8	Manzer
1970	Apr-Jun	Puget Sound	53	-	14	-	-	-	-	+	-	2	4	10	16	46	-	4	4	Kaczynski et al. (1973)
1964-66	Apr-Jun	Traitors Cove	170	+	3	3	-	-	-	+	1	1	+	-	8	77	-	6	1	Bailey et al. (1975)
1976	Jul-Sep	Strait of Georgia	22	2	-	-	-	-	-	32	11	7	37	-	-	8	-	-	3	Healey (1980)
Percentage index of relative importance																				
1978	Apr-May	Hood Canal	55	2	-	2	-	-	-	+	+	1	+	+	-	91	-	-	5	Simenstad et al. (1980)
1975	Mar-Jun	Auke Bay North	206	-	39	-	-	-	-	+	-	+	-	-	-	56	-	-	5	Landingham (1982)
		Auke Bay South	247	-	11	+	-	-	-	+	-	8	-	-	-	76	-	-	5	"
Percentage composition in number																				
1976	May	Departure Bay	78	-	-	-	-	-	-	-	-	-	+	-	17	82	-	-	1	Godin (1981a)
		Hammond Bay	87	-	-	-	-	-	-	-	-	-	3	-	28	62	-	-	7	"
Percentage composition in weight																				
1963	Nov	Hecate Strait	255	2	-	1	-	5	5	-	-	50	7	-	-	19	-	-	11	LeBrasseur & Barner (1964)
1965	Sep-Oct	Bering Sea	-	6.9	-	-	-	-	83.2	0.1	4.1	-	4.3	-	-	0.1	-	-	1.3	Andrievskaya (1968)
1964-65	Sep-Oct	Sea of Okhotsk	-	-	-	-	-	-	0.1	-	-	1.8	69.3	-	-	27.3	-	-	1.5	"
1966-67	Aug-Oct	Off west Kamchatka	-	20.1	-	-	42.4	-	29.0	-	3.2	3.2	0.3	-	-	1.8	-	-	-	Birman (1969a)
1966	Aug	Sea of Okhotsk, east Sakhalin coast	-	3.6	-	-	-	-	-	+	25.5	0.3	18.5	-	-	52.1	-	-	-	Andrievskaya (1970)
		Sea of Okhotsk, off east Sakhalin	-	-	-	-	5.9	-	-	-	-	2.3	57.7	-	-	34.1	-	-	-	"
1967	Sep	Sea of Okhotsk, off west Kamchatka	-	-	-	-	0.3	-	-	-	-	2.3	88.7	-	-	0.1	-	-	-	"
		Sea of Okhotsk, west Kamchatka coast	-	19.1	-	-	40.2	-	27.6	+	3.1	3.1	0.3	-	-	1.6	-	-	5.0	"

1974	May	Iturup coast	385	1.1	9.9	-	-	-	-	-	-	-	2.8	2.8	-	73.0	-	-	10.4	Shershnev et al. (1982)
	June		18	2.4	1.0	-	-	-	-	+	+	+	2.6	1.0	-	89.5	-	-	3.5	″
1977	Jul		523	3.3	0.2	-	-	-	-	2.4	0.8	3.2	2.2	1.2	-	82.8	-	-	3.9	″
	Aug		214	0.2	7.0	-	-	-	-	3.8	4.7	11.5	-	-	-	72.0	-	-	0.8	″
1968	Jul	Southwest Sakhalin	-	-	57.1	-	-	-	-	0.9	-	11.7	0.6	24.5	-	1.1	-	-	4.1	″
	Aug		-	10.8	-	-	-	-	-	23.3	16.5	-	13.3	12.7	-	1.7	-	-	21.7	″
Average number of prey animals found per stomach																				
1966	Apr	Strait of Georgia	155	+	1.7	-	-	-	-	0.7	0.1	+	0.6	-	-	26.4	-	-	0.2	Barraclough (1967a, 1967b)
1966	Jun	Strait of Georgia	8	-	-	-	-	-	-	1.1	-	-	2.6	-	-	7.3	1.9	-	-	Barraclough (1967c)
1966	Jul	Strait of Georgia	58	8.6	0.4	27.3	3.8	-	-	9.6	8.5	0.5	31.7	-	11.7	8.0	0.2	+	1.7	Barraclough & Fulton (1967)
1967	May	Strait of Georgia	2	0.5	-	15.0	-	-	-	5.5	1.0	-	23.0	-	12.5	3.0	-	3.5	-	Robinson et al. (1968a)
1967	Jun	Strait of Georgia	3	0.7	-	-	-	-	-	4.3	0.3	-	-	-	0.3	0.7	-	9.3	0.7	Robinson et al. (1968b)
1966	Jun	Saanich Inlet	19	0.6	2.6	135.7	10.3	-	-	16.9	0.5	1.8	21.0	-	-	80.1	-	-	-	Barraclough & Fulton (1968)
	Jul		10	-	0.9	-	0.2	-	-	1.0	-	1450.2	14.1	-	-	153.3	-	-	-	″
1968	Apr	Saanich Inlet	2	0.5	0.5	-	-	-	-	-	0.5	-	-	-	-	-	-	-	-	Barraclough et al. (1968)
	May		6	-	28.9	-	-	-	-	5.0	0.3	1.5	-	-	-	37.7	-	-	-	″
	Jun		217	0.1	253.5	9.2	-	-	-	2.9	2.9	59.6	2.4	-	3.6	66.1	0.6	0.4	1.8	″
	Jul		227	0.2	398.8	21.3	-	-	-	0.6	3.3	17.0	56.0	-	+	88.1	+	9.2	+	″
1977	May	Prince William Sound	20	-	-	-	-	-	+	+	1	-	-	-	-	54.5	-	-	+	Cooney et al. (1978)
	Jun		20	-	-	88.5	-	-	1	-	+	1	-	-	-	80.5	1	8.5	+	″

Source: Adapted from Tagaki et al. (1981)

pink salmon fry consumed mean rations of 23.6% and 39.8% of dry body weight during 12-hour feeding periods after they had been starved for 24 and 72 hours, respectively. Juvenile pink salmon apparently have a lower hunger threshold than other salmonids. This is probably related to a greater average feeding rate and daily food ration and to the relatively higher growth rate during early marine life, than those of other salmon species. Godin's (1981b) data suggested that juvenile pink salmon have the ability to feed continuously at a rate that balances the gastric evacuation rate. With a full stomach, less than 15% of the contents need be evacuated for spontaneous feeding to resume. Simenstad et al. (1980), in a diel collection of wild pink fry, noted that they maintained greater stomach fullness throughout the day than chum fry.

Mortensen (1983) studied daily rations with first-feeding pink fry at variable plankton densities in static and flowing water at different temperatures. He found that daily rations ranged from 15.4% to 18.2% of body weight per day at 8°C with intermediate and high plankton densities (2,031 and 10,154 mg/m^3). Feeding rates were somewhat higher in flowing than in static water. At 12°C, Mortensen estimated that daily rations ranged from 28.4% to 37.3% of body weight per day at the two plankton densities and water conditions.

In summary, food consumption rates for pink salmon fry and juveniles have been estimated to range from as low as 1.52% to as high as 37.3% of body weight per day, depending on factors such as size of fry, season, water temperature, locality, and density of available food organisms.

Early Marine Growth and Survival

The growth rates of juvenile pink salmon during their early sea life have been estimated to range from about 3.5% to 7.6% of body weight per day (LeBrasseur and Parker 1964; LeBrasseur 1969; Phillips and Barraclough 1978; Healey 1980). LeBrasseur and Parker (1964) estimated that Bella Coola River pink salmon fry in Burke Channel (British Columbia) grew 0.87 mm/d during the first 30 days in salt water. In the lower Strait of Georgia, Healey (1980) found that juvenile pinks captured in the Gulf Islands area increased from 36 mm in fork length in May to 198 mm in October, about 0.97 mm/d. Hurley and Woodall (1968) suggested that juvenile pink salmon increased about 1 mm/d in length in the first two to three months in the same area. Most pink salmon, generally larger fish, began leaving the Strait of Georgia in June and July at about 100 mm in length (Phillips and Barraclough 1978; Healey 1980). Healey (1980) calculated instantaneous daily growth rates for two groups of juvenile pink salmon caught in these waters. He estimated growth in the nearshore period at about 7.6% per day and in the pelagic period at about 4.5% per day. The average over the length interval of 35–100 mm was about 5.7% per day.

Although juvenile pink salmon exhibit highly developed schooling behaviour, which is generally considered an anti-predator defense (Major 1978), mortality is probably greater during this time than in subsequent periods. Parker (1965, 1968) estimated that average daily losses of Burke Channel-Bella Coola stocks of pink salmon juveniles varied from 2% to 4% in the first 40 days of sea life, producing overall mortality estimates between 55% and 77%. Parker believed that these losses were largely due to predation by juvenile coho salmon. He demonstrated in experimental tanks that these predators had a strong bias for smaller individual pink salmon in the prey population (Parker 1971).

In certain years, Karpenko (1982b) found that the predominant food of juvenile coho salmon were migrant pink and chum fry in Bering Sea waters along the eastern Kamchatka coast. Juvenile sockeye and chinook salmon, *Oncorhynchus tshawytscha*, in this area also fed on pink and chum juvenile migrants. In 1978, stomachs of 22 chinook salmon all contained pink fry (from 1 to 6/stomach) (Karpenko 1982b).

Parker (1971) suggested that rapid growth by young pink salmon fry is the best defense against marine predators and that selective predation pressures strengthen the high innate growth capacity of this species. Vernon (1962) speculated that other factors such as a large turbid stream and estuary, like those of the Fraser River, afforded protection against predation. In mixed pink and chum fry schools, the more numerous pink fry may provide buffer protection for chum fry. Hargreaves and LeBrasseur (1985) found that yearling coho salmon selectively prey upon pink salmon fry in marine enclosures, even when chum fry are both significantly smaller and more abundant than pink fry. Bailey et al. (1983) described a similar buffer benefit for pink fry occurring in mixed schools in early sea life with juvenile Pacific sandfish, *Trichodon trichodon*.

Offshore Migration

Behavioural Changes

With growth, the behaviour of pink salmon fry changes from nearshore to offshore areas. The exact size when the shift to deeper water away from shore occurs varies somewhat in different areas. LeBrasseur and Parker (1964) found that 4.5-5.5 cm was the critical size for offshore migration, whereas Cooney et al. (1978) observed that juveniles began to migrate when they were 6.0-7.0 cm in length. The larger juveniles are the first to migrate to open coastal waters (Sakagawa 1972; Healey 1982; Karpenko 1987).

The migration of juvenile pinks offshore coincides with physiological changes. LeBrasseur and Parker (1964) found a reduction in the growth rate of juvenile pinks when they reached 6-8 cm in length and suggested that this might signify a change similar to the parr-smolt transformation in other Pacific salmon. Elsewhere, Parker and LeBrasseur (1974) considered the initial shore-oriented estuarine pre-smolt period of early sea life to be separate from subsequent coastal offshore periods. The pink salmon "smolt" stage was identified by a shift to lower lipid levels (Parker and Vanstone 1966) and by the formation of smaller schools of 500-2,000 fish that abandoned shoreline habitats and actively migrated seaward (Parker and LeBrasseur 1974). Takagi et al. (1981) speculated that some of these changes coincide with changes in feeding habits and stomach contents.

Juvenile pink salmon arrive in coastal waters along both continents from major fry production areas through many pathways and entry points. The scope of migration routes travelled by juveniles from natal streams to coastal waters is highly varied and includes (1) long open straits with large inland waters often interspersed with islands, such as Tatar Strait, Puget Sound, Strait of Georgia, and Cook Inlet; (2) complex interconnecting fiords and channels common to much of central British Columbia, southeastern Alaska, parts of Prince William Sound, and Kodiak Island; and (3) relatively open areas, generally with more or less direct access to major seas, bays, or open ocean, such as the Alaska Peninsula, Bristol Bay, and much of the Far East.

Migrations of North American Juvenile Pink Salmon

The migration of juvenile pink salmon from the lower Strait of Georgia apparently occurs in July, based on declining catches in those waters (Barraclough and Phillips 1978; Healey 1980) and catches made seaward in the outer Juan de Fuca Strait (Sakagawa 1972, Hartt 1980). Juvenile pinks are still present in Juan de Fuca Strait in August (Hartt 1980), and Healey (1980, 1982) found some in the lower Strait of Georgia as late as October in 1976.

Juvenile pink salmon found in central Puget Sound in October and November are probably a resident stock that feeds and matures in Puget Sound without going to sea (Hartt and Dell 1986). Jensen (1956) tagged residual immature pinks in the Tacoma Narrows area of Puget Sound, and subsequent returns suggested that the Stillaguamish River may be the origin of the resident pinks.

According to Martin (1966), most migration within southeastern Alaska followed routes leading to major and minor summer schooling areas before juveniles migrated into open ocean waters. Beginning in late July and August during 1964 and 1965, he observed major migrations of juvenile pinks into the Gulf of Alaska from Sumner Strait, Chatham Strait, Peril Strait, and an area between west Prince of Wales and Noyes islands. Martin also noted that pink salmon juveniles from the Nass and Skeena rivers, and probably from more southerly regions, migrate north through Dixon

Entrance in August into coastal waters. This observation is consistent with those of Manzer (1956), who caught juvenile pink salmon in Dixon Entrance in September and found none along beaches of the Chatham Sound area after mid-August.

Tagging of juvenile salmon in the eastern North Pacific Ocean suggests that young pink salmon from the Fraser River and Puget Sound leave these waters relatively early and migrate rapidly northward along the coasts of British Columbia and southeastern Alaska (Figure 23) (Hartt and Dell 1986; Hartt 1980; Takagi et al. 1981). Two fish tagged before mid-August at about 58°N off southeastern Alaska had migrated almost 1,800 km from their likely point of entry into the ocean through the Juan de Fuca Strait (Hartt 1980). In addition, 35 pink salmon tagged as juveniles off southeastern Alaska and northern British Columbia in even-numbered years (Figure 23) and 14 tagged in odd-numbered years were all recovered as adults a year later, somewhat southward from the original capture areas (Hartt 1980; Takagi et al. 1981).

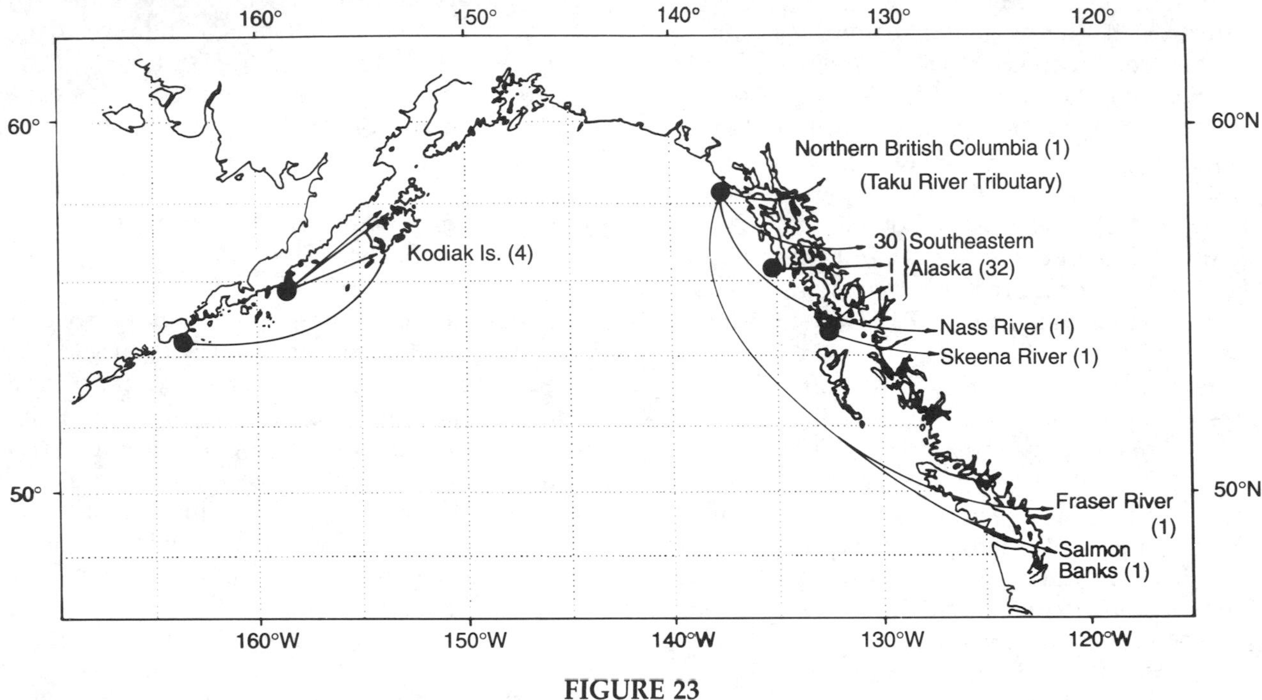

FIGURE 23

Tagging and recovery locations of 41 pink salmon tagged as juveniles from July to October in even-numbered years and recovered a year later (in odd-numbered years) as adults. The directions of the arrows do not indicate the routes of migrations. (From Takagi et al. 1981)

This information, combined with coastal purse seine catches of tagged juvenile pink salmon, indicates that a band of migrating fish progresses northward along the North American coastline from the Juan de Fuca Strait to Yakutat during the period July through October (Figure 24) (Hartt 1980). Stocks from southern areas tend to enter the ocean first, followed by those in more northerly and westerly locations later in the season. This basic migration pattern was found by Hartt (1980) to be similar for juvenile pink, chum, and sockeye salmon.

Examination of catch summaries shown in Figure 24C-D indicates that there is a second concentration of juvenile pink salmon during August to October along the south-central Alaska and Alaska Peninsula coastline. Stocks in the south-central region, including Prince William Sound, Cook Inlet, and the Alaska Peninsula, likely comprise a significant part of these catches. The directions of migration of these two bands of juvenile pinks, as inferred from paired seine sets held open in opposite directions and from recovery patterns of tagged fish (Figure 23), suggest that fish along the

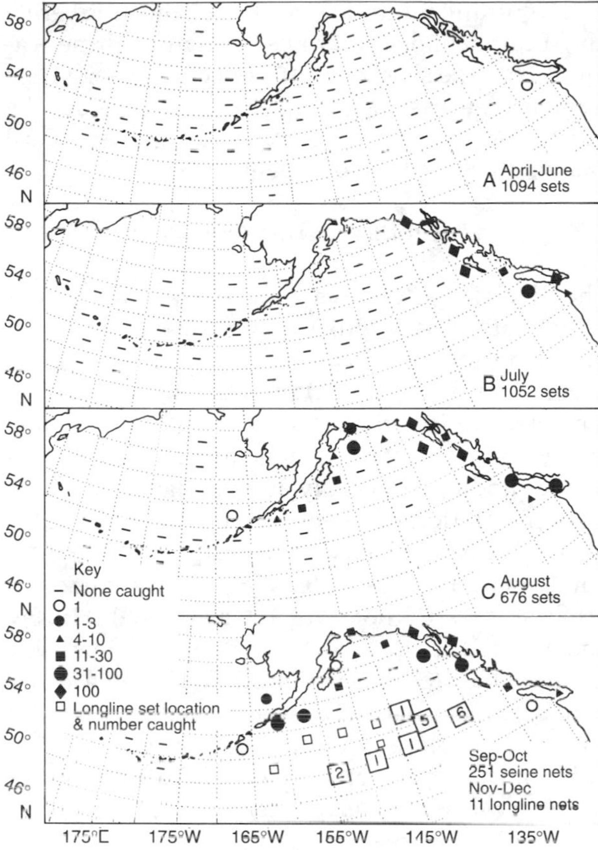

FIGURE 24

Catch per seine set of age .0 pink salmon by area and by time period; 3,073 sets, 1956–70 (from Hartt 1980). Also catches during Canadian offshore longline operations at 11 stations 21 November–7 December, 1966. (Based on data from the Fisheries Research Board of Canada 1969)

eastern Gulf of Alaska migrate in a northwesterly direction, whereas those along the south-central Alaska–Alaska Peninsula coastline migrate in a southwesterly direction (Hartt 1980). A third, smaller and less well defined, group of North American pink salmon in coastal waters involves western Alaska stocks from Bristol Bay and the Aleutian Islands. Small numbers of these were caught north of the Alaska Peninsula (Figure 24D).

Along the British Columbia and southeastern Alaska coastline, the band of juvenile salmon is up to 1,800 km long and often less than 40 km wide, generally paralleling the continental shelf (Hartt 1980). The times and locations at which juvenile pinks in the coastal bands disperse offshore into the open Gulf of Alaska and North Pacific Ocean are unknown.

Not all juvenile pink salmon migrate along the coastline in the temporal sequence described by Hartt (1980), as is indicated by two observations. First, as shown in Figure 24D, longline catches of age .0 pink salmon in the central Gulf of Alaska in late November-early December 1966 indicate that some juveniles likely moved offshore to the high seas early in the year in order to be in that location at that time. This movement could involve specific stocks that do not participate in the coastal migration pattern. Second, some juvenile pink salmon, in some years at least, remain in protected inshore coastal areas late into the fall, as suggested by trawl catches in November 1963 in Hecate Strait (LeBrasseur and Barner 1964). These fish may have entered the sea too late to participate in the general coastal migration pattern described by Hartt, they may represent stocks or groups of fish that make less extensive ocean migrations, they may have yet to leave coastal waters, or their behaviour may reflect a different pattern of ocean migration.

Hartt (1980) considered the possibility that juvenile salmon might re-enter inside passages after migrating along parts of the coast. He suggested that the lack of juveniles west of the Queen Charlotte Islands might indicate that they migrated inside through Hecate Strait rather than along the outer coast in that area.

Migrations of Asian Juvenile Pink Salmon

Data on the distribution of juvenile pink salmon in Asian coastal waters are available from small-meshed gillnet catches (Ishida 1966; Takagi et al. 1981), from beach seines and dip-net sampling (Karpenko 1983), and from trawls (Karpenko and Safronov 1985; Shuntov 1989). Juvenile pink salmon tend to remain in coastal waters in Asia, as in North America, at least until August. Birman (1969a) found small schools more than 168 km offshore in the Sea of Okhotsk in August. By September, some were found throughout the central Sea of Okhotsk region between Sakhalin Island and the western Kamchatka coast. Birman (1968) pointed out that the migration of juvenile salmon from northern to southern regions in the Sea of

Okhotsk is not confined to a narrow coastal band but covers a wide front across the open sea (Figure 25). Sea of Okhotsk catches in August and September (1969–72) by K. Shimazaki, as reviewed by Takagi et al. (1981), were generally supportive of Birman's observation, although Shimazaki's samples apparently were concentrated more in coastal waters in the northern and western areas, with fewer samples in the central Sea of Okhotsk. Shimazaki noted that larger fish tended to be further offshore.

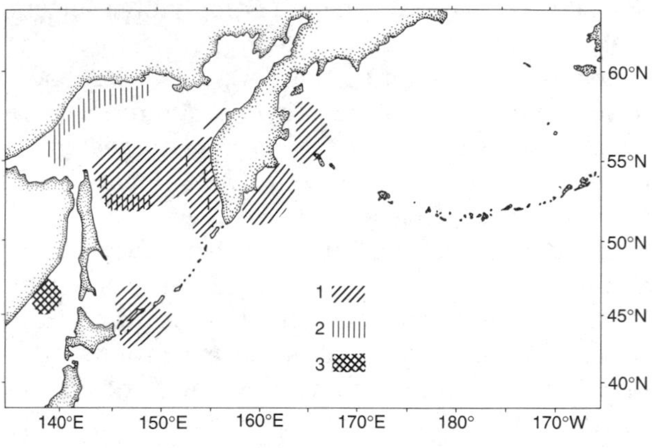

FIGURE 25

Some areas of observations on distribution of age .0 pink salmon in Asian waters: *1*, Birman 1969a; *2*, Shimazaki 1977, as reported by Takagi et al. 1981; *3*, Dubrovskaia 1934

Along the eastern Kamchatka coast, Birman (1969a) found juvenile pink salmon distributed through September only in waters west of 170°E. This pattern might indicate a southward moving coastal band similar to that described by Hartt (1980) in certain North American waters.

Birman (1969a) reported concentrations of age .0 pink salmon both east and west of the southern Kuril Islands in December. Off southwestern Iturup Island juvenile pinks average 26.2 cm in length. Shuntov (1989) documented catch rates of age .0 pink salmon in Pacific waters off southeastern Kamchatka and the southern Kuril Islands to be from 100 to 300 fish per hour of towing with a large otter trawl in November and December. Because southern Kuril Islands passes are known to be important migration routes for both juvenile and adult salmon, juveniles present in these waters late in the season could conceivably originate from many areas, including stocks throughout the greater Sea of Okhotsk region, local southern Kuril Islands stocks, or Sea of Japan stocks.

Orientation during Migration

The diversity of pathways to coastal waters suggests that juvenile pink salmon may use a variety of orientation cues to assist them in reaching the ocean. McInerney (1964) hypothesized that they may use salinity gradients, whereas Healey (1967) suggested that celestial cues may be used for direction-finding, especially on clear days. Sun-compass orientation has been demonstrated in adult pink salmon (Churmasov and Stepanov 1977; Stepanov et al. 1979) as well as in juvenile salmon of other species (Groot 1965; Hasler 1966). Hurley and Woodall (1968) found that pink salmon fry smaller than 60 mm in length selected warmer temperatures (12.0°–13.5°C) in a gradient and that larger fry selected slightly cooler water (9.5°–11.0°C). They speculated that offshore migration of larger fry may relate to cues from temperature selection preference.

Martin (1966) suggested that migration routes of juvenile pinks in southeastern Alaska correspond to net surface-water transport towards the Gulf of Alaska. Different migration routes in some areas in 1964 and 1965 were associated with changes in prevailing winds and, presumably, differences in wind-driven currents.

Diets of Juveniles in Coastal Waters

When juvenile pink salmon move between July to November from nearshore into coastal waters, 5–50 km offshore, there is a gradual shift in foods eaten to larger size zooplankton. In North American waters, the principal food items are larvaceans, copepods, euphausiids, arrowworms, and amphipods, whereas in Asian waters, young pink salmon prey on larval fishes, arrowworms, pteropods, euphausiids, amphipods, and mysids (Table 18).

High-seas Phase

Knowledge of the general offshore distribution of pink salmon in the North Pacific Ocean is derived from high-seas fishing and research activities conducted primarily since the early 1950s. Much information has come from western and central North Pacific Ocean waters, particularly the Japanese mothership and land-based fisheries in those areas. Beginning in 1955, large-scale co-operative investigations by Canada, Japan, and the United States, under the auspices of the International North Pacific Fisheries Commission (INPFC), expanded research activities into all offshore waters of the North Pacific Ocean north of about latitude 40°N. Information on the oceanic phase of pink salmon has been compiled by Manzer et al. (1965), Neave et al. (1967), and Takagi et al. (1981) as part of INPFC activities.

Gillnets were the major high-seas research gear used before 1960; later, longlines and purse seines were also employed. During the period 1961–71, high-seas research covered all months of the year, with emphasis on specific times, areas, and species. Throughout all regions, INPFC high-seas data on Pacific salmon distribution and abundance are reported and analysed on the basis of 2° longitude by 5°-latitude areas. Details about numbers of vessels, type and amount of fishing gear, and analyses of data available pertaining to pink salmon in the North Pacific Ocean from 1961 to 1971 are reviewed by Takagi et al. (1981).

Much of the understanding about pink salmon ocean migrations is based on high-seas tagging of age .1 fish early in their second year of marine life (April-June) and on the subsequent recovery of tagged fish, usually in coastal waters or spawning streams. Tag recovery information, plus supportive data from other stock-identifying features such as scale and meristic characters, along with seasonal distribution and abundance data, allow probable migration routes to be inferred for different regional stock groups.

Behaviour

Behaviour of pink salmon on the high seas, in terms of schooling, feeding, vertical distribution, and diel movements, are derived from information from offshore catches. Fishing is concentrated mostly in surface waters where gillnets and longlines sample to about 6 m depth and purse seines to about 36 m depth.

Information on vertical distribution patterns of pink salmon in the ocean is limited. Manzer and LeBrasseur (1959) and Manzer (1964) studied vertical distribution with sunken gillnets to about 61 m depth in the central Gulf of Alaska. In 1959, during May and July, most pink salmon (63%) were caught at night at the 12–14 m depth, with fewer caught at the 0–12 m depth (30%) and 24–37 m depth (7%). In July and August, all pinks were caught in the 10–12 m depth. Triple-tiered gillnet shackles in the western Gulf of Alaska from April to June in 1969 caught more pink salmon at the 16–23 m depth than at shallower depths (French et al. 1971). In general, however, most catches of pinks are concentrated in the upper 10 m depth of water. Takagi et al. (1981) suggested that vertical distribution patterns may change seasonally, with pink salmon remaining in deeper ocean water at night early in the year and then moving into surface waters at night as the season progresses.

Diel vertical movements of pink salmon are likely related to feeding patterns. Studies in oceanic areas with age .1 pink salmon indicate that feeding is principally diurnal, with peak activity occurring near dusk or at night (Shimazaki and Mishima 1969; Ueno et al. 1969; Pearcy et al. 1984). Takagi (1971) fished simultaneously with gillnets and longlines during four diel periods: morning, daytime, evening, and nighttime. He found that, although both fishing gears caught salmon effectively around sunrise and sunset, longlines were not effective in darkness, and gillnets were not effective in daylight. Vertical diel movements of food organisms, and of salmon feeding on specific prey as it becomes available during certain diel periods, likely account in part for stratification of some foods in stomachs (Allen and Aron 1958; Ueno 1968). Shimazaki and Mishima (1969) suggested that pink salmon failed to feed on abundant amphipods during periods of maximum darkness because they could not see them. However, Pearcy et al. (1984) found that pink salmon in the Gulf of Alaska, caught in gillnets set for 2 hours each over a 24 hour period, fed extensively on dense concentrations of euphausiids during maximum darkness. The nighttime catches were concentrated in the upper 2 m of gillnets and coincided with diel vertical movements of euphausiids into surface water layers (Pearcy et al. 1984).

Schooling of pink salmon in open ocean waters appears to be less structured and more loosely organized than in coastal waters or rivers. Analyses of catches in high-seas gillnets and of echogram images, as reviewed by Takagi et al. (1981), suggested that shoals tend to be composed of a few tens of individuals and, most often, of pairs of fish. Royce et al. (1968) suggested that particular stocks of salmon have no tendency to school as a group in the ocean. They noted that different species, age groups, and sizes often are caught in the same sets of fishing gear. Discrete schools likely do not redevelop until maturing fish migrate into coastal waters.

Diet of Age .1 Pink Salmon

Food found in stomachs of age .1 pink salmon tends to consist of larger prey than that eaten by juveniles (Table 19). Fish, squid, euphausiids, and amphipods are identified as major diet items in many stomach-content studies in different areas. Other important foods include pteropods, decapod larvae, and copepods. As with juveniles, the relative importance of specific foods eaten varies greatly with the time and area.

Diet studies summarized in Table 19, along with other information on foods eaten by age .1 pink salmon, are reviewed in detail by Takagi et al. (1981). From available knowledge on ocean diets of pink salmon, these authors concluded that the relative feeding index is low in winter, reflecting reduced feeding, and that the index increases in spring and summer, reflecting active feeding. Feeding is again reduced as gonadal maturation progresses and as maturing fish migrate into coastal waters and prepare to move into streams. Takagi et al. (1981) also noted that feeding indices apparently were lower with increased abundance of pink salmon. This was most evident in the eight-year study by Ito (1964), who found that food shortages in 1957 occurred concomitantly with high pink salmon abundance. Density-dependent ocean growth apparently occurs among many Asian stocks of pink salmon where the size of fish is smaller in big years (Kaganovsky 1949; Dvinin 1952; Semko 1954). However, notable exceptions to this pattern are the stocks from the Primore Region that reside in the Sea of Japan (Birman 1956).

According to the review by Takagi et al. (1981), pink salmon from the Sea of Japan were smaller in even years than in odd years when even-year populations were more abundant before 1962. Population size has reversed to odd-year abundance since 1962, and body size has changed inversely with small fish in odd years and large fish in even years.

Identity of Regional Stock Groups in the Ocean

Much research on pink salmon in the ocean has been directed at identifying migration patterns of regional groups of stocks and their areas of intermingling. The principal methods used in these stock separation studies were tag recovery patterns, and analysis of scale and meristic characteristics. In addition, biochemical techniques were employed using horizontal starch gel electrophoresis and serological and morphometric studies.

Neave et al. (1967) summarized much of the early effort to separate stocks on the high seas on the basis of scale and meristic characters. They noted that Asian pinks usually could be separated by scale and meristic features from North American stocks from southeastern Alaska and southward, but not from North American stocks from the Alaska Peninsula, Aleutian Islands, and western Alaska. The western Gulf of Alaska and central Pacific Ocean regions south of the Aleutian Islands and Alaska Peninsula became an important region of attention because eastern Kamchatka and western and central Alaska pink salmon stocks were known to mingle extensively in these ocean waters. According to Neave et al. (1967) tag recoveries indicate an overlap between Kamchatka and western Alaska pink salmon of about 1,600 km (177°E to 155°W) in an east-west direction.

Takagi et al. (1981) reviewed most recent studies on the use of scales and meristic characters to separate high-seas stocks of pink salmon through

TABLE 19

Information on the stomach contents of age .1 pink salmon sampled from various areas in the North Pacific Ocean and adjacent waters

Year	Month	Area	Number examined	Percentage of empty stomachs	Average amount of food per fish	Percentage composition in weight (or in volume)							Source
						Fishes	Squids	Euphausiids	Amphipods	Copepods	Pteropods	Others	
1958	May	Eastern N. Pacific, Coastal Domain	94	4	8.2 g	29.9	0.4	21.4	8.2	1.0	9.6	29.5	LeBrasseur (1966)
		Eastern N. Pacific, Alaskan Stream Domain	10	10	177.0 g	95.5	-	2.2	0.6	-	-	1.7	"
	Jun	Eastern N. Pacific, Subarctic Domain	47	15	5.5 g	10.8	74.6	-	14.6	-	-	-	"
		Eastern N. Pacific, Transitional Domain	16	31	15.9 g	2.2	-	43.1	34.3	4.4	6.9	9.1	"
1964	Jan-Feb	Eastern N. Pacific	24	38	52.4 cc	(1.0)	(+)	(1.0)	(97.0)	(0.0)	(0.0)	(1.0)	Manzer (1968)
1952	Jun-Jul	Aleutian waters	97	2	9.8 g	8.8	5.4	65.9	10.6	1.7	+	7.6	Maeda (1954)
1956			1061	-	16.4 g	44.8	28.9	11.0	5.7	1.7	6.2	1.7	Ito (1964)
1957			175	-	6.5 g	2.7	0.6	36.5	7.8	30.0	-	22.4	"
1958	May	47°N-57°N, 154°E-177°W	356	-	15.0 g	19.3	41.7	18.6	10.9	5.4	1.1	3.0	"
1959		Western and central N. Pacific	204	-	9.5 g	16.0	0.7	37.3	13.7	16.2	3.9	12.2	"
1960		including Okhotsk and Bering seas	176	-	16.6 g	16.2	57.0	11.3	6.5	1.8	0.7	6.1	"
1961	Aug		178	-	10.4 g	21.8	1.1	52.4	1.7	6.8	5.8	10.4	"
1962			196	-	12.0 g	19.0	8.2	21.8	13.2	9.5	13.4	14.9	"
1963			198	-	12.8 g	10.1	1.5	46.4	12.6	17.7	5.4	6.3	"
1964	Jun-Jul	Sea of Okhotsk	254	45	2.0 g	7.3	26.9	0.4	24.0	+	31.6	9.8	Takeuchi (1972)
1965	May-Jul	Western N. Pacific, 49°N-58°N, 164°E-175°W	290	22	6.7 g	17.7	2.2	35.2	4.0	5.0	3.8	32.1	"
1966	Apr-Jul	Western N. Pacific, 40°N-52°, 147°E-172°E	447	15	11.6 g	6.2	21.6	22.8	14.6	16.0	7.8	11.0	"
1966	Jun-Jul	Central Bering Sea	107	12	24.6 g	6.5	91.3	0.7	1.2	-	-	0.3	Kanno & Hamai (1971)
		Northern and eastern Bering Sea	79	8	10.2 g	53.2	0.0	16.4	27.5	-	-	2.9	"
1955	Jun	Sea of Okhotsk, off west Kamchatka	46	35	21.2 cc	(13.7)	(6.7)	(0.3)	(56.9)	(0.0)	(10.1)	(12.2)	Allen & Aron (1958);
		Off east Kamchatka	136	13	18.8 cc	(20.5)	(1.7)	(7.6)	(49.0)	(1.0)	(3.5)	(16.7)	Aron (1956)
	Aug	Western N. Pacific, 48°N-50°N, 165°E-175°E	111	4	23.6 cc	(17.0)	(3.1)	(20.2)	(12.5)	(41.7)	(5.5)	(0.0)	"

(continued on next page)

Table 19 (continued)

Year	Month	Area	Number examined	Percentage of empty stomachs	Average amount of food per fish	Percentage composition in weight (or in volume) Fishes	Squids	Euphausiids	Amphipods	Copepods	Pteropods	Others	Source
					8.2 g	29.9							
1955	Jun-Jul	Aleutian waters southward to Attu Is.	-	-	10.2‰	10.0	-	39.0	+	29.0	13.0	+	Andrievskaya (1957)
		Komandorskiy Is. waters	-	-	10.2‰	60.0	+	27.0	-	-	10.0	+	″
		Off east Kamchatka	-	-	2.9‰	35.0	-	+	45.0	+	13.0	+	″
		Off Cape Shipunskii	-	-	6.9‰	14.0	+	48.0	30.0	+	6.0	+	″
		Gulf of Kamchatka	-	-	-	20.0	-	25.0	40.0	-	-	+	″
		North Kuril Is. waters	-	-	2.5‰	45.0	+	24.0	18.0	-	6.0	+	″
		(Sum of the above six areas)	-	-	-	30.0	+	27.0	23.0	5.0	9.0	6.0	″
1940	Aug	Estuary of Kolpakova R., Kamchatka	12	-	-	85.0	-	-	+	-	-	+	Synkova (1951)
	Aug	Avacha Bay, Kamchatka	16	-	-	92.0	-	+	+	+	+	+	″
	Jul	Gulf of Kronotskiy, Kamchatka	43	-	-	85.0	-	+	+	-	+	+	″
	Aug		-	-	-	80.0	-	+	-	+	-	+	″
1948	Jul	North Kuril Is. waters	64	-	-	45.0	-	+	+	-	+	+	″
	Aug		-	-	-	78.0	-	+	+	-	+	+	″
1948	Jun-Jul	Datta Bay, Primore	400	97	0.01–0.69‰	87.8	-	4.6	4.8	-	+	2.8	Pushkareva (1951)
1949	Jun-Aug	Datta Bay, Primore	833	88	-	50.2	1.5	11.1	6.2	-	0.1	30.9	″
1949	Jul-Aug	Proster Gulf, South Kuril Is.	125	74	0.2–2.3‰	38.6	0.1	52.3	7.6	+	-	1.4	″
1965	Mar	Offshore, Sea of Japan	79	-	10.8 g	0.4	-	0.2	95.8	-	-	3.6	Fukataki (1967)
	Apr		229	-	8.3 g	0.7	1.8	34.3	54.4	2.2	0.0	6.6	″
	May		277	-	9.8 g	0.2	12.0	43.3	37.8	0.4	-	6.3	″
	(Mar-May)		585	-	9.3 g	0.5	6.5	33.4	52.6	1.0	0.0	6.0	″

Source: Adapted from Takagi et al. (1981)

the early 1970s, including those by Pearson (1966a, 1969) and Bilton (1966, 1971). Pearson (1966a, 1969) examined 27 scale characters and found that the number and spacing of circuli in the first-year zone, and the width from the centre of the focus to the thirtieth circulus were useful for separating some stocks in some years. Bilton (1971) reported that the number of circuli in the first ocean zone and the width of the first year's ocean growth from scale focus to the annulus could be used to correctly separate over half the samples of Fraser River and Aleutian Islands pink salmon according to their place of origin. These features, especially the average interval between circuli, were also used to identify percentages of scales from British Columbia and Alaska in the eastern and central Gulf of Alaska. In some years, most fish east of 140°W had British Columbia-type scales. Bilton (1966) concluded that the number and spacing between circuli on pink salmon scales during the first ocean year was not a fixed genetic character but could vary in different environments. He based this conclusion on a comparison of scales from Glendale River pink salmon from central British Columbia transplanted to Newfoundland, with other scales from several Pacific stocks of the same year, including the Glendale River.

Takagi et al. (1981) recognized six ocean migration patterns of pink salmon from regional stock groups throughout the North Pacific. These groups include (1) Washington and British Columbia (Figure 26); (2) southeastern, central, and southwestern Alaska (Figure 27); (3) western Alaska (Figure 28); (4) eastern Kamchatka north to Anadyr Gulf (Figure 29); (5) the greater Sea of Okhotsk basin and the Pacific Ocean side of Hokkaido and the Kuril Islands (Figure 30); and (6) the Sea of Japan (Figure 31).

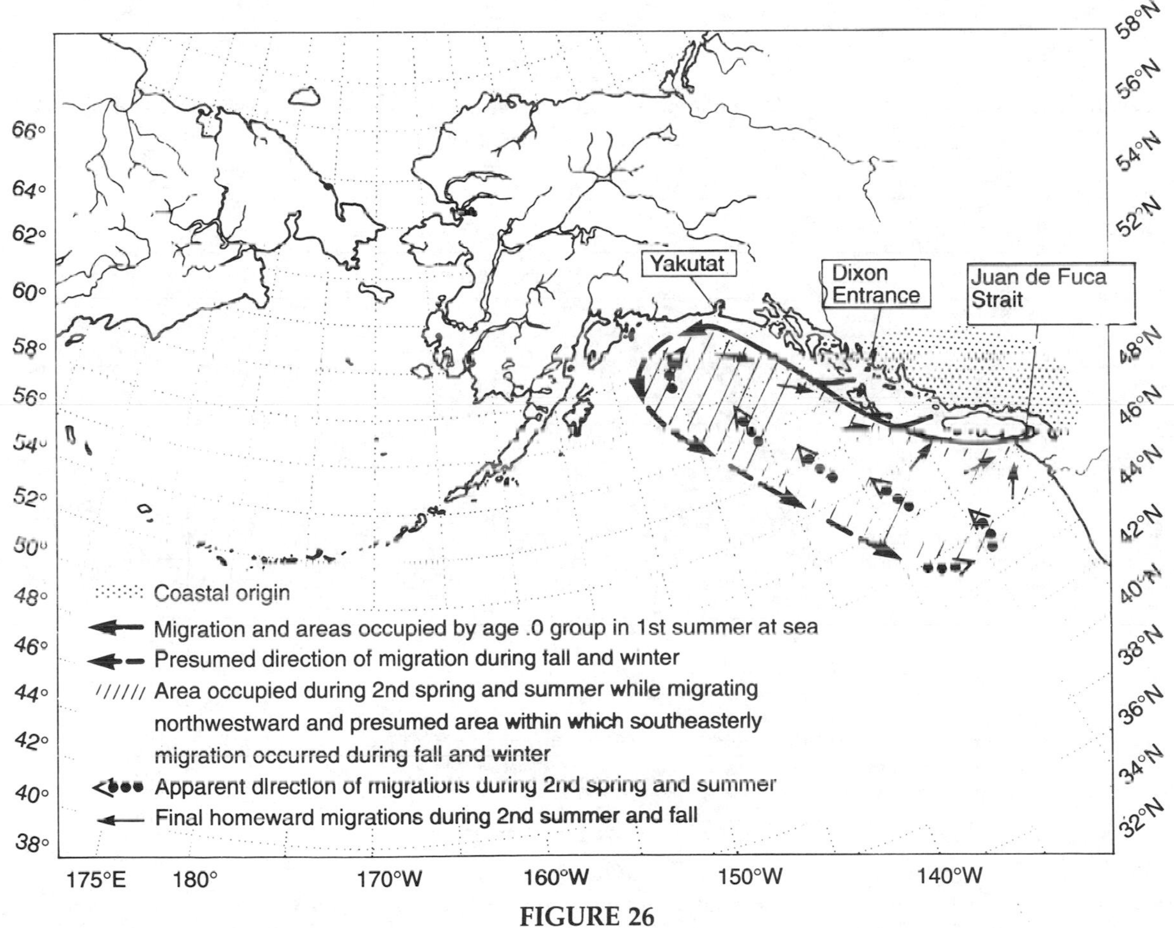

FIGURE 26

Diagram of ocean migration of pink salmon originating in Washington and British Columbia. (From Takagi et al. 1981)

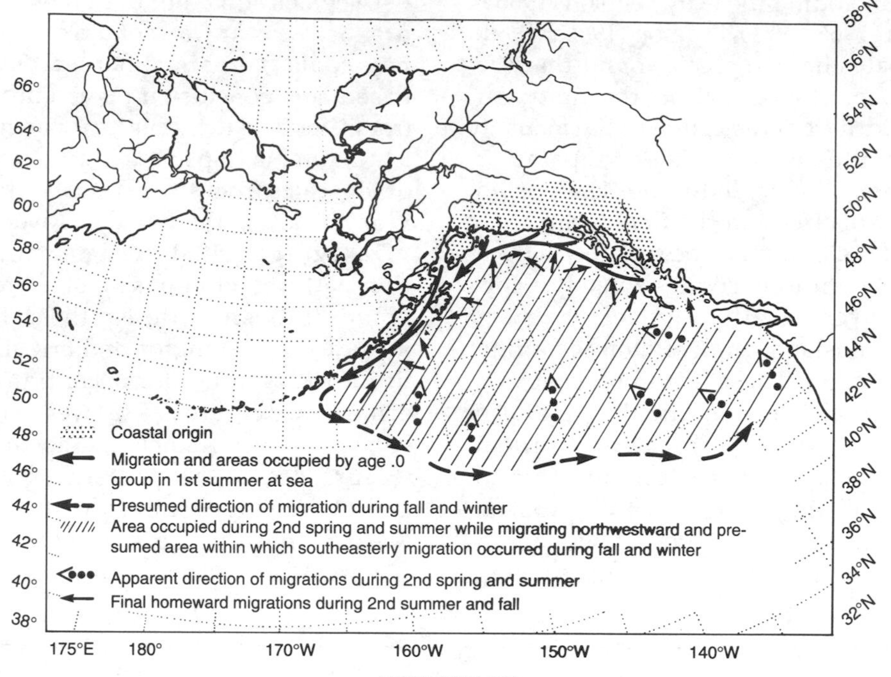

FIGURE 27

Diagram of ocean migration of pink salmon originating in southeastern, central, and southwestern Alaska. (From Takagi et al. 1981)

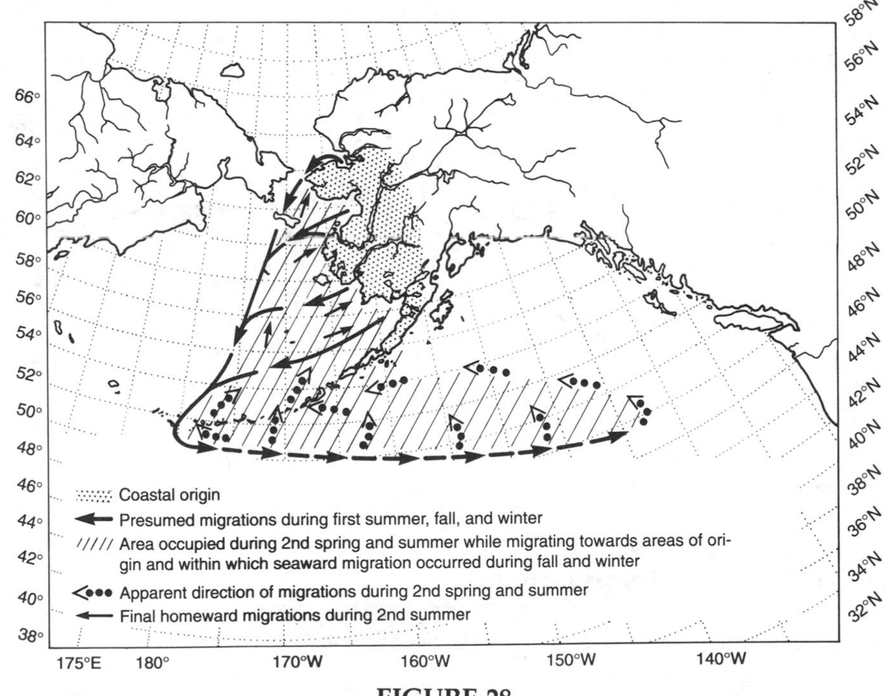

FIGURE 28

Diagram of ocean migration of pink salmon originating in western Alaska (north side of Alaska Peninsula to Kotzebue Sound, including St. Lawrence Island). (From Takagi et al. 1981)

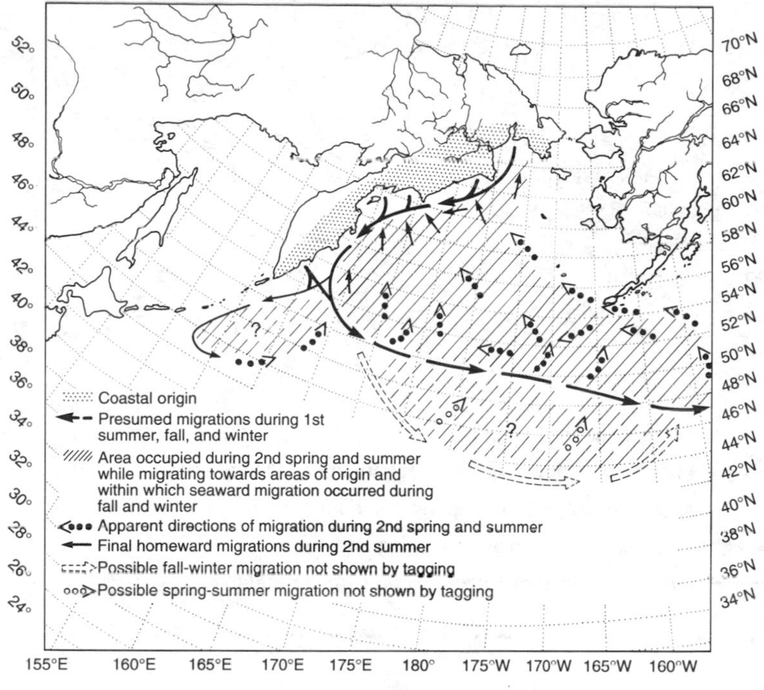

FIGURE 29

Diagram of ocean migration of pink salmon originating in eastern Kamchatka and northward to the Anadyr Gulf. (From Takagi et al. 1981)

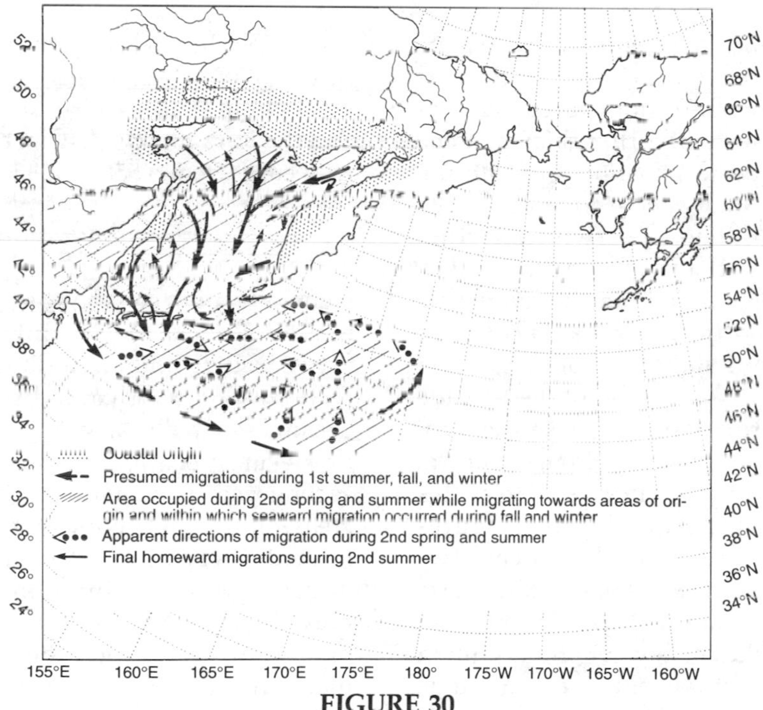

FIGURE 30

Diagram of ocean migration of pink salmon originating in western Kamchatka, northern Okhotsk coast, eastern Sakhalin, Kuril Islands, and Hokkaido. (From Takagi et al. 1981)

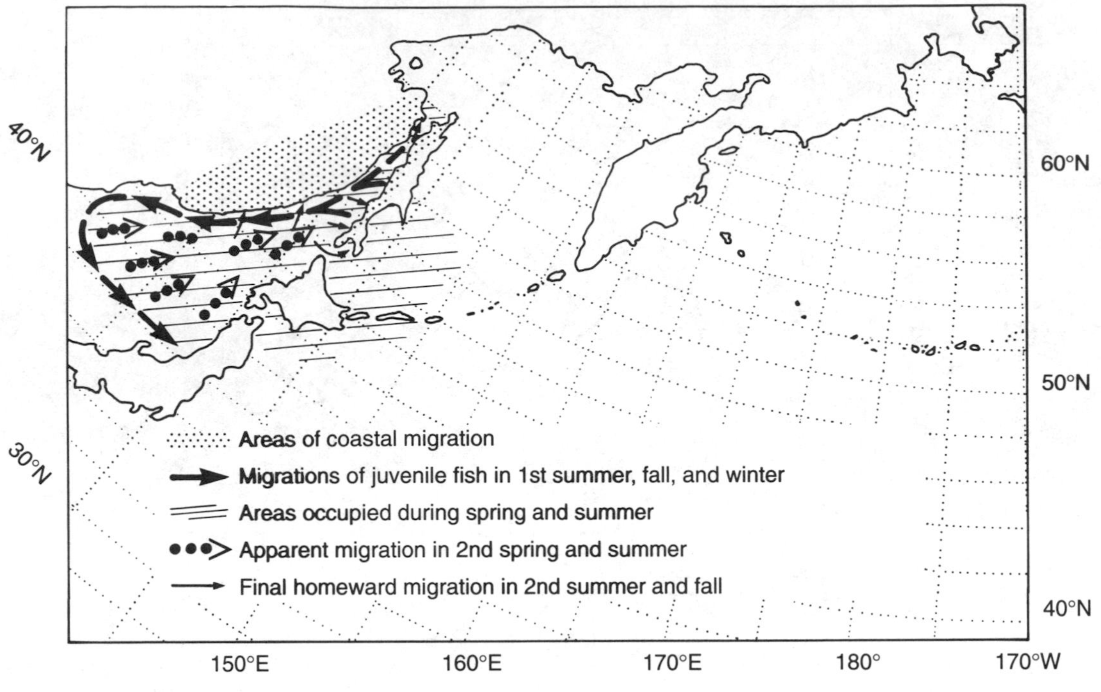

FIGURE 31

Diagram of ocean migration of pink salmon originating in the Amur River, Primore, and western Sakhalin. (From Takagi et al. 1981)

The marine migrations of regional stock groups appear to be similar for odd-year and even-year cycles. Although these migrations often occur at the same time and place with other salmon, some features tend to be species specific. Pink, chum, and sockeye salmon usually occur together on the high seas, but pink salmon overwinter farther south in warmer waters than sockeye (Royce et al. 1968). Neave et al. (1967) noted that, as favourable temperatures extend northward, pink salmon from southerly spawning populations in both Asia and North America migrate northward of the latitude of their ultimate destination and make the final approach to their spawning stream from the north.

A common feature of the ocean migration routes of the six regional stock groups shown in Figures 26–31 is that they all occur in a broad counter-clockwise circular pattern. The same ocean area is covered more than once at the beginning and end of the migration. Unlike sockeye or chum salmon which mature at more than one ocean-life age, pink salmon make only one circuit of their migration route. The areas covered in the oceanic migration of pink salmon involve considerable distances. Migration distances are least for stocks along the Sea of Japan. According to Royce et al. (1968), stocks from southeastern Alaska, British Columbia, and the Karaginskiy district of eastern Kamchatka may cover from 5,556–7,408 km while at sea.

Maturation

Generally, pink salmon start their final homeward migration from May to July. This is a period of active feeding and migration involving significant increases in growth and distances covered in the ocean. Essentially, all growth in pink salmon in the first year at sea and during the first part of the second year is somatic. A three- to four-fold increase in weight occurs after the winter annulus is formed (Ricker 1964). During the final phases of marine life, as maturation advances, much of growth becomes gonadal. Final maturation in pinks, however, as in other Pacific salmon, takes place after feeding has ceased, generally in fresh water, so that terminal gonadal growth and development is derived from body nutrients. According to Eniutina (1972), gonadal weights in Im River pink salmon reached a maximum of 15% and 25% of body weight, respectively, for males and females; mean values were about half these levels.

Maturation and marine growth schedules are interrelated so that early maturing females, even though they may have a high rate of growth in the final marine period, have smaller gonads and are less fecund (Grachev 1971).

Coastal Migration

Tag recovery patterns reveal that maturing pink salmon in the same ocean area at the start of their return migration may travel to widely different final destinations (Figure 32) and that fish present over broad distances within ocean regions may migrate to the same coastal location (Figure 33). The homeward migration of pink salmon from the high seas may involve a more or less direct route to the final spawning area, or it may include considerable migration paralleling the coastline (Neave 1964). An example of the latter is provided by the stocks from southern British Columbia and Washington that may make lengthy migrations southward along the coastline before moving to inshore waters (Figure 26). Pink salmon stocks that take a

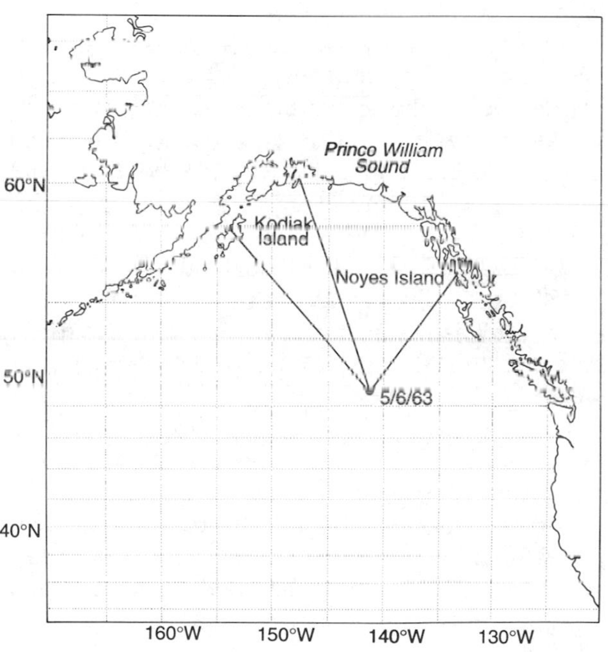

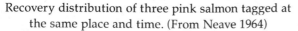

FIGURE 32

Recovery distribution of three pink salmon tagged at the same place and time. (From Neave 1964)

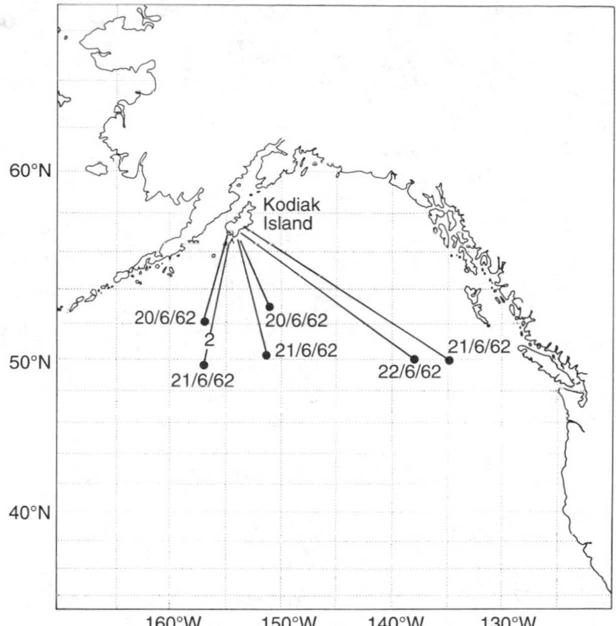

FIGURE 33

Ocean localities from which pink salmon, tagged on or near the same date, travelled to Kodiak Island, Alaska. (From Neave 1964)

more direct route from spring feeding areas to spawning areas include the central Alaska stocks (Figure 27) and the Karaginskiy stocks of eastern Kamchatka (Figure 29). The final migration of Karaginskiy District stocks is remarkably direct (Figure 34) (Royce et al. 1968).

Neave (1964) argued that coastal movements of maturing fish should be viewed relative to their total migration pattern and the effective use of favourable environments. Lengthy coastal migrations should not be interpreted as meaning that stocks are off course or searching for cues to make landfalls.

Concentrations of adult pink salmon develop in specific North American and Asian coastal waters, after extended high-seas migrations, with remarkable consistency from year to year. Arrival timing may vary slightly depending on the abundance of a particular brood, unusual oceanographic features, or characteristics of odd-year and even-year abundance patterns in the region. The overall consistency of these patterns forms the basis for coastal fisheries along both continents.

Coastal concentrations of adults often consist of

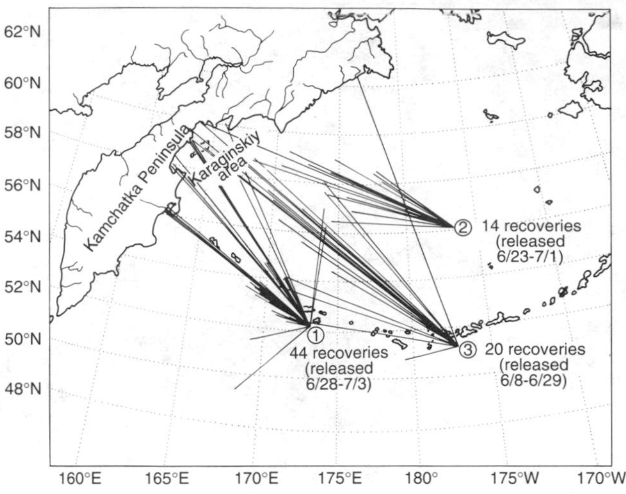

FIGURE 34

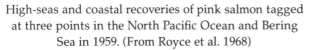

High-seas and coastal recoveries of pink salmon tagged at three points in the North Pacific Ocean and Bering Sea in 1959. (From Royce et al. 1968)

both local stocks near the end of directed migration and other more remote stocks with considerable migration distances remaining. Along the south side of the Alaska Peninsula, for example, relatively large numbers of pink salmon return in July and August to spawn in streams flowing into local bays, whereas other stocks that spawn in Bristol Bay and other western Alaska districts migrate through the same coastal waters (see Figures 27 and 28), but earlier, primarily in June (Thorsteinson 1959). Slightly offshore and westward, large numbers of eastern Kamchatka pink salmon (Figure 29) migrate through eastern Aleutian Islands passes in late June and early July (Hartt 1966; Royce et al. 1968) at the time when local southern Alaska Peninsula stocks start to appear to the east. Hartt (1962) noted the lack of southern Alaska Peninsula recoveries from eastern Aleutian pink salmon tagging. Although these two areas are in reasonably close proximity, stocks of pink salmon in adjacent waters, migrating homeward at about the same time, originate from different continents.

Migration Rates

Minimum rates of travel for pink salmon at sea can be determined based on lapsed time and distance for tagged individuals observed at separate locations. These are "straight line" estimates of minimum average migration rates between two points and do not necessarily reflect the actual rate. Hartt (1980) estimated that two juvenile pinks from the Fraser River and the Puget Sound area had migrated from 17.2 to 19.8 km/d during their first two to three months at sea. Neave (1964) estimated rates of travel during the outbound migration leg for two groups of pink salmon captured in the ocean in 1962. One group of three fish, tagged in the eastern Pacific (west and southwest of the Oregon-Washington coast) in April and recovered in Cook Inlet or Prince William Sound, travelled at least 5.6–9.3 km/d for about nine months to reach where they were initially caught and tagged. The other group, four fish, tagged in the western Gulf of Alaska in May, had travelled at least 9.3 km/d from eastern Kamchatka. Both of the foregoing groups would need to travel from two to three times the indicated outbound rates to return to the appropriate coastal waters during their remaining ocean life.

Hartt (1966) provided estimates of rates of travel for pink salmon during the last month at sea ranging from 43.5 to 60.2 km/d. These estimates are for eastern Kamchatka stocks tagged in Aleutian Islands passes or the Bering Sea in 1959 (Figure 34). The average rate of travel for fish recovered at sea and in coastal waters was 54.3 and 45.0 km/d, respectively, whereas the average time at liberty between tagging and recapture for the same groups was 12.0 and 27.8 days. Kaganovsky (1949) reviewed earlier migration rates for tagged pink salmon and reported these average rates: for 7 fish tagged off the east coast of Hokkaido, 7.6 km/d; for 23 fish tagged in the northern Kuril Islands, 28.5 km/d; for 9 fish tagged in the central and southern Kuril Islands, 38 km/d; and for 2 fish tagged off western Kamchatka, 65 km/d. Hartt (1966) mentioned one tagged pink salmon that travelled 463 km in 6 days (77.2 km/d), an unusually rapid rate for a fish at liberty for such a short time, since there is normally a noticeable one- to two-day delay in migration due to the effects of capture and tagging. Kaganovsky (1949) recorded the migration of a tagged pink salmon at liberty 3 days that travelled at a rate of 74 km/d.

Data reviewed by Kaganovsky (1949) and Hartt (1966) suggest a seasonal increase in the apparent rate of travel. However, as Hartt (1966) pointed out, this apparent seasonal increase in rate does not necessarily indicate a faster swimming speed, but may be due to later migrating fish following a more direct route to their coastal destination.

Growth and Mortality

Ocean growth of pink salmon is a matter of considerable interest because, although the species has the shortest life span, it also is among the fastest growing Pacific salmon. Entering the estuary as fry at or slightly above 3 cm in length, maturing adults return to the same area 14–16 months later ranging in length from 45 to 55 cm.

Winter annulus formation in pink salmon may start as early as November and is completed by January (Birman 1960; Pearson 1966b; Bilton and Ludwig 1966), somewhat earlier than in other Pacific salmon species. From the time of annulus formation until subsequent capture in nearshore areas, it is estimated that pink salmon increase their weight two- to six-fold (Ricker 1964). The average fork length of pink salmon at annulus formation at the end of the first year's growth has been estimated for several North American and Asian stocks to range from 27.0 to 33.9 cm, with North American stocks, on average, larger than Asian stocks (Ricker 1964).

Oceanic growth in pink salmon, in addition to a true winter annulus, often produces an annulus-like supplementary check on scales, usually from July to September during the first summer. This "fry ring," "fingerling ring," or "false check" has been studied in Asian stocks from the Sea of Japan and southern Sea of Okhotsk (Miyaguchi 1959; Eniutina 1962) and Kuril Islands (Ivankov 1965), and in North American stocks from British Columbia (Bilton and Ricker 1965). This growth phenomenon develops when pink salmon are from 10 to 18 cm long and is generally thought to be associated with environmental changes or shifts in diet, perhaps as fish move offshore into more oceanic waters. In some years, along both continents, the supplemental check is found in up to one-third of the scales studied. Some Soviet scientists have argued that this check is a true annulus and that pink salmon possessing it mature at ages older than two years (Vedensky 1954; Lapin 1963, 1971).

Pink and chum salmon have similar early ocean life histories and might be expected to have similar rates of ocean growth. Ricker (1964) noted that, in spite of much variation between years within stocks, pinks grew faster than chums at comparable age and size. Giving locality samples equal weight from North America and Asia for both species, Ricker found that the average weights of pink and chum salmon at first annulus were 307 and 258 g, respectively. He further pointed out that because chum salmon annulus formation is later than that for pink salmon (April versus January), the difference in growth between the two species is even greater. Finally, Ricker observed that, although comparisons in second-year growth are limited because pink salmon mature before the growing season is over, the average length of pink salmon was greater when growth stopped (July-August) than that of chums at their second annulus the following April.

Growth rates of pink salmon have been examined in detail by LeBrasseur and Parker (1964), Ricker (1964), Sakagawa (1972), and Takagi et al. (1981). These studies provide estimates of growth curves (Figure 35). Although these curves, based on a wide variety of data sets and analyses, vary considerably in detail, they give a broad overview of the range of lengths that may be found during the ocean life of pink salmon.

Size of Adults

The average body weight of pink salmon caught in coastal fisheries in North American and Asian fishing districts, 1962–71, is reviewed in Table 20. On both continents, the average size is smallest at the northern end and largest at the southern end of the range. In general, for roughly comparable latitudes, fish tend to be larger in North America than

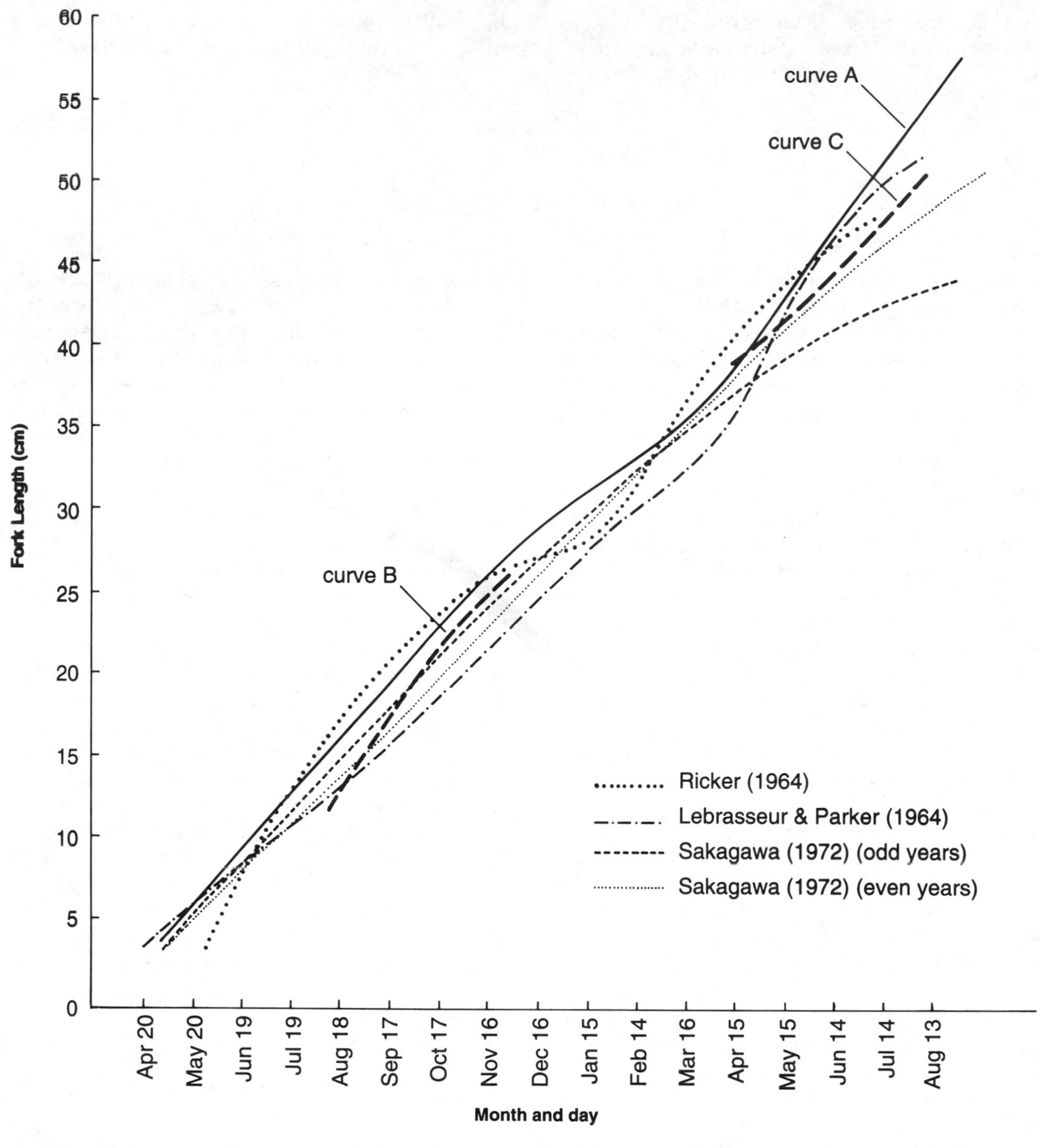

FIGURE 35

Comparison of growth curves of pink salmon. Curve *A* is a general growth curve estimated from plots of pink salmon lengths sampled from North American waters, 1962–67. Curve *B* is from age .0 fish data from the Sea of Okhotsk (adapted from Birman 1969a). Curve *C* is estimated from age .1 fish data sampled with longlines by Canadian, Japanese, and U.S. research vessels 1962–71. (From Takagi et al. 1981)

TABLE 20
Average body weight, in kilograms, of adult pink salmon caught in coastal areas of North America and Asia, 1962–71

Year	Coastal areas in North America													
	Arctic	Bristol Bay	Alaska Pen.	Chignik	Kodiak	Cook Inlet	Prince Wm. Sd.	S.E. Alaska	Northern BC	Southern BC	Fraser R. BC	Wash.	Ore.	Calif.
1962	1.09	1.45	1.36	1.22	1.36	1.45	1.36	1.77	1.81	1.88	–	2.09	–	–
1963	1.04	1.50	1.59	1.63	1.50	1.54	2.04	1.41	2.15	2.32	2.41	2.26	2.08	2.63
1964	1.09	1.36	1.41	1.45	1.50	1.59	1.77	1.72	1.73	1.79	–	–	–	–
1965	1.18	1.36	1.45	1.50	1.72	1.63	1.50	1.77	1.84	2.58	2.91	2.84	2.34	2.98
1966	1.13	1.41	1.59	1.54	1.77	1.63	1.72	2.00	1.97	1.80	–	2.15	–	–
1967	1.63	1.41	1.50	1.50	1.91	1.77	2.04	2.04	2.23	2.41	2.53	2.44	2.09	2.73
1968	1.59	1.45	–	–	1.54	1.36	1.63	1.50	1.28	1.31	–	1.76	–	–
1969	1.63	1.54	–	1.95	1.91	1.77	1.77	1.95	2.07	2.61	2.85	2.69	2.08	2.24
1970	1.45	1.41	1.81	1.63	1.63	1.72	1.81	1.77	1.77	1.77	–	2.35	–	–
1971	1.41	1.77	1.54	1.68	1.77	1.59	1.63	1.68	1.88	2.13	2.34	2.33	2.50	3.33
Average														
Odd-numbered	1.38	1.52	1.52	1.65	1.76	1.66	1.80	1.77	2.03	2.41	2.61	2.51	2.22	2.78
Even-numbered	1.27	1.42	1.54	1.46	1.56	1.55	1.66	1.75	1.71	1.71	–	2.09	–	–

Year	Coastal areas in Asia										
	N.E. Kamchatka	East Kamchatka	S.W. Kamchatka	N.W. Kamchatka	N.E. Okhotsk	N.W. Okhotsk	East Sakhalin	South Kuril	West Sakhalin	Primore	Amur River
1962	1.22	–	1.71	1.59	1.52	1.74	1.71	1.60	1.92	2.06	1.78
1963	1.38	1.36	1.81	1.50	1.38	1.47	1.48	1.20	1.66	2.11	1.76
1964	1.17	–	1.48	1.17	1.31	1.43	1.52	1.64	1.43	1.91	1.82
1965	1.21	–	1.48	1.48	1.31	1.34	1.56	1.43	1.66	1.90	1.77
1966	1.03	–	1.81	–	1.26	1.38	1.72	1.40	1.67	1.81	1.79
1967	1.34	1.25	1.47	1.88	1.14	1.22	1.44	1.27	1.73	1.94	1.65
1968	1.10	–	1.92	1.81	1.37	1.36	1.82	1.64	1.63	1.68	1.72
1969	1.36	1.33	2.25	1.79	1.15	1.21	1.44	1.31	1.41	1.75	1.54
1970	–	–			1.34	1.28	1.48	1.61	1.23	1.77	1.52
1971		1.41	2.03	2.14	1.68	1.52	1.38	1.25	1.20	2.22	1.99
Average:											
Odd-numbered	1.32	1.34	1.81	1.76	1.33	1.35	1.46	1.29	1.63	1.98	1.74
Even-numbered	1.13	–	1.70	1.52	1.30	1.41	1.65	1.58	1.58	1.85	1.73

Source: From Takagi et al. (1981)

in Asia. In Asia, average weights ranged from 1.13 kg in even-numbered years in northeastern Kamchatka to 1.98 kg in odd-numbered years in Primore. In North America, average weights ranged from 1.27 kg in even-numbered years in the Arctic to 2.78 kg in odd-numbered years in California (Table 20). The California data are based on rather small catches, from a troll fishery, of late-season pink salmon migrating to more northerly spawning areas. The largest North American pink salmon that are identifiable within an assemblage of specific stocks are from the Fraser River.

Neave (1966) noted that in British Columbia significant differences in weight of pink salmon are recognizable (1) between fish caught in different areas, (2) between even- and odd-year fish in the same area, and (3) between fish of either genetic line in different years. In North America there is a pronounced tendency for odd-year fish to be larger than even-year fish (Table 20). There is also a gradient in the difference between lines that intensifies latitudinally from north to south and this is especially evident throughout British Columbia. Larger odd-year pinks in British Columbia have persisted since the early days of the fishery (Hoar 1951; Godfrey 1959). Unlike some regions in Asia, where size of fish varies inversely with size of run (Kaganovsky 1949; Semko 1954), the phenomenon of larger odd-year pinks versus smaller even-year fish in British Columbia is independent of run size. Godfrey (1959) showed that fish are larger and catches greater (and presumably run size) in odd than in even years for several fishing districts. He concluded that odd-year pinks in British Columbia have a larger "basic" size than even-year pinks from the same area of origin. Ricker (1972) indicated that this "strongly suggests some genetic difference between the two lines in this region in respect to characters affecting rate of growth."

Ricker et al. (1978) found a significiant decrease in size of both odd- and even-year pink salmon in British Columbia since about 1951. They attributed the primary cause to selective genetic effects of fisheries, especially trolling, that tend to remove larger, faster growing individuals from populations.

In southeastern Alaska, Skud (1958) found that the timing of runs is earlier when pink salmon are large. In the Kuril Islands, Ivankov (1968a) found that pink salmon were "far smaller every year at the beginning of the run than at its end . . . both in the coastal zone and in the rivers."

Size in adults, then, is a highly variable aspect of pink salmon biology. Part of this variability can be attributed to latitude and continent of origin, to stock-specific genetics of even- and odd-year runs, and to annual differences in oceanic conditions. In some cases, run strength or size of population can influence adult size of pink salmon.

Ocean Mortality

Parker (1962a, 1962b) and Ricker (1964) have considered natural mortality rates of maturing pink salmon during the pelagic and returning coastal life history stages. Using long-term Hooknose Creek observations with assumptions on mean fecundity and equilibrium population levels, Parker (1962a) estimated total instantaneous mortality rates for pink and chum salmon at $i = 6.746$ and 7.146, respectively. The difference in the two species, $i = 0.40$, was attributed to the average mortality for two extra ocean years in chum salmon. While at sea and during all other life stages, mortality rates were assumed to be the same for both species. Thus, the total instantaneous mortality rate for pink salmon during the marine stage of life was $i = 0.40$ for 2 yr or 0.20/yr or 0.17 for the 10-month pelagic period (Parker 1962a). A second similar estimate was derived by Parker from marked Hooknose Creek pink and chum salmon fry after correcting for high mortality rates of marked fish. The instantaneous monthly rate was estimated at $i = 0.017$/mo for the pelagic period. Ricker (1964) used a value of $i = 0.02$/mo for pink salmon in the period from annulus formation to the final month in coastal waters for fish in the open sea $>$30 cm long.

In coastal waters, Parker (1962a) assigned a natural instantaneous mortality rate of $i = 0.07$ or 0.035/mo for a two-month period, whereas Ricker (1964) used a coastal rate of $i = 0.07$ for one month. The assumption that the natural instantaneous mortality rate of maturing salmon increases in coastal waters is based on greater concentrations of fish and assumed predation, particularly from marine mammals.

The overall natural marine mortality rate of pink salmon before inshore fishing equals 95.7% of the fry that entered the estuary. Adding a 65% fishing exploitation rate on pink salmon that reach coastal waters increases the total marine mortality to 98.5%. Based on return per spawner statistics for several Alaska stocks, McNeil (1980) estimated total marine mortality for pink salmon, including coastal fisheries, at 96.6%. The prevailing belief is that the greatest part of marine mortality occurs during the initial inshore and coastal period, before fish move to oceanic waters. Parker (1962a) presented monthly mortality rates for all life history stages of pink salmon that illustrate the importance of the early marine period in the schedule of total mortality (Figure 36).

Measured marine survivals for pink salmon from migrant fry to adults returning to streams are reviewed in Table 21 for five North American

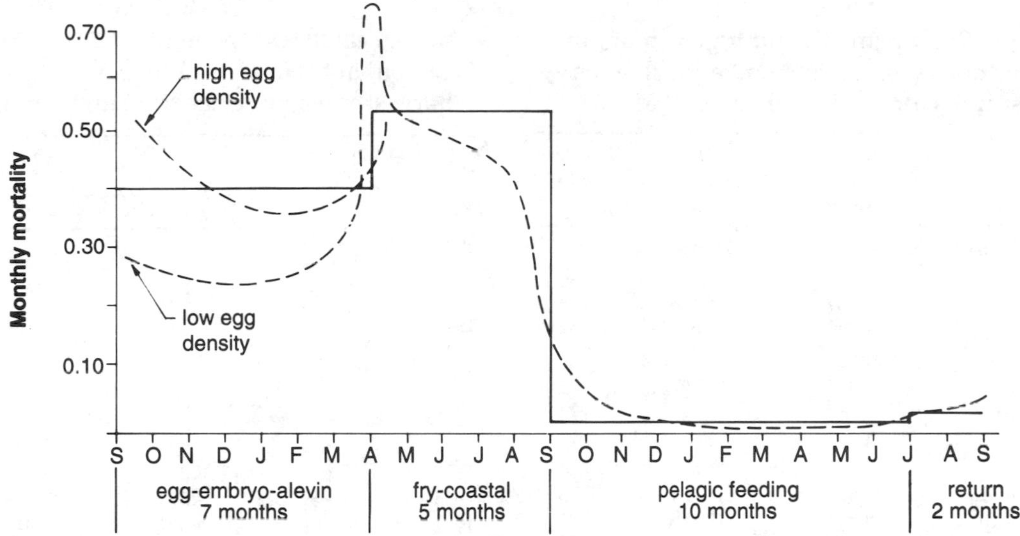

FIGURE 36

Major life history phases of pink salmon and suggested monthly instantaneous mortality rates for each phase. (From Parker 1962)

stocks. The range of mean values, 1.7%–4.7%, include the effects of coastal fisheries. The data series for Hooknose Creek, according to Parker (1964), may reflect major differences in fishing intensity on that particular stock. The effects of coastal fisheries on McClinton Creek, Sashin Creek, and Auke Creek data are also unknown and may vary widely. Only Fraser River runs reviewed in Table 21 have separate estimates available for both catch and escapement. These data (Table 22) show that the coastal harvest of Fraser River pink salmon since 1961 has ranged from 42% to 86% per year and averaged 68% for 13 broods. A comparable data series of harvest rates is available for the Prince William Sound district in Alaska for the years after 1945. These data (Table 23) show that estimated harvest rates ranged from 34% to 91% and averaged 75.6% over thirty-three years. The above calculations for Prince William Sound excluded data for 1954, 1955, 1959, and 1972, when severely depressed runs restricted fishing to token levels. No estimates on marine survival of natural spawning populations of pink salmon in Prince William Sound are available.

TABLE 21

Some reported marine survivals for pink salmon from migrant fry to adult return

Region	Stream	Inclusive brood years	*N*	Survivals (%) Mean	Range	Source
Northern BC	McClinton Cr.	1930–40	6	1.7	0.3–6.8	Pritchard (1948b)
Central BC	Hooknose Cr.	1947–60	14	1.8	0.2–5.2	Parker (1964)
Southern BC	Fraser R.	1961–79	12	3.1*	0.8–5.4	IPSFC (1986)
S.E. Alaska	Sashin Cr.	1940–65	17†	3.2	0.5–17.5	Ellis (1969)
		1966–78	13	4.7	0.2–23.1	Vallion et al. (1981)
	Auke Cr.	1971–81	11	2.9	0.2–6.7	Taylor (1983)

Notes: *Fraser River data include known levels of fishery harvest on those stocks for each year; other data include only adults returning to streams.

†Estimates from escapements less than 1,000 adults excluded because of errors inherent in sampling small populations

TABLE 22

Annual catch, escapement, and total run of pink salmon, in thousands of fish, and annual fishery harvest rates for the Fraser River, 1961–85

Year	C Catch	E Escapement	T Total run	Harvest rate C/T
1961	794	1 094	1 888	0.42
1963	3 524	1 953	5 477	0.64
1965	1 129	1 191	2 320	0.49
1967	11 137	1 831	12 968	0.86
1969	2 399	1 529	3 928	0.61
1971	7 963	1 804	9 767	0.82
1973	5 035	1 754	6 789	0.74
1975	3 527	1 367	4 894	0.72
1977	5 855	2 388	8 243	0.71
1979	10 843	3 561	14 404	0.75
1981	14 197	4 488	18 685	0.76
1983	10 714	4 632	15 346	0.70
1985	12 403	6 461	18 864	0.66
Mean				0.68

Source: Data calculated from IPSFC (1986), Tables 8 and 9

Marine Predators

Thirty-four marine mammal species commonly occur with salmon in waters of the North Pacific and contiguous seas, and fifteen are known to prey on salmon (Fiscus 1980). Other marine predators capable of feeding on adult salmon include several sharks and a few other fishes, such as large Pacific halibut, *Hippoglossus stenolepis*. Much of the available information on marine mammal-adult salmon interactions comes from estuarine and nearshore coastal waters usually associated with fisheries for salmon. Few data specifically identify pink salmon as prey because marine mammal food studies often group all salmonids in one category due to difficulty in identifying species from remains in stomach contents. Because pink salmon are the most abundant salmon in North Pacific waters, it is likely that they comprise a significant portion of the salmonids eaten by marine mammals.

Marine mammals known to prey on free-swimming pink salmon, as reviewed by Fiscus (1980), include the harbour seal, *Phoca vitulina*, based on observations in southeastern Alaska; the northern fur seal, *Callorhinus ursinus*, off the Washington coast; the Pacific whitesided dolphin, *Lagenorhynchus obliquidens*, off the Columbia River mouth; and the humpback whale, *Megaptera novaeangliae*, in the Kuril Islands.

TABLE 23

Annual catch, escapement, and total run of pink salmon, in thousands of fish, and annual fishery harvest rates in Prince William Sound, 1945–81

Year	C Catch	E Escapement	T Total run	Harvest rate C/T
1945	11 632	1 699	13 331	0.87
1946	8 025	942	8 967	0.89
1947	8 078	816	8 894	0.91
1948	2 461	447	2 908	0.85
1949	6 089	642	6 721	0.91
1950	1 851	400	2 251	0.82
1951	803	631	1 434	0.56
1952	2 161	295	2 456	0.88
1953	1 996	393	2 389	0.84
1954	12	1 934	1 946	0.01
1955	27	727	754	0.03
1956	4 526	1 304	5 830	0.78
1957	649	130	779	0.83
1958	6 289	858	7 147	0.88
1959	– *	601	601	0.00
1960	1 842	1 348	3 190	0.58
1961	2 297	2 204	4 501	0.51
1962	6 742	2 019	8 761	0.77
1963	5 296	1 345	6 641	0.80
1964	4 207	1 845	6 052	0.70
1965	2 460	976	3 436	0.72
1966	2 697	1 300	3 997	0.67
1967	2 626	1 227	3 853	0.68
1968	2 451	1 084	3 535	0.69
1969	4 828	1 128	5 956	0.81
1970	2 810	952	3 762	0.75
1971	7 311	2 115	9 426	0.78
1972	55	607	662	0.08
1973	2 054	1 230	3 286	0.63
1974	449	859	1 308	0.34
1975	4 453	1 618	6 071	0.73
1976	3 019	857	3 876	0.78
1977	4 513	1 693	6 206	0.73
1978	2 914	1 056	3 970	0.74
1979	15 630	2 927	18 557	0.84
1980	14 161	1 576	15 737	0.90
1981	20 558	2 912	23 470	0.86
Mean				0.76

Source: Data for 1945–71 are from Fredin et al. (1974); catch for 1972–81 from Alaska Department of Fish and Game catch statistics; escapement for 1972–75 from Pirtle (1977), for 1976, 1978, and 1979 from Pirtle and McCurdy (1980a, 1980b, 1980c), for 1977 from McCurdy and Pirtle (1980), and for 1980–81 from McCurdy (1984).
Note: *<500 fish

Homing and Straying

Homing of salmonids and the biological mechanisms that make it work have been studied extensively in recent decades. A number of in-depth assessments of present understandings of this phenomenon have been made by several researchers (Hasler 1966; Harden Jones 1968; Hasler et al. 1978; Brannon 1982; Hasler and Scholz 1983; Stabell 1984). Studies have focused on spawning stream selection by returning adults and how the "home" stream is identified, whether it be a short coastal creek flowing directly to sea or a distant upriver tributary of a major river system.

Homing to the parent stream is preceded by some form of natal area recognition by returning adults. A memory imprint of unique stream odours by juveniles before migrating seaward from natal areas (Hasler 1966; Brannon 1982; Hasler and Scholz 1983) or recognition of stock-specific pheromones (Nordeng 1971, 1977; Solomon 1973; Stabell 1984) are likely mechanisms used in homing. Pink salmon are Type A salmon (Hikita 1962) that migrate quickly at emergence to downstream nursery areas, and imprinting in this species would have to occur fast, possibly involving different processes from those used in freshwater rearing species that migrate seaward at later stages. Brannon (1982) suggested that imprinting in Type A salmon might occur as early as hatching. Regardless of whether homing is accomplished principally by imprinting or by pheromone recognition or by some combination of both, a highly developed olfactory acuity plays a major role in how most salmon return to their natal stream.

As migration onto spawning grounds nears, pink salmon in coastal areas begin milling in the vicinity of streams. At this time, schools may consist of individuals that originated from separate natal streams. In the process of searching and probing, presumably a part of normal homing behaviour, individual fish may be "carried" (Neave 1966) or decoyed into the wrong group of fish and eventually stray into and spawn in a non-natal stream. With advancing maturity, the focus on homing or straying must be superseded by a focus on spawning (Pritchard 1941). Only when a fish has spawned can a judgement be made as to whether it has homed or strayed. Otherwise, there is still the possibility for the fish to migrate elsewhere.

An example of searching and probing is provided by Jones and Thomason (1983), who tagged adult pink salmon with multi-coloured Petersen disk tags at or very near the mouth of 12 streams in southeastern Alaska and then released the fish at the point of capture. Beach seining and tagging all took place within a three-week period. Each stream was assigned a characteristic identifying tag. The numbers of fish tagged ranged from 288 to 505 per stream. Seining was usually within the intertidal zone of the streams or in freshwater pools at the stream mouth. Subsequent spawning distribution was determined through extensive surveys of several hundred streams in the district. Extensive movement of tagged pink salmon into streams at locations other than those where they were initially tagged and released was documented. For example, 50 tagged fish were recovered in Goat Creek, a stream where no tagged fish were released. These fish originated from three nearby streams, 4.8–9.7 km from Goat Creek. In streams of the Rudyard Bay area, more tags were recovered from "stray" pink salmon than in streams where they were initially tagged (Jones and Thomason 1983). An important factor to remember in this type of study, however, is that the true natal origin of fish originally schooled at the stream mouth is unknown.

Returns to Parent Streams

Davidson (1934) and Pritchard (1939, 1941, 1948b) marked fry from natal streams and observed marked adults returning to the same stream two years later at Duckabush River (Hood Canal, Puget Sound), at Snake Creek (Etolin Island, southeastern Alaska), and at McClinton Creek (Masset Inlet, British Columbia). These results demonstrated a strong tendency for pink salmon to return to parent streams. Allowing for differences in fin marks used and for the extent of searching for marks in other streams, these and other studies summarized in Table 24 suggest that homing precision in pink salmon, although pronounced, is somewhat dependent on the proximity of other spawning streams. Populations in isolated streams showed little or no straying. Those in close proximity to others exhibited a low level of "wandering" between streams.

TABLE 24

Summaries of some studies that measure, in part, homing precision of marked pink salmon fry from indigenous runs, including stream of origin, brood year, numbers of marked fry released, fin marks used, and marked adult recoveries in various locations

		Fry released		Marked adult recoveries				
Stream	Brood	*N* (1000's)	Mark*	Parental stream	Nearby streams	Bay fishery near parent stream	Distant fisheries	Source
Duckabush Cr.	1929	50	Ad+D	8 (80)	2	–	–	Davidson (1934)
Snake Cr.	1930	50	Ad+D	54 (100)	0	–	–	Davidson (1934)
McClinton Cr.	1930	186	Ad	96	?	22	66†	Pritchard (1939)
	1932	108	BV	2941	?	324	11	Pritchard (1939)
	1934	86	Ad+BV	35	?	?	–	Pritchard (1939)
	1938	179	BV	781 (89)	93	234	2	Pritchard (1941, 1948b)
Morrison Cr.	1941	101	BV	865 (99)	1	–	–	Pritchard (1944b)
Herman Cr.	1950	95	D+V	72 (97)	2	–	–	Ricker (1972)
Old Tom Cr.	1950	123	D+V	162 (100)	0	–	–	Ricker (1972)

Notes: Numbers in parentheses are percentages of stream recoveries of marked adults from parental stream.
**Ad* = adipose fin; *D* = dorsal fin; *V* = ventral fin; *BV* = both ventrals
†Probably naturally missing adipose fins

Data summarized in Table 24 indicate that the extent of straying in natural populations of pink salmon is about 10% (excluding the Duckabush River study in which a total of only ten marked fish was found). Ricker (1962) suggested that straying of pink salmon, although variable, "probably rarely exceeds 10% for indigenous stocks to judge by experiments for other salmonids." Helle (1966) found that 91% of the recovered adult pink salmon, displaced 5 km from Olsen Creek, returned to that stream a second time.

Biochemical Evidence

Further evidence suggesting relatively low straying rates in pink salmon comes from starch gel electrophoresis research. In the Prince William Sound district, spawning occurs in several hundred streams, and in early- and late-run components (Noerenberg 1963; Helle 1970; Roys 1971). Seeb and Wishard (1977) and Nickerson (1979) examined populations in 16 streams for genetic differences and reported on ten polymorphic gene loci. A total of 37 subpopulations were identified. Nickerson (1979) interpreted these data as evidence of "little or no straying" between populations. He suggested that differences in early and late spawners in Humpback Creek were such that "even minor straying would tend to homogenize gene frequencies in a few generations."

Other biochemical analyses of genetic variation in North American pink salmon populations include those of Aspinwall (1974), Johnson (1979), Utter et al. (1980), McGregor (1982), and Beacham et al. (1985). Johnson (1979) suggested that there were two, possibly more, subpopulations within each of the two largest even-year spawning streams – Karluk River and Red River – on Kodiak Island. Although electrophoretic studies did not directly measure rates of straying in pink salmon, they demonstrated, in general, considerable genetic variation within populations, including fluctuations in allelic frequencies. However, there are too few similarities to define different populations within a brood line over broad geographic areas (Utter et al. 1980). Cluster analyses of allelic frequencies from 21 odd-year and 4 even-year stocks in southern British Columbia and Puget Sound indicated that Fraser River, Canadian non-Fraser, and Puget Sound stocks were reasonably distinctive (Beacham et al. 1985). In northwest Alaska, Gharrett et al. (1988) found apparent homogeneity within stock groups from the Aleutian Islands, Norton Sound, and Bristol Bay collections. However, substantial heterogeneity existed among pink salmon from these three geographical regions, and from Kodiak Island.

Studies in Asia have identified genetic differentiation in early- and late-run components of pink salmon in the same geographical regions (Altukhov et al. 1983), suggesting fidelity in run timing and homing. However, Glubokovskii and Zhivotovskii (1986) argued that the population structure of pink salmon within the same line "is fairly labile" due to "intensity of exchange by individuals between populations." They propose a concept of fluctuating stocks, with periodic divergence from homing and migrational fidelity in pink salmon allowing fairly large-scale intermixing among normally discrete populations.

Genetic differences between odd- and even-year spawners within a stream are generally much greater than those between streams within the same line over broad areas (Aspinwall 1974; Johnson 1979; McGregor 1982; Beacham et al. 1985). The extent of this difference has led to the suggestion that odd- and even-year brood lines of pink salmon in some regions occupied different refugia during the last glaciation (Aspinwall 1974).

Extent of Straying

Although Horrall (1981) stated that pink salmon "tend to stray at a higher rate than other salmon," he presented no evidence beyond the expansion of the accidentally transplanted fish into the Great Lakes. Gharrett et al. (1988) suggested that the extent of even-year pink salmon straying in the Aleutian Islands may exceed that from other geographic regions. This suggestion, based on apparent genetic homogeneity of Aleutian populations, postulated that homing fidelity to small, isolated, and relatively undifferentiated islands may be more difficult than homing to larger islands or to mainland areas. Unfortunately, few studies have documented natural rates of homing and straying in endemic populations of Pacific salmon. Most information is derived from transplanted or hatchery fish.

Studies of Far East pink salmon frequently showed more extensive migrations into upstream and often marginal spawning areas in years when runs are large (Kaganovsky 1949; Semko 1954; Dvinin 1952; Eniutina 1972). Similar straying within a large North American river is inferred by re-invasion of the upper Fraser River, where greater upstream migration occurs when runs are large (Vernon 1962).

Some transplanted pink salmon stray extensively, at least initially, based on data from the Kola Peninsula (Isaev 1961; Berg 1961, 1967; Bjerknes and Vaag 1980a), from Newfoundland (Lear 1975, 1980), and from the Great Lakes (Kwain and Lawrie 1981; Emery 1981). Some marked Tlell River fry transplanted 48 km to McClinton Creek (Graham Island, northern British Columbia), in an attempt to establish an odd-year run in the latter stream, were later recovered as adults over 600 km distant in Fraser River and Puget Sound fisheries (Pritchard 1939). Whether rates of straying in progeny following initial transplants are greater than those in indigenous stocks, or, for that matter, in other species of salmon, has not been determined.

Not all pink salmon transplants, however, show extensive straying. In the Far East, Sano and Kobayashi (1953) transplanted eyed pink salmon eggs from the Shibetsu River in northern Hokkaido to the Yurappu River in southern Hokkaido. In spring, they released marked fry into the recipient transplant stream. Although recoveries of marked adults were in coastal fisheries, 22 of 23 stream recoveries (96%) were from the recipient stream.

Pink salmon straying has been associated with the transfer of eggs between USSR hatcheries (Rukhlov 1978; Rukhlov and Lyubayeva 1980; Roukhlov et al. 1982). In 1977, 797 marked adults, representing 15% of the stream recoveries that year from the Lesnaya Hatchery on southeastern Sakhalin Island, were recovered in the Kurilka River on Iturup Island. Another 238 marked pinks (4.5%) were recovered from the Naiba River near the parent stream, and the remainder of the stream recoveries were from the parent stream. Larger numbers of marked adults were caught in coastal fisheries along Iturup and Sakhalin islands (Rukhlov and Lyubayeva 1980). The Lesnaya Hatchery stream and the Kurilka River are separated by about 500 km across the southern Sea of Okhotsk. Almost 8% of the stream recoveries of marked adults in 1979 from the Sokolov Hatchery on Sakhalin came from Iturup Island streams (105 in the Kurilka River, 48 in the Reidovaya River) (Roukhlov et al. 1982). Although transfer of eggs between hatcheries is "practiced fairly widely in Sakhalin fish hatcheries" (Rukhlov and Lyubayeva 1980), nowhere is it explicitly stated that eggs were moved between Sakhalin and Kuril islands.

Dominance and Cycles

A prominent feature of pink salmon in many areas is a two-year cycle dominance in either odd- or even-numbered years. Other patterns of population fluctuation have been noted in some stocks of pink salmon, including four-year and twelve-year cycles. To quantify disparity in pink salmon, Ricker (1962) suggested that (1) dominance exists when one line is at least twice the size of the other, and this condition persists for at least four generations; (2) weak, moderate, marked, or extreme dominance can be based on geometric increases in ratios of abundance between lines with extreme dominance assigned to any ratio >16:1; and (3) complete dominance occurs when the "off-year" line is missing.

Neave (1953) believed that the factors affecting dynamics of pink salmon populations occurred in fresh water. He based this belief on extreme variations in freshwater environments, depensatory predation on migrant fry in streams, and the total separation of odd- and even-year lines in fresh water. Neave felt that freshwater and ocean survival were largely independent of each other and that higher fry production in fresh water will, on average, produce larger returns of adults. On interaction between lines, Neave (1953) succinctly noted that "large runs in both even and odd years are not fundamentally incompatible." Throughout much of its range, there is no clearcut pattern of line dominance in pink salmon populations. Neave (1952) suggested that the initial cause or production of disparity in population size is different from that which maintains it; the former requires survival rates of the two lines to differ, whereas the latter requires that they be the same. He concluded that the most probable cause initiating disparity was a disastrous decrease (or conceivably a dramatic increase) in survival rates of one line over the other when both were formerly on more equal footing, probably at a relatively low level of abundance. In this scheme, a single-year catastrophic event sets into motion a subsequent series of events leading to line dominance. Examples include a dramatic sequence in the Amur River district beginning in 1913 that led to extreme dominance by the even-year line within five generations (Neave 1952; Ricker 1962) and drought conditions in northern British Columbia in 1925 leading to a long-term decline in odd-year runs in that area (Wickett 1958b).

Odd- or Even-Year Dominance

Ricker (1962) reviewed in detail relevant features of the dominance of one line over the other and pink salmon abundance. From available published data he developed fourteen points and eight hypotheses related to observed variations in abundance cycles. Principal characteristics of line dominance covered in the first eight of Ricker's fourteen points are summarized as follows. Dominance (1) is highly variable from extreme to none, (2) can occur with large or small runs, (3) tends to be consistent over wide areas, (4) existed before commercial fishing began, (5) can be intensified by fishing, (6) can shift between lines, (7) can spread to adjacent areas, and (8) may disappear. Ricker's remaining points consider depensatory fry predation, poor survival of transplants, density-dependent ocean growth, and genetic aspects of morphological and growth differences in stocks.

Hypotheses 1 and 2 in Ricker's analyses are based on depensatory predation of fry or fingerlings in either fresh water or salt water where small populations suffer disproportionately large losses. Although this predation in either area could maintain some of the observed disparity, and although depensatory predation in fresh water is well documented, neither freshwater nor marine predation has been shown to initiate dominance or change it once established.

Wertheimer (1980) argued that marine mortality of pink salmon at Sashin Creek, Alaska, was compensatory rather than depensatory. He examined available long-term marine survival data for this stream (Table 21) and concluded that high survivals were rare when fry abundance was high, the opposite of what would be expected if depensatory marine mortality were important. Implicit in Wertheimer's argument was that predation did not remain at a fixed level, and larger fry numbers caused an increased aggregation of marine predators with higher losses.

Hypothesis 3 suggests that there is a direct suppressive interaction between lines, with depensatory predation by returning adults of the dominant line on seaward migrants of the weak line. Cannibalism is a popular notion related to line dominance in pink salmon (Barber 1979; Donnelly 1983), and in some areas there is ample overlap of seaward and homeward migration patterns between lines for it to occur. Pink salmon do feed heavily on fishes in their final year (Table 19). However, since Ricker (1962) advanced this hypothesis years ago, there are still few substantive data to support it. Sakagawa (1972) analysed coastal purse seine catches of pink salmon from Cape Flattery (Washington) to Yakutat Bay (Alaska) between 1965 and 1968, including seine sets with both juveniles and adults, and concluded that cannibalism was unimportant in mortality of pink salmon.

Hypothesis 4 suggests that there is a suppressive interaction between lines based on the persistence of dead eggs in gravel from a large run fouling the intragravel environment for the adjacent-line smaller run. Dead eggs can persist in streambeds between years (Hunter 1959; McNeil et al. 1964); however, there are also a host of intragravel scavengers that remove residual organic debris from many spawning gravels. In many regions where dominance is weak or absent, substantial runs spawn in the same gravel in subsequent years. Ricker (1962) suggested that perhaps between-line spawning ground contamination is more likely to happen in some types of gravel than in others.

Donnelly (1983) found evidence of density-dependent between-line interaction for 1952 through 1979 broods on Kodiak Island and suggested that predation or gravel medium were the most likely connections (Ricker's hypotheses 3 and 4). Although pink runs in the Kodiak area showed only weak to moderate even-year dominance in the 1960s, Donnelly's analysis does establish a possible suppressive interaction between the two lines.

Hypotheses 5 and 6 of Ricker's analysis of line dominance relate to fishing. In developmental days, fishing operated in a depensatory manner, catching a larger percentage of small runs than of large runs, thus intensifying differences in spawner abundance. Hypothesis 6 involves density-independent fishing that may still favour line inequality if a constant percentage is caught each year. These hypotheses together might explain many, but not all, features of dominance, for dominance was well developed in some areas before fishing started.

Hypotheses 7 and 8 consider possible food competition between lines at sea, and separation of the same stock between lines during migration or while at sea. Based on extensive INPFC tagging data on pink salmon (summarized by Takagi et al. 1981, and see Figures 29–34), there is no indication of different migration patterns for adjacent-year brood lines in any region. Ricker (1962) constructed a possible scenario to accomplish possible food competition between lines at sea but discounted it as "rather far-fetched." However, based on feeding patterns of pink salmon and their dependence on the hyperiid amphipod, *Parathemisto japonica*, in the Sea of Okhotsk, Andrievskaya (1970) identified a likely between-line interaction in age .0 and age .1 fish. An abundance of age .1 pink salmon feeding on sexually mature *P. japonica* in June and July could limit larval amphipods, a major food item of age .0 pink salmon (see Table 18) in the same waters in August to October (Andrievskaya 1970). Willette (1985) found that the marine survival of both odd- and even-year brood pink salmon in Prince William Sound, Alaska, was significantly correlated with ocean temperatures occurring only during odd years.

In his treatise on line dominance in pink salmon, Ricker (1962) suggested that there is no unique reason for the phenomenon beyond the two-year cycle of the species, that not one but many factors interact to develop and maintain dominance, and that the focus in this phenomenon should be on an individual case basis. The remaining mystery, according to Ricker, is why, in some stocks, line dominance does not and has not occurred.

Do variable ages at maturity in pink salmon account for some observed cyclic phenomena in this species? Kaganovsky (1949) suggested that a considerable part of the run along western Kamchatka in 1935 and 1947 might have come from age .2 pink salmon returning in their third year of life. He based the suggestion on unfavourable growth or on large numbers of pinks at sea. In either case, the presumed result was that part of the respective broods did not mature until a year later than normal. Vedensky (1954) and Lapin (1963, 1971), based on their interpretations of supplementary scale checks, believed three- or four-year-old pink salmon caused between-line shifts in stock abundance. However, other Soviet scientists (Birman 1960; Eniutina 1962, 1972; Ivankov 1965, 1968c) have argued against this interpretation and maintained that the supplemental scale check is not a true annulus.

Ivankov (1968c) found that sclerites on pink salmon scales are deposited on a regular temporal basis, averaging about 2.7 per month, independent of growth rate, and that age can be determined from sclerite counts. He indicated that pink salmon maturing at age .2 should have 60 or more sclerites on their scales.

Natural three-year-old pink salmon do occur, although rarely. Two such fish, a mature female from the Skeena River (Anas 1959) and an immature male from a high-seas longline catch (Turner and Bilton 1968), both had clear annular rings consistent with their time of capture, and sclerite counts similar to those suggested by Ivankov (1968c). Dvinin (1952) also documented the oceanic collection of an immature female starting its third year.

Three-year-old pink salmon have been documented from transplanted freshwater populations on the Kola Peninsula (Bakshtansky 1962) and in the Great Lakes (Kwain and Chappel 1978; Wagner and Stauffer 1980). In the Great Lakes, three-year-old fish are responsible for developing an even-year run from an odd-year transplant. It is of interest to note that the parent 1955-brood Lakelse River pink salmon stock responsible for the present Great Lakes populations of this species is from lineage similar to that of a naturally occurring three-year-old pink female reported by Anas (1959).

Kwain (1982) raised slow-growing pink salmon in the laboratory that did not mature until three and four years of age. All were females except for one three-year-old fish. Other laboratory research with pink salmon includes the precocious maturing of both one-year-old males (Funk and Donaldson 1972; Donaldson et al. 1972; MacKinnon and Donaldson 1976) and females (Funk et al. 1973). These accounts, along with those of naturally occurring one-year-old males in both Asian (Ivankov et al. 1975; Hikita 1984) and North American (Foster et al. 1981) runs, demonstrate that, biologically, pink salmon have a broader scope of maturation ages than usually thought. None of the above variations in ages at maturity, however, can be realistically construed to influence, to any extent, the observed cyclic and dominance patterns of abundance for endemic pink salmon populations.

Other Cyclic Patterns

Smirnov (1947) noted four-year cycles in Amur region pink salmon, as well as the normal two-year patterns. Because chum salmon in this region mature principally at four years of age, Smirnov implied an interaction between pink and chum cyclic patterns. Birman (1956) suggested a cyclic relation between growth patterns of Primore and Amur district pinks and chums in years of high and low abundance based on their common use of the Sea of Japan. Birman (1969b, 1976) discussed four-year cycles of pink salmon abundance in western Kamchatka stocks and suggested that they are derived from climatic effects, perhaps related to periods of solar activity or cyclic patterns in ocean currents. Donnelly (1983) found a twelve-year cycle in pink salmon catch data from Kodiak Island but could not identify causal factors. Gallagher (1979) found a cyclic interaction between pink and chum salmon in Puget Sound in which the presence and absence of pink salmon affects the abundance and age of returning chum salmon. The study suggested that the pink salmon's influence on chum salmon takes place in the early marine environment. Smoker (1984) developed a model to test various factors in these Puget Sound data and found that a genetic component for the maturation age of chum was necessary to maintain biennial cyclicity between pink and chum salmon stocks. Low heritability of maturation age in chum salmon would not permit biennial cycles in numbers of chums unless one pink salmon line was completely absent.

Concluding Remarks

Biologically, pink salmon represent a specialized part of the evolutionary spectrum present in extant Pacific salmon. As a species within the genus, *O. gorbuscha* occurs widely throughout the North Pacific Ocean and Bering Sea north of about 40°N latitude and is uniquely characterized by several factors. First, it is least dependent on fresh water, a point of either specialized or simplified development depending on freshwater or marine origin. Second, it is the smallest in adult size, has the highest rate of growth, is the least fecund, and is the most abundant Pacific salmon in both North America and Asia. Third, it has a short, fixed two-year life span so that even- and odd-year brood lines represent genetically distinct populations even within the same stream. There are rare exceptions to the two-year life cycle. However, other than the non-endemic even-year run in the Great Lakes, these exceptions have no bearing on the ecology or population dynamics of the species.

Although major populations occasionally occur in large river systems like the Fraser or Amur rivers, the pink salmon is principally a species of shorter coastal streams. In some regions of North America, with high levels of population abundance, significant amounts of spawning and fry production occur in the upper intertidal portions of streams. Pink salmon spawn in fast flowing riffles, generally in substantial groups on the same riffle. Breeding males develop a pronounced dorsal hump which is likely a result of the spawning environment.

Emergent pink salmon fry migrate quickly to sea and exhibit strong schooling behaviour as a primary defense against predation. Schooling behaviour lessens during the pelagic marine stage of the life cycle, then again becomes prounounced as maturing fish approach coastal waters and ascend spawning streams.

Pink salmon have been widely transplanted throughout the world. With few exceptions, these transplants have failed both within and outside the natural geographic range of the species. Reasons for the success or failure of pink salmon transplants are largely unknown but likely relate to specific genetic features of the stocks involved.

Six major population groups of pink salmon, from broad geographic regions, have somewhat discrete migration patterns in the ocean. These population groups, three each from North America and Asia, include (1) Puget Sound and British Columbia stocks; (2) southeastern Alaska, central Alaska, and Alaska Peninsula stocks; (3) Bristol Bay and western Alaska stocks; (4) eastern Kamchatka and Anadyr stocks; (5) western Kamchatka, Sea of Okhotsk and eastern Sakhalin Island stocks; and (6) western Sakhalin Island, Amur, and Primore stocks (Sea of Japan).

The population biology of pink salmon revolves around the two-year life cycle. A phenomenon of cycle dominance between odd- and even-year brood lines within specific regions is common. Dominance can be weak or strong, complete or non-existent. It can also shift between brood lines. With complete dominance, the "off-year" line is absent while non-dominance is characterized by similar population strength between odd- and even-year runs. Although many causes for dominance and its various characteristics in pink salmon populations have been proposed, none satisfactorily explains the event. Pink salmon populations can be resilient, rebounding from weak to strong run strength within regional stock groups in one or two generations. Genetically, pink salmon are more similar within odd- or even-year brood lines across broad geographic regions than across brood lines within the same stream. It has been suggested, for some geographic areas, that present odd- and even-year pink salmon populations arose from separate glacial refuges during late Pleistocene times.

Acknowledgments

I am grateful to Paula Johnson, Librarian at the Auke Bay Laboratory, for her capable assistance in helping me obtain much of the scientific literature on the biology of pink salmon, including many English translations of foreign-language articles. Mrs. Johnson has assembled an extensive reference list on this species that can be provided to anyone interested by writing her at P.O. Box 210155, Auke Bay, Alaska 99821.

References

Alderdice, D.F., J.O.T. Jensen, and F.P.J. Velsen. 1984. Measurement of hydrostatic pressure in salmonid eggs. Can. J. Zool. 62:1977–1987

Alderdice, D.F., and F.P.J. Velsen. 1978. Relation between temperature and incubation time for eggs of chinook salmon *(Oncorhynchus tshawytscha)*. J. Fish. Res. Board Can. 35:69–75

Alderdice, D.F., W.P. Wickett, and J.R. Brett. 1958. Some effects of temporary exposure to low dissolved oxygen levels on Pacific salmon eggs. J. Fish. Res. Board Can. 15:229–249

Allen, G.H., and W. Aron. 1958. Food of salmonid fishes of the western North Pacific Ocean. U.S. Fish Wildl. Serv. Spec. Sci. Rep. Fish. 237:11 p.

Altukhov, Y.P., E.A. Salmenkova, V.T. Omel'chenko, and V.N. Efanov. 1983. Genetic differentiation and population structure in the pink salmon of the Sakhalin-Kurile region. Sov. J. Mar. Biol. 9:98–102

Anas, R.E. 1959. Three-year-old pink salmon. J. Fish. Res. Board Can. 16:91–94

Andrew, F.J., and G.H. Geen. 1960. Sockeye and pink salmon production in relation to proposed dams in the Fraser River system. Int. Pac. Salmon Fish. Comm. Bull. 11:259 p.

Andrievskaya, L.D. 1957. The food of Pacific salmon in the northwestern Pacific Ocean, p. 64–75. *In*: Materialy po biologii morskovo perioda zhizni dalnevostochnykh lososei. Vsesoyuzny Nauchno-Issledovatelskii Institut Morskovo Rybnovo Khozyaistva i Okeanografii, Moscow. (Transl. from Russian; Fish. Res. Board Can. Transl. Ser. 182)

——. 1968. Feeding of Pacific salmon fry in the sea. Izv. Tikhookean. Nauchno-Issled. Inst. Rybn. Khoz. Okeanogr. 64:73–80 (Transl. from Russian; Fish. Res. Board Can. Transl. Ser. 1423)

——. 1970. Feeding of Pacific salmon juveniles in the Sea of Okhotsk. Izv. Tikhookean. Nauchno-Issled. Inst. Rybn. Khoz. Okeanogr. 78:105–115 (Transl. from Russian; Fish. Res. Board Can. Transl. Ser. 2441)

Andriyashev, A.P. 1954. Fishes of the northern seas of the USSR. Izdatel'stvo Akademii Nauk SSSR, Moscow. (Transl. from Russian; Israel Program for Scientific Translations, Jerusalem, 1964)

Anonymous. 1938. Pink runs coming later in southeastern Alaska. Pac. Fisherman 36:22–23

Anonymous. 1951. Drought brings death to salmon. Fish. Res. Board Can. Prog. Rep. Pac. Coast Stn. 88:72

Armstrong, R.H. 1970. Age, food, and migration of Dolly Varden smolts in southeastern Alaska. J. Fish. Res. Board Can. 27:991–1004

Armstrong, R.H., and P.C. Winslow. 1968. An incidence of walleye pollock feeding on salmon young. Trans. Am. Fish. Soc. 97:202–203

Aro, K.V. 1979. Transfers of eggs and young of Pacific salmon within British Columbia. Fish. Mar. Serv. (Can.) Tech. Rep. 861:147 p.

Aro, K.V., and M. P. Shepard. 1967. Pacific Salmon in Canada, p. 225–327. *In*: Salmon of the North Pacific Ocean. Part IV. Spawning populations of North Pacific salmon. Int. North Pac. Fish. Comm. Bull. 23

Aron, W.I. 1956. Food of salmonid fishes of the

North Pacific Ocean. C. Food of pink salmon *Oncorhynchus gorbuscha*. D. Preliminary comparative study of the feeding behavior of the pink, sockeye and chum salmons. Univ. Wash. Dep. Oceanogr. Fish. Rep. 4:10 p.

Aspinwall, N. 1974. Genetic analysis of North American populations of pink salmon, *Oncorhynchus gorbuscha*, possible evidence for the neutral mutation-random drift hypothesis. Evolution 28:295-305

Astafeva, A.V. 1964. Concerning the fauna of pink salmon nests in east Murman rivers. Tr. Murm. Morsk. Biol. Inst. 5:148-153. (Transl. from Russian; Fish. Res. Board Can. Transl. Ser. 579)

Atkinson, C.E., J.H. Rose, and T.O. Duncan. 1967. Pacific salmon in the United States, p. 43-223. *In*: Salmon of the North Pacific Ocean. Part IV. Spawning populations of North Pacific salmon. Int. North Pac. Fish. Comm. Bull. 23

Baggerman, B. 1960. Salinity preference, thyroid activity and the seaward migration of four species of Pacific salmon *(Oncorhynchus)*. J. Fish. Res. Board Can. 17:295-322

Bailey, J.E. 1969. Alaska's fishery resources - the pink salmon. Fish Leafl. (U.S.) 619:8 p.

Bailey, J.E., and D.R. Evans. 1971. The low-temperature threshold for pink salmon eggs in relation to a proposed hydroelectric installation. Fish. Bull. (U.S.) 69:587-593

Bailey, J.E., J.J. Pella, and S.G. Taylor. 1976. Production of fry and adults of the 1972 brood of pink salmon, *Oncorhynchus gorbuscha*, from gravel incubators and natural spawning at Auke Creek, Alaska. Fish. Bull. (U.S.) 74:961-971

Bailey, J.E., S.D. Rice, J.J. Pella, and S.G. Taylor. 1980. Effects of seeding density of pink salmon, *Oncorhynchus gorbuscha*, eggs on water chemistry, fry characteristics, and fry survival in gravel incubators. Fish. Bull. (U.S.) 78:649-658

Bailey, J.E., B.L. Wing, and J.H. Landingham. 1983. Juvenile Pacific sandfish, *Trichodon trichodon*, associated with pink salmon, *Oncorhynchus gorbuscha*, fry in the nearshore area, southeastern Alaska. Copeia 1983(2):549-551

Bailey, J.E., B.L. Wing, and C.R. Mattson. 1975. Zooplankton abundance and feeding habits of fry of pink salmon, *Oncorhynchus gorbuscha*, and chum salmon, *Oncorhynchus keta*, in Traitors Cove, Alaska, with speculations on the carrying capacity of the area. Fish. Bull. (U.S.) 73:846-861

Bakshtansky, E.L. 1962. A pink salmon from a lake. Nauchno-Tekh. Biul. Polyarn. Nauchno- Issled. Inst. Morsk. Rybn. Khoz. Okeanogr. 4(22):46-47. (Transl. from Russian; Fish. Res. Board Can. Transl. Ser. 451)

——. 1964. Effect of predators on the young of *Oncorhynchus gorbuscha* (Walb.) and *Oncorhynchus keta* (Walb.) in the White and Barents Seas. Vopr. Ikhtiol. 4(30):136-141. (Transl. from Russian; Fish. Res. Board Can. Transl. Ser. 507)

——. 1965. The impact of the environmental factors on survival of the Far Eastern young salmon during the acclimatization of the latter in the northwest part of the USSR, p. 477-479. *In*: ICNAF Environmental Symposium, FAO, Rome, 1964. Int. Comm. Northwest Atl. Fish. Spec. Publ. 6

——. 1970. Downstream migrations of pink and red salmon and causes of their delay in the streams of the Kola Peninsula, p. 129-143. *In*: Fish culture in natural waters. Tr. Vses. Nauchno-Issled. Inst. Morsk. Rybn. Khoz. Okeanogr. 74. (Transl. from Russian; Fish. Res. Board Can. Transl. Ser. 1765)

——. 1980. The introduction of pink salmon into the Kola Peninsula, p. 245-259. *In*: J. Thorpe (ed.). Salmon ranching. Academic Press, New York, NY

Bams, R.A. 1969. Adaptations of sockeye salmon associated with incubation in stream gravels, p. 71-87. *In*: T.G. Northcote (ed.). Symposium on Salmon and Trout in Streams. H.R. MacMillan Lectures in Fisheries. Institute of Fisheries, University of British Columbia, Vancouver, BC

——. 1970. Evaluation of a revised hatchery method tested on pink and chum salmon fry. J. Fish. Res. Board Can. 27:1429-1452

——. 1972. A quantitative evaluation of survival to the adult stage and other characteristics of pink salmon (*Oncorhynchus gorbuscha*) produced by a revised hatchery method which simulates optimal natural conditions. J. Fish. Res. Board Can. 29:1151-1167

——. 1976. Survival and propensity for homing as affected by presence or absence of locally adapted paternal genes in two transplanted populations of pink salmon (*Oncorhynchus gorbuscha*). J. Fish. Res. Board Can. 33:2716-2725

——. 1979. Gravel incubation studies - continued, p. 417-418. *In*: J.C. Mason (ed.). Proceedings of the 1978 Northeast Pacific Pink and Chum Salmon Workshop. Pacific Biological Station,

Nanaimo, BC

Barber, F.G. 1979. Disparity of pink salmon runs, a speculation. Fish. Mar. Serv. (Can.) MS Rep. 1504:7 p.

Barraclough, W.E. 1967a. Data record. Number, size, and food of larval and juvenile fish caught with a two boat surface trawl in the Strait of Georgia, April 25-29, 1966. Fish. Res. Board Can. MS Rep. Ser. 922:54 p.

——. 1967b. Data record. Number, size and food of larval and juvenile fish caught with an Isaacs-Kidd trawl in the surface waters of the Strait of Georgia, April 25-29, 1966. Fish. Res. Board Can. MS Rep. Ser. 926:79 p.

——. 1967c. Data record. Number, size composition and food of larval and juvenile fish caught with two-boat surface trawl in the Strait of Georgia, June 6-8, 1966. Fish. Res. Board Can. MS Rep. Ser. 928:58 p.

Barraclough, W.E., and J.D. Fulton. 1967. Data record. Number, size composition and food of larval and juvenile fish caught with two-boat surface trawl in the Strait of Georgia, July 4-8, 1966. Fish. Res. Board Can. MS Rep. Ser. 940:82 p.

——. 1968. Data record. Food of larval and juvenile fish caught with a surface trawl in Saanich Inlet during June and July 1966. Fish. Res. Board Can. MS Rep. Ser. 1003:78 p.

Barraclough, W.E., and A.C. Phillips. 1978. Distribution of juvenile salmon in the southern Strait of Georgia during the period April to July 1966-1969. Fish. Mar. Serv. (Can.) Tech. Rep. 826:47 p.

Barraclough, W.E., D.G. Robinson, and J.D. Fulton. 1968. Data record. Number, size composition, weight, and food of larval and juvenile fish caught with two-boat surface trawl in the Saanich Inlet, April 23-July 21, 1968. Fish. Res. Board Can. MS Rep. Ser. 1004:305 p.

Basham, L., and L. Gilbreath. 1978. Unusual occurrence of pink salmon (*Oncorhynchus gorbuscha*) in the Snake River of southeastern Washington. Northwest Sci. 52:32-34

• Beacham, T.D., and C.B. Murray. 1986. Comparative developmental biology of pink salmon, *Oncorhynchus gorbuscha*, in southern British Columbia. J. Fish Biol. 28:233-246

Beacham, T.D., R.E. Withler, and A.P. Gould. 1985. Biochemical genetic stock identification of pink salmon (*Oncorhynchus gorbuscha*) in southern British Columbia and Puget Sound. Can. J. Fish. Aquat. Sci. 42:1474-1483

Berdyshev, G.D., S.I. Baranova, and G.K. Korotayev. 1969. Change in nucleic acid contents of organs and tissues of *Oncorhynchus gorbuscha* (Walb.) at different stages of its spawning migration. Probl. Ichthyol. 9:112-119

Berg, L.S. 1948. Freshwater fishes of the U.S.S.R. and adjacent countries. Vol. 1. Opred. Faune SSSR 27:466 p. (Transl. from Russian; Israel Program for Scientific Translations, Jerusalem. 1963)

Berg, M. 1961. Pink salmon (*Oncorhynchus gorbuscha*) in northern Norway in the year 1960. Acta Borealia A Sci. 17:1-23

——. 1977. Pink salmon, *Oncorhynchus gorbuscha* (Walbaum), in Norway. Inst. Freshwater Res. Drottningholm Rep. 56:12-17

Berg, R.E. 1979. External morphology of the pink salmon, *Oncorhynchus gorbuscha*, introduced into Lake Superior. J. Fish. Res. Board Can. 36:1283-1287

Bigelow, H.B., and W.C. Schroeder. 1953. Fishes of the Gulf of Maine. 1st rev. Fish. Bull. Fish Wildl. Serv. 53:577 p.

Bilton, H.T. 1966. Characteristics of scales from pink salmon (*Oncorhynchus gorbuscha*) transplanted from the Glendale River in central British Columbia to the North Harbour River in Newfoundland. J. Fish. Res. Board Can. 23:939-940

——. 1971. Identification of major British Columbia and Alaska runs of even-year and odd-year pink salmon from scale characters. J. Fish. Res. Board Can. 29:295-301

Bilton, H.T., and S.A.M. Ludwig. 1966. Times of annulus formation on scales of sockeye, pink, and chum salmon in the Gulf of Alaska. J. Fish. Res. Board Can. 23:1403-1410

Bilton, H.T., and W.E. Ricker. 1965. Supplementary checks on the scales of pink salmon (*Oncorhynchus gorbuscha*) and chum salmon (*O. keta*). J. Fish. Res. Board Can. 22:1477-1489

Birman, I.B. 1950. Parasitism of salmon of the genus *Oncorhynchus* by the Pacific lamprey. Izv. Tikhookean. Nauchno-Issled. Inst. Rybn. Khoz. Okeanogr. 32:158-160. (Transl. from Russian; Fish. Res. Board Can. Transl. Ser. 290)

——. 1956. Concerning the causes of a peculiarity of the pink salmon. Zool. Zh. 35:1681-1684. (Transl. from Russian; Fish. Res. Board Can. Transl. Ser. 142)

———. 1960. On the period of formation of the annual rings on the scales of Pacific salmon and the rate of growth of pinks. Dokl. Akad. Nauk SSSR 132:1187–1190. (Transl. from Russian; Fish. Res. Board Can. Transl. Ser. 327)

———. 1968. Migration of Pacific salmon in the Okhotsk Sea. Izv. Tikhookean. Nauchno-Issled. Inst. Rybn. Khoz. Okeanogr. 64:35–42. (Transl. from Russian; Fish. Res. Board Can. Transl. Ser. 1420)

———. 1969a. Distribution and growth of young Pacific salmon of the genus *Oncorhynchus* in the sea. Probl. Ichthyol. 9:651–666

———. 1969b. Solar activity and periodic fluctuations in the abundance of salmon. Tr. Vses. Nauchno-Issled. Inst. Morsk. Rybn. Khoz. Okeanogr. 67:171–189. (Transl. from Russian; Fish. Res. Board Can. Transl. Ser. 1561)

———. 1976. Minor cycles in the abundance dynamics of salmon. J. Ichthyol. 16:364–372

Bjerknes, V., and A.B. Vaag. 1980a. The status of pink salmon in north Norway. Int. Counc. Explor. Sea. C.M. 1980/M:16:11 p.

———. 1980b. Migration and capture of pink salmon, *Oncorhynchus gorbuscha* Walbaum in Finnmark, North Norway. J. Fish Biol. 16:291–297

Boyd, F.C. 1964. Return of pink salmon to Robertson Creek shows promise of success. Can. Fish Cult. 32:59–62

Brannon, E.L. 1982. Orientation mechanisms of homing salmonids, p. 219–227. *In*: E.L. Brannon and E.O. Salo (eds.). Proceedings of the Salmon and Trout Migratory Behavior Symposium. School of Fisheries, University of Washington, Seattle, WA

Brickell, D.C. 1971. Oxygen consumption by dead pink salmon eggs in salmon spawning beds. M.Sc. thesis. University of Alaska, Fairbanks, AK. 53 p.

Briggs, J.C. 1953. The behavior and reproduction of salmonid fishes in a small coastal stream. Calif. Dep. Fish Game Fish Bull. 94:62 p.

Carey, W.E., and D.L.G. Noakes. 1981. Development of photo behavioural responses in young rainbow trout, *Salmo gairdneri* Richardson. J. Fish Biol. 19:285–296

Cavender, T.M., and R.R. Miller. 1972. *Simlodonichthys rastrosus*, a new Pliocene salmonid fish from western United States. Bull. Mus. Nat. Hist. Univ. Oregon 18:1–44

Chebanov, N.A. 1980. Spawning behavior of the pink salmon, *Oncorhynchus gorbuscha*. J. Ichthyol. 20:64–73

———. 1982. Spawning behavior of the humpbacked salmon *Oncorhynchus gorbuscha* with different ratios of sexes at the spawning ground. Sov. J. Ecol. 13:53–61

Chernenko, Ye.V. 1969. Evolution and cytotaxonomy of the family Salmonidae. Probl. Ichthyol. 9:781–788

Churmasov, A.V., and A.S. Stepanov. 1977. Sun orientation and guideposts of the humpback salmon (*Oncorhynchus gorbuscha*). Sov. J. Mar. Biol. 3:363–369

Claire, E.W., and R.W. Phillips. 1968. The stonefly *Acroneuria pacifica* as a potential predator on salmonid embryos. Trans. Am. Fish. Soc. 97:50–52

Coburn, A., and P. McCart. 1967. A hatchery release tank for pink salmon fry with notes on behavior of the fry in the tank and after release. J. Fish. Res. Board Can. 24:77–85

Collins, J.J. 1975. Occurrence of pink salmon (*Oncorhynchus gorbuscha*) in Lake Huron. J. Fish. Res. Board Can. 32:402–404

Cooney, R.T., D. Urquhart, R. Nevé, J. Hilsinger, R. Clasby, and D. Barnard. 1978. Some aspects of the carrying capacity of Prince William Sound, Alaska for hatchery released pink and chum salmon fry. Alaska Sea Grant Rep. 78–4; Univ. Alaska Inst. Mar. Resourc. IMS R78–3:98 p.

Cooper, A.C. 1965. The effect of transported stream sediments on the survival of sockeye and pink salmon eggs and alevins. Int. Pac. Salmon Fish. Comm. Bull. 18:71 p.

———. 1977. Evaluation of the production of sockeye and pink salmon at spawning and incubation channels in the Fraser River system. Int. Pac. Salmon Fish. Comm. Prog. Rep. 36:80 p.

Cordone, A.J., and D.W. Kelly. 1961. The influences of inorganic sediment on the aquatic life of streams. Calif. Fish Game 47:189–228

Corkum, L.D., and P.J. McCart. 1981. A review of the fisheries of the Mackenzie Delta and nearshore Beaufort Sea. Can. MS Rep. Fish. Aquat. Sci. 1613:55 p.

Craig, P., and L. Haldorson. 1986. Pacific salmon in the North American Arctic. Arctic 39:2–7

Craig, P.C. 1984a. Fish resources, p. 117–131. *In*: J. Truett (ed.) The Barrow Arch environment and possible consequences of planned offshore oil and gas development: proceedings of a synthesis meeting, Girdwood, Alaska, 30 October-1 November 1983. National Oceanic and Atmospheric Administration, Ocean Assessments Division, Alaska Office, Anchorage, AK

——. 1984b. Fish use of coastal waters of the Alaskan Beaufort Sea: a review. Trans. Am. Fish. Soc. 113:265-282

• Crisp, D.T. 1981. A desk study of the relationship between temperature and hatching time for the eggs of five species of salmonid fishes. Freshwater Biol. 11:361-368

Cunningham, P.B. 1976. Arctic salmon studies. Technical Report for period July 1, 1975 to June 30, 1976, Anadromous Fish Conservation Act, project. No. AFC-55-2. Alaska Department of Fish and Game, Juneau, AK. 48 p.

Dangel, J.R., and J.F. Jewell. 1975. Southeastern Alaska pink and chum salmon pre-emergent fry data file, 1963-75. Alaska Dep. Fish Game ADF&G Tech. Data Rep. 21:226 p.

Davidson, F.A. 1933. Temporary high carbon dioxide content in an Alaska stream at sunset. Ecology 14:238-240

——. 1934. The homing instinct and age at maturity of pink salmon (*Oncorhynchus gorbuscha*). Bull. Bur. Fish. (U.S.) 48:27-29

——. 1935. The development of the secondary sexual characters in the pink salmon (*Oncorhynchus gorbuscha*). J. Morphol. 57:169-183

Davidson, F.A., and L.S. Christey. 1938. The migrations of pink salmon (*Oncorhynchus gorbuscha*) in the Clarence and Sumner Straits regions of southeastern Alaska. Bull. Bur. Fish. (U.S.) 48:643-666

Davidson, F.A., and S.J. Hutchinson. 1943. Weather as an index to abundance of pink salmon. Pac. Fisherman 41:21-29

Davidson, F.A., and A.E. Vaughan. 1939. Cyclic changes in time of southeast Alaska pink salmon runs, part 2. Pac. Fisherman 37:40-42

Davidson, F.A., and E. Vaughan. 1941. Relation of population size to marine growth and time of spawning migration in the pink salmon (*Oncorhynchus gorbuscha*) of southeastern Alaska. J. Mar. Res. 4:231-246

Davidson, F.A., E. Vaughan, S.J. Hutchinson, and A.L. Pritchard. 1943. Factors influencing the upstream migration of the pink salmon (*Oncorhynchus gorbuscha*). Ecology 24:149-168

Dawson, C.E. 1964. A bibliography of anomalies of fishes. Gulf Res. Rep. 1:308-399

——. 1966. A bibliography of anomalies of fishes. Supplement 1. Gulf Res. Rep. 2(2):169-176

Dill, L.M. 1967. Studies on the early feeding of sockeye salmon alevins. Can. Fish Cult. 39:23-34

——. 1969. The sub-gravel behaviour of Pacific salmon larvae, p. 89-99. *In*: T.G. Northcote (ed.). Symposium on Salmon and Trout in Streams. H.R. MacMillan Lectures in Fisheries. Institute of Fisheries, University of British Columbia, Vancouver, BC

Dill, L.M., and T.G. Northcote. 1970. Effects of some environmental factors on survival, condition, and timing of emergence of chum salmon fry (*Oncorhynchus keta*). J. Fish. Res. Board Can. 27:196-201

Dill, P.A. 1982. Behavior of alevin influencing distribution in gravel, p. 61-70. *In*: E.L. Brannon and E.O. Salo (eds.). Proceedings of the Salmon and Trout Migratory Behavior Symposium. School of Fisheries, University of Washington, Seattle, WA

Disler, N.N. 1953. Ecological and morphological characteristics of the development of the Amur autumn chum salmon *Oncorhynchus keta* (Walb.). Tr. Soveshch. Ikhtiol. Kom. Akad. Nauk SSSR 1:354-362. (Transl. from Russian; *In*: Pacific salmon: selected articles from Soviet periodicals, p. 34-41. Israel Program for Scientific Translations, 1961)

Donaldson, E.M., J.D. Funk, F.C. Withler, and R.B. Morley. 1972. Fertilization of pink salmon (*Oncorhynchus gorbuscha*) ova by spermatozoa from gonadotropin-injected juveniles. J. Fish. Res. Board Can. 29:13-18

Donnelly, R.F. 1983. Factors affecting the abundance of Kodiak Archipelago pink salmon (*Oncorhynchus gorbuscha*, Walbaum). Ph.D. thesis. University of Washington, Seattle, WA. 157 p.

Dubrovskaia, N.V. 1934. The life cycle of the pink salmon of Primor'ye. Rybn. Khoz. 1-2:65-68 (Transl. from Russian; Fish. Res. Board Can. Transl. Ser. 542)

Dvinin, P.A. 1952. The salmon of south Sakhalin. Izv. Tikhookean. Nauchno-Issled. Inst. Rybn. Khoz. Okeanogr. 37:69-108 (Transl. from Russian; Fish. Res. Board Can. Transl. Ser. 120)

Dyagilev, S.Ye., and N.B. Markevich. 1979. Different times of maturation of the pink salmon, *Oncorhynchus gorbuscha*, in even and uneven years as the main factor responsible for different acclimatization results in the northwestern USSR. J. Ichthyol. 19:30–44

Dymond, J.R. 1940. Pacific salmon in the Arctic Ocean. Proc. 6th Pac. Sci. Congr. 1939(3):435

Earp, B.J., and R.L. Schwab. 1954. An infestation of leeches on salmon fry and eggs. Prog. Fish-Cult. 16:122–124

Elliott, S.T., and R. Bartoo. 1981. Relation of larval *Polypedilum* (Diptera: Chironomidae) to pink salmon eggs and alevins in an Alaskan stream. Prog. Fish-Cult. 43:220–221

Ellis, C.H., and R.E. Noble. 1959. Even year–odd year pink salmon. Wash. Dep. Fish. Annu. Rep. 69:36–43

Ellis, R.J. 1969. Return and behavior of adults of the first filial generation of transplanted pink salmon, and survival of their progeny, Sashin Creek, Baranof Island, Alaska. U.S. Fish Wildl. Serv. Spec. Sci. Rep. Fish. 589:13 p

——. 1970. Alloperla stonefly nymphs: predators or scavengers on salmon eggs and alevins? Trans. Am. Fish. Soc. 99:677–683

Emery, L. 1981. Range extension of pink salmon (*Oncorhynchus gorbuscha*) into the lower Great Lakes. Fisheries 6:7–10

Eniutina, R. 1954. Morphobiological and morphometric characteristics of the pink salmon of the Amur and Iski rivers. Izv. Tikhookean. Nauchno-Issled. Inst. Rybn. Khoz. Okeanogr. 41:333–336. (Transl. from Russian; Fish. Res. Board Can. Transl. Ser. 289)

Eniutina, R.I. 1954. Local stocks of pink salmon in the Amur Basin and neighbouring waters. Vopr. Ikhtiol. 2:139–143. (Transl. from Russian; Fish. Res. Board Can. Transl. Ser. 284)

——. 1962. Concerning the supplementary ring in the nuclear portion of the scales of pink salmon, *Oncorhynchus gorbuscha* (Walb.). Vopr. Ikhtiol. 2:740–742. (Transl. from Russian; Fish. Res. Board Can. Transl. Ser. 510)

——. 1972. The Amur pink salmon (*Oncorhynchus gorbuscha*): a commercial and biological survey. Izv. Tikhookean. Nauchno- Issled. Inst. Rybn. Khoz. Okeanogr. 77:3–126. (Transl. from Russian; Fish. Res. Board Can. Transl. Ser. 3160)

Fabricius, E. 1963. Aquarium observations on the spawning behavior of the char, *Salmo alpinus*. Inst. Freshwater Res. Drottningholm Rep. 34:14–48

Fabricius, E., and K.-J. Gustafson. 1954. Further aquarium observations on the spawning behavior of the char, *Salmo alpinus*. Inst. Freshwater Res. Drottningholm Rep. 35:58–104

Fiscus, C.H. 1980. Marine mammal-salmonid interactions: a review, p. 121–132. *In*: W.J. McNeil and D.C. Himsworth (eds.). Salmonid ecosystems of the North Pacific. Oregon State University Press, Corvallis, OR

Fisheries Research Board of Canada. 1969. Progress in 1967 in Canadian research on problems raised by the protocol. I. Results of high seas fishing and tagging. Int. North Pac. Fish. Comm. Annu. Rep. 1967:29–53

Foerster, R.E. 1935. Inter-specific cross-breeding of Pacific salmon. Proc. Trans. R. Soc. Can. Ser. 3 29(5):21–33

——. 1968. The sockeye salmon. Bull. Fish. Res. Board Can. 162:422 p.

Foerster, R.E., and A.L. Pritchard. 1941. Observations on the relation of egg content to total length and weight in the sockeye salmon (*Oncorhynchus nerka*) and the pink salmon (*O. gorbuscha*). Proc. Trans. R. Soc. Can. Ser. 3 35(5).51–60

Forrester, C.R. 1961. A note on the practical use of a salmon repellent. Can. Fish Cult. 30:61

Foster, R.W., C. Bagatell, and H.J. Fuss. 1981. Return of one-year-old pink salmon to a stream in Puget Sound. Prog. Fish-Cult. 43:31

Frame, G.W. 1974. Black bear predation on salmon at Olsen Creek, Alaska. Z. Tierpsychol. 35:23–38

Fredin, R.A. 1980. Trends in North Pacific salmon fisheries, p. 59–119. *In*: W.J. McNeil and D.C. Himsworth (eds.). Salmonid ecosystems of the North Pacific. Oregon State University Press, Corvallis, OR

Fredin, R.A., R.L. Major, R.G. Bakkala, and G.K. Tanonaka. 1977. Pacific salmon and the high seas salmon fisheries of Japan (Processed report). Northwest and Alaska Fisheries Center, National Marine Fisheries Service, Seattle, WA. 324 p.

Fredin, R.A., S. Pennoyer, K.R. Middleton, R.S. Roys, S.C. Smedley, and A.S. Davis. 1974. Information on recent changes in the salmon fisheries of Alaska and the condition of the stocks, p. 37–42. *In*: Additional information on the exploitation, scientific investigation, and management of salmon stocks on the Pacific coast of Canada and the United States in relation to the abstention provisions of the North Pacific Fisheries Convention. Int. North Pac. Fish. Comm. Bull. 29

French, R., R. Bakkala, J. Dunn, and D. Sutherland. 1971. Ocean distribution, abundance, and migration of salmon. Int. North Pac. Fish. Comm. Annu. Rep. 1969:89-102

Frolenko, L.A. 1970. Feeding of chum and pink salmon juveniles migrating downstream in the main spawning rivers of the northern coast of the Sea of Okhotsk. Izv. Tikhookean. Nauchno-Issled. Inst. Rybn. Khoz. Okeanogr. 71:179-188. (Transl. from Russian; Fish. Res. Board Can. Transl. Ser. 2416)

Fukataki, H. 1967. Stomach contents of the pink salmon, *Oncorhynchus gorbuscha* (Walbaum), in the Japan Sea during spring season of 1965. Bull. Jpn. Sea Reg. Fish. Res. Lab. 17:49-66. (In Japanese, English summary)

Funk, J.D., and E.M. Donaldson. 1972. Induction of precocious sexual maturity in male pink salmon (*Oncorhynchus gorbuscha*). Can. J. Zool. 50:1413-1419

Funk, J.D., E.M. Donaldson, and H.M. Dye. 1973. Induction of precocious sexual development in female pink salmon (*Oncorhynchus gorbuscha*). Can. J. Zool. 51:493-500

Gallagher, A.F., Jr. 1979. An analysis of factors affecting brood year returns of wild stocks of Puget Sound chum (*Oncorhynchus keta*) and pink salmon (*Oncorhynchus gorbuscha*). M.Sc. thesis. University of Washington, Seattle, WA. 152 p.

Gerke, R.J., and V.W. Kaczynski. 1972. Food of juvenile pink and chum salmon in Puget Sound, Washington. Wash. Dep. Fish. Tech. Rep. 10:27 p.

Gharrett, A.J., C. Smoot, and A.J. McGregor. 1988. Genetic relationships of even-year northwestern Alaskan pink salmon. Trans. Am. Fish. Soc. 117:536-545

Gharrett, A.J., and M.A. Thomason. 1987. Genetic changes in pink salmon (*Oncorhynchus gorbuscha*) following their introduction into the Great Lakes. Can. J. Fish. Aquat. Sci. 43:787-792

Girsa, I.I., V.N. Zhuravel', and Yu.Ye. Lapin. 1980. Salinity preferences of juvenile whitefish, *Coregonus lavaretus*, cisco, *Coregonus sardinella marisalbi*, and the pink salmon *Oncorhynchus gorbuscha*, from the White Sea Basin. J. Ichthyol. 20:138-148

Glubokovskii, M.K., and L.A. Zhivotovskii. 1986. Population structures of pink salmon: systems of fluctuating stocks. Sov. J. Mar. Biol. 12:92-97

Godfrey, H. 1959. Variations in annual average weights of British Columbia pink salmon, 1944-1958. J. Fish. Res. Board Can. 16:329-337

Godfrey, H., W.R. Hourston, J.W. Stokes, and F.C. Withler. 1954. Effects of a rock slide on Babine River salmon. Bull. Fish. Res. Board Can. 101: 100 p.

Godfrey, H., W.R. Hourston, and F.C. Withler. 1956. Babine River salmon after removal of the rock slide. Bull. Fish. Res. Board Can. 106:41 p.

Godin, J.G.J. 1980a. Temporal aspects of juvenile pink salmon (*Oncorhynchus gorbuscha* Walbaum) emergence from a simulated gravel redd. Can. J. Zool. 58:735-744

——. 1980b. Ontogenetic changes in the daily rhythms of swimming activity and of vertical distribution in juvenile pink salmon (*Oncorhynchus gorbuscha* Walbaum). Can. J. Zool. 58:745-753

——. 1981a. Daily patterns of feeding behavior, daily rations, and diets of juvenile pink salmon (*Oncorhynchus gorbuscha*) in two marine bays of British Columbia. Can. J. Fish. Aquat. Sci. 38: 10-15

——. 1981b. Effect of hunger on the daily pattern of feeding rates in juvenile pink salmon, *Oncorhynchus gorbuscha* Walbaum. J. Fish Biol. 19:63-71

——. 1982. Migrations of salmonid fishes during early life history phases: daily and annual timing, p. 22-49. *In*: E.L. Brannon and E.O. Salo (eds.). Proceedings of the Salmon and Trout Migratory Behavior Symposium. School of Fisheries, University of Washington, Seattle, WA.

Golovanov, I.S. 1982. Natural reproduction of pink salmon, *Oncorhynchus gorbuscha* (Salmonidae) on the northern shore of the Okhotsk Sea. J. Ichthyol. 22:32-39

Grachev, L.Ye. 1971. Alteration in the number of oocytes in the pink salmon (*Oncorhynchus gorbuscha* (Walbaum)) in the marine period of life. J. Ichthyol. 11:199-206

Graybill, J.P. 1979. Role of depth and velocity for nest site selection by Skagit River pink and chum salmon, p. 391-392. *In*: J.C. Mason (ed.). Proceedings of the 1978 Northeast Pacific Pink and Chum Salmon Workshop. Pacific Biological Station, Nanaimo, BC

Greeley, J.R. 1933. The spawning habits of brook, brown and rainbow trout and the problem of egg predators. Trans. Am. Fish. Soc. 62:239-248

Groot, C. 1965. On the orientation of young sockeye salmon (*Oncorhynchus nerka*) during their seaward migration out of lakes. Behaviour Suppl. 14:98 p.

——. 1982. Modifications on a theme-a perspective on migratory behavior of Pacific salmon, p. 1-21. *In*: E.L. Brannon and E.O. Salo (eds.). Proceedings of the Salmon and Trout Migratory Behavior Symposium. School of Fisheries, University of Washington, Seattle, WA

Groot, E.P., and D.F. Alderdice. 1985. Fine structure of the external egg membrane of five species of Pacific salmon and steelhead trout. Can. J. Zool. 63:552-566

Hallock, R.J., and D.H. Fry, Jr. 1967. Five species of salmon, *Oncorhynchus*, in the Sacramento River, California. Calif. Fish Game 53:5-22

Hanavan, M.G., and B.E. Skud. 1954. Intertidal spawning of pink salmon. Fish. Bull. Fish Wildl. Serv. 56:167-185

Harden Jones, F.R. 1968. Fish migration. St. Martin's Press, New York, NY. 325 p.

Hargreaves, N.B., and R.J. LeBrasseur. 1985. Species selective predation on juvenile pink (*Oncorhynchus gorbuscha*) and chum salmon (*O. keta*) by coho salmon (*O. kisutch*). Can. J. Fish. Aquat. Sci. 42:659-668

Harry, G.Y., Jr., and J.M. Olson. 1963. Straying of pink salmon to Sashin Creek, Little Port Walter Bay. U.S. Bur. Comm. Fish. Biol. Lab. Auke Bay Alaska MS Rep. MR 63-3:9 p.

Hartt, A.C. 1962. Observations of pink salmon in the Aleutian Island area, 1956-1960, p. 123-133. *In*: N.J. Wilimovsky (ed.). Symposium on Pink Salmon. H.R. MacMillan Lectures in Fisheries. Institute of Fisheries, University of British Columbia, Vancouver, BC

——. 1966. Migrations of salmon in the North Pacific Ocean and Bering Sea as determined by seining and tagging, 1959-1960. Int. North Pac. Fish. Comm. Bull. 19:141 p.

——. 1980. Juvenile salmonids in the oceanic ecosystem: the critical first summer, p. 25-57. *In*: W.J. McNeil and D.C. Himsworth (eds.). Salmonid ecosystems of the North Pacific. Oregon State University Press, Corvallis, OR

Hartt, A.C., and M.B. Dell. 1986. Early oceanic migrations and growth of juvenile Pacific salmon and steelhead trout. Int. North Pac. Fish. Comm. Bull. 46:105 p.

Hasler, A.D. 1966. Underwater guidepost: homing of salmon. University of Wisconsin Press, Madison, WI. 155 p.

Hasler, A.D., and A.T. Scholz. 1983. Olfactory imprinting and homing in salmon: investigations into the mechanism of the imprinting process. Springer-Kerlag, Berlin, Germany. 134 p.

Hasler, A.D., A.T. Scholz, and R.M. Horrall. 1978. Olfactory imprinting of homing in salmon. Am. Sci. 66:347-355

Hayes, F.R. 1942. The hatching mechanism of salmon eggs. J. Exp. Zool. 89:357-373

——. 1949. The growth, general chemistry and temperature relations of salmonid eggs. Q. Rev. Biol. 24:281-308

Hayes, F.R., I.R. Wilmot, and D.A. Livingstone. 1951. The oxygen consumption of the salmon egg in relation to development and activity. J. Exp. Zool. 116:377-395

Healey, M.C. 1967. Orientation of pink salmon (*Oncorhynchus gorbuscha*) during early marine migration from Bella Coola River system. J. Fish. Res. Board Can. 24:2321-2338

——. 1978. The distribution, abundance and feeding habits of juvenile Pacific salmon in Georgia Strait, British Columbia. Fish. Mar. Serv. (Can.) Tech. Rep. 788:49 p.

——. 1980. The ecology of juvenile salmon in Georgia Strait, British Columbia, p. 203-229. *In*: W.J. McNeil and D.C. Himsworth (eds.). Salmonid ecosystems of the North Pacific. Oregon State University Press, Corvallis, OR

——. 1982. The distribution and residency of juvenile Pacific salmon in the Strait of Georgia, British Columbia, in relation to foraging success, p. 61-69. *In*: B.R. Melteff and R.A. Neve (eds.). Proceedings of the North Pacific Aquaculture Symposium. Alaska Sea Grant Rep. 82-2

Heard, W.R. 1964. Phototactic behaviour of emerging sockeye salmon fry. Anim. Behav. 12:382-388

——. 1972. Spawning behavior of pink salmon on an artificial redd. Trans. Am. Fish. Soc. 101:276-283

——. 1975. Studies on 1966 brood year pink salmon, *Oncorhynchus gorbuscha*, and survival of their progeny in intertidal spawning channels, Lovers Cove Creek, Alaska. Auke Bay Lab., U.S. Natl. Mar. Fish. Ser. MS. Rep.-File MR-F 117:39 p.

——. 1978. Probable case of streambed overseeding: 1967 pink salmon, *Oncorhynchus gorbuscha*, spawners and survival of their progeny in Sashin Creek, southeastern Alaska. Fish. Bull. (U.S.) 76:569–582

Helle, J.H. 1966. Behavior of displaced adult pink salmon. Trans. Am. Fish. Soc. 95:188–195

——. 1970. Biological characteristics of intertidal and fresh-water spawning pink salmon at Olsen Creek, Prince William Sound, Alaska, 1962- 63. U.S. Fish Wildl. Serv. Spec. Sci. Rep. Fish. 602: 19 p.

Helle, J.H., R.S. Williamson, and J.E. Bailey. 1964. Intertidal ecology and life history of pink salmon at Olsen Creek, Prince William Sound, Alaska. U.S. Fish Wildl. Serv. Spec. Sci. Rep. Fish. 483: 26 p.

Herrmann, R.B. 1959. Occurrence of juvenile pink salmon in a coastal stream south of the Columbia River. Res. Briefs Fish Comm. Oreg. 7(1):81 p.

Hikita, T. 1962. Ecological and morphological studies of the genus *Oncorhynchus* (Salmonidae) with particular consideration on phylogeny. Sci. Rep. Hokkaido Salmon Hatchery 17:1–95. (In Japanese)

——. 1965. Some cases of the anomalous coloration of the pink salmon, *Oncorhynchus gorbuscha* (Walbaum), caught in the North Pacific Ocean. Sci. Rep. Hokkaido Salmon Hatchery 19:75–77. (In Japanese)

——. 1984. Further records on the small pink salmon (*Oncorhynchus gorbuscha*) caught in Hokkaido, Japan. Sci. Rep. Hokkaido Salmon Hatchery 38:83–88. (Transl. from Japanese; available Auke Bay Laboratory, Northwest and Alaska Fisheries Center)

Hikita, T., and T. Terao. 1967. An example of the pink salmon, *Oncorhynchus gorbuscha* (Walbaum), from Chitose River. Sci. Rep. Hokkaido Salmon Hatchery 21:77–78. (In Japanese)

Hikita, T., and Y. Yokohira. 1964. Biological study on hybrids of the salmonid fishes: a note of F_1 hybrids between chum (*Oncorhynchus keta*) and pink salmon (*Oncorhynchus gorbuscha*). Sci. Rep. Hokkaido Salmon Hatchery 18:57–64. (Transl. from Japanese; Fish. Res. Board Can. Transl. Ser. 1064)

Hoar, W.S. 1951. The chum and pink salmon fisheries of British Columbia 1917–1947. Bull. Fish. Res. Board Can. 90:46 p.

——. 1956. The behaviour of migrating pink and chum salmon fry. J. Fish. Res. Board Can. 13:309–325

——. 1958. The evolution of migratory behaviour among juvenile salmon of the genus *Oncorhynchus*. J. Fish. Res. Board Can. 15:391–428

——. 1969. Comments on the contribution of physiological and behavioural studies to the ecology of salmon in streams, p. 177–180. *In*: T.G. Northcote (ed.). Symposium on Salmon and Trout in Streams. H.R. MacMillan Lectures in Fisheries. Institute of Fisheries, University of British Columbia, Vancouver, BC

——. 1976. Smolt transformation: evolution, behavior, and physiology. J. Fish. Res. Board Can. 33:1233–1252

Hoar, W.S., M.H.A. Keenleyside, and R.G. Goodall. 1957. Reactions of juvenile Pacific salmon to light. J. Fish. Res. Board Can. 14:815–830

Horrall, R.M. 1981. Behavioral stock-isolating mechanisms in Great Lakes fishes with special reference to homing and site imprinting. Can. J. Fish. Aquat. Sci. 38:1481–1496

Hoshiai, G., and R. Sato. 1973. Record of the pink salmon *Oncorhynchus gorbuscha*: new record in the Akka River, Honshu, Japan. Jpn. J. Ichthyol. 20:125–126. (In Japanese, English summary)

Hourston, W.R., and D. MacKinnon. 1956. Use of an artificial spawning channel by salmon. Trans. Am. Fish. Soc. 86:220–230

Hunter, J.G. 1949a. Natural propagation of salmon in the central coastal area of British Columbia. II. The 1948 run. Fish. Res. Board Can. Prog. Rep. Pac. Coast Stn. 79:33–34

——. 1949b. Occurrence of hybrid salmon in the British Columbia commercial fishery. Fish. Res. Board Can. Prog. Rep. Pac. Coast Stn. 81:91–92

——. 1959. Survival and production of pink and chum salmon in a coastal stream. J. Fish. Res. Board Can. 16:835–886

Hurley, D.A., and W.L. Woodall. 1968. Responses of young pink salmon to vertical temperature and salinity gradients. Int. Pac. Salmon Fish. Comm. Prog. Rep. 19:80 p.

Hutton, K.E. 1967. Characteristics of the blood of adult pink salmon at three stages of maturity. Fish Bull. (U.S.) 66:195–202

Ievleva, M.Ya. 1951. Morphology and rate of embryonic development of Pacific salmon. Izv. Tikhookean. Nauchno- Issled. Inst. Rybn. Khoz. Okeanogr. 34:123–130. (Transl. from Russian; *In*:

Pacific salmon: selected articles from Soviet periodicals, p. 236–244. Israel Program for Scientific Translations, Jerusalem, 1961)

International North Pacific Fisheries Commission (INPFC). 1979. Historical catch statistics for salmon of the North Pacific Ocean. Int. North Pac. Fish. Comm. Bull, 39:166 p.

——. 1961–85. Statistical Yearbooks, 1960–81. International North Pacific Fisheries Commission, Vancouver, BC

International Pacific Salmon Fisheries Commission (IPSFC). 1958. Annual report, 1957. International Pacific Salmon Fisheries Commission, New Westminister, BC. 44 p.

——. 1968. Annual report, 1967. International Pacific Salmon Fisheries Commission, New Westminister, BC. 51 p.

——. 1979. Annual report, 1978. International Pacific Salmon Fisheries Commission, New Westminister, BC. 54 p.

——. 1982. Annual report, 1981. International Pacific Salmon Fisheries Commission, New Westminister, BC. 48 p.

——. 1983. Annual report, 1982. International Pacific Salmon Fisheries Commission, New Westminister, BC. 46 p.

. 1986. Annual report, 1984. International Pacific Salmon Fisheries Commission, New Westminister, BC. 54 p.

Isaev, A.I. 1961. Acclimatization of Pacific salmon in the Barents and White Seas. Vopr. Ikhtiol. 1(1):46–51. (Transl. from Russian; Fish. Res. Board Can. Transl. Ser. 361)

Ishida, T. 1966. Pink salmon in the Far East, p. 29–39. *In*: Salmon of the North Pacific Ocean. Part III. A review of the life history of North Pacific salmon. Int. North Pac. Fish. Comm. Bull. 18

——. 1967. Pink salmon in the Far East, p. 9–22. *In*: Salmon of the North Pacific Ocean. Part IV. Spawning populations of North Pacific salmon. Int. North Pac. Fish. Comm. Bull. 23

Ito, J. 1964. Food and feeding habits of Pacific salmon (genus *Oncorhynchus*) in their oceanic life. Bull. Hokkaido Reg. Fish. Res. Lab. 29:85–97. (Transl. from Japanese; Fish. Res. Board Can. Transl. Ser. 1309)

Ivankov, V N. 1965. On the age structure of pink salmon populations (*Oncorhynchus gorbuscha* (Walb.)). Vopr. Ikhtiol. 5:662–667. (Transl. from Russian; Fish. Res. Board Can. Transl. Ser. 705)

——. 1967. Local humpback-salmon schools of the Kurile Islands. Rybn. Khoz. 43:62–66. (Transl. from Russian; Israel Program for Scientific Translations, Jerusalem, 1968)

——. 1968a. Pacific salmon in the region of Iturup Island (Kuril Islands). Izv. Tikhookean. Nauchno- Issled. Inst. Rybn. Khoz. Okeanogr. 65:49–74. (Transl. from Russian; Fish. Res. Board Can. Transl. Ser. 1999)

——. 1968b. Influence of ocean driftnet fishing on the structure of spawning schools of pink salmon. Izv. Tikhookean. Nauchno-Issled. Inst. Rybn. Khoz. Okeanogr. 65:263–265. (Transl. from Russian; Fish. Res. Board Can. Transl. Ser. 1452)

——. 1968c. A method of determining the age of pink salmon (*Oncorhynchus gorbuscha*). Izv. Tikhookean. Nauchno- Issled. Inst. Rybn. Khoz. Okeanogr. 65:75–79. (Transl. from Russian; Fish. Res. Board Can. Transl. Ser. 1440)

Ivankov, V.N., and V.L. Andreyev. 1969. Fecundity of Pacific salmon (genus *Oncorhynchus* spp.). Probl. Ichthol. 9:59–66

Ivankov, V.N., Yu.A. Mitrofanov, and V.P. Bushuyev. 1975. An instance of the pink salmon (*Oncorhynchus gorbuscha*) reaching maturity at an age of less than 1 year. J. Ichthyol. 15:497–499

Jensen, H.M. 1956. Migratory habits of pink salmon found in the Tacoma Narrows area of Puget Sound. Wash. Dep. Fish. Fish. Res. Pap. 1:21–24

Jensen, J.O.T., and D.F. Alderdice. 1983. Changes in mechanical shock sensitivity of coho salmon (*Oncorhynchus kisutch*) eggs during incubation. Aquaculture 32:303–312

Johnson, K.R. 1979. Genetic variation in populations of pink salmon (*Oncorhynchus gorbuscha*) from Kodiak Island. M.Sc. thesis. University of Washington, Seattle, WA. 94 p.

Jones, D. 1978. Pink salmon stock predictions – S.E. Alaska. Technical report for period July 1, 1977 to June 30, 1978, Anadromous Fish Conservation Act Project AFC-59-1. Alaska Department of Fish and Game, Juneau, AK. 27 p.

Jones, D., and J. Dangel. 1982. Southeastern Alaska 1981 brood year pink (*Oncorhynchus gorbuscha*) and chum (*O. keta*) escapement surveys and pre-emergent fry program. Alaska Dep. Fish Game ADF&G Tech. Data Rep. 80:209 p.

Jones, J.D., and G. Thomason. 1983. Southeast

Alaska pink salmon secondary tagging and escapement enumeration studies, 1982, p. 1-13. *In*: Final report-1982 salmon research conducted in southeast Alaska by the Alaska Department of Fish and Game in conjunction with joint U.S.-Canada interception investigations. Commercial Fisheries Division, Alaska Department of Fish and Game, Juneau, AK.

Jones, J.W., and J.N. Ball. 1954. The spawning behaviour of brown trout and salmon. Anim. Behav. 2:103-114

Jones, J.W., and G.M. King. 1949. Experimental observations on the spawning behaviour of Atlantic salmon (*Salmo salar* Linn.). Proc. Zool. Soc. Lond. 119:33-48

Kaczynski, V.W., R.J. Feller, and J. Clayton. 1973. Trophic analysis of juvenile pink and chum salmon (*Oncorhynchus gorbuscha* and *O. keta*) in Puget Sound. J. Fish. Res. Board Can. 30:1003-1008

Kaeriyama, M., and T. Hikita. 1984. Southern extreme in the range of pink salmon, *Oncorhynchus gorbuscha*, along the Pacific Coast of Honshu, Japan. Sci. Rep. Hokkaido Salmon Hatchery 38:79-82. (In Japanese, English summary)

Kaganovsky, A.G. 1949. Some problems of the biology and population dynamics of pink salmon. Izv. Tikhookean. Nauchno-Issled. Inst. Rybn. Khoz. Okeanogr. 31:3-57. (Transl. from Russian; *In* Pacific salmon: selected articles from Soviet periodicals, p. 127-183. Israel Program for Scientific Translations, Jerusalem, 1961).

Kamyshnaia, M.S. (i.e., Kamyshnaya, M.S.). 1961. On the biology of the hybrid between chum and pink salmon: *Oncorhynchus keta* (Walbaum), infras. *autumnalis* Berg x *O. gorbuscha* (Walbaum) - family Salmonidae. Nauchn. Dokl. Vyssh. Shk. Biol. Nauk 4:29-33. (Transl. from Russian; Fish. Res. Board Can. Transl. Ser. 403)

Kamyshnaya, M.S. 1967. Downstream migration and behaviour of the young of introduced pink salmon. Rybn. Khoz. 43:9-12. (Transl. from Russian; Fish. Res. Board Can. Transl. Ser. 833)

Kamyshnaya, M.S., and M.I. Shatunovsky. 1969. Lipid composition in the humpback salmon in their natural habitat rivers and in acclimatized regions. Vestn. Mosk. Univ. Ser. VI Biol. Pochvoved. 24(2):33-37. (Transl. from Russian; Fish. Res. Board Can. Transl. Ser. 1417)

Kanid'yev, A.N., G.M. Kostyunin, and S.A. Salmin. 1970. Hatchery propagation of the pink and chum salmons as a means of increasing the salmon stocks of Sakhalin. J. Ichthyol 10:249-259

Kanno, Y., and I. Hamai. 1971. Food of salmonid fish in the Bering Sea in summer of 1966. Bull. Fac. Fish. Hokkaido Univ. 22:107-128. (In Japanese, English summary)

Karpenko, V.I. 1979. Feeding habits of juvenile Pacific salmon in the coastal waters of Kamchatka. Sov. J. Mar. Biol. 5:398-405

——. 1982a. Diurnal feeding rhythm of young salmon during the initial stage of marine life. J. Ichthyol. 22:131-134

——. 1982b. Biological peculiarities of juvenile coho, sockeye, and chinook salmon in coastal waters of east Kamchatka. Sov. J. Mar. Biol. 8:317-324

——. 1983. Influence of environmental factors on qualitative parameters of juvenile far-eastern salmon of the genus *Oncorhynchus* (Salmonidae) in Kamchatkan waters of the Bering Sea. J. Ichthyol. 23:93-101

——. 1987. Growth variation of juvenile pink salmon, *Oncorhynchus gorbuscha*, and chum salmon *Oncorhynchus keta*, during the coastal period of life. J. Ichthyol. 27:117-125

Karpenko, V.I., and V.V. Maksimenkov. 1988. Preliminary data on the interactions between Pacific salmon and herring during early ontogeny. J. Ichthyol. 28:136-140

Karpenko, V.I., and L.V. Piskunova. 1984. Importance of macroplankton in the diet of young salmons of the genus *Oncorhynchus* (Salmonidae) and their trophic relationships in the southwestern Bering Sea. J. Ichthyol. 24:98-106

Karpenko, V.I., and S.G. Safronov. 1985. Juvenile pink salmon, *Oncorhynchus gorbuscha*, from the coastal waters of the Okhotsk Sea. J. Ichthyol. 25:54-157

Kassov, E.G., M.S. Lazarev, and L.V. Polikashin. 1960. Pink salmon in the basins of the Barents and White seas. Rybn. Khoz. 36:20-25. (Transl. from Russian; Fish. Res. Board Can. Transl. Ser. 323)

Keenleyside, M.H.A., and H.M.C. Dupuis. 1988. Comparison of digging behavior by female pink salmon (*Oncorhynchus gorbuscha*) before and after spawning. Copeia 1988:1092-1095

Khorevin, L.D., V.A. Rudnev, and A.P. Shershnev.

1981. Predation on juvenile pink salmon by predatory fishes during the period of their seaward migration on Sakhalin Island. J. Ichthyol. 21:47–53

Kingsbury, A.P. 1979. Pink and chum forecasting in Alaska, p. 55–101. *In*: J.C. Mason (ed.). Proceedings of the 1978 Northeast Pacific Pink and Chum Salmon Workshop. Pacific Biological Station, Nanaimo, BC

Kirkwood, J.B. 1962. Inshore-marine and freshwater life history phases of pink salmon, *Oncorhynchus gorbuscha* (Walbaum), and the chum salmon, *O. keta* (Walbaum), in Prince William Sound, Alaska. Ph.D. thesis. University of Kentucky, Louisville, KY. 300 p.

Kitahara, T. 1983. Behavior of carotenoids in the chum salmon (*Oncorhynchus keta*) during anadromous migration. Comp. Biochem. Physiol. B Comp. Biochem. 76:97–101

Kobayashi, H. 1964. Biological study of hybrids of the salmonid fishes. Cytological observation of fertilization in the cross between chum salmon and pink salmon. Sci. Rep. Hokkaido Salmon Hatchery 18:67–71. (Transl. from Japanese; Fish. Res. Board Can. Transl. Ser. 1050)

Kobayashi, T. 1968a. Some observations on the natural spawning ground of chum and pink salmon in Hokkaido. Sci. Rep. Hokkaido Salmon Hatchery 22:7–13. (In Japanese, English summary)

——. 1968b. A note of the seaward mirgation of pink salmon fry. Sci. Rep. Hokkaido Salmon Hatchery 22:1–6. (Transl. from Japanese; Can. Transl. Fish Aquat. Sci. 5136)

Kobayashi, T., and S. Harada. 1966. Ecological observations on the salmon of Nishibetsu River. II. Movements, growth and feeding habits of pink salmon fry, *Oncorhynchus gorbuscha* (Walbaum), during seaward migration. Sci. Rep. Hokkaido Salmon Hatchery 20:1–10. (Transl. from Japanese; Fish. Res. Board Can. Transl. Ser. 785)

Koo, T.S.Y. 1962. Differential scale characters among species of Pacific salmon, p. 127–135. *In*: T.S.Y. Koo (ed.). Studies of Alaska red salmon. Univ. Wash. Publ. Fish. New Ser. 1

Kron, T.M., and H.J. Yuen. 1978. Zooplankton abundance and the feeding habits and migration of juvenile pink salmon, Tutka Bay, Alaska, 1975. Alaska Department of Fish and Game, FRED Division, Anchorage, AK. 25 p.

Krupitskiy, Yu.G., and A.F. Ustyugov. 1977. The pink salmon, *Oncorhynchus gorbuscha*, in the rivers of the north of Krasnoyarsk territory. J. Ichthyol. 17:320–322

Kuhlmann, F.W. 1977. Arctic anadromous fish investigations. Completion report for period July 1, 1974 to June 30, 1977. Anadromous Fish Conservation Act Project AFC-55. Alaska Department of fish and Game, Juneau AK. 72 p.

Kusnetzov, I.I. 1928. Some observations on spawning of the Amur and Kamtchatka salmons. Izv. Tikhookean. Nauchno-Issled. Inst. Rybn. Khoz. Okeanogr. 2:1–124. (Transl. from Russian; Fish. Res. Board Can. Transl. Ser. 22)

Kwain, W. 1978. Pink salmon are here to stay. Ont. Fish Wildl. Rev. 17:17–19

——. 1982. Spawning behavior and early life history of pink salmon (*Oncorhynchus gorbuscha*) in the Great Lakes. Can. J. Fish. Aquat. Sci. 39:1353–1360

Kwain, W., and J.A. Chappel. 1978. First evidence for even-year spawning pink salmon, *Oncorhynchus gorbuscha*, in Lake Superior. J. Fish. Res. Board Can. 35:1373–1376

Kwain, W., and A.H. Lawrie. 1981. Pink salmon in the Great Lakes. Fisheries 6:2–6

Lagler, K.F., and A.T. Wright. 1962. Predation of the Dolly Varden, *Salvelinus malma*, on young salmons, *Oncorhynchus* spp., in an estuary of south eastern Alaska. Trans. Am. Fish. Soc. 91:90–93

Landingham, J.H. 1982. Feeding ecology of pink and chum salmon fry in the nearshore habitat of Auke Bay, Alaska. M.Sc. thesis. University of Alaska, Juneau, AK. 132 p.

Lapin, Yu.E. 1963. On the age and population dynamics of the Pacific pink salmon *Oncorhynchus gorbuscha* (Walb.). Vopr. Ikhtiol. 3:243–255. (Transl. from Russian; Fish. Res. Board Can. Transl. Ser. 509)

Lapin, Yu.Ye. 1971. New data relating to direct determination of the age of the pink salmon (*Oncorhynchus gorbuscha* (Walbaum)) by marking. J. Ichthyol. 11:9–59

Lear, W.H. 1975. Evaluation of the transplant of Pacific pink salmon (*Oncorhynchus gorbuscha*) from British Columbia to Newfoundland. J. Fish. Res. Board Can. 32:2343–2356

——. 1980. The pink salmon transplant experiment in Newfoundland, p. 213–243. *In* J.E. Thorpe (ed.). Salmon ranching. Academic Press,

New York, NY

Lear, W.H., and F.A. Day. 1977. An analysis of biological and environmental data collected at North Harbour River, Newfoundland, during 1959–1975. Fish. Mar. Serv. (Can.) Tech. Rep. 697:61 p.

LeBrasseur, R.J. 1966. Stomach contents of salmon and steelhead trout in the northeastern Pacific Ocean. J. Fish. Res. Board Can. 23:85–100

——. 1969. Growth of juvenile chum salmon (*Oncorhynchus keta*) under different feeding regimes. J. Fish. Res. Board Can. 26:1631–1645

LeBrasseur, R.J., and W. Barner. 1964. Midwater trawl salmon catches in northern Hecate Strait, November 1965. Fish. Res. Board Can. MS Rep. Ser. 176:10 p.

LeBrasseur, R.J., and R.R. Parker. 1964. Growth rate of central British Columbia pink salmon (*Oncorhynchus gorbuscha*). J. Fish. Res. Board Can. 21:1101–1128

Levanidov, V.Ya. 1964. Salmon population trends in the Amur basin and means of maintaining the stocks, p. 35–40. *In*: E.N. Pavlovskii (ed.). Lososevoe khozyaistvo dal'nego vostoka. Izdatel'stvo Nauka, Moscow, USSR. (Transl. from Russian; Univ. Wash. Fish. Res. Inst. Circ. 227:35–40)

Levanidov, V.Ya, and I.M. Levanidova. 1957. Food of downstream migrant young summer chum salmon and pink salmon in Amur tributaries. Izv. Tikhookean. Nauchno-Issled. Inst. Rybn. Khoz. Okeanogr. 45:3–16. (Transl. from Russian; *In*: Pacific Salmon: selected articles from Soviet periodicals, p. 269–284. Israel Program for Scientific Translations, Jerusalem, 1961)

Lindberg, J.M., and P. Brown. 1982. Continuing experiments in salmon ocean ranching in southern Chile. Int. Counc. Explor. Sea C.M. 1982/M:21:6 p.

Lysaya, N.M. 1951. Changes in the blood composition of salmon during the spawning migration. Izv. Tikhookean. Nauchno-Issled. Inst. Rybn. Khoz. Okeanogr. 35:47–60. (Transl. from Russian; *In*: Pacific Salmon: selected articles from Soviet periodicals, p. 199–215. Israel Program for Scientific Translations, Jerusalem, 1961)

MacKinnon, C.N., and E.M. Donaldson. 1976. Environmentally induced precocious sexual development in the male pink salmon (*Oncorhynchus gorbuscha*). J. Fish. Res. Board Can. 33:2602–2605

MacKinnon, D. 1960. A successful transplant of salmon eggs in the Robertson Creek spawning channel. Can. Fish Cult. 27:25–31

——. 1963. Salmon spawning channels in Canada, p. 108–110. *In*: R.S. Croker (ed.). Report of Second Governor's Conference on Pacific Salmon. State Printing Plant, Olympia, WA

Maeda, H. 1954. Ecological analyses of pelagic shoals. I: analysis of salmon gill-net association in the Aleutians. 1: quantitative analysis of food. Jpn. J. Ichthyol. 3:223–231

Major, P.F. 1978. Predator-prey interactions in two schooling fishes, (*Caranx ignobilis*) and (*Stolephorus purpureus*). Anim. Behav. 26:760–777

Manzer, J.I. 1956. Distribution and movement of young Pacific salmon during early ocean residence. Fish. Res. Board Can. Prog. Rep. Pac. Coast Stn. 106:24–28

——. 1964. Preliminary observations on the vertical distribution of Pacific salmon (genus *Oncorhynchus*) in the Gulf of Alaska. J. Fish. Res. Board Can. 21:891–903

——. 1968. Food of Pacific salmon and steelhead trout in the northeast Pacific Ocean. J. Fish. Res. Board Can. 25:1085–1089

——. 1969. Stomach contents of juvenile Pacific salmon in Chatham Sound and adjacent waters. J. Fish. Res. Board Can. 26:2219–2223

Manzer, J.I., T. Ishida, A.E. Peterson, and M.G. Hanavan. 1965. Salmon of the North Pacific Ocean. Part V: offshore distribution of salmon. Int. North Pac. Fish. Comm. Bull. 15:452 p.

Manzer, J.I., and R.J. LeBrasseur. 1959. Further observations on the vertical distribution of salmon in the Northeast Pacific. Fish. Res. Board Can. MS Rep. Ser. 689:9 p.

Manzer, J.I., and M.P. Shepard. 1962. Marine survival, distribution and migration of pink salmon off the British Columbia coast, p. 113–122. *In*: N.J. Wilimovsky (ed.). Symposium on Pink Salmon. H.R. MacMillan Lectures in Fisheries. Institute of Fisheries, University of British Columbia, Vancouver, BC

Marr, D.H.A. 1963. The influence of surface contour on the behaviour of trout alevins, *S. trutta* L. Anim. Behav. 11:412

Martin, J.W. 1966. Early sea life of pink salmon, p. 111–125. *In*: W.L. Sheridan (ed.). Proceedings of the 1966 Northeast Pacific Pink Salmon Workshop. Alaska Dep. Fish Game Inf. Leafl. 87

Mathisen, O.A. 1962. The effect of altered sex ratios

on the spawning of red salmon, p. 141–245. *In*: T.S.Y. Koo (ed.). Studies of Alaska red salmon. Univ. Wash. Publ. Fish. New Ser. 1

McCurdy, M.L. 1984. Prince William Sound general districts pink (*Oncorhynchus gorbuscha*) and chum (*O. keta*) salmon aerial and ground escapement surveys and pre-emergent alevin index surveys, brood years 1980 through 1983. Alaska Dep. Fish Game ADF&G Tech. Data Rep. 116: 116 p.

McCurdy, M.L., and R.B. Pirtle. 1980. Prince William Sound general districts 1977 brood year pink (*Oncorhynchus gorbuscha*) and chum salmon (*O. keta*) aerial and ground escapement surveys and consequent egg deposition and pre-emergent fry index programs. Alaska Dep. Fish Game ADF&G Tech. Data Rep. 52:60 p

McDaniel, T.R. 1981. Evaluation of pink salmon (*Oncorhynchus gorbuscha*) fry plants at Seal Bay Creek, Afognak Island, Alaska. Alaska Dep. Fish Game Inf. Leafl. 193:9 p.

McDaniel, T.R., T. Kohler, and D.M. Dougherty. 1984. Results of pink salmon (*Oncorhynchus gorbuscha*) fry transplants to Hobo Creek, Prince William Sound, Alaska. Alaska Dep. Fish Game FRED Rep. 33:15 p.

McDonald, J. 1960. The behaviour of Pacific salmon fry during their downstream migration to freshwater and saltwater nursery areas. J. Fish. Res. Board Can. 17:655–676

McDonald, J.G. 1960. A possible source of error in assessing the survival of Pacific salmon eggs by redd sampling. Can. Fish Cult. 26:27–30

McGregor, A.J. 1982. A biochemical genetic analysis of pink salmon (*Oncorhynchus gorbuscha*) from selected streams in northern southeast Alaska. M.Sc. thesis. University of Alaska, Juneau, AK. 93 p.

McInerney, J.E. 1964. Salinity preference: an orientation mechanism in salmon migration. J. Fish. Res. Board Can. 21:995–1018

McLarney, W.O. 1964. The coast range sculpin, *Cottus aleuticus*: structure of a population and predation on eggs of the pink salmon *Oncorhynchus gorbuscha*. M.Sc. thesis. University of Michigan, Ann Arbor, MI. 83 p.

——. 1967. Intra-stream movement, feeding habits, and population of coast range sculpin, *Cottus aleuticus*, in relation to eggs of the pink salmon, *Oncorhynchus gorbuscha*, in Alaska. Ph.D. thesis. University of Michigan, Ann Arbor, MI. 131 p.

McNeil, W.J. 1962. Mortality of pink and chum salmon eggs and larvae in southeast Alaska streams. Ph.D. thesis. University of Washington, Seattle, WA. 270 p.

——. 1964a. Redd superimposition and egg capacity of pink salmon spawning beds. J. Fish. Res. Board Can. 21:1385–1396

——. 1964b. A method of measuring mortality of pink salmon eggs and larvae. Fish. Bull. (U.S.) 63:575–588

——. 1966a. Distribution of spawning salmon in Sashin Creek, southeastern Alaska, and survival of their progeny. U.S. Fish Wildl. Serv. Spec. Sci. Rep. Fish. 538:12 p.

——. 1966b. Effect of the spawning bed environment on reproduction of pink and chum salmon. Fish. Bull. (U.S.) 65:495–523

——. 1967. Randomness in distribution of pink salmon redds. J. Fish. Res. Board Can. 24:1629–1634

——. 1968. Migration and distribution of pink salmon spawners in Sashin Creek in 1965, and survival of their progeny. Fish. Bull. (U.S.) 66:575–586

——. 1980. Vulnerability of pink salmon populations to natural and fishing mortality, p. 147–151. *In*: W.J. McNeil and D.C. Himsworth (eds.). Salmonid ecosystems of the North Pacific. Oregon State University Press, Corvallis, OR

McNeil, W.J., and W.H. Ahnell. 1964. Success of pink salmon spawning relative to size of spawning bed materials. U.S. Fish Wildl. Serv. Spec. Sci. Rep. Fish. 469:15 p.

McNeil, W.J., S.C. Smedley, and R.J. Ellis. 1969. Transplanting adult pink salmon to Sashin Creek, Baranof Island, Alaska, and survival of their progeny. U.S. Fish Wildl. Serv. Spec. Sci. Rep. Fish. 587:9 p.

McNeil, W.J., R.A. Wells, and D.C. Brickell. 1964. Disappearance of dead pink salmon eggs and larvae from Sashin Creek, Baranof Island, Alaska. U.S. Fish Wildl. Serv. Spec. Sci. Rep. Fish. 485:13 p.

McPhail, J.D., and C.C. Lindsey. 1970. Freshwater fishes of northwestern Canada and Alaska. Bull. Fish. Res. Board Can. 173:381 p.

Merrell, T.R. 1962. Freshwater survival of pink salmon at Sashin Creek, p. 59–72. *In*: N.J. Wilimovsky (ed.). Symposium on Pink Salmon. H.R.

MacMillan Lectures in Fisheries. Institute of Fisheries, University of British Columbia, Vancouver, BC

Milovidova-Dubrovskaya, N.V. 1937. Material on biology and fishery of Primorsky (Maritime) gorbuscha (pink salmon). Izv. Tikhookean. Nauchno- Issled. Inst. Rybn. Khoz. Okeanogr. 12:101–114. (Transl. from Russian: available Northwest and Alaska Fisheries Center)

Miyaguchi, K. 1959. On the so-called "fry zone" appearing in the scale of pink salmon *Oncorhynchus gorbuscha* (Walbaum). Bull. Hokkaido Reg. Fish. Res. Lab. 20:104–108. (In Japanese, English summary)

Mortensen, D.G. 1983. Laboratory studies on factors influencing the first feeding of newly emerged pink salmon (*Oncorhynchus gorbuscha*) fry. M.Sc. thesis. University of Alaska, Juneau, AK. 108 p.

Moyle, P. 1966. Feeding behavior of the glaucous-winged gull on an Alaskan salmon stream. Wilson Bull. 78:175–190

Munro, J.A. 1936. Dipper (*Cinclus mexicanus unicolor*) eating salmon fry (*Oncorhynchus gorbuscha*) in British Columbia. Condor 38:120 p.

Munro, J.A., and W.A. Clemens. 1937. The American merganser in British Columbia and its relation to the fish population. Biol. Board Can. Bull. 55:50 p.

Murphy, M.L. 1985. Die-offs of pre-spawn adult pink salmon and chum salmon in southeastern Alaska. N. Am. J. Fish. Manage. 5:302–308

• Murray, C.B., and T.D. Beacham. 1986. Effect of varying temperature regimes on the development of pink salmon (*Oncorhynchus gorbuscha*) eggs and alevins. Can. J. Zool. 64:670–676

Neave, F. 1948. Fecundity and mortality in Pacific salmon. Proc. Trans. R. Soc. Can. Ser. 3 42(5):97–105

——. 1952. "Even-year" and "odd-year" pink salmon populations. Proc. Trans. R. Soc. Can. Ser. 3 46(5):55–70

——. 1953. Principles affecting the size of pink and chum salmon populations in British Columbia. J. Fish. Res. Board Can. 9:450–491

——. 1955. Notes on the seaward migration of pink and chum salmon fry. J. Fish. Res. Board Can. 12:369–374

——. 1958. The origin and speciation of *Oncorhynchus*. Proc. Trans. R. Soc. Can. Ser. 3 52(5):25–39

——. 1962. The observed fluctuations of pink salmon in British Columbia, p. 3–14. *In*: N.J. Wilimovsky (ed.). Symposium on Pink Salmon. H.R. MacMillan Lectures in Fisheries. Institute of Fisheries, University of British Columbia, Vancouver, BC

——. 1964. Ocean migrations of Pacific salmon. J. Fish. Res. Board Can. 21:1227–1244

——. 1965. Transplants of pink salmon. Fish. Res. Board Can. MS Rep. 830:23 p.

——. 1966. Pink salmon in British Columbia, p. 71–79. *In*: Salmon of the North Pacific Ocean. Part III. A review of the life history of North Pacific salmon. Int. North Pac. Fish. Comm. Bull. 18

Neave, F., T. Ishida, and S. Murai. 1967. Salmon of the North Pacific Ocean. Part VI. Pink salmon in offshore waters. Int. North Pac. Fish. Comm. Bull. 22:33 p.

Neave, F., and W.P. Wickett. 1953. Factors affecting the freshwater development of Pacific salmon in British Columbia. Proc. 7th Pac. Sci. Congr. 1949(4):548–556

——. 1955. Transplantation of pink salmon into the Fraser Valley in a barren year. Fish. Res. Board Can. Prog. Rep. Pac. Coast Stn. 103:14–15

Nickerson, R.B. 1979. Separation of some pink salmon (*Oncorhynchus gorbuscha* Walbaum) subpopulations in Prince William Sound, Alaska by length-weight relationships and horizontal starch gel electrophoresis. Alaska Dep. Fish Game Inf. Leafl. 181:36 p.

Nicola, S.J. 1968. Scavenging by *Alloperla* (Plecoptera: Chloroperlidae) nymphs on dead pink (*Oncorhynchus gorbuscha*) and chum *(O. keta)* salmon embryos. Can. J. Zool. 46:787–796

Nikolsky, G. V. 1963. The ecology of fishes. Academic Press, New York, NY. 352 p.

Noerenberg, W.A. 1963. Salmon forecast studies on 1963 runs in Prince William Sound. Alaska Dep. Fish Game Inf. Leafl. 21:17 p.

Noerenberg, W.H. 1971. Earthquake damage to Alaskan fisheries, p. 170–193. *In*: The great Alaska earthquake of 1964: biology. Natl. Acad. Sci. NAS Publ. 1604

Noltie, D.B. 1987. Incidence and effects of sea lamprey (*Petromyzon marinus*) parasitism on breeding pink salmon (*Oncorhynchus gorbuscha*) from the Carp River, eastern Lake Superior. Can. J. Fish. Aquat. Sci. 44:1562–1567

Nordeng, H. 1971. Is the local orientation of anadromous fishes determined by pheromones. Nature 233:411-413

——. 1977. A pheromone hypothesis for homeward migration in anadromous salmonids. Oikos 28:155-159

Nunan, P.J. 1967. Pink salmon in Lake Superior. Ont. Fish Wildl. Rev. 6:9-14

O'Brien, R.N., S. Visaisouk, R. Raine, and D.F. Alderdice. 1978. Natural convection: a mechanism for transporting oxygen to incubating eggs. J. Fish. Res. Board Can. 35:1316-1321

Olifirenko, L.N. 1970. Biochemical differences of quality within populations of the pink salmon and the masu salmon [*Oncorhynchus*] during the spawning migration. J. Ichthyol. 10:46-55

Olson, J.M., and W.J. McNeil. 1967. Research on pink salmon at Little Port Walter, Alaska, 1934-64. U.S. Fish Wildl. Serv. Data Rep. 17:301 p.

Ouchi, A., and M. Kuroiwa. 1963. The aberrant body color found in the pink salmon *Oncorhynchus gorbuscha*. Bull. Jpn. Sea. Reg. Fish. Res. Lab. 11:125-127. (In Japanese)

Parker, R.R. 1962a. A concept of the dynamics of pink salmon populations, p. 203-211. *In*: N.J. Wilimovsky (ed.). Symposium on Pink Salmon. H.R. MacMillan Lectures in Fisheries. Institute of Fisheries, University of British Columbia, Vancouver, BC

——. 1962b. Estimations of ocean mortality rates for Pacific salmon (*Oncorhynchus*). J. Fish. Res. Board Can. 19:561-589

——. 1964. Estimation of sea mortality rates for the 1960 brood-year pink salmon of Hook Nose Creek, British Columbia. J. Fish. Res. Board Can. 21:1019-1034

——. 1965. Estimation of sea mortality rates for the 1961 brood-year pink salmon of the Bella Coola area, British Columiba. J. Fish. Res. Board Can. 22:1523-1554

——. 1968. Marine mortality schedules of pink salmon of the Bella Coola River, central British Columbia. J. Fish. Res. Board Can. 25:757-794

——. 1971. Size selective predation among juvenile salmonid fishes in a British Columbia inlet. J. Fish. Res. Board Can. 28:1503-1510

Parker, R.R., and R.J. LeBrasseur. 1974. Ecology of early sea life, pink and chum juveniles, p. 161-171. *In*: D.R. Harding (ed.). Proceedings of the 1974 Northeast Pacific Pink and Chum Salmon Workshop. Department of the Environment, Fisheries, Vancouver, BC

Parker, R.R., and W.E. Vanstone. 1966. Changes in chemical composition of central British Columbia pink salmon during early sea life. J. Fish. Res. Board Can. 23:1353-1384

Pavlov, I.S. 1959. Experiments on the hybridization of Pacific salmon. Rybn. Khoz. 35:23-24. (Transl. from Russian; Fish. Res. Board Can. Transl. Ser. 263)

Pearcy, W., T. Nishiyama, T. Fujii, and K. Masuda. 1984. Diel variations in the feeding habits of Pacific salmon caught in gill nets during a 24-hour period in the Gulf of Alaska. Fish. Bull. (U.S.) 82:391-399

Pearson, R.E. 1966a. Report on the investigations by the United States for the International North Pacific Fisheries Commission-1964: pink salmon scale studies. Int. North Pac. Fish. Comm. Annu. Rep. 1964:118-124

——. 1966b. Number of circuli and time of annulus formation on scales of pink salmon (*Oncorhynchus gorbuscha*). J. Fish. Res. Board Can. 23:747-756

——. 1969. Investigations by the United States for the International North Pacific Fisheries Commission-1967: scale studies of pink salmon. Int. North Pac. Fish. Comm. Annu. Rep. 1967: 112-115

Pennoyer, S. 1970. Pink salmon run trends and 1970 forecast central region of Alaska, p. 9-12. *In*: C.E. Walker (ed.). Report of the 1970 Northeast Pacific Salmon Workshop. Department of Fisheries and Forestry, Vancouver, BC

Persov, G.M. 1963. The "potential" and "final" fecundity of fish through the example of the pink salmon *Oncorhynchus gorbuscha* (Walb.) acclimatized in White and Barents Sea Basins. Vopr. Ikhtiol. 3(28):490-496. (Transl. from Russian; *In*: Recent developments in Soviet ichthyology, p. 19-23. U.S. Joint Publications Research Service, Washington, DC)

Peterman, R.M., and M. Gatto. 1978. Estimation of functional responses of predators of juvenile salmon. J. Fish. Res. Board Can. 35:797-808

Phillips, A.C., and W.E. Barraclough. 1978. Early marine growth of juvenile Pacific salmon in the Strait of Georgia and Saanich Inlet, British Columbia. Fish. Mar. Serv. (Can.) Tech. Rep. 830: 19 p.

Phillips, R.W., and E.W. Claire. 1966. Intragravel movement of the reticulate sculpin, *Cottus perplexus* and its potential as a predator on salmonid embryos. Trans. Am. Fish. Soc. 95:210–212

Pirtle, R.B. 1977. Historical pink and chum salmon estimated spawning escapements from Prince William Sound, Alaska streams, 1960–1975. Alaska Dep. Fish Game ADF&G Tech. Data Rep. 35:332 p.

Pirtle, R.B., and M.L. McCurdy. 1980a. Prince William Sound general districts 1976 pink (*Oncorhynchus gorbuscha*) and chum salmon (*O. keta*) aerial and ground escapement surveys and consequent brood year egg deposition and pre-emergent fry index programs. Alaska Dep. Fish Game ADF&G Tech. Data Rep. 51:62 p.

——. 1980b. Prince William Sound general districts 1978 brood year pink (*Oncorhynchus gorbuscha*) and chum salmon (*O. keta*) aerial and ground escapement surveys and pre-emergent fry index program. Alaska Dep. Fish Game ADF&G Tech. Data Rep. 53:43 p.

——. 1980c. Prince William Sound general districts 1979 brood year pink (*Oncorhynchus gorbuscha*) and chum salmon (*O. keta*) aerial and ground escapement surveys and pre-emergent fry index program. Alaska Dep. Fish Game ADF&G Tech. Data Rep. 54:44 p.

Popov, A.V., and A.I. Zotin. 1961. The relationship between the time of hatching of salmon and whitefish embryos and certain environmental factors. Rybn. Khoz. 37:22–28. (Transl. from Russian; Fish. Res. Board Can. Transl. Ser. 4092)

Poy, A. 1970. On the behaviour of larvae of teleost fishes at hatching. Ber. Dtsch. Wiss. Komm. Meeresforsch. 21:377–392. (Transl. from German; Fish. Res. Board Can. Transl. Ser. 3519)

Pritchard, A.L. 1934a. Do caddis fly larvae kill fish? Can. Field-Nat. 48:39 p.

——. 1934b. Note on the water ousel, *Cinclus mexicanus*. Can. Field-Nat. 48:53 p.

——. 1934c. Was the introduction of the muskrat in Graham Island, Queen Charlotte Islands, unwise? Can. Field-Nat. 48:103 p.

——. 1936a. Factors influencing the upstream spawning migration of the pink salmon, *Oncorhynchus gorbuscha* (Walbaum). J. Biol. Board Can. 2:383–389

——. 1936b. Stomach content analysis of fishes preying upon the young of Pacific salmon during fry migration at McClinton Creek, Masset Inlet, British Columbia. Can. Field-Nat. 50:104–105

——. 1937. Variation in the time of run, sex proportions, size and egg content of adult pink salmon *(Oncorhynchus gorbuscha)* at McClinton Creek, Masset Inlet, B.C. J. Biol. Board Can. 3:403–416

——. 1938. Transplantation of pink salmon *(Oncorhynchus gorbuscha)* into Masset Inlet, British Columbia, in the barren years. J. Fish. Res. Board Can. 4:141–150

——. 1939. Homing tendency and age at maturity of pink salmon *(Oncorhynchus gorbuscha)* in British Columbia. J. Fish. Res. Board Can. 4:233–251

——. 1941. The recovery of marked Masset Inlet pink salmon during the season of 1940. Fish. Res. Board Can. Progr. Rep. Pac. Biol. Stn. Pac. Fish Exp. Stn. 48:13–17

——. 1944a. Physical characteristics and behaviour of pink salmon fry at McClinton Creek, B.C. J. Fish. Res. Board Can. 6:217–227

——. 1944b. Return of two marked pink salmon *(Oncorhynchus gorbuscha)* to the natal stream from distant places in the sea. Copeia 1944(2): 80–82

——. 1948a. Efficiency of natural propagation of the pink salmon *(Oncorhynchus gorbuscha)* in McClinton Creek, Masset Inlet, B.C. J. Fish. Res. Board Can. 7:224–236

——. 1948b. A discussion of the mortality in pink salmon *(Oncorhynchus gorbuscha)* during their period of marine life. Proc. Trans. R. Soc. Can. Ser. 3 42(5):125–133

Pushkareva, N.F. 1951. Food of pink salmon at the end of the marine stage of migration. Izv. Tikhookean. Nauchno-Issled. Inst. Rybn. Khoz. Okeanogr. 35:33–39. (Transl. from Russian; *In*: Pacific salmon: selected articles from Soviet periodicals, p. 191–198. Israel Program for Scientific Translations, Jerusalem, 1961)

Pyefinch, K.A. 1962. Capture of pink salmon on the Scottish coast. Ann. Biol. 17(1960):238

Reed, R.J. 1967. Observation of fishes associated with spawning salmon. Trans. Am. Fish. Soc. 96:62–67

Regnart, R. 1976. Arctic-Yukon-Kuskokwim region: a developing fishery, p. 78–88. *In*: G.K. Gunstrom (ed.). Proceedings of the 1976 Northeast Pacific Pink and Chum Salmon Workshop. Alaska Department of Fish and Game, Juneau, AK

Regnart, R.I., and M.F. Geiger. 1974. Status of commercial and subsistence salmon fisheries in the western Alaska region from Cape Newenham to Cape Prince of Wales, p. 143–157. *In*: Additional information on the exploitation, scientific investigation, and management of salmon stocks on the Pacific coasts of Canada and the United States in relation to the abstention provisions of the North Pacific Fisheries Convention. Int. North Pac. Fish. Comm. Bull. 29

Rice, S.D., and J.E. Bailey. 1980. Survival, size, and emergence of pink salmon, *Oncorhynchus gorbuscha*, alevins after short- and long-term exposures to ammonia. Fish. Bull. (U.S.) 78:641–648

Ricker, W.E. 1962. Regulation of the abundance of pink salmon populations, p. 155–206. *In*: N.J. Wilimovsky (ed.). Symposium on Pink Salmon. H.R. MacMillan Lectures in Fisheries. Institute of Fisheries, University of British Columbia, Vancouver, BC

——. 1964. Ocean growth and mortality of pink and chum salmon. J. Fish. Res. Board Can. 21:905–931

——. 1972. Hereditary and environmental factors affecting certain salmonid populations, p. 27–160. *In*: R.C. Simon (ed.). The stock concept in Pacific salmon. H.R. MacMillan Lectures in Fisheries. Institute of Fisheries. University of British Columbia, Vancouver, BC

Ricker, W.E., H.T. Bilton, and K.V. Aro. 1978. Causes of decrease in size of pink salmon *(Oncorhynchus gorbuscha)*. Fish. Mar. Serv. (Can.) Tech. Rep. 820:93 p.

Ricker, W.E., and K.H. Loftus. 1968. Pacific salmon move east. Fish. Counc. Can. Annu. Rev. 43: 37–39

Ricker, W.E., and J.I. Manzer. 1974. Recent information on salmon stocks in British Columbia, p. 1–24. *In*. Additional information on the exploitation, scientific investigation, and management of salmon stocks on the Pacific coasts of Canada and the United States in relation to the abstention provisions of the North Pacific Fisheries Convention. Int. North Pac. Fish. Comm. Bull. 29

Rimsh, E.Ya. 1977. Results of acclimatization of hump-back salmon in the Baltic, p. 97–99. *In*: Proceedings of Soviet/American Aquaculture Conference, Tallin, October 1977

Robinson, D.G., W.E. Barraclough, and J.D. Fulton. 1968a. Data record. Number, size compositon, weight and food of larval and juvenile fish caught with a two-boat surface trawl in the Strait of Georgia, May 1-4, 1967. Fish. Res. Board Can. MS Rep. Ser. 964:105 p.

——. 1968b. Data record. Number, size composition, weight, and food of larval and juvenile fish caught with a two-boat surface trawl in the Strait of Georgia, June 5–9, 1967. Fish. Res. Board Can. MS Rep. Ser. 972:109 p.

Rockwell, J., Jr. 1956. Some effects of sea water and temperature on the embryos of the Pacific salmon, *Oncorhynchus gorbuscha* (Walbaum) and *Oncorhynchus keta* (Walbaum). Ph.D. thesis. University of Washington, Seattle, WA. 416 p.

Rogers, D.E. 1984. Trends in abundance of northeastern Pacific stocks of salmon, p. 100–127. *In*: W.G. Pearcy (ed.). The influence of ocean conditions on production of salmonids in the North Pacific: a workshop. Oreg. State Univ. Sea Grant Coll. Program ORESU-W-83001

Roppel, P. 1982. Alaska's salmon hatcheries, 1891–1959. Alaska Historical Comm. Stud. Hist. 20: 299 p.

Rounsefell, G.A. 1957. Fecundity of North American Salmonidae. Fish. Bull. Fish Wildl. Serv. 57:451–468

——. 1958. Factors causing decline in sockeye salmon of Karluk River, Alaska. Fish Bull. Fish Wildl. Serv. 58:83–169

Roukhlov, F.N. (i.e., Rukhlov, F.N.), O.S. Ljubaeva, and L.D. Khorevin. 1982. Effectiveness of pink salmon reproduction at the hatcheries of the Sakhalin region, p. 119–122. *In*: B.R. Melteff and R.A. Nevé (eds). Proceedings of the North Pacific Aquaculture Symposium. Alaska Sea Grant Rep. 82-2

Royce, W.F. 1962. Pink salmon fluctuations in Alaska, p. 15–23. *In*: N.J. Wilimovsky (ed.). Symposium on Pink Salmon. H.R. MacMillan Lectures in Fisheries. Institute of Fisheries, University of British Columbia, Vancouver, BC

Royce, W.F., L.S. Smith, and A.C. Hartt. 1968. Models of oceanic migrations of Pacific salmon and comments on guidance mechanisms. Fish. Bull. (U.S.) 66.441–462

Roys, R.S. 1968. Forecast of 1968 pink and chum salmon runs in Prince William Sound. Alaska Dep. Fish Game Inf. Leafl. 116:50 p.

——. 1971. Effect of tectonic deformation on pink salmon runs in Prince William Sound, p. 220–

237. *In*: The Great Alaska Earthquake of 1964: biology. Natl. Acad. Sci. NAS Publ. 1604

Rukhlov, F.N. 1969. Materials characterizing the texture of bottom material in the spawning grounds and redds of the pink salmon (*Oncorhynchus gorbuscha* (Walbaum)) and the autumn chum (*Oncorhynchus keta* (Walbaum)) on Sakhalin. Probl. Ichthyol. 9:635-644

——. 1978. On mixing of local groups of Sakhalin humpback salmon on migration ways and at spawning grounds, p. 115-116. *In*: Biology of salmons: abstracts. Pacific Research Institute of Fisheries and Oceanography, Vladivostok, USSR

Rukhlov, F.N., and O.S. Lyubayeva. 1980. The results of marking of young pink salmon, *Oncorhynchus gorbuscha*, in the Sakhalin fish hatcheries in 1976. J. Ichthyol. 20:110-118

Sakagawa, G.T. 1972. The dynamics of juvenile salmon, with particular emphasis on pink salmon (*Oncorhynchus gorbuscha*), during their early marine life. Ph.D. thesis. University of Washington, Seattle, WA. 352 p.

Sano, S., and T. Kobayashi. 1953. On the returning of pink salmon, *Oncorhynchus gorbuscha* (Walbaum), to the Yurappu River. Sci. Rep. Hokkaido Fish Hatchery 8:1-9. (In Japanese, English summary)

Schumacher, R.E., and E. Eddy. 1960. The appearance of pink salmon, *Oncorhynchus gorbuscha* (Walbaum), in Lake Superior. Trans. Am. Fish. Soc. 89:371-373

Schumacher, R.E., and J.G. Hale. 1962. Third generation pink salmon, *Oncorhynchus gorbuscha* (Walbaum), in Lake Superior. Trans. Am. Fish. Soc. 91:421-422

Scofield, N.B. 1916. The humpback and dog salmon taken in San Lorenzo River. Calif. Fish Game 2: 41 p.

Seeb, J., and L. Wishard. 1977. Genetic characterization of Prince William Sound pink salmon populations. Prepared for Alaska Department of Fish and Game by Pacific Fisheries Research. 21 p.

Semko, R.S. 1954. The stocks of west Kamchatka salmon and their commerical utilization. Izv. Tikhookean. Nauchno- Issled. Inst. Rybn. Khoz. Okeanogr. 41:3-109. (Transl. from Russian; Fish. Res. Board Can. Transl. Ser. 288)

Senn, H., and R.M. Buckley. 1978. Extended freshwater rearing of pink salmon at a Washington hatchery. Prog. Fish-Cult. 40:9-10

Seymour, A.H. 1956. Effects of temperature upon chinook salmon. Ph.D. thesis. University of Washington, Seattle, WA. 127 p.

Sheridan, W.L. 1960. Frequency of digging movements of female pink salmon before and after egg deposition. Anim. Behav. 8:228-230

• ——. 1962a. Relation of stream temperatures to timing of pink salmon escapements in southeast Alaska, p. 87-102. *In*: N.J. Wilimovsky (ed.). Symposium on Pink Salmon. H.R. MacMillan Lectures in Fisheries. Institute of Fisheries, University of British Columbia, Vancouver, BC

——. 1962b. Waterflow through a salmon spawning riffle in southeastern Alaska. U.S. Fish Wildl. Serv. Spec. Sci. Rep. Fish. 407:20 p.

Shershnev, A.P., V.M. Chupakhin, and V.A. Rudnev. 1982. Ecology of juvenile pink salmon, *Oncorhynchus gorbuscha* (Salmonidae), from Sakhalin and Iturup islands during the marine period of life. J. Ichthyol. 22:90-97

Shershnev, A.P., and A.I. Zhul'kov. 1979. Features of the downstream migration of young pink salmon and some indices of the efficiency of reproduction of the pink salmon, *Oncorhynchus gorbuscha*, from Pritornaya River. J. Ichthyol. 19:114-119

Shimazaki, K., and S. Mishima. 1969. On the diurnal changes of the feeding activity of salmon in the Okhotsk Sea. Bull. Fac. Fish. Hokkaido Univ. 20:82-93. (In Japanese, English abstract)

Shmidt, P.Yu. 1950. Fishes of the Sea of Okhotsk. Tr. Tikhookean. Kom. Akad. Nauk SSSR 6:392 p. (Transl. from Russian: Israel Program for Scientific Translations, Jerusalem, 1965)

Shuman, R.F. 1950. On the effectiveness of spermatozoa of the pink salmon (*Oncorhynchus gorbuscha*) at varying distances from point of dispersal. Fish. Bull. Fish Wildl. Serv. 51:359-363

Shuntov, V.P. 1989. Distribution of juvenile Pacific salmon, *Oncorhynchus*, in the Okhotsk Sea and adjoining Pacific waters. J. Ichthyol. 29:155-164

Simenstad, C.A., W.J. Kinney, S.S. Parker, E.O. Salo, J.R. Cordell, and H. Buechner. 1980. Prey community structure and trophic ecology of outmigrating juvenile chum and pink salmon in Hood Canal, Washington: a synthesis of three years' studies, 1977-1979: final report. Univ. Wash. Fish. Res. Inst. FRI-UW-8026:113 p.

Simon, R.C. 1963. Chromosome morphology and

species evolution in the five North American species of Pacific salmon *(Oncorhynchus)*. J. Morphol. 112:77-97

Simon, R.C., and R.E. Noble. 1968. Hybridization in *Oncorhynchus* (Salmonidae). I. Viability and inheritance in artificial crosses of chum and pink salmon. Trans. Am. Fish. Soc. 97:109-118

Skud, B.E. 1958. Relation of adult pink salmon size to time of migration and freshwater survival. Copeia 1958:170-176

Smirnov, A.G. 1947. Condition of stocks of the Amur salmon and causes of the fluctuations in their abundance. Izv. Tikhookean. Nauchno-Issled. Inst. Rybn. Khoz. Okeanogr. 25:33-51. (Transl. from Russian; *In*: Pacific salmon: selected articles from Soviet periodicals, p. 66-86. Israel Program for Scientific Translations, Jerusalem, 1961)

Smirnov, A.I. 1953. Some characteristics of the interspecific hybrid between autumn chum salmon and pink salmon (*Oncorhynchus keta* (Walbaum) infraspecies *autumnalis* Berg x *O. gorbuscha* (Walbaum), family Salmonidae). Dokl. Akad. Nauk. SSSR 91:409-412. (Transl. from Russian; Fish. Res. Board Can. Transl. Ser. 957)

——. 1954. The effect of mechanical agitation on developing eggs of the pink salmon *Oncorhynchus gorbuscha* (Walbaum). Salmonidae. Dokl. Akad. Nauk SSSR 97:365-368. (Transl. from Russian; Fish. Res. Board Can. Transl. Ser. 231)

——. 1959. The functional importance of the prespawning changes in the skin of salmon (as exemplified by the genus *Oncorhynchus*). Zool. Zh. 38:731-741. (Transl. from Russian; Fish. Res. Board Can. Transl. Ser. 348)

——. 1963. The fertilization of the eggs and spermatozoa of pink salmon *(Oncorhynchus gorbuscha* (Walbaum)) when held in water. Nauchn. Dokl. Vyssh. Shk. Biol. Nauk 3:37-41. (Transl. from Russian; Fish. Res. Board Can. Transl. Ser. 544)

——. 1975. The biology, reproduction and development of the Pacific salmon. Izdatel'stvo Moskovskogo Universiteta, Moscow, USSR. (Transl. from Russian; Fish. Res. Board Can. Transl. Ser. 3861)

Smirnov, A.I., and M.S. Kamyshnaya. 1965. The biology of young pink salmon in relation to certain questions of their propagation and acclimatization. Zool. Zh. 44:1813-1824. (Transl. from Russian; Fish. Res. Board Can. Transl. Ser. 782)

Smith, H.D., A.H. Seymour, and L.R. Donaldson. 1966. The salmon resource, p. 861-876. *In*: N.J. Wilimovsky and J.N. Wolfe (eds.). Environment of the Cape Thompson region, Alaska. United States Atomic Energy Commission, Division of Technical Information, Oak Ridge, TN

Smith, S. 1957. Early development and hatching, p. 323-359. *In*: M.E. Brown (ed.). The physiology of fishes. Vol. 1: Metabolism. Academic Press, New York, NY

Smoker, W.W. 1984. Genetic effect on the dynamics of a model of pink *(Oncorhynchus gorbuscha)* and chum salmon *(O. keta)*. Can. J. Fish. Aquat. Sci. 41:1446-1453

Soin, S.G. 1953. The development of unfertilized salmon eggs. Rybn. Khoz. 5:55-58 (Transl. from Russian; Fish. Res. Board Can. Transl. Ser. 4557)

——. 1954. Pattern of development in summer chum, masu and pink salmon. Tr. Soveshch. Ikhtiol. Kom. Akad. Nauk SSSR 4:144-155. (Transl. from Russian; *In*: Pacific salmon: selected articles from Soviet periodicals. Israel Program for Scientific Translations, Jerusalem, 1961)

——. 1956. Respiratory significance of the carotinoid pigment in the eggs of salmonids and other representatives of the Clupeiformes. Zool. Zh. 35:1362-1369. (Transl. from Russian; Fish. Res. Board Can. Transl. Ser. 4538)

Solomon, D.J. 1973. Evidence for pheromone-influenced homing by migrating Atlantic salmon *Salmo salar* (L.). Nature 244(5413):23

Solovjova, V.K. 1976. Acclimatization of pink salmon in the Baltic Sea. Rev. Trav. Inst. Peches. Marit. 40:747

Stabell, O.B. 1984. Homing and olfaction in salmonids: a critical review with special reference to the Atlantic salmon. Biol. Rev. 59:338-388

Stepanov, A.S., A.V. Churmasov, and S.A. Cherkashin. 1979. Migration direction finding by chum [sic] salmon according to the sun. Sov. J. Mar. Biol. 5:92-99

Stuart, T.A. 1953. Spawning migration, reproduction and young stages of loch trout *(Salmo trutta* L.). Freshwater Salmon Fish. Res. 5: 39 p.

Sullivan, C.R. 1980. Forests and fishes in southeastern Alaska. Fisheries 5:2-8

Synkova, A.I. 1951. Food of Pacific salmon in Kamchatka water. Izv. Tikhookean. Nauchno-Issled. Inst. Rybn. Khoz. Okeanogr. 34:105-121. (Transl. from Russian; *In*: Pacific salmon: selected articles from Soviet periodicals, p. 216–235. Israel Program for Scientific Translations, Jerusalem, 1961)

Taft, A.C. 1938. Pink salmon in California. Calif. Fish Game 24:197–198

Tagmaz'yan, Z.I. 1971. Relationship between the density of the downstream migration and predation of young pink salmon (*Oncorhynchus gorbuscha* (Walb.)). J. Ichthyol. 11:984–987

Tait, J.S. 1960. The first filling of the swim bladder in salmonids. Can. J. Zool. 38:179–187

Takagi, K. 1971. Information on the catchable time period for Pacific salmon obtained through simultaneous fishing by longlines and gillnets. Bull. Far Seas Fish. Res. Lab. 5:177–194. (In Japanese, English summary)

Takagi, K., K.V. Aro, A.C. Hartt, and M.B. Dell. 1981. Distribution and origin of pink salmon *(Oncorhynchus gorbuscha)* in offshore waters of the North Pacific Ocean. Int. North Pac. Fish. Comm. Bull. 40:195 p.

Takeuchi, I. 1972. Food animals collected from the stomachs of three salmonid fishes *(Oncorhynchus)* and their distribution in the natural environments in the northern North Pacific. Bull. Hokkaido Reg. Fish. Res. Lab. 38:1–119. (In Japanese, English summary)

Tautz, A.F., and C. Groot. 1975. Spawning behavior of chum salmon *(Oncorhynchus keta)* and rainbow trout *(Salmo gairdneri)*. J. Fish. Res. Board Can. 32:633–642

Taylor, S.G. 1980. Marine survival of pink salmon fry from early and late spawners. Trans. Am. Fish. Soc. 109:79–82

——. 1983. Vital statistics on juvenile and adult pink and chum salmon at Auke Creek, northern southeastern Alaska. Auke Bay Lab., U.S. Natl. Mar. Fish. Serv. MS Rep.-File MR-F 152:35 p.

Terhune, L.D.B. 1958. The Mark VI groundwater standpipe for measuring seepage through salmon spawning gravel. J. Fish. Res. Board Can. 15:1027–1063

Thorpe, J.E. 1982. Migrations in salmonids with special reference to juvenile movements in freshwater, p. 86–97. *In*: E.L. Brannon and E.O. Salo (eds.). Proceedings of the Salmon and Trout Migratory Behavior Symposium. School of Fisheries, University of Washington, Seattle, WA

Thorsteinson, F.V. 1959. Pink salmon migrations along the Alaska Peninsula, p. 1–7. *In*: Alaska fisheries briefs. U.S. Fish Wildl. Serv. Circ. 59

——. 1962. Herring predation on pink salmon fry in a southeastern Alaska estuary. Trans. Am. Fish. Soc. 91:321–323

Thorsteinson, F.V., J.H. Helle, and D.G. Birkholz. 1971. Salmon survival in intertidal zones of Prince William Sound streams in uplifted and subsided areas, p. 194–219. *In*: The great Alaska earthquake of 1964: biology. Natl. Acad. Sci. NAS Publ. 1604

Tinbergen, N. 1951. The study of instinct. Oxford University Press, London. 228 p.

Todd, I., and P.A. Larkin. 1971. Gillnet selectivity on sockeye *(Oncorhynchus nerka)* and pink salmon *(O. gorbuscha)* of the Skeena River system, British Columbia. J. Fish. Res. Board Can. 28:821–842

Triplett, E., and J.R. Calaprice. 1974. Changes in plasma constituents during spawning migration of Pacific salmons. J. Fish. Res. Board Can. 31: 11–14

Turner, C.E., and H.T. Bilton. 1968. Another pink salmon *(Oncorhynchus gorbuscha)* in its third year. J. Fish. Res. Board Can. 25:1993–1996

Ueno, M. 1968. Food and feeding behavior of Pacific salmon. I: the stratification of food organisms in the stomach. Bull. Jpn. Soc. Sci. Fish. 34:315–318

Ueno, M., S. Kosaka, and H. Ushiyama. 1969. Food and feeding behavior of Pacific salmon. II. Sequential change of stomach contents. Bull. Jpn. Soc. Sci. Fish. 35:1060–1066

Utter, F.M., D. Campton, S. Grant, G. Milner, J. Seeb, and L. Wishard. 1980. Population structures of indigenous salmonid species of the Pacific northwest, p. 285–304. *In*: W.J. McNeil and D.C. Himsworth (eds.). Salmonid ecosystems of the North Pacific. Oregon State University Press, Corvallis, OR

Vallion, A.C., A.C. Wertheimer, W.R. Heard, and R.M. Martin. 1981. Summary of data and research pertaining to the pink salmon population at Little Port Walter, Alaska, 1964–80. NWAFC Processed Rep. 81-10:102 p.

Vasilenko-Lukina, O.V. 1962. On the biology of Primorsky pink salmon, *Oncorhynchus gorbuscha* (Walbaum). Vopr. Ikhtiol. 2:604–608. (Transl. from Russian; Univ. Wash. Fish. Res. Inst. FRI Circ. 197)

Vaux, W.G. 1968. Intragravel flow and interchange of water in a streambed. Fish. Bull. (U.S.) 66:479–489

Vedensky, A.P. 1954. Age of pink salmon and the pattern of their fluctuations in abundance. Izv. Tikhookean. Nauchno- Issled. Inst. Rybn. Khoz. Okeanogr. 41:111–195. (Transl. from Russian; Israel Program for Scientific Translations, Jerusalem)

Velsen, F.P.J. 1980. Embryonic development in eggs of sockeye salmon, *Oncorhynchus nerka*. Can. Spec. Publ. Fish. Aqua. Sci. 49:17 p.

Vernon, E.H. 1958. An examination of factors affecting the abundance of pink salmon in the Fraser River. Int. Pac. Salmon Fish. Comm. Prog. Rep. 5:49 p.

——. 1962. Pink salmon populations of the Fraser River system, p. 53–58. *In*: N.J. Wilimovsky (ed.). Symposium on Pink Salmon. H.R. MacMillan Lectures in Fisheries. Institute of Fisheries, University of British Columbia, Vancouver, BC

——. 1966. Enumeration of migrant pink salmon *(Oncorhynchus gorbuscha)* fry in the Fraser River estuary. Int. Pac. Salmon Fish. Comm. Bull. 19: 83 p.

Vernon, E.H., A.C. Hourston, and G.A. Holland. 1964. The migration and exploitation of pink salmon runs in and adjacent to the Fraser River Convention Area in 1959. Int. Pac. Salmon Fish. Comm. Bull. 15:296 p.

Volovik, S.P. 1966. On releasing young pink and chum salmon from fish hatcheries. Rybn. Khoz. 42(4):13–14. (Transl. from Russian; Fish. Res. Board Can. Transl. Ser. 713)

Volovik, S.P., and O.F. Gritsenko. 1970. Effects of predation on the survival of young salmon in the rivers of Sakhalin, p. 193–209. *In*: Biological foundations of the fishing industry and regulations of marine fisheries. Tr. Vses. Nauchno Issled. Inst. Morsk. Rybn. Khoz. Okeanogr. 71. (Transl. from Russian; Fish. Res. Board Can. Transl. Ser. 1716)

Wagner, W.C., and T.M. Stauffer. 1980. Three-year-old pink salmon in Lake Superior tributaries. Trans. Am. Fish. Soc. 109:458–460

Walker, C.E., and D.B. Lister. 1971. Results for three generations from transfers of pink salmon *(Oncorhynchus gorbuscha)* spawn to the Qualicum River in 1963 and 1964. J. Fish. Res. Board Can. 28:647–654

Ward, F.J. 1959. Character of the migration of pink salmon to Fraser River spawning grounds in 1957. Int. Pac. Salmon Fish. Comm. Bull. 10:70 p.

Wells, R.A., and W.J. McNeil. 1970. Effect of quality of the spawning bed on growth and development of pink salmon embryos and alevins. U.S. Fish Wildl. Serv. Spec. Sci. Rep. Fish. 616:6 p.

Wertheimer, A. 1980. Evidence of compensatory mortality during the marine life of Sashin Creek pink salmon, p. 93–107. *In*: A.P. Kingsbury (ed.). Proceedings of the 1980 Northeast Pacific Pink and Chum Salmon Workshop. Alaska Department of Fish and Game, Anchorage, AK

Wickett, W.P. 1954. The oxygen supply to salmon eggs in spawning beds. J. Fish. Res. Board Can. 11:933–953

——. 1957. The development of measurement and quality standards for water in the gravel of salmon spawning streams. Proc. Alaska. Sci. Conf. 8:95–99

——. 1958a. Adult returns of pink salmon from the 1954 Fraser River planting. Fish. Res. Board Can. Prog. Rep. Pac. Coast Stn. 111:18–19

——. 1958b. Review of certain environmental factors affecting the production of pink and chum salmon. J. Fish. Res. Board Can. 15:1103–1126

——. 1959a. Observations on adult pink salmon behaviour. Fish. Res Board Can. Prog. Rep. Pac. Coast Stn. 113:6–7

——. 1959b. Effects of siltation on success of fish spawning, p. 16–17. *In*: Proceedings of the Fifth Symposium, Pacific Northwest, on Siltation its Sources and Effects on the Aquatic Environment. U.S. Public Health Service, Portland, OR

——. 1959c. Note on the behaviour of pink salmon fry. Fish Res. Board Can. Prog. Rep. Pac. Coast Stn. 113:8–9

——. 1962. Environmental variability and reproduction potentials of pink salmon populations in British Columbia, p. 73–86. *In*: N.J. Wilimovsky (ed.). Symposium on Pink Salmon. H.R. MacMillan Lectures in Fisheries. Institute of Fisheries, University of British Columbia, Vancouver, BC

——. 1975. Mass transfer theory and the culture of fish eggs, p. 419–434. *In*: W.A. Adams, G. Greer, G.S. Kell, J.E. Desnoyers, K.B. Oldham, G. Atkinson, and J. Walkley (eds.). Chemistry and physics of aqueous gas solutions. Electrochemical Society, Princeton, NJ

Willette, T.M. 1985. The effect of ocean temperatures on the survival of the odd- and even-year pink salmon (*Oncorhynchus gorbuscha*) originating from Prince William Sound, Alaska. M.Sc. thesis. University of Alaska, Fairbanks, AK. 115 p.

Williams, I.V., and P. Gilhousen. 1968. Lamprey parasitism on Fraser River sockeye and pink salmon during 1967. Int. Pac. Salmon Fish. Comm. Prog. Rep. 18:22 p.

Williamson, R.B. 1974. Futher captures of Pacific salmon in Scottish waters. Scott. Fish. Bull. 41:28–30

Withler, F.C. 1982. Transplanting Pacific salmon. Can. Tech. Rep. Fish. Aquat. Sci. 1079:27 p.

Withler, F.C., and R.B. Morley. 1982. Use of milt from on-year males in transplants to establish off-year pink salmon (*Oncorhynchus gorbuscha*) runs. Can. Tech. Rep. Fish. Aquat. Sci. 1139:49 p.

Woodhead, P.M.J. 1957. Reactions of salmonid larvae to light. J. Exp. Biol. 34:402–416

Yastrebkov, A.A. 1966. Effect of egg size upon size and growth rate of pink salmon larvae. Tr. Murm. Morsk. Biol. Inst. 12 (16):45–53.. (Transl. from Russian; Fish. Res. Board Can. Transl. Ser. 1822)

Yefanov, V.N., and V.M. Chupakhin. 1982. The dynamics of some population indices of pink salmon, *Oncorhynchus gorbuscha*, from Iturup (Kuril Islands). J. Ichthyol. 22:42–50

Zotin, A.I. 1958. The mechanism of hardening of the salmonid egg membrane after fertilization or spontaneous activation. J. Embryol. Exp. Morphol. 6:546–568

Life History of Chum Salmon

CONTENTS

HEINE

12

PLATE 12. Chum salmon fry. *Photograph by S.L. Schroder*

13

PLATE 13. Adult chum salmon during spawning act, Big Beef Creek, Washington. *Photograph by G. Duker*

PLATE 11 (*previous page*). Chum salmon life history stages. *Painting by H. Heine*

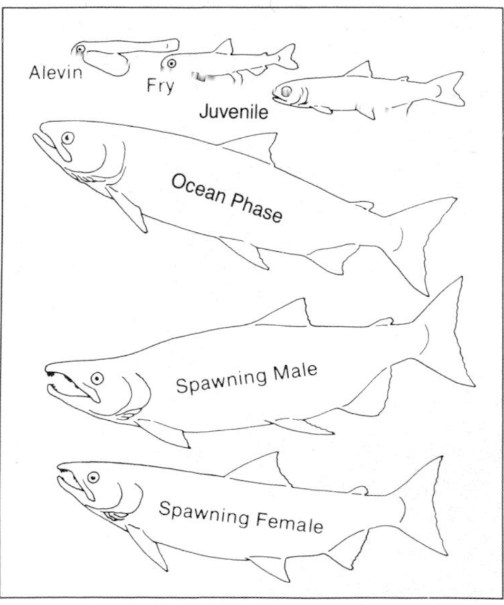

LIFE HISTORY OF CHUM SALMON

(*Oncorhynchus keta*)*

E.O. Salo†

Introduction

CHUM SALMON, *Oncorhynchus keta* (Walbaum), have the widest natural geographic distribution of all Pacific salmon species (Bakkala 1970; Fredin et al. 1977), ranging in Asia from Korea to the Arctic coast of the USSR and west to the Lena River (Laptev Sea), and in North America from Monterey, California, to the Arctic coast and east to the Mackenzie River (Beaufort Sea). Historically, they may have constituted up to 50% of the annual biomass of the seven species of Pacific salmon in the North Pacific Ocean.

Chum salmon are semelparous and anadromous. They spawn successfully in streams of various sizes, and the fry migrate directly to the sea soon after emergence. The immatures distribute themselves widely over the North Pacific Ocean, and the maturing adults return to the home streams at various ages, usually at two through five years, and in some cases at up to seven years (Bigler 1985). All die after spawning. With individuals reported to be up to 108.8 cm in length and 20.8 kg in weight (Anonymous 1928), chum is second only to chinook salmon (*O. tshawytscha*) in size. Spawning fish are characterized by the calico nuptial coloration, particularly evident in the dominant males, and the metamorphosis of the head with its prominent canine-like teeth. The eggs are comparatively large and the alevins are large and mobile.

The valid scientific name for chum salmon is *Oncorhynchus keta* (Walbaum) (Jordan and Gilbert 1882), and the type specimen was described by Walbaum (1792) from the Kamchatka River under the name *Salmo keta*. The derivation of the word "keta" is from the language of the Nanai, who live in the Khabarovsk and Primore regions of the USSR and between the Sungari and Ussuri rivers of the People's Republic of China. This language is a subdialect of the Amur people, and "keta" literally means fish. Vernacular names include dog salmon and calico salmon in the United States and Canada, and there are at least nine names, varying among and within areas, in the USSR. In Japan more than ten provincial names are used for chum salmon, with the name "gila" reserved for the late run of bright silvery fish with deciduous scales.

Tchernavin (1939), Hoar (1958), and Neave (1958) assumed that the family Salmonidae had a freshwater origin and that the Pacific salmon species diverged from the trout genus *Salmo*, with the main trend in evolution towards greater adaptation to marine life. In this view, masu (*O. masou*) and coho (*O. kisutch*) salmon are closer to the ancestral form than chum and pink salmon (*O. gorbuscha*), both of which migrate to sea shortly after emergence. This hypothesis has been supported by others (Tsuyuki and Roberts 1966). Behnke (1979) concluded that the genus *Oncorhynchus* was derived from the evolutionary line leading to the subgenus *Parasalmo* after its divergence from the other *Salmo* species group. This places the various Pacific salmon and the western trout species closer to each other than either of them are to the brown trout (*Salmo trutta*) or Atlantic salmon (*S. salar*).

*Contribution 814, School of Fisheries, University of Washington, Seattle, Washington 98115

†Fisheries Research Institute, University of Washington, Seattle, Washington 98195

Recently, the western trouts have been removed from the genus *Salmo* and placed in the genus *Oncorhynchus* (Smith and Stearly 1989).

Utter et al. (1973) and Miller and Brannon (1982) concluded that chum salmon are not as highly specialized as either pink salmon or sockeye salmon (*O. nerka*). This accounts for the chum salmon's more versatile behaviour in both freshwater and marine environments. This versatility is limited because no freshwater residents or landlocked forms have been reported; however, chum have been reared in captivity to maturity in fresh water (R.L. Burgner, Fisheries Research Institute, University of Washington, Seattle, Washington, pers. comm.).

Day (1887), Regan (1911), and Thorpe (1982), on the other hand, considered the Salmonidae to have a marine origin and that its evolutionary development has been towards greater freshwater adaptation. Thorpe (1982) argued that the occurrence of landlocked populations provides evidence of evolutionary advancement in juvenile life histories (juvenilization) and that the trend in salmonids appears to be away from dependence on the sea. In this respect, the least advanced species among the Pacific salmon would be the chum salmon and the closely related pink salmon, which both have short freshwater and extensive marine life stages. The arguments presented so far in the literature appear to favour a freshwater origin of Pacific salmon; thus, chum salmon can be considered one of the more advanced species among the Pacific salmon.

Common to virtually every region of the chum salmon's area of distribution is the occurrence of early and late returning stocks to the natal stream. Berg (1934) separated Asian chum salmon into seasonal races - summer and autumn - and classified "autumn chums" as the infraspecies *autumnalis*. His justifications for separating autumn from summer chum salmon were (1) later entrance into spawning streams, (2) less developed reproductive products at time of entry into these streams, (3) a later spawning period, (4) larger size, and (5) greater fecundity. Although Berg's classification has been supported by other investigators (Lovetskaya 1948; Grigo 1953; Birman 1956; Hirano 1958; Sano 1966), it has not been widely used. In North America the only true summer chum salmon may be in the Yukon River, where the summer chum have the distinguishing characteristics of the Asian summer chum. From western Alaska south to British Columbia and Washington, there are runs referred to as "summer" chum, which spawn from June to early September; these chum are characterized by large body size, older age composition, and high fecundity, and are probably early autumn chum (T. Beacham, Pacific Biological Station, Nanaimo, British Columbia, Canada, pers. comm.).

In general, early-run spawners spawn in main stems of streams, while the late spawners seek out spring water that has more favourable temperatures through the winter. The timing of the runs varies from north to south, as does age at maturity and absolute (and, probably, relative) fecundity. This temporal and spatial partitioning may have originated in stocks spawning in inland streams as opposed to those spawning in coastal or island streams. In recent times, the early- and late-running stocks have adapted to rivers with appropriate characteristics, regardless of geographical location of the river basin. In this study, the summer and autumn chum runs are considered as different stocks which vary in a number of morphological, physiological, and behavioural characteristics.

Distribution of Spawning Stocks

Range of Spawning Stocks

On the Asian continent, spawning chum salmon range from the Naktong River in Korea and the Nagasaki and Fukuoka prefectures of Kyushu Island of Japan (Atkinson et al. 1967) in the south, to the rivers emptying from Siberia into the Arctic Ocean as far west as the Lena River (Laptev Sea) in the north (Figure 1). Historically, chum salmon were also present in the Komandorskiy Islands (Smirnov 1975) and the area of present-day northeastern China (K. Chew and L. Donaldson, School of Fisheries, University of Washington, Seattle, Washington, pers. comm.).

The southernmost mainland Asian runs of commercial importance are in the Amur River. Exploitable runs exist on Sakhalin Island, the Kuril Islands, and the continental streams emptying into the Sea of Okhotsk. Chum salmon are also abundant on the Kamchatka Peninsula as well as in the Anadyr River, which flows into the Bering Sea.

On the North American continent, chum salmon range from the San Lorenzo River in Monterey, California, in the south (Scofield 1916), to Arctic coast streams in the north, as far east as the Mackenzie River system (Dymond 1940; Wynne-Edwards 1952), and west to Attu Island in the Aleutian Islands (Holmes 1982) (Figure 1). Atkinson et al. (1967) noted that chum and pink salmon occur in the Colville area (Beaufort Sea), which is approximately 71°N latitude and 152°W longitude, and that they probably also spawn in the Tunagoruk and the Usuktuk rivers, which are at approximately the same latitude as the Lena River in Asia. The northernmost large runs of commercial importance are in Kotzebue Sound (Noatak and Kobuk rivers). Substantial runs have been reported for the Yukon River by Buklis and Barton (1984), where spawning takes place as far as 2,800 km from the sea. Runs of significant numbers occur from Kotzebue Sound to Tillamook Bay, Oregon, in streams that range greatly in size (Henry 1953, 1954). As recently as the 1940s chum salmon were abundant in the Columbia River; however, in 1982 the total run was only about 1,000 fish. Hallock and Fry (1967) reported spawning populations of chum salmon in the Sacramento River as far as 322 km upstream, but, at present, they are

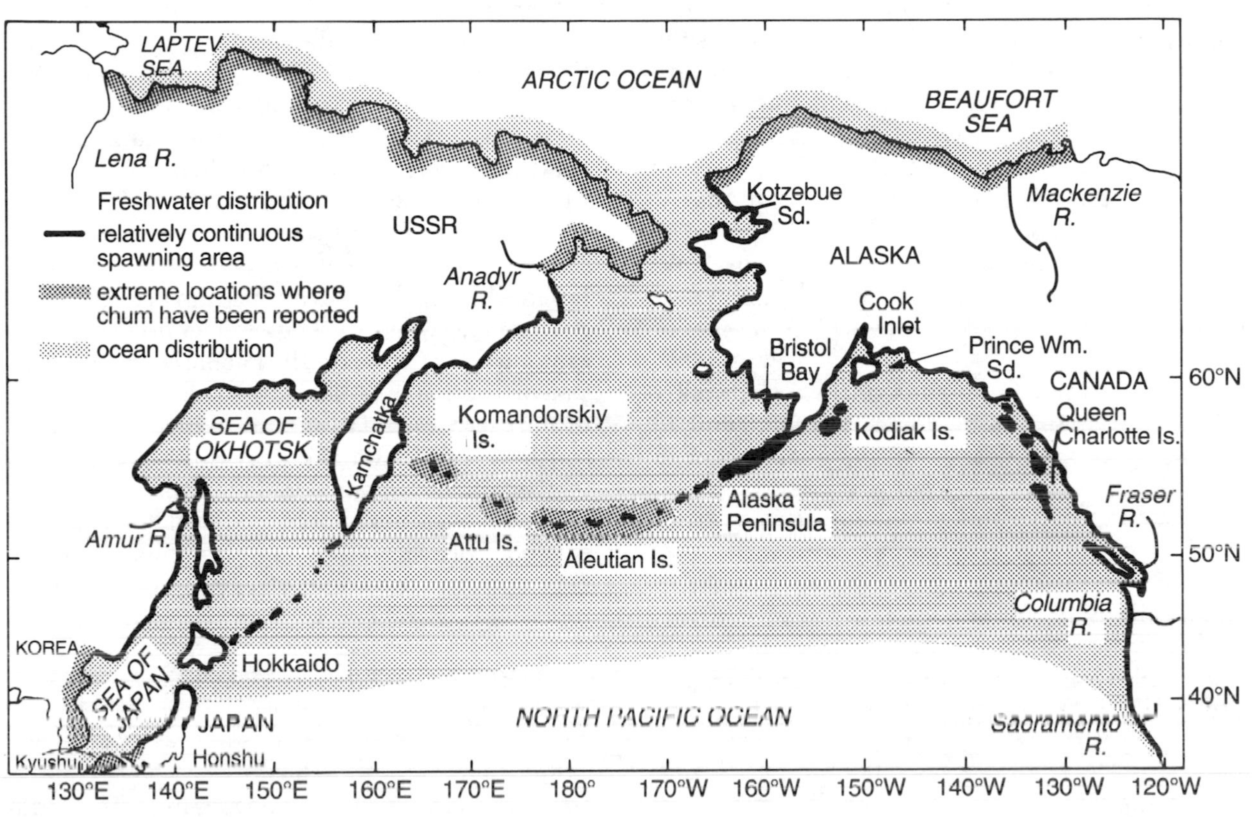

FIGURE 1

Generalized freshwater and ocean distribution of chum salmon. (From Neave et al. 1976)

only occasionally seen in northern California.

Distribution and Run Timing in Asia

There are two major groups of chum salmon in Asia: the summer chum, native to Kamchatka, the Okhotsk coast, the Amur River, and the east coast of Sakhalin; and the autumn chum, native to Japan, the west coast of Sakhalin, the southern Kuril Islands, and the Amur River. The Amur River is the only major river with both summer and autumn chum. Generally, the northern runs migrate upstream in June, July, and August, with the peak of the runs occurring progressively later farther south. The runs are principally in July and August in Kamchatka, in September and October in Sakhalin, in September through November in Hokkaido, and during October and November in Honshu.

USSR. The chum salmon runs of the Olyutorskiy and Anadyr districts begin in June and end in September in most years, and the peaks occur in July and August.

Chum salmon ascend to spawn in west Kamchatka from June to September, with peak runs occurring in July in some years and August in others. The major chum salmon streams in this region are the Bolshaya, Icha, and Kikhchik rivers (Semko 1954), all of which are in the southern half of the peninsula. Although there are about eighteen large streams on the west coast of Kamchatka, more than 80% of the chum salmon catch is from six streams from the Icha River area and southward (Sano 1966). The twelve streams north of the Icha River, which have an average stream length of 210 km, produce relatively small numbers of chum.

The migration time in the Okhotsk District is similar to that in western Kamchatka (June to September), and more than 80% of the chum salmon arrive in August. The major runs are in the Okhota and Kukhtuy rivers.

Summer chum occur in the Amur River and on the east coast of Sakhalin. Autumn chum are also present in the Amur River and along the west coast of Sakhalin, as well as in the Primore region (Sano 1966). Summer chum migrate upstream shortly after their appearance in coastal waters in July and August. Autumn chum migrate from August to early October, with about 90% of autumn fish appearing in coastal waters during the month of September. The Amur populations are more numerous than those in other areas of this region, although the summer runs have become depressed to a level too low to support fisheries exploitation.

There are two peaks in the abundance of chum salmon in the nearshore waters of Sakhalin and the Kuril islands, one in July and August, and the other in September and October. The former group is not large and is probably destined mainly for the Amur River and the Okhotsk District, although there are a few summer chum populations in the Tym and Poronai rivers of Sakhalin. The autumn chum ascend the rivers shortly after their appearance in the nearshore waters and are considered to be related to those of northern Japan (Hirano 1953).

Japan. Historically, autumn chum occurred along the east and west coasts of Hokkaido and Honshu ranging from Chiba on the Pacific coast and Nagasaki on the Sea of Japan side. Again, the migrations of northern populations are a little earlier than southern ones, and the peak runs occur in September and October in Hokkaido and mostly October and November in areas south of Hokkaido. The existence of a late-arriving (mid-January and early February) run to various streams in Japan was also reported by Sano (1964).

Distribution and Run Timing in North America

Alaska. The large runs of chum salmon in Kotzebue Sound support the northernmost commercial chum salmon fishery and occur in two modes. The migration into the Kobuk River, which receives 25% of the run, is principally in July, and the run into the Noatak River, which produces 75% of the escapement, is in August (Bigler and Burwen 1984).

Summer chum salmon enter the Yukon River in early May, and the run overlaps with the autumn chum salmon run in June and July, although, for management purposes, 15 July is considered the end of the summer run. The autumn chum spawn from September through November in spring-fed streams and sloughs. Summer chum are more abundant, not as large, and arrive in fuller nuptial coloration than the autumn chum (Buklis 1981; Buklis and Barton 1984).

Chum salmon in the Kuskokwim River arrive in

late August and September, and the major runs in the Togiak, Nushagak, and Kvichak rivers in Bristol Bay begin in mid- to late June, peak the first week of July, and end in late July (D. Rogers, Fisheries Research Institute, University of Washington, Seattle, Washington, pers. comm.).

Small runs of chum salmon enter numerous streams of the Alaska Peninsula in July and August, especially on the south side of the peninsula. At the same time, substantial runs occur on the north side of Kodiak Island and lesser runs on the south side. There are small runs in the Aleutian Islands chain, at least as far west as Attu Island (Holmes 1982).

In central Alaska, a substantial run of chum salmon occurs in the Susitna River and in about fifteen short streams along the northern portion of Cook Inlet (Atkinson et al. 1967). The peaks of the runs are primarily in late July. In southern Cook Inlet, the runs occur in July and August in the Kenai River and in streams north of Kachemak Bay to Resurrection Bay. Numerous runs enter Prince William Sound destined for Port Wells, the Valdez Arm, and Port Fidalgo areas, as well as the lesser arms and bays.

For chum salmon native to southeastern Alaska, the peak of nearshore abundance was established, for 1984, as the first two weeks in August (Clark and Weller 1986). The mid-point of the catches varied from 5 August in northern areas to 19 August in the south. The median of the escapements varied from late August to mid-September (Clark and Weller 1986).

Although many runs remain to be catalogued for southeastern Alaska, information on abundance and timing is available for some. In the Yakutat area, chum salmon spawn in the East River (near the mouth of the Alsek River) in October (J.H. Helle, National Marine Fisheries Service, Auke Bay, Alaska, pers. comm.). An exceptionally late and large autumn run occurs in the Chilkat River (near Haines) where unusual upwellings of warm water keep portions of the river ice-free throughout the winter (Cline 1982). This phenomenon has provided sustenance for up to 3,500 bald eagles through the winter. Runs are not numerous in the Icy Strait area except for one large run in Excursion Inlet (July and August). Smaller runs in August and September are present in the eastern part of southeastern Alaska, particularly in Stephens Passage, but more numerous and larger runs occur in the western district of southeastern Alaska. Several large runs arrive in August and September. There are numerous small runs in the Wrangell district.

Farther down the coast, many small runs spawn from July to September in the mainland streams and in the island streams all the way to the Ketchikan area (J.H. Helle, National Marine Fisheries Service, Auke Bay, Alaska, pers. comm.). Runs can be particularly strong on Prince of Wales Island where spawning occurs mainly in September and October. Recently, the runs in southeastern Alaska have been considerably enhanced by releases of large numbers of juveniles from private and public aquaculture enterprises.

British Columbia. In British Columbia, chum salmon spawn in over 800 streams. The most productive 58 streams produce only 50% of the total and less than 13 have large runs (Aro and Shepard 1967).

Runs in the northern part of British Columbia are earlier than those in the south. In the Queen Charlotte Islands, fisheries adjacent to the spawning grounds are in August and September, whereas chum salmon bound for streams on the northern mainland pass through the inshore fishing areas mainly in July and August. Along the central part of the British Columbia coast, peak catches are made in August, mostly in the Bella Bella and Bella Coola areas.

In the south, spawning takes place principally from October to January: October in the Chehalis area (Fraser River) and November to January in the main stem of the Fraser, Chilliwack, Vedder, and Harrison rivers. Peak spawning in streams of Johnstone Strait and the Strait of Georgia varies between early October in the northern rivers (Knight Inlet) and late December in some southern streams (Cowichan River). The peak of the catches north of Vancouver Island ranges from mid July to mid September, and is in October in southern British Columbia (Beacham 1984b).

Washington and Oregon. The pattern of broad distribution holds for the state of Washington also, although the spawning areas may be relatively farther upstream because of the comparatively numerous moderate to large rivers in the Puget

Sound region. In the north, spawning occurs in the upper Nooksack, central and upper Skagit, Stillaguamish, Skykomish, and Snohomish rivers (Atkinson et al. 1967). Large runs occur in the Skagit River system, the Hood Canal system (where artificial propagation is the primary source), the Nisqually River system, and the Grays Harbor and Willapa Harbor areas. The spawning areas are near salt water in the coastal rivers of the Olympic Peninsula and in the smaller and shorter lowland streams of the Puget Sound area.

There are often variations in timing within the early and late autumn runs. In some Puget Sound streams a more or less distinct "middle run" may occur (Koski 1975). With runs beginning in early September and continuing in some streams as late as March, a spawning period of four to five months is common for southern Puget Sound.

Chum salmon are limited to the lower part (300 km) of the Columbia River, with more runs on the Washington than on the Oregon side. There are October runs in the Washougal, Lewis, Kalama, Cowlitz (Washington), and Sandy (Oregon) river systems. Historically, there were chum salmon in the Toutle River, but none have been seen since the eruption of Mt. St. Helens in 1980. Chum salmon spawn in the Abernathy, Elokomin, and Grays River areas on the Washington side of the Columbia River and nearby in some of the smaller Oregon streams.

Chum salmon populations in the coastal area of Oregon are small. The principal runs enter Tillamook Bay in October and November. Smaller runs of chum salmon are found south of Tillamook Bay at Netarts Bay and in the Nestucca River. Chum salmon are present in very small numbers in the Yaquina and Siuslaw Rivers, and Coos Bay.

SPAWNING MIGRATION

Homing and Straying

The precision of homing and the degree of straying of chum are not well known and only incidental references are available. Returns of adults that were marked as juveniles indicate that the homing tendency of these fish is strong. For two seasons, Salo and Noble (1952) surveyed streams adjacent to and near Minter Creek in Washington and noted no strays, as determined from marked individuals.

For many years, chum salmon fry were released from a nearby hatchery into Walcott Slough, near Brinnon, on Hood Canal, Washington, and the adults returned, apparently unerringly, to a trapping device on the slough, where no natural run existed (Wolcott 1978). At Big Beef Creek, Washington, adult chum salmon return to a trap at the outlet of a spawning channel from which they emigrated as fry, although an alternate trap on the mainstream is available (E.O. Salo, unpublished data).

Tagging of adults near the mouths of streams may give erroneous results if the assumption is made that all fish captured are native to those particular streams; nevertheless, for management purposes these types of studies can be valuable. By tagging mature chum salmon in Skagit Bay (Washington), Eames et al. (1981) estimated a straying rate of 14%. Hiyama et al. (1967) determined that adult chum salmon with their olfactory organs occluded showed no ability to return to their parent stream, whereas fish that were blinded did.

The strong homing tendency of mature chum salmon leads to generally uniform migration patterns from year to year and contributes to stock isolation (Beacham 1984b; Beacham et al. 1985), which, in turn, forms the basis for efficient utilization of the stream by spawning stocks. For example, in the Noatak River the stocks spawning above and below Kelly River differ electrophoretically, suggesting that they are genetically distinct (Davis and Olito 1986).

Rate of Migration during River Approach

Pacific salmon characteristically go through morphological and physiological changes prior to spawning. Specifically, they change from a saltwater to a freshwater physiological state and from a feeding and growing to a reproductive state. Whitish and very mushy meat, called "hottchare" in Japan, is commonly observed in pre-spawning chum salmon, frequently while the salmon are still in the estuary. This deteriorating muscle condition is due to the lysing of proteins (Konagaya 1983). Also, the scales of chum salmon become embedded early in the spawning migration and the integument thickens markedly so that the skin is often marketed as "salmon leather."

The approach by chum salmon to the estuaries of their natal streams is usually fairly rapid, but varies from stream to stream. Lyamin (1949) traced the migration of tagged chum salmon as they approached and ascended the Bolshaya River and determined that they had migrated 1,200 km in 15 days (80 km/d), whereas those approaching another river had travelled only 300 km in 15 days (20 km/d). Shmidt (1947) (quoted by Lyamin 1949) reported that the "maximum speed" of travel during river approach is from 43 to 63 km/d. A travel time of 21 days is used by Anderson and Beacham (1983) for chum salmon on their migration from Johnstone Strait to the mouth of the Fraser River, a distance of 300 km.

Seven chum salmon tagged at the north end of Whidbey Island, Puget Sound, travelled at a mean time of 21 days (range 5–32 d) for distances that varied from 70 to 106 km. In the same year, 59 tagged fish were recaptured in Skagit Bay, a distance of only 15 km, and they had travelled, on the average, seven days (range 1–21 d) (Barker 1979).

Once chum salmon arrive at the mouth of their natal stream they may spend several days "milling" before ascending the stream (Hunter 1959; Koski 1975). The period of milling becomes shorter as the spawning season progresses. Eames et al. (1981) reported that some fish remained in Skagit Bay 21 days after tagging. Usually chum enter the stream when ripe and in full spawning coloration (Fiscus 1969; Koski 1975).

Stimuli for Stream Entry

Once near the mouth of the stream, chum are stimulated to move upstream by an increase in stream runoff of almost any magnitude. However, late in the season, high water is not essential for a timely ascent (Salo and Noble 1952; Hunter 1959).

The "summer" runs, whether truly summer or early autumn runs, respond to high flows caused by spring and summer snow melt, whereas the autumn runs arrive at a time when fall rains occur. Once past the tidal currents chum salmon travel upstream at a slower rate, compared to what Lyamin (1949) described as "their impetuous entry into the river."

The first chum salmon enter Japanese streams when temperatures drop to 15°C and most enter when the temperatures are 10°–12°C. The peak of migration generally occurs when the temperatures range between 7° and 11°C. Helle (1960) noted an absence of chum salmon in a glacially-fed stream in Alaska until the water cleared up, even though spawning was taking place in adjacent streams.

Rate of Stream Migration and Instream Orientation

Chum salmon are large, strong swimmers and are capable of swimming in currents of moderate to high velocities. The maximum swimming speed recorded is 3.05 m/s or 67% of the maximum burst speed of 4.6 m/s (Powers and Orsborn 1985). They are not leapers and usually are reluctant to enter long-span fish ladders. Thus, they are generally found below the first barrier of any significance in a river.

In the Bolshaya River the migration rate is about 14 km/d. In the Anadyr River the first fish arrive on 3–5 July and they migrate at a rate of 40–50 km/d for a period of 10 days. This brings them to the halfway point of the river (Lyamin 1949). As they approach the spawning grounds, their speed increases.

Autumn run chum in the Yukon River migrate close to the river banks (Buklis 1981; Buklis and Barton 1984). The stocks destined for the upper Yukon and Porcupine rivers move mostly along the north bank of the Yukon River in the Galena-Ruby area between the 850 and 930 km section of the river (about halfway to the spawning grounds). The Tanana River stocks, which have a shorter distance to migrate, follow about five days later and swim mostly along the south bank.

Age Composition and Sex Ratio

The age composition of the spawners often varies

over the spawning season. In Big Beef Creek, a lowland stream, the early run has a higher proportion of three-year-olds than the late run (E.O. Salo, unpublished data). Henry (1954) reported that older fish appeared later in the run than younger fish at Tillamook Bay, Oregon. However, in Minter Creek, a lowland Puget Sound stream, the older chum salmon returned before the younger fish (Salo and Noble 1953). Trends similar to that at Minter Creek have been reported for Fraser River and other British Columbian (Beacham and Starr 1982; Beacham 1984a), central Alaskan (Helle 1979), and southeastern Alaskan stocks (Clark and Weller 1986). The runs in these streams are comprised of three-, four-, and five-year-old (or age 0.2, 0.3, and 0.4)[1] chum salmon, with four-year-olds being dominant. The five-year-olds complete their migration earlier than the more numerous four-year-old fish, which are typically well represented throughout the run. The three-year-olds return later than the five-year-olds, and the ratio of three-year-olds to four-year-olds increases to the end of the spawning run.

In general, males predominate early and females late in the run, with the overall ratio of males to females approaching 1:1 for the entire period (Bakkala 1970). Mattson et al. (1964) reported that male to female ratios on the spawning beds of Traitors Creek, Alaska, varied from a high of 3.56:1 to a low of 1:1.34 during a single season. The ratio was generally between 1.25:1 and 1.70:1 during the early part of the run and stabilized to approximately 1:1 during the peak spawning period. In Minter Creek the ratio was 1.7:1 early in the season and equalized during the peak after which females predominated (1:1.2) (Salo and Noble 1953). Semko (1954) reported similar changes in the sex ratios within each age group in the Bolshaya River, USSR. Beacham and Starr (1982) did not observe changes in the sex ratio during the season in the Fraser River chum salmon run. However, they were sampling a mixture of stocks which may have individually exhibited this trait. Later, Beacham (1984a) pooled stocks from southern British Columbia and reported that males were more abundant than females at three and five years of age, but less abundant at four years of age.

Spawning

The spawning behaviour of chum salmon has been described by Sano and Nagasawa (1958), Tautz and Groot (1975), Duker (1977), Helle (1981), and Schroder (1973, 1982). Basically, it consists of a combination of nest-building activities by the female and courtship display by the male, leading to deposition of fertilized eggs in the nest. Immediately following egg deposition, the female fills the nest pocket with gravel and digs a new nest in front of the first one. The females are resident and usually build four to six nests in succession in one place. When all the eggs are buried they defend the redd (the combined pockets of covered eggs) until death. The males are transitory and move from one spawning female to another. Once attracted to a female the male will use physical force to exclude rivals.

Nest Site Selection

Selection of the nest site by the female involves searching for preferred hydrological and geophysical features, such as water odour, depth and velocity, gravel composition, and the presence of cover. Chum salmon prefer to spawn immediately above turbulent areas or where there is upwelling.

The female explores potential nest-building sites by "nosing" with the head pointed down towards the gravel substrate. With pectoral, ventral, and anal fins fully extended, and while moving her head from side to side, she swims slowly upstream.

1 The age designation system used in this chapter is the European formulation used for Atlantic salmon and now widely employed for Pacific salmon. In this system the winters spent in fresh water and salt water are indicated and separated by a decimal point.

After a nosing bout, she usually swims back and digs in the area just passed over (Schroder 1982). Prior to final nest site selection the female may nose and perform exploratory digging over a fairly large area. Once a suitable nest site has been located, digging movements become more concentrated into an increasingly smaller area (Tautz and Groot 1975). Much of this activity may be visually mediated (Duker 1977).

The primary tactic used by the female is to search for an unoccupied space without fighting. Once a spawning territory is established, the female attempts to protect as much space as possible. Territorial evictions of already established females are rare events, even under relatively high (0.6 females/m^2) spawner densities (Schroder 1982). Only weaker, spawned-out females are evicted.

Nest Construction

In digging, the female turns her body on its side and performs a series of four to six flexures, slapping her tail on the gravel substrate. During each dig the pectoral fins are held perpendicular to the body surface and appear to function as brakes to stop her from shooting forward. After a dig, the female normally turns and swims back to the rear part of the nest. Besides turning and circling, the female also "weaves" over the nest in tight circles or figure-eight movements (Schroder 1982).

Digging is initially performed by fanning out from a downstream position of the nest to create a general impression in the gravel of several square metres. As nest construction progresses, digging occurs more and more in the center of the nest. This results in a cone-shaped hollow in the gravel of about 20–40 cm deep, with a porous layer of stones around the bottom portion.

While the female is digging, the male courts her. His principal courtship activities consist of "quivering," which is a quick approach towards the female accompanied by a high frequency low amplitude undulation of the body, and "crossing over," in which the male swims from side to side over the caudal peduncle region of the female (Tautz and Groot 1975). The first courting movement a male performs after locating a female is often a quiver.

After the nest develops a centralized depression, the female lowers her tail and mid-body and "probes" the substrate with her anal fin extended. When contact is made with the gravel she reverses the movement and returns to the original position. She may then weave, dig, or perform another probe. As the nest nears completion the female decreases her tendency to turn and circle after each dig and increases the frequency of probes; in response, the male increases his performance of crossing-over and quivering activities (Tautz and Groot 1975). Probing and quivering usually occur in a predictable order. As soon as the female initiates a probe, the male, while quivering, approaches her from his position behind and to one side of her. After completing the probe the female comes up again and the male drifts back to his original courtship position. The angle of the probes becomes more and more pronounced as the nest develops and when it reaches about 20° the nest is complete.

Duker (1982) described a model for the prespawning phase of chum salmon involving orientation and species recognition, and the sensory systems utilized to locate and choose conspecific mates (Figure 2). Of the potential auditory cues available in the noisy lotic environment, only the sounds of digging by the female appear to provide information on the location of reproductively active females. Tactile cues involving physical contact in reproductive behaviour appear to be limited to the female's interaction with the gravel substrate. Visual cues, however, are important because the body coloration clearly distinguishes the species, especially the conspicuous pigment patterns found in the mouths of the spawners. Females appear to respond more actively to the external body coloration of males than do males to the coloration and configuration of females. The black and white pigmentation pattern inside the mouth is species specific, and is very evident as the courting pair gape simultaneously at spawning (Schroder 1981; Duker 1982) (Plate 13). Duker suggested that non-visual cues are perceived by both sexes during their short-range courtship interactions and that combinations of visual and nonvisual cues contribute to the process of identification. He also concluded that olfaction does not play a major role in mate selection in guiding males to active female conspecifics. However, the response of chum salmon males to heterospecific females suggests that long-range olfactory cues may be

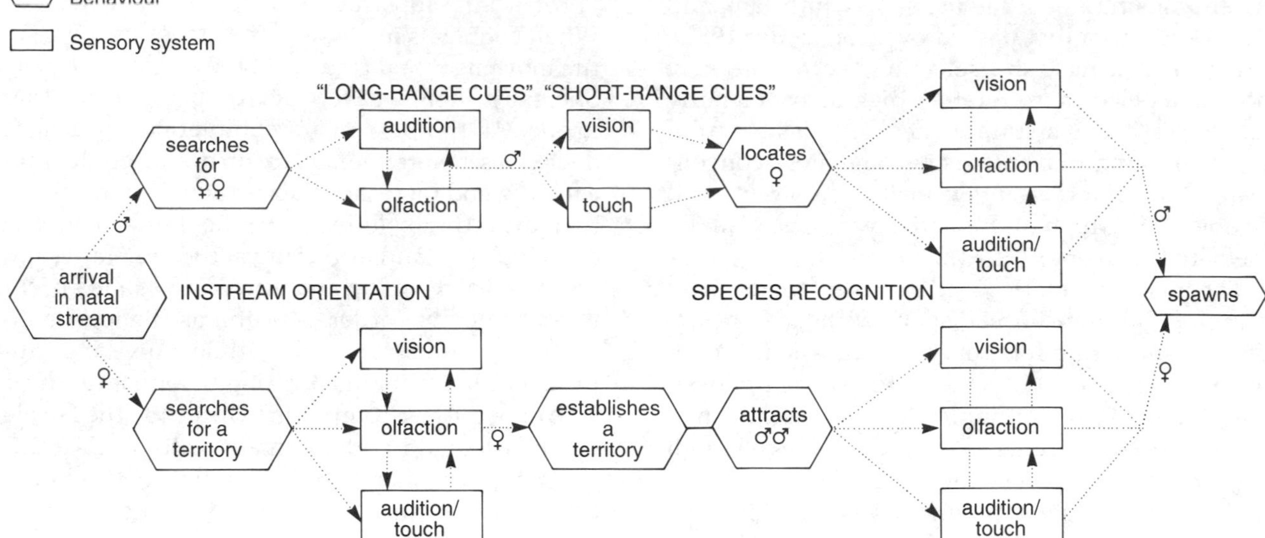

FIGURE 2

A model illustrating the behaviour of Pacific salmon and the sensory systems utilized by these fish to locate and choose conspecific mates. (From Duker 1982)

important in separating chum from coho salmon in sympatric situations.

Mating and Covering of Eggs

When the female is ready to deposit her eggs, she will move into the nest and "crouch," which looks like a probe but with mouth agape. When she starts to crouch, the male immediately moves forward and lies next to her assuming a similar posture (Plate 13). At this point the reproductive products are released with both partners vibrating their caudal peduncle and anal fin. Sometimes the female will perform several crouches in succession before releasing the eggs. In each crouch she is followed by the male. The spawning act lasts, on average, 10 seconds (Schroder 1982).

Within seconds of egg deposition, the female starts to cover the eggs with gravel. She moves upstream, turns sideways and, while laying her tail on the gravel, gently flexes her body two or three times. The first few "covering digs" do not move any gravel but drive the eggs into the gravel interstices (Tautz and Groot 1975; Schroder 1982). Subsequent digs become more vigorous as the nest pocket fills with gravel, and after 15-30 minutes, when the nest is completely closed, the female returns to normal nest digging. This results in the construction of the second nest in front of the first one.

Schroder (1982) reported that most females (>80%) completed spawning 30-40 hours after starting their first nest in the Big Beef Creek experimental spawning channel. He also noted that about 35% of the eggs were deposited in the first nest and that the last few nests contained only one-half to one-quarter the number of eggs of the first one.

The male may stay around for a little while after mating but generally moves on to find another female in the process of nest construction. Males remain sexually active for 10-14 days. The difference in duration of sexual activity between sexes increases the likelihood of intrasexual competition, especially when the ratio of sexually-active males to sexually active females is above unity.

Male chum salmon use physical force, that is, open mouth rushes, bites, and body blocks (Figure 3), to compete with other males for spawning opportunities and, occasionally, battles occur (Schroder 1973, 1981). The ability of the male to maintain a mate, or obtain one, depends on his relative size. Large and dominant males have a greater chance of obtaining a mate, whereas small or weak males spend more time searching for po-

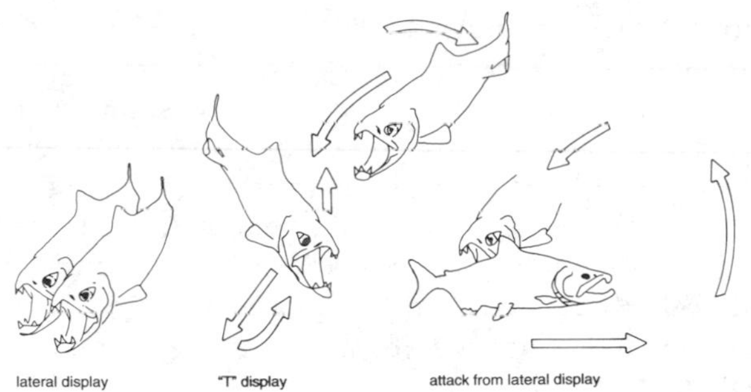

FIGURE 3
Ritualized fighting displays used by male chum salmon. (From Schroder 1981)

tential mates and attempt to avoid conflicts (Schroder 1982). Subdominant or satellite chum males will adopt the strategy of positioning themselves downstream from a courting pair. From this position the subdominant male continuously approaches the female and attempts to fertilize some of her eggs when she crouches with the dominant male. Schroder (1982) found that as competition for females increased, the occurrence of satellite males rose. At male-to-female ratios of less than one, the percentage of males employing the subdominant male strategy approached zero, but increased to as high as 30% when the ratio equalled three. Using electrophoretic techniques to estimate the gametic contributions of alpha (dominant) and satellite males, Schroder (1982) concluded that satellite males can make significant gametic contributions (up to 25%). He also observed from analysis of video tapes that the closer the male is to the female the more eggs he is able to fertilize. Extreme competition makes participation of some satellite males difficult because the female is obscured by competitors. Schroder (1982) also found that mate selection by males was not influenced by size of the female but by her "attractiveness" as expressed by her behaviour in terms of nosing, turning, digging, and weaving. Under extremely high spawning densities, courtship and territorial defenses break down and "mass spawning" occurs (E.O. Salo, unpublished data; D. E. Rogers, Fisheries Research Institute, University of Washington, Seattle, Washington, pers. comm.).

Nest and Redd Characteristics

Burner (1951) and Helle (1979) reported that the average depth of chum salmon nests is 21.5 cm (range 7.5–43 cm), not including the depth of the eggs. Bruya (1981), on the other hand, found that chum salmon nests have a mean depth of 42.5 cm and that high survivals to emergence (mean 84%) resulted from egg depositions in gravel depths ranging from 20 to 50 cm. He concluded that a minimum of 30 cm is essential and a depth of 40 cm is optimal. Premature emergence of fry occurs in nests of less than 20 cm deep and rises to 80% at nest depths of 10 cm.

The size composition of gravel selected for spawning by chum salmon in Hokkaido averages 25% for gravel less than 0.5 cm in diameter, 5% for gravel from 0.6 to 3.0 cm, and 30% for gravel greater than 3.1 cm (Sano 1959). Redds of chum salmon in the tributaries of the Columbia River consisted of gravel of which 13% was larger than 15 cm, 81% was 15 cm or less, and 6% was silt and sand (Burner 1951).

Rukhlov (1969) described the spawning gravels of chum salmon of six rivers in Sakhalin in terms of silts, sand, gravel, and "shingle." He noted that the percentage of fines and sand was less in the nests than in the surrounding gravels (11.5% versus 14.7%) and that when the proportion of sand was 22% or greater, the survival of the eggs was less than 50%. Because of the predominance of pink salmon, the percentage of sand in the gravel was less in odd years. Of the six rivers, chum inhabited mainly those with significant groundwater and higher base flows.

In measurements of over a thousand redds in the state of Washington, Johnson et al. (1971) noted that, although chum salmon spawned in velocities ranging from 0.0 to 167.6 cm/s, 80% spawned in velocities between 21.3 and 83.8 cm/s, with the mean being 50.3 cm/s. The water depth over 80% of the redds ranged from 13.4 to 49.7 cm, and the distribution of the depths was highly skewed with a mean of 27.1 cm.

Water velocities selected by autumn chum in Hokkaido are from 10 to 20 cm/s (Sano and Nagasawa 1958), whereas summer chum in the My River spawn in velocities of 10–100 cm/s (Soin 1954). Water depths range from 20 to 110 cm in Hokkaido and from 30 to 100 cm in the My River. Artificial spawning channels for chum salmon are typically regulated to have flows of about 20 cm/s. Flows over Japanese-type incubation "keeper" channels, which utilize systems of screens to retain the eggs until hatching, have flows of about 50 cm/min. Bams (1982) found that the developmental rate was increased and larger fry were produced at higher (66 cm/min) than at lower flows (33 cm/min). Chum salmon, although the second largest in size of the Pacific salmon, have adapted to spawning in waters of lesser depths and velocities than the pink salmon and some of the other species in the genus. Typically, summer chum spawn in deeper waters and higher velocities than autumn chum (Soin 1954; Sano and Nagasawa 1958).

Post-Spawning Longevity and Egg Retention

Koski (1975) defined post-spawning longevity of the chum salmon as the elapsed time in hours after they had been placed in the spawning channel at Big Beef Creek (which was very soon after arrival into the stream) until their observed death. The average longevity of females and males combined for an early stock was 8.8 days and 10.5 days for 1968 and 1969, respectively. The late stock averaged 11.2 days and 15.2 days for the same years. Koski suggested that the colder water temperatures in 1969 may have been partly responsible for the increased longevity. The males lived longer than the females, and Koski (1975) and Schroder (1977) observed that there was no significant change in longevity with different spawning densities.

The number of eggs retained by the females after spawning vary considerably. The longer the female delays spawning and the higher the spawner density, the greater the egg retention (Schroder 1981). Although spawning generally occurs during a period of falling temperatures, prolonged cold water temperatures increase spawning time and egg retention. Spermatazoa are produced for as long as 26 days (Koski 1975), which allows males to spawn with more than one female (Smirnov 1975).

Fecundity

Fecundities of chum salmon reported in the literature are not reliable because of the uncertainty in data collection. Individual measurements may be comparable, but it is not certain how representative the samples are for the reported geographical regions and rivers of origin. This is particularly true of samples collected in commercial catches. Nevertheless, when both absolute fecundity (number of eggs/female) and relative fecundity (number of eggs/cm of length) are considered, similarities and differences among regions can be noted.

In Asia, individual absolute fecundities ranged from 909 to 7,779, and annual means ranged from 1,800 to 4,297 eggs per female. In North America, the reported fecundity of individuals ranged from 2,018 to 3,977, and annual means ranged from 2,107 to 3,629 eggs per female. Generally, the northern stocks in Asia have a higher relative fecundity than the southern stocks. This trend is apparent for both summer and autumn chum in the USSR (Kulikova 1972) (Table 1). These differences are masked, to some extent, when absolute fecundities are considered because of body size differences among stocks. In North America, there is a weak trend for northern stocks of autumn chum to have lower absolute and relative fecundities than the southern stocks (Table 2; Figure 4).

The reasons for the latitudinal trends are not obvious but are probably related to survival rates decreasing from south to north in Asia and north to south in North America. Information is lacking on latitudinal differences in fecundity by age and size, along with their relative survival rates.

Fecundity and Stream Size

Races in small, short streams tend to be less fecund than stocks from longer streams (Kayev 1983). Beacham (1982) reported lower fecundities for stocks on Vancouver Island and the Queen Charlotte Islands than for a few mainland stocks. Ex-

TABLE 1
Length, absolute fecundity, relative fecundity, and egg diameter of female chum salmon (age 0.3), in the USSR, 1966

Location	Fork length mean (cm)	Absolute fecundity	Relative fecundity (No. egg/cm)	egg diameter (cm)
Summer chum				
N Anadyr R. (Bering Sea)	65.5 ± 1.10	3160 ± 191.80	48.2	.71 ± .01
Bolshaya R. (Kamchatka)	61.5 ± 0.80	2490 ± 93.53	40.5	-
Taui R.	62.7 ± 0.65	2770 ± 85.30	45.0	-
Kukhtuy	62.2 ± 0.37	2850 ± 110.15	45.0	.75 ± .02
Iski R.	61.5 ± 0.92	2510 ± 83.00	40.8	.79 ± .02
S Amur R.	52.2 ± 0.59	2200 ± 55.20	39.8	.79 ± .02
Autumn chum				
N Amur R.	64.5 ± 0.71	3450 ± 716.30	53.5	.70 ± .03
Naiba R.	65.5 ± 0.93	2805 ± 67.42	42.8	.93 ± .01
Kalininka R. (W. Sakhalin)	67.3 ± 0.76	2720 ± 115.30	40.4	.90 ± .03
S Kurilka R. (S. Kuril)	69.3 ± 0.84	2600 ± 89.32	37.5	.95 ± .01

Source: Adapted from Kulikova (1972)
Notes: Rivers are listed north (N) to south (S); 100 fish sampled in each river

TABLE 2
Relative fecundity (eggs/cm of fork length) of North American autumn chum salmon

Location	Eggs/cm	Sample size	Source
N Yukon R. (Delta R, AK)	41.2	14	Trasky (1974)
Skeena R. (BC)	45.2	54	Beacham (1982)
Pallant Cr. (BC)	40.2	58	Beacham (1982)
Mathers Cr. (BC)	41.5	14	Beacham (1982)
Tlupana R. (BC)	41.0	26	Beacham (1982)
Little Qualicum R. (BC)	43.7	33	Beacham (1982)
Big Qualicum R. (BC)	44.1	577	Beacham (1982)
Squamish R. (BC)	43.9	61	Beacham (1982)
Fraser R. (BC)	44.9	222	Beacham and Starr (1982)
Fraser R. (Harrison R., BC)	46.9	15	Beacham (1982)
S Big Beef Cr. (WA)	46.3	-*	Koski (1975)

Notes: Locations are listed north (N) to south (S).
*Unreported

ceptions include Big Beef Creek, which is a small stream with high fecundity levels (Koski 1975).

Beacham (1982) suggested that differences in fecundity among streams may result from high exploitation rates. Kayev (1983) found that chum salmon inhabiting short streams on the Kuril Islands, with good groundwater and favourable estuarine and marine conditions, were less fecund and had greater fry survival.

Differences in Seasonal Runs

In Asia, the Amur River autumn stocks are more fecund (relatively and absolutely) than summer stocks (Lovetskaya 1948; Birman 1956; Svetovidova 1961; Sano 1966; Kulikova 1972) (Tables 1 and 3; Figure 4). Early-run chum salmon at Big Beef Creek are more fecund (50 eggs/cm) than the late-run chum (46.0 eggs/cm) (Koski 1975), which is the opposite of the situation seen in the distinct seasonal races of the Amur River. Data from Andersen (1983) and Trasky (1974) showed a slight difference in relative fecundity between summer and autumn chum of the Yukon River (summer, 43.5 eggs/cm, $N = 23$; autumn, 41.2 eggs/cm, $N = 24$). Although the samples are small, the differences are similar to those presented by Koski (1975) for the early-run and late-run chum at Big Beef Creek.

Egg Diameter

In Asia, egg diameter increases from north to south (Table 1). Southern stocks incubate at higher temperatures, and the higher metabolic rates require a greater supply of energy; also, if the northern stocks are more fecund, egg size may be limited by egg number. However, egg diameter is correlated with female size and dependent upon spawning time (Beacham and Murray 1986; T. Beacham, Pacific Biological Station, Nanaimo, British Columbia, pers. comm.). Although the size of eggs of chum salmon has no effect on hatching time, exogenous yolk absorption ("button-up"), or emergence from the redd, larger eggs produce alevins that are longer and have greater amounts of yolk than those produced from smaller eggs. The differences are maintained through the alevin to the newly emerged fry stage (Beacham et al. 1985). Whether southern stocks encounter more potential predators than northern stocks is not known, and

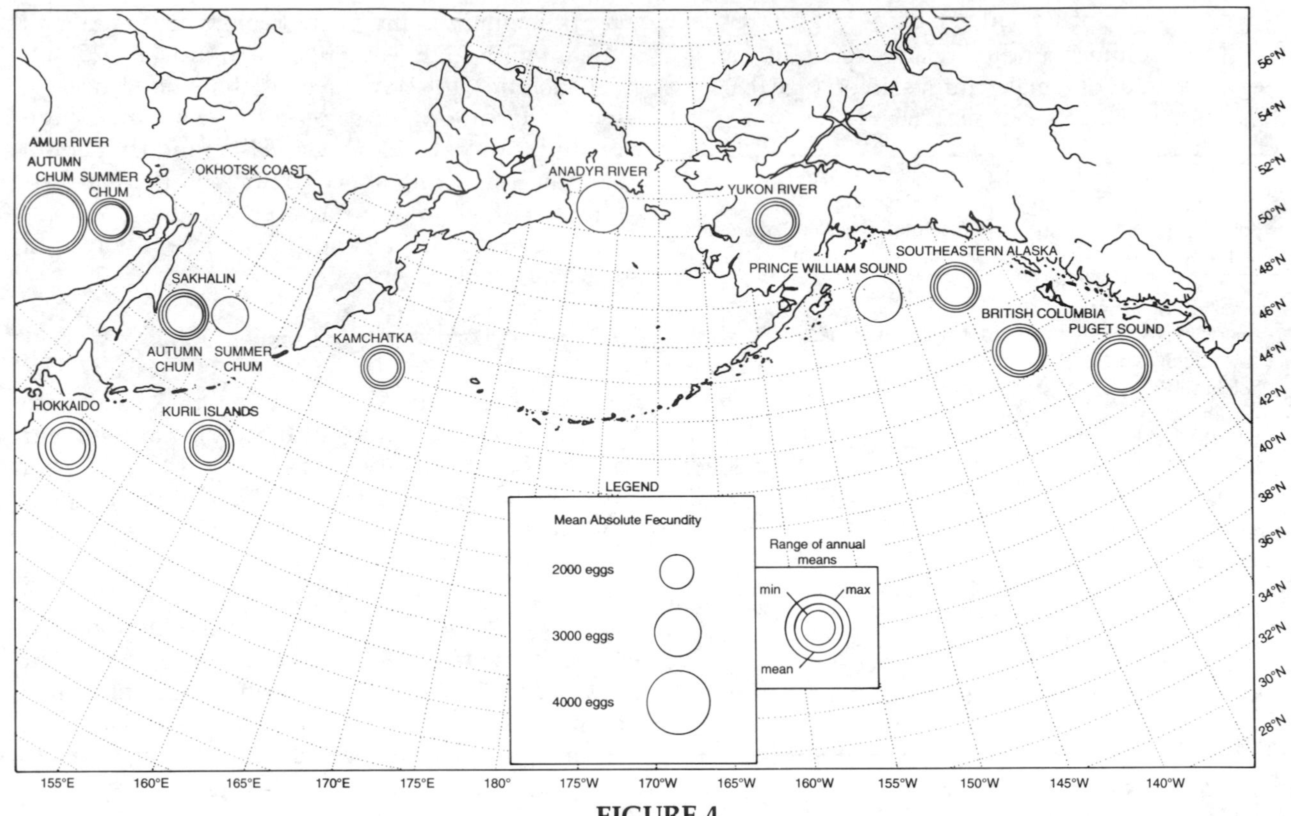

FIGURE 4

Absolute fecundity of chum salmon in Asia and North America

TABLE 3

Relative fecundity (eggs/cm of fork length) of Amur River system chum salmon

	Summer chum			Autumn chum			
Year	My	Ul	Beshenaya	Amgun	Kur	Khor	Bira
	→	→	→	→	→	→	→
1946	–	–	–	–	–	46.0	56.3
1947	–	–	–	–	–	55.7	55.2
1948	–	–	–	–	–	56.7	57.1
1949	–	–	38.8	50.4	–	53.8	59.3
1950	37.2	–	40.1	52.0	61.1	–	55.4
1951	39.2	37.7	39.5	50.2	–	54.7	58.5
1952	39.5	37.5	40.0	53.3	–	52.3	57.2
1953	37.5	35.5	39.2	53.5	–	56.5	54.0
1954	39.5	39.0	39.3	–	–	–	–
1955	42.5	38.5	47.5	–	–	–	–
Average	39.6	37.7	39.5	51.7	61.1	53.7	56.5

Source: Data for summer chum from Svetovidova (1961); data for autumn chum from Birman (1956)
Note: Arrows indicate increasing distance from estuary.

whether this is related to the northern fish being smaller is also conjectural. The relationships between prey availability and predator gape are documented but these have not been related to the north-south change in fry size.

Incubation and Emergence

Incubation and emergence and the "quality" and "fitness"[2] of the emerging fry are affected by stream flow, water temperature, dissolved oxygen, gravel composition, spawning time, spawner density, and genetic characteristics (Figure 5) (Koski 1975, 1981; Bams 1982, 1983; Beacham and Murray 1986).

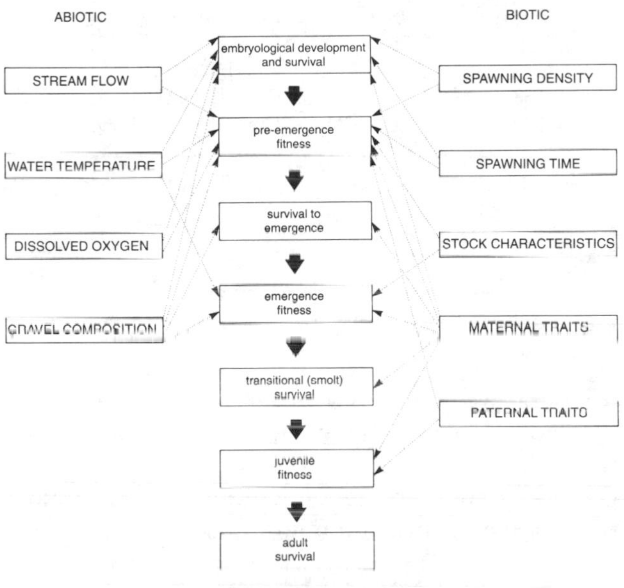

FIGURE 5

Diagram of the interrelationships of the abiotic and biotic factors in the salmonid incubation environment which affect fitness and survival. (From Koski 1975)

Water Temperatures and Temperature Units

The incubation time of eggs is prolonged by lower temperatures, and the time of hatching and emergence varies among stocks because of differences in the number of temperature units (TU's)[3] required for hatching and development. These variations may be genetically controlled, and the differences between the temperature units required for hatching and emergence of early and late stocks result in a tendency towards similar emergence times (Koski 1975).

In many Asian and North American streams, the late-running stocks select areas with springs having water temperatures generally above 4°C. This protects the eggs from freezing and results in more or less consistent times of emergence from year to year. Nikolskii (1952) pointed out that during severe winters the redds of autumn chum near the outflow of groundwaters were less affected by freezing than redds of the summer chum in the main stream. Adaptation to intertidal spawning also allows for compensation for environmental extremes because warmer marine waters cover the spawning areas during each tidal cycle. Thorsteinson (1965) reported that a 3.2 m tidal differential caused an 8°C change in the temperature of an intertidal redd in Olsen Creek.

Low water temperatures (near or at freezing) during spawning and incubation can account for significantly high mortalities of salmonid eggs and alevins (Smirnov 1947; McNeil 1962, 1966; Levanidov 1964; Sano 1967). In the state of Washington, a drop in water temperature below 2.5°C inhibits nest construction and spawning by chum salmon (Schroder 1973). Schroder et al. (1974) found significantly higher mortalities of chum salmon eggs, alevins, and fry when the eggs were incubated in water temperatures below 1.5°C during the early stage of development (before blastopore closure).

Chum salmon eggs require about 400 to 600 TU's to hatch and about 700–1,000 TU's for yolk absorption (Table 4). Values vary among stocks and among individuals within stocks. At Big Beef Creek, a range of 30 days occurred between completion of emergence of the early and late stocks in a single year, and a difference of 60 days occurred between years (Figure 6). The Susitna River (Alaska) and the Amur River autumn chum stocks have very low TU requirements (Table 4), which are stock-specific adaptations to low incubation temperatures.

2 For definitions of "quality" and "fitness" see Bams (1983) and Koski (1975, 1981). They discuss the deviation of these terms from the genetic definitions (relative reproductive success).

3 TU = the average number of degrees above 0°C during a 24-hour period.

TABLE 4
Accumulated temperature units required for hatching and yolk absorption for chum salmon eggs

Stock	Temperature units (C°-days) to hatching	Days	Mean temp. (C°)	Temperature units (C°-days) to yolk absorption	Days	Mean temp. (C°)	Source
Amur R., USSR (autumn)	408–420	122–128	3.4[1]			3.4[1]	Disler (1954)[5]
Susitna R., AK[6]	292	173	1.7[2]	623	284	2.2[2]	Wangaard & Burger (1983)
	447	123	3.6[2]	728	250	2.9[2]	
	489	106	4.6[2]	847	215	3.9[2]	
	473	117	4.0[3]	860	213	4.0[3]	
BC (specific stock not reported)	510–589	52–61	9.7–9.8[3]			9.7–9.8[3]	Alderdice et al. (1958)[5]
Big Beef Cr. WA (early)				1060[4]	166	-	Koski (1975)
Big Beef Cr. (late)				933[4]	146	-	Koski (1975)
Skagit R., WA	453	114	4.0[2]	867	155	4.7[2]	Graybill et al. (1979)
	623	86	7.2[3]	1124	157	7.2[3]	Graybill et al. (1979)
	565	118	4.8[3]	958	200	4.8[3]	Graybill et al. (1979)
	509	182	2.8[3]	909	325	2.8[3]	Graybill et al. (1979)

Notes: *1* Water temperatures varied less than 1°C
2 Mean of variable stream water temperature
3 Constant water temperature
4 C°-days to emergence
5 Cited by Bakkala (1970)
6 Values are averages based on Appendix Table 1 of Wangaard and Burger (1983)

Although the number of TU's required for hatching and yolk absorption is generally less at low than at high temperatures (Table 4), these compensations are not sufficient to offset all the effects of temperature on development. In warm years, hatching and development is accelerated. Some compensation for annual variation in temperatures within stocks is provided for by the number of temperature units required, which is partially a function of the temperature regime experienced. Beacham and Murray (1986), working with early, middle, and late spawning Fraser River stocks, and constant temperatures of 4°, 8°, and 12°C, found that time of spawning had no effect on hatching time of alevins. Thus, development rates of eggs from fertilization to hatching provided no evidence for stock adaptation of Fraser River chum salmon to their time of spawning. On the other hand, timing of emergence of the fry within each incubation temperature was dependent on the relative timing of the stocks. Early-spawning stocks had heavier eggs, and fry from these stocks had later emergence times.

In summary, differences in TU's required for emergence of fry vary within stocks depending on temperature regimes (weather) and among stocks depending on long-range temperature characteristics (climate) of the incubation environment. Generally, there is a tendency towards a common time for emergence and migration.

Dissolved Oxygen

A number of authors (Wickett 1954; Coble 1961; Phillips and Campbell 1962; McNeil 1966) have shown that the survival of salmonid eggs and alevins is directly related to the intragravel dissolved oxygen content. Wickett (1954) calculated that the lethal level (minimum) for chum salmon is 1.67 mg/l. Koski (1975) found that the survival rate decreased rapidly when the concentration of oxygen dropped below 2 mg/l. Alderdice et al. (1958) conducted experiments on the survival of various-aged chum salmon eggs after seven days of exposure to low dissolved oxygen levels and found that (1) eggs were most sensitive to hypoxia between 100 and 200 C degree-days, and compensated for reduced oxygen availability by reducing the oxygen demand and rate of development (increase in TU requirements); (2) very low oxygen levels at

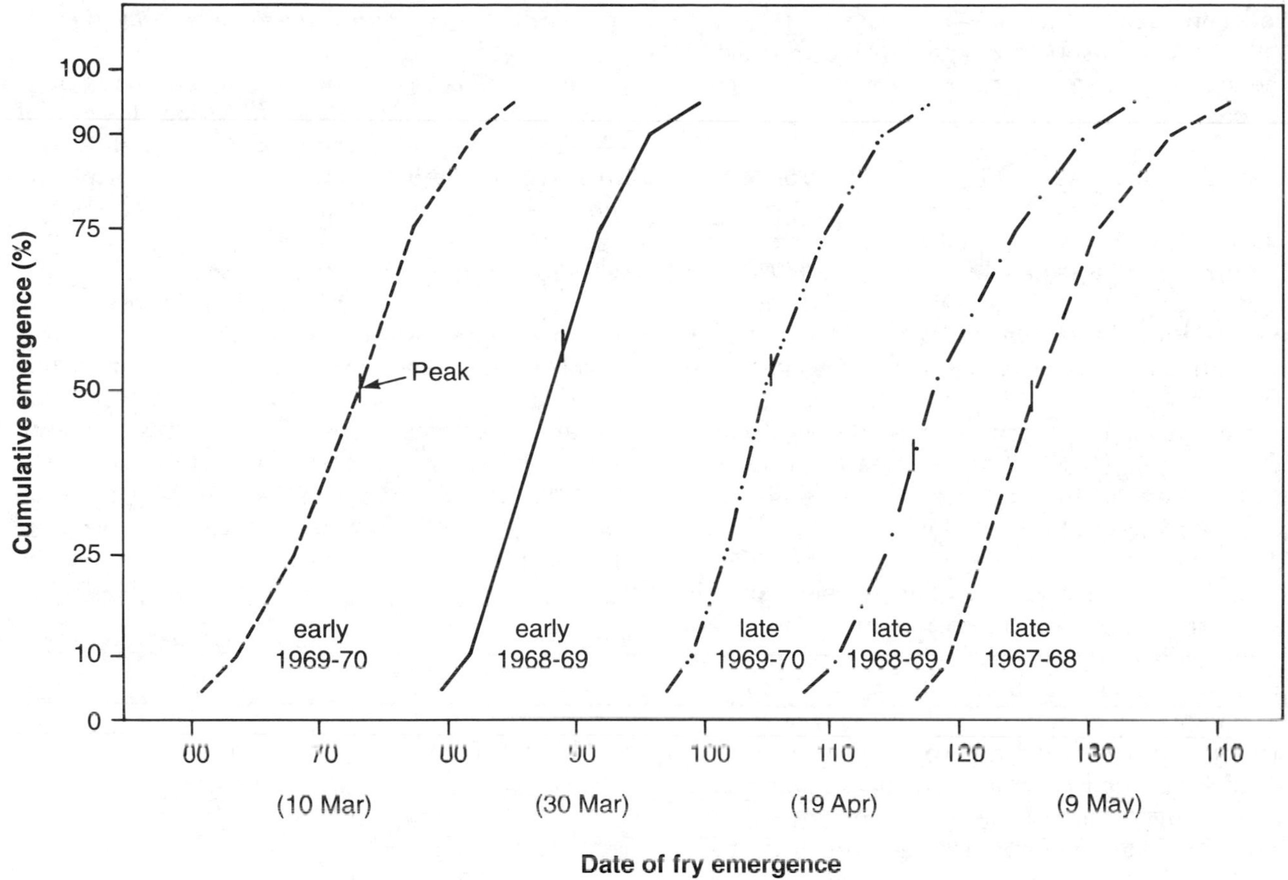

FIGURE 6

Annual variation in the timing of fry emergence for early and late stocks of chum salmon from the Big Beef Creek channels (Julian calendar and the calendar year date are given). (From Koski 1975)

early incubation stages cause monstrosities; (3) estimated median lethal levels rose slowly from fertilization to hatching; and (4) oxygen consumption per egg rose from fertilization to hatching, although the consumption per gram of larval tissue declined from a high at about the time of blastopore closure. Under experimental conditions, the incipient median lethal level for dissolved oxygen rose with development from approximately 0.4 ppm in early development to 1.0–1.4 ppm prior to hatching. The calculated critical oxygen levels varied from 0.72 ppm at 4 C degree-days to 7.1 ppm at 452 C degree-days. Fast and Stober (1984) corroborated the increasing oxygen requirements for developing salmon embryos and determined that chum salmon have lower oxygen requirements than either coho salmon or steelhead (*O. mykiss*), reflecting a lower metabolic demand.

Koski (1975) found no significant differences in the size of chum salmon fry emerging from redds with prolonged minimum dissolved oxygen concentrations (less than 6.0 mg/l). Alderdice et al. (1958) reported that eggs subjected to low dissolved oxygen levels just prior to hatching, hatch prematurely at a rate dependent on the degree of hypoxia, and Koski (1975) noted that the number of days to initial emergence was greater at prolonged low dissolved oxygen concentrations (less than 3.0 mg/l).

Bams and Lam (1983) concluded that deteriorating water quality in a Japanese-style keeper channel (upstream to downstream) measurably reduced larval development rate, growth rate, and yolk conversion efficiency. The main effective factor was low dissolved oxygen which stimulated the pre-emergence of fry during unfavourable conditions.

Condition Factor, Egg Size, and Fry Size

The coefficient of condition, K_D (=$10.\sqrt[3]{\text{Weight(mg)}}$/ Length (mm)), was developed by Bams (1970) as a comparative index of the condition of salmon alevins and fry. Bams reported a K_D of 1.92 at tissue resorption for chum from Hooknose Creek, British Columbia. The K_D at this stage for Big Beef Creek chum fry was 1.89 for those incubated in 10 cm of gravel, and 1.83 for those incubated in deeper gravel (Bruya 1981). Although there are differences among stocks (Abbasov and Polyakov 1978; Koski 1981; Beacham and Murray 1986), Bruya (1981) felt that the difference between Bams' results and his were due to the effects of preservation in formalin. Incubation environments with low oxygen can produce alevins with high indices of development (K_D factors) (Alderdice et al. 1958). Also, large eggs produce alevins with higher K_D values than do smaller eggs, so the actual factor of development is affected by the size of the egg (Beacham and Murray 1986). Late-spawning Fraser River stocks have smaller eggs and shorter times from fertilization to fry emergence than early-spawning stocks (Beacham and Murray 1986).

The need to optimize the development of eggs and alevins in a biologically efficient manner has been recognized by a number of investigators (Disler 1953; Brannon 1965; Bams 1967, 1969; Poon 1970; Emadi 1973; Mathews and Senn 1975; McNeil and Bailey 1975). The results of these studies are being applied in chum salmon hatcheries in the Soviet Union, Japan, and North America.

Chum salmon alevins incubated on unaltered screen substrates at production levels of 8,000 eggs per tray in water with and without sediment were found to be significantly smaller, by as much as 0.1 g/fry in weight and 4.9 mm in length, than fry incubated on artificial substrates (B. Snyder, Big Beef Creek Fish Research Station, University of Washington, Seattle, Washington, pers. comm.). Also, at the time of "ponding" of the former fry, they were less developed and had a higher condition coefficient (K_D) than fry incubated on artificial substrates.

Gravel Composition and Spawner Density

Sediment affects the survival of salmonids in at least three ways: (1) direct suffocation of eggs and alevins, (2) reduced intragravel water flow and dissolved oxygen content, and (3) a physical barrier to emergence (Koski 1966, 1975; Gibbons and Salo 1973; Iwamoto et al. 1978). Using substrates with four different levels of intragravel sediments, Koski (1966, 1975) determined the rates of survival to emergence in experimental channels. The survival to emergence was highest (63%) in gravel containing 11%-30% sand. For each 1.0% increase in sand there was a 1.26% decrease in survival to emergence (Koski 1981) (Figure 7). Koski assumed that the amount of fines in the spawning gravel is, in essence, an index of the "living space" available for the developing eggs and alevins, reflecting the

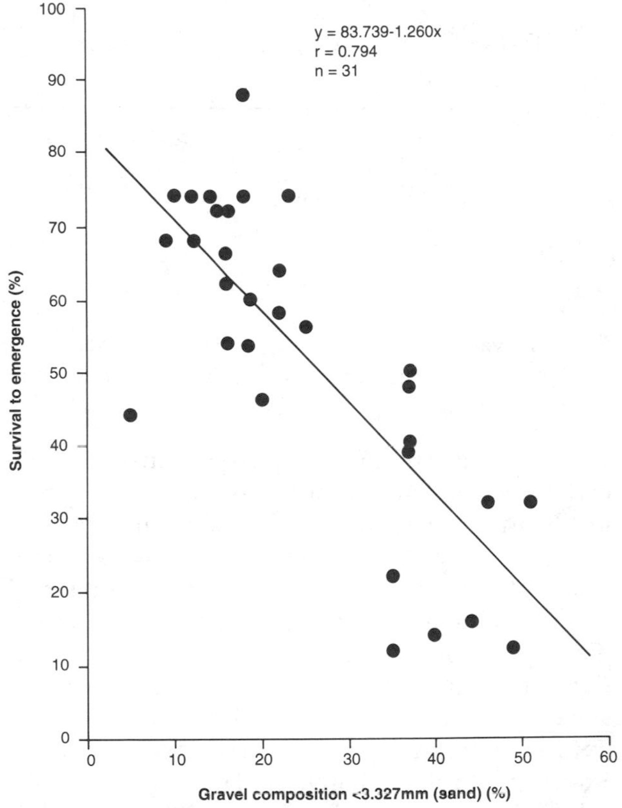

FIGURE 7
Relationship between the percentage of sand (fines <3.327 mm, ⩾0.105 mm) in the gravel and the rate of survival to emergence of chum salmon (1968 and 1969). (From Koski 1975)

percentage of voids within the gravel.

In an experiment in the incubation channels at Robertson Creek, British Columbia, Dill and Northcote (1970) determined that the survival of chum salmon from planting of eggs to emergence of fry was higher in large gravel (5.1–10.2 cm) than in small gravel (1.0-3.8 cm). Gravel size, depth of egg burial (20.3 and 30.5 cm), or density of eggs (50 and 100) had no effect on the condition coefficient or timing of emergence.

Using survival from egg deposition to downstream migration of the fry as a criterion, Schroder (1973) determined that the optimum spawning density of chum salmon in controlled flow channels of Big Beef Creek was 1.7 m^2/female (0.6 females/m^2). Thorsteinson (1965) found that the optimum density of spawners in the intertidal areas of Olsen Creek, Alaska, was between 2 and 3 females/m^2, with no increase in successful egg deposition at five or more females per square metre.

Effects of Salt Water and Other Water Quality Factors

Bailey (1964) and Thorsteinson (1965) determined that chum salmon spawning in the intertidal areas of Prince William Sound, Alaska, was widespread and that the intertidal spawners were discrete stocks with an extended spawning period. Although chum rarely spawn in the lower intertidal areas where the pink salmon spawn, significant changes in temperature, salinity, and dissolved oxygen do occur with each rise and fall of the tide. Survival of the eggs and alevins decreases from the upper to the lower areas. Hashimoto (1971) concluded that treatment with a salt solution lowered the incubation rate of chum eggs and increased the occurrence of dead eggs and abnormal hatching. Although the period from fertilization to the beginning of hatching was reduced, the number of days from the first hatching until the last hatching increased. In eggs that hatched prematurely, the chorion was digested by the hatching enzyme when the moisture in the perivitelline liquid was dehydrated.

Land uplift and subsidence associated with the large earthquake in Alaska in March 1964 caused major ecological changes in the intertidal areas of Prince William Sound. The effects of these changes in land level on behaviour and survival of intertidal spawning chum salmon are discussed by Thorsteinson et al. (1971).

Relationships of Eggs and Fry with Benthic Fauna

Nicola (1968) discussed a case of potential mutualism between the normal predaceous stonefly nymphs of the genus *Alloperla* and developing chum salmon embryos in the Harris River, Alaska. Analyses of the numbers of missing and dead embryos in relation to the numbers of stonefly nymphs in containers of buried eggs supported the hypothesis that stonefly nymphs are scavengers and not predators. Nicola noted that fungus was not found in any of the containers. Because of the biological oxygen demand, the removal of dead eggs was probably beneficial. The absence of saprophytes, along with the disappearance of eggs, was also noted by Vibert (1956).

Behaviour of Alevins in the Gravel Environment

Under conditions of adequate velocity, dissolved oxygen, and darkness, alevins move downward through the gravel substrate. Fast and Stober (1984) showed that chum alevins made more successful migrations through 20 cm of gravel than either coho or chinook alevins. The head/body thickness ratio allowed chum alevins (71.5 mg) on the day after hatching to pass through a number 7 sieve (2.80 mm mesh size). This adaptation allows the relatively large egg and alevin to use gravels with small interstitial spaces. The factor of robustness became evident when steelhead alevins, which are smaller than chum fry, migrated through number 7 and number 8 (2.36 mm) meshes, but in fewer numbers than the chum fry.

Fast and Stober (1984) also showed that chum alevins are photonegative from day 6 to 25 after hatching. After this time there is a rapid reversal to photopositive behaviour corresponding with the onset of emergence (Figure 8). The early photonegative behaviour is believed to be an adaptation for predator avoidance by keeping the alevins in the relative safety of the gravel until they have developed sufficiently to survive upon emergence.

Disler (1953) reported that alevins of autumn chum in the Amur River feed prior to yolk absorption; however, it is not known whether this con-

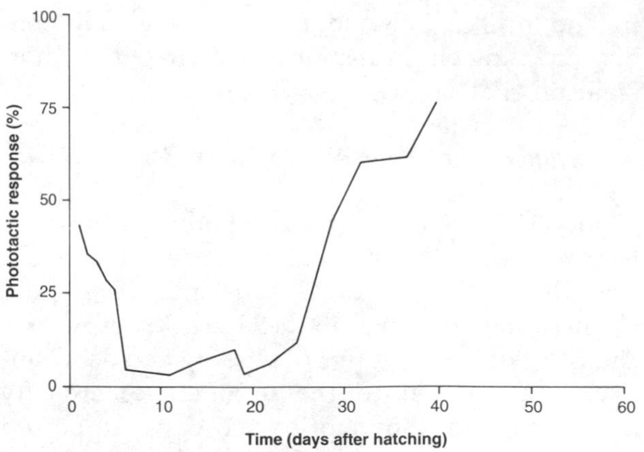

FIGURE 8
Photobehaviour of chum salmon alevins from hatching to yolk absorption. Each data point represents the mean percentage of 30 alevins in the light compartment of four separate light-dark choice tests. (From Fast and Stober 1984)

tributes to the growth of the alevins or if it is important in the development of their feeding behaviour.

Downstream Migration of Chum Salmon Fry

General Pattern of Downstream Migration

Chum salmon fry (Plate 12) typically emerge during nighttime hours and promptly migrate downstream to estuarine waters where they linger until they make the transition to waters of higher salinity. In the shorter rivers the migration is over in about 30 days, whereas in the longer rivers the migration is prolonged. Migration timing varies from early spring to midsummer by latitude, length of stream, timing of spawning of parental stocks, and interactions with other species, particularly pink salmon.

The movement downstream generally starts in the early nighttime hours and ceases during the middle of the night. In the early morning hours there is some aggregation or schooling leading to minimal to moderate downstream movement until broad daylight. As the lengths of the streams and their flows increase, variations (some of which are extreme) of the typical patterns occur. Migration during daylight hours, in well-lit areas, is not uncommon in some of the Asian and northern North American streams. Deviations from nocturnal migrations also occur with increased turbidity of the stream. Regardless of the variations in migration patterns, for most of their lives chum salmon are obligatory ocean dwellers (Hoar 1958).

The migration of chum fry has been described variously as either a displacement by the current after loss of orientation during darkness (Neave 1955) or as an active migration downriver where the fry are oriented with respect to the river flow (Hoar 1958). Combinations of the two have also been reported. The problem is that responses of migratory fry in the field are commonly quite different from those in test streams (Neave 1955). The migration-by-displacement principle received support from Kobayashi and Ishikawa (1964). Others (Semko 1954; Kostarev 1970; Iwata 1982a, 1982b) described active migrations, often with lingering and feeding, particularly among the larger fry. Chum salmon probably combine the elements of displacement and active swimming, and the behaviour varies with the relative strength of the orienting factors, such as current, temperature, and visual reference points (Hoar 1958).

Seasonal and Diel Timing

In both Asia and North America the seasonal migration of chum fry is progressively earlier from north to south with the duration of the migration tending to be longer in the southern streams. Migrations in larger rivers are generally of longer duration than those in shorter streams.

The migration in Kamchatka is from early April through June with the peak in late April and early May (Semko 1954), whereas in the Okhotsk area the migration is from May to July with the peak in late June (Volobuyev 1984). In Hokkaido the migration lasts from March to June depending upon the river, with the peaks varying from April to late May (Kobayashi and Ishikawa 1964; Kobayashi et al. 1965; Sano 1966; Kobayashi and Kurohagi 1968).

A pronounced north to south pattern is evident in North America. Chum fry migration in the Yukon and Noatak River systems is from ice break-up in late spring until autumn, with the principal outmigration in June and July (Martin et al. 1986). The Noatak River has a less-defined peak than the Yukon River (Merritt and Raymond 1983). Farther south, in Olsen Creek (Prince William Sound) and the Taku River, the migrations are in May and June with peaks in mid-May and early May, respectively (Kirkwood 1962; Meehan and Siniff 1962). In the Skeena River system, the chum fry migration extends from mid-March to mid-April (McDonald 1960), and in Hooknose Creek the peak of migration occurs in late April and early May (Hunter 1959).

Fraser River chum fry move downstream from February to June (Todd 1966) with the majority migrating between mid-March and the end of April (Beacham and Starr 1982). In the Nooksack and Skagit rivers (Washington) the migrations are from April to June (Tyler and Bevan 1964; Davis 1981). In Minter Creek (southern Puget Sound), the migration extends from late January to late April (Salo and Noble 1954). At Big Beef Creek the migration lasts from February to June and has two peaks, in April and May (Koski 1975). The Satsop and Humptulips rivers, which empty into Grays Harbor, Washington, have migrations that peak in late April (Brix 1981).

Although the diel migrations of chum fry are typically described as nocturnal, there are some extreme variations from this nighttime pattern. The literature is not representative enough by region to establish "norms" and their variations.

Nocturnal migrations have been described by Volobuyev (1984) for the Okhotsk region, by Kobayashi and Ishikawa (1964) for the Ishikari River (Hokkaido), by Hunter (1959) for Hooknose Creek (British Columbia), by Davis (1981) for the Skagit River (Washington), by Koski (1975) for Big Beef Creek (Washington), and by E.O. Salo and C.H. Ellis (unpublished data, Washington Department of Fisheries) for Minter Creek (Washington). Diurnal and nocturnal migrations have been recorded by Rosly (1972) for the Amur River, by Kostarev (1970) for the Ulkhan River, by Semko (1954) for the Karymaiskiy Spring areas (West Kamchatka), by Meehan and Siniff (1962) for the Taku River, by McDonald (1960) for the Kispiox River (British Columbia), and by Todd (1966) for the Fraser River.

Following their initial emergence and movement, chum fry prefer well-lit areas, and more than 50% of them are often found in exposed locations. However, when the light intensity exceeds 500 foot candles they seek deeper, less illuminated areas. Some chum fry can be found at nearly all natural light intensities (Hoar 1958). On the other hand, Kobayashi (1960) noted, as did C.H. Ellis and E.O. Salo (unpublished data), that active nocturnal migrations were terminated by bright moonlight.

Kostarev (1970) found that the main body (66%) of fry on the Ulkhan River migrated on clear, cloudless days; whereas on days of variable cloudiness, 33.9% migrated; and on cloudy days, only 0.1% showed definite seaward movement. Peak migration occurred during morning hours with good illumination and high water temperatures. On bright days the migrating fry stayed close to the surface, whereas during evening hours or cloud cover conditions the fry were in deeper water. There was no definite peak on cloudy days, although the main body of fry still migrated during early morning hours.

McDonald (1960) reported that, in general, chum salmon fry migration in the Skeena River takes place nocturnally where the migration distances are short, but that some daylight movement occurs where travel distances are longer. The time of maximum migration of chum fry in the lower Fraser River happens during the early afternoon at the beginning of the season and becomes progressively earlier in the day as the season advances (Todd 1966).

Behaviour during Migration

Chum fry do not school as strongly as pink and sockeye fry. Their schools are not compact and if left undisturbed for some time, the individuals tend to scatter. Chum fry apparently lack a pronounced hiding behaviour whether schooled or not. When one individual approaches another the mutual recognition is generally evident. This can take the form of either attraction or agonistic behaviour. Vestigial territorial behaviour may be observed in relatively constricted areas (Newman 1956; Hoar 1958). Older chum fry near the end of their downstream migration feed more and school less if left undisturbed.

According to Hoar (1958), chum fry respond consistently and positively to currents at all times of the day. He noted that at 0600 h, 70% were in the current; at 1200 h, 50%; and in the evening (1800-2000 h), about 40%. In general, they respond to changes in flow by heading into the current as long as they can maintain their position.

The chum fry of the Karymaiskiy Spring areas, western Kamchatka, migrate to sea shortly after emergence but move slowly in the currents and tarry on the spawning grounds, some reaches of the river, at log jams, and in flooded valley areas (Semko 1954). Also, they feed intensively in fresh water. Most of the fry in the Ulkhan River migrate in a narrow band in the main channel of the river, but some that are in the shoals orient head-upstream and those in calm waters swim head-downstream (Kostarev 1970).

In the streams of Hokkaido, chum fry move seaward in small schools facing upstream, staying near shore, and avoiding strong currents. They migrate actively when the stream temperature rises to about 15°C and leave the coastal area when the temperature exceeds 17°C (Mihara 1958).

In the Fraser River, chum are generally distributed across the entire river throughout the season. As the season progresses, more fry tend to migrate near the surface in the top metre of water. This change in migratory behaviour was not as pronounced in even years (1962 and 1964) as in odd years (1963 and 1965), when pink fry were not present (Todd 1966). Hunter (1959) and K.L. Fresh and S.L. Schroder (Washington Department of Fisheries, Olympia, Washington, pers. comm.) found that in a smaller stream the chum migrated in the stronger currents in the middle of the stream, and Kobayashi and Ishikawa (1964) observed, in a Japanese stream, that most of the movement of fry was close to the bank and was not affected by current velocity.

The larger the river, the greater are the influences of variation in flow, turbidity, and temperature within and between years. Rosly (1972) found that fingerlings in the Amur River were in better condition due to increased feeding opportunities during years of low flows than during years of high flows. This relationship was reflected in the catches of autumn chum four years later with higher survival resulting from low flow years.

In the Otsuchi River, Iwata (1982a) observed that upon release at about 1 g, chum fry formed schools within 3-5 minutes, and when in slow moving water of less than 20-30 cm/s, they started to move downstream along the bank with their heads directed seaward. However, in currents of 30-80 cm/s their heads were oriented upstream and they were displaced by water flow. In shallow, rapid flows (80-120 cm/s) their position became random due to strong turbulence. Thus, the fry showed either positive or negative rheotaxis dependent upon the flow.

While migrating, chum fry are attracted by shade or darkness of waterweed communities (Figure 9). When the density of fish becomes high in the shaded areas they continue to move downstream. When the fish reach sea water they respond strongly to the mixed water and either turn back to fresh water or swim in the upper layer of lower salinity.

Feeding

It is not clear to what extent chum fry feed as they migrate down the larger rivers because only a few cases have been documented.

Chum fry in the Ulkhan River begin feeding early as "they spend a long time gaining weight in the spawning beds" (Kostarev 1970). Lingering as late as June, their basic food consists of the larvae

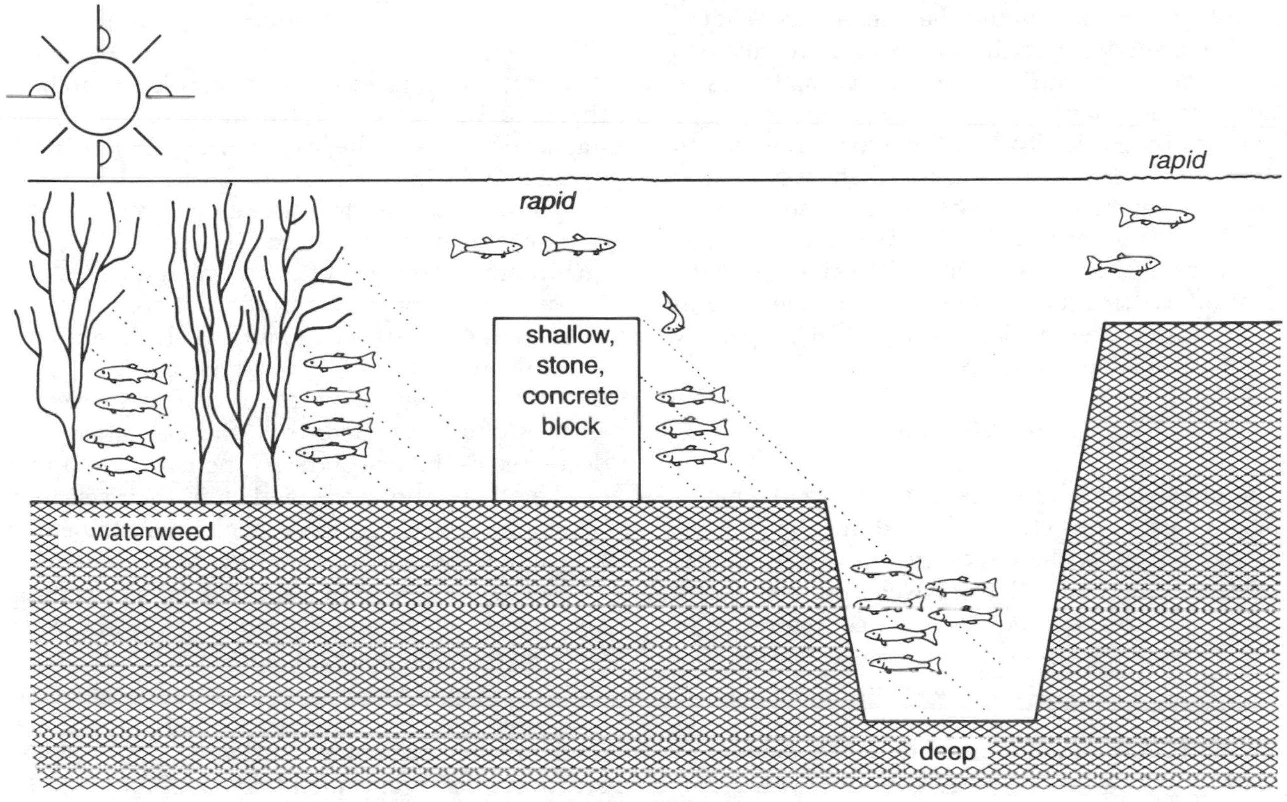

FIGURE 9

Schematic representation of chum salmon fry in fresh water showing areas of lingering in the shade (negative phototaxis). Fry migrated downstream in flatbottom, sunny areas. The fish in the rapids exhibit negative rheotaxis. (From Iwata 1982a)

and chrysalis of chironomids, mayfly larvae, *Trichoptera*, and other insects. The chum salmon migrating downstream in the rivers of the Sea of Okhotsk coast of Sakhalin and in the Lovetskaya River, Sea of Japan, feed much more intensively than pink salmon, and the intensity of their feeding increases towards the end of the downstream movement; the feeding spectrum also appears to be broader than that of the pink salmon (Frolenko 1970). The staples in the freshwater diet are benthic chironomids, and *Ephemeroptera* and *Plecoptera* larvae.

The chum fry released from the Chitose River hatchery in Hokkaido gradually increase in size as they migrate downstream. Their growth rate is low in March and April, the time of the normal migration (Kobayashi and Ishikawa 1964), and increases in May and June for both the fry from late or early releases that are already in the lower reaches of the river. Chironomids are the most abundant of the benthic invertebrates in their diet. Similar findings were reported by Mayama (1976) on the Anabetsu River. According to Kobayashi and Abe (1977), cited by Iwata (1982a), a small number of fry released from a hatchery were observed feeding in a river 1.5 months after release. Kaeriyama (1986) documented feeding and growth of the "river type" of chum fry. Delayed migrations of actively feeding fry have been observed in several Washington streams (Salo and Noble 1954; Tyler and Bevan 1964).

Downstream migrating chum salmon in the Yukon River range in size from 29 to 107 mm, with the majority being less than 70 mm. Differences in time of emergence were cited by Martin et al. (1986) as contributing to the variance; however, this implies that some of the groups feed in the river.

Kobayashi (1960) noted that feeding accelerated after sunset when light intensity is reduced to about zero lux during both active and passive downstream migration. Daytime feeding could be induced by artificially supplying food. The growth rate of fry that fed during the night was greater than that of fry fed during daytime (also reported by S. Schroder and K. Fresh, Washington Department of Fisheries, Olympia, Washington, pers. comm.). Other reports of chum salmon feeding in fresh water are in Baggerman (1960), Sparrow (1968), and Bailey et al. (1975).

Smoltification

The physiological changes associated with smoltification are not as clearly defined in chum salmon as in species with longer freshwater residence. Iwata (1982b) noted that fry weighing from 0.4 to 1.8 g had a plasma sodium (Na^+) concentration of 130–140 mmol/l. Upon reaching the estuarine area (salinity 2–25 ppt), the Na^+ level increased slightly to 150–160 mmol/l. The osmoregulatory ability of chum fry decreased gradually when they were reared for an extended period in fresh water.

Chum salmon achieve maximum ATPase activities at sizes 48–55 mm (0.9–1.2 g). After introduction into brackish water and, later, sea water, Na-K^+ ATPase specific activities increase rapidly after a short period of decline (3–24 hours) (H. Fuss, Washington Department of Fisheries, Olympia, Washington, pers. comm.). The increase is greatest in fish smaller than 45 mm. The basal levels in the larger fish are higher prior to introduction into brackish water.

Predation

Although there has been considerable research on the predation of chum fry in fresh water, the impact on total survival is not clear. Hunter (1959) found that during the course of a 2.6 km journey to sea, chum and pink fry mortalities averaged 45% (ranging from 22.6% to 85.5%, 1948–57). Other quantitative estimates of chum salmon losses due to predation vary in magnitude and are indicative of sampling differences (Table 5). In a series of experiments with Big Beef Creek chum salmon, Beall (1972) found that coho yearlings selected smaller chum fry and that the predation rate decreased as the size of the fry increased. Sculpins, on the other hand, were random in their selection of prey by size. No selection of chum fry (by either length or weight) by coho salmon or rainbow trout (*O. mykiss*) was found in a subsequent study at Big Beef Creek (Fresh and Schroder 1987). Chum fry with some yolk reserves appeared to be more vulnerable, however. Significant predation by coho and trout was limited to larger individuals, and predation by sculpins was insignificant. Fresh and Schroder (1987) concluded that satiation of predators can be reached in small streams by controlled release of fry. In Japan, Hiyama et al. (1972) released small (36 mm) and large (50 mm) chum fry into a small coastal river where, previously, losses were estimated to be 50%. Although the fate of all chum fry in their three-day experiment was not described, a greater percentage of the larger chum juveniles was recovered at a weir 3 km downstream.

TABLE 5

Estimates of freshwater mortality by predation for chum salmon

Location	Major predator species	Time period of estimate	Estimate (%)		Source
			Mean	Range	
Nile Cr. BC	–	Annual	47	35–62	Neave (1953)
Karymaiskiy Sp., USSR	Coho, charr	Annual	37	2–68	Semko (1954)
Hooknose Cr., BC	Coho, cottids	One week	58	33–85	Hunter (1959)
Big Beef Cr., WA	Coho, trout	48 hours	22	5–60	Fresh & Schroder (1987)
Big Beef Cr.*	Coho, cottids	24 h +	47–56†		Beall (1972)
Hooknose Cr.‡	Coho, cottids	Annual	45	23–85	Hunter (1959)

Source: *In a test aquarium and test stream gauntlet
†47% for those incubated in gravel, 56% for those incubated in open troughs
‡Includes pink salmon

Chum-Pink Salmon Interactions

Chum fry change their vertical distribution in the water column during downstream migration in years when pink salmon fry are present (Todd 1966). Semko (1954) noted that in 1944 (pink salmon present), 41.5% of the chum fry migrated in the light and 58.5% in the dark; whereas in 1945 (pink salmon absent), 1.2% migrated in the light and 98.8% in the dark. Somewhat the same pattern was repeated in 1946 when 29.2% migrated in the light and 70.8% in the dark, whereas in 1947, 1.8% migrated in the light and 98.2% in the dark. Thus, chum fry show a greater tendency to migrate during daylight hours in years when pink fry are present in the river system than when they are absent.

Early Sea Life

Chum salmon are second only to chinook salmon in dependence upon estuaries, and they may choose either the upper or lower estuaries, depending on the relative productivity of each. This selective use of habitats of differing salinities is made possible by the euryhaline tolerance of the fry (Kubo 1953; Baggerman 1960; Congleton et al. 1982).

Time of Entry into Sea Water

The timing of entry of juvenile chum salmon into sea water is commonly correlated with the warming of the nearshore waters and the accompanying plankton blooms. A model of optimal annual mean time of downstream migration and entry into the estuary, maximizing early-marine survival was developed by Walters et al. (1978). The model included parameters of (1) production of zooplankton, (2) the rations and growth of young salmon, (3) survival related to body size, and (4) timing of arrival of fry into the ocean. The predicted optimal mean time for saltwater entrance coincided closely with the known peak abundance of chum fry in the Fraser River estuary.

The median date of downstream migration and entry into sea water is directly related to latitude, with a variance from 31 March at 46°N to as late as 20 June at 57°N - 59°N (Godin 1982). The migrations occur from May to June in western Alaska and from April to June in southeastern Alaska. In Oregon, Washington, and southern British Columbia, migrations occur from February through May, and are earlier to the south. Variations in time of entry into estuaries, which can affect early marine survival, are caused by fluctuations in weather and stream runoff patterns.

Outmigrations from North American and mainland Asian streams are correlated in a broad sense with the warming of nearshore waters (March through June). In Honshu and Hokkaido, seaward movements occur earlier (March - May) and peak in late April, apparently as an adaptation to avoid the approaching warm (14°C) Tsushima Current (Irie 1985). Usually the juveniles have moved offshore by mid-June (Figures 10 and 11).

Martin et al. (1986) found that juvenile chum salmon of the Yukon River did not utilize the nearshore habitat of the delta because the outmigrants were widely distributed and occurred more frequently in the offshore waters than in the coastal habitats. The fish were dispersed by the large river plume, and the smaller fry (36.8–43.8 mm), which Martin et al. (1986) surmised to be summer-run stock, were particularly vulnerable to dispersion.

The behaviour patterns at entry and during estuarine residency appear to be consistent among the more typical North American estuaries (Healey 1982a). The young chum salmon spend up to three weeks rearing in the estuaries of the Fraser and Nanaimo rivers and occupy tidal creeks and sloughs high in the delta area (Healey 1982a). Their movements in and out of the estuary are correlated with the tides (Congleton et al. 1982). In Hood Canal, Washington, the initial distribution of the juveniles is widespread after entry into salt water,

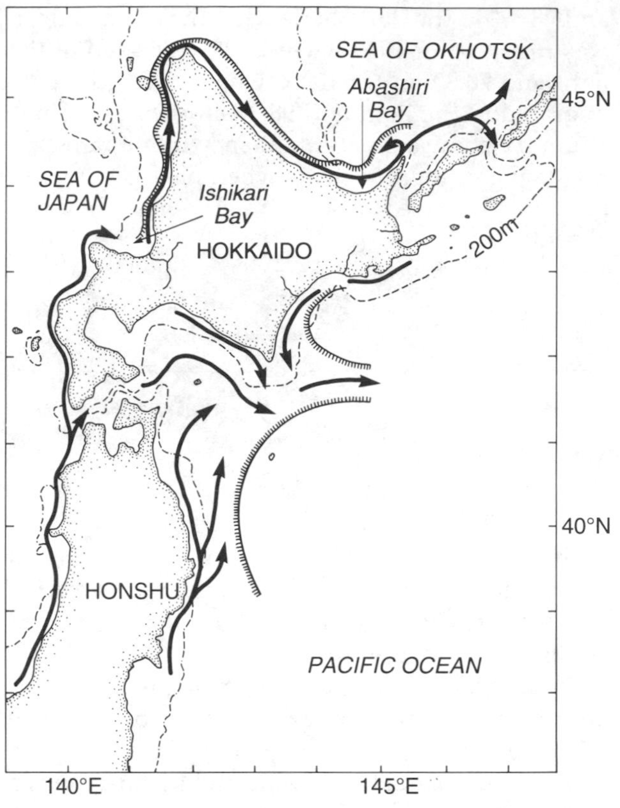

FIGURE 10

Estimated migration routes and distribution areas of young salmon in their early marine life. ➔ = migration route, ▥▥▥ = distribution area. (Adapted from Irie 1985)

and then changes to a distribution closely oriented to the shoreline until the fish are large enough to move offshore. Distributions resemble loose aggregations and the fish scatter after dark (Schreiner 1977; Bax 1983a).

In Hood Canal, the juveniles (<40 mm fork length) enter sea water in February and March and migrate rapidly (7–14 km/d), which suggests that food availability is low. Later in spring (May, June), as the epibenthic and neritic zooplankton increase, the migration rates decline (3–5 km/d) (Bax 1983a). As food resources decline in summer, the juveniles move offshore and the diet changes to pelagic and nectonic organisms (Simenstad and Salo 1982). This suggests that the pattern of outmigration of chum fry relates to availability of preferred prey organisms.

The migration of chum fry is a combination of active and passive movements. Correcting for the residual tidal outflow, Bax (1982) calculated that the active migration rate of chum fry is 7 km/d (8 cm/s) early in the season for migrants ranging from 42 to 61.5 mm in length. This corresponds to 1.9–1.3 body lengths (L)/s, which approximates the optimum (energetic) speed of 1 L/s, as derived by Weihs (1975) and Trump and Leggett (1980). Hatchery-reared and natural populations of chum juveniles emigrate out of Hood Canal at different rates at different times of the year. Only the slowest recorded rate approximates the optimum rate of 1 L/s. The highest recorded rate is 3.5 times as high as the optimum rate, and, if this rate is due to active swimming, it is bioenergetically inefficient (Bax 1982). During the outmigration, most of the juvenile chum salmon were on the down-current side of spits, irrespective of tide stage, indicating a selection of areas with lower velocities. Balchen (1976) described fish movement at any one time as a "simple minded process of maximizing comfort."

In Hood Canal the emigration rates of each size group decrease as the season progresses; also, the rates decrease as the size of the fish increases (Bax 1982). Migration rates for different size classes are positively correlated with the estimated residual surface outflow and are dependent on the density of pink and chum salmon juveniles of the same size class (Figure 12).

The migration rate of chum salmon, therefore, changes with the season, the size of the fish, and with the density of outmigrants. Emigration rates change from year to year (Cooney et al. 1978; Healey 1979). Mason (1974) suggested that juvenile chum salmon use a repertoire of behavioural responses to delay seaward movement. Fry released from hatcheries on Hood Canal, Washington, showed varied but delayed migration rates with later release dates (Figure 13) (C.R. Simenstad, J.R. Cordell, and R.C. Wissmar, University of Washington, Seattle, Washington, pers. comm.).

Feeding and Offshore Movement

As the chum salmon fry leave the rivers they begin feeding in estuaries and shallow nearshore marine habitats on epibenthic and neritic food resources (Okada and Taniguchi 1971; Mason 1974; Feller and Kaczynski 1975; Healey 1980; Simenstad et al. 1982). The food web supporting juvenile salmonids in the estuarine habitat appears to be detritus-

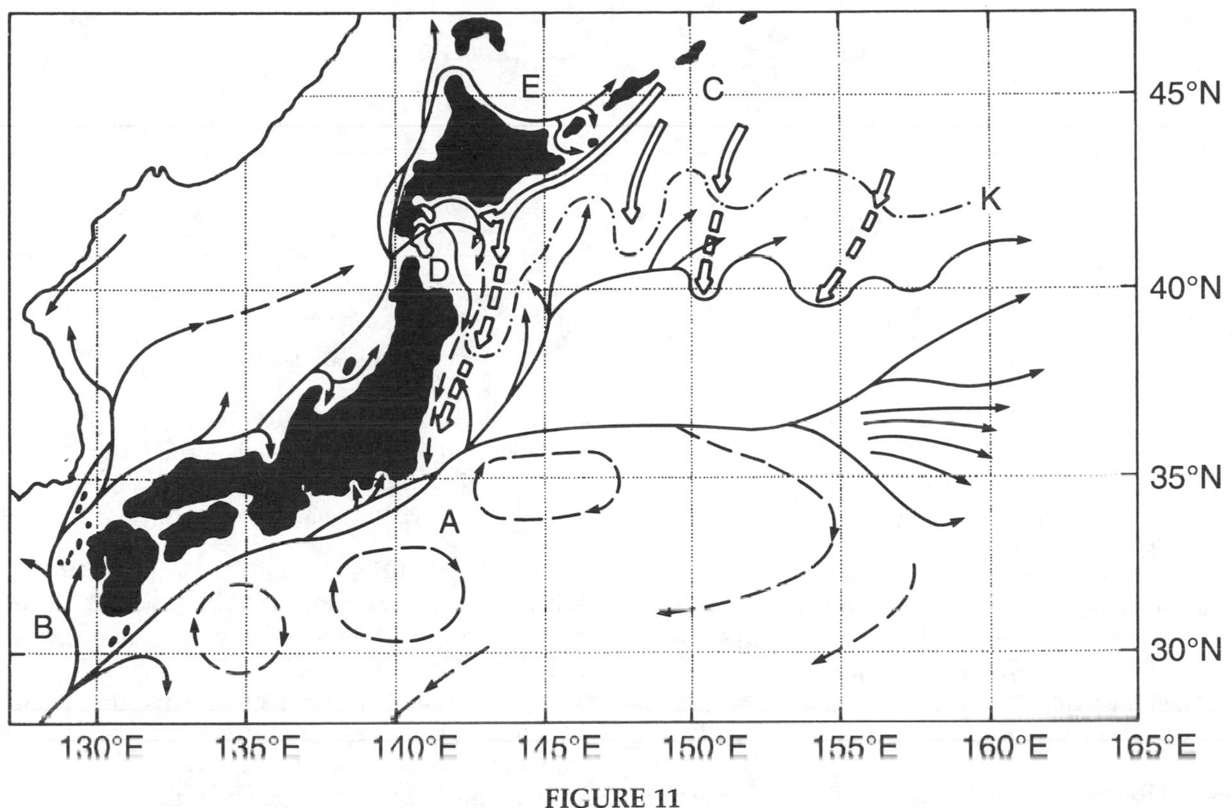

FIGURE 11

A schematic of the upper structure of currents and waters around Japan. *A* = Kuroshio Current; *B* = Tsushima Warm Current; *C* = Oyashio Current (cold); *D* = Tsugaru Warm Current; *E* = Soya Warm Current; *K* = boundary of the cold and the warm water mass. (Adapted from Irie 1985)

based. In terms of grams of carbon/m²/year, the Nanaimo River estuary, for example, produces 0.9 g of macrophytic algae and 73 g of *Zostera marina*. The marshland produces about 560 g/m²/y (mainly Carex), but the river discharges over 2,000 g/m²/y (Naiman and Sibert 1979; Healey 1982a). Thus, a prime rearing area combines high carbon input from freshwater sources with adequate wetland and intertidal areas for conversion of the carbon to forms available to the chum salmon (Simenstad et al. 1982).

In the nearshore waters of Puget Sound and British Columbia, the diet of chum juveniles is dominated by harpacticoid copepods and gammarid amphipods (Kaczynski et al. 1973; Simenstad et al. 1980; Simenstad and Salo 1982; Simenstad et al. 1982). Some investigators found that the early diet consists almost exclusively of harpacticoid copepods (Kaczynski et al. 1973; Sibert et al. 1977; Healey 1980). A single harpacticoid, *Harpacticus uniremis*, is the major prey item in the Nanaimo River estuary (Sibert 1979; Healey 1980) and Auke Bay, Alaska (Landingham 1982). Dipterans, primarily chironomids, are the predominant prey for chum salmon in high salt marsh habitats in the Skagit River (Congleton 1979) and in estuaries of Vancouver Island (Mason 1974). Diptera are also the predominant prey in Kotzebue Sound, Alaska (Merritt and Raymond 1983), whereas in the Fraser River estuary the diet consists of harpacticoid copepods, chironomid larvae and pupae, adult insects, and amphipods (Levy and Northcote 1981).

Diel changes in feeding activity and diet composition of juvenile chum salmon in relation to their euryhaline tendencies are evident in salt marsh habitats where they feed successfully on freshwater, estuarine, and marine organisms during a tidal cycle (Congleton 1979). The feeding was most intense at high tides when the marsh was sub-

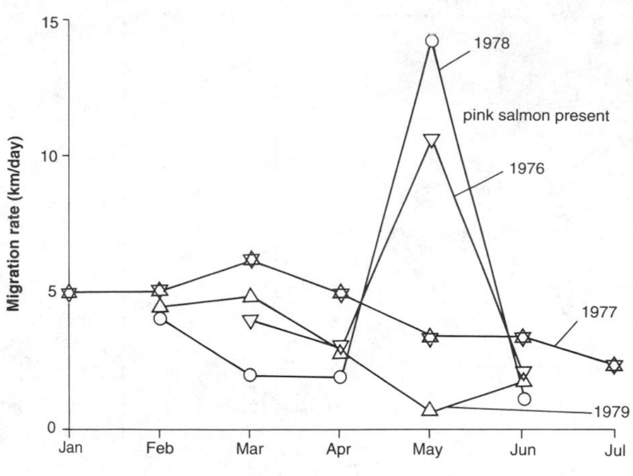

FIGURE 12

The most common migration rates and the rates at which distinguishable proportions of the 45-59 mm size group of juvenile chum salmon (intercepted at Bangor Annex, Hood Canal, Washington), migrated from 1976-79. The points are joined to emphasize consistent monthly patterns. Each point represents a migration rate which produced a minimum value of the least squares statistic. (Adapted from Bax 1983a)

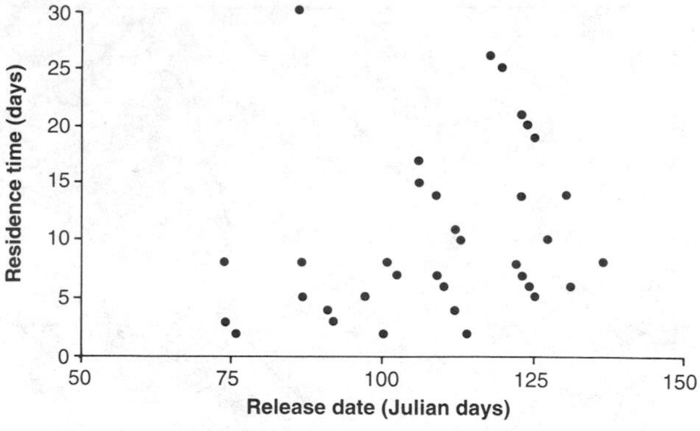

FIGURE 13

Relationship between release date and residence time of juvenile chum salmon monitored at a distance of 20 km from release point

merged for four to five hours. During this time the water was from 0.3 to 1.0 m deep. Chum fry exploit both the freshwater and marine food webs and, by so doing, are exposed to marked daily fluctuations (0-27 ppt) in salinity (Mason 1974).

The movement offshore generally coincides with the decline of inshore prey resources and is normally at the time when the fish have grown to a size that allows them to feed upon larger neritic organisms and avoid predators. This transition takes place in Puget Sound at 45-55 mm fork length (Simenstad et al. 1980; Simenstad and Salo 1982). In Prince William Sound, Alaska, the offshore migration is at 60 mm fork length (Cooney et al. 1978).

Chum salmon tend to be size (larger) and taxa selective with respect to the available planktonic assemblage (Healey 1980; Simenstad et al. 1982; Cordell 1986). The most important planktonic organisms eaten are calanoid copepods, hyperiid amphipods, larvaceans, euphausiids, chaetognaths, decapod larvae, and fish larvae (Okada and Taniguchi 1971; Healey 1980; Simenstad et al. 1980; Simenstad and Salo 1982; Simenstad et al. 1982). Selectivity for large prey is apparent in both the epibenthic and neritic feeding and is attributed to (1) visual perception, (2) active selection associated with functional morphology (gape), and (3) optimization of bioenergetics of foraging (Simenstad and Salo 1982; Cordell 1986).

In the Strait of Georgia, chum juveniles occupy the shallow waters until early June when they move to waters 20-40 m deep. The movement out of the strait is in July, with the larger fish tending to move first (Healey 1980). Healey believes that, although some of the movements could be innate and related to age and ontogeny, occupation of a particular habitat is in some instances related to foraging success.

Kaeriyama (1986) divided the migrating fry and fingerlings into three groups: (1) the "river" type, which remain in the river until they are large enough to migrate offshore as the Oyashio Current approaches; (2) the "foraging migration" type, which migrate to the coastal region (February-March) where they forage until the Oyashio Current approaches; and (3) the "escape foragers," which migrate to low salinity shoreline and estuarine waters where they forage until the Oyashio Current retreats in June-July.

The distribution of chum salmon in nearshore and offshore waters of northern Hokkaido (Abashiri Bay) is closely related to water temperature and the distribution of zooplankton in May and June. As the warmer currents move near shore, the plankton populations, which consist primarily of oceanic cold-water species, move farther offshore (Figure 14). The increased demand for food in the warmer waters contributes to forcing the fish

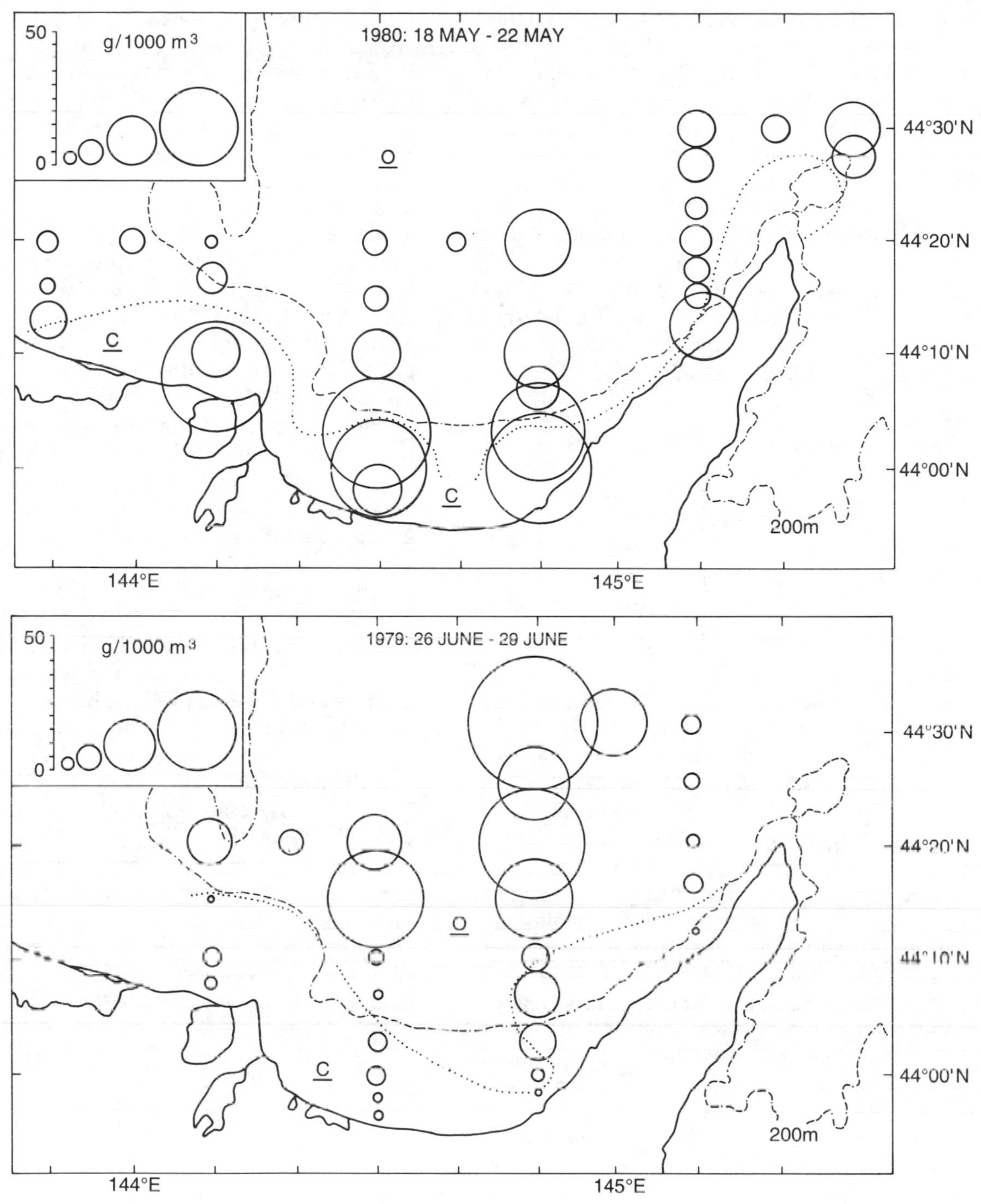

FIGURE 14

Distribution of zooplankton and water masses in Abashiri Bay. Density of zooplankton is related to diameter of circle. Dotted lines indicate the boundary of two water masses. O = surface water of the Sea of Okhotsk, C = coastal water mass. (Adapted from Irie 1985)

north and offshore (Irie 1985). Prior to the changes in temperature, the smaller fish (30-50 mm fork length) feed on small zooplankton in the near-coastal areas whereas the larger chum salmon (50-110 mm) feed on the larger zooplankton in the offshore water masses. The relationship between size of juvenile chum salmon and the characteristic of the water mass in which they are found is presented schematically in Figure 15. In some Oregon estuaries, juvenile chum salmon of a similar size leave regardless of their time of entry (Bax 1983a), and larger juveniles entering the estuary later in the season emigrate faster than those entering earlier.

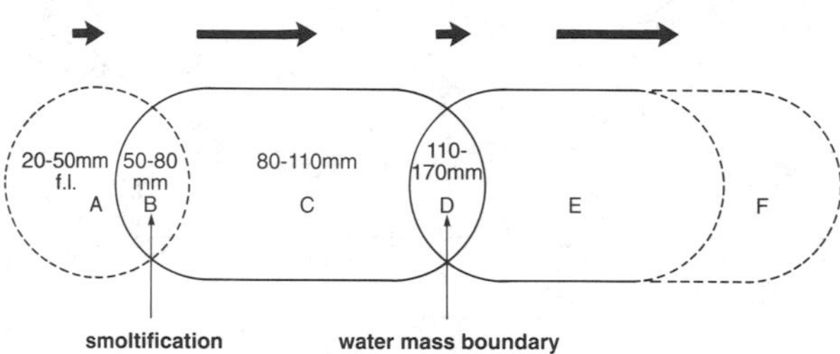

FIGURE 15

A schematic of the offshore migration pattern of juvenile chum salmon by size in relation to water mass. *A* = fresh water, *B* = brackish water, *C* = coastal water, *D* = cold or warm water with lower salinity, *E* = offshore cold water, *F* = subarctic current water, ➜ = migration course. (Adapted from Irie 1985)

Although most juvenile chum salmon migrate rapidly from fresh water to shallow nearshore marine habitats after emergence from the gravel beds, some may remain up to a year in fresh water in large northern rivers. In these rivers their diet consists primarily of insects, with dipterans and plecopterans as the most common prey in the Noatak River (Kotzebue Sound, Alaska) (Merrit and Raymond 1983), and Chironomidae the most common in the Amur River system (Levanidov and Levanidova 1957).

Survival Rates

Early marine residence is considered to be a critical time in the marine life history of chum salmon (Bax 1983a), especially because of possible food limitations at time of entry (Gunsolus 1978; Helle 1979; Gallagher 1979; Simenstad and Salo 1982). Bax (1983a) and Mason (1974) pointed out that although early-marine mortality in nearshore areas may be high, the factors controlling the numbers of adults returning (year-class strength) may occur in the offshore or open ocean regimes. Each locale has a variance in the median date of entry of chum fry into marine waters that may contribute to varying annual rates of survival. Bax (1983a) assumed that the emigration rate, by affecting the time at which the juveniles migrate from Puget Sound and reach the coastal water masses, may be an important factor influencing overall survival. Walters et al. (1978), quoted in Godin (1982), showed survival rates from 0.15 for 15 March and 15 July seawater entries to 0.25 for 15 May entries. In Hood Canal the average daily mortality of juvenile chum salmon was 31% and 46% over a two- and four-day period (Bax 1983b), and the survival of another group was 42%, two days after release (Whitmus 1985). Whitmus (1985) found that the size of a juvenile chum salmon influenced its emigration rate. Although a large percentage of fish from different-sized cohorts resided in Hood Canal for up to four weeks, the larger fish were the first to migrate and did not suffer the higher rates of mortality experienced by smaller fish. Smaller chum salmon, in the prey size range of 43-63 mm fork length, were chosen by coho (112-130 mm fork length) in large saltwater enclosures (Hargreaves and LeBrasseur 1986).

The recent increases in survival rates (about 2%) of chum salmon released from Japanese hatcheries have been facilitated by feeding the fry so that they are 70 mm in length and 3 g in weight by the time the coastal waters reach 12°-13°C. At that size, the fish are able to move north and east out to sea.

Ocean Distribution

Research by member nations of the International North Pacific Fisheries Commission (INPFC)[4] since 1955 has contributed greatly to the understanding of the distribution and movements of chum salmon during the high-seas phase (summarized by Manzer et al. 1965; Shepard et al. 1968; Neave et al. 1976; Hartt and Dell 1986). Based on this ongoing research, a conceptual "working model" of chum salmon migrations throughout the North Pacific Ocean and adjacent seas has been advanced.

4 INPFC was established in 1953 by the International Convention for the High Seas Fisheries of the North Pacific Ocean, a treaty between Japan, Canada, and the United States

General Offshore Distribution

Some general features of the offshore distribution of Asian and North American chum salmon (all ages and maturity groups combined) are illustrated in Figure 16. The eastward extension (to at least 140°W) of Asian chum salmon shows a more distant migration than the North American chum salmon, which are not commonly found west of 175°E. The westernmost occurrence of a North American chum salmon, as indicated by tagging, was a Yukon River fish tagged in July at 60°09′N, 174°30′E near the USSR coast (Neave et al. 1976). Recent analyses of scale patterns by Japanese scientists provisionally indicate a broader overlap of

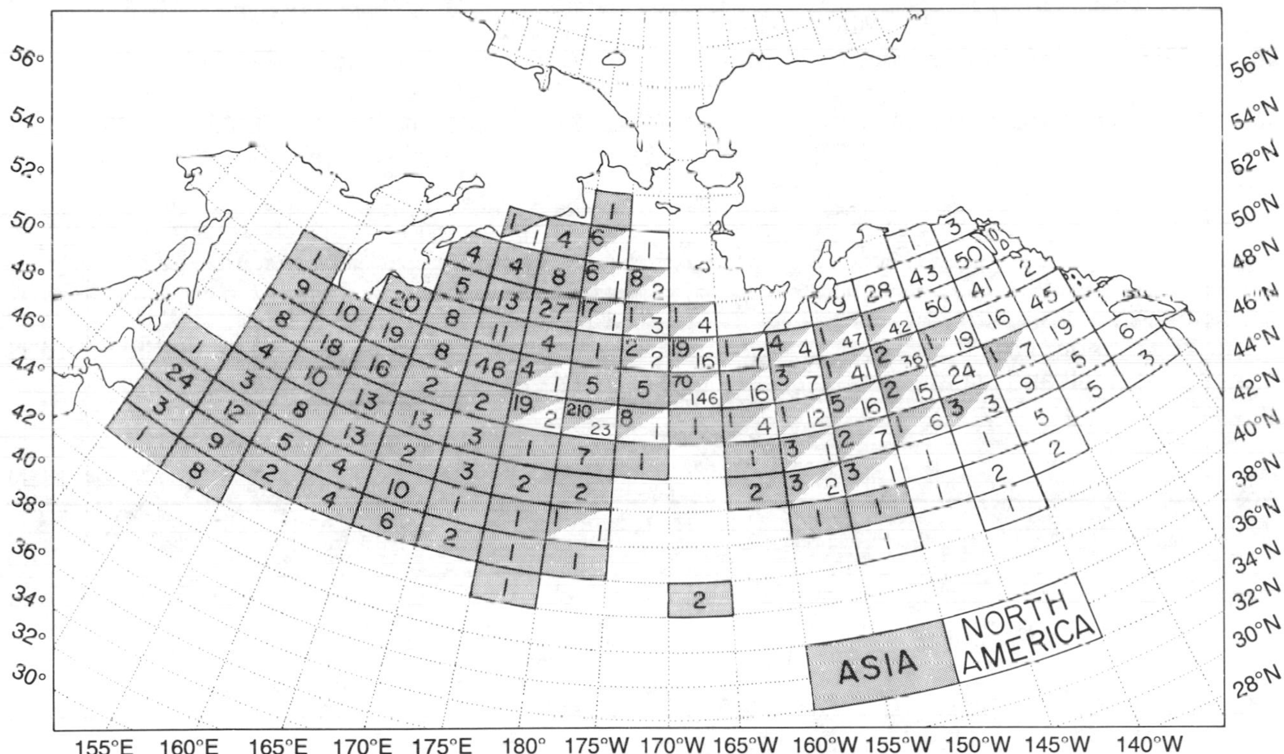

FIGURE 16

Recoveries of chum salmon tagged on the high seas by participants in INPFC-related research (1956–84), summarized by continent of origin and 2°×5° sector of release, for all ages combined. (Figure provided by Colin Harris, Fisheries Research Institute, University of Washington, Seattle, WA)

North American and Asian chum salmon than do tagging studies (Colin Harris, Fisheries Research Institute, Seattle, Washington, pers. comm.; Ishida et al. 1983).

Offshore Distribution of Juveniles

The distribution of juvenile Asian chum salmon in the Sea of Okhotsk changes from coastal waters in early summer to 160 km or more offshore by August and extends across the Sea of Okhotsk from Sakhalin to the Kamchatka coast by September. Movement into the North Pacific Ocean begins in October and lasts until December. The offshore migration, which starts from Hokkaido in June, places the juvenile chum salmon (age 0.0) well into the western North Pacific Ocean by July.

North American chum salmon (age 0.0) occur together with juvenile sockeye and pink salmon along the coast of North America in a narrow band that broadens to about 36 km offshore in southeastern Alaska (Hartt 1980). Hartt inferred from catch-per-unit-effort data, directional seine sets, and tag returns, that stocks from southern locations entered the ocean first, followed by stocks from more northern and western locations as the season progressed. The juveniles apparently migrate northerly, westerly, and southwesterly along the coastal belt of the Gulf of Alaska (Figure 17). Chum and sockeye juveniles tend to remain nearshore, whereas juvenile coho and chinook salmon and steelhead trout occur in both coastal as well as offshore catches, indicating a more diverse migratory pattern.

In the southeastern Bering Sea, substantial numbers of juvenile chum salmon are only 460–555 km from their estuaries of origin by September, indicating a less extensive and slower migration than the Gulf of Alaska stocks during their first summer at sea (Figure 17). The direction of migration of these stocks is variable and appears to be strongly influenced by tidal currents. A net movement to the southwest is evident, and densities are highest along the north side of the Alaska Peninsula, within 93–110 km of the shore. The full southwestward extent of this migration in September is unknown (Hartt 1980).

Although a "working model" of these juvenile

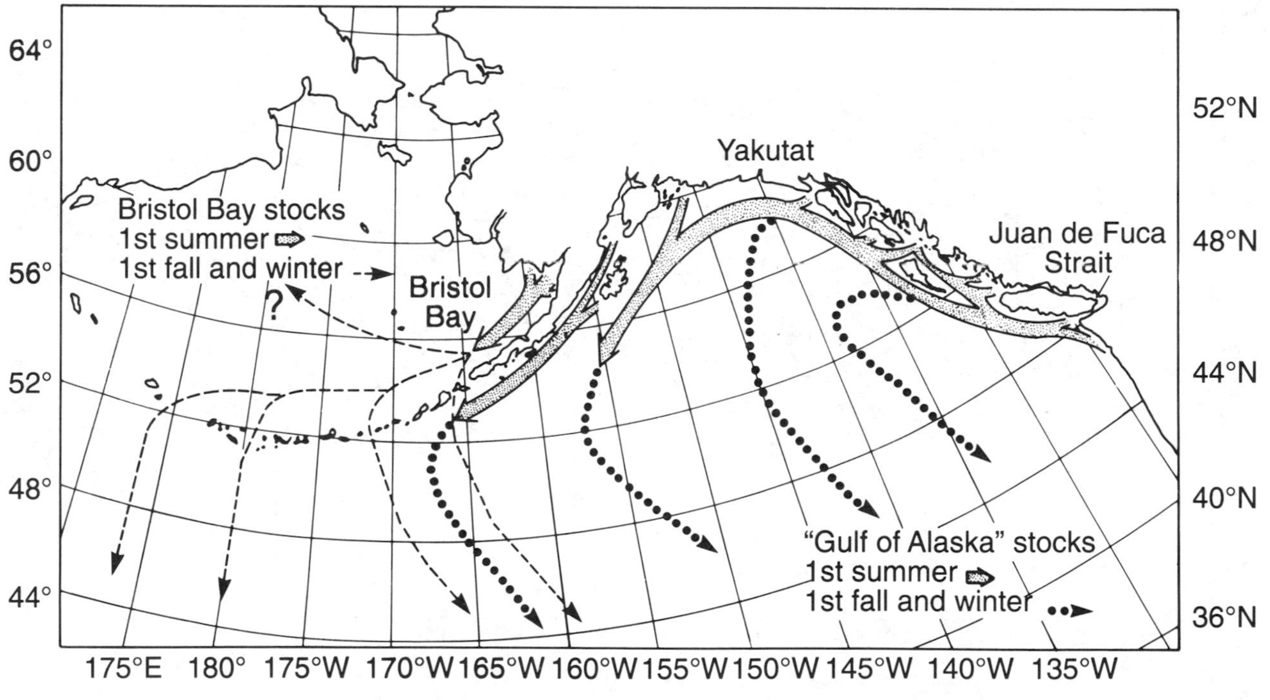

FIGURE 17

Oceanic migration patterns of some major stocks of North American sockeye, chum, and pink salmon during their first summer at sea, plus probable migrations during their first fall and winter. (From Hartt 1980)

chum salmon migrations during their first summer at sea is possible, many questions remain unanswered. It is unclear to what extent chum salmon reside in inside and nearshore waters prior to their embarkation on the coastal belt migration. Hartt and Dell (1986) reported that some juvenile chum salmon are still present in northern Hecate Strait and in central Puget Sound during November. Residual chum salmon juveniles in Puget Sound averaged 23 cm in November, indicating good growth conditions. Although Puget Sound fish were clearly local, those sampled in Hecate Strait may have migrated from elsewhere and re-entered inside waters while migrating northward (Hartt 1980). Jensen (1956) tagged age 0.1 (30–40 cm) chum salmon in central Puget Sound in June, and the recoveries indicated that they moved northward to the Strait of Georgia and the west coast of Vancouver Island shortly after release.

Another unclear aspect of juvenile chum salmon migration is the pattern of offshore dispersion from the coastal belt in fall and winter, or in the following spring (Jensen 1956). At present, migrations during this period are inferred primarily from the distribution of age 0.1 chum sampled in the spring. Late migrants (e.g., the residuals of Puget Sound) may not make the extended northwest migration along the coastal belt depicted in Figure 17 but, rather, may proceed more directly offshore from their point of ocean entry (Hartt 1980).

Offshore Distribution of Immature Chum Salmon

Fish of age 0.1 (after 1 January) and older are referred to as "immatures" throughout their ocean residence until the year they return to their spawning streams; during this final year at sea, they are referred to as "maturing." Immature chum salmon from Asia and North America form two concentrations that overlap in the eastern North Pacific Ocean after the first winter but appear to migrate independently in the following spring and summer (Figures 18 and 19) (Neave et al. 1976; Fredin et al. 1977). Asian chum salmon from the larger of the two groups are distributed widely throughout the Bering Sea and the central and western North Pacific Ocean. During late summer, fall, and winter the migration of this group is generally southward and extends into waters well south of 50°N in the winter (Figure 18A). During this period, fish in the western North Pacific are distributed farther south than fish in the central North Pacific Ocean. This east-west cline in the southern range coincides with differences in the sea surface temperature across the northern Pacific Ocean.

Tagging experiments have indicated some partitioning by age groups within this distribution. Apparently, by the end of their first year at sea, the majority of Asian chum salmon (age 0.1) are distributed only as far east as the central Aleutian Islands and do not intermingle significantly with age 0.1 fish of North American origin.[5] Asian chum salmon of age 0.2, however, show a more pronounced overlap in distribution with North American fish of the same age. Increased intermingling between Asian and North American stocks is believed to result from an eastward and southeastward extension of Asian chum salmon into the northeastern Pacific Ocean during their second and third winters at sea (Neave et al. 1976).

By spring, and continuing through early summer, Asian chum salmon not destined to mature that year are moving northward (Figure 18B). Most move into the Bering Sea, but a considerable number remain in the western and central North Pacific Ocean. This migration is led by fish of age 0.2 and older in May, followed by fish of age 0.1 in late June. In July, a rapid movement of immature fish into the Bering Sea occurs, and abundance declines in the central and western North Pacific Ocean. The distribution of these stocks is believed to reach its most northern location in late August (Neave et al. 1976).

By comparison, immature chum of North American origin are more restricted in their high-seas distribution (Figures 19A and 19B). By late fall or winter, most of the newly arriving western Alaska chum salmon have left the Bering Sea to join immatures from more southerly locations. Migration during late summer, fall, and winter occurs in a broad southeasterly fashion, primarily south of 50°N and east of 155°W in the Gulf of Alaska (Figure 19A).

During spring and early summer, immature

5 Very few age 0.1 fish have been sampled throughout the North Pacific and the Bering Sea in winter, so the winter distribution is not well known. By the time age 0.1 chum salmon appear in catches in the spring a broad distribution is indicated (Neave et al. 1976).

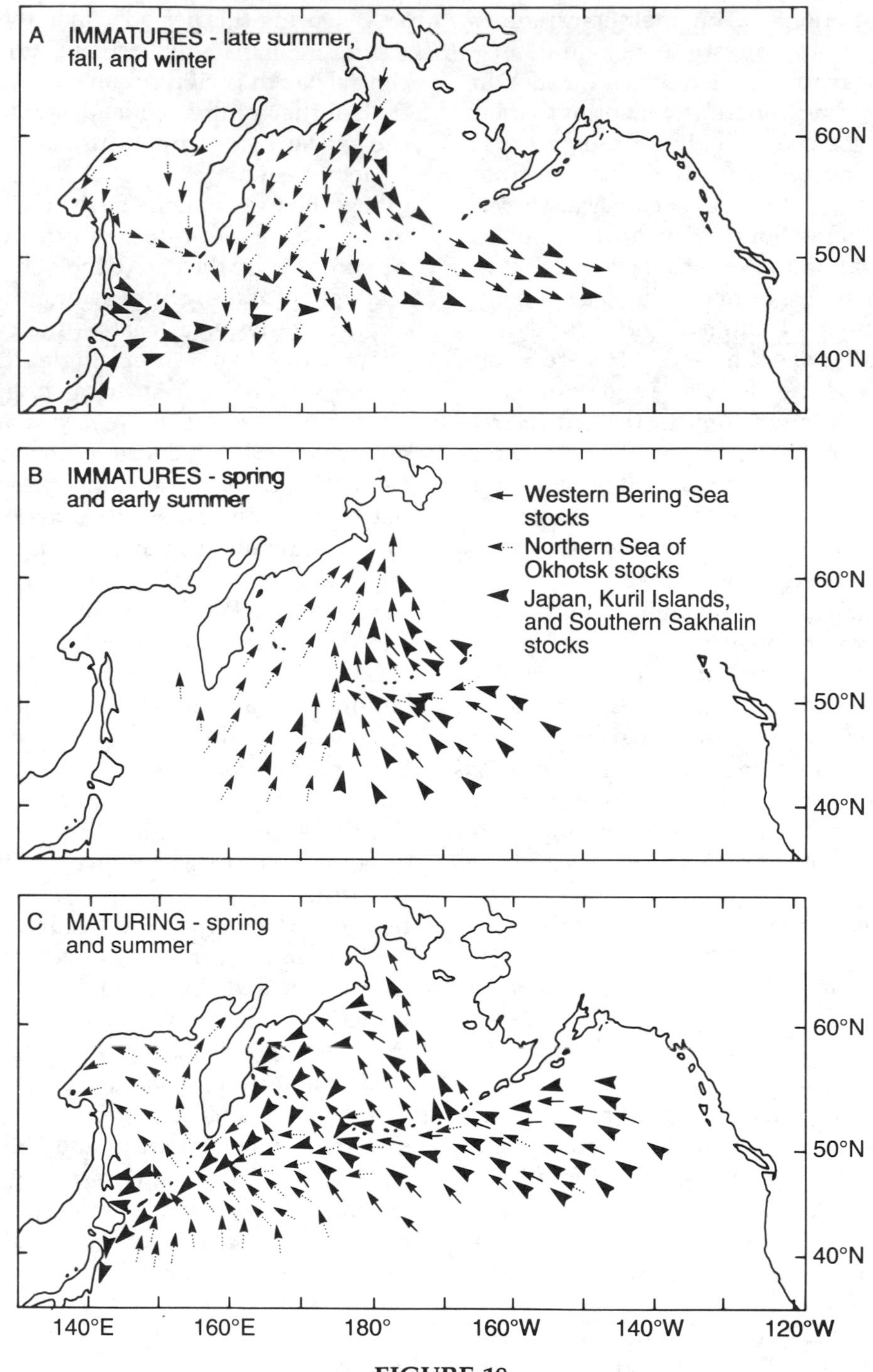

FIGURE 18

Model of migration of Asian chum salmon. (From Fredin et al. 1977)

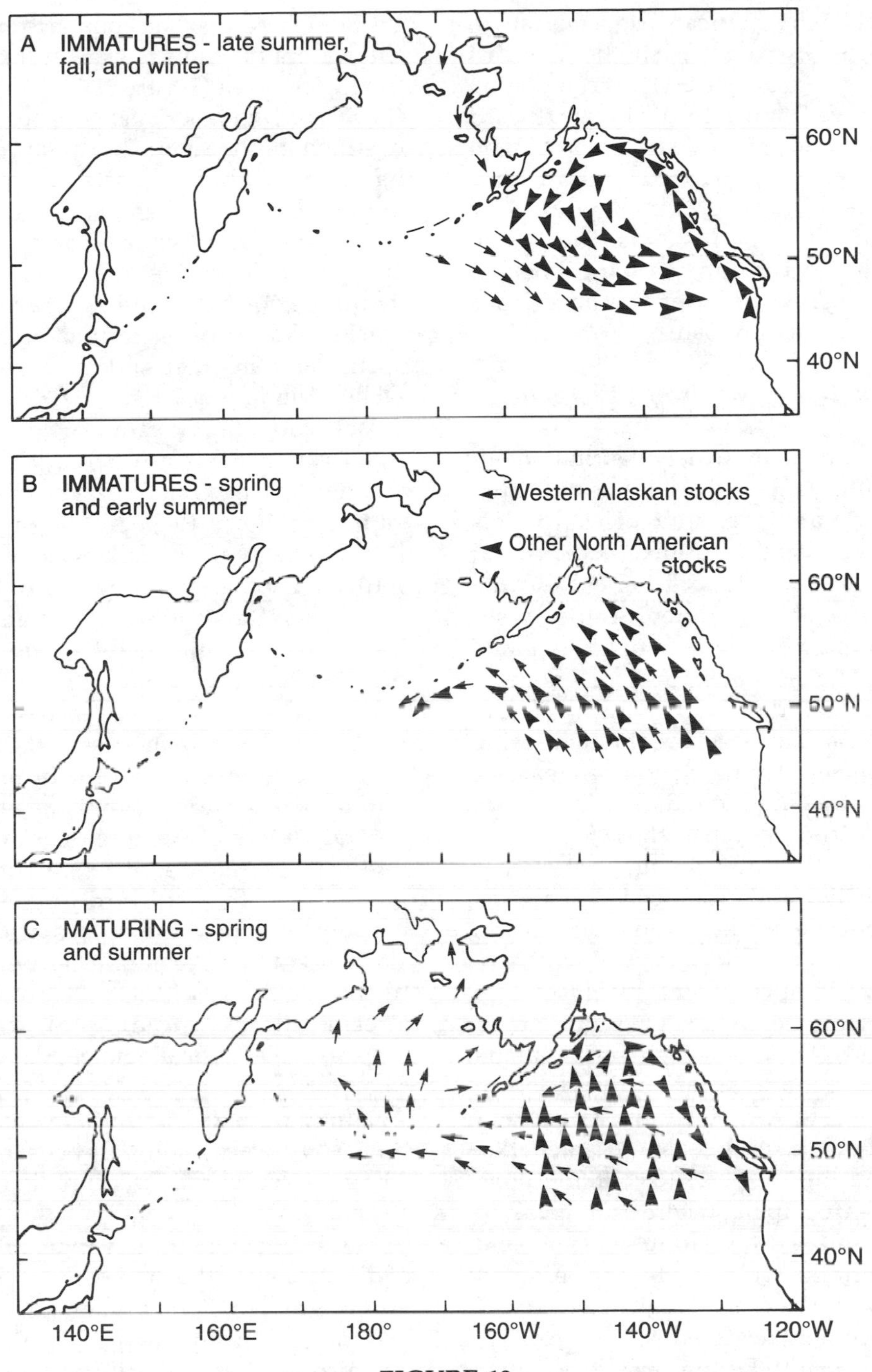

FIGURE 19

Model of migration of North American chum salmon. (From Fredin et al. 1977)

chum salmon of North American origin migrate to the north and west, with older fish in the vanguard followed by younger ones (age 0.1) just completing their first year at sea (Figure 19B). This northward movement lags behind that of Asian chum salmon in spring, and is not as pronounced until late June. A minor component of this group migrates as far west as the central Aleutian Islands. There is no evidence to indicate that any chum salmon of North American origin re-enter the Bering Sea prior to returning as mature fish.

Offshore Distribution of Maturing Chum Salmon

Maturing chum salmon are widely distributed offshore, but the abundance is variable as indicated by differences in catch per unit of effort. This variability is partially due to mixed stocks and seasonal migrations (C. Harris, Fisheries Research Institute, University of Washington, Seattle, Washington, pers. comm.). The pattern of formation of aggregations and the movement inshore is not well known, but the abundance of maturing chum salmon declines gradually offshore through spring and summer as the intermingled stocks depart for their respective spawning grounds.

For some Asian stocks, such as those from Japan and eastern Kamchatka, the spawning migration is particularly extensive and prolonged. Distributed widely throughout the northeastern Pacific Ocean in April and May, these stocks migrate northwestward or westward in spring, and travel well into the Bering Sea before heading southward to their streams of origin (Figure 18C). Chum salmon bound for Japan are found as far north as 64°N in the Bering Sea in late July; by August and September most are off the east coast of Kamchatka. The southern migration appears to continue along the Kuril Islands to Japan, where runs peak in October and November. Most chum salmon bound for eastern Kamchatka enter the Bering Sea early (June) and approach the coastal regions by July.

Asian chum salmon destined for the Anadyr River in the northwestern Bering Sea also occupy a wide part of the Gulf of Alaska (primarily west of 149°W) in April and May. These stocks enter the Bering Sea in large numbers in June and follow a migratory pattern similar to the early movements of the eastern Kamchatka stocks.

Chum salmon returning to coastal streams of the Sea of Okhotsk, by comparison, are confined to the waters off Kamchatka by the spring of their final year at sea (Figure 18C). Peak migration into the Sea of Okhotsk appears to occur earlier for fish destined for the Amur River and the northern Sakhalin and Okhotsk coasts, as compared to fish bound for western Kamchatka. The former stocks enter the Sea of Okhotsk primarily in June and subsequently reside offshore for one or two months prior to ascending the rivers. The latter stocks tend to be distributed farther to the east in spring and summer and migrate into the Sea of Okhotsk in July or later.

Maturing chum salmon destined for North American streams are found widely distributed throughout the Gulf of Alaska during spring and summer (Figure 19C). Although some fish move south of 50°N in the fall or winter preceding the return for spawning, the majority remain in more northerly waters and shift eastward sometime during winter. Chum salmon bound for western Alaska begin their homeward migration from diverse locations as far east as near the British Columbia coast and as far west as the central Aleutian Islands. The primary movement through the Aleutian passes occurs in June, and runs peak in the coastal waters of western Alaska in July. Stocks destined for coastal streams throughout the Gulf of Alaska migrate northward from May to July. Migration rates of these stocks are believed to be slower than those of the western Alaska stocks due to the proximity of these fish to their spawning streams. Stocks heading for central Alaskan streams move basically north, whereas stocks destined for southeastern Alaska, British Columbia, Washington, and Oregon move north, then east, then southward along the coast to their home streams (Neave et al. 1976; Fredin et al. 1977; Hartt 1980).

As the rate of migration appears to be a function of distance from the more-or-less common feeding grounds, Asian chum salmon migrate the greatest distance, followed, in range of migration, by Oregon, Washington, and British Columbia stocks. The stocks from western Alaska have the shortest distance to travel to their rearing areas.

Feeding in the Ocean

The rapid growth of salmon in the sea attests to

their intensive and effective feeding behaviour. Salmon continue to feed up to the time of their spawning migration, with greater variances in the fullness of the stomach as they approach the coastal waters. This variance was thought by Birman (1960) to be a function of availability of food, but it is also related to cessation of feeding as the gonads mature.

Birman (1960) reported that during their last summer at sea in 1955 (June-July) up to 40% of the food of chum salmon was made up of pteropods, with euphausiids being second (17%), followed by salps (10%) and miscellaneous (10%). At the same time, the food of pink salmon consisted of fish (30%), euphausiids (27%), and hyperiids (23%). In May-June of 1956, euphausiids made up 60% of the chum salmon's diet, with pteropods being second (17%), and, by August of 1956, fish constituted 52% of their diet.

In a study in the Bering sea in the summer of 1966, Kanno and Hamai (1971) found that there were great differences in the composition of food organisms among three species of salmon (chum, pink, and sockeye) and among the areas that the fish were caught. In chum salmon there were noticeable differences in the dominant organisms in the diet among four areas of the Bering Sea: copepods in the western, pteropods in the southern, euphausiids in the eastern (including Bristol Bay), and amphipods in the northern Bering Sea. There were also differences between chum and sockeye salmon and between chum and pink salmon. Although squid was the dominant organism in the diet of sockeye and pink salmon, it was absent in the stomachs of chum salmon. The order of incidence of food organisms in the stomachs of chum salmon was (1) amphipod, euphausiid, pteropod, and copepod; (2) fish; and (3) squid larvae.

Age, Growth, and Survival

Chum salmon from different regions differ in growth rate, age at maturity, and size at maturity. These differences result from genetic variability among stocks and habitat heterogeneity in the nearshore and North Pacific Ocean environments.

In the marine environment, Asian chum grow faster (greater instantaneous growth) than North American fish; however, North American chum are usually larger at each stage of their marine life. The faster growth rate of Asian chum may be a genetic (bioenergetic) adaptation to relatively poor habitat conditions in the western North Pacific Ocean. LaLanne (1971) and Ricker (1964) suggested that the growing season is longer for North American fish, and studies of chum salmon scales indicated that growth of these fish begins earlier and ends later in the Gulf of Alaska than in the western North Pacific Ocean. Ricker (1964), based on Kobayashi's (1959) data, concluded that, on average, chum salmon in the western Pacific Ocean complete their annulus about 1 May, whereas February or March is the likely time of annulus completion for fish in the Gulf of Alaska (Bilton and Ludwig 1966). Although most of the growth occurs from May through October, with the bulk of it occurring in June and July (Koo 1961), some growth takes place in the winter and early spring. Little data exist on the precise time of seasonal decrease of growth (beginning of annulus formation), but Bilton and Ludwig (1966) indicated that this may occur in November in the Gulf of Alaska. Koo (1961) found that growth decreases about the end of September for chum salmon located south of the Aleutians. These dates likely vary from year to year with general temperature trends in the North Pacific Ocean.

An example of the greater ultimate growth for the North American chum salmon is illustrated in the schematic growth diagrams (Figures 20 and 21). In the model, summer growth for the Amur River fish is assumed to begin on 1 May and end 30 September for immature fish, with the final year of growth ending on 1 September. This is the peak of the autumn chum run in the Amur River, according to Ricker (1964). For calculation of growth, both immature and maturing Minter Creek fish are as-

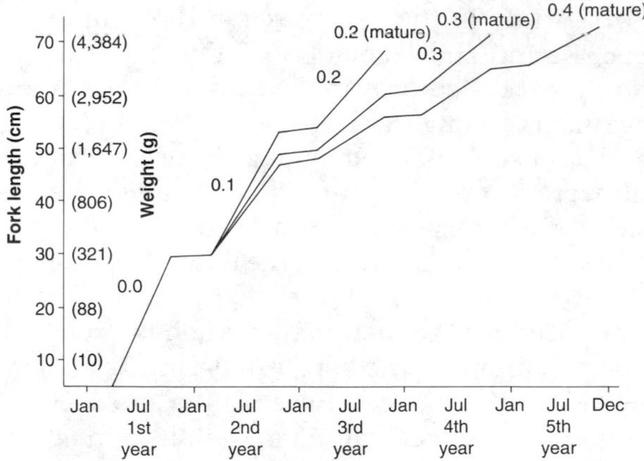

FIGURE 20

Schematic growth of Minter Creek chum salmon

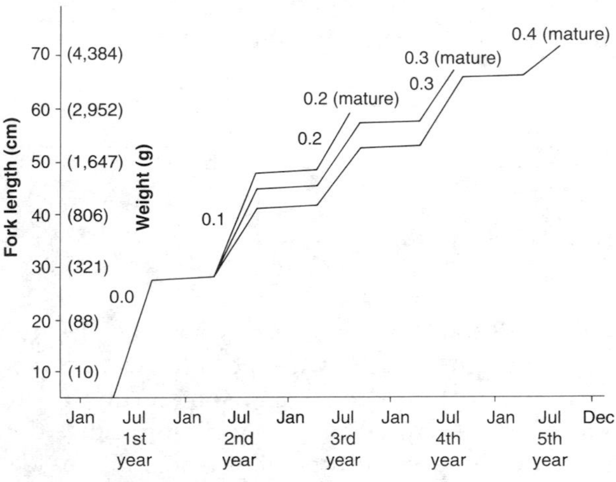

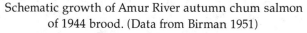

FIGURE 21

Schematic growth of Amur River autumn chum salmon of 1944 brood. (Data from Birman 1951)

sumed to begin growing 15 March and to continue until 1 November. These figures are not intended to show true growth, which is best represented by a version of the logarithmic equation[6] within each growing season.

Bilton and Ludwig (1966) and Ricker (1964) (based on Kobayashi's 1959 data) indicated that younger fish begin growing earlier in the spring. Ricker (1964) concluded that the average beginning date for a chum salmon in its fourth year was 25 April, and 5 May for a fish in its fifth year. Male and female chum salmon grow at similar rates until the final year of life, when males outgrow the females (Bakkala 1970; E.L. Salo, unpublished data). Summer chum grow slower and have a shorter final growing season than autumn chum (Ricker 1964; LaLanne 1971).

First-Year Growth

Growth during the first year in the marine environment differs by stock, largely due to environmental differences. For example, northern stocks often grow slower than southern stocks (Bakkala 1970; Kovtun 1981). Growth is similar in the first year for all fish of a given stock regardless of age at maturity (Helle 1979); however, the youngest maturing fish sometimes grow slightly faster (Kaev 1981; Kovtun 1981). Because of the longer growing season, North American fish appear to grow faster than Asian fish in the first year (Ricker 1964). Also, latitudinal bias in these data likely exaggerates the difference.

Growth of chum salmon fry is exponential during their nearshore residence, with averages from 4% to 6% body weight/d (Phillips and Barraclough 1978; Whitmus and Olsen 1979; Healey 1980; Congleton et al. 1982) and ranges from 3.5% (Fedorenko et al. 1979) to 6.7% (Prinslow et al. 1980). Higher values (8.6%) were noted for the first few days of feeding (Bax and Whitmus 1981). Under experimental conditions (floating fibreglass tubs), growth rates varied from 2.2% to 5.7% with daily rations from 3% to 17% body weight, respectively (LeBrasseur 1969). Back-calculating from otolith microstructures, Volk et al. (1984) estimated that growth rates of juvenile chum salmon, exposed to different rations of natural prey, are as high as 3.4% body weight/d. Optimal ration appears to be between 15% (LeBrasseur 1969; Healey 1980) and 20% (Simenstad et al. 1980) body weight/d. Daily change in length is on average 1 mm/d (Healey 1980; Congleton et al. 1982) with a range from 0.75 mm/d (Healey 1980) to 1.7 mm/d (Whitmus and Olsen 1979). Sampling with a beach seine in Hood Canal (Washington), Bax (1982) observed differences in lengths of chum juveniles from 30 mm to over 90 mm in a 20-week period.

Assuming that size-dependent daily ration requirements varied between 15% and 25% body

6 $l_t = L_\infty (1 - e^{-k(t - t_0)})$
L = maximum size towards which the fish is approaching
K = a measure of the rate at which length (l) approaches L_∞
From Dickie (1971), after Ricker (1964)

weight/d, Simenstad and Salo (1982) estimated that the carrying capacity of juvenile chum salmon in Hood Canal for a two-week period ranged from 0.03 to 0.65 fish/m² in shallow sublittoral habitats and from 0.01 to 0.07 fish/m² in neritic habitats. In recent years, releases of hatchery chum salmon into Hood Canal may have resulted in densities of 5 fish/m² for a linear distance of 520 m, whereas densities as low as 1 fish/m² may have the potential to deplete the preferred prey resources and cause depressed growth within just a few days (Simenstad et al. 1980). On the other hand, under more "normal" conditions, the growth rate of chum salmon during the first few weeks of ocean life is very rapid, suggesting that food is not directly limiting growth at this time (Healey 1982c).

The size of chum salmon sampled during their first summer at sea varies with time at sea. Prior to entering the open ocean in July, fish from North America (Washington) and Asia (Hokkaido) are about 10–12 cm in length (Sano 1966; Hartt et al. 1970). Peterson et al. (1982) reported an average length of 124 mm for chum salmon during mid- to late June, off the coast of Oregon. By August and September, chum salmon ranged in size from 13 to 20 cm along coastal areas of the Gulf of Alaska, and between 13 and 18 cm in the Bering Sea (Shepard et al. 1968). Juvenile chum salmon in the Sea of Okhotsk ranged from 11 to 18 cm in early August, 14 to 25 cm in late August and September, and from 16 to 24 cm in October. By September they were about 200 km offshore (Birman 1969). W. Pearcy (Oregon State University, Corvallis, Oregon, pers. comm.) found that 9–15 cm chum salmon were distributed within 7 km of the coasts of Washington and Oregon during May 1981. In August they ranged from 17 to 27 cm and were about 17–35 km offshore from these states. The calculated length for Minter Creek chum salmon at the end of the first year was nearly 30 cm, and the length for Amur River fish was estimated to be about 28 cm (Birman 1951).

Growth in the Second through Penultimate Years

Growth decreases with increasing age for both Asian and North American chum salmon (Ricker 1964; LaLanne 1971), and during the second year of marine life differences in growth among stocks and among the age groups from a year class become apparent. However, some Asian stocks show compensatory growth during their second year at sea (Table 6). The younger maturing fish from a year class generally grow faster than the older maturing fish (Figures 20 and 21) (Helle 1979; Kaev 1981; Kovtun 1981). These differences in growth rates among age groups continue until maturity. However, there is considerable overlap in the growth rates of the individuals of the different age groups during this period (Kaev 1981; Kovtun 1981). Amur River fish average 44 cm at the end of the second year, 55.2 cm at the end of the third year, and 65 cm at the end of the fourth year. The faster growing fish mature in their third year at about 60 cm (Figure 21). Maturing three-year-old Minter Creek chum are about 70 cm in length.

TABLE 6

Growth rates of four-year-old (.3+) Amur, Sakhalin, and South Kuril chum salmon

Region and year of observations	Size by years (cm)				Growth increments (cm)			
	1	2	3	4	1	2	3	4
S. Sakhalin, 1952	29.1	44.0	57.1	69.3	29.1	14.9	13.1	12.2
Amur, Ozerpakh, 1948	25.7	44.0	55.2	62.6	25.7	18.3	11.2	7.4
Amur, Yuzhnaya Gavan, 1948	27.8	44.1	55.2	67.6	27.8	16.3	11.1	12.4
Iturup Is. (S. Kuril Is.), 1962	29.3	46.3	60.7	72.3	29.3	17.0	14.4	11.6

Source: From Ivankov (1968); data for Amur chum salmon, from Lovetskaya (1948)

LaLanne (1971) deduced that maturing and immature chum salmon of equal age grow at a similar rate; however, this conclusion is likely inaccurate because of sampling of mixed stocks, the pooling of data from different years, and the technique used to calculate growth.

Final Marine Year

The final growing season is important in determining the size at maturity. Ricker (1964) demonstrated a great increase in weight during this period for four stocks of North American chum salmon (123% - 156% increase from the end of the second year to the end of the third year for a fish maturing at age 0.2). The increase is smaller, 73% - 92% and 46% - 52%, for chum salmon maturing at ages 0.3 and 0.4, respectively, reflecting a decreasing growth rate with age (Ricker 1964). The increase is less for Asian stocks, 88% (one stock) for age 0.1 to age 0.2, 62% - 84% (three stocks) for age 0.2 to age 0.3, and 48% (one stock) for age 0.3 to age 0.4. These lesser growth rates in the ultimate year are likely due to the early maturation of many of the Asian stocks and to differences in productivity between the western and eastern North Pacific Ocean.

Helle (1979) found for one stock that mean length at maturity of four-year-old male (age 0.3) chum salmon was positively correlated with sea surface temperature, mean air temperature, and mean dew point; and negatively correlated with mean cloud cover. He concluded that between 40% and 70% of the variation in adult length at maturity was associated with variability in these environmental parameters during the final growing season.

Age at Maturity

Chum salmon mature after two to six years of age (0.1–0.5) with more than 95% in the three- to five-year (0.2 - 0.4) age group (Table 7). In most areas, four-year-olds predominate, ranging from 60% to 90%. Age composition varies according to latitude in both North America and Asia. Fish in the north tend to have a higher proportion of four- and five-year-olds (0.3 and 0.4) than fish in the south, with three-year-olds (0.2) being proportionally greater in the south than in the north. Generally, therefore, there is a shift in age composition from predominantly four- and five-year-olds in the north to predominantly three- and four-year-olds in the south.

TABLE 7
Age composition at maturity of chum salmon from Asia and North America

Location	Years	Age composition (%)					Source
		0.1	0.2	0.3	0.4	0.5	
ASIA							
USSR							
Anadyr R.	1953, 1956, 1962	0	19.2	59.6	21.1	0.1	Ostroumov (1967)
Kamchatka							
East coast	1959	0	0.8	88.6	9.4	1.4	Sano (1966)
S.W. coast	1959	0	1.7	94.9	3.4	0	Sano (1966)
Bolshaya R.	1932–51	0	4.1	56.0	38.7	1.2	Semko (1954)
N.W. coast	1959	0	0	88.6	11.4	0	Sano (1966)
Okhotsk	1959	0	2.2	86.8	7.9	3.1	Sano (1966)
	1945–64	0	4.0	53.3	37.8	5.1	Kostarev (1967)
Northern coast	1956–65	0	2.4	57.8	37.9	1.9	Klokov (1973)
Amur River							
Summer chum	1927–30	0	14.7	76.3	8.6	0.4	Lovetskaya (1948)
	1947–51	0	8.5	86.6	4.9	0	Birman (1967)
	1959	0	3.5	86.0	10.5	0	Sano (1966)
Autumn chum	1927–33	0	4.9	67.1	26.4	1.6	Lovetskaya (1948)
	1946–51	0	11.5	76.7	11.7	0.1	Birman (1967)
	1959	0	10.7	81.8	7.2	0.3	Sano (1966)
Sakhalin	1959	0	9.0	84.0	7.0	0	Sano (1966)
South coast							
Summer chum	1948	0	0	94.8	5.2	0	Dvinin (1952)
Autumn chum	1948	0	0	74.0	26.0	0	Dvinin (1952)
Kuril Is.							
S. Kuril Is.	1956–67	0.2	16.7	56.4	26.5	0.2	Ivankov & Andreyev (1971)
Japan							
Hokkaido		0.4	22.0	62.6	14.7	0.3	Sano (1960)
Tokoro R.	1950–57	0.3	33.4	59.8	6.3	0.2	Bakkala (1970)
	1959	0.4	53.7	44.1	1.8	0	Sano (1966)

(continued on next page)

TABLE 7 (continued)

Location	Years	Age composition (%) 0.1	0.2	0.3	0.4	0.5	Source
Nishibetsu R.	1950–57	0.1	27.1	61.9	11.0	0	Bakkala (1970)
	1959	0	0	62.5	37.5	0	Sano (1966)
Tokachi R.	1950–57	0.4	18.3	57.6	23.0	0.6	Bakkala (1970)
	1959	0	2.0	80.7	17.3	0	Sano (1966)
Ishikari R.	1950–57	3.2	30.8	52.4	13.3	0.3	Bakkala (1970)
	1958	0	40.5	57.1	2.4	0	Sano (1966)
Yurappu R.	1950–57	0.1	15.0	68.0	17.0	0	Bakkala (1970)
	1958	0.6	27.4	70.8	1.2	0	Sano (1966)
Honshu							
Sea of Japan coast	1959	6.9	61.5	31.6	0	0	Sano (1966)
Pacific coast	1959	1.3	11.2	79.8	7.7	0	Sano (1966)
NORTH AMERICA							
Alaska							
Kotzebue Sd.	1962–81	0	17.1	64.3	18.2	0.5	Bird (1982)
Yukon R.	1920	0	3.3	68.1	28.6	0	Gilbert (1922)
	1955–57	0	10.5	66.2	22.7	0.6	Birman (1967)
	1961–65	0	9.1	77.6	13.2	0.1	Regnart et al. (1966)
Emmonak	1973–82	0	12.0	72.7	15.1	0.2	Buklis & Barton (1984)
Delta R.	1973–82	0	14.0	77.1	8.8	0.2	Buklis & Barton (1984)
Nenana	1974–82	0	20.6	68.7	10.7	0	Buklis & Barton (1984)
Toklat R.	1974–82	0	34.4	57.7	7.9	0	Buklis & Barton (1984)
Rampart	1974–79	0	23.8	66.7	9.5	0	Buklis & Barton (1984)
Sheenjek R.	1974–82	0	15.5	59.6	24.6	0.3	Buklis & Barton (1984)
Commercial catch	1973–82						
Summer chum		0	8.8	70.5	20.4	0.4	Buklis & Barton (1984)
Autumn chum		0	12.0	72.7	15.1	0.2	Buklis & Barton (1984)
Alaska Pen.	1951–57	0	9.3	75.2	15.5	0	Thorsteinson et al. (1963)
Cook Inlet	1956–57	–	–	91.7	–	–	Birman (1967)
Kodiak Is.	1948–57	0	4.7	67.2	28.1	0	Thorsteinson et al. (1963)
Prince Wm. Sd.	1952–58	0	9.2	73.8	17.0	0	Thorsteinson et al. (1963)
Olsen Creek	1959–77	0	9.0	62.5	28.0	0.5	Helle (1979)
S.E. Alaska	1961, 1963	0	5.7	89.6	4.7	0	Mattson & Hobart (1962); Mattson et al. (1964)
British Columbia							
	1916–17, 1928–35, 1940–42	0.1	21.4	43.3	31.7	3.5	Pritchard (1943)
Queen Charlotte Is.	1940–42	0	0.2	16.4	71.8	11.6	Pritchard (1943)
	1955	–	–	21.2	78.4	–	Birman (1967)
	1981–85	0	42.4	53.2	4.3	0.1	Beacham & Murray (1987)
Mainland							
North	1957–72	0	13.4	73.8	12.8	0	Ricker (1980)
	1981–85	0	8.4	83.0	8.4	0.2	Beacham & Murray (1987)
Central	1981–85	0	49.6	46.1	4.1	0.2	Beacham & Murray (1987)
South	1981–85	0	20.6	68.9	10.5	0	Beacham & Murray (1987)
Vancouver Is.							
East	1916–17	0	55.8	43.6	0.6	0	Pritchard (1943)
	1981–85	0.2	22.9	72.2	4.7	0	Beacham & Murray (1987)
West	1933–35	0	16.2	47.4	35.0	1.4	Pritchard (1943)
	1941	0	6.5	46.0	44.7	2.8	Pritchard (1943)
Johnstone St.	1958–72	0	30.2	66.8	3.0	0	Ricker (1980)
	1930	0	31.0	67.7	1.3	0	Pritchard (1943)
Juan de Fuca St.	1981–85	0	47.6	51.0	1.4	0	Beacham & Murray (1987)
Fraser R.	1981–85	0	28.1	61.4	10.5	0	Beacham & Murray (1987)

(continued on next page)

TABLE 7 (continued)

Location	Years	Age composition (%) 0.1	0.2	0.3	0.4	0.5	Source
Washington							
Puget Sd.							
Bellingham	1910	0	53.5	44.8	1.7	0	Gilbert (1922)
Admiralty Inlet	1935	0	38.2	52.9	8.9	0	Rounsefell & Kelez (1938)
North Sd.	1957–68	0	26.1	72.8	1.1	0	Pratt (1974)
	1969–83	0	19.7	72.4	7.9	0	Ames (pers. comm.)*
Hood Canal	1963–70	0	57.1	42.2	0.7	0	Ames (pers. comm.)*
	1971–83	0	36.2	61.1	2.7	0	Ames (pers. comm.)*
	1965–76†	0	35.7	62.3	2.0	0	Wolcott (1978)
South Sd.	1962–70	0	60.3	39.2	0.5	0	Ames (pers. comm.)*
	1971–83	0	40.9	56.8	2.3	0	Ames (pers. comm.)*
Bellingham–Samish bays	1957–67	0	33.1	64.3	2.6	0	Pratt (1974)
Skagit Bay	1954–70	0	29.8	67.9	2.3	0	Pratt (1974)
Everett	1963–70	0	22.9	75.1	2.0	0	Pratt (1974)
Admiralty Inlet, Point No Point	1961–70	0	50.2	49.1	0.7	0	Pratt (1974)
Seattle	1962–70	0	60.0	39.4	0.6	0	Pratt (1974)
Minter Cr.	1938–55	0	49.1	49.3	1.6	0	Salo (unpubl. data)
Grays Harbor	1969–82	0	36.2	62.3	1.5	0	WDF (pers. comm.)‡
Columbia R.	1914	0	70.5	28.7	0.8	0	Marr (1943)
Oregon							
Tillamook Bay	1947–61	0	52.7	46.4	0.9	0	Oakley (1966)
	1947–49	0	10.2	89.2	0.6	0	Henry (1954)

Notes: *J. Ames, Washington Department of Fisheries, Olympia, WA
†Late run fish only, mid-October and later
‡Washington Department of Fisheries, Olympia, WA

A few chum stocks show an alternation of dominance between three- and four-year-olds (four-year-olds and five-year-olds in some Asian stocks) that may be related to the presence of dominant year classes of pink salmon (see section on chum-pink salmon interactions). The mean age at maturity is negatively correlated with growth during the second year of marine life but not with growth in the first year (Helle 1979) and is also negatively correlated with abundance of the brood (Helle 1979; Beacham and Starr 1982).

The relationships of growth and brood abundance to mean age at maturity indicate that the environment and competition can override the genetic factors that control age at maturity. Alternatively, genetics may control the "basic" rate of growth and thresholds of size with maturity. Helle's (1979) correlation with growth during the second year suggests that this may be the time that determines whether a fish returns as a three-year-old. Although Helle (1979) did not correlate the return of four-year-olds with third year growth or the return of five-year-olds with fourth year growth, his data provide some evidence that growth during the third or fourth year can influence age at maturity. For example, about 58% of the Olsen Creek 1956 brood returned as three-year-olds. In an average brood year, 15% of the males and 9% of the females are three-year-olds. The 1956 brood experienced a warm first year (1957) and their second marine year was the El Niño year of 1958 which led to good growth. In 1960, no five-year-old fish returned, and this is the only year with no five-year-olds recorded. Four-year-olds were dominant (nearly 100%). It is possible that good conditions in 1958 (the third year of life for these potential five-year-old fish) may have led to their maturing early in 1959 as four-year-olds. Data are not adequate to determine whether the second year at sea is the pivotal year for all age groups at all times or whether the penultimate year can be an important determinant, or if perhaps it is dependent on both environment and genetics.

Size at Maturity and Age Composition

Within its range, the average size of chum salmon (all ages combined) increases from north to south in both Asia and North America (Table 8, Figure 22). Fish in southern areas have a longer growing period and mature at a younger age than northern populations. Also, within age groups there is an

TABLE 8
Average weights of chum salmon from Asia and North America

Location	Years	Mean weight (kg)	Range of annual means	Source
ASIA				
USSR				
Anadyr R.		3.2		Kaganovsky (1933)
Kamchatka				
N.E. coast	1965–69	3.74	(3.21–4.04)	INPFC (1979)
East coast	1958–69	3.65	(3.22–4.07	INPFC (1979)
S.W. coast	1958–69	3.52	(2.97–4.26)	INPFC (1979)
N.W. coast	1958–69	3.45	(3.05–4.61)	INPFC (1979)
Sea of Okhotsk				
Northern coast	1960–69	3.61	(3.20–4.31)	INPFC (1979)
Okhotsk coast	1958–69	3.52	(3.01–4.01)	INPFC (1979)
Amur R.				
Summer chum	1958–69	2.52	(2.31–2.79)	INPFC (1979)
Autumn chum	1958–69	4.37	(3.83–5.06)	INPFC (1979)
Sakhalin				
West coast	1963–69	3.47	(3.33–3.76)	INPFC (1979)
East coast	1963–69	3.41	(2.86–3.72)	INPFC (1979)
Kuril Is.	1955–67	4.04	(3.68–4.50)	Ivankov & Andreyev (1971)
Japan				
Coastal	1962–76	3.19	(2.96–3.44)	INPFC (1979)
NORTH AMERICA				
Alaska	1960–76			
Arctic, Yukon, & Kuskokwim rivers		3.17	(2.77–3.63)	INPFC (1979)
Bristol Bay		3.01	(2.68–3.40)	INPFC (1979)
Alaska Pen. & Aleutian Is.		3.10	(2.63–3.36)	INPFC (1979)
Chignik		3.34	(2.95–3.99)	INPFC (1979)
Kodiak Is.		3.67	(3.08–4.97)	INPFC (1979)
Cook Inlet, Resurrection Bay		3.41	(2.99–3.95)	INPFC (1979)
Prince Wm. Sd. & Copper and Bering rivers		3.73	(3.00–4.45)	INPFC (1979)
Southeastern		4.27	(3.76–4.94)	INPFC (1979)
British Columbia	1951–75			
Northern (areas 1-10)		5.30	(4.29–6.43)	Ricker (1980)
Southern (areas 11-27)		5.00	(4.20–5.71)	Ricker (1980)
Fraser R. (area 29)		5.32	(4.40–6.07)	Ricker (1980)
Washington				
Puget Sd.	1960–70	4.47	(3.98–4.97)	Pratt (1974)
Willapa Harbor*	1968–81	4.87	(4.15–5.59)	
Coastal and Grays Harbor*	1968–81	5.33	(4.90–6.00)	
Columbia R.	1938–76	5.58	(4.90–6.40)	INPFC (1979)

Note: *Calculated from pound and numbers data in "1981 Fisheries Statistical Report," Washington Department of Fisheries, Olympia, WA (net fishery only)

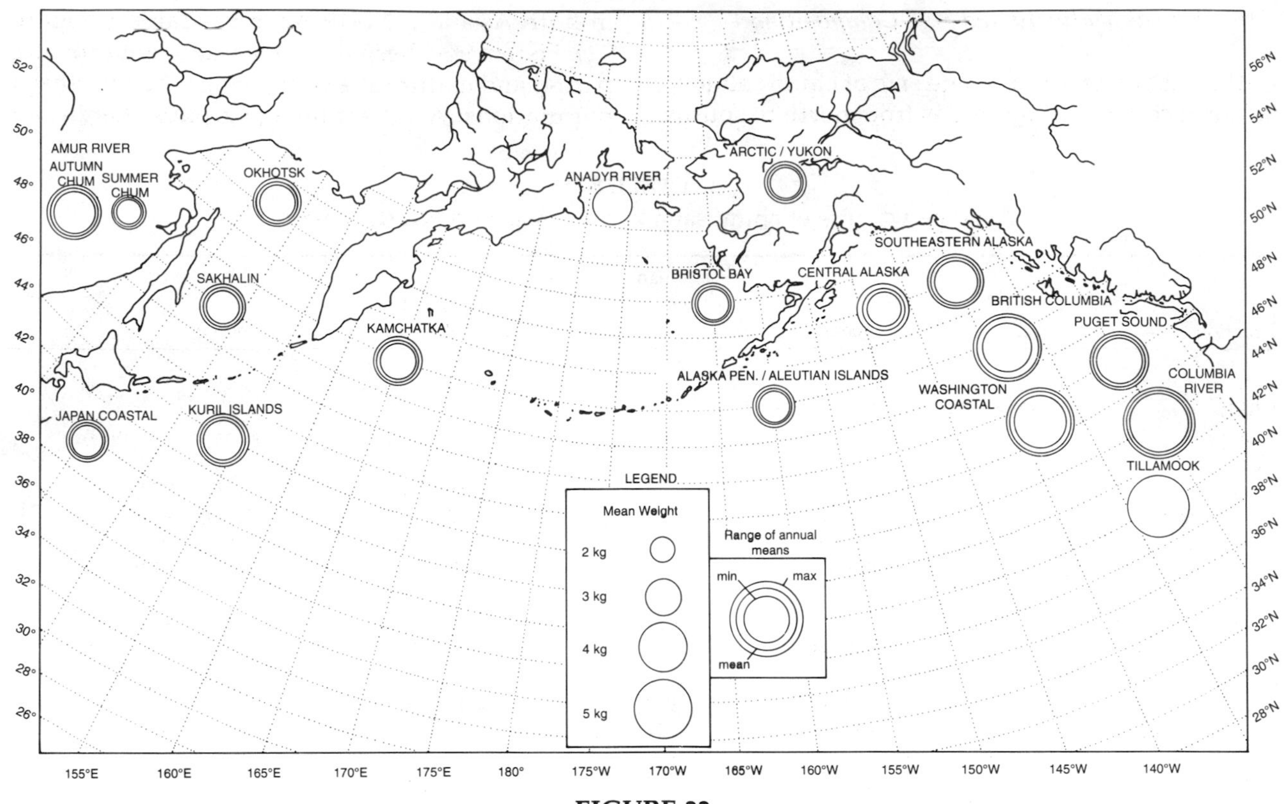

FIGURE 22
Mean weight of chum salmon

increase in size from north to south (see age 0.3 chum in Table 9). The Asian fish are generally older (Figure 23) and smaller than the North American stocks (Figure 22) at a given latitude. Considerable spatial and temporal variability in size within regions is superimposed upon this broad latitudinal trend. For fish of a given age and sex, Ricker (1980) found no significant differences in size of chum salmon from northern and southern British Columbia, although the overall size tended to be larger but more variable in northern British Columbia (Table 9). The generally decreasing trend in mean weight from north to south was attributed to the relatively greater number of age 0.2 chum salmon in southern catches. In Alaska, the average weight of chum salmon is less than in British Columbia, and generally increases from north to south (Table 9). A notable exception in this trend are chum salmon from Kotzebue Sound, which exceed the size of chum salmon from Olsen Creek (Helle 1984). Interestingly, the returns to Kotzebue Sound comprise a relatively greater proportion of younger fish as compared to other northern Alaskan areas (Table 7). This reversion to younger maturing fish in the far north was also mentioned by M.L. Frey (College of Fisheries, University of Washington, Seattle, Washington, pers. comm.) (see also Figure 23). In Puget Sound, age 0.3 chum salmon from northern areas tend to be larger than age 0.3 chum from southern areas (Pratt 1974) (Table 9). Fish harvested in northern Puget Sound are primarily from Fraser River stocks, and are predominantly age 0.3. In southern harvest areas, age composition is less stable, with age 0.3 chum dominant in some years, and age 0.2 chum in other years. There has been a consistently greater proportion of younger fish in southern Puget Sound since the early 1960s (Table 7).

Considerable temporal variability in chum salmon size is also evident within localized regions. A general decrease in the size of British Columbia chum salmon at a given age, along with a general decrease in ocean temperature between 1951–75 was reported by Ricker (1980). More re-

TABLE 9

Mean length and weight of age 0.3 chum salmon at maturity from Asia and North America

Location	Years	Fork length (cm) female	male	both	Weight (kg) female	male	both	Source
ASIA								
USSR								
Anadyr R.				60.2				Ostroumov (1967)
Kamchatka								
East coast	1959	59.8	63.5		2.83	3.49		Sano (1966)
S.W. coast	1959	61.7	65.2		2.79	3.29		Sano (1966)
Bolshaya R.	1932–49			61.3			2.88	Semko (1954)
	1951–60			62.1			2.90	Petrova (1964)
N.W. coast	1959	63.0	68.3		2.93	3.74		Sano (1966)
Okhotsk	1959	59.1	62.6		2.70	3.26		Sano (1966)
				62.3				Kostarev (1967)
Amur R.								
Summer chum	1959	55.5	57.7		2.21	2.57		Sano (1966)
				59.0				Lovetskaya (1948)
Autumn chum	1959	66.1	69.4		3.40	4.16		Sano (1966)
	1925–49			66.8				(?)
				67.3				Lovetskaya (1948)
Sakhalin	1959	64.7	64.9		3.28	3.68		Sano (1966)
Kuril Is.	1967	71.5	74.4	73.3	4.02	4.14	4.09	Ivankov & Andreyev (1971)
Japan								
Hokkaido								
Nemuro district	1959	70.3	73.1		3.78	3.97		Sano (1966)
Okhotsk coast	1959	69.1	70.7		3.99	4.45		Sano (1966)
Pacific coast	1959	73.3	76.7		4.49	5.62		Sano (1966)
S.W. coast	1958	77.6	77.2		5.37	5.52		Sano (1966)
Japan Sea coast	1959	68.7	70.4		4.13	4.26		Sano (1966)
Honshu								
Pacific coast	1959	75.7	76.4		4.55	5.18		Sano (1966)
Sea of Japan	1959	73.9	75.9		4.79	5.24		Sano (1966)
NORTH AMERICA								
Alaska								
Kotzebue Sd.	1962–65*	64.3	70.0					Regnart et al. (1966)
Yukon R.	1920	62.0	67.1					Gilbert (1922)
	1962–65*	60.3	66.8					Regnart et al. (1966)
Alaska Pen.	1951–57*	63.6	69.3					Thorsteinson et al. (1963)
Kodiak Is.	1948–57*	65.7	71.3					Thorsteinson et al. (1963)
Prince Wm. Sd.	1952–58	67.2	71.9					Thorsteinson et al. (1963)
Olsen Cr.	1959–78†	66.3	71.1					Helle (1979)
S.E. Alaska								
Traitors Cove	1961*	70.8	76.6					Mattson & Hobart (1962)
	1963*	72.1	77.0					Mattson et al. (1964)
East R.	1963*	67.1	72.2					Mattson et al. (1964)
Yakutat	1961*	62.0	72.3					Mattson & Hobart (1962)
Lynn Canal	1961*	71.1	74.8					Mattson & Hobart (1962)
Icy St.	1961*	67.4	72.6					Mattson & Hobart (1962)
Portland Canal	1961*	73.4	78.8					Mattson & Hobart (1962)
British Columbia‡								
Northern								
Nass (area 3)	1957–72	73.2	76.8					Ricker (1980)
Skeena (area 4)	1957–72	74.5	78.1					Ricker (1980)

(continued on next page)

TABLE 9 (continued)

Location	Years	Fork length (cm) female	Fork length (cm) male	Fork length (cm) both	Weight (kg) female	Weight (kg) male	Weight (kg) both	Source
Ogden–Principe (area 5)	1957–72	74.2	77.8					Ricker (1980)
Whale Channel (area 6)	1946			70.3				Ricker (1980)
	1948			68.9			4.57	Ricker (1980)
	1958–72	74.7	78.7					Ricker (1980)
Bella Bella (area 7)	1947			68.9			4.16	Ricker (1980)
	1958–72	73.2	74.2					Ricker (1980)
Bella Coola (area 8)	1946			78.9				Ricker (1980)
	1947			71.6				Ricker (1980)
	1948			73.9			5.66	Ricker (1980)
	1958–72	75.9	78.8					Ricker (1980)
Rivers Inlet (area 9)	1946			74.9				Ricker (1980)
	1947			74.6			5.25	Ricker (1980)
	1958–70	74.5	78.3					Ricker (1980)
Smith Inlet (area 10)	1959–70	74.6	78.0					Ricker (1980)
Southern								
Upper Johnstone Str.	1945			70.4			4.61	Ricker (1980)
(area 12)	1948			69.4			4.07	Ricker (1980)
	1950			72.5				Ricker (1980)
	1953			77.6				Ricker (1980)
	1958–72	74.8	76.5					Ricker (1980)
Lower Johnstone St.	1945			69.8			4.61	Ricker (1980)
(area 13)	1950			72.6				Ricker (1980)
	1958–72	74.8	76.0					Ricker (1980)
St. of Georgia (areas 14–18)								
Nanaimo (area 17)	1916	70.8	74.9		5.14	5.88		Ricker (1980)
	1917	73.1	76.3		5.33	6.09		Ricker (1980)
Little Qualicum	1917	76.3	73.4		5.53	6.28		Ricker (1980)
(area 17)	1978	73.3						Beacham (1982)
Chemainus (area 18)	1917	73.1	74.6		4.77	5.25		Ricker (1980)
	1978	72.3						Beacham (1982)
Areas 14–18	1960–72	75.8	76.3					Ricker (1980)
Juan de Fuca St.	1946			73.0				Ricker (1980)
(area 20)	1948			73.9			5.34	Ricker (1980)
	1958–72	72.9	74.7					Ricker (1980)
West Vancouver Is. (areas 23–26)								
Barkley Sd. (area 23)	1946			70.3				Ricker (1980)
Areas 23–26	1959–63	73.5	74.3					Ricker (1980)
Fraser R. (area 29)	1950			73.3				Ricker (1980)
	1957–72	75.4	77.2					Ricker (1980)
Washington (Puget Sd.)								
Northern Sd.	1964			78.3				Pratt (1974)
	1970			75.7				Pratt (1974)
Bellingham	1910	70.4	76.0					Bakkala (1970)
Southern Sd.	1964			72.9				Pratt (1974)
	1970			70.9				Pratt (1974)
Big Beef Cr.								
Early Run	1968–69	69.8	77.5					Koski (1975)
	1969–70	69.8	76.6					Koski (1975)
Late Run	1968–69	72.1	77.2					Koski (1975)
	1969–70	71.2	76.9					Koski (1975)
Discovery Bay to Tacoma	1963–66	71.2	74.3					Pratt (1974)
	1970	69.6	72.4		4.43	4.97		Pratt (1974)
Columbia R.	1914	74.8	80.6					Marr (1943)

(continued on next page)

TABLE 9 (continued)

Location	Years	Fork length (cm) female	male	both	Weight (kg) female	male	both	Source
Oregon								
Tillamook Bay	1947	73.2	79.7		4.76	5.63		Henry (1954)
	1949	70.9	76.9		4.67	5.86		Henry (1954)
	1959	72.1	80.0					Bakkala (1970)

Notes: *Mid-eye-fork of tail length (MEFT) converted to tip of snout-fork of tail length (TSFT) using the following equations (developed from equations of Helle 1979):

$$♀\ \text{TSFT} = 7.9948 + 1.0706\ \text{MEFT}$$
$$♂\ \text{TSFT} = 132.8937 + .9285\ \text{MEFT}$$

†Mid-eye-hypural length (MEHP) converted to tip of snouth-fork of tail length using the regression equations of Helle (1979):

$$♀\ \text{TSFT} = 49.148 + 1.123\ \text{MEHP}\ (r^2 = .91)$$
$$♂\ \text{TSFT} = 132.669 + 1.038\ \text{MEHP}\ (r^2 = .94)$$

‡Post-orbital-hypural length converted to tip of snout-fork length by using a factor of 1.25 (Ricker 1980). Number in parentheses indicates British Columbia statistical areas.

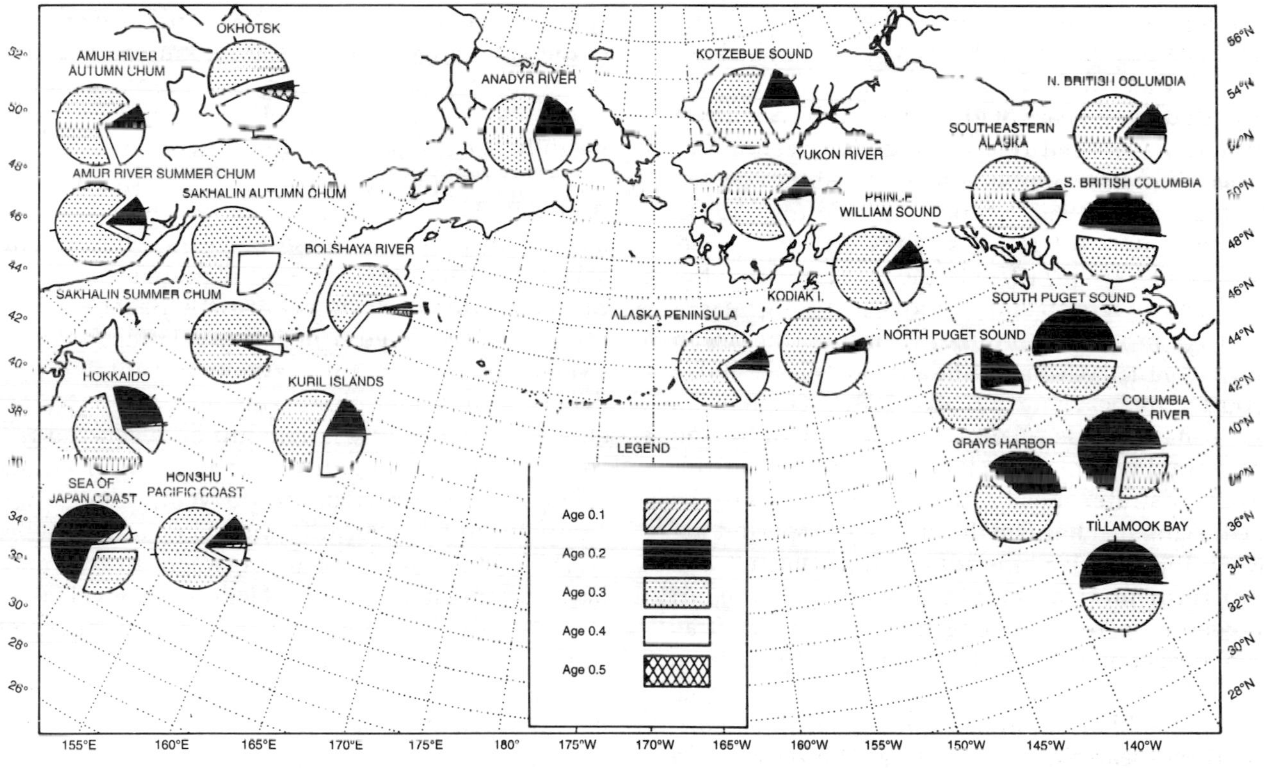

FIGURE 23
Age composition of chum salmon at maturity

cently, small increases in the mean weight of chum salmon from British Columbia (all ages combined) were noted, possibly due in part to selection by gillnets, which remove more of the smaller fish (Ricker 1984). Selective removal of smaller fish tends to increase progeny size within an age group, but also favours survival of older ages, which have slower growth rates (Ricker 1984).

Similarly, a decrease in the average length of age 0.3 chum for Puget Sound was evident for fish sampled in 1964 and 1970 (Pratt 1974). Pratt also noted that reduced average weights during odd years suggested competition for food between chum and pink salmon during the pre-maturation period for years that coincide with returning pinks.

Relationship between Abundance and Size of Adults

Although there are no studies that provide direct evidence of density-dependent growth in chum salmon, there are several reports indicating lesser growth during years of greater abundance. Soviet authors have cited relationships between abundance of chum salmon stocks and size of adults (Birman 1951, 1960; Semko 1954; Petrova 1964); however, none of these studies are conclusive. Conversely, Helle (1979) found no correlation between mean length and abundance during year of return or brood year, but he did find a higher mean age at maturity from abundant year classes (also reported by Beacham and Starr 1982). A high mean age at maturity is related to slow growth during the second year of ocean life and, although difficult to demonstrate, density-dependent factors may be the linking mechanism. In the six-year period between 1979 and 1984 the adult chum salmon (0.3 in age) returning to the Ohkawa River (Honshu) showed a decrease in length (Y. Ishida, Far Seas Fish Research Laboratory, Shimizu, Japan, pers. comm.). Also, the numbers of fish returning as four-year-olds that showed slow growth as two- and three-year-olds is increasing. This decrease in length may be due to density dependency associated with the massive releases from Japanese hatcheries. Ricker (1980) found a positive correlation between mean weight and catch for the major areas of British Columbia; however, he concluded that this positive correlation may be an artifact caused by variable year-class strength.

Density-dependent growth is difficult to show for chum salmon because the final year in the ocean is so important in determining the final size of the adults (Ricker 1964; Helle 1979). Therefore, comparisons of the size of adults produced by different year classes may not be an accurate index of density-dependent growth. Rather, they may reflect the interactions of genetically-determined size thresholds and environmental differences encountered during the final year of life. Also, density-independent factors (weather) will determine the carrying capacity of the marine environment for chum salmon. For this reason, an index of competition, such as abundance of fish, may not accurately express the differences in intensity of competition among years.

Little is known about the carrying capacity of the North Pacific Ocean or even the density of prey and the optimal ration for chum salmon. Neave (1961) calculated that the maximum density of chum salmon was about 180 kg/km^2 during the late 1930s when chum salmon were very abundant. It is not known whether densities of this magnitude can lead to slower growth.

Density dependence may not always be reflected in the size of returning adult chum salmon, as mortality may increase in years when chum salmon are abundant because of higher predation rates resulting from slower growth. Beacham and Starr (1982) showed a negative relationship between survival of chum salmon and total abundance of pink and chum salmon fry. They also showed an inverse relationship between the return/spawner ratio and the abundance of chum salmon of the previous brood year, suggesting competition between adjacent year classes. Helle (1979) found no relationship between survival and brood abundance; however, Helle's "survival" was based on return/spawner ratios, which incorporated freshwater mortality. Also, it is difficult to compare results when the effects of the fishery (size of catch) are not clear.

Survival

Chum salmon experience differential losses during each stage of their life history. The magnitude of survival is a reflection of complex interactions between biota and environment at each stage.

The survival of chum salmon eggs from spawning to emergence varies widely among streams and can vary by factors as high as twenty from year to year in a particular stream (Table 10). Parker (1962) reported a range of 1%-22% survival over fourteen years in Hooknose Creek, British Columbia, which was attributed to radical changes in flow (scouring and freezing). Higher survivals with less variability are obtained from controlled streams, as in the Big Qualicum River, where survival to emergence averaged 11.2% prior to flow stabilization and 24.9% after stabilization (Table 10). Egg-to-fry survival was further improved to 74% when a spawning channel was built.

TABLE 10

Survival of chum salmon in early stages of development in natural and artificial environments

Location	No. of years sampled	Method of measuring survival	Survival[1] Range (%)	Mean (%)	Source
Natural stream environment					
Disappearance Cr. AK	2	-	8.7-16.9	12.8	Wright (1964)
Big Qualicum R., BC	4	Downstream migrant fry counts	5.0-17.0	11.2	Lister & Walker (1966)[2]
Nile Cr., BC	4	Downstream migrant fry counts	0.1- 7.0	1.5	Wickett (1952)[2]
Hooknose Cr., BC	14	Downstream migrant fry counts	1.0-22.0	8.5	Parker (1962)[2]
Karymaiskiy Sp., Bolshaya R., USSR	7	Downstream migrant fry counts	0.7- 4.2	2.4	Semko (1954)[2]
Khor R., USSR	-	Examination of redds at hatching	25.0-30.0	-	Levanidov (1964)
Five tributaries of the Amur R., USSR	7	-	2.0-12.0	-	Levanidov (1964)[2]
Iski R. (Amur Trib.), USSR					
1940		Examination of redds	54.3-85.9[3]	71.6[3]	Smirnov (1947)[4]
1941			3.3-17.3	6.8[3]	
Memu R., Japan	3	Downstream migrant fry counts	16.2-34.4	27.6	Nagasawa & Sano (1961)[2]
Controlled stream environment					
Abernathy Cr. spawning channel, WA	1	Downstream migrant fry counts	-	82.1	Bur. Commerc. Fish.[5]
Jones Cr. spawning channel, BC	1	Downstream migrant fry counts	-	30.0	Trade News (1956)[2]
Nile Cr. (natural stream protected from floods), BC	4	Downstream migrant fry counts	3.4-11.8	7.5	Wickett (1952)[2]
Big Qualicum R. (natural stream with controlled flow), BC	2	Downstream migrant fry counts	24.5-25.2	24.9	Lister & Walker (1966)[2]
Big Qualicum R. spawning channel, BC	6	Downstream migrant fry counts	64.2-85.7	74.0	Paine (1974)
Big Beef Cr., WA	3	Downstream migrant fry counts	25.6-57.9[6]	-	Koski (1975)

Source: Adapted from Bakkala (1970)
Notes: *1* Percentage survival calculated from potential egg deposition
2 Cited in Bakkala (1970)
3 Ranges and means of several areas in the Iski River
4 Cited in Sano (1966)
5 Bakkala (1970)
6 Survival range was 7.2%-88.4% in individual experimental units.

The survival of fry to maturity also varies among regions and between years, with the average reported survival ranging from 0.3% to 3.2% for wild chum salmon, whereas hatchery-produced chum have an even greater variance (Table 11). The survival of chum salmon from the Fraser River and Puget Sound frequently varies on an odd- and even-year basis (see section on chum-pink salmon interactions). However, most of the variability in marine survival is related to ocean conditions (e.g., temperature, cloud cover, and salinity). Blackbourn (1985) reported a negative correlation be-

TABLE 11

Survival of chum salmon from egg to fry and fry to adult under natural and hatchery conditions

Location	Date	Freshwater survival (egg to fry)		Marine survival (fry to adult)		Source
		Mean (%)	Range (%)	Mean (%)	Range (%)	
Natural streams						
USSR						
Five tributaries of the Amur R.						
Summer chum	1955–58, 1960	6.1	1.3–13.1	2.5[2]	1.5–3.2[2]	Levanidov (1964)[1]
British Columbia						
Fraser River	1961–79	14.2	5.7–35.4	1.2[2,4,5]	0.3–2.7[2,5]	Beacham & Starr (1982)
Hooknose Cr.		7.8	–	2.8[2]	–	Parker (1962)[1]
Washington						
Minter Cr.	1938–54 (N = 10)	9.1	2.8–16.9	1.9[3]	1.4–2.4[3]	Salo (unpub. data)
Big Beef Cr. (spawning channel)	1967–69 (broods)	–	25.6–58.9	–	0.5–2.6[3]	Koski (1975)
Walcott Slough	1916–18	–	–	0.8[6]	–	Wolcott (1978)
Hatchery production						
Japan						
Hokkaido	1962–77	80	–	2.0[2]	0.5–2.7[2]	Hiroi (1985)
Honshu	1962–77	80	–	1.0[2]	0.3–2.5[2]	Hiroi (1985)
USSR						
S.W. Sakhalin	1964–78	–	–	0.3[2]	0.1–0.7[2]	Roukhlov (1982)
S.E. Sakhalin	1964–78	–	–	0.3[2]	0.01–1.8[2]	Roukhlov (1982)
U.S.						
Alaska					–	
Washington						
Hood Canal	1916–69	–	–	0.34[7]	0.25–0.43	Wolcott (1978)
Hood Canal		–	–	–	0.50–2.70[8]	
Hood Canal	1966–71	–	–	–	0.96–3.0	Wolcott (1978)

Notes: *1* Cited by or calculated from values in Bakkala (1970)
2 Does not include fishing mortality
3 Includes fishing mortality
4 Survival varies on odd and even cycle
5 1961–74
6 Minimum estimate
7 Unfed fry
8 Fed fry

tween salinity (indexed by rainfall or river discharge) during the first summer of ocean residence and marine survival for seven stocks of chum in Washington and British Columbia.

The marine survival rate of hatchery-produced chum salmon in Japan has recently increased to over 2% due to artificial feeding of the fry before release. Similar and even greater rates have been obtained in North American hatcheries, but not as consistently. Reportedly, hatchery-reared chum salmon from the Soviet Union have a lower survival rate (Table 11), although these data are incomplete.

Most of the mortality suffered by chum salmon in the marine environment occurs within the first few months of life. Parker (1962) suggested a survival of 5.4% for the first five months of marine life. Bax (1983a), in separate sequential experiments, estimated average daily rates of mortality to be between 31% and 46% over a two- and four-day period, respectively. Later, Whitmus (1985), working in the same area of Hood Canal (Washington), estimated the survival of one marked group to be 42% over a two-day period. He also found the emigration and the survival rate to be size-dependent. The rates estimated by Bax (1983a) and Whitmus (1985) are an order of magnitude higher than those estimated for pink salmon over a 40-day period subsequent to saltwater entry (Parker 1968).

Healey (1982b) also determined that the mortality of juvenile salmon during early sea life is probably size-dependent. Significant mortality by size occurred in the time period that the fish were laying down scale circuli numbers 2-4 and was size-selective over the size range 45–55 mm fork length. Chum salmon lay down the first scales at about 40 mm and they are completely scaled by the time they reach 50 mm. Although no particular mortality could be identified, the size range corresponded with the size at which the chum salmon moved from shallow water to the pelagic habitat.

ABUNDANCE

Productivity of the North Pacific Ocean

The annual catch of North Pacific Ocean chum salmon from 1925 through 1981 averaged over 41 million fish (range 18–83 million). In this period the mean Asian catch was 26.5 million and the North American mean was 11.5 million (Figure 24). In the thirty-year period after 1952 the mean annual harvest of chum salmon was 24.9 million for Japan, 6.9 million for the USSR, 6.9 million for the United States, and 2.4 million for Canada (Figures 25 and 26).

The catch trends by area for North America and Asia are shown in Figure 27. The ratio of Asian to North American chum, based on catches since 1930, has been as high as 3.5:1 in 1936 and as low as nearly 1:1 in 1952–53. However, over the years, Asian fish have constituted the bulk of the production. The decade with the greatest total catch of chum salmon was 1934-43, with an annual average of 57.5 million fish (Figure 24). If the fish had an average weight of 3.66 kg, the average catch would have been 210,600 t. The peak year was 1936, when 83 million fish weighing 304,000 t were harvested. If one assumes a catch to escapement ratio of 3:1, the production would have been 111 million mature fish or 407,000 t. If the catch to escapement ratio was 2:1, the annual production would have been 126 million mature fish or 462,000 t.

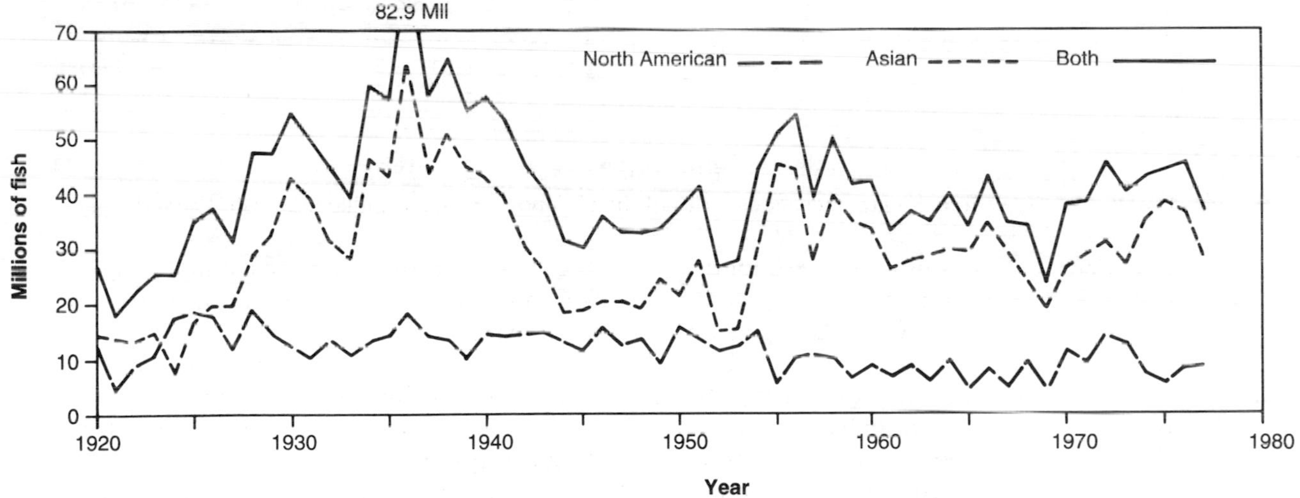

FIGURE 24

Total catch of chum salmon of Asian and North American origins by commercial fisheries of the USSR, Japan, U.S., and Canada, in millions of fish, 1920–77

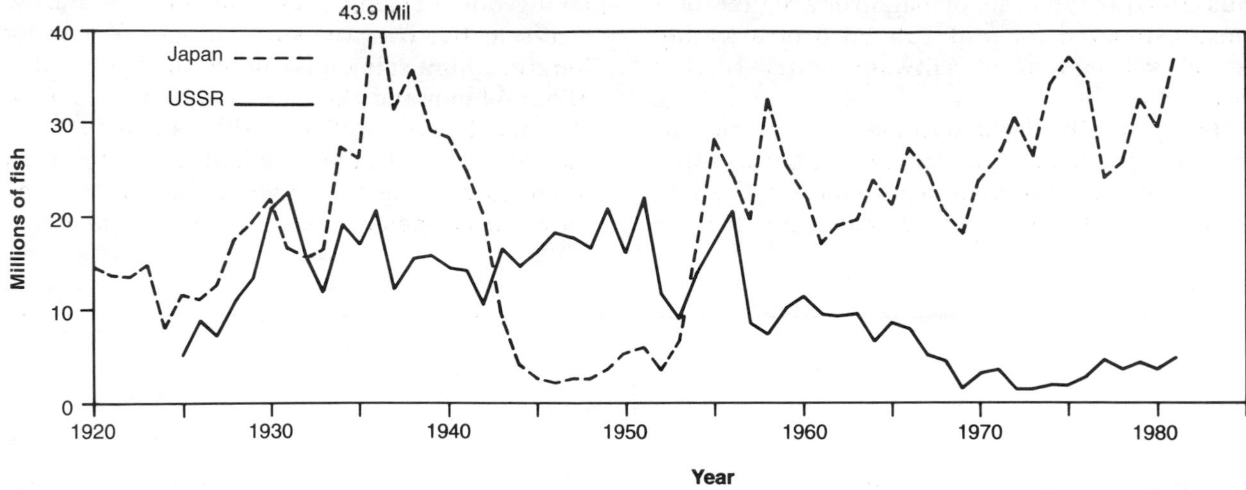

FIGURE 25
Catch of chum salmon by commercial fisheries of Japan and the USSR, in millions of fish, 1920–81

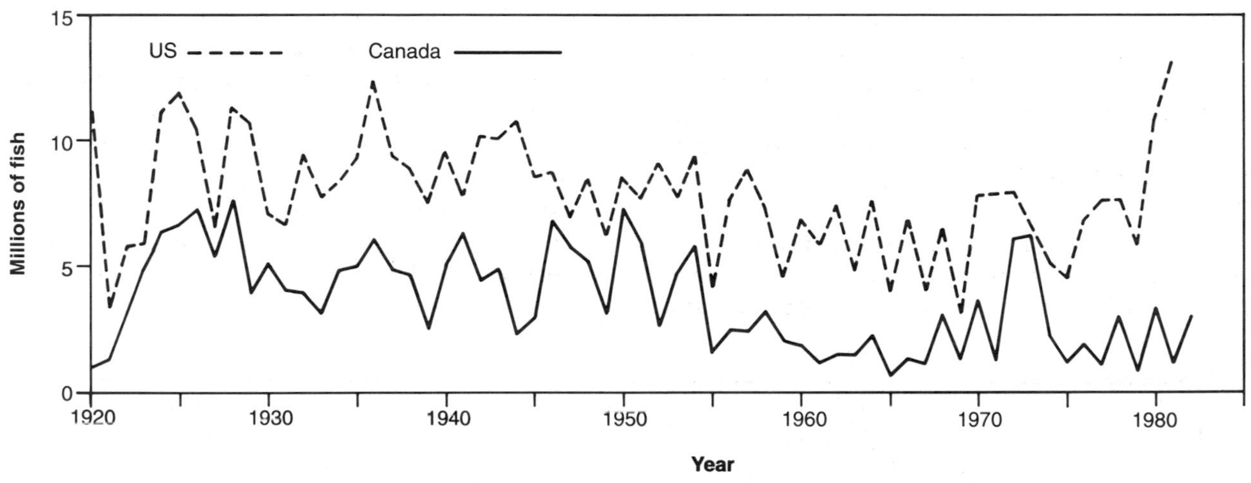

FIGURE 26
Catch of chum salmon by commercial fisheries of the U.S. and Canada, in millions of fish, 1920–81

G. Grette (Pentec Environmental, Edmonds, Washington, pers. comm.) estimated the age composition for the maturing chum for 1936 by using the age composition for Asian and North American chum salmon separately and then estimating the average numbers in each age group by incorporating a factor based on the proportion of chum salmon produced on each continent. The number of immature fish was determined by working backwards from the catch and age composition of the maturing stock and by multiplying by the inverse of the survival rate. Using the estimated age composition for the catch and for a production of 126 million mature fish, the total biomass (immature and mature) of chum salmon supported by the North Pacific Ocean was estimated to be between 860,000 and 1,300,000 t. The biomass estimates vary with assumptions on survival rates and growth from time of entry into the ocean to the time of maturity (ages 0.2, 0.3, and 0.4). Then, by using the available data on the age and size at maturity for the Asian and North American stocks, the total

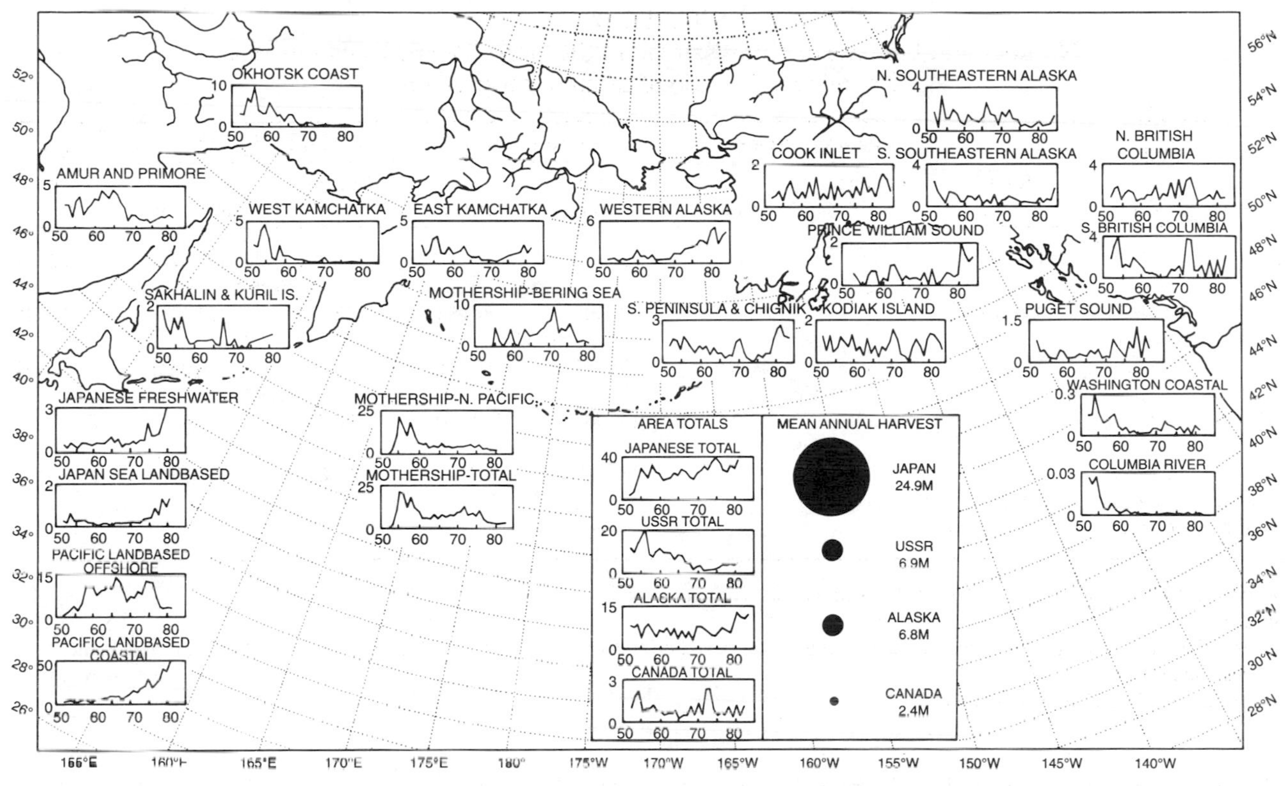

FIGURE 27

Commercial harvest of chum salmon, 1952–84, in millions of fish, and mean annual catch by Japan, USSR, Alaska, and Canada

annual biomass was calculated to be 1,300,000 t (Table 12). If the catch to escapement ratio was 1:1 the total biomass would be 1,600,000 t and the harvest of mature fish would be 604,000 t.

Neave (1961) estimated the annual biomass of chum salmon for the years 1936–39 to be 510,000 t for the mature fish, and 845,000 t for the immatures, for a total of 1,355,000 t. This is about equal to the lower estimate calculated above using a 2:1 catch ratio. According to Neave (1961), chum salmon contributed 47% of the total biomass of the six species of Pacific salmon.

Asian Catch

Japan caught the major portion of Asian salmon from 1933 to 1942 (average annual catch 28.4 million versus 15.1 million for the USSR), although the production at this time was primarily from the Asian mainland (Figure 27). The catch by Japan dropped dramatically during the war and immediate postwar years, 1943–53. After the resumption of distant-water fishing, the Japanese catch of chum salmon averaged 22.8 million from 1955 to 1969. During this period the USSR catch averaged 9.1 million but plummeted to 1.5 million in 1969 (Figure 27). In the 1970s the Japanese developed a massive and very successful hatchery program, and the nearshore and terminal catch rose to over 48 million fish (152,600 t) in 1985, while the far seas catch diminished from 6 million fish in 1965 to 2.5 million fish in 1981, primarily due to international restraints (Figure 27). Correspondingly, the coastal sea catch rose from 16.4 to 34 million fish in the same period.

Birman (1960) developed a relationship for the long-term fluctuations in abundance of Amur River autumn chum salmon, pink salmon, and south Sakhalin herring (*Clupea harengus pallasi*). Relating the warming and cooling of the waters in the Asiatic region of the Kuroshio Current to the abundance of the three species of fishes, he con-

TABLE 12

Number and biomass of chum salmon in the North Pacific Ocean in 1936, assuming a 2:1 ratio of catch to escapement

Year class	Millions of fish	Total tonnes	Mature 3-y-olds (millions of fish)	Mature 3-y-olds (tonnes)	Mature 4-y-olds (millions of fish)	Mature 4-y-olds (tonnes)	Mature 5-y-olds (millions of fish)	Mature 5-y-olds (tonnes)	Age in 1936
1931	27.72	127 512					27.72	127 512	5 (22%)†
							(0.85)*		
1932	114.51	400 786			81.90	286 650	32.61	114 135	4 (65%)†
					(0.85)*		(0.8)*		
1933	153.49	414 423	16.38	44 226	96.351	260 145	40.76	110 052	3 (13%)†
			(0.85)*		(0.8)*		(0.8)*		
1934	190.66	285 990	19.27	28 905	120.44	180 660	50.95	76 425	2
			(0.8)*		(0.8)*		(0.8)*		
1935	238.33	71 499	24.09	7 227	150.55	45 165	63.69	19 107	1

Total annual biomass = 1 300 210 tonnes
Total mature stock = 458 388 tonnes

Source: Calculated by G. Grette, Pentec Environmental, Edmonds, WA

Notes:

*Proportion surviving; partly based on Parker (1962); size based on Ricker (1964). Log w = –2.22 + 3.2 log L and using $\bar{w}$ = 3.6k and 13:65:22 age ratio

†Age composition of maturing stock

cluded that, synchronously, chum salmon and herring thrived during the warmer years and pink salmon were more abundant during the colder years.

Earlier, Birman (1957) had correlated the cyclic rise and fall in abundance of chum salmon and herring with the eleven-year cycle of sunspot activity which, in turn, he associated with warm and cool climatic periods. Wolcott (1978) also compared cyclic sunspot activities to the marine survival rates of Walcott Slough chum salmon and concluded that the relationship was strong. Others, however, felt that the reductions were due to overfishing as well as to natural causes. The catches of chum salmon in 1961–65 in the continental Okhotsk region, as well as those from western and eastern Kamchatka, were produced by the 1955–60 year classes. Kostarev (1982) stated that "in spite of the favorable conditions for natural production (in these winter periods), it was impossible to compensate for the significant spawner deficiency in these periods."

North American Catch

The U.S. catch, predominantly Alaskan, averaged about 8.7 million fish between 1920 and 1951, while the Canadian catch averaged about 4.7 million fish (Figure 27). In the decade 1955–65 the Canadian, Washington, and Oregon catches dropped significantly, indicating extremely poor marine survival rates (Figures 27 and 28). This was very noticeable at Minter Creek, Washington, where records were kept of wild outmigrant juveniles as well as the hatchery contribution for the 1950–61 year classes. No unusual freshwater influences were detected during this period. The period 1950-55 was characterized by falling temperatures and low sunspot activity (Wolcott 1978). This was followed by a decade of warmer than normal northeastern Pacific Ocean temperatures (Chelton 1984). During this period, the central Alaskan stocks showed a definite odd-even year relationship (Figure 29). From 1975 to 1984, again a period of relatively warm ocean temperatures, the western and central Alaskan stocks thrived. Catches increased without significant changes in escapement (Bigler 1985). Confounding the analysis, to some extent, was the resurgence of the Japanese mothership fishery from 1955 to 1960.

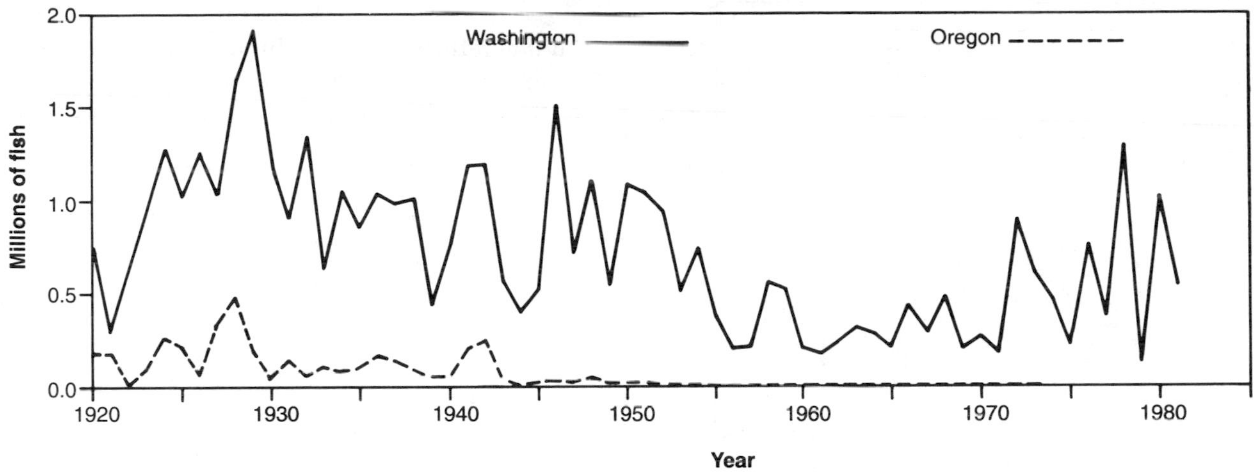

FIGURE 28
Catch of chum salmon by commercial fisheries in the states of Washington and Oregon, in millions of fish, 1920-81

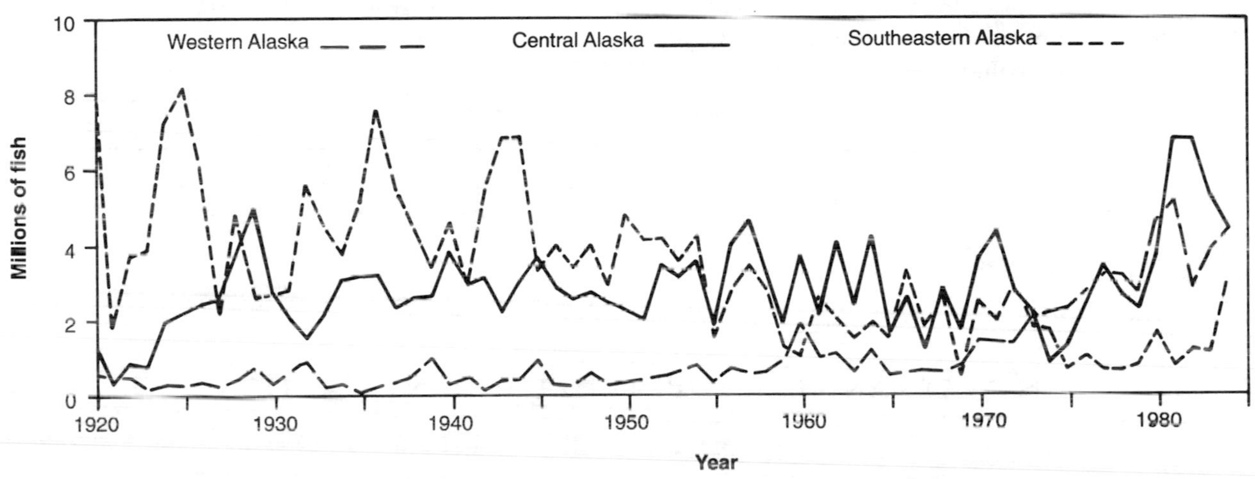

FIGURE 29
Catch of chum salmon by commercial fisheries in Alaska, in millions of fish, 1920-84

The Japanese Hatchery Program

Other than fisheries, perhaps the greatest human influence on the status of chum salmon stocks has been the phenomenal concentration of hatcheries on Hokkaido and Honshu where 262 rivers are managed almost entirely by artificial propagation. Before 1966, the fry were released without supplemental feeding, and the rate of return averaged about 1%. The larger the fry, the sooner they migrate offshore, avoiding the warm currents (Figures 10 and 11). Since feeding of the fry became an established practice in the late 1960s, the rate of return has been consistently about 2% and, on occasion, nearly 3% (Figure 30) (Shirahata 1985).

The number of chum fry released annually from 1982 to 1985 was over 2 billion (300 t), and adult returns in 1984, 1985, and 1986 exceeded 40 million fish (Table 13), for a greater than 2% survival from time of release. The mean survival for fish released from 1976 to 1980 was 2.6%. To put these numbers in perspective in the calculations of the biomass of chum salmon in the North Pacific Ocean as indicated in Table 12, the combined estimate of fry emanating from Asia and North America was 4 billion in 1936.

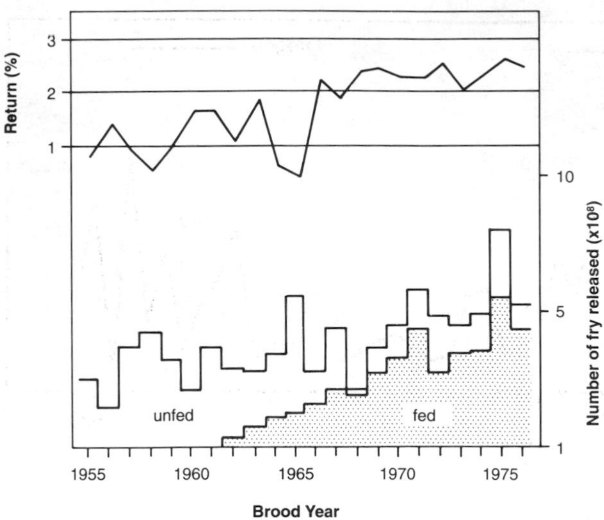

FIGURE 30
Chum salmon fry released from Hokkaido sea ranches, 1955-76, and return (%) related to fed and unfed fry. (From Mahnken et al. 1986)

TABLE 13
Annual returns of adult chum salmon to Hokkaido and Honshu, 1966-86

Year	Total (Thousands)
1966	4 442
1967	5 012
1968	2 513
1969	4 620
1970	5 851
1971	8 548
1972	7 884
1973	9 175
1974	10 772
1975	17 686
1976	10 419
1977	12 559
1978	16 208
1979	24 028
1980	22 418
1981	29 904
1984	37 928
1985	48 086
1986	48 014

Source: From Hiroi (1985) for the years 1966-81; for the years 1984-86, Hiroi (pers. comm.)

Cycles and Chum-Pink Salmon Interaction

Some chum salmon stocks exhibit definite and quite regular even- and odd-year variations in behaviour, age at maturity, size, marine survival, and abundance (Rounsefell and Kelez 1938; Smirnov 1947; Lovetskaya 1948; Noble 1955). These patterns appear to be related to the presence of pink salmon which have strong biennial cycles of abundance. In years when pink salmon juveniles are abundant, the feeding rates of juvenile chum salmon are lower and growth rates are less (Ivankov and Andreyev 1971). As mentioned earlier, the diets of chum salmon may also change. These variations probably result from interspecific competition and from responses that evolved to minimize this competition. Gallagher (1979) suggested that Gause's "exclusion principle," which states that two species cannot occupy the same niche at the same time, is the proper paradigm in which to consider the chum-pink salmon interaction. No matter what the proximal causes may be in the observed cycles in chum salmon stocks, competition must still be viewed as the ultimate cause.

The effects of chum-pink salmon competition vary with the life history phase of the fish and the environment in which it occurs. Chum and pink salmon often spawn in the same reaches of the river, and the possibility exists for density-dependent effects on the deposited eggs resulting from redd superimposition. During years when spawners of both species are abundant, the effects can be both positive and negative. Excessive movement of the gravel can be detrimental. On the other hand, as noted earlier, during years of pink salmon dominance the spawning gravels are cleaner. The deterioration of large numbers of salmon carcasses can cause oxygen deficiencies, as can large clumps of dead and dying eggs. However, density-independent factors associated with weather and climate (e.g., stream flow, scouring, freezing) are very important in determining egg-to-fry survival in these species. More frequently than not, these factors override the density-dependent effects in fresh water and control survival to the fry stage. Variations in temporal and spatial behaviour of downstream migrants can occur in a cyclic pattern, as noted earlier.

Recent research indicates that there are density-dependent effects in the marine environment for some salmonid stocks (Peterman 1978; Rogers 1980; Beacham and Starr 1982; McGie 1984). For chum and pink salmon, direct and indirect evidence indicates that density-dependent effects are present in the early marine environment (Birman 1960). In the southern Kuril Islands, feeding rates of chum and pink salmon juveniles were lower in years when juveniles were abundant (Ivankov and Shershnev 1967, 1968; both cited by Ivankov and Andreyev 1971). Chum fry in the Strait of Georgia near the Fraser River estuary were found to be larger in years when pink salmon were not present (Phillips and Barraclough 1978; cited by Beacham and Starr 1982). Survival of chum fry to adulthood varies with environmental conditions, but for some stocks an even-odd year pattern is evident. Fraser River chum salmon have a higher survival rate during even "non-pink" years than during "pink" years (Figure 31). A similar pattern of survival is present for chum salmon from Hoodsport Hatchery at Hood Canal, Washington (Gallagher 1979). The period of life history at which these patterns are formed is not clear, but it is probably during the early stages at age 0.0 to 0.1.

The possibility also exists for chum-pink salmon competition to occur later in their marine life. Beacham and Starr (1982) suggested that marine survival of chum salmon is influenced by the abundance of adjacent year classes. Such competition would occur in the ocean, not the nearshore or estuarine environment. Also, some chum salmon stocks that do not compete with pink salmon in the early marine environment show odd and even cycles. In Tillamook Bay, Oregon, chum salmon stocks show an alternating age at maturity, yet no pink salmon are present in this area. The same patterns are evident for Willapa Harbor and Grays Harbor on the coast of Washington (S.L. Schroder, Washington Department of Fisheries, Olympia, Washington, pers. comm.).

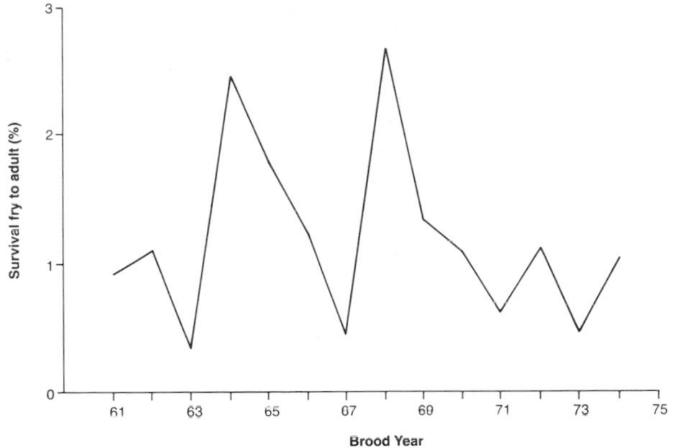

FIGURE 31
Percentage survival from fry to adult for Fraser River chum salmon for brood years 1961–74.
(Data from Beacham and Starr 1982)

Andrievskaya (1966) reported an even-odd cycle shift in the diet of maturing chum salmon in western Kamchatka. In even years (low pink salmon abundance), chum and pink salmon ate similar prey, but during odd years (high pink salmon abundance), the chum diet consisted of prey of lower nutritional quality. Pratt (1974) reported that Puget Sound chum salmon were smaller during odd "pink" years than during even "non-pink" years (Figure 32). It is possible that Puget Sound chum salmon compete with pinks in a way similar to that reported by Andrievskaya (1966).

The catch of chum salmon in the northeast Pacific Ocean exhibits a distinct even-odd year relationship with higher catches during even years (Figure 33). Also, North American pink salmon stocks show an alteration of dominant cycles, changing in roughly ten- to fifteen-year periods (Figure 33). Near the time of these changes, there are indications of breaks in the odd-even chum salmon patterns, although they are not precise or regular (1954–58 for chum salmon, 1954–62 for pink salmon).

Pink salmon are present nearly exclusively in odd years in the streams of Puget Sound and tribu-

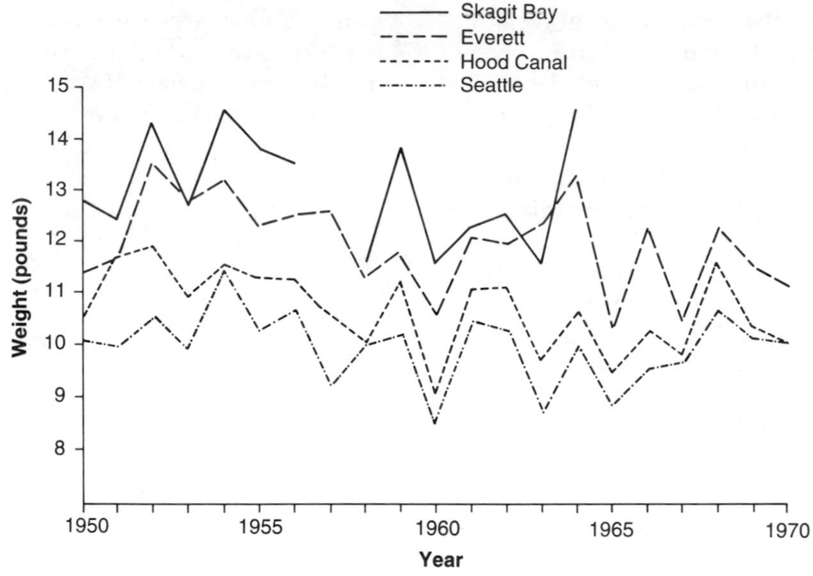

FIGURE 32

Average purse seine chum weights from selected areas in Puget Sound, 1950–70. (From Pratt 1974)

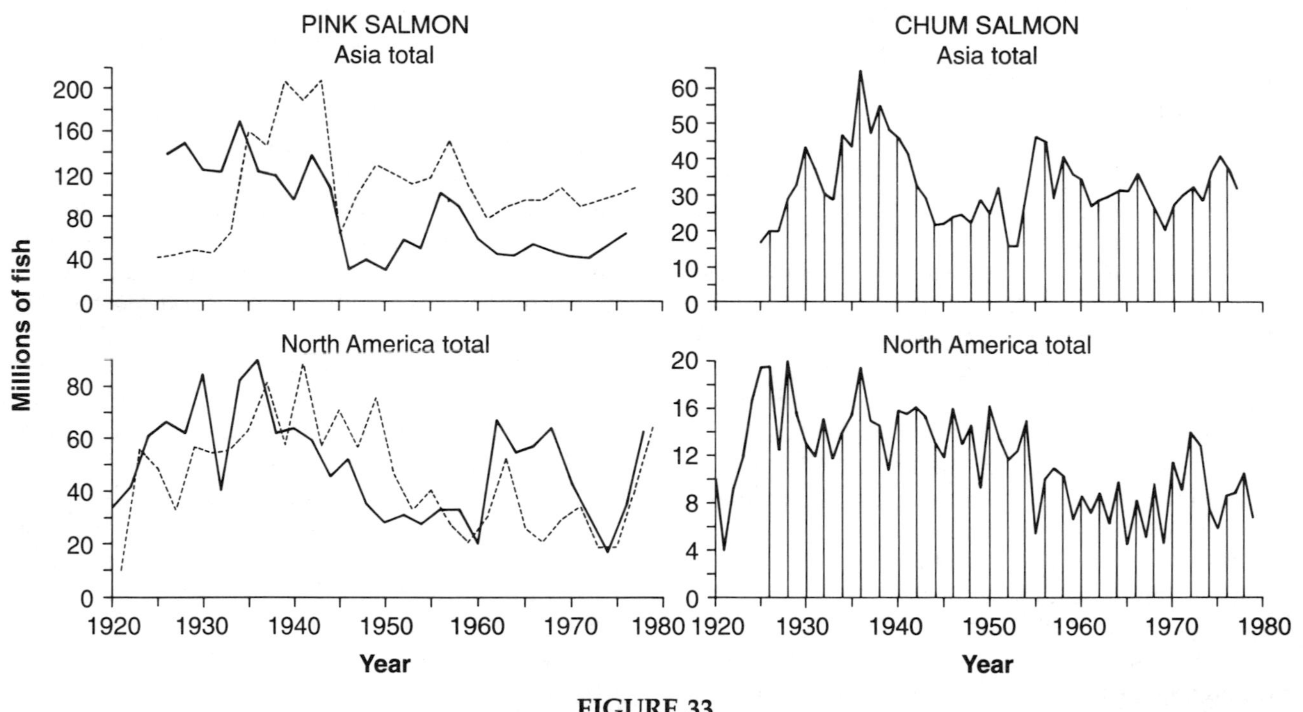

FIGURE 33

Total Asian (1920–77) and North American (1925–79) pink and chum salmon catches. For pink salmon graphs, odd-year data are plotted by the dotted line and even-year data are plotted by the solid line. (Adapted from Asian and North American pink and chum salmon catch statistics, 1980; Proceedings of Pink and Chum Salmon Workshop of 1979; and C.K. Harris, Fisheries Research Institute, University of Washington, Seattle, WA, pers. comm.)

taries of the Fraser River. In this region, chum salmon are more abundant in even years than during odd years (Gallagher 1979). This biennial cycling of chum salmon is manifested in two underlying cycles: 1) survival from fry to adult is higher for even-year than for odd-year broods; and 2) regular alternations occur in age at maturity between even- and odd-year broods (Gallagher 1979). For Puget Sound stocks, odd-year broods return in roughly a 50:50 ratio of three-year-olds to four-year-olds (Figure 34). Three-year-olds comprise about 35% of even-year broods while four-year-olds make up about 65%. A similar pattern exists for Fraser River stocks, although the values vary slightly (Figure 35) (Gallagher 1979; based on data from Bilton 1973). The net result of these alternations is that chum salmon put more reproductive effort into even "non-pink" years than into "odd-pink" years. Gallagher (1979) concluded that these alternations represent a genetic adaptation which allows chum salmon to minimize competition with pink salmon in the early marine environment. This tendency for odd-year brood chum salmon to return at a younger age is even more noteworthy because, due to competition, they are smaller juveniles and, therefore, might be expected to return older as more time is needed to reach critical size at maturity.

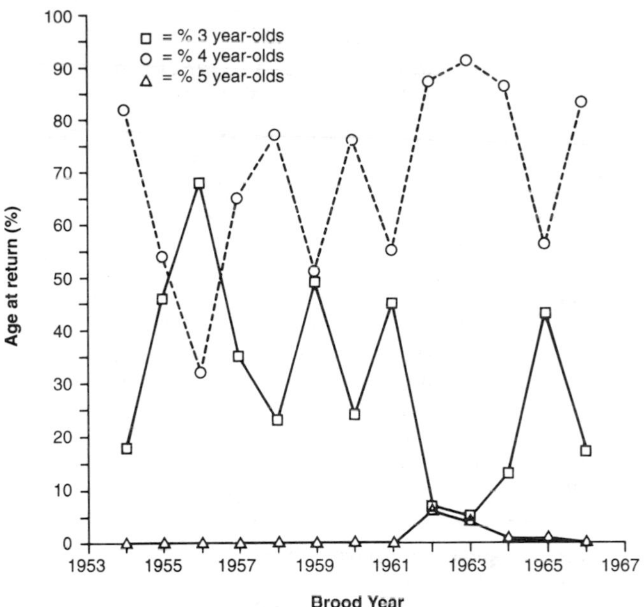

FIGURE 35

Age at return of the total chum salmon brood to the Fraser River, BC, as a percentage of three-, four-, and five-year-olds, 1954–66. (From Bilton 1973, cited in Gallagher 1979)

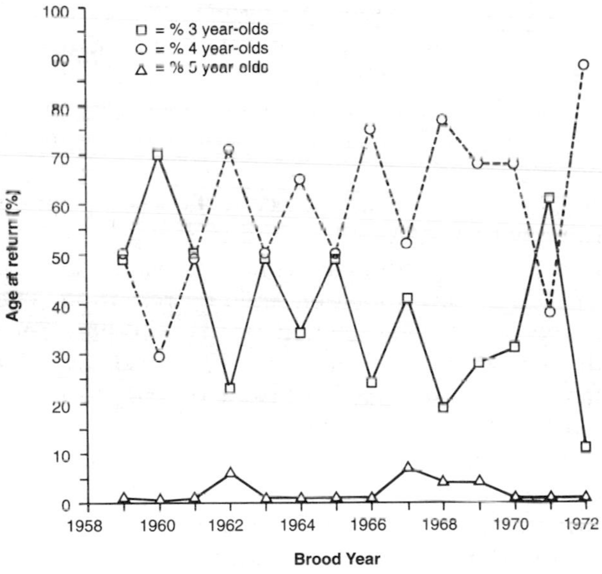

FIGURE 34

Age at return of the total chum salmon brood to Puget Sound as a percentage of three-, four-, and five-year-olds, 1959–72. (From Gallagher 1979)

Smoker (1984) developed a model of the chum-pink salmon interaction based on an adaptation of the Ricker curve to two interacting stocks. He concluded that environmental influences alone could not account for the type of fluctuations in abundance and age at maturity of Puget Sound chum salmon stocks and that a fairly strong genetic mechanism must also be present.

Fluctuations in age at maturity also occur in some stocks which coexist with both even- and odd-year pinks. Chum in Olsen Creek, Alaska, show a more complicated pattern than do Puget Sound stocks, although this pattern appears to agree with Gallagher's conclusion that more reproductive effort is concentrated into odd years which coincide with the subdominant pink year (Gallagher 1979). A regular pattern exists in age at maturity for Bolshaya River chum (Table 14). However, in this region (western Kamchatka), the dominant pink cycles periodically alternate between odd and even cycles; and the presence of a regular pattern in Bolshaya River chum cannot be easily

TABLE 14
Age composition of Bolshaya River chum salmon for brood years 1937–46

Brood year	Catch from brood*	Age composition (%)			
		0.2	0.3	0.4	0.5
1937	462 000	0.4	62.8	36.4	0.4
1938	25 000	19.3	68.9	11.8	0
1939	259 000	4.6	26.8	67.6	1.1
1940	129 000	1.5	84.2	10.1	4.2
1941	1 115 000	2.6	33.8	63.7	0
1942	257 000	1.0	89.9	1.8	7.2
1943	1 703 000	9.0	17.3	71.7	2.0
1944	294 000	0.3	60.8	38.9	0
1945	1 342 000	0.9	60.2	35.3	3.6
1946	360 000	1.0	29.7	69.2	– †

Source: Adapted from Sano (1966); based on data from Semko (1954)

Notes: *Values rounded to nearest 1,000; includes only catches in the Bolshaya River area

†Data unavailable, percentage was assumed to be zero

explained by a genetic model because there does not appear to be time for adaptation to the alternating pink cycles. Table 14 indicates that even-year broods produce fewer chum than odd-year broods (assuming uniform escapement) in the Bolshaya River. Based on this, we might expect the effect of selection to be an increase in spawning effort in the odd brood years at the expense of even brood years. However, the opposite is true. The population is putting more effort into low-survival even broods than into high-survival odd broods. However, the even broods do coincide with the subdominant pink salmon cycle, which conforms to Gallagher's postulate of competitive exclusion between pink and chum salmon. Consequently, the apparent differential mortality appears to be acting in the opposite direction to produce this effect. Also, the pattern of age alteration appears to break down in the 1945 and 1946 brood years and assumes a direction in agreement with selective pressure.

Other stocks show odd-even variations in age at return, but often the data are presented in terms of catch year (Table 15). These data defy analysis without data on the relative strength of the year classes.

Available data limit the determination of the relative contributions of genetics and environment to these fluctuations in age at maturity. Gallagher (1979) noted reversals in these alternatives which occurred simultaneously in Puget Sound, Fraser River, and Olsen Creek stocks. This strongly suggests an effect of the environment interacting with or overriding heredity. Such genetic control must be very malleable and sensitive to be influenced by environmental conditions and by density-dependent interactions in the ocean.

Gallagher (1979) and Smoker (1984) considered the early marine environment to be the most important site of chum-pink salmon competition in Puget Sound chum stocks. Limiting the scope of competition to this life history phase leads to two implicit assumptions for Puget Sound stocks: (1) local chum stocks compete with local pink stocks; and (2) odd-brood chum salmon compete with pink salmon, whereas even-brood chum salmon do not. But Puget Sound chum salmon mature at three or four years of age and pink salmon mature after two years. Because of this age overlap, a Puget Sound chum salmon from an even brood year could compete with a Puget Sound pink salmon in the North Pacific Ocean. More likely is the possibility of competition with pink salmon or chum salmon from other areas. The paths of migration of chum and pink salmon and their ocean distribution provide an opportunity for chum-pink and chum-chum competition among distant stocks.

Table 16 provides a theoretical model for considering the possibilities for competition (intraspecific or interspecific) for Puget Sound chum salmon stocks. The model assumes effects of competition to be density-dependent and to result in slowed growth. Competition in the home stream is limited to redd superimposition and egg retention. The effects of reduced growth depend on the life stages during which it occurs. Slow growth during the first year of life may lead to increased mortality. During the first year, Puget Sound chum salmon interact with local chum and pink salmon and with regional stocks as they migrate northward up the British Columbia coast. From the end of their first year to the end of their penultimate year, as they feed in the North Pacific Ocean, they will interact with pink and chum salmon from Washington, Canada, and, to some extent, with those from Alaska and Asia. During this time, density-dependent growth influences age at maturity and may

TABLE 15
Yearly variation in age composition of chum salmon populations

Area and year	No. of fish sampled	Age composition (%) 0.2	0.3	0.4	0.5	Source
NORTH AMERICA						
Alaska						
Kotzebue Sd.						
1962	68	7.3	63.3	28.0	1.4	Regnart et al. (1966)
1963	255	32.6	47.4	18.8	1.2	"
1964	463	55.7	42.5	1.8	0	"
1965	480	2.7	92.3	5.0	0	"
Yukon R.						
1961	97	4.1	75.3	20.6	0	Regnart et al. (1966)
1962	915	1.9	69.3	28.8	0	"
1963	650	6.0	83.3	10.2	0.5	"
1964	268	33.2	63.0	3.7	0	"
1965	486	0.2	97.3	2.5	0	"
Prince Wm. Sd.						
1952	187	23.5	47.1	29.4	0	Thorsteinson et al. (1963)
1953	819	8.4	76.4	15.1	0	"
1954	100	45.0	45.0	10.0	0	"
1955	55	10.9	81.8	7.3	0	"
1956	617	11.0	86.2	2.8	0	"
1957	218	6.9	72.0	21.1	0	"
1958	141	15.6	76.6	7.8	0	"
British Columbia						
Nootka						
1933	160	14.4	24.4	59.4	1.8	Pritchard (1943)
1934	124	16.9	73.3	9.0	0.8	"
1935	186	17.2	44.6	36.6	1.6	"
1941	518	9.1	50.6	39.6	0.7	"
Oregon						
Tillamook Bay						
1947	65	32.3	66.2	1.5	0	Oakley (1966)
1949	287	4.9	94.7	0.4	0	"
1950	481	76.2	22.5	1.3	0	"
1959	310	51.2	48.0	0.8	0	"
1960	92	68.2	30.8	1.0	0	"
1961	123	83.4	16.0	0.6	0	"
ASIA						
Sea of Okhotsk coast						
1957	-	1.4	63.1	9.8	25.7	Kondo et al. (1965)
1958	-	6.8	25.2	68.0	0	"
1959	-	1.9	86.0	9.5	2.6	"
1960	-	0.3	42.1	57.0	0.6	"
1961	-	1.2	32.9	63.0	2.8	"
West Kamchatka coast						
1957	-	0	68.2	23.0	8.8	Kondo et al. (1965)
1958	-	19.0	58.6	22.4	0	"
1959	-	0.6	91.7	7.7	0	"
1960	-	0.2	59.8	39.7	0.3	"
1961	-	0	37.6	59.2	3.1	"
East Kamchatka coast						
1957	-	5.0	72.5	21.5	1.0	Kondo et al. (1965)
1958	-	9.0	75.6	15.4	0	"
1959	-	0.8	83.7	13.7	1.8	"
1960	-	1.0	41.4	54.8	2.8	"
1961	-	0.8	51.1	44.7	3.4	"

Source: From Bakkala (1970)

TABLE 16
Theoretical location and effect of competition (chum/chum and chum/pink) for Puget Sound chum salmon stocks

	Location of Competition					
Stock/effect	Home stream	Nearshore estuary	BC coast	North Pacific	BC coast	Puget Sound
	First year →			End of first year to end of penultimate year	Final year →	
Local stocks	+	+	minimal*	0	minimal*	+
Major effect of competition	↓Survival to migrant fry†	↓Survival to adult	↓Survival to adult	-	↓Size at return‡	↓Size at return‡
Regional stocks (WA, OR, BC)	0	0	+	minimal*	+	0
Major effect of competition	-	-	↓Survival to adult	↑Age of maturation‡	↓Size at return‡	-
North Pacific stocks	0	0	0	+	0	0
Major effect of competition	-	-	-	↑Age of maturation‡	-	-

Notes: ↑ = increased effect, ↓ = decreased effect; + = competition, 0 = no competition
*The magnitude of effect will depend on the size of the local stock compared to the size of the regional stocks or size of regional stock compared to size of North Pacific stock.
†Effect only for chums on chums or chums on pinks, not pinks on chums, because pinks spawn before chums. Effect is due to redd superimposition and egg retention.
‡Survival to adult may also decrease due to predation.

influence survival. To define these interactions, further studies similar to Peterman's (1984) approach to sockeye salmon interactions are needed.

G. Grette (Pentec Environmental, Edmonds, Washington, pers. comm.) hypothesized that as Asian, Canadian, Washington, and southeastern Alaskan chum salmon have a moderate overlap of feeding grounds, and as competition is not strategically advantageous, a pattern of avoidance has developed. Both Asian and southern North American stocks are influenced by pink salmon, so competition at that time is minimized by conserving reproductive energies to off-year cycles. This temporal avoidance is not evident for western Alaskan dominant even-year chum salmon, which may be present in large numbers as age 0.1 fish in the same feeding areas as are northern even-year pink salmon. The adaptations in these instances may be limited to a shift in age at maturity to put more reproductive effort into the "off-years."

Interactions between chum and pink salmon are evident, and interactions between chum, pink, and sockeye salmon are implied. The partitioning of the oceanic feeding areas among the various age groups of chum and sockeye salmon, along with the variable input of odd- and even-year pink salmon, suggests that the dynamics and plasticity of the chum and sockeye salmon are "fitted in" the rigid, structured life history of the pink salmon. Perhaps it is necessary for one of the three species to be structured. It is increasingly apparent that the occurrence of odd- and even-year cycles has been brought about by oceanic phenomena. Concurrently, the suggested eleven-year periodicity of sunspots and the resultant cooling and warming of the oceans superimposes environmental effects that override the genetic mechanisms affecting odd- and even-year behaviour.

Concluding Remarks

In spite of our inability to define precisely the patterns of evolution of the salmonids, more specifically, the species in the genus *Oncorhynchus*, the different adaptive features and degrees of specialization of the seven species of salmon are quite evident. Whether *O. keta* is the most primitive in a geneological line or whether it evolved as a species between the sockeye-pink salmon and the chinook-coho salmon is presently not clear.[7] However, we can identify the remarkable adaptations of the chum salmon, which, in total, constitute an effective strategy for survival.

Chum salmon have maintained subtle but complex patterns of diversity in behaviour that have led to accommodation to a wide spectrum of environmental conditions. This has resulted in the chum salmon having the widest geographical distribution of the Pacific salmon and, prior to the influences of harvest, probably the greatest biomass of any of the salmon species in the Pacific Ocean. The adaptations that brought about these successes include: an ability to spawn successfully in streams of various sizes in a number of systems; the separation of runs by time and space, maintained by well-developed homing and migratory behaviour, thus making for efficient utilization of stream and ocean pasturage; exhibition of broad patterns of phototaxis and rheotaxis; retention of some aggressive but vestigial territorial behaviour as fry in fresh water, whether schooled or separated as individuals; the ability to feed in fresh water (perhaps as long as a year) when necessary or advantageous (otherwise their residence in fresh water is short); alteration of patterns of diel outmigration from streams with large numbers of pink and chum salmon outmigrants; the loss of (or never attaining) obligatory schooling, or the maintenance of loose schools which break up when advantageous; the alteration of diet on the high seas during intense competition with pink salmon or with chum salmon cohorts; the maturation and return as adults of cohorts (and siblings) at various ages, thus increasing fecundity by increased size, and distributing the gene pool over several years; a genetic adaptation to alter the age at return to increase spawning potential to coincide with the subdominant pink salmon cycles; indications of spatial and temporal (odd-even year) partitioning of ocean feeding areas as strategies to minimize intra- and interspecific competition; and the maintenance of a broad and plastic genetic base which can be overridden by environmental factors when strategically advantageous.

Under certain conditions the chum salmon has proven to be a highly successful animal for artificial propagation (aquaculture), and, in recent times, the number of chum salmon in the North Pacific has reached historic proportions.

7 Smith and Stearley (1989) indicated that chum salmon lie between the coho-chinook and pink-sockeye groupings on a phylogenetic tree of the salmonids. (*Editors*)

Acknowledgments

Many thanks to Drs. Kees Groot and Leo Margolis of the Pacific Biological Station, Nanaimo, British Columbia, first, for the offered opportunity to gather the lore of chum salmon, and, second, for their continual support. Once again I relied upon the studies of past graduate students of the University of Washington's Fisheries Research Station at Big Beef Creek, especially Drs. K Koski (National Marine Fisheries Service, Auke Bay, Alaska) and Steve Schroder (Washington Department of Fisheries, Olympia, Washington). I am especially indebted for direct contributions by Glenn Grette (Pentec Environmental, Edmonds, Washington) in the areas of age, growth, and abundance, and Tom Jagielo (Washington Department of Fisheries, Olympia, Washington) for population statistics. I am also indebted to T. Moritz of the East Asia Library, Marino Kraabel Lundin of the Suzzalo Library, and Janet Ebaugh of the Department of Botany at the University of Washington, Seattle, Washington. Gary Duker, formerly of the University of Washington, and currently Chief of Publications for the Northwest and Alaska Fisheries Science Center, National Marine Fisheries Service, Seattle, Washington, contributed throughout. The review of chum salmon by Dr. Richard Bakkala was drawn upon heavily. Critical reviews were made by Drs. Colin Harris (Fisheries Research Institute, Seattle, Washington), Terry D. Beacham (Pacific Biological Station, Nanaimo, British Columbia), John H. Helle (National Marine Fisheries Service, Auke Bay, Alaska), and Gary Duker. Much gratitude to Ms. Carol Sisley for processing the many versions. Finally, heartfelt thanks to my wife, Helen, for her patient support.

References

Abbasov, G.S., and G.D. Polyakov. 1978. Evaluation of feeding conditions and the biological state of a fish population by means of the condition factor. J. Ichthyol. 18:404-417

Alderdice, D.F., W.P. Wickett, and J.R. Brett. 1958. Some effects of temporary exposure to low dissolved oxygen levels on Pacific salmon eggs. J. Fish. Res. Board Can. 15:229-249

Andersen, F.M. 1983. Upper Yukon test fishing studies, 1982. Alaska Department of Fish and Game, Division of Commercial Fisheries, Fairbanks, AK. 24 p.

Anderson, A.D., and T.D. Beacham. 1983. The migration and exploitation of chum salmon stocks of the Johnstone Strait-Fraser River study area, 1962-1970. Can. Tech. Rep. Fish. Aquat. Sci. 1166:125 p.

Andrievskaya, L.D. 1966. Food relationships in the Pacific salmon in the sea. Vopr. Ikhtiol. 6: 84-90. (Transl. from Russian; available Northwest and Alaska Fisheries Center, Seattle, WA)

Anonymous. 1928. Record chum caught off Quadra. Pac. Fisherman 1928(Oct.):13

Aro, K.V., and M.P. Shepard. 1967. Pacific salmon in Canada, p. 225-327. *In*: Salmon of the North Pacific Ocean. Part IV. Spawning populations of North Pacific salmon. Int. North Pac. Fish. Comm. Bull. 23

Atkinson, C.E., J.H. Rose, and T.O. Duncan. 1967. Pacific salmon in the United States, p. 43-224. *In*: Salmon of the North Pacific Ocean. Part IV. Spawning populations of North Pacific salmon. Int. North Pac. Fish. Comm. Bull. 23

Baggerman, B. 1960. Salinity preference, thyroid activity and the seaward migration of four species of Pacific salmon (*Oncorhynchus*). J. Fish. Res. Board Can. 17: 295-322

Bailey, J.E. 1964. Intertidal spawning of pink and chum salmon at Olsen Bay, Prince William Sound, Alaska. U.S. Fish Wildl. Serv. Biol. Lab. Auke Bay MS Rep. MR 64-6:23 p.

Bailey, J.E., B.L. Wing, and C.R. Mattson. 1975. Zooplankton abundance and feeding habits of fry of pink salmon, *Oncorhynchus gorbuscha*, and chum salmon, *Oncorhynchus keta*, in Traitors Cove, Alaska, with speculations on the carrying capacity of the area. Fish. Bull. (U.S.) 73:846-861

Bakkala, R.G. 1970. Synopsis of biological data on the chum salmon, *Oncorhynchus keta* (Walbaum) 1792. FAO Fish. Synop. 41; U.S. Fish. Wildl. Serv. Circ. 315:89 p.

Balchen, J.G. 1976. Principles of migration in fishes. Prepared for Norges Teknisk Naturvitenskapelige Forskningsrad og Norges Fiskeriforskningsrad. Selsk. Ind. Tekh. Forsk. Nor. Tekh. Hoegsk. Rapp. STF48 A76045; Tekh. Notat 81:33 p.

Bams, R.A. 1967. Differences in performance of naturally and artificially propagated sockeye salmon migrant fry, as measured with swimming and predation tests. J. Fish. Res. Board Can. 24:1117–1153

——. 1969. Adaptations of sockeye salmon associated with incubation in stream gravels, p. 71–88. *In*: T.G. Northcote (ed.). Symposium on Salmon and Trout in Streams. H.R. MacMillan Lectures in Fisheries. Institute of Fisheries, University of British Columbia, Vancouver, BC

——. 1970. Evaluation of a revised hatchery method tested on pink and chum salmon fry. J. Fish. Res. Board Can. 27:1429–1452

——. 1982. Experimental incubation of chum salmon (*Oncorhynchus keta*) in a Japanese-style hatchery system. Can. Tech. Rep. Fish. Aquat. Sci. 1101:65 p.

——. 1983. Early growth and quality of chum salmon (*Oncorhynchus keta*) produced in keeper channels and gravel incubators. Can. J. Fish. Aquat. Sci. 40:449–505

Bams, R.A., and C.N.H. Lam. 1983. Influences of deteriorating water quality on growth and development of chum salmon (*Oncorhynchus keta*) larvae in a Japanese-style keeper channel. Can. J. Fish. Aquat. Sci. 40:2098–2104

Barker, M. 1979. Summary of salmon travel time from tagging studies, 1950–1974. Washington Department of Fisheries, Olympia, WA. 30 p.

Bax, N.J. 1982. Seasonal and annual variations in the movement of juvenile chum salmon through Hood Canal, Washington, p. 208–218. *In*: E.L. Brannon and E.O. Salo (eds.). Proceedings of the Salmon and Trout Migratory Behavior Symposium. School of Fisheries, University of Washington, Seattle, WA

——. 1983a. The early marine migration of juvenile chum salmon (*Oncorhynchus keta*) through Hood Canal - its variability and consequences. Ph.D. thesis. University of Washington, Seattle, WA. 196 p.

——. 1983b. Early marine mortality of marked juvenile chum salmon (*Oncorhynchus keta*) released into Hood Canal, Puget Sound, Washington, in 1980. Can. J. Fish. Aquat. Sci. 40:426–435

Bax, N.J., and C.J. Whitmus. 1981. Early marine survival and migratory behavior of juvenile salmon released from the Enetai Hatchery, Washington, in 1980. Univ. Wash. Fish. Res. Inst. FRI-UW-8109:48 p.

Beacham, T.D. 1982. Fecundity of coho salmon (*Oncorhynchus kisutch*) and chum salmon (*O. keta*) in the northeast Pacific Ocean. Can. J. Zool. 60:1463–1469

——. 1984a. Age and morphology of chum salmon in southern British Columbia. Trans. Am. Fish. Soc. 113:727–736

——. 1984b. The status of the chum salmon fishery and stocks of British Columbia. (Submitted to annual meeting International North Pacific Fisheries Commission, Vancouver, BC, October 1984.) Department of Fisheries and Oceans, Pacific Biological Station, Nanaimo, BC. 29 p.

Beacham, T.D., and C.B. Murray. 1986. Comparative developmental biology of chum salmon (*Oncorhynchus keta*) from the Fraser River, British Columbia. Can. J. Fish. Aquat. Sci. 43:252–262

——. 1987. Adaptive variation in body size, and developmental biology of chum salmon (*Oncorhynchus keta*) in British Columbia. Can. J. Fish. Aquat. Sci. 44:244–261

Beacham, T.D., and P. Starr. 1982. Population biology of chum salmon, *Oncorhynchus keta*, from the Fraser River, British Columbia. Can. J. Fish. Aquat. Sci. 43:252–262

Beacham, T.D., F.C. Withler, and R.B. Morley. 1985. Effect of egg size on incubation time and alevin and fry size in chum salmon (*Oncorhynchus keta*) and coho salmon (*O. kisutch*). Can. J. Zool. 63:847–850

Beacham, T.D., R.E. Withler, and A.P. Gould. 1985. Biochemical genetic stock identification of chum salmon (*Oncorhynchus keta*) in southern British Columbia, 1985. Can. J. Fish. Aquat. Sci. 43:437–448

Beall, E.P. 1972. The use of predator-prey tests to assess the quality of chum salmon *Oncorhynchus keta* fry. M.Sc. thesis. University of Washington, Seattle, WA. 105 p.

Behnke, R.J. 1979. Monograph of the native trouts of the genus *Salmo* of western North America. U.S. Fish and Wildlife Service, Fishery Resources, Denver, CO. 163 p.

Berg, L.S. 1934. Vernal and hiemal races among anadromous fishes. Izv. Akad. Nauk SSSR Otd. Mat. Estestv. Nauk Ser. 7:711–732. (Transl. from Russian; J. Fish. Res. Board Can. 16(1959):515–537)

Bigler, B. 1985. Kotzebue Sound chum salmon (*Oncorhynchus keta*) escapement and return data,

1962–1984. Alaska Dep. Fish Game ADF&G Tech. Data Rep. 149:112 p.

Bigler, B., and D. Burwen. 1984. Migratory timing and spatial entry patterns of chum salmon (*Oncorhynchus keta*) in Kotzebue Sound. Alaska Dep. Fish Game Inf. Leafl. 238:23 p.

Bilton, H.T. 1973. The relation between the catches of three-year and four-year-old chum salmon in consecutive years. Fish. Res. Board Can. Tech. Rep. 413:45 p.

Bilton, H.T., and S.A.M. Ludwig. 1966. Times of annulus formation on scales of sockeye, pink, and chum salmon in the Gulf of Alaska. J. Fish. Res. Board Can. 23:1403–1410

Bird, F. 1982. Preliminary forecast model for Kotzebue Sound, Alaska, chum salmon (*Oncorhynchus keta*). Alaska Dep. Fish Game Inf. Leafl. 203:27 p.

Birman, I.B. 1951. Qualitative characteristics of the stocks and the dynamics of abundance of the autumn chum salmon in the Amur River. Izv. Tikhookean. Nauchno-Issled. Inst. Rybn. Khoz. Okeanogr. 35:17–31. (Transl. from Russian; Fish. Res. Board Can. Transl. Ser. 349)

——. 1956. Local stocks of autumn chum salmon in the Amur Basin. Vopr. Ikhtiol. 7:158–173. (Transl. from Russian; Fish. Res. Board Can. Transl. Ser. 349)

——. 1957. The Kuro-Sio and the abundance of Amur autumn chum salmon (*Oncorhynchus keta* (Walbaum) infrasp. *autumnalis*). Vopr. Ikhtiol. 8:3–7. (Transl. from Russian; Fish. Res. Board Can. Transl. Ser. 265)

——. 1960. New information on the marine period of life and the marine fishery of Pacific salmon, p. 151–164. *In*: Trudy Soveshchaniia po biologicheskim osnovam okeanicheskovo rybolovstva, 1958. Tr. Soveshch. Ikhtiol. Kom. Akad. Nauk SSSR 10. (Transl. from Russian; Fish. Res. Board Can. Transl. Ser. 357)

——. 1967. Interspecific relationships of Pacific salmon at sea. Izv. Tikhookean. Nauchno-Issled. Inst. Rybn. Khoz. Okeanogr. 57:3–24. (In Russian)

——. 1969. Distribution and growth of young Pacific salmon of the genus *Oncorhynchus* in the sea. Probl. Ichthyol. 9:651–666

Blackbourn, D.J. 1985. The 'salinity' factor and the marine survival of Fraser River pinks and other stocks of salmon and trout, p. 67–75. *In*: B.G. Shepherd (rapp.). Proceedings of the 1985 Northeast Pacific Pink and Chum Salmon Workshop. Department of Fisheries and Oceans, Vancouver, BC

Brannon, E.L. 1965. The influence of physical factors on the development and weight of sockeye salmon embryos and alevins. Int. Pac. Salmon Fish. Comm. Prog. Rep. 12:26 p.

Brix, R. 1981. Data report of Grays Harbor juvenile salmon seining program, 1973–1980. Wash. Dep. Fish. Prog. Rep. 141:78 p.

Bruya, K.J. 1981. The use of different gravel depths to enhance the spawning of chum salmon, *Oncorhynchus keta*. M.Sc. thesis. University of Washington, Seattle, WA. 86 p.

Buklis, L.S. 1981. Yukon and Tanana River fall chum salmon tagging studies, 1976–1980. Alaska Dep. Fish Game Inf. Leafl. 194:40 p.

Buklis, L.S., and L.H. Barton. 1984. Yukon River fall chum salmon biology and stock status. Alaska Dep. Fish Game Inf. Leafl. 239:67 p.

Burner, C.J. 1951. Characteristics of spawning nests of Columbia River salmon. Fish. Bull. Fish Wildl. Serv. 61:97–110

Chelton, D.B. 1984. Commentary: Short-term climatic variability in the northeast Pacific Ocean, p. 87–99. *In*: W.G. Pearcy (ed.). The influence of ocean conditions on the production of salmonids in the North Pacific: a workshop. Oreg. State Univ. Sea Grant Coll. Program ORESU-W-83-001

Clark, J.E., and J.L. Weller. 1986. Age, sex, and size of chum salmon (*Oncorhynchus keta* Walbaum) from catches and escapements in southeastern Alaska, 1984. Alaska Dep. Fish Game ADF&G Tech. Data Rep. 168:288 p.

Cline, D. 1982. Council grounds of the eagles. Alaska Fish Tales Game Trails 1982–83 (Winter):22–25

Coble, D.W. 1961. Influence of water exchange and dissolved oxygen in redds on survival of steelhead trout embryos. Trans. Am. Fish. Soc. 90:469–474

Congleton, J.L. 1979. Feeding patterns of juvenile chum in the Skagit River salt marsh, p. 141–150. *In*: S.J. Lipovsky and C.A. Simenstad (eds.). Gutshop '78: fish food habits studies: proceedings of the Second Northwest Technical Workshop. Wash. Sea Grant Program WSG-WO-79-1

Congleton, J.L., S.K. Davis, and S.R. Foley. 1982. Distribution, abundance and outmigration timing of chum and chinook salmon fry in the Skagit salt marsh, p. 153–163. *In*: E.L. Brannon and E.O. Salo (eds.). Proceedings of the Salmon and Trout Migratory Behavior Symposium. School of Fisheries, University of Washington, Seattle, WA

Cooney, R.T., D. Urquhart, R. Nevé, J. Hilsinger, R. Clasby, and D. Barnard. 1978. Some aspects of the carrying capacity of Prince William Sound, Alaska, for hatchery released pink and chum salmon fry. Alaska Sea Grant Rep. 78-4; Univ. Alaska Inst. Mar. Resourc. IMS R78-3:98 p.

Cordell, J.R. 1986. Structure and dynamics of an epibenthic harpactacoid assemblage and the role of predation by juvenile salmon. M.Sc. thesis. University of Washington, Seattle, WA. 186 p.

Davis, R.H., Jr., and C. Olito. 1986. Preliminary investigations of genetic structure of chum salmon, *Oncorhynchus keta*, in the Noatak and Kobuk River drainages of northwestern Alaska. Alaska Dep. Fish Game FRED Rep. 62:23 p.

Davis, S.K. 1981. Determination of body composition, condition, and migration timing of juvenile chum and chinook salmon in the lower Skagit River, Washington. M.Sc. thesis. University of Washington, Seattle, WA. 97 p.

Day, F. 1887. British and Irish Salmonidae. Williams and Norgate, London. 298 p.

Dickie, L.M. 1971. Mathematical models of growth. Chpt. 5 addendum (p. 126–130) *In*: W.E. Ricker (ed.). Methods for assessment of fish production in fresh waters. 2nd. ed. Int. Biol. Programme IBP Handbk. 3

Dill, L.M., and T.G. Northcote. 1970. Effects of some environmental factors on survival, condition, and timing of emergence of chum salmon fry (*Oncorhynchus keta*). J. Fish. Res. Board Can. 27:196–201.

Disler, N.N. 1953. Ecological and morphological characteristics of the development of the Amur autumn chum salmon, *Oncorhynchus keta* (Walb). Tr. Soveshch. Ikhtiol. Kom. Akad. Nauk SSSR 1:354–362. (Transl. from Russian. *In*: Pacific salmon: selected articles from Soviet periodicals, p. 33–41. Israel Program for Scientific Translations, Jerusalem, 1961)

——. 1954. Development of autumn chum salmon in the Amur River, p. 129–143. *In*: Trudy Soveshchaniia po voprosam lososevogo khozyaistva dal'nego vostoka, 1953. Tr. Soveshch. Ikhtiol. Kom. 4. (Transl. from Russian; Israel Program for Scientific Translations, Jerusalem, 1963)

Duker, G.J. 1977. Nest site selection by chum salmon (*Oncorhynchus keta*) in a spawning channel. M.Sc. thesis. University of Washington, Seattle, WA. 113 p.

——. 1982. Instream orientation and species recognition by Pacific salmon, p. 286–295. *In*: E.L. Brannon and E.O. Salo (eds.). Proceedings of the Salmon and Trout Migratory Behavior Symposium. School of Fisheries, University of Washington, Seattle, WA

Dvinin, P.A. 1952. The salmon of south Sakhalin. Izv. Tikhookean. Nauchno-Issled. Inst. Rybn. Khoz. Okeanogr. 37:69–108. (Transl. from Russian; Fish. Res. Board Can. Transl. 120)

Dymond, J.R. 1940. Pacific salmon in the Arctic Ocean. Proc. 6th Pac. Sci. Cong. 1939(3):435

Eames, M., T. Quinn, K. Reidinger, and D. Harding. 1981. Northern Puget Sound 1976 adult coho and chum tagging studies. Wash. Dep. Fish. Tech. Rep. 64:217 p.

Emadi, H. 1973. Yolk-sac information in Pacific salmon in relation to substrate, temperature, and water velocity. J. Fish. Res. Board Can. 30:1249–1250

Fast, D.E., and Q.J. Stober. 1984. Intragravel behavior of salmonid alevins in response to environmental changes. Final report for Washington Water Research Center and City of Seattle. Univ. Wash. Fish. Res. Inst. FRI-UW-8414:103 p.

Fedorenko, A.Y., F.J. Fraser, and D.T. Lightly. 1979. A limnological and salmonid resource study of Nitinat Lake: 1975–1977. Fish. Mar. Serv. (Can.) Tech. Rep. 839:86 p.

Feller, R.J., and J.W. Kaczynski. 1975. Size selective predation by juvenile chum salmon on epibenthic prey in Puget Sound. J. Fish. Res. Board Can. 32:1419–1429

Fiscus, G. 1969. 1968 Admiralty Inlet chum salmon tagging. Wash. Dep. Fish. Annu. Rep. 78:13–19

Fredin, R.A., R.L. Major, R.G. Bakkala, and G. Tanonaka. 1977. Pacific salmon and the high seas salmon fisheries of Japan. (Processed report.) Northwest and Alaska Fisheries Center, National Marine Fisheries Service, Seattle, WA. 324 p.

Fresh, K.L., and S.L. Schroder. 1987. Influence of the abundance, size and yolk reserves of juvenile chum salmon (*Oncorhynchus keta*) on predation by freshwater fishes in a small coastal stream. Can. J. Fish. Aquat. Sci. 44:236-243

Frolenko, L.A. 1970. Feeding of chum and pink salmon juveniles migrating downstream in the main spawning rivers of the northern coast of the Sea of Okhotsk. Izv. Tikhookean. Nauchno-Issled. Inst. Rybn. Khoz. Okeanogr. 71:179-188. (Transl. from Russian; Fish. Res. Board Can. Transl. Ser. 2416)

Gallagher, A.F., Jr. 1979. An analysis of factors affecting brood year returns in the wild stocks of Puget Sound chum (*Oncorhynchus keta*) and pink salmon (*Oncorhynchus gorbuscha*). M.Sc. thesis. University of Washington, Seattle, WA. 152 p.

Gibbons, D.R., and E.O. Salo. 1973. Annotated bibliography of the effects of logging on fish of the western United States and Canada. U.S. For. Serv. Gen. Tech. Rep. PNW-10:143 p.

Gilbert, C.H. 1922. The salmon of the Yukon River. Bull. Bur. Fish. (U.S.) 38:317-332

Godin, J.G.J. 1982. Migrations of salmonid fishes during early life history phases: daily and annual timing, p. 22-50. *In*: E.L. Brannon and E.O. Salo (eds.). Proceedings of the Salmon and Trout Migratory Behavior Symposium. School of Fisheries, University of Washington, Seattle, WA

Graybill, J.P., R.L. Burgner, J.C. Gislason, P.E. Hoffman, K.H. Wyman, R.G. Gibbons, K.W. Kurko, Q.J. Stober, T.W. Fagman, A.P. Stayman, and D.M. Eggers. 1979. Assessment of the reservoir-related effects of the Skagit project on downstream fishery resources of the Skagit River, Washington. Final report for City of Seattle, Department of Lighting. Univ. Wash. Fish. Res. Inst. FRI-UW-7905:602 p.

Grigo, L.D. 1953. Morphological differences between the summer and the autumn chum salmon *Oncorhynchus keta* (Walbaum), *O. keta* (Walbaum) infraspecies *autumnalis* Berg. Dokl. Akad. Nauk SSSR 92:1225-1228. (Transl. from Russian; *In*: Pacific salmon: selected articles from Soviet periodicals, p. 13-17. Israel Program for Scientific Translations, Jerusalem, 1961)

Gunsolus, R.T. 1978. The status of Oregon coho and recommendations for managing the production, harvest, and escapement of wild and hatchery reared stocks. Oregon Department of Fish and Wildlife, Columbia Region, Portland, OR. 59 p.

Hallock, R.J., and D.H. Fry, Jr. 1967. Five species of salmon, *Oncorhynchus*, in the Sacramento River, California. Calif. Fish Game 53:5-22

Hargreaves, N.B., and R.J. LeBrasseur. 1986. Size selectivity of coho (*Oncorhynchus kisutch*) preying on juvenile chum salmon (*O. keta*). Can. J. Fish. Aquat. Sci. 43:581-586

Hartt, A.C. 1980. Juvenile salmonids in the oceanic ecosystem - the critical first summer, p. 25-57. *In*: W.J. McNeil and D.C. Himsworth (eds.). Salmonid ecosystems of the North Pacific. Oregon State University Press, Corvallis, OR

Hartt, A.C., and M.B. Dell. 1986. Early oceanic migrations and growth of juvenile Pacific salmon and steelhead trout. Int. North Pac. Fish. Comm. Bull. 46:105 p.

Hartt, A.C., M.B. Dell, and L.S. Smith. 1970. Investigations by the United States for the International North Pacific Fisheries Commission-1968: tagging and sampling. Int. North Pac. Fish. Comm. Annu. Rep. 1968:68-79

Hashimoto, S. 1971. The effect of treatment with salt water on the development of chum salmon, *Oncorhynchus keta* (Walbaum), eggs. I. Removing the unfertilized and undeveloped eggs from the trays using salt water. Sci. Rep. Hokkaido Salmon Hatchery 25:45-51. (In Japanese, English abstract)

Healey, M.C. 1979. Detritus and juvenile salmon production in the Nanaimo Estuary. I. Production and feeding rates of juvenile chum salmon (*Oncorhynchus keta*). J. Fish. Res. Board Can. 36:488-496

——. 1980. The ecology of juvenile salmon in Georgia Strait, British Columbia, p. 203-229. *In*: W.J. McNeil and D.C. Himsworth (eds.). Salmonid ecosystems of the North Pacific. Oregon State University Press, Corvallis, OR

——. 1982a. Juvenile Pacific salmon in estuaries: the life support system, p. 315-341. *In*: V.S. Kennedy (ed.). Estuarine comparisons. Academic Press, New York, NY

——. 1982b. Timing and relative intensity of size-selective mortality of juvenile chum salmon (*Oncorhynchus keta*) during early sea life. Can. J. Fish. Aquat. Sci. 39:952-957

——. 1982c. The distribution and residency of juvenile Pacific salmon in the Strait of Georgia, British Columbia, in relation to foraging success, p. 61-69. *In*: B.R. Melteff and R.A. Nevé (eds.).

Proceedings of the North Pacific Aquaculture Symposium. Alaska Sea Grant Rep. 82-2

Helle, J.H. 1960. Characteristics and structure of early and late spawning runs of chum salmon, *Oncorhynchus keta* (Walbaum), in streams of Prince William Sound, Alaska. M.Sc. thesis. University of Idaho, Moscow, ID. 53 p.

———. 1979. Influence of marine environment on age and size at maturity, growth, and abundance of chum salmon, *Oncorhynchus keta* (Walbaum), from Olsen Creek, Prince William Sound, Alaska. Ph.D. thesis. Oregon State University, Corvallis, OR. 118 p.

———. 1981. Significance of the stock concept in artificial propagation of salmonids in Alaska. Can. J. Fish. Aquat. Sci. 38:1665-1671

———. 1984. Age and size at maturity of some populations of chum salmon in North America, p. 126-143. *In*: Proceedings of the Pacific Salmon Biology Conference (USSR, USA, Canada, Japan), Yuzhno-Sakhalinsk, USSR, 1978. Tikhookeanskii Nauchno-Issledovatel'skii Institut Rybnogo Khoziaistva i Okeangrafii, Valadivostok, USSR

Henry, K.A. 1953. Analysis of factors affecting the production of chum salmon (*Oncorhynchus keta*) in Tillamook Bay. Fish Comm. Oreg. Contrib. 18:37 p.

———. 1954. Age and growth study of Tillamook Bay chum salmon (*Oncorhynchus keta*). Fish Comm. Oreg. Contrib. 19:28 p.

Hirano, Y. 1953. An outline of the results of the tagging experiments on Pacific salmon. Hokkaido Prefectural Fisheries Experiment Station, Kushiro, Japan. 134 p. (In Japanese)

———. 1958. Migration route of the autumn salmon of Hokkaido. Hokusuishi Geppo 15:349-353

Hiroi, O. 1985. Hatchery approaches in the artificial chum salmon enhancement, p. 45-53. *In*: C.J. Sindermann (ed.). Proceedings of the Eleventh U.S.-Japan Meeting on Aquaculture, Salmon Enhancement, Tokyo, Japan, October 1982. NOAA Tech. Rep. NMFS 27

Hiyama, Y., Y. Nose, M. Shimizu, T. Ishihara, H. Abe, R. Sato, M. Takashi, and T. Kajihara. 1972. Predation of chum salmon fry during the course of its seaward migration: II: Otsuchi River investigation 1964 and 1965. Bull. Jpn. Soc. Sci. Fish. 38:223-229

Hiyama, Y., T. Taniuchi, K. Suyama, K. Ishioka, R. Sato, T. Kajihara, and T. Maiwa. 1967. A preliminary experiment on the return of tagged chum salmon to the Otsuchi River, Japan. Bull. Jpn. Soc. Sci. Fish. 33:18-19

Hoar, W.S. 1958. The evolution of migratory behaviour among juvenile salmon of the genus *Oncorhynchus*. J. Fish. Res. Board Can. 15:391-428

Holmes, P.B. 1982. Aleutian Islands salmon stock assessment study. Special report to the Alaska Board of Fisheries. Alaska Department of Fish and Game, Anchorage, AK. 83 p.

Hunter, J.G. 1959. Survival and production of pink and chum salmon in a coastal stream. J. Fish. Res. Board Can. 16:835-886

International North Pacific Fisheries Commission (INPFC). 1979. Historical catch statistics for salmon of the North Pacific Ocean. Int. North Pac. Fish. Comm. Bull. 39:166 p.

Irie, T. 1985. The migration and ecology of young salmon in their early marine life, p. 55-65. *In*: C.J. Sindermann (ed.). Proceedings of the Eleventh U.S.-Japan Meeting on Aquaculture, Salmon Enhancement, Tokyo, Japan, October 1982. NOAA Tech. Rep. NMFS 27

Ishida, Y., S. Ito, and K. Takagi. 1983. An analysis of scale patterns of Japanese hatchery reared chum salmon in the North Pacific. (Submitted to the annual meeting of the International North Pacific Fisheries Commission, Anchorage, Alaska, November 1983.) Fisheries Agency of Japan, Tokyo, Japan. 8 p.

Ivankov, V.N. 1968. Pacific salmon in the region of Iturup Island (Kuril Islands). Izv. Tikhookean. Nauchno-Issled. Inst. Rybn. Khoz. Okeanogr. 65:49-74. (Transl. from Russian; Fish. Res. Board Can. Transl. Ser. 1999)

Ivankov, V.N., and V.L. Andreyev. 1971. The South Kuril chum (*Oncorhynchus keta* (Walb.)): ecology, population structure and the modeling of the population. J. Ichthyol. 11:511-524

Ivankov, V.N., and A.P. Shershnev. 1967. The biology of the South Kuril pink and chum salmon in the initial period of life in the sea. *In*: Abstracts of the First Scientific Conference on Problems of Navigation and Study of the Pacific Ocean and Use of the Resources of the Far Eastern Seas. Vladivostok, USSR. (In Russian)

———. 1968. The biology of young pink and chum salmon in the sea. Rybn. Khoz. 1968(4):16-17. (In Russian)

Iwamoto, R.N., E.O. Salo, M.A. Madej, and R.L.

McComas. 1978. Sediment and water quality: a review of the literature including a suggested approach for water quality criteria; with summary of workshop and conclusions and recommendations, by E.O. Salo and R.L. Rulifson. U.S. Environ. Protect. Agency EPA 910/9-78-048:150 p.

Iwata, M. 1982a. Downstream migration and seawater adaptability of chum salmon (*Oncorhynchus keta*) fry, p. 51-59. *In*: B.R. Melteff and R.A. Nevé (eds.). Proceedings of the North Pacific Aquaculture Symposium. Alaska Sea Grant Rep. 82-2

——. 1982b. Transition of chum salmon fry into salt water, p. 204-207. *In*: E.L. Brannon and E.O. Salo (eds.). Proceedings of the Salmon and Trout Migratory Behavior Symposium. School of Fisheries, University of Washington, Seattle, WA

Jensen, H.M. 1956. Recoveries of immature chum salmon tagged in southern Puget Sound. Wash. Dep. Fish. Fish. Res. Pap. 1(4):32

Johnson, R.C., R.J. Gerke, D.W. Heiser, R.F. Orrell, S.B. Mathews, and J.G. Olds. 1971. Pink and chum salmon investigations, 1969: supplementary progress report. Washington Department of Fisheries, Fisheries Management and Research Division, Olympia, WA. 66 p.

Jordan, D.S., and C.H. Gilbert. 1882. Synopsis of the fishes of North America. U.S. Natl. Mus. Bull. 16:1018 p.

Kaczynski, V.W., R.J. Feller, and J. Clayton. 1973. Trophic analysis of juvenile pink and chum salmon (*Oncorhynchus gorbuscha* and *O. keta*) in Puget Sound. J. Fish. Res. Board Can. 30:1003-1008

Kaeriyama, M. 1986. Ecological study on early life of the chum salmon, *Oncorhynchus keta* (Walbaum). Sci. Rep. Hokkaido Salmon Hatchery 40:31-92. (In Japanese, English summary)

Kaev, A.M. 1981. Age and growth of chum salmon, *Oncorhynchus keta*, from Iturup Island. J. Ichthyol. 21:54-60

Kaganovsky, A.G. 1933. Commercial fishes of the Anadyr River and the Anadyr Estuary. Vestn. Dal'nevost. Fil. Akad. Nauk SSSR 1933 (1/3):137-139. (Transl. from Russian; Fish. Res. Board Can. Transl. Ser. 282)

Kanno, Y., and I. Hamai. 1971. Food of salmonid fish in the Bering Sea in summer of 1966. Bull. Fac. Fish. Hokkaido Univ. 22:107-127. (In Japanese, English abstract)

Kayev, A.M. 1983. Some questions on the factors influencing abundance of the autumn chum salmon, *Oncorhynchus keta* (Salmonidae), from Sakhalin and Iturup Island. J. Ichthyol. 22: 39-46

Kirkwood, J.B. 1962. Inshore-marine and freshwater life history phases of the pink salmon, *Oncorhynchus gorbuscha* (Walbaum) and the chum salmon, *O. keta* (Walbaum) in Prince William Sound, Alaska. Ph.D. thesis. University of Kentucky, Louisville, KY. 300 p.

Klokov, V.K. 1973. The dynamics of chum salmon abundance in various northern coastal areas of the Okhotsk Sea. Izv. Tikhookean. Nauchno-Issled. Inst. Rybn. Khoz. Okeanogr. 86:66-80. (Transl. from Russian; Fish. Mar. Serv. (Can.) Transl. Ser. 3380)

Kobayashi, T. 1959. Age determination of chum salmon in the northern Pacific Ocean during the early parts of fishing season. Sci. Rep. Hokkaido Salmon Hatchery 13:1-10. (In Japanese, English summary)

——. 1960. An ecological study of the salmon fry, *Oncorhynchus keta* (Walbaum). VI: Note on the feeding activity of chum salmon fry. Bull. Jpn. Soc. Sci. Fish. 26:577-580. (In Japanese, English abstract)

Kobayashi, T., and S. Abe. 1977. Studies on the Pacific salmon in the Yurappu River and Volcano Bay. 2: seaward migration and growth of salmon fry (*Oncorhynchus keta*) and the return of marked adults. Sci. Rep. Hokkaido Salmon Hatchery 31:1-11. (In Japanese, English summary)

Kobayashi, T., S. Harada, and S. Abe. 1965. Ecological observations on the salmon of Nishibetsu River. I: the migration and growth of chum salmon fry, *Oncorhynchus keta* (Walbaum). Sci. Rep. Hokkaido Salmon Hatchery 19:1-10. (In Japanese, English summary)

Kobayashi, T., and Y. Ishikawa. 1964. An ecological study on the salmon fry, *Oncorhynchus keta* (Walbaum). VIII: the growth and feeding habit of the fry during seaward migration. Sci. Rep. Hokkaido Salmon Hatchery 18:7-12. (In Japanese, English abstract)

Kobayashi, T., and T. Kurohagi. 1968. A study of the ecology of chum salmon fry, *Oncorhynchus keta* Walbaum, in Abashiri Lake and its protection. Sci. Rep. Hokkaido Salmon Hatchery 22:37-71. (In Japanese, English summary)

Konagaya, S. 1983. Enhanced protease activity in

the muscle of chum salmon during spawning migration with reference to softening or lysing phenomenon of the meat. Bull. Tokai Reg. Fish. Res. Lab. 109:41–55. (In Japanese, English abstract)

Kondo, H., Y. Hirano, N. Nabayama, and M. Miyake. 1965. Offshore distribution and migration of Pacific salmon (genus *Oncorhynchus*) based on tagging studies (1958–1961). Int. North Pac. Fish. Comm. Bull. 17:213 p.

Koo, T.S.Y. 1961. Circulus growth and annulus formation on scales of chum salmon (*Oncorhynchus keta*). (Presented to ICES Salmon and Trout Committee Symposium on Scale Reading, Copenhagen, October 1961.) Univ. Wash. Coll. Fish. Contrib. 137:12 p.

Koski, K V. 1966. The survival of coho salmon (*Oncorhynchus kisutch*) from egg deposition to emergence in three Oregon coastal streams. M.Sc. thesis. Oregon State University, Corvallis, OR. 84 p.

——. 1975. The survival and fitness of two stocks of chum salmon (*Oncorhynchus keta*) from egg deposition to emergence in a controlled-stream environment at Big Beef Creek. Ph.D. thesis. University of Washington, Seattle, WA. 212 p.

——. 1981. The survival and quality of two stocks of chum salmon (*Oncorhynchus keta*) from egg deposition to emergence, p. 330–333. *In*: R. Lasker and K. Sherman (eds.). The early life history of fish: recent studies: the second ICES symposium, Woods Hole, April 1977. Rapp. P.-V. Reun. Cons. Int. Explor. Mer 178

Kostarev, V.L. 1967. The age and growth of the Okhotsk chum. Izv. Tikhookean. Nauchno-Issled. Inst. Rybn. Khoz. Okeanogr. 61:173–181 (In Russian)

——. 1970. Quantitative calculation of Okhotsk keta juveniles. Izv. Tikhookean. Nauchno-Issled. Inst. Rybn. Khoz. Okeanogr. 71:145–158. (Transl. from Russian; Fish. Res. Board Can. Transl. Ser. 2589)

——. 1982. Natural reproduction of the Far East chum (*Oncorhynchus keta* Walb.), p. 71–75. *In*: B.R. Melteff and R.A. Nevé (eds.). Proceedings of the North Pacific Aquaculture Symposium. Alaska Sea Grant Rep. 82-2

Kovtun, A.A. 1981. Age and growth of chum salmon, *Oncorhynchus keta*, from Sakhalin. J. Ichthyol. 21:61–69

Kubo, I. 1953. On the blood of salmonid fishes of Japan during migration. I. Freezing point of blood. Bull. Fac. Fish. Hokkaido Univ. 4:138–139

Kulikova, N.I. 1972. Variability and speciation in the chum salmon (*Oncorhynchus keta* (Walb.)). J. Ichthyol. 12:185–197

LaLanne, J.J. 1971. Marine growth of chum salmon. Int. North Pac. Fish. Comm. Bull. 27:71–91

Landingham, J.H. 1982. Feeding ecology of pink and chum salmon fry in the nearshore habitat of Auke Bay, Alaska. M.Sc. thesis. University of Alaska, Juneau, AK. 132 p.

LeBrasseur, R.J. 1969. Growth of juvenile chum salmon (*Oncorhynchus keta*) under different feeding regimes. J. Fish. Res. Board Can. 26:1631–1645

Levanidov, V.Ya. 1964. Salmon population trends in the Amur basin and means of maintaining the stocks, p. 35–40. *In*: E.N. Pavlovskii (ed.). Lososevoe khozyaistvo dal'nego vostoka. Izdatel'stvo Nauka, Moscow. (Transl. from Russian; Univ. Wash. Fish. Res. Inst. Circ. 227)

Levanidov, V.Ya., and I.M. Levanidova. 1957. Food of downstream migrant young summer chum salmon and pink salmon in Amur tributaries. Izv. Tikhookean. Nauchno-Issled. Inst. Rybn. Khoz. Okeanogr. 45:3–16. (Transl. from Russian; *In*: Pacific salmon: selected articles from Soviet periodicals, p. 269–284. Israel Program for Scientific Translations, Jerusalem, 1961)

Levy, D.A., and T.G. Northcote. 1981. The distribution and abundance of juvenile salmon in marsh habitats of the Fraser River Estuary. Westwater Res. Cent. Univ. Br. Col. Tech. Rep. 25:117 p.

Lister, D.B., and C.E. Walker. 1966. The effect of flow control on freshwater survival of chum, coho, and chinook salmon in the Big Qualicum River. Can. Fish Cult. 37:3–25

Lovetskaya, E.A. 1948. Data on the biology of the Amur chum salmon. Izv. Tikhookean. Nauchno-Issled. Inst. Rybn. Khoz. Okeanogr. 27:115–137. (Transl. from Russian; *In*: Pacific salmon: selected articles from Soviet periodicals, p. 101–125. Israel Program for Scientific Translations, Jerusalem, 1961)

Lyamin, K.A. 1949. Results of tagging Pacific salmon in the Gulf of Kamchatka. Izv. Tikhookean. Nauchno-Issled. Inst. Rybn. Khoz. Okeanogr. 29: 173–176. (Transl. from Russian; *In*: Pacific salmon: selected articles from Soviet periodicals, p. 184–190. Israel Program for Scientific Translations, Jerusalem, 1961)

Mahnken, C.V.W., D.M. Damkaer, and V.G. Wespestad. 1986. Perspectives of North Pacific salmon sea ranching, p. 186–216. *In*: International fish farming: new perspectives for growth. Proceedings of the Second International Fish Farming Conference, March 1983. Janssen Services, Chislehurst, England

Manzer, J.I., I. Ishida, A.E. Peterson, and M.G. Hanavan. 1965. Salmon of the North Pacific Ocean. Part V. Offshore distribution of salmon. Int. North Pac. Fish. Comm. Bull. 15:452 p.

Marr, J.C. 1943. Age, length and weight studies of three species of Columbia River salmon (*Oncorhynchus keta*, *O. gorbuscha* and *O. kisutch*). Stanford Ichthyol. Bull. 2:157–197

Martin, D.J., D.R. Glass, C.J. Whitmus, C.A. Simenstad, D.A. Milward, E.C. Volk, M.L. Stevenson, P. Nunes, M. Savvoie, and R.A. Grotefendt. 1986. Distribution, seasonal abundance, and feeding dependencies of juvenile salmon and non-salmonid fishes in the Yukon River Delta. U.S. Dep. Comm. NOAA OCSEAP Final Rep. 55(1988):381–770

Mason, J.C. 1974. Behavioral ecology of chum salmon fry (*Oncorhynchus keta*) in a small estuary. J. Fish. Res. Board Can. 31:83–92

Mathews, S.B., and H.B. Senn. 1975. Chum salmon hatchery rearing in Japan, in Washington. Wash. Sea Grant Publ. WSG-TA 75-3:24 p.

Mattson, C.R., and R.A. Hobart. 1962. Chum salmon studies in southeastern Alaska, 1961. U.S. Fish Wildl. Serv. Biol. Lab. Auke Bay MS Rep. MR 62-5:32 p.

Mattson, C.R., R.G. Rowland, and R. A. Hobart. 1964. Chum salmon studies in southeastern Alaska, 1963. U.S. Fish Wildl. Serv. Biol. Lab. Auke Bay MS Rep. MR 64-8:22 p.

Mayama, H. 1976. Aquatic fauna of the Anebetsu River during downstream migration of chum salmon fry. Sci. Rep. Hokkaido Salmon Hatchery 30:55–73. (In Japanese, English Abstract)

McDonald, J. 1960. The behaviour of Pacific salmon fry during their downstream migration to freshwater and saltwater nursery areas. J. Fish. Res. Board Can. 17:655–676

McGie, A.M. 1984. Commentary: evidence for density dependence among coho salmon stocks in the Oregon production index area, p. 37–39. *In*: W.G. Pearcy (ed.). The influence of ocean conditions on the production of salmonids in the North Pacific: a workshop. Oreg. State Univ. Sea Grant Coll. Program ORESU-W-83-001

McNeil, W.J. 1962. Mortality of pink and chum salmon eggs and larvae in southeast Alaska streams. Ph.D. thesis. University of Washington, Seattle, WA. 270 p.

———. 1966. Effect of the spawning bed environment on reproduction of pink and chum salmon. Fish. Bull. (U.S.) 65:495–523

McNeil, W.J., and J.E. Bailey. 1975. Salmon ranchers manual. (Processed report.) Auke Bay Fisheries Laboratory, Northwest and Alaska Fisheries Center, National Marine Fisheries Service, Auke Bay, AK. 95 p.

Meehan, W.R., and D.B. Siniff. 1962. A study of the downstream migration of anadromous fishes in the Taku River, Alaska. Trans. Am. Fish. Soc. 91:399–407

Merritt, M.F., and J.A. Raymond. 1983. Early life history of chum salmon in the Noatak River and Kotzebue Sound. Alaska Dept. Fish Game FRED Rep. 1:56 p.

Mihara, T. 1958. An ecological study on the salmon fry, *Oncorhynchus keta*, in the coastal waters of Hokkaido. Sci. Rep. Hokkaido Salmon Hatchery 13:1–14. (Transl. from Japanese; Fish. Res. Board Can. Transl. Ser. 226)

Miller, R.J., and E.L. Brannon. 1982. The origin and development of life history patterns in Pacific salmonids, p. 296–309. *In*: E.L. Brannon and E.O. Salo (eds.). Proceedings of the Salmon and Trout Migratory Behavior Symposium. School of Fisheries, University of Washington, Seattle, WA

Nagasawa, A., and S. Sano. 1961. Some observations on the downstream chum salmon fry (*O. keta*) counted in the natural spawning ground at Memu Stream, 1957–1959. Sci. Rep. Hokkaido Salmon Hatchery 16:107–125. (In Japanese, English summary)

Naiman, R.J., and J.R. Sibert. 1979. Detritus and juvenile salmon production in the Nanaimo estuary. III. Importance of detrital carbon to the estuarine ecosystem. J. Fish. Res. Board Can. 36:504–520

Neave, F. 1953. Principles affecting the size of pink and chum salmon populations in British Columbia. J. Fish. Res. Board Can. 9:450–491

———. 1955. Notes on the seaward migration of pink and chum salmon fry. J. Fish. Res. Board

Can. 12:369-374

——. 1958. The origin and speciation of *Oncorhynchus*. Proc. Trans. R. Soc. Can. Ser. 3 52(5):25-39

——. 1961. Pacific salmon: ocean stocks and fishery developments. Proc. 9th Pac. Sci. Congr. 1957(10):59-62

Neave, F., T. Yonemori, and R. Bakkala. 1976. Distribution and origin of chum salmon in offshore waters of the North Pacific Ocean. Int. North Pac. Fish. Comm. Bull. 35:79 p.

Newman, M.A. 1956. Social behaviour and interspecific competition in two trout species. Physiol. Zool. 29:64-81

Nicola, S.J. 1968. Scavenging by *Alloperla* (Plecoptera: Chloroperlidae) nymphs on dead pink (*Oncorhynchus gorbuscha*) and chum (*O. keta*) salmon embryos. Can. J. Zool. 46:787-796

Nikolskii, G.V. 1952. The type of dynamics of stocks and the character of spawning of the chum *Oncorhynchus keta* (Walb.) and the pink salmon *Oncorhynchus gorbuscha* (Walb.) in the Amur River. Dokl. Akad. Nauk SSSR 86:873-875. (Transl. from Russian; *In*: Pacific salmon: selected articles from Soviet periodicals. p. 9-12. Israel Program for Scientific Translations, Jerusalem, 1961)

Noble, R.E. 1955. Minter Creek Biological Station progress report. Part 1: chum salmon. Washington Department of Fisheries, Olympia, WA. 9 p.

Oakley, A.L. 1966. A summary of information concerning chum salmon in Tillamook Bay. Res. Briefs Fish. Comm. Oreg. 12:5-21

Okada, S., and A. Taniguchi. 1971. Size relationship between salmon juveniles in shore waters and their prey animals. Bull. Fac. Fish. Hokkaido Univ. 22:30-36

Ostroumov, A.G. 1967. Some materials on the biology of the chum of the Anadyr' River. Izv. Tikhookean. Nauchno-Issled. Inst. Rybn. Khoz. Okeanogr. 57:80-92. (In Russian)

Paine, J. 1974. The Big Qualicum River artificial spawning channel for chum salmon, p. 72-78. *In*: D.R. Harding (ed.). Proceedings of the 1974 Northeast Pacific Pink and Chum Salmon Workshop. Department of the Environment, Fisheries, Vancouver, BC

Parker, R.R. 1962. A concept of the dynamics of pink salmon populations, p. 203-211. *In*: N.J. Wilimovsky (ed.). Symposium on Pink Salmon. H.R. MacMillan Lectures in Fisheries. Institute of Fisheries, University of British Columbia, Vancouver, BC

——. 1968. Marine mortality schedules of pink salmon of the Bella Coola River, central British Columbia. J. Fish. Res. Board Can. 25:757-794

Peterman, R.M. 1978. Testing for density dependent marine survival in Pacific salmonids. J. Fish. Res. Board Can. 35:1434-1450

——. 1984. Interaction among sockeye salmon in the Gulf of Alaska, p. 187-199. *In*: W.G. Pearcy (ed.). The influence of ocean conditions on the production of salmonids in the North Pacific: a workshop. Oreg. State Univ. Sea Grant Coll. Program ORESU-W-83001

Peterson, W.T., R.D. Brodeur, and W.G. Pearcy. 1982. Food habits of juvenile salmon in the Oregon coastal zone, June 1979. Fish. Bull. (U.S.) 80:841-851

Petrova, Z.I. 1964. Condition of the stocks of salmon of the Bol'shaya River, p. 36-42. *In*: E.N. Pavlovskii (ed.). Lososevoe khozyaistvo dal'nego vostoka. Izdatel'stvo Nauka, Moscow. (Abstract transl. from Russian; Univ. Wash. Fish. Res. Inst. Circ. 227)

Phillips, A.C., and W.E. Barraclough. 1978. Early marine growth of juvenile Pacific salmon in the Strait of Georgia and Saanich Inlet, British Columbia. Fish. Mar. Serv. (Can.) Tech. Rep. 830:19 p.

Phillips, R.W., and H.J. Campbell. 1962. The embryonic survival of coho salmon and steelhead trout as influenced by some environmental conditions in gravel beds. Pac. Mar. Fish. Comm. Annu. Rep. 14(1961):60-73

Poon, D.C. 1970. Development of a streamside incubator for culture of Pacific salmon. M.Sc. thesis. Oregon State University, Corvallis, OR. 84 p.

Powers, P.D., and J.F. Orsborn. 1985. Analysis of barriers to upstream fish migration: an investigation of the physical and biological conditions affecting fish passage success at culverts and waterfalls. Final project report, part 4 of 4, project 82-14. U.S. Department of Energy, Bonneville Power Administration, Division of Fish and Wildlife, Portland, OR. 120 p.

Pratt, D.C. 1974. Age, sex, length, weight, and scarring of adult chum salmon (*Oncorhynchus keta*) harvested by Puget Sound commercial net fisheries from 1954 to 1970: supplemental progress report, Marine Fisheries Investigation,

Puget Sound Commercial Net Fisheries. Washington Department of Fisheries, Olympia, WA. 78 p.

Prinslow, T.E., C.J. Whitmus, J.J. Dawson, N.J. Bax, B.P. Snyder, and E.O. Salo. 1980. Effects of wharf lighting on outmigrating salmonids, 1979. Final report. Univ. Wash. Fish. Res. Inst. FRI-UW-8007:137 p.

Pritchard, A.L. 1943. The age of chum salmon taken in the commercial catches in British Columbia. Fish. Res. Board Can. Prog. Rep. Pac. Coast Stn. 54:9–11

Regan, C. 1911. The freshwater fishes of the British Isles. Methuen, London. 287 p.

Regnart, R.I., P. Fridgen, and M. Geiger. 1966. Age, sex and size compositions of arctic and subarctic Alaska stocks of chum and sockeye salmon. (Submitted to the annual meeting International North Pacific Fisheries Commission, Vancouver, BC, November 1966.) Alaska Department of Fish and Game, Division of Commercial Fisheries, Juneau, AK. 9 p.

Ricker, W.E. 1964. Ocean growth and mortality of pink and chum salmon. J. Fish. Res. Board Can. 21:905–931

——. 1980. Changes in the age and size of chum salmon (*Oncorhynchus keta*). Can. Tech. Rep. Fish. Aquat. Sci. 930:30 p.

——. 1984. Trends in size of British Columbia salmon 1975–1982. (Submitted to the annual meeting International North Pacific Fisheries Commission, Vancouver, BC, November 1984). Department of Fisheries and Oceans, Pacific Biological Station, Nanaimo, BC

Rogers, D.E. 1980. Density-dependent growth of Bristol Bay sockeye salmon, p. 267–283. *In*: W.J. McNeil and D.C. Himsworth (eds.). Salmonid ecosystems of the North Pacific. Oregon State University Press, Corvallis, OR

Rosly, Yu.S. 1972. The scale structure of the Amur chum (*Oncorhynchus keta* (Walb.)) as an indicator of growth and living conditions in the freshwater stage. J. Ichthyol. 12:483–497

Roukhlov, F.N. (i.e., Rukhlov, F.N.). 1982. The role of salmon production and the perspectives of its development in the Sakhalin region, p. 99–103. *In*: B.R. Melteff and R.A. Nevé (eds.). Proceedings of the North Pacific Aquaculture Symposium. Alaska Sea Grant Rep. 82–2

Rounsefell, G.A., and G.B. Kelez. 1938. The salmon and salmon fisheries of Swiftsure Bank, Puget Sound, and the Fraser River. Bull. Bur. Fish. (U.S.) 48:693–823

Rukhlov, F.N. 1969. Materials characterizing the texture of bottom material in the spawning grounds and redds of the pink salmon (*Oncorhynchus gorbuscha* (Walbaum)) and the autumn chum (*Oncorhynchus keta* (Walbaum)) on Sakhalin. Probl. Ichthyol. 9:635–644

Salo, E.O., and R.E. Noble. 1952. Chum salmon. Part I, p. 1–10. *In*: Minter Creek Biological Station progress report (Sept.-Dec. 1952). Washington Department of Fisheries, Olympia, WA. 33 p.

——. 1953. Chum salmon upstream migration, p. 1–9. *In*: Minter Creek Biological Station progress report, September through October 1953. Washington Department of Fisheries, Olympia, WA. 14 p.

——. 1954. Downstream migration of chum salmon, 1954, p. 1–3. *In*: Minter Creek Biological Station progress report, March 1954. Washington Department of Fisheries, Olympia, WA. 7 p.

Sano, S. 1959. The ecology and propagation of genus *Oncorhynchus* found in northern Japan. Sci. Rep. Hokkaido Salmon Hatchery 14:21–90. (In Japanese, English abstract)

——. 1960. Ecological study (particularly on spawning) and propagation of the Hokkaido salmon. Hokkaido Salmon Hatchery Data 108. (In Japanese)

——. 1964. A study of the silver colored chum salmon during their upstream migration: preliminary report I. J. Shimonoseki Univ. Fish. 1:39–43. (In Japanese)

——. 1966. Chum salmon in the Far East, p. 4–58. *In*: Salmon of the North Pacific Ocean. Part III. A review of the life history of North Pacific salmon. Int. North Pac. Fish. Comm. Bull. 18

——. 1967. Chum salmon in the Far East, p. 23–42. *In*: Salmon of the North Pacific Ocean. Part IV. Spawning populations of North Pacific salmon. Int. North Pac. Fish. Comm. Bull. 23

Sano, S., and A. Nagasawa. 1958. Natural propagation of chum salmon, *Oncorhynchus keta*, in Memu River, Tokachi. Sci. Rep. Hokkaido Salmon Hatchery 12:1–19. (Partial transl. from Japanese; Fish. Res. Board Can. Transl. Ser. 198)

Schreiner, J.V. 1977. Salmonid outmigration studies in Hood Canal, Washington. M.Sc. thesis. University of Washington, Seattle, WA. 91 p.

Schroder, S.L. 1973. Effects of density on the spawning success of chum salmon (*Oncorhynchus keta*) in an artificial spawning channel. M.Sc. thesis. University of Washington, Seattle, WA. 78 p.

———. 1977. Assessment of production of chum salmon fry from the Big Beef Creek spawning channel. Completion report. Univ. Wash. Fish. Res. Inst. FRI-UW-7718:77 p.

———. 1981. The role of sexual selection in determining overall mating patterns and mate choice in chum salmon. Ph.D. thesis. University of Washington, Seattle, WA. 274 p.

———. 1982. The influence of intrasexual competition on the distribution of chum salmon in an experimental stream, p. 275-285. *In*: E.L. Brannon and E.O. Salo (eds.). Proceedings of the Salmon and Trout Migratory Behavior Symposium. School of Fisheries, University of Washington, Seattle, WA.

Schroder, S.L., KV. Koski, B.P. Snyder, K.J. Bruya, G.W. George, and E.O. Salo. 1974. Big Beef Creek studies, p. 26-27. *In*: Research in fisheries 1973. Univ. Wash. Coll. Fish. Contrib. 390

Scofield, N.B. 1916. The humpback and dog salmon taken in San Lorenzo River. Calif. Fish Game 2:41

Semko, R.S. 1954. The stocks of West Kamchatka salmon and their commercial utilization. Izv. Tikhookean. Nauchno-Issled. Inst. Rybn. Khoz. Okeanogr. 41:3-109. (Transl. from Russian; Fish. Res. Board Can. Transl. Ser. 288)

Shepard, M.P., A.C. Hartt, and T. Yonemori. 1968. Salmon of the North Pacific Ocean. Part VIII. Chum salmon in offshore waters. Int. North Pac. Fish. Comm. Bull. 25:69 p.

Shirahata, S. 1985. Strategy in salmon farming in Japan, p. 91-95. *In*: C.J. Sindermann (ed.). Proceedings of the Eleventh U.S.-Japan Meeting on Aquaculture, Salmon Enhancement, Tokyo, Japan, October 1982. NOAA Tech. Rep. NMFS 27

Shmidt, P.Yu. 1947. Migrations of fish. 2nd ed. Izdatel'stvo Akademiia Nauk SSSR, Moscow. 361 p. (In Russian)

Sibert, J. 1979. Detritus and juvenile salmon production in the Nanaimo estuary. 2: meiofauna available as food for juvenile salmon. J. Fish. Res. Board Can. 36:497-503

Sibert, J., T.J. Brown, M.C. Healey, B.A. Kask, and R.J. Naiman. 1977. Detritus-based food webs: exploitation by juvenile chum salmon (*Oncorhynchus keta*). Science 196:649-650

Simenstad, C.A., K.L. Fresh, and E.O. Salo. 1982. The role of Puget Sound and Washington coastal estuaries in the life history of Pacific salmon: an unappreciated function, p. 343-364. *In*: V.S. Kennedy (ed.). Estuarine comparisons. Academic Press, New York, NY

Simenstad, C.A., W.J. Kinney, S.S. Parker, E.O. Salo, J.R. Cordell, and H. Buechner. 1980. Prey community structure and trophic ecology of outmigrating juvenile chum and pink salmon in Hood Canal, Washington: a synthesis of three years' studies, 1977-1979. Final report. Univ. Wash. Fish. Res. Inst. FRI-UW-8026:113 p.

Simenstad, C.A., and E.O. Salo. 1982. Foraging success as a determinant of estuarine and nearshore carrying capacity of juvenile chum salmon (*Oncorhynchus keta*) in Hood Canal, Washington, p. 21-37. *In*: B.R. Melteff and R.A. Nevé (eds.). Proceedings of the North Pacific Aquaculture Symposium. Alaska Sea Grant Rep. 82-2

Smirnov, A.G. 1947. Condition of stocks of the Amur salmon and causes of the fluctuations in their abundance. Izv. Tikhookean. Nauchno-Issled. Inst. Rybn. Khoz. Okeanogr. 25:33-51. (Transl. from Russian; *In*: Pacific salmon: selected articles from Soviet periodicals, p. 66-85. Israel Program for Scientific Translations, Jerusalem, 1961)

Smirnov, A.I. 1975. The biology, reproduction and development of the Pacific salmon. Izdatel'stvo Moskovogo Universiteta, Moscow, USSR. 335 p. (Transl. from Russian; Fish. Mar. Serv. (Can.) Transl. Ser. 3861)

Smith, G.R., and R.F. Stearley. 1989. The classification and scientific names of rainbow and cutthroat trouts. Fisheries (Bethesda) 14:4-10

Smoker, W.W. 1984. Genetic effect on the dynamics of a model of pink (*Oncorhynchus gorbuscha*) and chum salmon (*O. keta*). Can. J. Fish. Aquat. Sci. 41:1446-1453

Soin, S.G. 1954. Pattern of development of summer chum, masu, and pink salmon. Tr. Soveshch. Ikhtiol. Kom. Akad. Nauk SSSR 4:144-155. (Transl. from Russian; *In*: Pacific salmon: selected articles from Soviet periodicals, p. 42-54. Israel Program for Scientific Translations, Jerusalem, 1961)

Sparrow, R.A.H. 1968. A first report of chum salmon fry feeding in fresh water of British Columbia. J. Fish. Res. Board Can. 25:599-602

Svetovidova, A.A. 1961. Local stocks of summer keta, *Oncorhynchus keta* (Walbaum), of the Amur Basin. Vopr. Ikhtiol. 17:14-23. (Transl. from Russian; Fish. Res. Board Can. Transl. Ser. 347)

Tautz, A.F., and C. Groot. 1975. Spawning behavior of chum salmon (*Oncorhynchus keta*) and rainbow trout (*Salmo gairdneri*). J. Fish. Res. Board Can. 32:633-642

Tchernavin, V. 1939. The origin of salmon. Salmon Trout Mag. 95:120-140

Thorpe, J.E. 1982. Migration in salmonids, with special reference to juvenile movements in freshwater, p. 86-97. *In*: E.L. Brannon and E.O. Salo (eds.). Proceedings of the Salmon and Trout Migratory Behavior Symposium. School of Fisheries, University of Washington, Seattle, WA

Thorsteinson, F.V. 1965. Some aspects of pink and chum salmon research at Olsen Bay, Prince William Sound. U.S. Fish Wildl. Serv. Biol. Lab. Auke Bay MS Rep. MR 65-3:30 p.

Thorsteinson, F.V., J.H. Helle, and D.G. Birkholz. 1971. Salmon survival in intertidal zones of Prince William Sound streams in uplifted and subsided areas, p. 194-219. *In*: The great Alaska earthquake of 1964: biology. Natl. Acad. Sci. NAS Publ. 1604

Thorsteinson, F.V., W.H. Noerenberg, and H.D. Smith. 1963. The length, age, and sex ratio of chum salmon in the Alaska Peninsula, Kodiak Island, and Prince William Sound areas of Alaska. U.S. Fish Wildl. Serv. Spec. Sci. Rep. Fish. 430:84 p.

Todd, I.S. 1966. A technique for the enumeration of chum salmon fry in the Fraser River, British Columbia. Can. Fish Cult. 38:3-35

Trasky, L.T. 1974. Yukon River anadromous fish investigations. Anadromous Fish Conservation Act completion report for period July 1, 1973 to June 30, 1974. Alaska Department of Fish and Game, Commercial Fisheries Division, Juneau, AK. 111 p.

Trump, C.L., and W.C. Leggett. 1980. Optimum swimming speeds in fish: the problems of currents. Can. J. Fish. Aquat. Sci. 37:1086-1092

Tsuyuki, H., and E. Roberts. 1966. Inter-species relationships within the genus *Oncorhynchus* based on biochemical systematics. J. Fish. Res. Board Can. 23:101-107

Tyler, R.W., and D.E. Bevan. 1964. Migration of juvenile salmon in Bellingham Bay, Washington, p. 44-55. *In*: Research in fisheries 1963. Univ. Wash. Coll. Fish. Contrib. 166

Utter, F.M., F.W. Allendorf, and H.O. Hodgins. 1973. Genetic variability and relationships in Pacific salmon and related trout based on protein variations. Syst. Zool. 22:257-270

Vibert, R. 1956. Laboratory methods for studying and increasing the survival in the wild of hatchery raised trout and salmon fry. Ann. Stn. Cent. Hydrobiol. Appl. 6:347-439. (In French, English summary)

Volk, E.C., R.C. Wissmar, C.A. Simenstad, and D.M. Eggers. 1984. Relationship between otolith microstructure and the growth of juvenile chum salmon (*Oncorhynchus keta*) under different prey rations. Can. J. Fish. Aquat. Sci. 41:126-133

Volobuyev, V.V. 1984. Spawning characteristics and ecology of young chum salmon, *Oncorhynchus keta*, in the Tauy River Basin (northern Okhotsk Sea). J. Ichthyol. 24:156-173

Walbaum, J.J. 1792. Petri Artedi Sueci Genera piscium in quibus systema totum ichthyologiae proponitur cum classibus, ordinibus, generum characteribus, specierum differentiis, observationibus plurimis: Ichthyologiae. pars. iii. Grypeswaldiae. 723 p.

Walters, C.J., R. Hilborn, R.M. Peterman, and M.J. Staley. 1978. Model for examining early ocean limitation of Pacific salmon production. J. Fish. Res. Board Can. 35:1303-1315

Wangaard, D.B., and C.B. Burger. 1983. Effects of various water temperature regimes on the egg and alevin incubation of Susitna River chum and sockeye salmon. U.S. Fish and Wildlife Service, Anchorage, Alaska. 43 p.

Weihs, D. 1975. An optimum swimming speed of fish based on feeding efficiency. Isr. J. Technol. 13: 163-167

Whitmus, C.J., Jr. 1985. The influence of size on the migration and mortality of early marine life history of juvenile chum salmon (*Oncorhynchus keta*). M.Sc. thesis. University of Washington, Seattle, WA. 69 p.

Whitmus, C.J., and S. Olsen. 1979. The migratory behavior of juvenile chum salmon released in 1977 from the Hood Canal hatchery at Hoodsport, Washington. Univ. Wash. Fish. Res. Inst. FRI-UW-7916:46 p.

Wickett, W.P. 1952. Production of chum and pink salmon in a controlled stream. Fish. Res. Board

Can. Prog. Rep. Pac. Coast Stn. 93:7–9

——. 1954. The oxygen supply to salmon eggs in spawning beds. J. Fish. Res. Board Can. 11:933–953

Wolcott, R.S.C., Jr. 1978. The chum salmon run at Walcott Slough. Special report to U.S. Department of the Interior, Fish and Wildlife Service, Reno, NV. 41 p.

Wright, A.T. 1964. Studies to determine optimum escapement of pink and chum salmon in Alaska. Part 9. *In*: A study of the carrying capacity of pink and chum salmon spawning areas in Alaska. Alaska Department of Fish and Game, Juneau, AK. 11 p.

Wynne-Edwards, V.C. 1952. Freshwater vertebrates of the Arctic and subarctic. Bull. Fish. Res. Board Can. 94:28 p.

Life History of Chinook Salmon

CONTENTS

HEINE

15

PLATE 15. Maturing chinook salmon holding in a pool below Stamp River falls, British Columbia. *Photograph by C. Groot*

PLATE 16. Maturing chinook salmon jumping three-metre falls during upstream migration, Stamp River, British Columbia. *Photograph by Marj Trim*

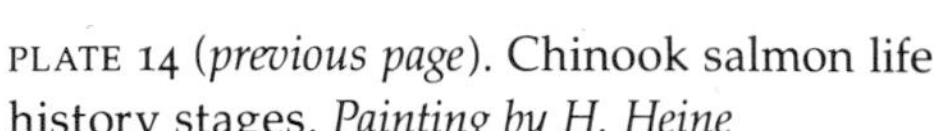

PLATE 14 (*previous page*). Chinook salmon life history stages. *Painting by H. Heine*

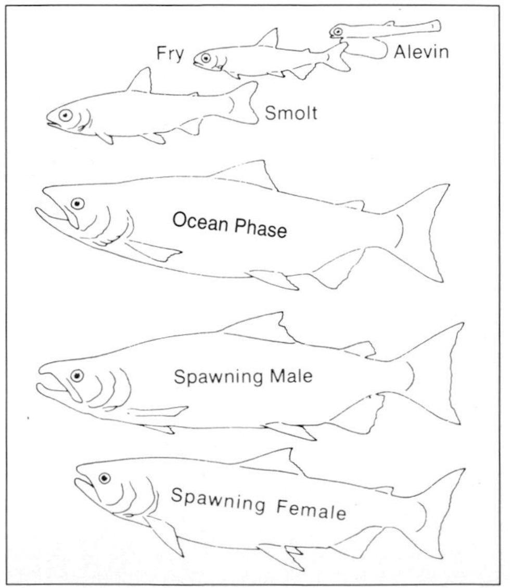

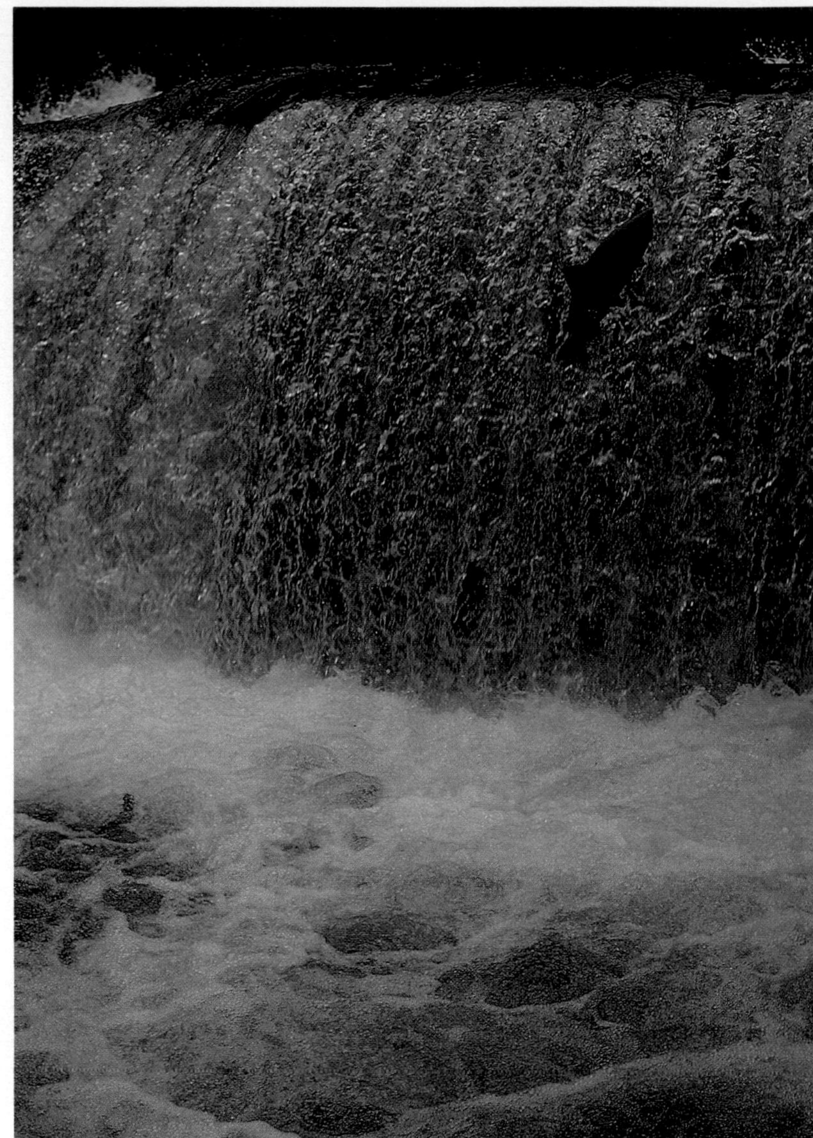

16

LIFE HISTORY OF CHINOOK SALMON
(*Oncorhynchus tshawytscha*)

M.C. Healey*

INTRODUCTION

THE GENUS *Oncorhynchus* dates at least from the Pliocene (Smith 1975) and probably originated from a stream- or lake-dwelling *Salmo*-like fish (Neave 1958). When the modern species evolved is uncertain, but the chinook salmon (*O. tshawytscha*), and the other Pacific salmon species, may have evolved as recently as 500,000 to 1,000,000 years ago (Neave 1958). A time span of one million years is extremely short for the differentiation of seven species. Nevertheless, Neave (1958) argued that there were good geological reasons for believing that such evolution had occurred. Most important, the restriction of *Oncorhynchus* species to the North Pacific suggested that their evolution postdated the faunistic connection between Atlantic and Pacific that existed during the late Pliocene. Thus, it seems possible that the species had undergone their complete elaboration during the Pleistocene. Recent investigation of mitochondrial DNA, however, suggests that the species may be two to three million years old (Thomas et al. 1986), which would be more in keeping with their elaboration during the Pliocene.

Morphologically, the chinook is distinguished from other *Oncorhynchus* species by its large size (adults may reach a weight of 45 kg), and by having small black spots on both lobes of the caudal fin, black pigment along the base of the teeth (Plate 14), and a large number of pyloric caeca (>100) (McPhail and Lindsey 1970; Hart 1973). Chinook also differ from the other species by their variable flesh colour, from white through various shades of pink to red.

Chinook fry and parr are distinguished by having large parr marks extending well below the lateral line (Plate 14). The adipose fin is normally unpigmented in the centre, but edged with black. The anal fin is usually only slightly falcate, and the leading rays do not reach past the posterior insertion of the fin when folded against the body. The anal fin has a white leading edge, but this is not set off by a dark pigment line as it is in coho salmon. Juvenile characteristics are highly variable, however, so that proper identification often requires meristic and pyloric caeca counts.

Within the species group, the chinook is most closely related to the coho (*O. kisutch*), with which it forms one subgrouping. Sockeye (*O. nerka*), chum (*O. keta*), and pink (*O. gorbuscha*) form a second subgrouping, and masu (*O. masou*) and amago (*O. rhodurus*), supposedly the most primitive, form a third subgrouping. According to Tsuyuki et al. (1965) and Tsuyuki and Roberts (1966), the probable evolutionary order of the species is: *masou*, *kisutch*, *tshawytscha*, *keta*, *nerka*, *gorbuscha*, although there is some question about the ordering of *keta* and *nerka*. Tsuyuki et al. (1965) and Tsuyuki and Roberts (1966) did not recognize *rhodurus* as a separate species, but see Kato, this volume. *Oncorhynchus rhodurus* is more primitive than *O. masou*.

The chinook, like all *Oncorhynchus* species, is anadromous and semelparous (i.e., dies after

*Department of Fisheries and Oceans, Biological Sciences Branch, Pacific Biological Station, Nanaimo, British Columbia V9R 5K6. Present address: Westwater Research Centre, University of British Columbia, Vancouver, BC V6T 1Z2

spawning once). Within this general life history strategy, however, chinook display a broad array of tactics that includes variation in age at seaward migration, variation in length of freshwater, estuarine, and oceanic residence, variation in ocean distribution and ocean migratory patterns, and variation in age and season of spawning migration.

An important objective of this chapter is to develop a conceptual model of the life history of chinook that can encompass this degree of variation. I shall argue that there must be two fundamental components of this model. The first component is racial (Healey 1983). (Here, "race" is used in the sense that Merrell (1981) used it, that is, to identify subdivisions of a population that are geographically separated to some degree and between which gene flow is reduced.) A large part of the variation in chinook life history apparently derives from the fact that the species occurs in two behavioural forms. One form, which has been designated "stream-type" (Gilbert 1913), is typical of Asian populations and of northern populations and headwater tributaries of southern populations in North America. Stream-type chinook spend one or more years as fry or parr in fresh water before migrating to sea, perform extensive offshore oceanic migrations, and return to their natal river in the spring or summer, several months prior to spawning (Figure 1). Occasionally, males of this form mature precociously without ever going to sea. The second form, which has been designated "ocean-type" ("sea-type" in Gilbert 1913), is typical of populations on the North American coast south of 56°N. Ocean-type chinook migrate to sea during their first year of life, normally within three months after emergence from the spawning gravel, spend most of their ocean life in coastal waters, and return to their natal river in the fall, a few days or weeks before spawning (Figure 1).

The second component of the life history model is tactical and encompasses variation within each race (Figure 1). This variation represents adaptation to uncertainties in juvenile survival and productivity within particular freshwater and estuarine nursery habitats. Briefly, chinook appear to have evolved a variety of juvenile and adult behaviour patterns in order to spread the risk of mortality across years and across habitats (e.g., Stearns 1976; Real 1980). By so doing, they avoid the potential disaster associated with high mortality in a particular year or habitat.

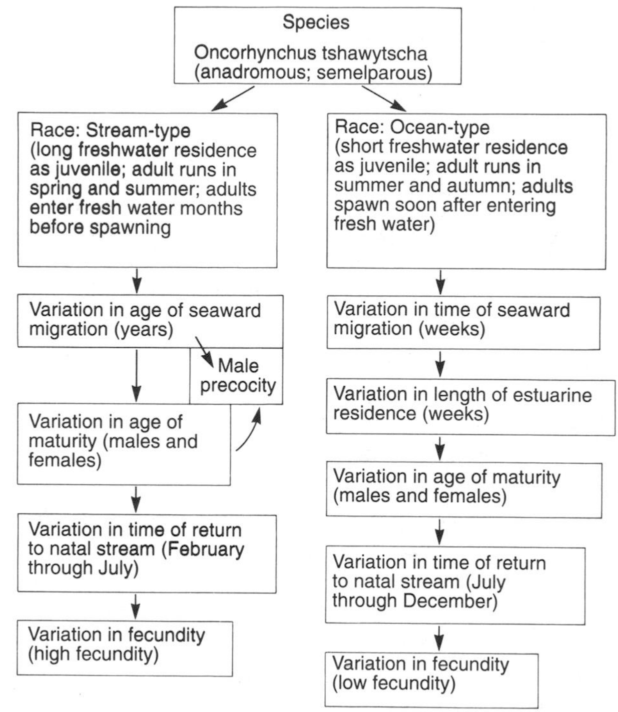

FIGURE 1

Life history structure of chinook salmon showing the division of the species into two races (ocean- and stream-type) and the range of tactical variation within each race

Relative Abundance

The chinook is a valuable commercial species. Fisheries for chinook are conducted by Japan on the high seas west of 175°E in the North Pacific Ocean (west of 175°W during 1955–77) and west of

175°W in the Bering Sea by means of gillnets. The Bering Sea fishery is to be phased out by 1994. Nationals of Canada, the United States, and the USSR fish for chinook in their coastal waters and rivers. In the eastern USSR, fishing is by trap net and gillnet. In the coastal waters of the United States and Canada, fishing is by gillnet, purse seine, and troll (hook and line). Annual recorded commercial landings averaged about 23,800 t between 1970 and 1979 (INPFC 1972–82). The largest catches are along the British Columbia coast, although substantial catches are also made in southeastern Alaska, in the Japanese mothership fishery, and along the coasts of Washington, Oregon, and California (Figure 2). In North America the chinook is also prized as a sport fish, and approximately one million are taken each year in sport fisheries (INPFC 1972–82; Department of Fisheries and Oceans, Vancouver, British Columbia, unpublished data). Chinook are, however, the third least abundant of the Pacific salmon, with only *O. rhodurus* and *O. masou* being less abundant.

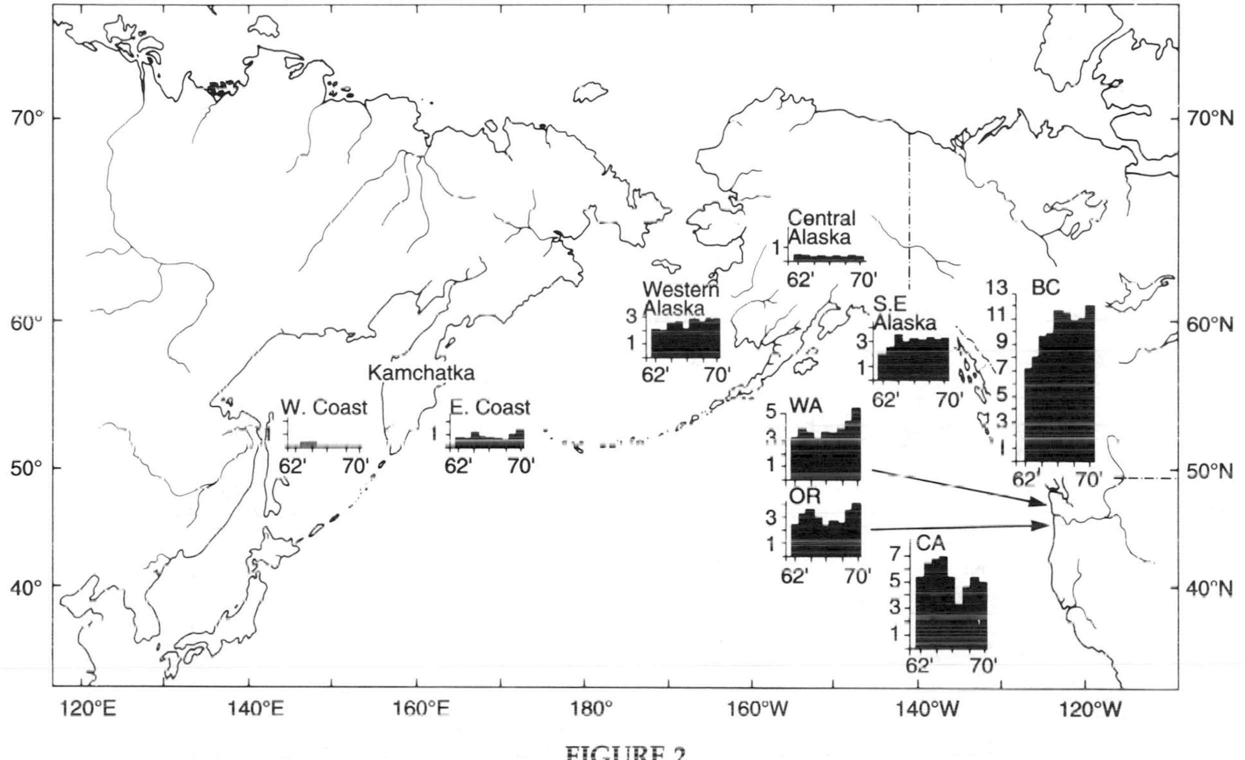

FIGURE 2

Map of the North Pacific Ocean with histograms showing the catch (hundreds of thousands of fish) of chinook in major coastal fisheries from 1962–70. (Adapted from Major et al. 1978)

Spawning Populations

Distribution and Abundance of Spawning Stocks

Spawning stocks of chinook are known to be distributed from northern Hokkaido to the Anadyr River on the Asian coast and from central California to Kotzebue Sound, Alaska, on the North American coast (Figure 3) (McPhail and Lindsey 1970; Major et al. 1978). Unconfirmed reports, however, suggest that chinook salmon may be distributed even further north and east on the Alaskan coast, and Hart (1973) cited an unpublished report of thirteen specimens from the Coppermine River (67°50′N, 115°00′W) in the Canadian Arctic. McLeod and O'Neil (1983) reported recovering a single specimen from the Liard River in the upper Mackenzie River drainage.

There are probably well in excess of a thousand spawning populations of chinook salmon on the North American coast (Atkinson et al. 1967; Aro and Shepard 1967) and an uncertain, but probably much lower, number on the Asian coast. Spawning occurs from near tidewater to over 3,200 km upstream in the headwaters of the Yukon River (Major et al. 1978). Individual spawning populations of chinook are relatively small, not exceeding a few tens of thousands. In British Columbia, where records have been kept on over three hundred chinook spawning populations for many decades, 80% of the populations have averaged fewer than one thousand spawners (Healey 1982a). Presumably, individual spawning populations are similarly small throughout the chinook's range.

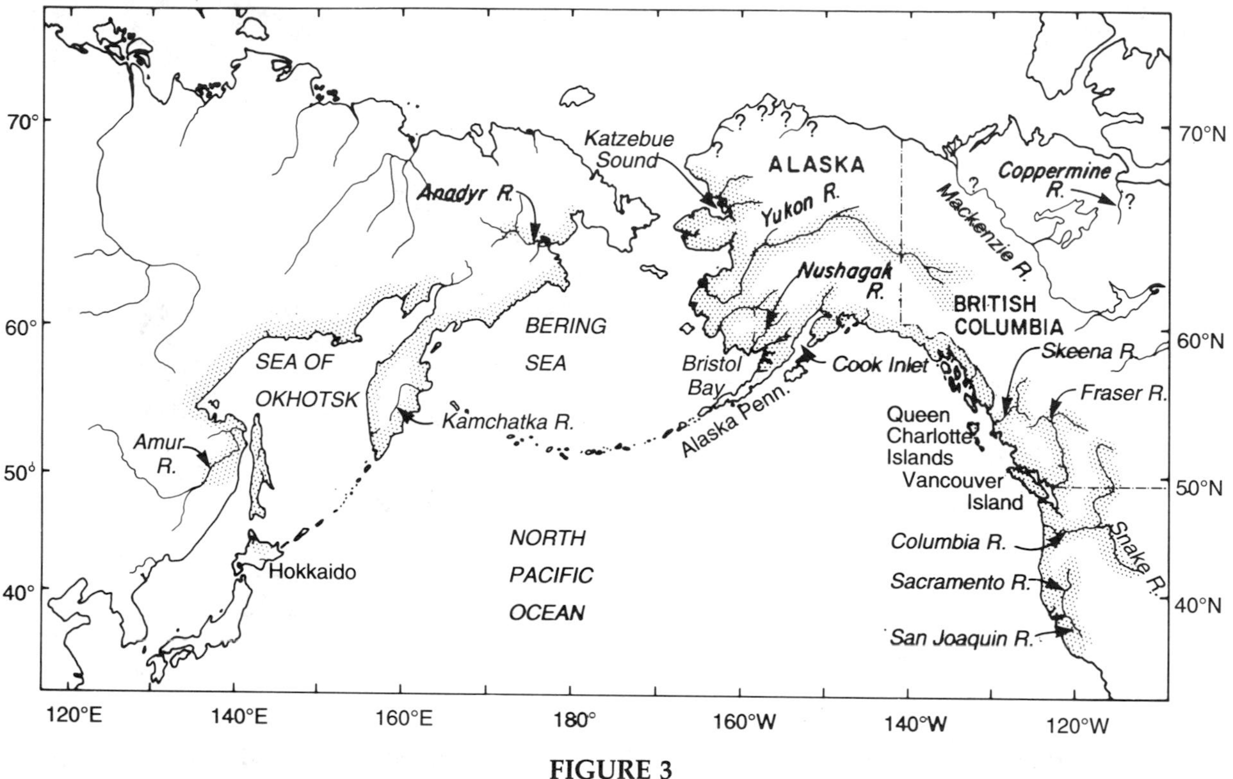

FIGURE 3

Map of the North Pacific Ocean and Bering Sea, showing the distribution of chinook spawning populations (stippled) and some of the landmarks referred to in the text. The distribution of chinook spawning populations north and east of Kotzebue Sound on the North American coast is unconfirmed (shown as question marks), except for a positive identification in the Mackenzie drainage.

The largest rivers tend to support the largest aggregate runs of chinook and also tend to have the largest individual spawning populations (Figure 4). This is not surprising, because larger rivers are likely to have more suitable spawning and rearing habitats than smaller rivers. What is somewhat surprising, however, is that major rivers at the northern and southern limits of the chinook's range support populations as large or larger than those in major rivers near the middle of the range. The Sacramento-San Joaquin River system, at the southern limit of the range of chinook, for example, had chinook runs of a million fish or more until the early part of this century (Clark 1929). These rivers

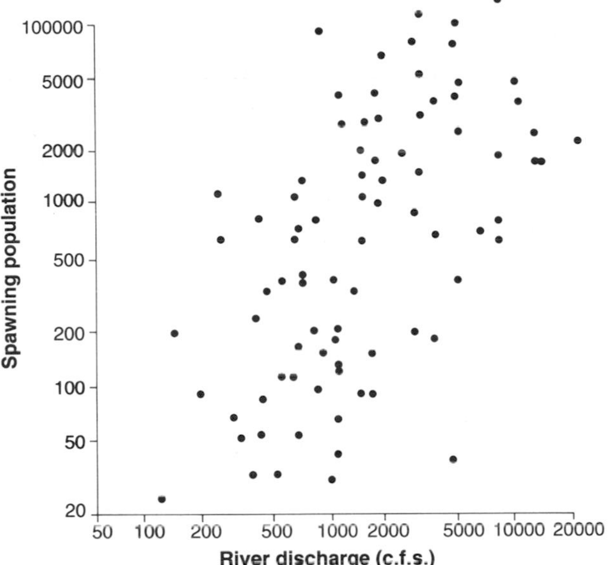

FIGURE 4

Relationship between average spawning population size (1952–76) and average river discharge for British Columbia populations of chinook salmon (both axes are on a logarithmic scale). These data are for individual spawning populations within each river system.

still have runs of several hundred thousand, even though habitat loss and water extraction have been so severe that the San Joaquin River was virtually unable to support chinook spawning, with escapement averaging less than 4,000 in the mid-1970s (Kjelson et al. 1982). Since 1980, however, escapements to the San Joaquin have recovered somewhat and escapement was 70,000 in 1985 (M. Kjelson, U.S. Fish and Wildlife Service, Stockton, California, pers. comm.). Run estimates for the Yukon and Nushagak rivers, near the northern limit of the range of chinook, are very uncertain, but catches of chinook in western Alaska and escapement estimates for these rivers indicate that runs to both rivers are probably on the order of 400,000 to 600,000 fish (Knudsen et al. 1983). By comparison, the Columbia River historically produced only about twice as many fish as the Sacramento-San Joaquin River system, and the Fraser River produced only about 200,000 chinook (Rich 1942; M.C. Healey unpublished data). Both the Columbia and Fraser rivers are near the centre of the chinook's range on the North American coast.

Chinook have been transplanted to a variety of locations outside their normal range. Transplants to the east coast of North America have been conducted for almost a century, initially with the hope of supplementing declining Atlantic salmon (*Salmo salar*) runs, but also as a means to "control" landlocked smelt (*Osmeridae*) and alewife (*Alosa pseudoharengus*) populations (Hoover 1936; Ricker and Loftus 1968). Most transplanted populations were maintained only by artificial propagation, if they were maintained at all. Recently, however, naturally spawning populations have become established in the Laurentian Great Lakes (Carl 1982). Both artificially and naturally produced chinook are now a valuable component of the sport fisheries in Lake Michigan and its tributary streams.

Chinook from California were successfully transplanted to New Zealand around the turn of the century. The first transplants to New Zealand were in the late 1800s, and intensive stocking began in 1901. By 1907, adults were returning to New Zealand hatcheries, and by 1910 it was possible for these hatcheries to export spawn to other parts of New Zealand and Tasmania. Chinook are now widespread in rivers along the east coast of the south island and occur in some rivers on the west and north coasts. Whether this has occurred as a result of deliberate planting or natural straying is not clear (Waugh 1980).

Although the original intent of the introduction of chinook to New Zealand was to create commercially exploitable runs of fish, established runs have provided only a low sustainable yield, and little commercial harvesting of salmon has occurred. Chinook are, however, an important sport fish. Limited suitable spawning and rearing areas in New Zealand streams, together with impoundments, water extraction for irrigation, and pollution, have served to limit smolt production (Waugh 1980). Poor smolt production is presumably an important reason for the relatively low productivity of chinook in New Zealand.

Chinook have also been transplanted to southern Chile. Apparently, landlocked populations have been established there, and the hope is that marine anadromous populations will also be successful in Chile. Some returns of anadromous chinook have been reported in sea ranching operations (Lindbergh 1982).

In this chapter, I shall concentrate on chinook within their natural range. This is not to discount the possible impact of successful transplants. The introduction of chinook and other Pacific salmon to the Great Lakes, for example, almost certainly means that they will ultimately invade Atlantic Ocean drainages, with unknown consequences for local fish fauna. It is the intent here, however, to emphasize what is known about the biology of the animal in its natural habitat.

Within the natural range of chinook salmon, stream- and ocean-type spawning populations are geographically separated to a considerable degree (Healey 1983). All Asian stocks are apparently stream-type (e.g., Knudsen et al. 1983). On the North American coast, chinook spawning populations are wholly or predominantly stream-type throughout Alaska. At the Alaska-British Columbia border, however, there is a rather abrupt shift in composition. From the Nass and Skeena rivers (56°N) (Figure 3) southward, ocean-type chinook dominate all runs except, perhaps, the Yakoun River on the Queen Charlotte Islands (Table 1). Stream-type chinook make an important contribution to runs to larger rivers south of 56°N (14%–48%) but are relatively scarce in smaller rivers (0%–12% of runs). Wherever they are sympatric with ocean-type fish, stream-type fish tend to be found in headwater spawning areas and ocean-type fish in downstream spawning areas (Rich 1925; Hallock et al. 1957; Healey and Jordan 1982). The geographic separation is not complete, however, as the behavioural types are sympatric on many spawning riffles.

TABLE 1

Occurrence of stream- and ocean-type chinook in spawning runs to rivers along the west coast of North America

River system	Approximate N. latitude	% of spawning runs	
		Stream	Ocean
Alaska			
Yukon	62°30′	100	0
Cook Inlet rivers	61°30′	97–99	1–3
Taku	58°30′	100	0
Stikine	56°40′	100	0
British Columbia			
Nass	55°20′	42	58
Skeena	54°20′	48	52
Kitimat	54°00′	12	88
Yakoun	53°30′	57	43
Bella Coola	52°25′	14	86
Docee (Rivers Inlet)	51°40′	3	97
Quinsam (Campbell)	50°00′	1	99
Big Qualicum	49°25′	0	100
Fraser	49°20′	34	66
Nanaimo	49°10′	5	95
Nitinat	48°50′	1	99
Chemainus	48°50′	0	100
Cowichan	48°50′	10	90
Washington/Oregon			
Columbia	46°10′	22	78
Sixes	42°50′	12	88
California			
Klamath	41°30′	14	86
Sacramento	38°00′	10–18	82–90

Source: From Healey (1983) with additional data from Clark (1929)
Note: The rivers are ordered in descending latitude.

Timing of Spawning Runs

Chinook salmon may return to their natal river mouth during almost any month of the year (Snyder 1931; Rich 1942; Hallock et al. 1957). There are, however, typically one to three peaks of migratory activity. The timing and the number of migratory peaks varies among river systems (Figure 5). For northern river systems (Kamchatka River: Vronskiy 1972; Yukon River: Brady 1983; Cook Inlet tributaries: Yancey and Thorsteinson 1963; Nass and Skeena rivers: Department of Fisheries and Oceans, Vancouver, British Columbia, unpubl. data), a single peak of migratory activity during June appears typical, although the run may extend from April to August (Figure 5). Particular spawning populations may return later in the season, however, and this may result in two peaks of migratory activity (e.g., Kenai Peninsula: Yancey and Thorsteinson 1963).

Further south, runs tend to occur progressively later. In the Bella Coola River, an early run in late May and June is followed by a second, but relatively smaller, run in August (Figure 5). In the Fraser River there are two runs of almost the same size: an early run peaking in July, and a late run peaking in September/October. The Fraser River has a third, smaller, run in August that corresponds in timing to the late run into the Bella Coola River (Figure 5) (Ball and Godfrey 1968a, 1968b; Fraser et al. 1982). A late August run domi-

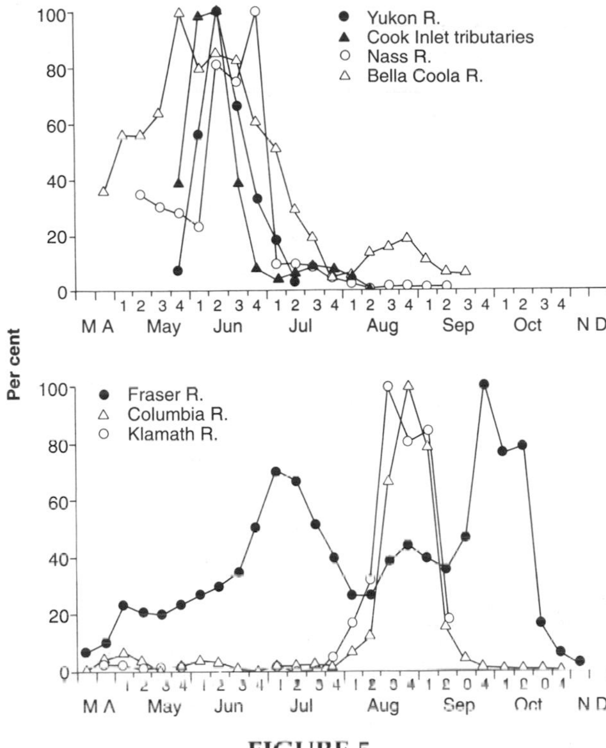

FIGURE 5

The timing of spawning runs to rivers throughout the North American range of chinook. The estimated abundance of fish during each quarter month is shown relative to the quarter monthly period with the greatest abundance of fish. Data sources are: Brady (1983) for the Yukon; Yancey and Thorsteinson (1963) for Cook Inlet tributaries; Canada Dept. of Fisheries and Oceans (unpublished data) for the Nass and Bella Coola rivers; Fraser et al. (1982) for the Fraser; Rich (1942) for the Columbia; and Snyder (1931) for the Klamath

nates chinook returns to both the Columbia and Klamath rivers. The Columbia River also has spring and summer runs, but these are numerically small relative to the late August run (Figure 5) (Rich 1942; Silliman 1950). The Klamath River has a spring run, but it, too, is relatively small (Figure 5) (Snyder 1931). Historically, the Klamath River spring run and the Columbia River summer run are believed to have been much larger, but they have since been decimated by habitat loss and overfishing.

The early run in the Fraser River may be analagous to the summer run in the Columbia River and to the spring runs in more northern rivers. The late Fraser River run appears unique in its timing, unless it is an analogue of the winter run into the Sacramento River which reaches Redding from late November through February, 350 km upstream (Slater 1963; Hallock and Fry 1967). The Sacramento River also has spring and fall runs of chinook, and a unique winter run. The winter and spring runs tend to overlap considerably in timing so that their separation is not precise during the run, but the two groups do separate on the basis of spawning area and time of spawning. Winter-run fish spawn mainly in May and June in the upper main stem of the Sacramento River. Spring-run fish, however, delay spawning until late August or September, and their spawning areas are in the upper reaches of the main stem of the Sacramento River and of the principal tributaries. The majority of chinook in the Sacramento River are fall-run fish that enter the river in September and October. These fish spawn shortly after entering the river in the middle and lower reaches (Hallock et al. 1957; Slater 1963; Hallock and Fry 1967; Kjelson et al. 1981). Thus, Chinook in the Sacramento River appear to exhibit typical stream-type (spring-run) and ocean-type (fall-run) behaviour. The winter-run fish are somewhat anomalous in that they have characteristics of both stream- and ocean-type races. They enter the river green and migrate far upstream. Spawning is delayed for some time after river entry. Young winter-run chinook, however, migrate to sea in November and December, after only four to seven months of river life. Without further information or details of the life history of these rare and interesting fish it is difficult to decide whether or not they fit the classification of stream- and ocean-type races.

Despite the wide variation in run timing within most rivers, spawning times tend to be similar among runs. Early-run fish normally delay spawning and spawn in the fall, about the same time as late-run fish. The winter run in the Sacramento River is an exception to this rule.

Detailed information on the racial composition of spawning runs is available for the Fraser and Columbia rivers (Rich 1925; Ball and Godfrey 1968a, 1968b). These data clearly demonstrate the alternation in timing of stream- and ocean-type races entering the two rivers (Figure 6). Stream-type chinook enter principally during the spring and early summer, and ocean-type chinook enter during the summer and fall. Given the fact that

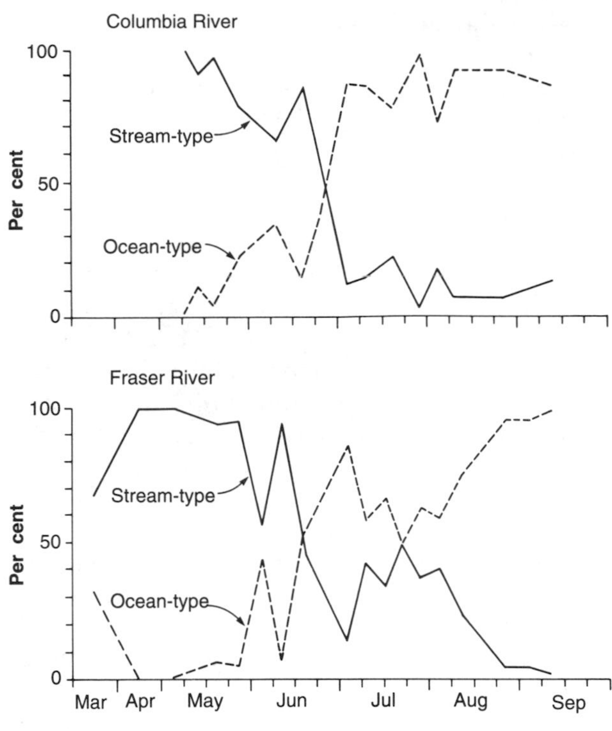

FIGURE 6

Seasonal changes in the percentage contribution of stream- and ocean-type chinook to the total run in the Columbia and Fraser rivers. (Data from Ball and Godfrey 1968a, 1968b, 1969, 1970; Rich 1942)

stream-type fish migrate to headwater tributaries of these two systems, their early entry to fresh water permits them to take advantage of peak summer flows to reach their spawning areas. Ocean-type fish, with shorter upriver migrations on average, can delay entry until after peak flows and thus take advantage of a slightly longer ocean feeding period. Stream-type fish, therefore, appear to suffer a double disadvantage. Not only do they lose feeding time in the sea, but they must also maintain their ion balance, without feeding, in the osmotically rigorous freshwater environment for several months before spawning. The adaptations required to achieve this suggest that there must be more than casual genetic separation between stream- and ocean-type chinook.

Timing of Spawning

The time at which chinook actually spawn is quite variable, ranging from May/June for some more northern populations and for the winter-run fish in the Sacramento River (Department of Fisheries and Oceans, Vancouver, British Columbia, unpublished data; Hallock et al. 1957; Slater 1963) to December/January for fall-run chinook in the Sixes and Elk rivers, Oregon (Reimers 1971; Burck and Reimers 1978), and in the Sacramento River (Hallock and Fry 1967). The trend throughout the range of chinook is to earlier spawning as one moves north (Figure 7), with northern populations tending to spawn from July to September and southern populations from November to January. Within river systems, however, individual populations may spawn at widely different times. In the Sacramento River, for example, spring-run fish spawn in August to September, fall-run fish in October to December, and winter-run fish not until May and June. Thus, there are (or were before the winter and spring runs were decimated) chinook spawning virtually every month of the year in the Sacramento River. In the Fraser River (Department of Fisheries and Oceans, Vancouver, British Columbia, unpublished data), median spawning dates for individual populations range from June to No-

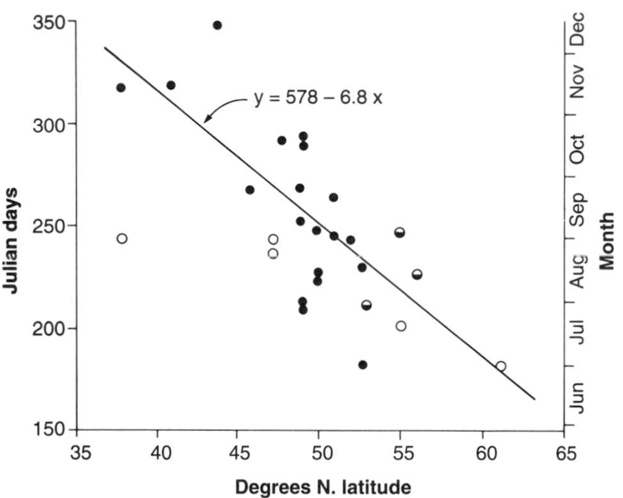

FIGURE 7

The relationship between the median spawning date and the latitude of the spawning population for stream- and ocean-type chinook. Symbols are: ○ = stream-type; ◒ = ocean-type; ● = mixed-type. The correlation for all data is significant ($r = -0.755$, $p < .01$). The races do not differ significantly ($p > .05$). The regression reported on the figure is the regression of Julian day on latitude.

vember, a span of six months. Spawning appears to be less protracted in northern rivers. In the Skeena River the range of median spawning dates among populations was 15 August to 25 September, and in the Kamchatka River the range among populations was 15 July to 15 August (Department of Fisheries and Oceans, Vancouver, British Columbia, unpublished data; Vronskiy 1972).

There is some suggestion that stream-type fish spawn earlier than ocean-type fish, at least in the central and southern parts of the chinook's range where the two types are sympatric. Burner (1951) noted that spring-run (stream-type) Columbia River chinook spawned in late August whereas summer- and fall-run fish (mainly ocean-type) spawned in late September. Hallock et al. (1957) also noted earlier spawning of spring-run fish, which presumably were mainly stream-type, relative to fall-run fish in the Sacramento River. Known populations of stream-type chinook in the Fraser River, however, do not obviously spawn earlier than ocean-type chinook(Department of Fisheries and Oceans, Vancouver, British Columbia, unpublished data). Northern populations, which are all stream-type, do spawn early, but this is not clearly related to race. If earlier spawning was mainly a function of race, then one would expect a discontinuity in the median date of spawning at about 56°N latitude, the latitude at which the composition of spawning populations shifts from predominantly ocean-type to predominantly stream-type. The trend to earlier spawning as one moves north is, however, continuous throughout the range of chinook, with no apparent discontinuity where population structure shifts to stream-type (Figure 7).

Redd Characteristics

The spawning beds chosen by chinook vary considerably in physical characteristics. Chinook will spawn in water depths from a few centimetres (Burner 1951; Vronskiy 1972) to several metres (Chapman 1943; Chapman et al. 1986). They will spawn in small tributaries two to three metres wide (Vronskiy 1972) and in the main stem of large rivers like the Columbia and Sacramento (Chapman 1943; Hallock et al. 1957).

Several authors have reported in some detail on the characteristics of chinook redds and spawning beds (Chapman 1943; Burner 1951; Briggs 1953; Vronskiy 1972; Neilson and Banford 1983). In addition, the expanding literature on instream flow requirements provides depth and velocity criteria for chinook spawning (Collings et al. 1972; Smith 1973; Bovee 1978). Comparison among observations by different authors is complicated by the fact that methods of measurement were different, and in some reports the methods were not given in sufficient detail to permit an assessment of comparability. The overriding impression is that, although there is good agreement among mean values for water depth and velocity in spawning beds, the range in depths and velocities that chinook find acceptable is very broad (Table 2). There is little agreement among observers about either the maximum or the minimum values for depth and velocity. For example, Burner (1951) observed chinook spawning in as little as 5 cm of water, whereas Collings et al. (1972) suggested that the minimum water depth for spawning is 30 cm. Maximum depths of spawning range over an order of magnitude, from 41 to >700 cm. Velocity minima range from 10 to 52 cm/s and maxima from 64.4 to 150.0 cm/s. There is apparently no agreement as to whether depth and velocity characteristics used in nest site selection differ between stream- and ocean-type chinook. Some authors assign greater depth and velocity preferences to stream-type and others to ocean-type chinook (Table 2).

In preparation for spawning, the female chinook digs a shallow depression in the gravel of the stream bottom by performing vigorous swimming movements on her side near the bottom. Gravel and sand thrown out of the depression accumulate in a mound, or tailspill, at the downstream margin of the depression. During the act of spawning the female deposits a group or "pocket" of eggs in the depression and then covers them with gravel. Over the course of one to several days, the female deposits four or five such egg pockets in a line running upstream, enlarging the spawning excavation in an upstream direction as she does so. The total area of excavation, including the tailspill, is here termed a "redd."

Four papers reported the size of the area of the redd excavation (Table 2). Burner (1951) observed a relatively narrow range of redd sizes in tributaries of the Columbia River, with stream-type chinook having the smaller redds. The redds of the stream-

TABLE 2
Summary of published information on water depth and velocity in chinook spawning beds, and area of the redd excavation

		Water depth (cm)		Water velocity (cm/s)		Redd area (m²)	
Source	Type*	Range	Mean	Range	Mean	Range	Mean
Chapman (1943)		30–460				2.4–4.0	
Burner (1951)	S	5–122	31			3.9–6.5	
Briggs (1953)	0	28–41	32	30–76			
Vronskiy (1972)	S	13–720	56†	30–150	61	4.0–15.0‡	
Collings et al. (1972)	S	45–52		52–68	54†		
	0	30–45		30–68			
Smith (1973)	S		31	21.7–64.4	43		
	0		38.9	18.6–80.5	49.7		
Bovee (1978)§	S	10–70	30	10–100	40		
	0	10–120	30	25–115	50		
Nelson & Banford (1983)	S			15–100	56	0.5–27.5	9.5
Chapman et al. (1986)	0	to 700		37–189	>100	2.1–44.8	17.0

Notes: *Separate values are reported for stream- and ocean-type chinook when available. *S* = stream-type (spring-run chinook, *0* = ocean-type (fall-run) chinook
†Geometric mean of the ranges given by Vronskiy (1972) for different tributaries
‡Maximum redd areas. These were calculated as the product of maximum and minimum length and breadth measurements given in Vronskiy (1972).
§Values taken from probability-of-use curves in Bovee's (1978) report

type fish also appeared to be in areas of coarser gravel and were often characterized by having a few large cobbles in the bottom of the excavation. Vronskiy (1972) gave only length and width measurements for stream-type chinook spawning mounds in the Kamchatka River. Maximum redd areas, inferred from the product of minimum and maximum measurements given by him, ranged from 4 to 15 m². Vronskiy (1972) also commented on the appearance of large cobbles in the bottom of chinook redds in the Kamchatka River but noted that in some tributaries chinook spawned in very fine gravel. Neilson and Banford (1983) reported a great range of sizes and large average size for redds of stream-type chinook in the Nechako River (British Columbia). Chapman et al. (1986) reported redd areas for the Hanford reach of the Columbia River that ranged from 2.1 to 44.8 m² and averaged 17.0 m². Neilson and Banford (1983) apparently estimated redd area as the product of maximum length and width of the redd, as I did from Vronskiy's data, whereas Burner (1951) and Chapman et al. (1986) took account of the oval shape of the redd. Thus, the measurements are not comparable, and Neilson and Banford's (1983) estimates of redd area, as well as those that I made from Vronskiy's data, will be up to twice as large as those made by Burner (1951) or Chapman et al. (1986).

Both Vronskiy (1972) and Neilson and Banford (1983) observed that the depth of the redd excavation was negatively correlated with water velocity in the spawning area. According to Vronskiy (1972), the higher mound in the tailspill of redds dug by chinook in low velocity water serves to improve subgravel irrigation of the eggs. Low velocity areas were also likely to be characterized by fine gravel that, presumably, could be dug into more easily by the fish, so that nests were deeper.

Briggs (1953) reported that chinook buried their eggs 20–36 cm deep (average 28 cm) in the gravel of two small streams in California. Vronskiy (1972) observed eggs buried from 10 to 80 cm deep in the gravel in the Kamchatka River, although he found few eggs below 50 cm. According to Vronskiy

(1972), the depth to which the eggs were buried was at least partly dependent on water flow, with the eggs being buried more deeply where flow was low. Chapman et al. (1986) found that the depth of gravel over eggs and embryos ranged from 10 to 33 cm and averaged 18.8 cm.

The range of depths and velocities within which chinook have been observed to spawn suggests that establishing meaningful minimum and maximum criteria for these factors is problematic. Although conventional wisdom states that chinook prefer deeper, faster rivers for spawning than the other *Oncorhynchus* species, measures of spawning area characteristics for other species do not confirm this wisdom (e.g., Burner 1951). Available measurements do not suggest that chinook avoid shallow water and low flows. Chinook may, however, spawn in water that is deeper and faster flowing than that used by other species because they are large enough to hold position in the faster current and to build a redd in the coarser gravel found there.

Minimum spawning depth is presumably governed by water depth needed for successful digging and spawning; however, Burner's (1951) observations suggest that this can be accomplished in as little as 5 cm of water. It is not clear why increasing depth should ever be a constraint, unless it is correlated with some other significant factor, such as velocity. Chapman (1943) observed Columbia River fall (ocean type) chinook spawning below Kettle Falls in water 15 ft (4.6 m) deep. In one spot, as many as thirty fish appeared to be spawning together in one large redd, although he also observed numerous individual redds and some with a few fish digging together. According to Indians whom he interviewed, this sort of main stem spawning was commonplace in the Columbia River when chinook were abundant and such spawning places had been important fishing places for them. Chapman (1943) considered such spawning to be unusual, however, and stated that, in his experience, chinook normally spawned at the head of a riffle in 0.3–1.2 m of water. Chapman (1943) speculated that the spawning below Kettle Falls may have been stimulated by high subgravel flow rates below the falls.

Other authors have emphasized the importance of subgravel flow in the choice of redd sites by chinook. Vronskiy's (1972) comment about the importance of the tailspill gravel mound in stimulating subgravel flow when redds are placed in low velocity water has already been mentioned. In addition, Vronskiy (1972), like Chapman (1943), observed that most redds (95%) were located at the head of a riffle, just before the crest of the rapid. Vronskiy (1972) attributed the attractiveness of this location to the high subsurface flows that occurred there. Other spawning occurred in pools below log jams where the log jam increased the rate of subgravel flow. In the Nechako River most chinook spawn on the upstream sides of large gravel dunes oriented across the river channel, presumably to take advantage of the subgravel flow stimulated by the dune (Russell et al. 1983).

Provided the condition of good subgravel flow is met, chinook apparently will spawn in water that is shallow or deep, slow or fast, and where the gravel is coarse or fine. The requirement for good subsurface flow is consistent with the probable incubation requirements of chinook relative to the other species. Chinook have the largest eggs (Rounsefell 1957) and, thus, their eggs have a small surface-to-volume ratio compared with the other species of Pacific salmon. Their eggs should, therefore, be more sensitive to reduced oxygen levels and require a more certain rate of irrigation. Silver et al. (1963) observed that the size of chinook at hatching was dependent on water velocity in the incubation apparatus even at velocities as high as 1,350 cm/h, and on oxygen concentration even near saturation levels, at least when the incubation temperature was about 11°C.

The apparent preference of chinook for spawning areas with high subgravel flow may explain their tendency to aggregate in particular locations for spawning and to ignore other, superficially similar, areas (Vronskiy 1972). Within areas of aggregation the distribution of spawning nests is not random but, rather, tends to an even distribution (Neilson and Banford 1983). Burner (1951) suggested that each spawning pair defends an area equal to about four times the area of its redd, and he recommended that the available spawning area be divided by four times the average redd area to arrive at an estimate of the maximum spawning population that the area could support. According to Burner's (1951) estimates of redd size, stream-type chinook require about 16 m^2 and ocean-type chinook require about 24 m^2 of gravel per spawning pair.

The chinook's apparent need for strong subsurface flow may mean that suitable chinook spawning habitat is more limited in most rivers than superficial observation might suggest, so that at high population density many chinook spawn in areas of low suitability, and their eggs consequently suffer high mortality. If this is the case, the continued high production of chinook in spite of greatly reduced spawning populations (Healey 1982a) becomes more understandable, since the apparent reduction in spawning populations will not have been accompanied by a corresponding reduction in fry production.

Length of Residence on the Redd

Chinook females in the Morice River, a tributary of the upper Skeena River drainage, spent between 4 and 18 days defending their redd after they began spawning (Neilson and Geen 1981). Chinook females in the Nechako River, a tributary of the upper Fraser drainage, spent between 6 and 25 days defending their redds (Neilson and Banford 1983). The average length of residence on the redds declined throughout the spawning period from about 14 days (early in the season) to about 5 days (late in the season) on the Morice River, and from about 15 days (early in the season) to about 4 days (late in the season) on the Nechako River. Both the Morice and Nechako river populations are mainly stream-type, although scale analysis indicates the presence of ocean-type chinook as well. As far as I am aware, these are the only measurements of residence on a redd for chinook. Apparently, no measurements exist for males, and, as Neilson and Geen (1981) pointed out, obtaining such measurements would be complicated by the fact that males are not faithful to a single redd.

Fecundity

The earliest measurements of fecundity in chinook salmon are those that McGregor (1922, 1923) reported for the Klamath and Sacramento rivers in California. McGregor's data indicated considerable variation in fecundity within each population but an even larger difference between populations. McGregor (1923) went so far as to propose that differences in fecundity would permit separation of Sacramento River and Klamath River chinook caught at sea. Healey and Heard (1984) summarized more recent data on the fecundity of 16 additional populations and confirmed the high intra- and interpopulation variation in chinook fecundity that McGregor (1923) had described. Fecundity of chinook females ranged from fewer than 2,000 eggs to more than 17,000 eggs. Fecundity was significantly correlated with female size in all but one of the populations examined to date. Size, however, explained only 50% or less of the variation in fecundity between individuals within a population (Figure 8). The slope of the fecundity on length relationship for chinook was also low (generally less than two in chinook compared with other fishes in which it is generally greater than three) (Healey and Heard 1984).

In several instances, fecundity had been measured for the same population over a number of years. Significant variation in average fecundity between years was evident in these populations, although the absolute interannual variation was less than that observed between years in other species (Healey and Heard 1984). The only other factor that appeared to contribute to within-population variation in fecundity was a small difference between fish of red and white flesh colours in the Fraser River population (Godfrey 1968a). Age apparently contributed nothing to variation in fecundity beyond that predicted by the difference in sizes between ages (Healey and Heard 1984).

Between-year and flesh colour variation in fecundity explained only a small additional amount of the within-population variation in fecundity, so that a great deal of individual variation remains to be explained. Healey and Heard (1984) speculated that this high variation may reflect an uncertain trade-off between egg size and egg number in the overall fitness of chinook populations. Unfortunately, there are no data on egg size in chinook sufficient to demonstrate that the more highly fecund fish within a size class also have smaller

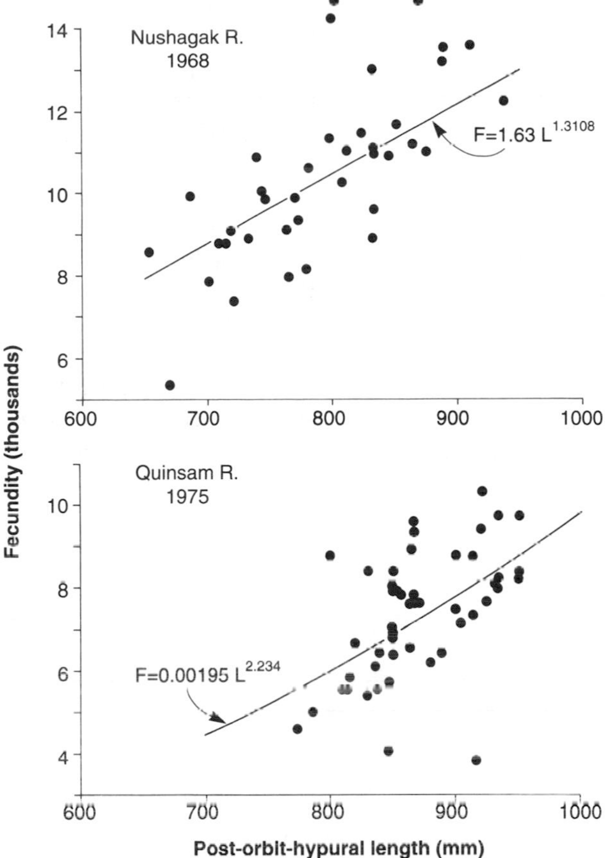

FIGURE 8

Two examples of the relationship between length and fecundity for chinook salmon. Nushagak River chinook are a stream-type race having high average fecundity, and Quinsam River chinook are an ocean-type race having low average fecundity. Although the fecundity on length regressions are significant in both instances, the amount of variation in fecundity attributable to variation in length is relatively small.

eggs. Egg size has been shown to increase with female size in Oregon chinook stocks (Nicholas and Hankin 1988).

Fecundity, therefore, appears to be less determined by body size in chinook than in other fishes. Partly, this may be due to the unresolved trade-off between egg size and egg number in chinook mentioned above, so that egg size varies more between chinook individuals than is usual for fishes. Also, it appears that in the trade-off between body size and fecundity among older chinook, body size was more critical, so that more energy is devoted to somatic growth and less to egg production than in other fishes (Healey and Heard 1984). More energy is devoted to individual eggs among these large fish as well (Nicholas and Hankin 1988).

Although, within populations, interannual variation in fecundity was significant, between-population variation was numerically greater (Healey and Heard 1984; Nicholas and Hankin 1988). Average fecundity at size varied approximately twofold between populations (4,347–9,427 at 740 mm post-orbit/hypural length). In general, fecundity increased from south to north in the chinook's range, contrary to Rounsefell's (1957) conclusion based on a few samples (Figure 9). The Sacramento River population, at the southern limit of the chinook's range, however, has an unusually high fe-

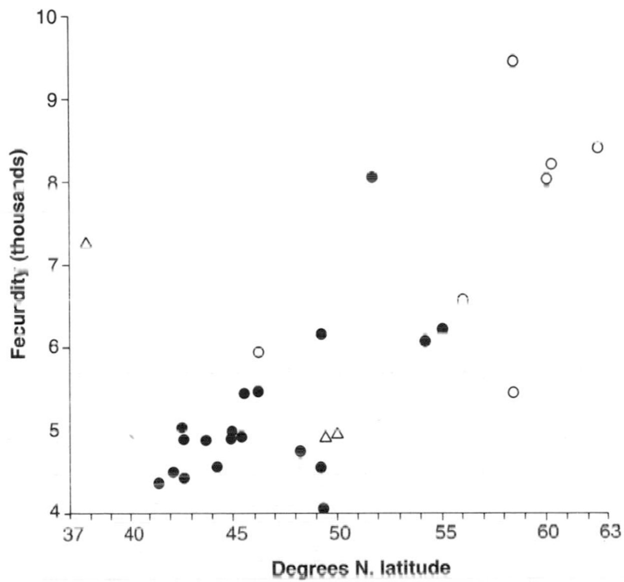

FIGURE 9

The relationship between average fecundity of chinook at 740 mm postorbit-hypural length and the latitude at which the population spawns (from data in Healey and Heard 1984; Nicholas and Hankin 1988). Symbols are: • = ocean-type populations having a common slope of fecundity on length regression; Δ = ocean-type populations having slopes of fecundity on length regression different from the above; o = stream-type populations having a common slope of fecundity on length regression. The overall correlation between fecundity and latitude is significant ($r = 0.663$, $p < .05$). If the populations with symbol Δ are left out of the regression, the relationship for ocean-type populations is also significant ($r = .642$, $p < .05$). Within the stream-type race, however, there is no relationship between latitude and fecundity.

cundity. This population, as well as three others (Quinsam River, Puntledge River, Rivers Inlet, all in British Columbia), differed from all other populations in the slope of the regression of fecundity on length (Healey and Heard 1984). Thus, it is difficult to compare the Sacramento River fish directly with most other populations. The high-fecundity populations near the northern limit of the chinook's range are all stream-type fish which spend a year in fresh water before going to sea, whereas the low-fecundity populations in the south are mainly ocean-type fish which go to sea during their first year of life. Thus, latitudinal differences in fecundity may partly reflect a racial difference between stream- and ocean-type chinook rather than a latitudinal cline. In the Columbia River, however, data are available on the fecundity of both stream- and ocean-type chinook. Although stream-type fish had a higher fecundity than ocean-type, the difference between the races was not statistically significant (Galbreath and Ridenhour 1966; Healey and Heard 1984). If the data are segregated into stream- and ocean-type life histories, there is still a latitudinal cline in fecundity within the ocean-type life history, provided the Sacramento, Puntledge, and Quinsam rivers and Rivers Inlet populations are excluded ($r = .642$, $p < .05$). The fecundity of stream-type populations alone is not significantly correlated with latitude, but, for all populations combined, the correlation is significant ($r = .663$, $p < .05$).

Spawning, Incubation, and Survival

Egg Deposition

The fecundity of females represents only the potential for production of the next generation. This potential is subject to successive losses that, in a stable population, ultimately result in an average production of one adult female spawner for each female spawner in the parent generation. Many of these successive losses have not been documented for chinook, except in an anecdotal way. They will be listed here in order of their occurrence, together with whatever estimates of loss are available in the literature. As before, the emphasis will be on evidence of variation among populations as adaptations to local environments.

Females that die unspawned on the spawning grounds, or that do not spawn all their eggs, represent an important potential loss in egg production. Such losses are generally lumped together in estimates of unspawned egg retention, and these estimates are seldom large (Chapman et al. 1986). Vronskiy (1972) reported that egg retention was generally about 0.6% of absolute fecundity for chinook; Major and Mighell (1969) reported it to be about 0.5% for Yakima River chinook; and Shepherd (1975) reported it to be 1.3% for Morice River chinook. Paine et al.(1975), however, reported an average of 11.9% egg retention for Big Qualicum River chinook, and Shepherd (1975) reported 25% egg retention for Bear River chinook with 9 of 47 females unspawned and 20% egg retention for Babine River chinook with 30 of 230 females unspawned. Fish in these latter two rivers, however, had been subject to harassment. In 1965 about 25% of adult chinook died without spawning in a spawning channel at Priest Rapids, Washington. This mortality was apparently caused by an infection of the gills with a protozoan of the genus *Dermocystidium* (Pauley 1967).

Ovarian disease may be a cause of reduced fecundity or egg deposition. A condition termed "bad eggs" has been known for Columbia River fall (ocean-type) chinook since the early 1940s. The condition is characterized by a number of signs, including a pus-like discharge from the vent and dead white eggs, either individually or in clusters, on the surface of the ovary or near the point of attachment. The condition may be present in one or both ovaries and occurred in 2%–20% of females returning to Columbia River hatcheries (Conrad

1965). Apparently, conditions of this type are not a problem in British Columbia hatcheries or known to be a problem in wild stocks (G. Hoskins, Department of Fisheries and Oceans, Pacific Biological Station, Nanaimo, British Columbia, pers. comm.).

Losses to eggs actually shed by females may occur in a variety of ways. Some eggs will be swept out of the redd during spawning and will be subject to heavy predation. Some eggs will not be fertilized. Some will not be buried deeply enough and will be accessible to vertebrate and invertebrate predators. Floods, siltation, freezing, desiccation, and disease can all take a toll. Poor gravel percolation or poor water quality can cause mortality of eggs. Finally, the quality of eggs laid and the embryos produced, which themselves depend on the genetic and phenotypic quality of the parents, may influence the survival of eggs. Few of these sources of egg and embryo loss have been quantified.

Opinions differ as to the quantity of eggs lost during spawning by being swept out of the redd. Vronskiy (1972) believed that there was a very high loss of eggs during spawning. He based this conclusion on the number of eggs recovered during complete excavation of redds. No more than 30% of the average fecundity of chinook females was ever recovered from a redd, and the average number of eggs recovered was about 12% of average fecundity. Briggs (1953), however, believed that very few eggs were lost during spawning. The difference in opinion between these authors may be due to the fact that Vronskiy (1972) made his observations at a spawning area where velocity was very high, whereas Briggs (1953) conducted his investigations on small streams with lower velocity. Unpublished observations of my own, on a high-velocity riffle in the Nanaimo River, British Columbia, indicated that few eggs were lost during spawning even in fast flowing water. Both Vronskiy (1972) and Briggs (1953) commented that trout, charr, and other small fish may dart into a redd and steal a few eggs while the female is spawning, but the loss due to this kind of predation was not quantified.

Most eggs deposited in redds appear to be fertilized. Briggs (1953) reported various studies that demonstrated that between 92% and 98% of eggs were successfully fertilized. Vronskiy (1972) reported fertilization in excess of 99% for Kamchatka River chinook.

Survival during Incubation

Information on mortality of fertilized eggs and agents of mortality comes from observations made of both artificially planted eggs and natural redds. Shelton (1955) investigated the survival to hatching and emergence of eggs planted at different depths in two sizes of gravel in artificial stream channels and subjected to several rates of water percolation through the gravel. He concluded that survival to hatching was greater than 97%, regardless of planting depth in the gravel or gravel size, provided the percolation rate was at least 0.001 ft/s (0.03 cm/s). Emergence was 13% or less, however, from small gravel and when percolation was less than 0.002 ft/s (0.06 cm/s). Eighty-seven per cent of fry emerged successfully from large gravel with adequate subgravel flows.

Alderdice and Velsen (1978) reviewed the available information on rate of egg development and temperature for chinook. Upper and lower temperatures for 50% pre-hatch mortality were 16°C and 2.5°–3°C, respectively, when the incubation temperature was constant. When incubation temperature varied with the ambient temperature, development rate and survival were better at low temperatures than when incubation temperature was constantly low. Presumably, this better performance reflected the development of greater low-temperature tolerance after initial cell division. Time to 50% hatch ranged from about 159 days at 3°C to 32 days at 16°C.

Alderdice and Velsen (1978) concluded that a log inverse form of Belehradek's equation (development rate = (temperature – C) exp b/K; where C, b, and K are constants) gave the best fit to the available data on egg development in relation to temperature. Even this model, however, underestimated development rate at low temperatures. For most practical purposes a simple thermal sum model (development time = $468.7/T$; where T is the average temperature during incubation) was adequate for predicting time to hatching.

Gangmark and Bakkala (1960) recorded percolation rate, temperature, and oxygen concentration near eggs planted in Mill Creek, California. Normal temperature fluctuations were not related to egg survival, but both percolation and oxygen concentration were. Mortality of eggs increased with decreasing percolation rate, being 2.9% at 4.0 ft/h

(0.034 cm/s) and nearly 40% at 0.5 ft/h (0.0042 cm/s). Mortality also increased rapidly at dissolved oxygen concentrations below 13 ppm, averaging 3.9% at 13 ppm and 37.9% at less than 5 ppm.

The survival of eggs in undisturbed natural redds appears to be quite good. Vronskiy (1972) reported survival of 97% to hatching, and Briggs (1953) reported 90% survival to the eyed stage and 82% to hatching. Vronskiy's estimate is based on the number of live and dead eggs and alevins recovered during excavation of total redds, whereas Briggs' estimates are based on samples from redds. Both authors attempted to correct for the disintegration of dead eggs, Briggs by planting dead eggs and observing the losses over time, and Vronskiy by counting egg shells as disintegrated eggs. Despite these corrections, the estimates of survival to hatching must be taken as maximum estimates for successful redds. Neither author dealt with losses due to scouring or siltation. Briggs (1953) did, however, observe some interesting instances of high mortality in redds due to attacks by an undescribed species of oligochaete worm. This is the only description of invertebrate predation on chinook eggs buried in a redd of which I am aware, although, under the right circumstances, eggs must be available to a variety of invertebrate predators.

Stream conditions during incubation can have a dramatic effect on the survival of eggs to hatching and emergence. In a series of experiments at Mill Creek, California, Gangmark and Broad (1955) and Gangmark and Bakkala (1960) demonstrated that flooding in Mill Creek was an important cause of high mortality of chinook eggs. Apart from loss of eggs washed out of the gravel by floods, mortality was associated with low oxygen concentrations in spawning gravel (<5 ppm) and poor percolation of water through spawning gravel, with increasing mortality below percolation rates of 4.0 ft/h (0.034 cm/s).

Temperature has seldom been implicated in any significant loss of eggs during incubation. Coombs and Burrows (1957) speculated that salmon spawning in cold headwater streams with temperatures below 40°F (4.4°C) would suffer high mortality, and, at the other extreme, Slater (1963) concluded that winter-run chinook spawn would suffer high mortality in some tributaries of the Sacramento River because of high water temperature during incubation.

Chinook normally begin spawning in late summer and may begin spawning when temperatures are near 16°C, the upper temperature limit for 50% egg mortality (Alderdice and Velsen 1978). As temperatures are falling rapidly at this time of year, however, the eggs are probably not exposed to near lethal temperatures for long. Similarly, temperatures may drop below 3°C part way through the incubation period in ice-covered rivers. The impact of seasonal exposure to extreme temperatures on survival and viability of eggs and alevins has not been studied systematically, but presumably the embryos are able to survive conditions such as these, which are typical in spawning rivers.

Adequate water percolation through the spawning gravels is essential for egg and alevin survival. There is no doubt that percolation is affected by siltation and that siltation in spawning beds can cause high mortality (Shaw and Maga 1943; Wickett 1954; Shelton and Pollock 1966). There appears to have been no systematic study of the interrelation between river discharge (velocity), sediment load, and survival of chinook spawn. Of particular significance to any assessment of the probable effects of siltation on survival is Shaw and Maga's (1943) observation that siltation resulted in greatest mortality when administered early in incubation. Thus, siltation during winter freshets in coastal rivers or during summer peak discharge in snow-fed rivers may have a greater effect on the amount of suitable spawning gravel than on the survival of previously deposited spawn.

Becker et al. (1982, 1983) investigated the effects of dewatering artificial chinook redds on survival and development rate of embryos at various stages of development. Becker et al. (1982) defined stage of development in terms of accumulated thermal units during incubation and studied four stages: cleavage eggs, incubated for 56–168 Celsius degree-days (DD), embryos (249–467 DD), eleutheroembryos (553–575 DD), and pre-emergent alevins (780–814 DD). Dewatering of redds is likely to occur in regulated rivers where discharge is varied to satisfy some domestic or industrial need but could also occur in natural rivers. Alevins were most sensitive to both periodic short-term dewatering and a prolonged single dewatering, surviving at less than 4% in periodic dewaterings of one hour or a single dewatering of six hours. Eleutheroembryos were less sensitive, and cleavage eggs and embryos least sensitive. In fact, embryos apparently suffered no ill effects from daily dewaterings of up to 22 hours over a 20-day period. The development rate was also reduced in those instances in which survival was affected but not in instances when survival was good. These results seem at variance with the observation of Silver et al. (1963) that chinook embryo development is highly sensitive to any reduction in oxygen concentration or percolation rate. Since the dewatered eggs and embryos remained damp, however, they probably suffered no shortage of oxygen. Elimination of metabolic waste products may have been a problem.

Emergence

Estimating survival to emergence poses significant problems with chinook, as some fish migrate downstream as fry whereas others rear for a variable length of time in the river before migrating downstream. Counts of downstream migrants, therefore, provide only a minimum estimate of the number of fry that emerged. In Mill Creek, 85%–100% of fertilized eggs deposited in plastic mesh bags in the gravel were lost prior to emergence. These losses were associated with floods in the creek. In a channel with controlled flow, the mortality of planted eyed eggs to emergence was only 40% (Gangmark and Bakkala 1960). In Fall Creek, California, Wales and Coots (1954) and Coots (1957) found a 68%–93% mortality from egg deposition to the emergent fry stage. The 93% mortality was associated with floods. From redd excavation, Gebhards (1961) estimated that only 42% of alevins would have emerged from a single redd. Although not well documented, it appears that emergence may be a difficult time for fry.

Apparently gravel conditions can influence the success of emergence. Shelton (1955) found that only 13% of hatched alevins emerged from experimental troughs in which eggs were planted in fine gravel compared with 80%–90% emergence in troughs with coarse gravel. Emergence from fine gravel was further influenced by the depth of planting and water velocity through the gravel. Greater emergence occurred when the eggs were planted near the surface and when water velocity was low.

Major and Mighell (1969) estimated that 5.4%–16.4% of spring chinook survived to migrate as yearling smolts from the potential egg deposition in the Yakima River, Washington. In the Cowichan River, British Columbia, 9.2% and 16.5% of potential egg deposition survived to migrate as fry and fingerlings (Lister et al. 1971), whereas in the Big Qualicum River survival from potential egg deposition to fry and fingerling migrants was 0.2%–7.0% prior to flow control and 12.0%–19.8% after flow control (Lister and Walker 1966; Paine et al. 1975). M.D. Bailey (Department of Fisheries and Oceans, Vancouver, British Columbia, pers. comm.) estimated that 4%–50% of potential egg deposition migrated as fry into the lower Fraser River, British Columbia, but cautioned that these estimates were based on poor data. Healey (1980b) estimated that about 12%–20% of potential eggs deposited migrated downstream as fry in the Nanaimo River, British Columbia. All these values are difficult to interpret because of uncertainty in estimates of both the potential eggs deposited and the numbers of fry produced. The values do suggest, however, that, barring serious floods, egg-to-fry survival in chinook is relatively good.

Summary of Egg-To-Fry Survival

Published estimates of the mortality rate between egg laying and fry emergence are so few and so variable that it is difficult to draw any firm generalizations (Table 3). In particular, there is little evidence that can corroborate or refute my assertions about variability among chinook populations. Under natural conditions, 30% or less of the potential eggs deposited resulted in emergent fry or fry and fingerling migrants in the systems studied (Table 3). When and how eggs die or are lost to the population is uncertain. Two features do bear some comment, however. Eggs properly buried in a redd that remains undisturbed, or which are artificially planted where subgravel percolation is good, apparently survive well. Floods, which scour the bottom or result in heavy siltation, are generally associated with high egg mortality, as is dewatering of redds (but see Becker et al. 1982, 1983). Flow control appears to result in a significant increase in average survival. These observations suggest that egg-to-fry and fingerling mortality probably occurs either at the time of spawning or as a result of

TABLE 3
Published estimates of mortality (%) of chinook to various development stages in fresh water (mean of ranges in parentheses)

River system	Eggs not spawned	Losses at spawning	Spawning to eyed stage	Spawning to alevin	Spawning to emergence	Spawning to fry/smolt	Remarks
Mill Cr. (CA)					85–100 (96)		Planted eggs, flooding channel
					40		Planted eggs, controlled flow
Fall Cr. (CA)					68–93 (85)		Natural spawning
Prairie Cr. (CA)		1.0	0–25.5 (10)	14–25 (18)			Natural spawning, redd sampling
Yakima (WA)	1.0					84–95 (89)	Stream-type, weir counts of smolts
Lemhi (ID)				27	58		Emergence trap over one redd
Cowichan (BC)						84–91 (87)	Ratio of fry/smolt migrants to eggs
Nanaimo (BC)						80–88 (84)	Ratio of fry/smolt migrants to eggs
Big Qualicum (BC)	12					93–100	Before flow control
						80–88	After flow control
Skeena System							
Bear R. (BC)	25						
Morice R. (BC)	1						
Babine R. (BC)	20						
Kamchatka (USSR)	1	88		1–6 (3)			Redd sampling

Source: See text for sources

redd disturbance due to floods. Losses at the time of spawning in particular bear further investigation, in view of Vronskiy's (1972) comment that high egg losses at spawning have been observed by Russian biologists for several species of Pacific salmon, and the International Pacific Salmon Fisheries Commission's observation that losses of eggs at spawning in Adams River sockeye were heavy when the spawning stock was large (T. Gjernes, Department of Fisheries and Oceans, Pacific Biological Station, Nanaimo, British Columbia, pers. comm.).

Major and Mighell's (1969) observations on spring chinook deserve comment, as their estimates of survival from egg deposition to smolt migrants are comparable to estimates of survival from egg deposition to fry and underyearling smolt migrants for fall chinook in other systems. It is tempting to assert that this is an example of the dichotomy between stream- and ocean-type chinook. There are, however, a number of possible sources of error in the data on which these estimates were based. Major and Mighell (1969) estimated potential egg deposition by counting redds and multiplying by the average fecundity of females in the Columbia River. In my view, this procedure is liable to give a biased estimate of potential eggs deposited because false redds may be counted as true redds, because redds may be missed, or because the fecundity of local stocks differs from that of the general population. The apparent high survival of spring chinook from egg to smolt may, therefore, be fortuitous. Egg-to-migrant survival was negatively correlated with redd count and potential egg deposition in the Yakima River, even excepting an unusual mortality of eggs in the upper Yakima River due to low flows in 1957 (Figure 10). Such a relationship is consistent with the notion that redd counts may have been inaccurate, so that variations in survival were more a reflection of error in redd counts than real variation in survival. It is also consistent, however, with

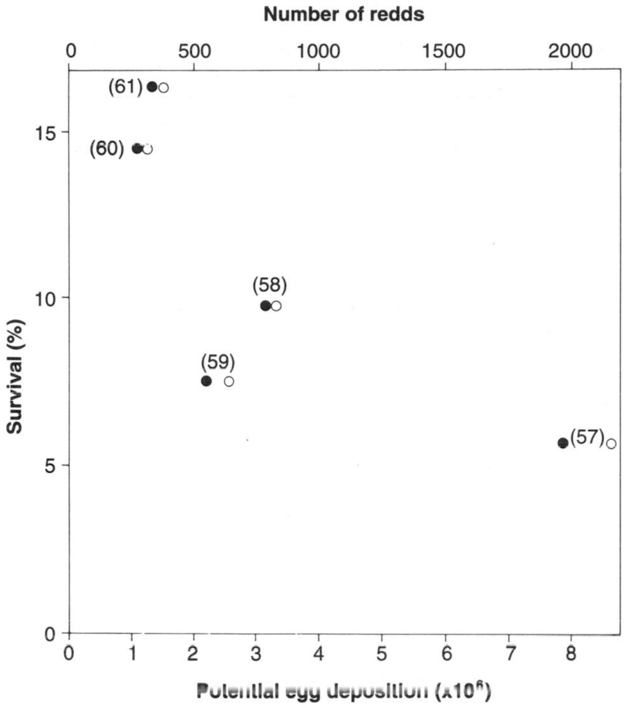

FIGURE 10
The relationship between the number of spring chinook redds counted and estimated egg-to-smolt survival (•), and between potential egg deposition and egg-to-smolt survival (o) for the Yakima River, Washington. Brood years are in brackets. (Data from Major and Mighell 1969)

the possibility suggested earlier that good redd sites are few in most rivers, and that fry and smolt production may be more related to the amount of good spawning area than to the number of spawners. A third possibility is that yearling smolt production is independent of spawner density over a wide range of spawner densities. Lister and Walker (1966) noted a rather small variance in underyearling smolt production compared with fry production in the Big Qualicum River and speculated that rearing habitat was a limiting factor in smolt production.

Freshwater Residence and Downstream Migration

Fry Migrations

Factors controlling the emergence of chinook fry from the spawning beds are not well studied. According to Reimers (1971), most emergence is at night, although up to 20% of fry emerged during the day in his experimental troughs. Emergence was reduced during a full moon. In Reimers' (1971) experiments, emergence peaked just after alevins reached maximum weight.

Upon emergence, fry swim, or are displaced, downstream. Thomas et al. (1969) found that fall chinook fry go through a period of reduced swimming ability just before the time of complete yolk absorption, and that this coincided with the time of peak downstream migration. They hypothesized that reduced swimming ability was the cause of downstream migration.

Downstream movement of fry occurs mainly at night, although small numbers may move during the day. Peak nightly catches of downstream migrants may occur before (Reimers 1971), at (Lister et al. 1971), or after midnight (Mains and Smith 1964). Differences in time of peak catch most likely reflect the distance of trapping sites below the main spawning area rather than differences in time of emergence from the gravel. Reimers (1971) also observed that downstream movement was inhibited by bright moonlight, so that, on moonlit nights, peak trap catch was shifted from before to after midnight.

Once started downstream, chinook fry may continue migrating downstream to the river estuary, or may stop migrating and take up residence in the stream for a period of time ranging from a few weeks to a year or more. What determines whether fry will hold and rear in the river, or migrate downstream to the estuary, is unknown. Kjelson et al. (1981) observed that peak catches of chinook fry in the Sacramento-San Joaquin delta often followed flow increases associated with storm runoff. They speculated that flow surges influence the numbers of fry that migrate from upper river spawning grounds to the delta. Healey (1980b) also observed that downstream movement of fry was correlated with river flow in the Nanaimo River. Reimers (1968) and Lister and Walker (1966), however, speculated that social interaction or density-dependent mechanisms may cause fry to be displaced downstream. Reimers (1968) observed lateral displays, chasing, nipping, fighting, fleeing, submission, and redirected aggression among juvenile fall chinook in stream tanks, where the agonistic behaviour of one or a few dominant fish apparently stimulated the downstream movement of subordinate fish. These same behaviours, with the exception of nipping and redirected aggression, also occurred in natural stream populations of chinook. Lister and Walker (1966) observed that, in the Big Qualicum River, fry migrants varied almost 100-fold in abundance between years, whereas fingerling migrants varied only about ten-fold in abundance. They concluded that available freshwater rearing area limited the number of fry that could reside in the river, and that the rest were displaced downstream. A similar conclusion could be drawn from Major and Mighell's (1969) observations on production of stream-type smolts in the Yakima River.

River discharge and intraspecific interaction may both play a role in stimulating downstream movement of chinook fry. Other factors, however, may also be important. Stein et al. (1972) observed that juvenile coho apparently were dominant to chinook and grew faster in sympatric groupings in stream troughs. Chinook were able to grow as rapidly as coho, however, when alone in the troughs. Stein et al. (1972) speculated that interaction with coho may influence the downstream movement of chinook. Recently, Taylor and Larkin (1986) demonstrated that stream-type chinook fry showed stronger positive rheotaxis and were more aggressive towards conspecifics and coho fry than were ocean-type chinook fry. Taylor (1988) confirmed for other stocks that stream-type chinook were more aggressive but not that they had stronger rheotaxis than ocean-type chinook. These behaviour patterns are consistent with the expected length of river residence of the two races and suggest that river residency and its associated behaviour patterns may be inherited.

Lister and Genoe (1970) reported habitat segregation among juvenile chinook and coho in the Big Qualicum River, as did Chapman and Bjornn (1969) and Everest and Chapman (1972) for chinook and steelhead (*Oncorhynchus mykiss*) in headwater tributaries of the Snake River, and Murphy et al. (1989) for chinook and coho in the Taku River. Everest and Chapman (1972) found that underyearling chinook in summer occurred over all substrate types, at all depths, and in water of all velocities (up to 1.2 m/s) studied, but that abundance generally declined with increasing substrate particle size, increasing depth, and increasing water velocity. In these studies, chinook were larger and emerged earlier than the associated coho and steelhead of the same brood year. Since the larger, older fish chose higher velocity habitats, there may have been little competition for space between the species. Habitat segregation in these studies seemed to be a mechanism for reducing competition rather than a result of competition.

A large downstream movement of chinook fry immediately after emergence is typical of most populations (e.g., Lister and Walker 1966; Bjornn 1971; Reimers 1971; Healey 1980b; Kjelson et al. 1982). The downstream migration of stream- and ocean-type chinook fry, when spawning grounds are well upstream, is probably a dispersal mechanism that helps distribute fry among the suitable rearing habitats. In the case of ocean-type populations that spawn close to tidewater, downstream migrant fry may be swept to the river estuary in a few hours. It has been hypothesized that migrant fry swept to the estuary represent those that are surplus to the carrying capacity of rearing habitat in the river (Lister and Genoe 1970). Often these fry represent the majority of the emergent population, however, and it seems doubtful that such a waste of reproductive potential would be adaptive (Figure 11). Furthermore, it is now known that estuaries provide important nursery habitat for recently emerged chinook fry (Northcote 1976; Healey 1980b, 1982b; Levy and Northcote 1982).

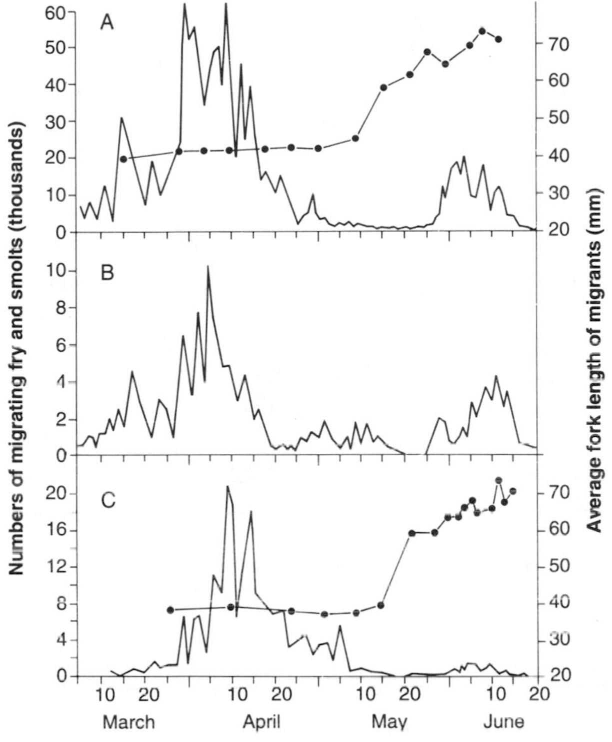

FIGURE 11

The temporal pattern in abundance of fry and underyearling smolts migrating seaward in three Vancouver Island rivers. *A* = Cowichan River, 1967; *B* = Big Qualicum River, 1967; *C* = Nanaimo River, 1980. Average fork length of migrants in the Cowichan and Nanaimo rivers is shown and demonstrates the rapid switch from fry (35–40 mm) to smolt (60-70 mm) migrants in late May.

Downstream migration of fry in the lower Nanaimo River, British Columbia, for example, appears not to be a consequence of limited rearing habitat in the river. In some years, few chinook remained to rear in the lower river after the fry migration, yet large numbers of fry consistently reared in the river estuary (Healey 1980b; Healey and Jordan 1982). It seems probable that the Nanaimo River estuary is the preferred rearing habitat for at least part of the Nanaimo River chinook population rather than a final refuge for displaced fry. Chinook in the Nitinat River, British Columbia, also migrate to the estuary in large numbers as fry, and most of the underyearling smolts from both the Nitinat and Nanaimo rivers are produced in the estuary rather than in the river (Healey 1982a). Similarly, large numbers of ocean-type fry migrate seaward in the Sacramento River, the Cowichan River (Figure 11), and the Fraser River (M.D. Bailey, Department of Fisheries and Oceans, Vancouver, British Columbia, pers. comm.) and rear in the river estuaries (Kjelson et al. 1981, 1982; Healey 1982a; Levy and Northcote 1982). It should be noted that the principal rearing areas in the Sacramento-San Joaquin and Fraser River estuaries are essentially fresh water whereas rearing areas in Vancouver Island estuaries range up to 20 ppm salinity.

In the Nanaimo River there are three geographically separated spawning areas, and fry from the two most downstream areas drift down to the estuary. Many fry from the middle spawning area, however, rear in the river and migrate to sea after six to eight weeks, and fry from the upper spawning area may spend up to a year in the river before migrating to sea (Healey and Jordan 1982). Analysis of polymorphic enzyme systems and body morphology suggested that the fry from the three spawning areas were genetically distinct and may be programmed to migrate seaward at different ages (Carl and Healey 1984). Clarke et al. (1989) found that stream-type chinook required a period of short day length before they would adapt to and grow well in sea water, whereas ocean-type chinook did not. From these and the observations of Taylor and Larkin (1986) and Taylor (1988), it appears that the downstream movement of fry after emergence may not be totally involuntary and that length of river residency may be dependent upon the fish's genotype. It is reasonable that stream- and ocean-type chinook would differ in this regard since length of freshwater residence is an important distinguishing characteristic for these races.

Downstream movement of fry is normally most intense between February and May (Figures 11 and 12), being earlier in more southern populations. Rich (1920), for example, observed a few fry as early as December in the lower Columbia River and in October and November in the Sacramento River. The timing of peak downstream migration

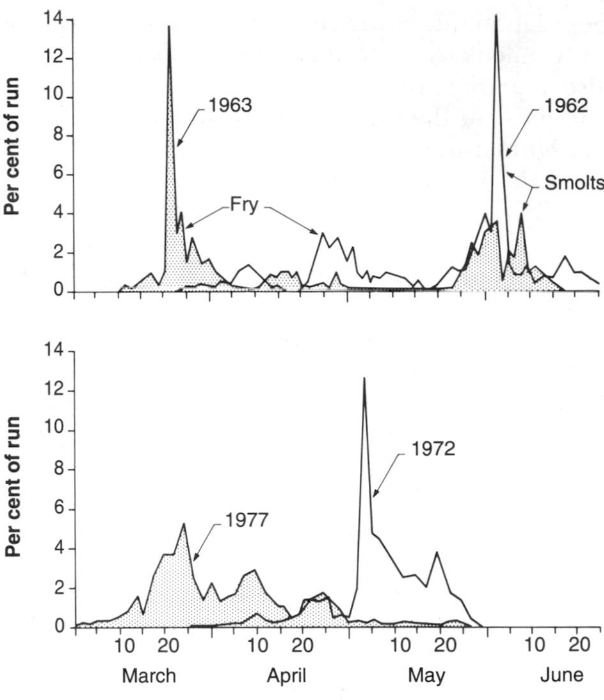

FIGURE 12

Examples of variation in downstream run timing of ocean-type chinook in the Big Qualicum River (*A*) and the Fraser River (*B*). Only fry migrants are shown for the Fraser, but both fry and smolt migrants are shown for the Big Qualicum. Note that large variation in fry migration timing is not paralleled by variation in smolt migration timing in the Big Qualicum.

can vary substantially from year to year in the same system. Time of peak downstream movement varied from the fourth week of March to the fourth week of April in the Big Qualicum River (Lister and Walker 1966), and from mid-March to early May near the mouth of the Fraser River between 1964 and 1977 (M.D. Bailey, Department of Fisheries and Oceans, Vancouver, British Columbia, pers. comm.) (Figure 12). The beginning and end of the run appear to vary less from year to year, so that, when the run peaks early, the temporal pattern has a negative skew, and when the run peaks late, the temporal pattern has a positive skew.

In addition to annual variation in the peak of the run, there is tremendous day-to-day variation in the abundance of downstream migrants (e.g., M.D. Bailey, Department of Fisheries and Oceans, Vancouver, British Columbia, pers. comm.; Healey and Jordan 1982). The causes of both annual and daily variation in run are not well understood. As noted earlier, Kjelson et al. (1981) speculated that downstream migration in the Sacramento River was stimulated by high discharge. Mains and Smith (1964) also suggested that peaks in downstream movement of chinook at Central Ferry on the Snake River were triggered by freshets. The greatest movement of chinook at this trapping location was at river temperatures between 4.5° and 13°C, but there appeared to be no relationship between migration and temperature. At Byer's Landing on the Columbia River, Mains and Smith (1964) found no relationship between the migration of chinook and either discharge or temperature. At Byer's Landing, temperatures ranged from 4.5° to 15.5°C during the run. In the Nanaimo River, daily variation in the downstream run of chinook fry was positively correlated with discharge while the run was increasing in 1975 and 1976, but not while the run was decreasing; yet variation in discharge was comparable over both the increasing and decreasing portions of the run. Greatest fry migration occurred at river temperatures of 6.0°-9.0°C in 1975 and 8.0°-11.0°C in 1976, but variation in the downstream run was not correlated with temperature in either year (Healey 1980b). Irving (1986) found that simulated freshets in experimental stream channels increased the numbers of fry moving downstream, provided water velocity at peak flows exceeded 25 cm/s.

In larger rivers, chinook fry migrate more at the edges of the river than in the high velocity water near the centre of the channel (Figure 13) and, when the river is deeper than about 3 m, they prefer the surface (Mains and Smith 1964; M.D. Bailey, Department of Fisheries and Oceans, Vancouver, British Columbia, pers. comm.; Healey and Jordan 1982). These observations provide further support for my earlier suggestion that downstream movement of fry is not simply a passive displacement controlled by water velocity, but that some active behaviour of the fry helps direct the migration.

Habitat Utilization in Fresh Water

The process by which chinook take up residence in a stream is not well studied. Reimers (1971) observed that, on the first night after emergence, virtually all fry drifted downstream in a stream

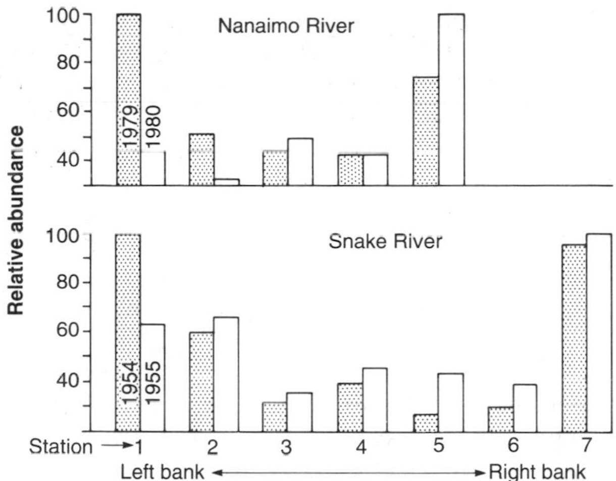

FIGURE 13

Lateral distribution of downstream migrating chinook fry in the Nanaimo and Snake rivers showing the tendency of the fry to concentrate near the river banks. Numbers of fry-per-unit-volume of water at each station across the river is shown as a percentage of the station with the greatest fry-per-unit-volume. For both rivers, stations were distributed with equal spacing from bank to bank.

trough. On each succeeding night, however, fry that were moved back upstream in the trough showed a stronger and stronger tendency to hold position in the trough. These observations suggest that stream residence develops over a number of days and that fry could be displaced quite far downstream before taking up residence. The tendency for all fry to drift downstream on the first night after emergence may partially explain why few fry were found rearing in the lower reaches of the Nanaimo River, where spawning occurs only a few kilometres from the estuary. This explanation is unsatisfactory, however, as fry were also observed drifting into the lower river from more upstream spawning areas, and presumably these could have occupied habitat in the lower river if they had been so inclined. Resident fry were relatively abundant in suitable habitat near the upstream spawning areas. Carl and Healey (1984) demonstrated that fry that migrated to the Nanaimo River estuary were genetically and morphologically different from those that reared in the river. Furthermore, Taylor and Larkin (1986) and Taylor (1988) showed that stream-type chinook fry displayed greater inter- and intraspecific aggression than ocean-type fry and that some stream-type stocks also displayed stronger positive rheotaxis. The behavioural mechanisms for holding position in the river after emergence thus appear to be better developed in some types of chinook fry than in others.

Lister and Genoe (1970) studied habitat segregation among juvenile fall (ocean-type) chinook and coho in the Big Qualicum River during the spring of 1967, when flow in the river was held constant at about 5.8 m^3/s. They examined three sites along the river, and each site was subdivided into two or three habitat types which differed in velocity, depth, and distance from shore. Chinook emerged during March and April, whereas coho emerged during May. Thus, chinook arrived in the study areas at least six weeks earlier than coho. Chinook were larger than coho at the time of emergence and, because they emerged earlier, grew even larger before the coho fry appeared in the study sites. Smaller fry of both species inhabited marginal areas of the river, particularly back eddies, behind fallen trees, undercut tree roots, or other areas of bank cover. As they grew larger, both species moved away from shore into midstream and higher velocity areas. Although the correlation between size of fish captured and velocity of sampling site within species was weak, chinook were always larger and more abundant than coho in high velocity subareas. Thus, there was important habitat segregation between chinook and coho in the Big Qualicum River, and this was mainly a consequence of the larger size and earlier emergence of the chinook salmon.

Chapman and Bjornn (1969) and Everest and Chapman (1972) reported qualitatively very similar observations on habitat segregation between stream-type chinook and steelhead in the Snake River. Juvenile chinook were most abundant where substrate particle size was small, velocity was low, and depth was shallow, but were found in small numbers in virtually every habitat investigated. Fish size was positively correlated with water velocity and depth for both species, but the species differed in size owing to differences in emergence timing and fry size between the species.

Murphy et al. (1989) sampled various habitat types in the lower Taku River for chinook, coho, and riverine sockeye. They found that chinook were mainly in riverine habitat and seldom in beaver ponds or off-channel sloughs. Velocity and turbidity were the principal factors associated with chinook distributions. Chinook were rare in still water or where velocity was greater than 30 cm/s. There was little overlap in chinook habitat with that of coho or sockeye. Thus, habitat segregation appears to provide a mechanism for reducing competition between cohabiting chinook and other stream salmonids, and the pattern of segregation is similar for stream- and ocean-type races.

The movement of fish offshore and into faster water represents a shift from predominantly sandy substrate to predominantly boulder and rubble substrate. Chapman and Bjornn (1969) suggested that chinook prefer finer substrates than steelhead of comparable size, but both species showed a strong preference for the rubble type of habitat. Any interpretation of substrate preferences is confounded by velocity preference, however, and needs further investigation.

Edmundson et al. (1968) reported limited day-to-day movement of young chinook in a stream aquarium, suggesting strong fidelity to a particular site. Reimers (1968) reported that juvenile chinook in the Sixes River were primarily solitary animals and displayed aggressive behaviour towards other chinook, suggesting the existence of defended areas in the stream, at least during the day. Chapman and his co-workers (Don Chapman Consultants 1989) observed temporary defence of feeding territories by chinook in the evening. As with habitat preferences, these observations need to be substantiated in other rivers and other situations.

Day and night distributions of chinook in streams may be quite different. Edmundson et al. (1968) and Don Chapman Consultants (1989) found that at night chinook moved inshore to quiet water over sandy substrates or into pools and that most settled to the bottom. With returning daylight, these fish returned to occupy the same riffle and glide areas that they had occupied on the previous day.

Fingerling Migrants

Fish that elect to hold in the river after emergence may migrate seaward almost any time of year. In the southern half of the chinook's range, many stream dwellers migrate seaward as fingerlings between April and June of their first year of life (Healey 1980b, 1982b; Kjelson et al. 1981, 1982) (Figure 11). Although following close on the heels of the fry migration, fingerling migrants are readily distinguished by their larger size. In all the rivers studied, a sharp change in size of fish accompanies the change from fry to fingerling migrants (Figure 11). Fry migrants normally range from 30 to 45 mm in fork length, although they have been recorded as small as 20 mm and as large as 55 mm (Mains and Smith 1964; Lister et al. 1971; Healey et al. 1977). Many fry migrants still have visible yolk and few have begun feeding, although those above 44 mm fork length may have some food in their stomachs. Fingerling migrants, on the other hand, normally range from 50 to 120 mm in fork length, and all have been actively feeding for some time (Mains and Smith 1964; Lister et al. 1971; M.C. Healey unpublished data).

The factors stimulating downstream movement of underyearling chinook are not known. Although it is well documented that June is a month of very active downstream migration for fingerlings, they are known to migrate downstream at other times of the year as well. The main fingerling migration tends to be earlier in the southernmost parts of the chinook's range (Kjelson et al. 1981) and is influenced by the presence of populations with unique spawning times (e.g., Slater 1963). In some Columbia River tributaries, juvenile chinook were found to be resident as late as October but were gone in November (Reimers and Loeffel 1967). Bjornn (1971) observed downstream movement of stream-type chinook fingerlings in the Lemhi River (Idaho) during the fall months. He proposed that this migration represented a redistribution of fish to more suitable wintering habitat. Fingerling migration is known to occur through August in the Fraser River (Northcote 1976; M.D. Bailey, Department of Fisheries and Oceans, Vancouver, British Columbia, pers. comm.) but is generally observed to be complete by the end of June in most other rivers sampled (Lister and Walker 1966; Lister et al. 1971; Healey and Jordan 1982). Reimers and Loeffel (1967) were able to relate extended residence in Columbia River tributaries to slow growth, and suggested that size was an important variable in determining when fish will move downstream. Nevertheless, downstream migrant fingerlings vary substantially in size, both within and between rivers (Table 4), so that some factor other than size must also play a role. Furthermore, it is known that chinook may move out of tributaries and into a river main stem, or simply relocate downstream with the approach of winter (Bell 1958; Chapman and Bjornn 1969; Park 1969). Presumably, as Bjornn (1971) suggested, suitable summer habitat may not be suitable winter habitat. The disappearance of chinook from some Columbia River tributaries in October or November (Reimers and Loeffel 1967), therefore, probably indicates relocation to an instream wintering area rather than seaward migration.

Fingerlings migrate downstream throughout the day, but the majority migrate at night (Mains and Smith 1964; Lister et al. 1971). In the Columbia and Snake rivers, fingerlings apparently preferred the shoreline during migration (Mains and Smith 1964). In the lower Fraser River and in the Nanaimo River, however, most fingerlings migrated in the fastest water near the centre of the river (M.D. Bailey, Department of Fisheries and Oceans, Vancouver, British Columbia, pers. comm.; Healey and Jordan 1982).

The rate of downstream migration of chinook fingerlings appears to be both time-and size-dependent and may also be related to river discharge and the location of the chinooks in the river. Cramer and Lichatowich (1978) observed that migrating spring chinook fingerlings in the Rogue River, Oregon, travelled downstream only about 0.3–5.0 km/d in the upper reaches of the river but travelled 6.1–24.0 km/d in the lower reaches during June-September. Rates of downstream migration in the Rogue River were also related to fish size and time of year. Larger chinook travelled downstream faster, and the rate of migration increased with the season. In 1975, a year of low and stable river flow, the rate of downstream migration was negatively correlated with discharge, whereas in 1976, when flows were higher and more variable, the rate of migration was positively correlated with discharge. Cramer and Lichatowich (1978) interpreted the negative correlation in 1975 to reflect a reduction in rearing habitat as discharge dropped and interpreted the positive correlation in 1976 to reflect a direct effect of discharge on the migration rate at higher discharge.

TABLE 4

Fork length (mm) of age 0.0 and 1.0 riverine smolts in various rivers and years

River	Year	Fork length		Source
		Mean	Range	
Age 0.0				
Sixes (OR)	1969	62.0	40-91	Reimers (1971)
Nitinat (BC)	1980	52.7	44-67.5	Healey (unpubl. data)
Cowichan (BC)	1966	77.3	63-98	Lister et al. (1971)
	1967	72.1	60-91	"
	1978	63.8	57-75	Healey (unpubl. data)
Nanaimo (BC)	1979	68.8	52-84	Healey & Jordan (1982)
	1980	63.5	49-76	"
Big Qualicum (BC)	1972	66.5		Paine et al. (1975)
Age 1.0				
Yakima (WA)	1959	125.5	105-170	Major & Mighell (1969)
	1960	124.6	105-170	"
	1961	127.0	105-170	"
	1962	134.0	90-170	"
	1963	132.6	90-160	"
Snake (OR)	1954	101.0	55-147	Mains & Smiths (1964)
	1955	101.0	55-147	"
	1957	68.4	45-105	Bell (1985)
	1958	67.9	50-95	"
Upper Columbia (OR)	1955	84.2	55-140	Mains & Smith (1964)
Taku (BC, AK)	1961	73.3	45-110	Meehan & Siniff (1962)
Crooked Cr. (AK)	1961	93.5	90-140	Waite (1979)

Growth of Fingerlings

Direct estimates of the growth of juvenile chinook in fresh water exist only for the Sacramento River (Kjelson et al. 1982). Inferences about growth in other river systems can be made from seasonal changes in the size of resident chinook or from the size of downstream migrants. These estimates must be viewed with caution, however, as the length of freshwater residence is not known precisely for either the downstream migrant fish or for those captured during river residence.

Tagged chinook fry in the upper Sacramento River grew an average of 0.33 mm/d over a period of 72 days. Fry that had migrated to the freshwater Sacramento-San Joaquin delta, however, grew significantly more quickly, averaging 0.53–0.86 mm/d in two years of observation (Kjelson et al. 1982). Chinook that migrated seaward as fingerling smolts in June of their first year of life averaged 52.7–77.3 mm fork length in four Vancouver Island rivers and one Oregon river (Table 4). Assuming that the length of river residence is indicated by the difference in run timing between the fry and fingerling smolt migrants, these fish spent an average of about 60 days in the river before migrating to sea. Growth rates thus ranged from a low of 0.21 mm/d in the Nitinat River in 1980 to a high of 0.62 mm/d in the Cowichan River in 1966. These rates are comparable with those based on tagged fish in the Sacramento River and delta.

Yearling Smolts

Stream-type chinook do not migrate to sea during their first year of life but delay migration until the spring following their emergence from the gravel and, in northern rivers, sometimes for an additional year as well (Healey 1983). As noted earlier, stream-type chinook characteristically return to their natal river in spring. Apparent exceptions to this pattern occur in some Oregon rivers, such as the Rogue, where spring-run adults produce underyearling smolts (Cramer and Lichatowich 1978; Nicholas and Hankin 1988). Chinook that overwinter in the larger rivers often move out of the tributary streams and into the river main stem, where they occupy deep pools or crevices between boulders and rubble during the winter. In the Nanaimo River, two lakes along the river main stem are also used as overwintering areas. A fall redistribution of fish, presumably from preferred summer habitat to preferred winter habitat, has been observed in some systems (Reimers and Loeffel 1967; Chapman and Bjornn 1969; Bjornn 1971; Carl and Healey 1984; Don Chapman Consultants 1989). Bjornn (1971) found that the number of downstream migrants in an experimental stream trough in the fall was related to the presence of suitable substrata for overwintering in the experimental stream trough.

Yearling smolts normally migrate seaward in the early spring, sometimes preceding the main migrations of fry and fingerlings and sometimes intermixed with them. In the Brownlee-Oxbow section of the Snake River, yearling smolts migrated downstream from April to June with peak numbers in May (Bell 1958). Bell also captured a few downstream migrant fry, mainly in May. In the Yakima River, Washington, yearling smolts migrated mainly in April and May, with the time of peak movement ranging from the third week of April to the second week of May (Major and Mighell 1969). Underyearling smolts were not abundant in the Yakima River until June. In the Taku River, yearling smolts migrated seaward from April to June, with peak movement in early May (Meehan and Siniff 1962). There were no underyearling migrants in the Taku River. The main period of yearling smolt outmigration from the Kasilof River on the Kenai Peninsula, however, was July (Waite 1979).

Bell (1958) related the peak in migration of yearling smolts to spring floods and increasing temperatures. As with the underyearling migration, however, there has been no systematic study of the factors triggering migration. Yearling migrants do appear to be less nocturnal than underyearlings. Meehan and Siniff (1962) could find no significant difference in the abundance of migrants between day and night in the Taku River, although, on average, more smolts moved at night. Major and Mighell (1969) also observed greater movement of yearling smolts during the night, but Bell (1958) found that the greatest movement was during the daylight hours.

Raymond (1968) found that the rate of downstream migration of yearling smolts in the Columbia and Snake rivers was positively correlated with discharge, but that rates of travel through free-flowing and impounded sections of these rivers were similar. At low discharge, the rate of migration was 21 km/d, whereas at moderate discharge it was 37 km/d. These rates of migration are considerably faster than those of underyearling smolts observed by Cramer and Lichatowich (1978). The rapid migration of smolts through impoundments on the Columbia River indicates that yearling smolts undertake a directed migration that is independent of river flows.

Growth of Yearling Smolts

Yearling smolts vary greatly in size. In the Yakima River, they ranged from 100 to 160 mm in fork length, and the average increased from 124.6 mm to 134.0 mm between 1959 and 1962 (Table 4). These sizes suggest an average growth rate of 0.25 mm/d, but the increase in size over time is perplexing. Major and Mighell (1969) could not explain the apparent increase in smolt size, but suggested that it could be due to differential growth of separate tributary spawning populations coupled with the differential contribution of these populations to the smolt run. This explanation is highly speculative.

Rich (1920) presented data for the Columbia River that suggest a growth rate of about 0.20 mm/d for spring chinook during the period March - September. During this period, the fish increased in length from 40.0 to 74.5 mm. Mains and Smith (1964) observed that yearling smolts in the Columbia River averaged about 84.2 mm fork length in 1955, for an annual growth rate of about 0.12 mm/d. In the Snake River, by comparison, Mains and Smith (1964) found yearling smolts to be about 101 mm in 1954 and 1955. Thus, growth in the Snake River was close to 0.17 mm/d. Bell (1958), on the other hand, found yearling smolts in the Snake River to be only about 68 mm fork length (growth = 0.077 mm/d) in May 1957 and 1958. In April 1958, however, Bell (1958) observed a second, larger size mode at 100–104 mm among the smolts captured in the Snake River. Fish in the larger mode were equivalent in size to those captured by Mains and Smith (1964). During May and June, 1958, Bell observed only one size mode, but the position of the mode changed from 70–74 mm in May to 85–89 mm in June. The modes in May and June appear to be a continuation of the smaller April mode, with appropriate growth during each month of about 0.33 mm/d.

In the Taku River the size range for yearling smolts was 50–105 mm with a mean of 73.3 mm (Meehan and Siniff 1962). Thus, in the Taku River the average growth rate was only about 0.09 mm/d. Loftus and Lenon (1977) recorded mid-eye to fork lengths of chinook smolts in the Salcha River (a tributary in the upper Yukon River drainage, Alaska) to be 55–86 mm with a mean of 73 mm. These smolts were slightly larger than those in the Taku River when the difference in length measurements was taken into account. In Crooked Creek, on the Kenai Peninsula, Alaska, yearling smolts averaged 93.5 mm fork length (Waite 1979). As these smolts did not emigrate until late July, however, they had the benefit of spring growth in the year of migration. Assuming their river residence time was about 430 days, the growth of Crooked Creek smolts averaged 0.124 mm/d.

In the Snake River, therefore, the average annual growth for smolts in the smaller mode was comparable to the growth in the Taku and Salcha rivers, whereas the fish in the larger mode had grown at a rate more comparable to that in the Yakima River. Smolts captured in the Columbia River and Crooked Creek were intermediate in size and growth rate.

The existence of two distinct size groups of fish in the same run (e.g., the Snake River) suggests that there may be important differences in microhabitat affecting chinook growth in rivers. If the increase in size of fish in the smaller mode in the Snake River represents growth, then the rate of growth during the spring months was 0.33 mm/d. Rich (1920) observed growth of 0.20 mm/d in the Columbia River. All these rates are slower than the growth rate observed for underyearling smolts.

Major and Mighell (1969) observed that, within the same year, the average size of yearling smolts migrating downstream in the Yakima River decreased with time, and suggested that the larger fish migrated first. Bell (1958), on the other hand, observed that the larger fish of one group of yearling smolts were caught later in the season in the Snake River, and Mains and Smith (1964) observed no systematic change in smolt size with time. These apparently conflicting results may simply reflect the unresolvable interactions of differences in stock growth and microhabitat as well as opportunities for spring growth during and prior to migration.

Summary of Freshwater Residence and Downstream Migration

Although there is some variation in timing, all populations of chinook appear to display similar migratory behaviour. At the time of emergence, there is an extensive downstream dispersal of fry, although some fry apparently are able to take up residence in the natal river at the spawning site. For populations that spawn close to tidewater, this downstream dispersal carries the fry to estuarine nursery areas, whereas in others it serves principally to distribute the fry among suitable freshwater nursery areas. Later in the spring, there appears to be a second dispersal that carries some populations to the sea or simply redistributes the population within the river system, presumably to more suitable summer rearing areas. For those populations that remain a year in fresh water there is a third late fall redistribution to suitable overwintering habitat, usually from the tributaries to the river main stem. Finally, in the spring there is a migration of yearling smolts to sea.

During the late spring and fall redistributions in fresh water, the population tends to shift into deeper water and to move seaward. These changes in habitat are consistent with the shorter term habitat changes observed by Lister and Walker (1966) and Chapman and Bjornn (1969), in which chinook moved into deeper, faster water as they grew in size. The redistributions may punctuate developmental stages as well as achieve more efficient utilization of freshwater nursery habitat. The tendency for redistribution to carry the fish downstream may be coincidental. Such a movement pattern may also be adaptive, however, by shortening the length of spring migration for yearling smolts, particularly for headwater spawning populations in larger rivers.

Mortality and Its Causes during Freshwater Residence

Rates of survival from fry to fingerling migrant stage and from fry to yearling migrant are unknown, with the exception of some recent data from the Sacramento River. Based on the ocean returns of chinook from the same brood year - marked and released as both fry and smolts in the Sacramento River at Red Bluff and in the Sacramento-San Joaquin River delta - survival from fry to smolt ranged from 3% to 34% for the 1980-82 year classes (M. Kjelson, U.S. Fish and Wildlife Service, Stockton, California, pers. comm.). Evidence from these and other releases of tagged fish in the Sacramento River system suggest that fry that rear in the upper river experience a higher survival to smolting than fry that rear in the delta (Kjelson et al. 1982; Brown 1986). Survival of smolts passing through the Sacramento-San Joaquin River delta was highly correlated with discharge of the Sacramento River.

Major and Mighell (1969) estimated that 5.4%-16.4% of potential egg deposition survived to migrate as yearling smolts in the Yakima River. Assuming even a high average egg-to-fry survival rate (30%), fry-to-smolt survival would have to have been about 30% to account for these rates. This seems too high a rate of survival to be generally true in other populations.

Even though estimates of fry and fingerling mortality rates are nonexistent, except for the Sacramento system, mortality is presumed to be heavy in all rivers. Mortality rates of 70%-90% among fry and fingerlings are recorded for other species of Pacific salmon (Foerster and Ricker 1941; Hunter 1959; Parker 1965), and, as these are similar to the losses of chinook observed in the Sacramento River system, it seems reasonable to suppose that chinook in other rivers suffer similar losses. Healey (1980b, 1982b, and unpublished data) could account for only about 30% or less of downstream migrant chinook fry in the estuaries of the Nanaimo and Nitinat rivers, suggesting that mortality was high during this stage of the chinook's life.

Predators are commonly implicated as the principal agent of mortality among fry and fingerlings of chinook and other species, and heavy losses due to predators have been documented in some instances (Foerster and Ricker 1941; Hunter 1959). Patten (1971), however, was only able to infer sculpin predation of 1%–4% among chinook fingerlings released from a hatchery in the Elokomin River, Washington, during 1962 and 1963. Most important predators were the prickly sculpin (*Cottus asper*) and the torrent sculpin (*C. rhotheus*). Less important were the reticulate sculpin (*C. bairdi*) and the coast range sculpin (*C. aleuticus*). However, the release of fingerlings occurred during a single night in 1962 (1.5 million) and during three nights in 1963 (2.5 million), so that the young chinook were available to the predators for only a brief period. Had the releases extended over two to three weeks, as in a natural fingerling migration, losses to predators might have approached 0.5 million (Patten 1971). Also, the large size of the fingerlings relative to the sculpins probably reduced the efficiency of sculpin predation. Although other quantitative estimates of the rate of predation on chinook are lacking, various authors have reported juvenile chinook in the stomachs of predatory fishes (Clemens and Munro 1934; Thompson 1959a, 1959b).

Other fish are generally considered to be the most important predators of juvenile salmon, but invertebrate predators have occasionally been observed to kill or injure juvenile salmon. Eisler and Simon (1961), for example, observed that *Hydra oligactus* caused high mortality among recently hatched chinook alevins in hatchery troughs. Exposure of alevins to 400 hydra in 500 ml of water for only five minutes was sufficient to cause mortality as high as 80%. Dead alevins showed symptoms comparable to white spot disease. Coho alevins exposed to as few as 20 hydra in 500 ml for several days suffered high mortality. Mortality of this magnitude due to hydra may be confined to hatcheries. Novotny and Mahnken (1971) observed a marine isopod, *Rocinella belliceps pugettensis*, attacking fry and fingerlings of chum, coho, and pink salmon in aquaria and in the field at night in Puget Sound. The isopod normally attached to the young fish on the side behind the dorsal fin. Young salmon so attacked would swim in a darting, erratic, and twisting manner, presumably attempting to dislodge the isopod. Even if the isopod attacks were not fatal, the erratic swimming of the fry could attract other predators.

Since the behaviour during comparable life history stages of populations which reside for different lengths of time in fresh water is essentially the same, it seems likely that they are exposed to similar patterns of instream mortality during these stages. The existence of different lengths of stream residence, however, suggests that, at least in the past, there must have been a survival advantage to protracted stream residence in some situations and not in others. The nature of that advantage is not immediately apparent, particularly in situations such as those encountered in the Nanaimo River, where different behaviour patterns coexist in a relatively small river system.

Food Habits in Fresh Water

The principal foods of chinook while rearing in fresh water appear to be larval and adult insects. Kjelson et al. (1982) found Cladocera, Diptera, Copepoda, and Homoptera to be the dominant foods of chinook fry in freshwater regions of the Sacramento-San Joaquin River delta. Chapman and Quistdorff (1938) found dipteran larvae, beetle larvae, stonefly nymphs, and leaf hoppers to be the most abundant diet items of young chinook (43–152 mm standard length) in tributaries of the Columbia River. Clemens (1934) found that young chinook in Shuswap Lake fed primarily on terrestrial insects, small crustaceans (mainly Cladocera), and chironomid larvae, pupae, and adults. Herrmann (1970) found young chinook in the lower Chehalis River feeding principally on crustaceans such as *Corophium*, and on immature and mature insects. The presence of *Corophium* in the diet of these fish suggests that they had been feeding in estuarine waters. Rutter (1902) found young chinook in the Sacramento River feeding mainly on larval and pupal insects. Loftus and Lenon (1977) found Diptera, Plecoptera, and Ephemeroptera to be the most important components of the diet of chinook smolts in the Salcha River, Alaska. In a more detailed study of the diet of chinook in fresh water, Becker (1973) found that insects constituted over 95% of their diet in all seasons. Adult Chironomidae were by far the most important dietary group, comprising 58%–63% of the diet. Following these, in order of importance, were: larval chironomids (17%–18%), Trichoptera adults (3%–5%), Notonectidae (3%–5%), and Collembola (1%–5%). Some seasonal variation was apparent, with Diptera declining in importance from 99% to 70% between March and May, then increasing again to 85% by July. Notonectids were most important in May, Trichoptera in June and July, and Collembola in April and May. In contrast to these results, Craddock et al. (1976) found crustacean zooplankton, especially Cladocera, to be important in the diet of chinook during July-August in the lower Columbia River. Insects predominated at other times of the year.

The importance of insects in the diet of chinook in fresh water indicates that chinook feed in the water column or at the surface on drifting food. Their basic diet is similar to that of coho, steelhead, and other stream-dwelling salmonids (Mundie 1969; Chapman and Bjornn 1969). Whether there is competition for food resources among the cohabiting species is not known; however, any such competition is presumably reduced by the habitat segregation among species described earlier.

Utilization of Estuarine Habitats

Fry Migrants

Many of the fry of ocean-type chinook that migrate downstream immediately after emerging from the spawning beds take up residence in the river estuary and rear there to smolt size. Recently emerged chinook fry are known to rear in the Sacramento and Columbia River estuaries (Rich 1920; Kjelson et al. 1981, 1982), in the Skagit River estuary (Congleton et al. 1981), in the Fraser River estuary (Dunford 1975; Goodman 1975; Levy and Northcote 1981; 1982; Levings 1982; Gordon and Levings 1984), the Nanaimo River estuary, the Campbell River estuary and other estuaries on the east coast of Vancouver Island (Healey 1980b, 1982b; Levings et al. 1986), and the Nitinat and Somass River estuaries on the west coast of Vancouver Island (Birtwell 1978;Healey 1982b).

In some instances, the salinity of the estuarine rearing habitat is low (e.g., Sacramento River: Kjelson et al. 1982; Fraser River: Levy and Northcote 1981, 1982) or is unknown, but observations on the Cowichan, Nanaimo, Courtenay, Campbell, and Nitinat River estuaries demonstrated that chinook fry will rear where salinity is commonly 15–20 ppm or more (Healey 1980b, 1982b; Levings et al. 1986). Thus, estuary rearing may be considered qualitatively different from rearing in the river channel further upstream. Although many chinook fry appear unable to survive immediate transfer to 30 ppm salinity, they are clearly able to survive transfer to 20 ppm or less, and osmoregulatory capability develops quickly in fry exposed to intermediate salinities (Weisbart 1968; Wagner et al. 1969; Clarke and Shelbourn 1985). I have transferred chinook fry directly from downstream migrant traps on the Nanaimo River into sea water of 32 ppm in the laboratory with no apparent short-term ill effects or retardation of growth compared with controls maintained in fresh water and brackish water of 15 ppm (M.C. Healey, unpublished data). Some chinook fry, therefore, appear to be able to tolerate immediate transfer to high salinity.

Rich (1920) reported observing chinook fry in the Columbia River estuary as early as December, and earlier still, in October and November, in the

Sacramento River estuary. During March and April, fry were abundant in the Columbia River estuary. Rich did not measure salinity at the capture sites but did observe that the smallest fry appeared to avoid brackish water and were consistently associated with freshwater inflows to the estuary. Without associated data on downstream migration of fry or estimates of relative abundance, little can be concluded from Rich's (1920) observations other than that healthy fry were present in the estuary. It is worth noting, however, that on the basis of a few samples collected irregularly from the Columbia, Sacramento, and other rivers, Rich (1920) hypothesized much of what we now know to be true about the early life history of chinook salmon.

More recently, Kjelson et al. (1981, 1982) have provided much more detailed observations on the Sacramento-San Joaquin River estuary. Fry arrive at the river delta mainly from January to March and reside there for about two months before migrating seaward. Most rearing occurs in freshwater habitats in the upper delta area, and the fry do not move into brackish water until they smoltify. Levy and Northcote (1981, 1982) described similar behaviour of chinook fry in freshwater marsh areas of the Fraser River delta. Fry migration to the delta was later in the Fraser River, however, it occurred predominantly in April and May.

Observations on a number of Vancouver Island estuaries and on the Fraser River estuary (Healey 1980b, 1982b; Levy and Northcote 1981, 1982; Levings 1982) showed that these estuaries were not only important nursery areas for chinook, but also that the distribution of chinook changed seasonally and tidally. At high tide, the young chinook were scattered along the edges of the marshes at the highest points reached by the tide. As the tide receded, the young chinook retreated into tidal channels and creeks that dissect the marsh areas and retain water at low tide. With the incoming tide, the chinook again dispersed along the edges of the marshes. On the Fraser River, Levy and Northcote (1982) found that chinook were among the last fish to vacate tidal channels in the marsh when the channels dried up at low tide.

The twice-daily pattern of migration from low-tide refuges to the marsh areas and back again was continued throughout the period of residence of fry in the estuaries. As the season progressed, however, the major concentration of young fish moved seaward through the delta area in Vancouver Island estuaries. This is partly due to the fact that larger fish appear to prefer deeper water, and that larger fish are able to osmoregulate in higher salinities. The redistribution of fish may, however, also be associated with increasing temperatures in shallow tidal channels, particularly at low tide. Healey (1980b) found that fry moved away from sampling stations where temperatures exceeded 20°–21°C. This seasonal, seaward movement was not apparent in the Fraser River estuary (Levy and Northcote 1981). Kjelson et al. (1982) reported that, in freshwater rearing areas of the Sacramento-San Joaquin River delta, fry distribution changed from day to night and with fish size. Fry were concentrated near shore in shallow water during the day but tended to move offshore at night. Larger fish also tended to be further offshore than smaller fry. During the day fry were concentrated in the upper 3 m of the water column but became more randomly distributed in the water column at night.

Fry remain in the estuarine nursery areas until they are about 70 mm fork length, after which they disperse to nearby marine areas. In the Fraser River estuary, peak abundance of fry in channels through the marsh was April and May. Juveniles were still abundant in major arms of the river in June (Dunford 1975; Goodman 1975; Levy et al. 1979). Most were gone from the river arms by July but remained abundant over the sand flats at the delta front (Roberts and Sturgeon banks) throughout August (Goodman 1975; Gordon and Levings 1984). In the Nanaimo River estuary and other Vancouver Island estuaries, fry were most abundant in April–June, but the time of peak abundance varied from year to year in accordance with changes in the timing of the downstream run of chinook fry (Healey 1980b, 1982b). Sasaki (1966) observed that young chinook salmon were most abundant in the Sacramento-San Joaquin River delta during April–June, similar to the timing observed in more northern deltas. However, Kjelson et al. (1981, 1982) observed that fry were most abundant in February and March in the Sacramento-San Joaquin River system, and that these were replaced by smolts from upriver in April to June.

The proportion of downstream migrant fry that

find a place to rear in the estuary is not well known. For both the Nanaimo and Nitinat River systems on Vancouver Island, only 30% or less of the estimated downstream migrants could be accounted for in the estuary. The fate of the remaining 70% is unknown, but it was unlikely that they reared elsewhere, despite their apparent ability to survive and grow in habitats with high salinity. Thorough sampling of other potential nursery areas in the vicinity of the Nanaimo River in 1976 and 1977 failed to reveal any significant numbers of chinook fry outside the estuary in April and early May (Healey 1980b, 1982b). Levy and Northcote (1981) estimated that the population of young chinook in the Ladner marsh complex of the Fraser River was approximately 305,000 during mid-May, 1979. Taking account of the fact that the Ladner marsh represented less than one-half the marsh area of the Fraser River delta, that some other important habitats had not been taken into consideration, and that the delta population would probably turn over several times owing to the arrival of new downstream migrants and the emigration seaward of fish which had completed their rearing in the delta, Levy and Northcote (1981) speculated that several millions of chinook probably reared in the delta. The population accounted for in the Fraser River delta is, nevertheless, small relative to the estimated 64 million chinook fry that migrated through the lower Fraser River during March–June 1979. The fate of the many millions of Fraser River chinook fry that do not rear in the delta is unknown. Thus, it appears that there may be high mortality of downstream migrant fry shortly after they complete their downstream migration. The agents of this mortality are unknown.

The residence time of cohorts of fry in estuarine habitats has been approximated by mark and recapture studies in the Nanaimo, Nitinat, Fraser, Skagit, and Sacramento-San Joaquin River estuaries (Healey 1980b; 1982b; Congleton et al. 1981; Levy and Northcote 1982; Kjelson et al. 1982). On the Nanaimo River estuary, recovery of marked fry suggested a maximum residence time of about 60 days. The average length of residence of fry, based on recaptures of marked fry and on rate of growth and maximum sizes of chinook in the inner estuary, was about 20–25 days (Healey 1980b). Residence times in Nitinat Lake were similar to those in the Nanaimo River estuary, the average in 1979 being about 21 days and in 1980 about 17 days. Levy and Northcote (1982) and Congleton et al. (1981) investigated residence in tidal channels through marshes and found much shorter residence times, about 8 days in the Fraser River marshes and about 3 days in a single channel of the Skagit River marsh. The relatively short residence times observed by Levy and Northcote (1982) and by Congleton et al. (1981) may reflect the limited area sampled relative to the range of habitat available to fry. In both instances, movement between tidal channels in the marsh (observed by Levy and Northcote 1982) and movement of cohorts of fry seaward with time (and thus out of the marsh but not out of the estuary) (Healey 1980b) may have contributed to shorter residence time estimates in these studies. The maximum residence time of chinook fry in the Sacramento-San Joaquin River delta was 64 days in 1980 and 52 days in 1981 (Kjelson et al. 1982). All the residence times so far calculated include the combined effects of migration and mortality.

Levy and Northcote (1981) investigated the relationship between occurrence and abundance of chinook fry in various marsh habitats according to the physical characteristics of the habitat. Chinook abundance was significantly correlated with twelve of twenty-two habitat characteristics. In a multiple regression analysis, however, only two characteristics (area of low tide refugia and elevation of tidal channel banks) explained significant amounts of variation in chinook catch. A large number of correlations among the habitat characteristics may have confounded the analysis, but the results suggest that young chinook prefer tidal channels with low banks and many subtidal refugia. Chinook, and associated fish species, also tended to be associated with larger tidal channels.

Fingerling and Yearling Migrants

The apparent movement of fry migrants away from the Nanaimo and Nitinat River estuaries in late May and June coincided with the downstream migration of fingerling smolts, so that the fingerling smolts took over the habitat vacated by the fry. In estuaries on the Oregon coast there appear to be few fry migrants, and the first chinook to enter these estuaries are fingerling smolts. (Reimers 1971; Reimers et al. 1979; Myers 1980; Myers and

Horton 1982). In the Sacramento-San Joaquin River estuary, as in the Vancouver Island estuaries, fingerling smolts replaced the fry in estuarine nursery areas (Kjelson et al. 1982). The fingerling smolts tend to occupy deeper water in the estuary and to remain there for varying periods. In the Nanaimo and Nitinat River estuaries, fingerling smolts were abundant during June and July but began to decline in abundance about mid-July and were rare after August, although a few occurred year-round in the outer estuary of the Nanaimo and other rivers (Healey 1982b) (Figure 14). In Oregon coastal estuaries, fingerling smolts appear to reside much longer–well into October (Reimers 1971; Myers 1980; Myers and Horton 1982; Nicholas and Hankin 1988) (Figure 14). The period of greatest abundance of fingerling smolts tends, however, to be June to August in Oregon estuaries. In the Sacramento-San Joaquin River estuary, fingerling smolts were most abundant from April to

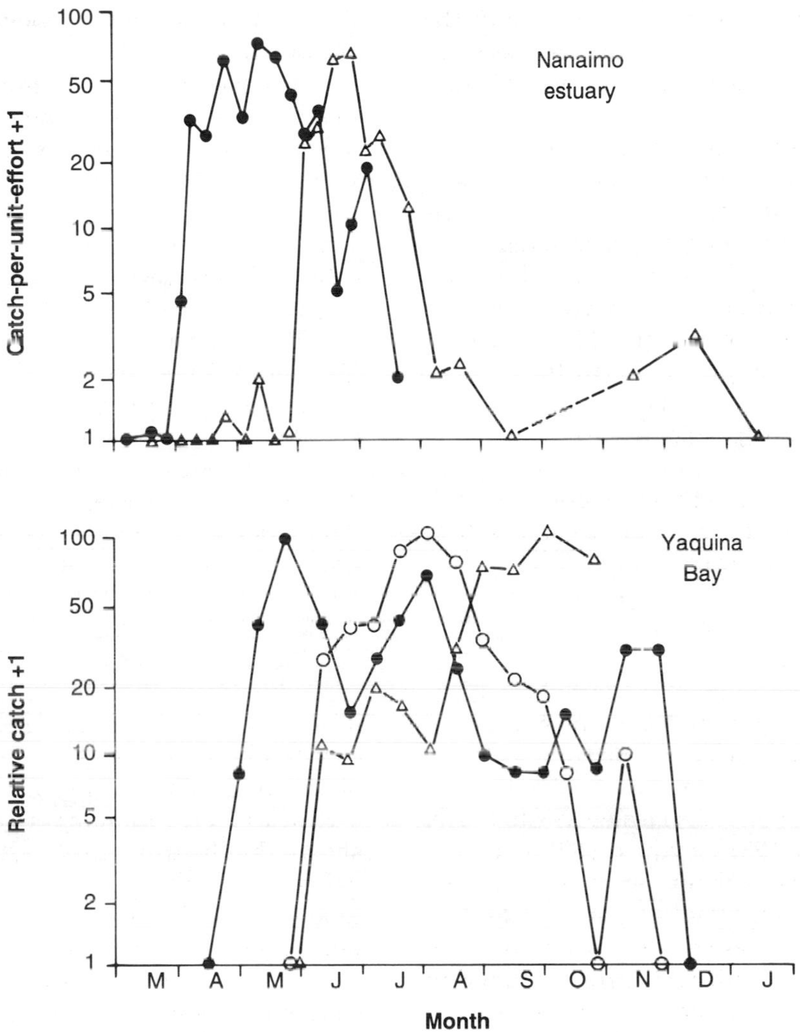

FIGURE 14

Abundance of juvenile chinook at various locations in the Nanaimo and Yaquina estuaries during March to January. Symbols are: • = inner estuary beaches; o = outer estuary beaches; Δ = outer estuary deep water. Note the different scales on the ordinates. (Adapted from Healey 1982b and Myers 1980)

mid-June but were scarce during summer months, apparently because of high water temperature in the delta and bays (Kjelson et al. 1982). There was a small secondary peak in smolt abundance in the fall, representing fish that had remained in cooler water upstream over the summer (Kjelson et al. 1982). It is possible that estuaries along the open coast from Washington to California provide important sheltered habitat for young fall chinook during the summer and autumn, provided temperatures in these esturies do not get too high. Sheltered habitat is much more common along the British Columbia coast, and there may be less stimulus for young chinook to remain in British Columbia river estuaries.

Chinook that migrate to sea as yearling smolts often do so together with the emergent fry and they, too, spend some time in the estuary of their natal stream. While the fry are concentrated in the delta area, yearling smolts occupy the delta front, so that there is no spatial conflict between the two life history types in the estuary. Not all downstream migrant yearlings remain in the estuary and some disperse to other nearshore areas adjacent to the river mouth. Also, it appears that the length of residence of yearling smolts in the estuary is relatively brief, as is their residence in sheltered coastal waters in general (Healey 1980b, 1982b, 1983; Levy and Northcote 1981).

Food Habits in Estuaries

The food habits of chinook in estuaries are documented for a number of British Columbia estuaries (Dunford 1975; Goodman 1975; Birtwell 1978; Sibert and Kask 1978; Fedorenko et al. 1979; Levy et al. 1979; Northcote et al. 1979; Healey 1980b, 1982b; Levy and Northcote 1981; Levings 1982), a number of Oregon estuaries (Reimers et al. 1978; Myers 1980; Bottom 1984), and the Sacramento-San Joaquin River estuary (Sasaki 1966; Kjelson et al. 1982). Diets vary considerably from estuary to estuary and from place to place within an estuary. Dunford (1975), Northcote et al. (1979), Levy et al. (1979), and Levy and Northcote (1981) reported that chironomid larvae and pupae were the most important diet items of ocean-type chinook in tidal channels throughout the Fraser River marshes. Of secondary importance were *Daphnia*, *Eogammarus*, *Corophium*, and *Neomysis*. Diets of chinook from the north arm of the Fraser River tended to be more restricted, with greater emphasis on *Neomysis*. Chinook captured in the main river channels or over Roberts and Sturgeon banks at the delta front were larger than those found in tidal channels; and juvenile herring, sticklebacks, and other small fish, as well as cumaceans, insects, and *Neomysis* were important in their diet (Dunford 1975; Goodman 1975; Northcote et al. 1979; Levy and Northcote 1981; Levings 1982). Yearling smolts (stream-type) were larger still and fed heavily on chum fry, as well as on *Eogammarus*, *Neomysis*, *Corophium*, and chironomids. Northcote et al. (1979) noted that chinook fry less than 50 mm long in the Fraser River marshes demonstrated an intermediate to long path food web dominated by benthic detritivores, but with significant input from other pathways such as herbiverous zooplankton and terrestrial insects. Larger chinook characteristically had long path food webs with multiple dietary compartments, including benthic detritivores, zooplankton, and fish.

Chinook fry (ocean-type) in the Nitinat River estuary fed mainly on adult insects, gammarids, crab larvae, and Cladocera, whereas those in the intertidal zone of the Nanaimo River estuary fed mainly on crab larvae, mysids, adult insects, and harpacticoid copepods. As was observed in the Fraser River estuary, chinook that had moved into deeper water in the Nanaimo River estuary began to feed heavily on fish (Healey 1980b). Insects and amphipods were the preferred diet items of ocean-type chinook in the Somass River estuary, except in some heavily industrialized areas where a paucity of other fauna resulted in chinook feeding on oligochaetes (Birtwell 1978).

Research on a number of Oregon estuaries has shown that benthic amphipods, particularly *Corophium* spp., and aquatic insects are the dominant food of juvenile fall chinook (Reimers et al. 1978; Bottom 1984). By contrast, Myers (1980) found that fishes, especially Engraulidae, Osmeridae, and Clupeidae, dominated the diet of juvenile wild chinook in the Yaquina estuary, Oregon.

Insects and Crustacea dominated the diet of young chinook in the Sacramento-San Joaquin River delta (Sasaki 1966; Kjelson et al. 1982). Sasaki (1966) found that chironomid larvae were important in the upstream areas of the delta, whereas *Neomysis* and *Corophium* were important in the lower delta. By contrast, Kjelson et al. (1982) found Cladocera, Copepoda, and Diptera were the most important foods in the upper and lower estuary and that amphipods and mysids constituted only a small percentage of the diet.

Seasonal changes in diet are typical, and presumably reflect seasonal changes in the abundance of prey organisms. In the Nitinat River estuary, chinook diet was dominated by insects and *Eogammarus* during April. Towards the end of April, however, larval herring became important and remained so until the end of June. About mid-May crab larvae began to comprise an important part of the diet and continued to be important until the end of June. Cladocera were not important until late May and were a significant diet item until late June (Figure 15).

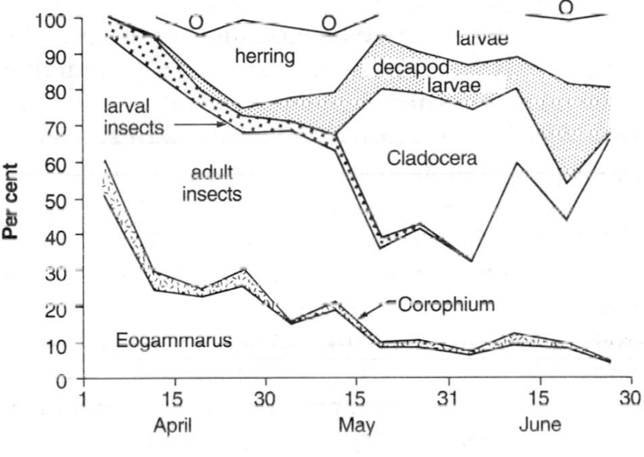

FIGURE 15

Seasonal changes in the diet of juvenile chinook in Nitinat Lake. At top of figure, *O* refers to other diet items. (From Healey 1982b)

Most studies of chinook diet are based on samples taken during daylight hours. Studies of the Sixes River estuary, however, indicate that juvenile chinook feed actively at night (Bottom 1984). Their principal foods in this estuary were the amphipods *Corophium* and *Eogammarus*, which are inactive during the day but migrate into the water column at night.

Sibert and Kask (1978) compared the diets of chinook among the Fraser, Cowichan, Nanaimo, and Campbell River estuaries, and found little correlation among the diets of fish from comparable physiographic regions of different estuaries or among the diets of fish from different physiographic regions of the same estuary. The same, or closely related organisms did tend to show up consistently in the diets, however, although their numerical contribution to diet varied widely between estuaries.

Chinook generally cohabit with other salmonids in estuaries, in particular with chum salmon. Although they often eat the same organisms, the correlation between their diets was weak in the Fraser and Nanaimo River estuaries (Dunford 1975; Sibert and Kask 1978). The diet of chinook also correlated poorly with the diet of cohabiting coho in the Nanaimo River estuary (Sibert and Kask 1978). The diet of chinook was, in fact, more similar to the diet of some non-salmonids in these estuaries (e.g., herring, *Clupea pallasi*; stickleback, *Gasterosteus aculeatus*; shiner perch, *Cymatogaster aggregata*; and sand lance, *Ammodytes hexapterus*) (Sibert and Kask 1978). Myers (1980), on the other hand, found that there was often considerable overlap in the diets of wild chinook and hatchery coho in Yaquina Bay, Oregon.

In general, chinook appear to be opportunistic feeders in estuaries. Comparison of their diet with that of other similar-sized salmonids in the same area suggests that chinook prefer slightly larger organisms and that larval and adult insects, as well as amphipods of various sorts, are their preferred prey in the intertidal regions of most estuaries. Dunford (1975) found that chinook were more efficient predators of chironomid larvae than chum and were able to capture and eat *Neomysis* that chum could not capture. In a mixed assemblage of Cladocera, chironomid larvae, and *Neomysis*, chinook fed preferentially on chironomids and Cladocera. As the chinook become larger and begin to inhabit deeper water, their dietary preference appears to shift to larval and juvenile fishes.

Growth in Estuaries

In the Columbia River estuary, young fall chinook (ocean-type) increased in length from 38 mm in April to 113 mm in October, an average daily increase of 0.44 mm, assuming that these fish were from the same cohort (Rich 1920). Chinook in the Sacramento River estuary increased in length by 0.48 mm/d between March and July, again assuming the same cohort was being sampled (Rich 1920). The probability that these samples were from the same cohort is rather small, however, as young chinook probably do not spend such a long time in the estuary, and their departure from estuaries appears to be size-related. Also, the "alternation" of different juvenile life history types in the estuary makes it unlikely that these growth estimates relate to a single behavioural type. The fish captured later are likely to have reared in the river for some weeks or months. These estimates of growth rate in estuaries must, therefore, be considered minimum estimates. Kjelson et al. (1982) estimated growth rates of fry tagged with coded-wire tags in the Sacramento-San Joaquin River estuary to be 0.86 mm/d in 1980 and 0.53 mm/d in 1981. These rates of growth are faster than those based on unmarked fry.

During 1978 and 1979, young-of-the-year chinook in the Fraser River estuary increased from 40 to 60–68 mm fork length between March and July (Levy and Northcote 1981). Their increase in length was slow until mid-May, rapid from mid-May until the end of June, and then slow again during July. From mid-May until the end of June they increased an average of 0.56 mm/d in 1978 and 0.39 mm/d in 1979.

The slow increase in length of chinook fry in the Fraser River estuary from March to mid-May was almost certainly due to the continued addition of recently emerged fry to the estuary population, since the main period of downstream migration of fry was from mid-March to mid-May in both years (Levy and Northcote 1981). The apparent slow growth of chinook in July may have been due to movement of larger chinook away from the beaches and into deeper water. Levy and Northcote (1982) observed that young chinook captured by purse seine in deep water in an old river channel on the Fraser River foreshore were significantly larger than those captured by beach seine along the margins of the channel. As noted earlier, Kjelson et al. (1982) observed a similar distribution of chinook in relation to size in the Sacramento-San Joaquin River estuary.

Reimers (1971) observed the growth rate of young fall chinook in the Sixes River estuary both by sampling the general population and by observing changes in the length of marked cohorts. Both sources of data gave similar results. Growth in the estuary from late April to early June was rapid. During this period, chinook increased in average length from 48 to 79 mm, an average of 0.9 mm/d. From June to August, growth in the estuary was poor, and the fish increased only 6 mm or 0.07 mm/d during this period. From September to November, growth was again rapid, averaging about 0.5 mm/d. Reimers (1971) hypothesized that poor growth during the June-to-August period was due to the large population of chinook in the estuary at this time, and that the increased rate of growth in September to November resulted from both a reduction in population size and better utilization of the whole estuary. By means of otolith microstructure, Neilson et al. (1985) confirmed that individual fish show a rapid growth until June in the Sixes River estuary, after which their growth slowed down. Neilson et al. (1985) suggested that the decline in growth rate after June resulted from a combination of high temperatures in the estuary that reduced growth efficiency and competition for food. Neilson and Geen (1986) also noted that chinook that entered the estuary at a large size remained large relative to members of the same cohort throughout their first year of ocean life.

The rate of growth of marked fry in the Nanaimo River estuary averaged 1.32 mm/d (4%–5% body weight/d) (Figure 16). From average length data for the general population, however, the rate of increase in length during April to June was only about 0.5 mm/d (Healey 1980b), or less than half the rate indicated by marked fish. Other estuaries on the east coast of Vancouver Island showed rates of growth based on average length data from 0.22mm/d in the Cowichan River estuary to 0.61 mm/d in the Courtenay River estuary (M.C. Healey, unpublished data) and 0.46–0.55 mm/d in the Campbell River estuary (Levings et al. 1986). Presumably, these are underestimates of the true growth rate in these estuaries. Average length in the general population of chinook in the Nitinat River estuary increased about 0.33 mm/d during the years 1975–77 (Fedorenko et al. 1979). In 1979, however, fish from a marked cohort increased about 0.62 mm/d (3% body weight/d) (Figure 16). Measurements based on increases in average length of the general population in estuaries, therefore, appear to underestimate the true growth rate by a factor of about two. The exception to this is Reimers' (1971) data for the Sixes River. Variation in growth rate among estuaries is also on the order of two times, as is evident from data of the Na-

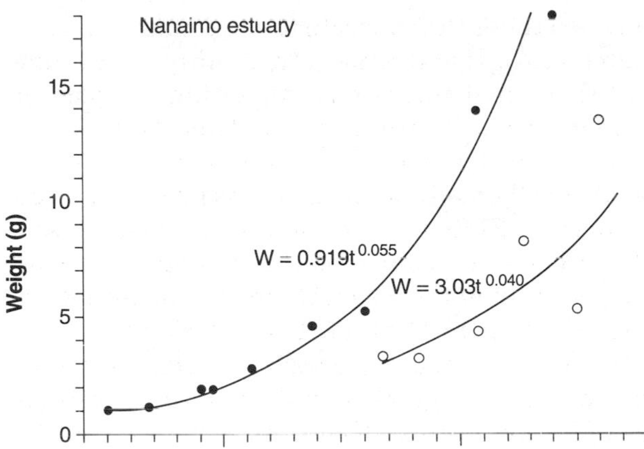

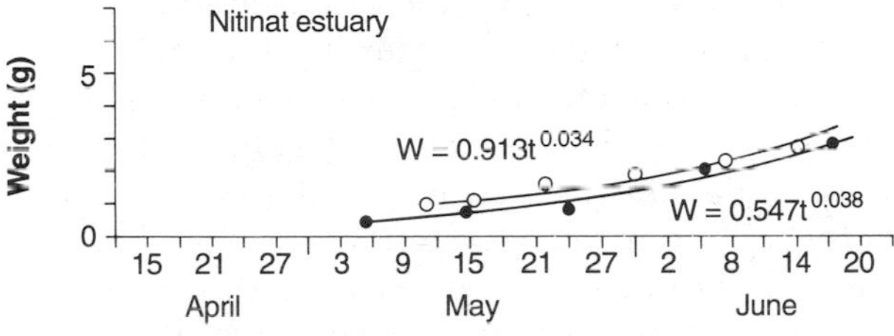

FIGURE 16

Growth of marked cohorts of chinook fry in the Nanaimo and Nitinat River estuaries. Closed and open circles refer to different tagged groups of fry. Regressions are the regression of log weight (W) on days (t) following marking.

naimo and Nitinat River estuaries. The available evidence suggests that variation in growth rates between estuaries and between years within an estuary is correlated with food supply (Healey 1982b; Neilson et al. 1985).

Considering the limited information available on rates of growth of young chinook, particularly their growth in fresh water, no substantial comparisons between freshwater and estuarine growth can be made. Kjelson et al. (1982), however, demonstrated by means of fry tagged both in the river and the estuary that fry grew more rapidly in the estuary. In many other instances, growth of chinook that rear in the river during their first spring appears slower than growth of those that migrate to the estuary. Reimers (1971) and Rich (1920) inferred this from the closer spacing of circuli on the scales of fish that had reared in the river, and suggested that circulus spacing could be used to distinguish between fish that reared to smolt size in the river and those that reared in the estuary. Reimers (1971) proposed that up to five, and Schluchter and Lichatowich (1977) proposed that up to seven, different juvenile life history patterns involving different periods of river and estuarine residence could be distinguished from patterns of scale growth in the Sixes River and the Rogue River, Oregon. These patterns of scale growth were presumed to reflect differences in growth rate between fry residing in the river and in the estuary. In other instances, there appears to be no substantial difference in growth between fish that rear to fingerling smolt size in the river and fish that rear in the estuary (Vancouver Island estuaries, M.C. Healey 1980b, 1982b; M.C. Healey, unpublished data). Fish from these systems that reared in the estuary and fish that reared in the river did not differ in the spacing of circuli on their scales.

Apparently egg size may influence rate of growth, at least in hatchery fry with abundant food. Fowler (1972) found that fry from large eggs were larger at hatching and maintained that advantage over 10–12 weeks after hatching. Rombough (1985) found that larger eggs produced alevins with greater maximum weight, but also that it took the alevins from larger eggs longer to reach their maximum weight after fertilization, compared with alevins from smaller eggs. Such variations in growth rate could have important implications for mortality if the larger fry outgrow potential predators more quickly. In Fowler's (1972) experiments, however, the more rapidly growing fry from larger eggs also had a higher mortality rate from unspecified causes. Thus, the advantage of faster growth may be offset by other unknown disadvantages associated with larger eggs.

Healey (1982b) estimated the production and food requirements of chinook and other salmon species in the Nanaimo and Nitinat River estuaries. Chinook were second to chum in production in the inner Nanaimo River estuary, producing about 200 kg during the spring and early summer, compared with 1,750 kg for chum. In the Nitinat River estuary, however, chinook dominated juvenile salmon production, contributing 774 kg of an estimated total production of 1,137 kg. These levels of production are small relative to the area of the two estuaries, being about 0.031 g/m^2 in the Nanaimo River estuary and 0.025 g/m^2 in the Nitinat River estuary. Food resources required to support this production were 0.093 g/m^2 in the Nanaimo River estuary, consisting of mainly benthic foods but including about 30% insects; and 0.078 g/m^2 in the Nitinat River estuary, of which about 40% were benthic organisms, 40% insects, and 20% plankton (Healey 1982b). The standing crops of food organisms in these estuaries appeared sufficient to support considerably greater production. Growth of chinook in the Nitinat River estuary was, however, considerably less than growth in the Nanaimo River estuary, and this was correlated with lower standing crops of food organisms in the Nitinat River estuary. Also, there was a positive correlation between abundance and growth rate in the Nanaimo River estuary and the fullness of chinook stomachs, suggesting that growth and production were positively related to food supply in this estuary (Healey 1982b).

Differences among stream- and ocean-type races are evident in their utilization of estuarine habitats. Ocean-type fish make extensive use of estuarine habitat, whereas stream-type fish spend little time in the estuary of their natal stream. Among the ocean-type races there is a further dichotomy between those that migrate to the estuary as fry in March or April and remain there until about June and those that migrate as fingerlings in May or June and remain until August or later. This dichotomy becomes blurred in Oregon, where a variety of juvenile behaviour patterns has been identified (Reimers 1971; Schluchter and Lichatowich 1977; Nicholas and Hankin 1988) but is apparent again further south in the Sacramento-San Joaquin River estuary (Kjelson et al. 1981, 1982). The period of estuarine occupancy by ocean-type chinook varies regionally, being greatest in the open coast estuaries of Washington to California, and least in the sheltered coastal estuaries of British Columbia. Estuaries apparently provide a rich feeding habitat for the smaller fry and fingerling migrants, and growth in estuarine habitats, although variable, is relatively rapid. Why stream-type chinook do not spend more time in estuarine habitats is not immediately apparent. The tendency for chinook to become piscivorous, however, as soon as their size permits, and the particular oceanic migratory pattern of stream-type chinook (Healey 1983), may preclude a longer estuarine residence.

Ocean Life

The distribution, seasonal abundance, and migratory behaviour of first ocean-year chinook salmon are described by Healey (1976, 1980a, 1980b) for the Strait of Georgia (British Columbia), by Miller et al. (1983) and Fisher et al. (1983, 1984) for the waters off the coasts of Washington and Oregon, and by Hartt (1980), Healey (1983), and Hartt and Dell (1986) for coastal and offshore waters from the Columbia River to the Bering Sea. The sampling device in all these studies was a small-mesh purse seine, so that the results pertain only to the surface waters (about 20 m deep).

Information on the distribution and migratory habits of older chinook derive from a wide variety of sources. Sampling of commercial troll catches from southeastern Alaska to California has provided information on the general coastal distribution of chinook and their distribution by age and race (Fry and Hughes 1951; Parker and Kirkness 1956; Milne 1964; Ball and Godfrey 1968a, 1968b, 1969, 1970; Wright et al. 1972; Argue and Marshall 1976; Nicholas and Hankin 1988). Sampling the catch of the Japanese mothership and landbased gillnet fisheries has provided information on the distribution of chinook in the western North Pacific Ocean. (Geographic distribution of these fisheries is shown in Figures 19 and 20.)

Research cruises by Canadian, U.S., and Japanese vessels have provided more extensive information on chinook salmon distribution throughout the North Pacific Ocean. Gear employed in research cruises included floating longlines and gillnets (Major et al. 1978). As with the seine sampling for first ocean-year chinook, samples taken on the high seas by gillnet and longline relate mainly to the surface waters. Chinook are probably underrepresented in these samples because of their tendency to be distributed deeper in the water column than the other species of Pacific salmon (Milne 1955; Taylor 1969; Argue 1970). Despite potential sampling biases and difficulties in comparing among gear types, the observations of all authors are consistent in that they indicate a major difference in the distribution and migratory behaviour of stream- and ocean-type chinook during their ocean life.

More specific information on the distribution of particular stocks or groups of stocks of chinook is derived from tagging and recapture. Immature and maturing salmon have been tagged in both coastal and high-seas fisheries, and tagged fish have been recaptured from a few weeks to several years after tagging (Kondo et al. 1965; Hartt 1966; Godfrey 1968b; Aro et al. 1971; Aro 1972, 1974). Comparatively few fish tagged in the ocean have been recaptured after a sufficient length of time, however, to permit a precise determination of migratory routes and timing. Considerably more detailed information is available from tagging of hatchery-produced smolts and noting their recovery in coastal troll and net fisheries (e.g., Cleaver 1969; Wahle and Vreeland 1977; Dahlberg 1982; Wertheimer and Dahlberg 1983, 1984; Dahlberg and Fowler 1985; Dahlberg et al. 1986; Nicholas and Hankin 1988). A wealth of this kind of information has been produced over the past decade as a result of the extensive application of coded-wire tagging in hatchery evaluation (Jefferts et al. 1963). Most of these data, however, remain completely unanalysed. Furthermore, the question of whether the behaviour of hatchery fish is comparable with that of wild stocks remains unresolved, although Healey and Groot (1987) found that wild and hatchery chinook from the east and west coasts of Vancouver Island had similar oceanic distributions.

Fish in their First Ocean Year

Observations in the Strait of Georgia indicated that, in these sheltered waters, young chinook began to disperse seaward from their natal estuary shortly after completing their downstream migration. Apparently, the first to do so were the stream-type smolts. In the Nanaimo River estuary, stream-type smolts were rarely captured in the inner estuary but were common in the outer estuary and in other nearshore sampling stations in the vicinity of Nanaimo in June and July of 1975 and 1976, but were rare after July (Healey 1980b) (Figure 17).

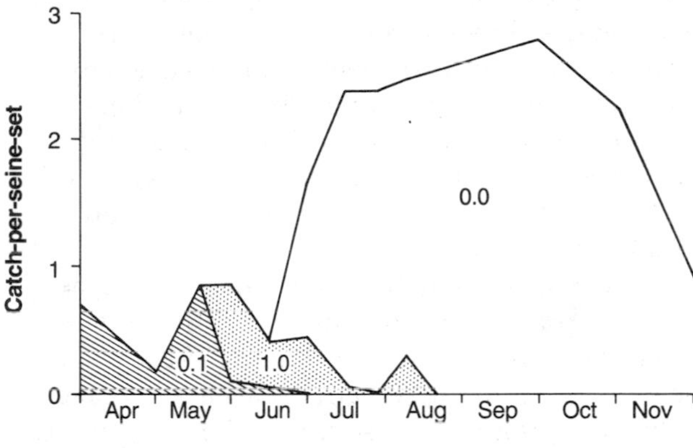

FIGURE 17

Seasonal occurrence of 0.0, 1.0, and 0.1-aged chinook in purse seine catches near Nanaimo, BC, during 1975-77. (From Healey 1980a)

Sampling from May to October, 1976, throughout the Gulf Islands (along the southeast coast of Vancouver Island) confirmed this seasonal pattern of abundance of stream-type smolts outside the Nanaimo area (Healey 1980a). As stream-type juveniles declined in abundance in the sheltered waters of the Strait of Georgia, ocean-type juveniles increased in abundance. In the Nanaimo area, ocean-type juveniles were about five times as abundant in purse seine samples as stream-type juveniles, and ocean-type fish remained abundant from July until November (Figure 17). Seine net catches were lower during the winter, but some chinook were present in surface waters and available to the seine throughout the year (Healey 1980b). In the Gulf Islands region of the Strait of Georgia, young chinook were rather constant in abundance from May to October, 1976. Samples taken throughout the Strait of Georgia in August to September, 1975 and 1976, revealed that chinook were most abundant in the region of the Fraser River plume and less abundant elsewhere. This distribution was much more pronounced in 1975 than 1976 (Healey 1980a).

The general conclusion from these observations was that stream-type chinook in their first ocean year are common in the surface waters of the Strait of Georgia only in the spring and early summer, but that ocean-type chinook in their first ocean year are abundant throughout the summer and autumn, and some remain in the surface waters of the strait throughout their first ocean year. There was no indication of an outmigration of young ocean-type chinook from the Strait of Georgia before November in 1976. Seine samples taken by the Snoquomish Indian band in Puget Sound in October and November, 1976, indicated an abundance of chinook in their first ocean year that was similar to that in the Strait of Georgia (Hartt and Dell 1986).

The estuaries along the open coasts of Washington, Oregon, and California afford the only sheltered water habitat in these regions. Here young ocean-type chinook remain in estuaries considerably longer than was observed in British Columbia estuaries (Reimers 1971; Myers 1980; Myers and Horton 1982) (Figure 14), although stream- and ocean-type chinook also occur in coastal waters outside the estuaries during the summer (Miller et al. 1983; Fisher et al. 1983, 1984). These observations suggest that the affinity of young ocean-type chinook for sheltered waters is general throughout the range of chinook, but that there is also some offshore movement of these fish.

Off the coasts of Washington and Oregon, Miller et al. (1983) found young chinook in abundance during sampling from 27 May to 7 June, 1980. Catches were greatest off the Columbia River mouth, north of this river, and in seine sets made within 20 km of shore. Catches of chinook in their first ocean year declined in July, presumably owing to high surface water temperatures at this time, and catches increased again in August–September when surface temperatures were lower. In the August–September sampling, young chinook were distributed about equally north and south of the Columbia River mouth but were still most abundant close to shore. Significantly, young spring chinook (stream-type) from Columbia River hatcheries were present only in the May–June samples, and then only in the northernmost sampling transects, indicating a rapid migration of these fish north from the river mouth. In 1982 and 1983 Fisher et al. (1983, 1984) sampled an area similar to that sampled by Miller et al. (1983) in 1980 and captured similar numbers of young chinook. During 1982, chinook were most common in samples taken in May and June but were rare in September. During 1983, catches were highest in May and September but were low in June. There was no clear trend in abundance from south to north in these samples; sometimes young chinook were more abundant south of the Columbia River and sometimes north. Nor was there any clear evidence from the samples of Fisher et al. (1983, 1984) that young chinook were more abundant close to the coast.

Miller et al. (1983) compared catches in seine sets held open towards the north or south. In May–June, 80% of chinook were captured in sets held open towards the south, indicating a significant northward dispersal at this time. During July and August–September sampling, however, the direction of the set had no effect on catch. These data indicated a northward dispersal of ocean-type chinook immediately on entry into the sea, but also indicated that movements were not directed throughout most of the summer. This is reminiscent of the summer residency of young ocean-type chinook in the Strait of Georgia.

Hartt (1980) and Hartt and Dell (1986) reported on an extensive series of samples taken by purse seine along the coast of North America from the Columbia River to Bristol Bay, throughout the Gulf of Alaska, along the Aleutian Islands chain, and into the central Bering Sea. Hartt and his co-workers made a total of 3,073 sets during the 15-year period, 1956–70, mostly during the summer months, but extending from April to October. The sampling was concentrated in coastal waters (Figure 18), with offshore seining principally in the spring and early summer. Most young chinook were caught near the coast, and catches overall were greatest from the Columbia River to southeastern Alaska. Smaller catches were made in central Alaska, along the Aleutian chain to Adak Island, and in Bristol Bay. No chinook in their first ocean year were captured in the central Gulf of Alaska or in the central Bering Sea. Greatest catches were in the June–August period, and there was an indication, although catches in all areas were small, that chinook appeared in the catches later in the north than in the south.

An important feature of Hartt's catches was that, by far, the majority of chinook were stream-type (245/253 or 97%)(Healey 1983). Of the eight ocean-type fish captured, six were captured off

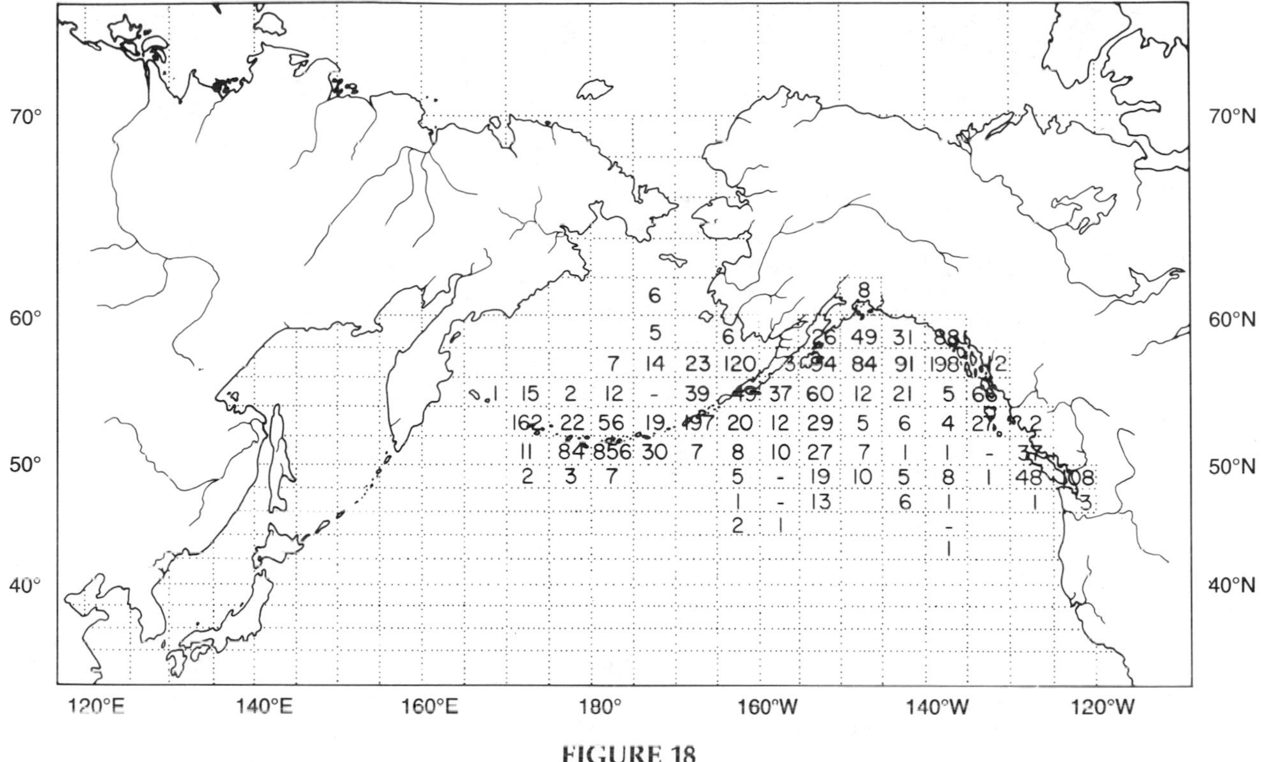

FIGURE 18

Numbers of small-mesh purse seine sets made by Hartt and his co-workers (Hartt and Dell 1986) between 1956 and 1970 in 2° × 5° geographic areas. These areas are the standard statistical recording areas of the International North Pacific Fisheries Commission (INPFC).

Cape Flattery and two off southeastern Alaska. Thus, it appeared that juvenile chinook in their first ocean year, living along the outer coast, at least from Cape Flattery and northward, were predominantly stream-type fish, whereas those in sheltered inside waters were predominantly ocean-type. The samples collected by Miller et al. (1983) and Fisher et al. (1983, 1984) suggest that ocean-type chinook were more common along the open coast south of Cape Flattery.

Fish in their Second and Subsequent Ocean Years

With the exception of organized coastal fisheries and the Japanese mothership and landbased fisheries (Figure 19 and 20), comparatively few chinook have been captured in the open ocean, but the distribution of captures has been very widespread (Figure 19). Chinook almost certainly occur further offshore south of 46°N on the North American coast than is currently indicated, probably even south of 40°N, considering the continued high production of chinook from Californian rivers. Sampling, however, has not been adequate to demonstrate this.

Manzer et al. (1965) reported on the freshwater ages of 847 chinook captured in the Japanese mothership fishery between 175°E and 175°W. These fish were all stream-type. The majority were probably of Asian and Alaskan origin, although recent returns of coded-wire tagged chinook showed that fish from British Columbia, Washington, and Oregon migrate as far west as 160°W–175°W longitude (Dahlberg 1982; Wertheimer and Dahlberg 1983, 1984; Dahlberg and Fowler 1985; Dahlberg et al. 1986). It is possible, therefore, that some of the fish in the sample reported by Manzer et al. (1965) originated from populations south of the Alaskan Panhandle. If so, none was an ocean-type fish. More recent sampling of the Japanese high-seas fisheries (Knudsen et al. 1983; Myers et al. 1984) revealed a similarly low contribution of

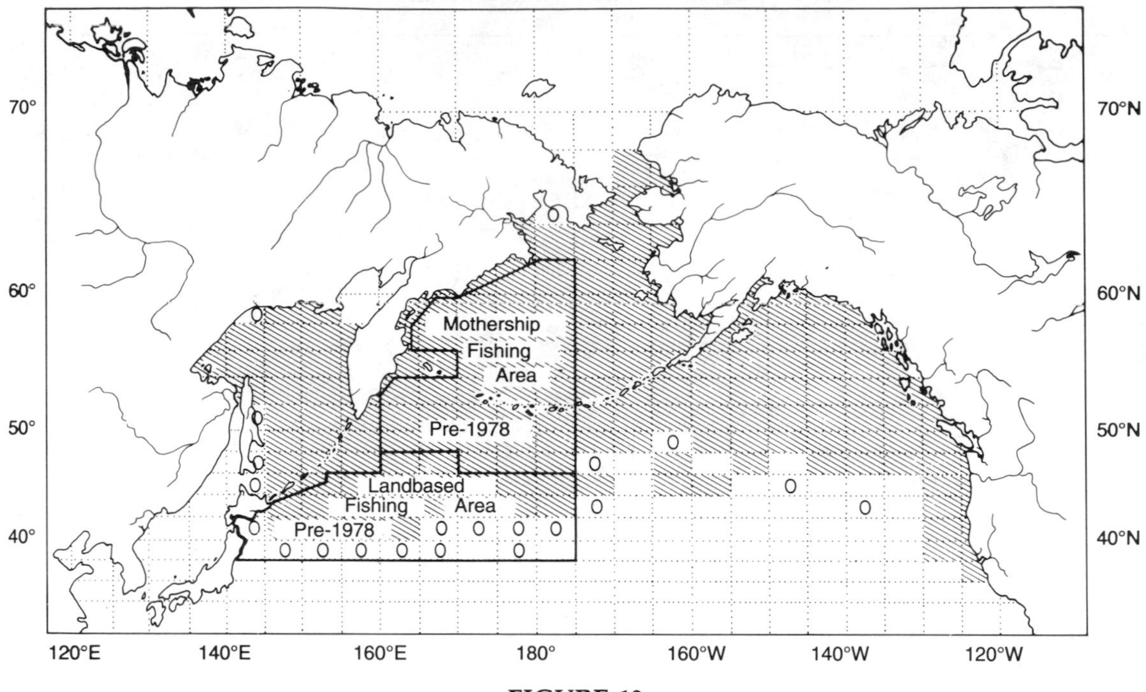

FIGURE 19

Known distribution of chinook in the North Pacific Ocean and Bering Sea based on captures in high-seas and coastal sampling (shaded 2° × 5° INPFC statistical areas). Areas in which sampling occurred but no chinook were captured are shown (0). The Japanese mothership and landbased fishing areas prior to 1978 are shown.

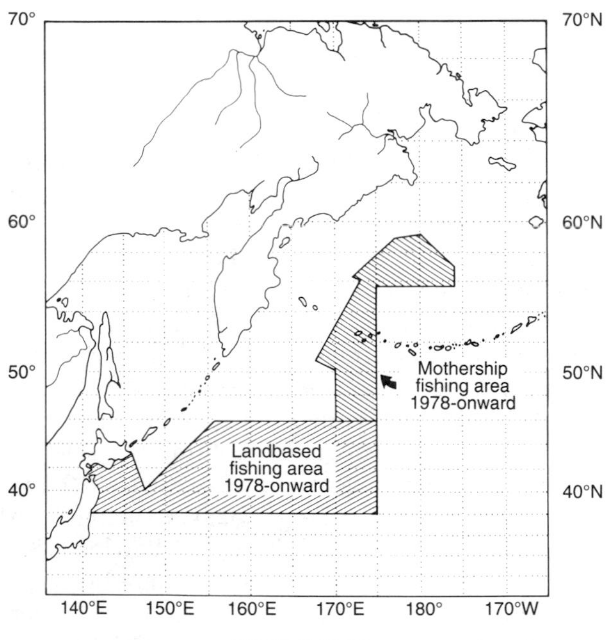

FIGURE 20

Japanese mothership and landbased fishing areas following renegotiation of agreements in 1978, which were in effect until 1986

ocean-type fish, although there are no absolute criteria for distinguishing age 0. and age 1. fish, and there was some disagreement between Japanese and U.S. ageing of the fish reported by Knudsen et al. (1983). The U.S. ageing team found some ocean-type fish (68/2779) and the Japanese team, none.

Further east, ocean-type fish do make a contribution to the high-seas population. Of 80 chinook captured by Canadian research vessels fishing with longlines north of 45°N and east of 170°W during 1961-67, 52 (65%) were stream-type and 26 (35%) were ocean-type (Healey 1983). These fish were almost certainly of North American origin, and were probably mainly from southeastern Alaska and rivers to the south. Since stream-type chinook appear to comprise no more than 25% of all spawning populations from the Sacramento River to southeastern Alaska, the percentage of stream-type fish in the Canadian research vessel catch on the high seas was significantly greater than expected, if stream- and ocean-type chinook have similar ocean distributions ($X^2 = 68.3$; $p <$ .001). The high percentage of stream-type fish in this sample was not a consequence of the fish being captured mainly in the northern Gulf of Alaska, as the catch was equally distributed north and south of 49°N. Furthermore, of the 28 ocean-type fish captured, 14 were captured north of 49°N, and 14 south of this latitude. Thus, there was no evidence of a higher proportion of ocean-type fish in the southern half of the sampling area, as would be expected if ocean-type fish were well distributed offshore, but they were concentrated in more southerly areas, in keeping with the distribution of their spawning populations (Healey 1983). It appears, therefore, that stream-type fish constitute a high proportion of the high-seas population regardless of latitude, although the proportion is lower in the eastern than in the central and western North Pacific Ocean.

Chinook are abundant in coastal waters along the coast of North America from southeastern Alaska to California throughout their ocean lives. In these coastal waters, the representation of stream- and ocean-type races is the opposite of that observed on the high seas, and ocean-type fish predominate (Healey 1983).

The proportion of stream-type chinook in the commercial troll fisheries of Alaska, British Columbia, and Washington in recent years has ranged from 3% to 4% in the Strait of Georgia to 25% in the Queen Charlotte Islands area of British Columbia (Table 5). The proportion off Washington and off southeastern Alaska was similar at 15%–16%. In recent years, therefore, stream-type fish have made up a relatively small proportion of the ocean troll catch, generally less than 20%, and significantly less than one would expect from the proportion of stream-type fish in the regional spawning populations.

There is evidence that stream-type chinook may, historically, have constituted a greater percentage of the coastal troll catch than they do at present. In the Strait of Georgia, stream-type fish were 28% of the catch during 1911–20 and declined to 3%–4% by 1961–70 (Table 5). Off the west coast of Vancouver Island, stream-type fish were 20% of the catch during 1921–30, declined to 3.9% during 1961–70, and increased to 9.0% during 1981–85 (Table 5). Off Washington and Oregon, stream-type chinook constituted more than 20% of the catch before 1950 but only 15% of the catch after 1960 (Table 5). These changes must be interpreted cautiously, however, as scale interpretation and fishing patterns may have changed over the years. For example, stream-type fish were a high percentage of the recent net catch in major river estuaries of British Columbia (Table 5) and, as noted earlier, maturing stream-type chinook return to the estuary of their spawning river early in the year. If historic troll fisheries were nearer river mouths, or were concentrated earlier in the year, then their catch of stream-type fish might naturally have been higher. On the other hand, habitat modification, particularly damming of the Columbia and

TABLE 5
Stream-type chinook (as a percentage) in coastal troll catch and river mouth gillnet catch during several decades

Fishery	Data source*	1911–20	1921–30	1941–50	1951–60	1961–70	1971–80	1981–85
S.E. Alaska troll	1				15.7			
North BC troll	2		23.0		5.6	20.6		25.0
Central BC troll	2					9.3		12.0
Van. Is. troll	2		20.0	12.4	10.9	3.9		9.0
Georgia St. troll	2	28.0	17.5		6.5	3.4		
Wash./Oreg. troll	3		22.0	28.4	34.4	15.3	12.8	
Fraser R. net	4					42.8	28.1	
Skeena R. net	4					48.1		
Nass R. net	4					46.2		

Notes: *1 Parker & Kirkness (1956)
2 Milne (1964); Ball & Godfrey (1968a, 1968b, 1969, 1970); Healey (1986)
3 Wright et al. (1972); Rich (1925); Van Hyning (1951)
4 Godfrey (1968a); Ball & Godfrey (1968a, 1968b); Fraser et al. (1982); Ginetz (1976)

Sacramento rivers, but also water extraction for irrigation and hydraulic mining (Kjelson et al. 1981, 1982), have probably had a much greater effect on stream-type chinook than ocean-type, owing to their longer upriver migrations and longer river residence, both as adults and as young. It is, therefore, quite conceivable that stream-type chinook were relatively more abundant a few decades ago. The possibility that the changes in percentage of stream-type fish in the troll catches represent a real reduction in the numbers of stream-type fish is further supported by the observation that spring and summer spawning runs (mainly stream-type fish) have declined drastically in several rivers in the southern half of the chinook's range (Snyder 1931; Rich 1942) and by the fact that the older maturing stream-type chinook are less able to support an intensive fishery than younger maturing ocean-type chinook (Hankin and Healey 1986).

The information on distribution of races indicates that stream-type chinook move offshore early in their ocean life, whereas ocean-type chinook remain in sheltered coastal waters. Stream-type fish maintain a more offshore distribution throughout their ocean life than do ocean-type. Although ocean-type fish are captured offshore in the eastern half of the North Pacific Ocean, they are much less common there than stream-type fish, whereas the reverse is true close to the coast. Since only stream-type chinook occur in western Alaska and in Asia, it is not surprising that catches in the western North Pacific Ocean are virtually all stream-type. Stream-type chinook are also common in river mouth fisheries of British Columbia (Table 5) and in early season catches in the Fraser and Columbia rivers (Figure 6). These observations suggest that maturing stream-type fish move rather quickly through the coastal troll fisheries to the river estuaries and so are available for only a relatively short time to the troll fisheries. Healey and Groot (1987) noted that maturing chinook returning to their native river travelled more than 45 km/d, or close to their optimal cruising speed, on a direct course towards the river. They appear to remain in the river estuaries for some time, however, and while there are vulnerable to the river mouth gillnet fisheries.

Ocean Distribution in Relation to Temperature

High-seas catches of chinook by research vessels generally have been too small to permit analysis of chinook distribution in relation to ocean temperature except for catches by Japanese research vessels in the western North Pacific Ocean. There, during 1962–70, chinook were encountered at temperatures ranging from 1° to 15°C (Major et al. 1978). There was no evidence for a preferred temperature within this range, except that, in April and May, relatively fewer chinook were encountered at temperatures below about 5°C. These data refer, of course, to Asian and Alaskan stream-type chinook. This may mean that chinook are indifferent to temperature over this range. The data may not be a true reflection of temperature selection (or lack thereof) by chinook, however, as the samples of fish and temperature measurements came only from surface waters. As will be demonstrated later, chinook are not concentrated at the surface. Surface catches may, therefore, not be representative of temperature selection occurring at depth.

Vertical Distribution of Chinook

Information on the depth distribution of chinook comes primarily from two sources, neither of which distinguished between stream- and ocean-type fish. Sampling was also conducted only during daylight hours. Taylor (1969) reported the depth of capture of immature chinook off the east and west coasts of Vancouver Island during survey cruises for Pacific herring. Most samples came from the west coast of the island, and these will be emphasized here. Chinook were captured in trawls fishing deeper than 60 fm (110 m), but most fish were captured above 40 fm (73 m) (Table 6). Chinook were not concentrated near the surface but were most abundant in the 30–40 fm (57–73 m) stratum. Chinook captured in trawls in the Strait of Georgia were more abundant near the surface, with most (51/53) captured between 10 and 20 fm (20–37 m) (Table 6). Argue (1970) reported a more extensive set of data based on troll sampling in Juan de Fuca Strait. Argue's (1970) samples only extended to 55 m, and most chinook (54%) were captured in the 48-55 m stratum (Table 6). Possibly, catches would have been as large at even greater depth. Both young chinook (younger than .2) and maturing chinook (older than .2) were captured more frequently at shallower depths than were older immature chinook (Table 6). There appeared

TABLE 6
Depth distribution of chinook (as a percentage of catch) captured off the west (W) and east (E) coasts of Vancouver Island during herring surveys (Taylor 1969), and in Juan de Fuca Strait during a troll fishery investigation (Argue 1970)

Depth		Taylor 1969		Argue 1970*				
		All ages		Immature				Maturing
		N = 194	*N* = 53	All ages	0.0+1.0	0.1	0.2	
fm	(m)	W	E	*N* = 150	*N* = 50	*N* = 76	*N* = 71	*N* = 19
0-5	(0-9)	23	2	3	2	10	0	11
6-10	(11-18)			3	7	8	6	17
11-15	(20-27)	7	96	7	14	21	21	44
16-20	(29-37)			17	21	29	10	18
21-25	(38-46)	7	2	16	32	16	25	0
26-30	(48-55)			54	24	14	38	10
31-35	(57-64)	46						
36-40	(66-73)							
41-45	(72-82)	6						
46-50	(84-91)							
>50	(>91)	10						

Note: *Argue's data are segregated by age and maturity

to be a seasonal change in depth distribution, with the average depth of capture dropping from 33 m in June and July to 41 m in August-October (Figure 21). Taylor (1969) did not indicate the freshwater ages of the fish captured. The fish in Argue's samples, however, were virtually all ocean-type.

Ocean Distribution in Relation to Area of Origin

There are two sources of information on oceanic distribution of chinook in relation to area of origin: tagging studies and analysis of scale patterns. Tagging studies include both the tagging of immature and maturing chinook in the ocean with subsequent recapture in the ocean or on the spawning beds, and the tagging of hatchery smolts at the time of seaward migration with subsequent recapture in the ocean. Scale pattern analysis attempts to determine the area of origin of chinook captured on the high seas from characteristic features of their scales. Neither source of information gives a

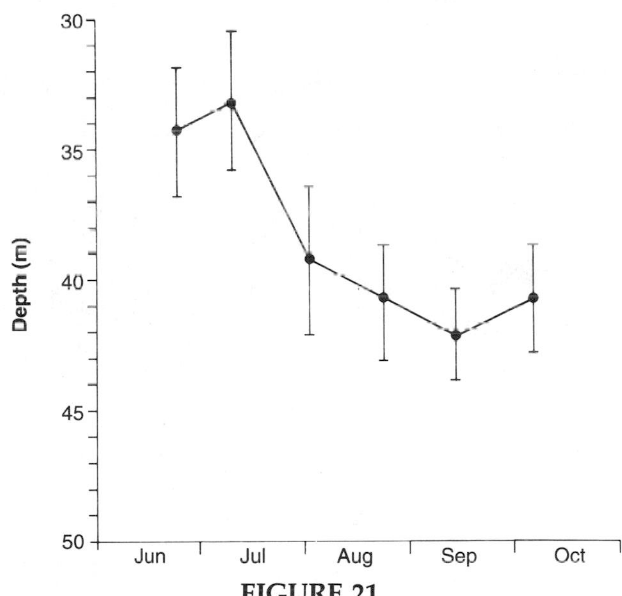

FIGURE 21
Seasonal changes in the average depth of capture of chinook in Juan de Fuca Strait. Bars show two standard errors around mean. (Data from Argue 1970)

very satisfactory picture of the ocean distribution of the many stocks of chinook salmon. The majority of ocean tagging of chinook has been conducted along the Pacific coast of North America from southeastern Alaska to California. Godfrey (1968b) summarized data from ocean tagging in this area during the period 1924–64. In addition, a few tags have been recovered from chinook tagged during high-seas fishing operations in the northern North Pacific Ocean and the Bering Sea (Major et al. 1978; Knudsen et al. 1983). The ocean distribution of chinook tagged as smolts and recaptured in coastal fisheries has been described by Cleaver (1969), Wahle and Vreeland (1977), Wahle et al. (1981), and Healey and Groot (1987).

A total of 21,566 chinook salmon were tagged in coastal waters during the 1924–64 period, and recaptures numbered 2,418 (Godfrey 1968b). Recoveries of tags came from recreational and commercial fisheries and from spawning ground surveys. Because of unequal sampling and recovery effort, the tag returns provide only a general indication of the stock composition in various regions of the coast. Some important generalizations are, however, possible. Most recaptures were made either in the area of tagging or to the south, except for the tagging off California, where it was only possible to recover tags in the area of tagging or to the north (Figure 22). When tagging was conducted in inside sheltered waters, most recaptures came from the area of tagging or immediately adjacent areas. When tagging was conducted in more open waters, however, such as off the west coast of southeastern Alaska or off the west coast of Vancouver Island, recaptures were more widely distributed geographically (Figure 22). Fish tagged off southeastern Alaska were recaptured as far south as Oregon coastal streams, whereas fish tagged off California were recaptured as far north as the Washington coast. These data suggest a northward dispersal of juveniles along the coast, followed by a southward homing migration of maturing adults. The more limited dispersal of fish tagged in inside waters is, however, perplexing. It may indicate the presence of "resident" stocks that do not undergo long-distance migrations. It may also be a reflection of a greater number of stream-type chinook in the groups tagged in outside waters, coupled with the apparent longer migrations of this race. Recent analysis of tag returns from hatchery and wild

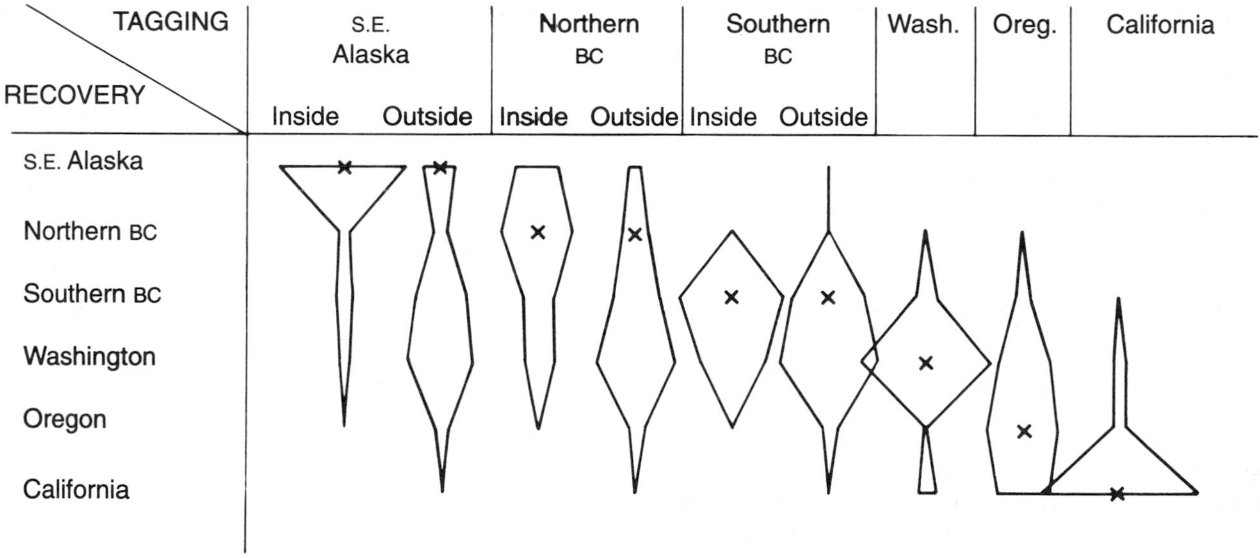

FIGURE 22

The distribution of recaptures from chinook tagged as immatures in coastal waters from southeastern Alaska to California. Tagging location in each diagram is marked (x). Proportion of recaptures in each region is indicated by the width of the diagram at that point. Inside and outside tagging areas for some regions refer to inside or outside coastal island chains.

chinook stocks on the east and west sides of Vancouver Island (Healey and Groot 1987) indicate that "inside" ocean-type stocks have a more restricted migration than "outside" ocean-type stocks.

Coastal Oregon stocks of chinook show diverse ocean migration patterns that do not correspond to the general northward migrating behaviour on other parts of the North American coast. Stocks that spawn in rivers on the central and northern parts of the Oregon coast (from the Elk River north) show the typical northward migration as immatures, and these stocks contribute to fisheries from Oregon to Alaska. Stocks that spawn in rivers on the southern part of the Oregon coast (from the Rogue River south), by contrast, disperse mainly south and contribute to fisheries off Oregon and northern California. One stock, the Umpqua River spring-run chinook, disperse both north and south from their natal river and are harvested from northern California to Alaska (Nicholas and Hankin 1988).

The distribution of particular stocks by age is revealed by returns of fish tagged from hatchery releases. Releases from Columbia River hatcheries dispersed mainly north of the Columbia River, although some were also captured south of the river mouth (Figure 23) (Cleaver 1969). Fish aged two years were caught mainly off the Washington coast; fish aged three years, off Washington and southern British Columbia; fish aged four years, off Washington and southern British Columbia but with a few returns from northern British Columbia

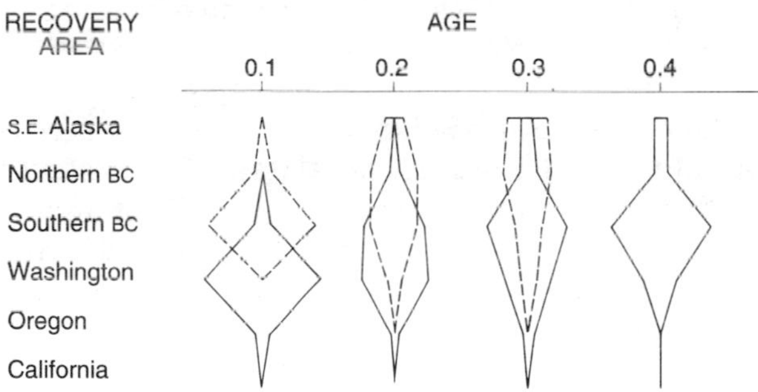

FIGURE 23

Distribution of recaptures by age for ocean-type chinook from Columbia River hatcheries (solid lines) and Robertson Creek, BC (dashed lines). Proportion of recaptures in each region is indicated by the width of the diagram.

and southeastern Alaska; and fish aged five years, mainly from southern British Columbia, but extending into northern British Columbia and southeastern Alaska (Figure 23). Chinook from the Robertson Creek Hatchery on the west coast of Vancouver Island also dispersed mainly north as they became older, but there was somewhat greater southward dispersal than was apparent for the Columbia River fish. Robertson Creek fish were clearly distributed further north than Columbia River fish (Figure 23). Robertson Creek fish, however, also dispersed further north than Big Qualicum River fish from eastern Vancouver Island (Healey and Groot 1987). It appears that there may, in some instances, be considerable differences in the ocean distribution of stocks from the same geographic area.

Wahle and Vreeland (1977) and Wahle et al. (1981) described the ocean distribution of fall (ocean-type) and spring (stream-type) chinook from various Columbia River hatcheries by means of recaptures in coastal fisheries. Fall chinook were recaptured mainly in British Columbia and Washington fisheries and, secondarily, in Columbia River estuary fisheries. Except for the Kalama Hatchery (Washington) release, there were no recoveries of fall chinook from southeastern Alaska and relatively few recoveries from south of the Columbia River (Figure 24). There were some interesting, and perhaps significant, differences in the contribution of different hatchery stocks to the different fisheries. For example, the proportion recovered in British Columbia fisheries ranged from 12% to 50%, and the proportion recovered in the Columbia River estuary ranged from 2% to 26% (Figure 24). Spring chinook, in general, had a wider distribution than fall chinook. For example, up to 20% were recaptured in southeastern Alaska, and all stocks contributed to the Alaskan catch. More spring chinook tended to be caught south of the Columbia River and in the river estuary as well. As with the fall chinook stocks, there were interesting differences in the apparent ocean distribution among spring chinook hatchery stocks.

To a large extent, the distribution of tag recoveries of various stocks has been determined by the location of intensive fisheries for chinook. This is clearly revealed by the recent recovery of tags from chinook captured incidentally in foreign trawl fisheries within the United States 200-mile conserva-

OCEAN-TYPE CHINOCK

HATCHERY
RECOVERY AREA
Spring Creek
Kalama
Elokomin
Oxbow
Grays River
Cascade
Klickitat
Big Creek
Bonneville
Little White Salmon
Alaska
British Columbia
Washington
Columbia River
Oregon
California

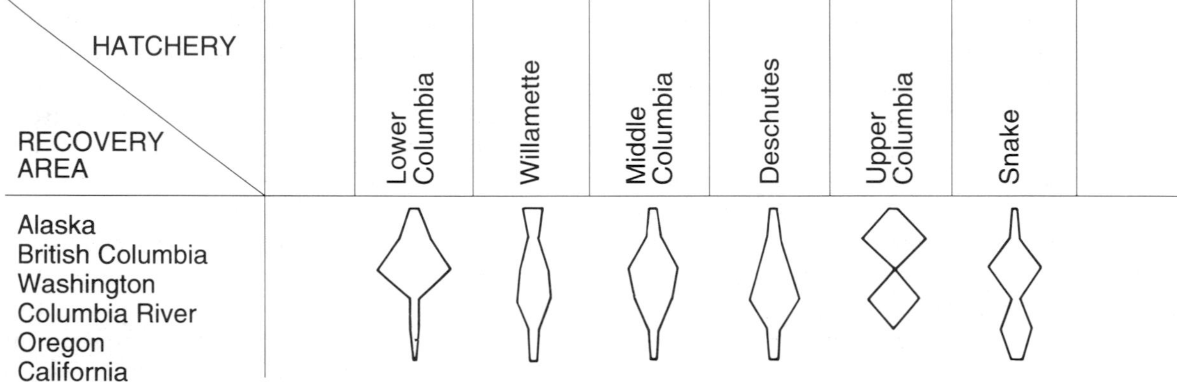

FIGURE 24

Distribution of recaptures of chinook from ten Columbia River hatcheries producing ocean-type chinook and six hatcheries producing stream-type chinook. Proportion of recaptures in each region is indicated by the width of the diagram.

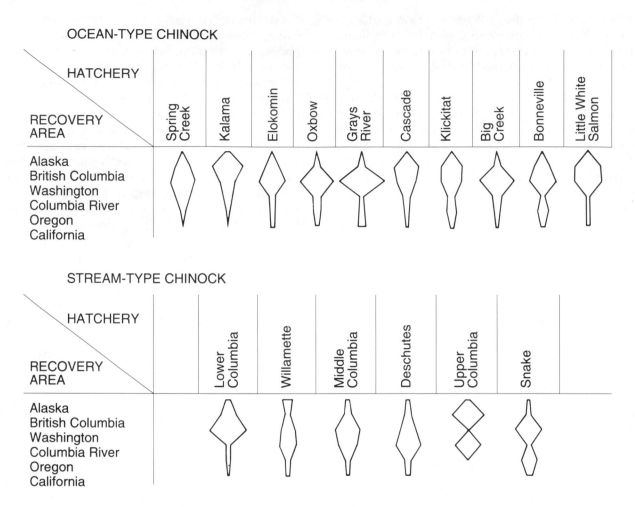

tion zone in Alaska (Dahlberg 1982; Wertheimer and Dahlberg 1983, 1984; Dahlberg and Fowler 1985; Dahlberg et al. 1986). Approximately 60,000 chinook have been examined by United States observers aboard foreign trawlers fishing mainly south of the Alaska Peninsula and in the southeastern Bering Sea, and these have yielded a total of 244 tagged chinook that originated from rivers ranging from central Alaska to California. The distribution of recaptures extends from northern southeastern Alaska westward through central Alaska, along the Alaska Peninsula and Aleutian Islands to about 175°W, and northward into the southeastern Bering Sea (Figure 25).

Of the recaptured chinook, 70 were from stocks in southeastern and central Alaska, 75 from stocks in British Columbia, 43 from stocks in Washington, 54 from stocks in Oregon, and one each from stocks in California and Idaho. The recaptures of Alaskan stream-type chinook in these coastal trawl fisheries provide evidence that at least some fish from northern stream-type stocks remain close to the coast for one to two years following their seaward migration, similar to the behaviour of ocean-type chinook from further south. The recaptures of chinook from stocks in British Columbia and further south represent definite range extensions for these stocks. Most of the recaptures were of ocean-

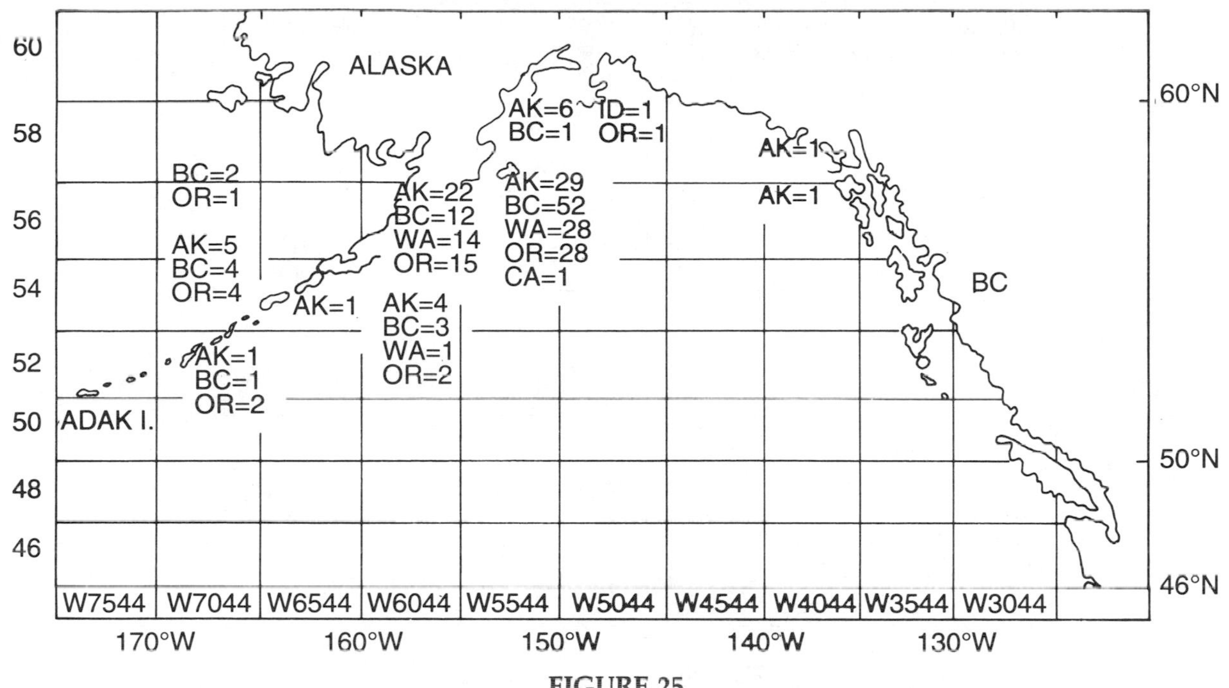

FIGURE 25

Numbers of tagged chinook from different coastal origins recovered as by-catch in the foreign trawl fleet fishing within the U.S. 200-mile zone off Alaska. Catches are shown within INPFC statistical areas.

type chinook, and the previous maximum extent of their range along the coast had been determined from recaptures in the troll fisheries of southeastern Alaska. These new recaptures extend the range of these stocks westward along the coast to nearly 175°W and, for British Columbian and Oregon chinook, northward into the Bering Sea (Figure 25).

Although the recaptures from incidental catch of chinook in foreign trawl fisheries indicate significant range extensions for numerous stocks of chinook, the incidence of tagged chinook among those landed is very low (0.4% overall, 0.3% if only ocean-type stocks are considered), indicating that only a small proportion of these stocks travels so far west and north. Furthermore, these recaptures, together with a single return to the Columbia River from chinook tagged south of Adak Island, appear to define the probable western limit of stocks from south of the Alaskan Panhandle at about 180°, because examination of almost 53,000 chinook captured in the Japanese mothership fishery operating within the United States 200-mile conservation zone west of 175°E produced no tagged chinook. Nor did examination of over 20,000 chinook captured by Japanese research vessels fishing in the northern North Pacific Ocean and Bering Sea produce any tagged chinook (Dahlberg 1982; Wertheimer and Dahlberg 1983, 1984; Dahlberg and Fowler 1985; Dahlberg et al. 1986).

The information presented to this point refers only to coastal distribution of particular stocks of chinook salmon. Information on the open ocean distribution is much more sketchy and speculative, as most of it is not based on tag returns from fish of known origin.

Approximately 2,099 chinook older than age 1. have been tagged and released on the high seas west of 150°W and in the Bering Sea between 1956 and 1984 (C. Harrison, Fisheries Research Institute, University of Washington, Seattle, Washington, pers. comm.). Thirty-three of these have been recaptured (Figure 26), and these recaptures provide some information on the ocean migrations of stream-type chinook. Recaptures up to 1972 were discussed by Major et al. (1978).

Twenty-one fish tagged in the central Bering Sea were recaptured within the Bering Sea and its coastal areas (Figure 26). Five had moved towards the west (northwest to southwest) from the

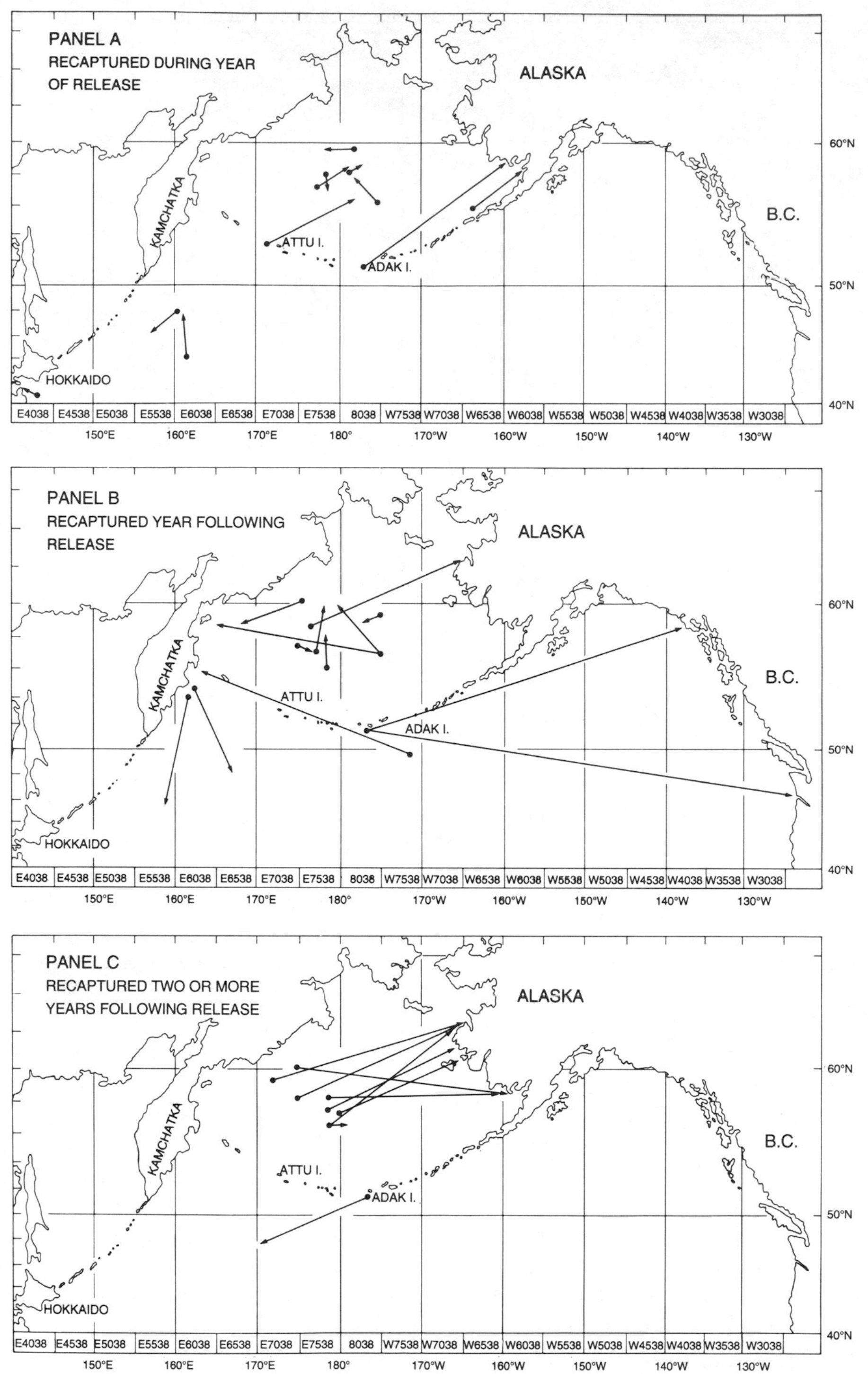

FIGURE 26

The location of tagging (•) and recapture (→) of chinook tagged on the high seas in the central Bering Sea, in the N.W. Pacific, and south of Adak Island in the Aleutian chain. *Panel A,* chinook recaptured in the year of tagging. *Panel B,* chinook recaptured one year after tagging. *Panel C,* chinook recaptured two or more years after tagging

point of release, and three a moderate distance north or northeast. Three of these were immature at the time of recapture and the other four were of undetermined maturity. Eight had moved east to river mouths in western Alaska, and these were all mature at the time of recapture, most being recaptured two years after tagging. The remaining five fish were recaptured only a short distance from their point of tagging. Scant as these results are, they suggest a north and westward movement of immatures and an eastward movement of matures, at least of those bound for western Alaskan rivers.

Of five chinook tagged and recaptured in the western North Pacific Ocean, three were recaptured south or southwest, one north, and one west of their tagging sites (Figure 26). Their maturity at the time of recapture was not determined. Four fish tagged south of Adak Island in the Aleutian chain were recaptured, three as maturing fish and one of undetermined maturity (Figure 26). Of the maturing fish, one was recaptured in Bristol Bay, one in southeastern Alaska, and one at the Columbia River mouth. The fourth fish was recaptured southwest of Adak Island. One other fish tagged south of the Aleutians was recaptured off the east coast of Kamchatka one year later. These returns reveal only that chinook from the northern North Pacific Ocean are from diverse origins and that fish from the Columbia River may migrate as far west as Adak Island.

The costliness, and the limitations, of ocean tagging have led investigators, particularly those in the United States, to explore other techniques for identifying the origin of chinook captured on the high seas. One of these techniques is the estimation of stock composition by discriminant analysis of scale measurements.

The earliest attempts to classify high-seas catches of chinook by this technique were described by Major et al. (1978). In these early studies only western Alaskan and Asian stock groupings were recognized. The analysis was expanded by Knudsen et al. (1983) and Myers et al. (1984) to include classification of central Alaskan, southeastern Alaskan/British Columbian, and Washington/Oregon/Californian stock groupings. The Washington/Oregon/Californian stock grouping was eliminated from further consideration by these authors when they found that it made an insignificant contribution to high-seas catches. Most recently, Ito et al. (1985, 1986) and Myers (1986) have investigated how altering the proportion of the different Asian stocks used as known standards in developing the discriminant functions affects the classification of unknown samples from the Japanese mothership and landbased fisheries. Myers et al. (1987) summarized findings from recent U.S. analyses.

The results of these analyses are controversial, particularly when they are used to estimate numbers of fish of different origins landed by a particular fishery. Before presenting some of the results of these analyses, the most important criticisms of the methodology will be summarized. There can be no doubt that the quantitative results of the discriminant function analysis are sensitive to the composition of the standards used to develop the discriminant functions, and to assumptions about which stocks or stock groupings contribute to the catch in any high-seas fishery. Because the analysis is sensitive to these factors, it is my view that it is inappropriate at this time to use the discriminant function analysis to estimate stock composition quantitatively. It is also my view, however, that the analyses are sufficiently consistent, particularly with respect to mixing proportions of some stocks, to provide a reason able basis for speculation about qualitative features of the high-seas distribution of chinook.

There are numerous technical difficulties involved in applying discriminant function analysis of scale features to the analysis of a sample of scales from fish of unknown origin. A set of scales appropriate for such an analysis must be compiled from known populations. Features of the scales that are unambiguous in terms of their measurement, and that differ between populations but are reasonably consistent within populations, must be discovered if classification by discriminant function analysis is to be successful. The scales of unknown origin that are to be classified must be comparable with those used to develop the original discriminant functions and must comprise an unknown mixture of only those stocks for which discriminant functions have been calculated. Failure to meet these and other technical requirements in the data can result in errors in classification of unknown seriousness. Myers et al. (1984) and Myers (1986) report rather careful screening of the data to minimize the possibility of such errors, such as ensuring that the scales used in the analysis were from the preferred area of the body, and that the discriminant functions were brood year specific. It is virtually impossible, however, to eliminate all such sources of error.

An important potential source of error in the classification of an unknown mixture of scales is the presence of scales from stocks not included in the standards used to develop the discriminant functions. Such scales will be classified into one or more of the stocks specified in the analysis and, if abundant, could greatly bias the estimate of stock composition.

The success in classifying the scales of known origin from which the discriminant functions were calculated provides one indicator of the magnitude of error that might occur in classifying a sample of unknown scales. It should be remembered, however, that there may be greater error in the classification of individual scales than in the estimate of overall stock composition.

The overall success in classifying fish of known origin by the discriminant functions developed from those fish ranged from 69.4% to 79.7% in the studies reported by Myers et al. (1984) and Myers (1986), and was 60.9% in the study reported by Ito et al. (1985). Although these are reasonably high rates of successful classification, there is still room for considerable error in quantitative estimation of stock composition. For individual stock groupings, the success in classifying known samples ranged from 58.5% to 95.5% for the Asian grouping, from 64.6% to 89.9% for the western Alaskan grouping, from 47.0% to 67.7% for the central Alaskan grouping, and from 70.5% to 83.5% for the southeastern Alaskan/British Columbian grouping (Myers et al. 1984, Ito et al. 1985; Myers 1986). The central Alaskan stock grouping was the grouping most likely to be misclassified in all analyses, and these fish tended to be misclassified as belonging either to the southeastern Alaskan/British Columbian or Asian stock groupings rather than to the western Alaskan grouping (Myers et al. 1984; Ito et al. 1985, 1986; Myers 1986). The directions and degree of misclassification of central Alaskan scales depended to a considerable extent on the stock proportions in scales used for the Asian standard (Myers 1986).

Considering the potential sources of bias and error it is not surprising that the quantitative estimates of stock composition produced by different analyses differ considerably. In Figure 27 the estimates from recent publications of stock composition of immature chinook captured in July in subareas of the Japanese mothership fishery are shown (Myers et al. 1984; Ito et al. 1986; Myers 1986). Despite the quantitative disagreement in the results of different analyses, there appears to be considerable qualitative agreement.

In the Bering Sea, all the analyses agree that western Alaskan (including the Canadian Yukon) chinook are the most abundant stock grouping. Next in abundance are Asian and central Alaskan chinook, which are each about half as abundant as western Alaskan chinook. The southeastern Alaskan/British Columbian stock grouping is rare in the Bering Sea (Figure 27).

In the western North Pacific Ocean, the greatest disagreement is over the relative contribution of central Alaskan and Asian stock groupings; the analyses of Myers et al. (1984) and Myers (1986) suggest that central Alaskan chinook often dominate in this area, whereas the analyses of Ito et al. (1985, 1986) suggest that Asian chinook dominate (Figure 27). This disagreement is not, as yet, technically resolvable. The very high percentage of central Alaskan chinook in the western North Pacific Ocean suggested by Myers et al. (1984) (sometimes as high as 100%) seems inconsistent with the apparent abundance of this species in spawning escapements in central Alaska. Myers et al. (1987), however, suggested that chinook may be more abundant in central Alaska than previously thought. Comparison of measured characteristics suggests similarity between Asian and central Alaskan scales in a number of features. It seems possible that there may be considerable misclassification of Asian scales as central Alaskan and vice versa, depending on the scale characters and the stocks used to calculate the discriminant functions. Leaving aside the relative contributions of Asian and central Alaskan chinook, all the analyses suggest that the chinook population of the western North Pacific Ocean is comprised mainly of stocks from Asia, western Alaska, and central Alaska, and that each of these stocks makes a substantial contribution to the population. Southeastern Alaskan/British Columbian chinook, although perhaps slightly more abundant than in the Bering Sea, still constitute only a small per-

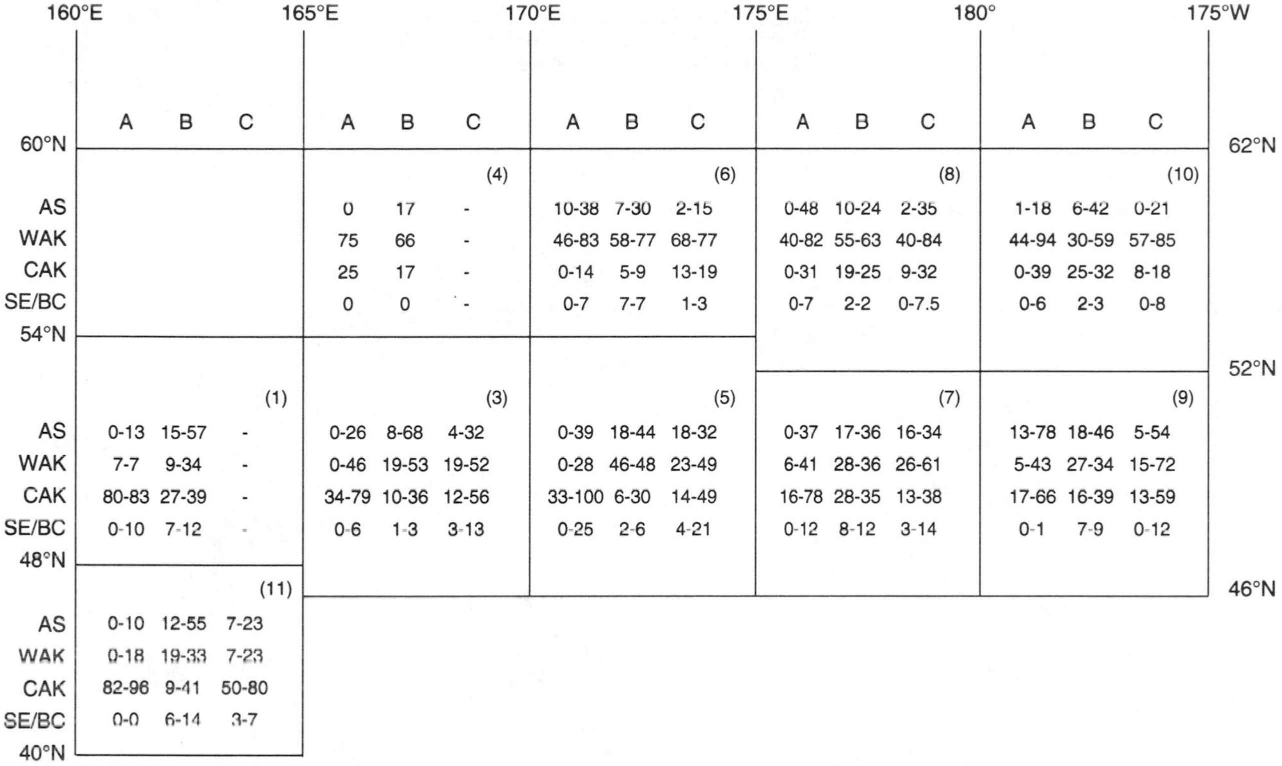

	160°E			165°E			170°E			175°E			180°			175°W
	A	B	C	A	B	C	A	B	C	A	B	C	A	B	C	
60°N																62°N
						(4)			(6)			(8)			(10)	
AS				0	17	-	10-38	7-30	2-15	0-48	10-24	2-35	1-18	6-42	0-21	
WAK				75	66	-	46-83	58-77	68-77	40-82	55-63	40-84	44-94	30-59	57-85	
CAK				25	17	-	0-14	5-9	13-19	0-31	19-25	9-32	0-39	25-32	8-18	
SE/BC				0	0	-	0-7	7-7	1-3	0-7	2-2	0-7.5	0-6	2-3	0-8	
54°N																52°N
			(1)			(3)			(5)			(7)			(9)	
AS	0-13	15-57	-	0-26	8-68	4-32	0-39	18-44	18-32	0-37	17-36	16-34	13-78	18-46	5-54	
WAK	7-7	9-34	-	0-46	19-53	19-52	0-28	46-48	23-49	6-41	28-36	26-61	5-43	27-34	15-72	
CAK	80-83	27-39	-	34-79	10-36	12-56	33-100	6-30	14-49	16-78	28-35	13-38	17-66	16-39	13-59	
SE/BC	0-10	7-12	-	0-6	1-3	3-13	0-25	2-6	4-21	0-12	8-12	3-14	0-1	7-9	0-12	
48°N																46°N
			(11)													
AS	0-10	12-55	7-23													
WAK	0-18	19-33	7-23													
CAK	82-96	9-41	50-80													
SE/BC	0-0	6-14	3-7													
40°N																

FIGURE 27

Estimates by various authors of stock composition of chinook captured in subdivisions of the Japanese mothership fishery based on discriminant function analysis of scale characteristics showing variation in results. AS = Asia, WAK = Western Alaska, CAK = Central Alaska, SE/BC = southeastern Alaska/BC. The numbers (1)–(11) represent subdivisions of the Japanese mothership fishing area. *A* = estimates of stock composition from Myers et al.'s (1984) discriminant analysis of scale characters; range of values for brood years 1971–77. *B* = estimates of stock composition from Ito et al.'s (1985, 1986) discriminant analysis of scale characters; range of values for different assumptions about Asian stock origins; brood year 1970. *C* = estimates of stock composition from Myers's (1986) discriminant analysis of scale characters; range of values for different assumptions about Asian stock origins; brood years 1973, 1974, and 1976

centage of the population in the western North Pacific Ocean (Figure 27).

Summary of Ocean Distribution of Chinook

Despite the many extensive and intensive investigations that have been conducted, our understanding of the ocean migratory and distribution patterns of chinook is still very sketchy. Information on the distribution of ocean-type populations has been derived almost exclusively from the intensive ocean troll fisheries off Oregon, Washington, and British Columbia. Nevertheless, the relatively small number of recaptures of southern British Columbian and Columbia River chinook in southeastern Alaska, where there is an active troll fishery, suggests that most ocean-type chinook do not disperse more than about 1,000 km from their natal river. Some go much further, of course, as revealed by the recaptures from south of the Aleutian chain and in the southeastern Bering Sea. All recaptures to date are from near the coast, and the dispersal of ocean-type populations offshore is revealed only in catches by Canadian research vessels fishing in the eastern North Pacific Ocean. The low catches of chinook in this part of the North Pacific Ocean, and the low proportion of ocean-type chinook among the few that were cap-

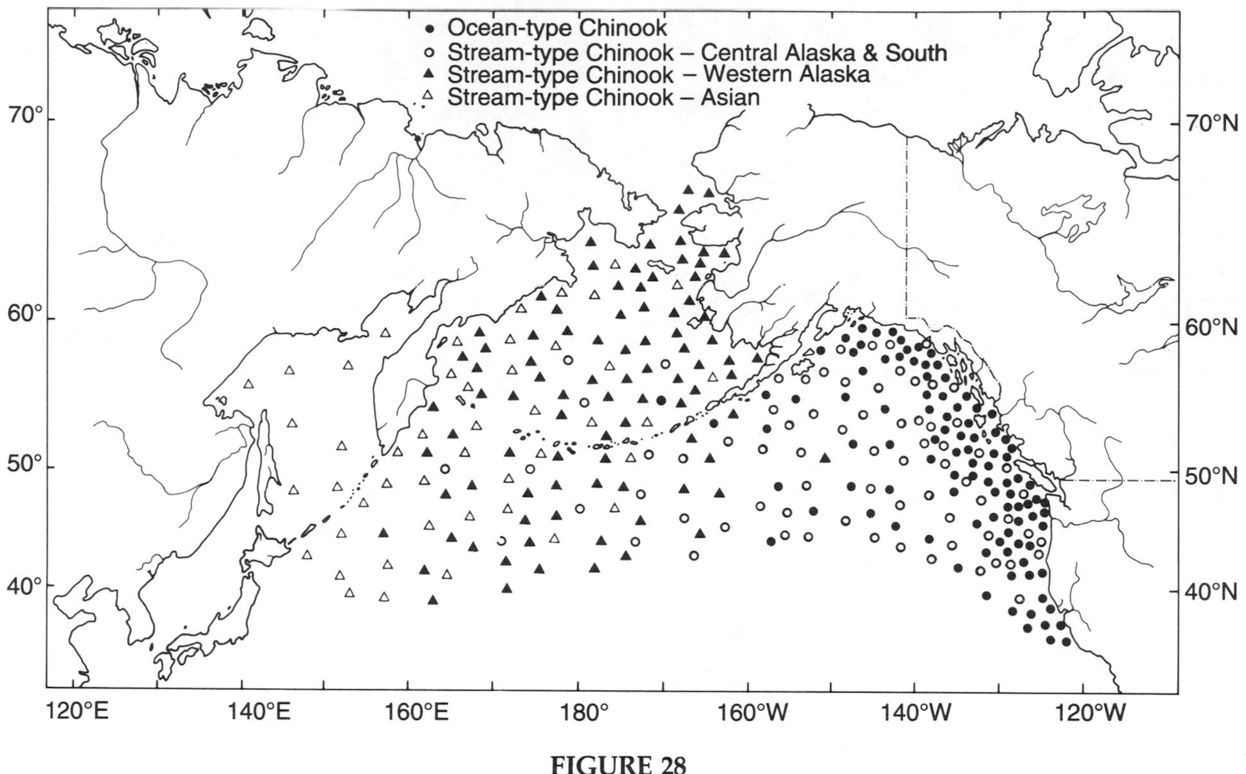

FIGURE 28

The likely ocean distributions and relative abundance in different parts of the North Pacific Ocean and Bering Sea of ocean- and stream-type chinook from various regions of the North American and Asian coasts. Stream-type chinook from western Alaska include Canadian Yukon River chinook. See text for details regarding the basis of these distributions and difficulties with the data.

tured, suggests that ocean-type chinook do not often wander far from shore (Figure 28) or that they are found at greater depths than those fished by the research vessel gear.

The dispersal of stream-type chinook appears to be much broader, as revealed by their contribution to both coastal and high-seas catches. Recaptures of tagged stream-type chinook from the Columbia River in coastal troll fisheries show greater dispersal of this race, both north and south of the river mouth, than of ocean-type populations. The high contribution of stream-type fish to the high-seas catches throughout the North Pacific Ocean and Bering Sea further indicates much greater offshore dispersal of this race (Figure 28).

The high-seas distribution of stream-type chinook from different regions of the North American coast and Asia is a matter of considerable debate. Nevertheless, I believe a few speculations may be made about such distributions, with the understanding that these will be subject to modification as more information becomes available.

Asian chinook are probably distributed throughout the Bering Sea but are also probably concentrated west of 180°. In the northern North Pacific Ocean they appear to be distributed at least as far east as 175°W, and probably further, and to be relatively more abundant in the western North Pacific Ocean than in the Bering Sea. The southern limit of their distribution is not known but is at least 40°N latitude and perhaps further south (Figure 28).

Western Alaskan chinook (including Candian Yukon chinook) are also distributed throughout the Bering Sea. Chinook are probably much more abundant in western Alaska than in Asia, so that chinook from western Alaska tend to dominate Bering Sea catches, even in the western half of the sea. Alaskan chinook are, nevertheless, probably relatively more abundant in the eastern and central than in the western Bering Sea. Western Alaskan chinook apparently also migrate south of the Aleutian chain into the North Pacific Ocean. The limits of their distribution there are not known, but they may occur as far westward as 160°E, and probably extend considerably east of 175°W. Western Alaskan chinook may also be distributed as far south as 40°N (Figure 28). I believe that western Alaskan chinook will ultimately be shown to have a very wide distribution in the western North Pacific Ocean, but they are probably no more abundant than Asian chinook in these waters (Knudsen et al. 1983; Myers et al. 1984) (Figure 27).

Central Alaskan chinook are also probably widely distributed in the central and western North Pacific Ocean and Bering Sea (Figure 28). As I noted earlier, the relative abundance of central Alaskan chinook in these waters is controversial. I suspect that they will ultimately be shown to be less abundant than Asian chinook in the Bering Sea and western North Pacific Ocean.

Southeastern Alaskan/British Columbian chinook appear to be relatively rare in both the Bering Sea and the western North Pacific Ocean, at least on the basis of scale analysis (Figure 27). Since virtually all chinook captured in the Bering Sea and western North Pacific Ocean are stream-type, any fish of the southeastern Alaskan/British Columbian stock grouping captured there must have been of the stream-type race. The southeastern Alaskan/British Columbian chinook, as well as those from Washington, Oregon, and California, are probably distributed mainly in the eastern North Pacific with the greatest concentrations over the continental shelf waters along the North American coast (Healey 1983).

Ocean Food Habits

Most data on food habits of chinook in the ocean are from samples taken in the commercial fishery and, therefore, relate to larger fish. Healey (1980a), however, reported on the diets of chinook of 10–30 cm fork length captured in the Strait of Georgia in late summer, 1975 and 1976. Small fish, particularly herring, pelagic amphipods, and crab megalopa made up between 70% and 92% of the diet, with fish being the largest single contributor at 28%–63% of the diet. Adult insects were also important in the diet of chinook captured in 1976 but not in 1975. There was evidence for regional variation in diet within the Strait of Georgia in that fish composed about 79% of stomach contents in the Gulf Islands region but only 37% in the Fraser River plume and central region of the Strait of Georgia. Pelagic amphipods were relatively unimportant in the Gulf Islands region but constituted 24% of stomach contents in the Fraser River plume and 14% of stomach contents in the central Strait of Georgia. Crab megalopa increased in importance from 11.5% of stomach contents in the Gulf Islands region to 46.4% of stomach contents in the central Strait of Georgia. Insect adults were only important in the Fraser River plume.

For larger chinook, diet information is available from southeastern Alaska south to the California coast. The data span a number of years and times of year and thus provide evidence for regional, annual, and seasonal changes in diet. Techniques of data collection and analysis have not been consistent among investigators, however, so that quantitative comparisons must be viewed with caution.

Reid (1961) described the diet of chinook captured in the troll fishery off southeastern Alaska in 1957 and 1958. Stomachs were sampled from mid-June to mid-September. The volume of stomach contents was measured, and organisms were identified taxonomically and counted. Thirteen different taxa were identified in the stomachs, but by far the most important diet item was herring (60%–68% of volume). There were some indications of differences in diet between years. For example, squid were about ten times as common in the diet in 1958 as in 1957, whereas fishes other than herring were much more important in 1957. There was also an indication that squid were more important in sheltered inside waters than in the more open water fishing areas. The importance of herring relative to other foods was related to size of chinook, increasing from less than 20% of diet for chinook less than 64 cm to more than 60% for chinook more than 100 cm in length.

Pritchard and Tester (1944) and Prakash (1962) described the diet of chinook salmon in British Columbian waters. In both studies, samples were collected from the commercial fisheries. Pritchard and Tester (1944) sampled from the whole coast during 1939-1941, whereas Prakash (1962) sampled only the west coast of Vancouver Island, the Juan de Fuca Strait, and the Fraser River estuary. Contribution of various taxonomic groupings to diet was estimated as a percentage of the total stomach volume in both studies.

Pritchard and Tester (1944) recorded 21 different taxonomic groupings in the diet of chinook during the three years of sampling. Fish of various sorts, but especially herring and sand lance (*Ammodytes hexapterus*), dominated the diet. Invertebrate taxa never exceeded 6% of the diet. Some between-year variation in diet was evident, most notably in the absence of pilchards (*Sardinops caerulea*) from the diet in 1939 and their relative importance in 1940 and 1941. This change appeared correlated with the strength of the pilchard stocks in southern British Columbia. Regional variation in diet was also apparent but chiefly in the alternation in relative importance of sand lance, herring, and pilchards. Off the northwest coast of the Queen Charlotte Islands, for example, herring and sand lance were almost equal in importance and pilchards were unimportant. On the northern mainland coast, herring were relatively important only in 1939 and were replaced by pilchards in 1940 and 1941. Off the northwest coast of Vancouver Island, sand lance predominated, constituting about 70% of the diet. Various invertebrates were also sometimes important in this region. On the southwest coast of Vancouver Island, herring predominated together with pilchards in 1940 and 1941. In 1939, sticklebacks were important (presumably replacing pilchards). In the northern Strait of Georgia, herring was the dominant food; and in the southern Strait of Georgia, pilchards were dominant. These regional differences in diet are unlikely to reflect patterns of dietary preference among regions. Rather, they probably reflect the relative abundance of potential prey between regions and years. Pritchard and Tester (1944) were of the opinion that chinook were largely opportunistic feeders. Apart from the chinook's apparent preference for fish as prey, this interpretation is probably correct.

Prakash (1962) found herring to be the most important diet item of chinook off southern Vancouver Island (72.5% of the diet). Larger herring were taken off the west coast of the Island than off the east coast (13–23 cm compared with 5–10 cm). As Reid (1961) noted in Alaska, large chinook tended to have a higher percentage of herring in their diets than did small chinook. Consequently, stomach samples from the west coast of Vancouver Island, which were from larger chinook, indicated a higher proportion of herring in the diet than stomach samples from the east coast of the island. Invertebrate food organisms (mainly euphausiids) dominated the diet of chinook from the east coast of Vancouver Island in May and June, but otherwise there was no evidence for significant seasonal changes in diet.

Robinson et al. (1982) reported on the stomach contents of 27 chinook captured in the Qualicum River area of the Strait of Georgia during April and May, 1981. They ranged from 29 to 72 cm in length. Their principal diet items were chum salmon fry, larval herring, sand lance, euphausiids, and adult herring. This is a rare instance in which predation by chinook upon the juveniles of another salmon species has been documented.

Silliman (1941) reported on the diet of chinook captured by trolling off the coasts of Washington and Vancouver Island in 1938. Stomach contents were measured gravimetrically. Herring dominated the diet of chinook captured off Vancouver Island (Pritchard and Tester 1944; Prakash 1962), but euphausiids dominated off Westport, Washington (43% of the diet). Fish of various sorts still made up most of the diet in the Westport area, but no one species overshadowed the rest.

Heg and Van Hyning (1951) described the diet of chinook captured by trolling off the Oregon coast in 1948–50 and of chinook captured by sports fishermen in 1950. Clupeids and clupeid remains dominated the diet, and anchovies (*Engraulis mordax*) were the most important single species. Invertebrates (predominantly euphausiids) constituted only 4%–5% of the diet.

Merkel (1957) reported on the diets of chinook captured in the ocean sport troll fishery off San Francisco. Sampling extended from February to November, and samples were adequate to permit monthly comparisons of diet as well as comparison by size of fish and location of capture. Northern anchovy (29.1%) and various juvenile rockfishes (22.5%) were the most important diet items, although euphausiids (14.9%) and herring (12.7%) also contributed significantly. Seasonal variations in diet were dramatic. Anchovies dominated the diet from August to November, rockfishes during June and July, and herring in February and March.

During April and May, invertebrate foods, especially euphausiids but including squid and crab megalopa, were dominant. The importance of invertebrate foods in April and May, but also to some extent in June, is reminiscent of Prakash's (1962) observation that invertebrates dominated the diet of chinook off southeastern Vancouver Island in May and June.

Diet was also related to location of capture (Merkel 1957). In water shallower than 20 fm (38 m), anchovies dominated (91%–92% of diet). In water deeper than 20 fm (38 m), rockfishes dominated (71%–74%); but a wider variety of organisms occurred in stomachs, and herring and euphausiids were significant diet items. Anchovies were less than 5% of the diet of chinook taken over deep water and rockfishes less than 1% of the diet of chinook taken in shallow water. As other authors have observed, small chinook fed more heavily on invertebrates than large chinook. Chinook less than 25 in. (63 cm) long captured near the Farallon Islands (California) had 85% invertebrates in their stomachs, whereas chinook more than 25 in. (63 cm) long had only 52% invertebrates in their stomachs.

Viewed together, the coastwide data on chinook diet suggest some regional trends (Figure 29). In general, the importance of herring and sand lance increases from south to north, whereas the importance of rockfishes and anchovies decreases. These trends probably reflect the relative abundance of the prey along the coast. Pilchards were important in the central parts of the chinook's range prior to the collapse of the pilchard stocks. More recent observations suggest that pilchards are not important in the diet of chinook at present, but that anchovies are more important further north than is indicated by the historic data. Euphausiids are important from time to time, and there is no clear geographic trend for their importance. Other prey items are incidental. The importance of fish in the diet of chinook is apparent in virtually all studies and, in fact, chinook appear to be the *Oncorhynchus* species most dependent on fish as food (Healey 1976).

Not only are seasonal and regional variations in diet composition apparent but so are seasonal and regional variations in feeding intensity. Healey (1982c) found that the weight of stomach contents of first ocean-year chinook in the Strait of Georgia ranged from 0.4%–1.73% of body weight between regions and years of sampling. Stomach contents tended to be highest in areas where catch was highest, suggesting that the young chinook were congregating in good feeding areas. For commercial-sized fish, Prakash (1962) found that stomach contents were greater off the west coast of Vancouver Island than off the east coast during May to September, and that contents tended to be greatest in August (Figure 30). The percentage of stomachs with food was uniformly high (greater than 80%) among troll-caught fish, except for samples from the west coast of Vancouver Island and the Juan de Fuca Strait in September. Chinook sampled from the Fraser River gillnet fishery had little in their stomachs, and the percentage of stomachs with food declined from about 50% in July to 0% in October. Stomach contents of chinook captured off the west coast of Vancouver Island and coastal Washington during April–October 1938 ranged from 0.1 to 63.7 g/fish (Silliman 1941). Contents increased during the sampling period off the west coast of Vancouver Island but decreased off the coast of Washington (Figure 30). Silliman (1941) found a significant relationship between salmon troll catch and the weight of fish in the diet of chinook. He interpreted this to reflect the relative attractiveness of troll lures, which mimic prey fishes, during periods when the chinook were feeding on fish. Finally, Merkel (1957) observed that stomach contents of chinook off California were least in spring and fall and greatest in early summer (Figure 30).

Although highly variable, the data on stomach contents suggest that chinook feed most actively in spring and summer. They also suggest that the best feeding periods are during July and August off southwest British Columbia but earlier, during April to June, from Washington to California. These differences in seasonal feeding patterns may, in turn, reflect the differences in diet along the coast described earlier (e.g., the greater importance of anchovy and rockfishes south of Washington and the greater importance of herring and sand lance north of Washington).

Ocean Growth

Healey (1980a) and Miller et al. (1983) provided data on the growth of chinook salmon during their

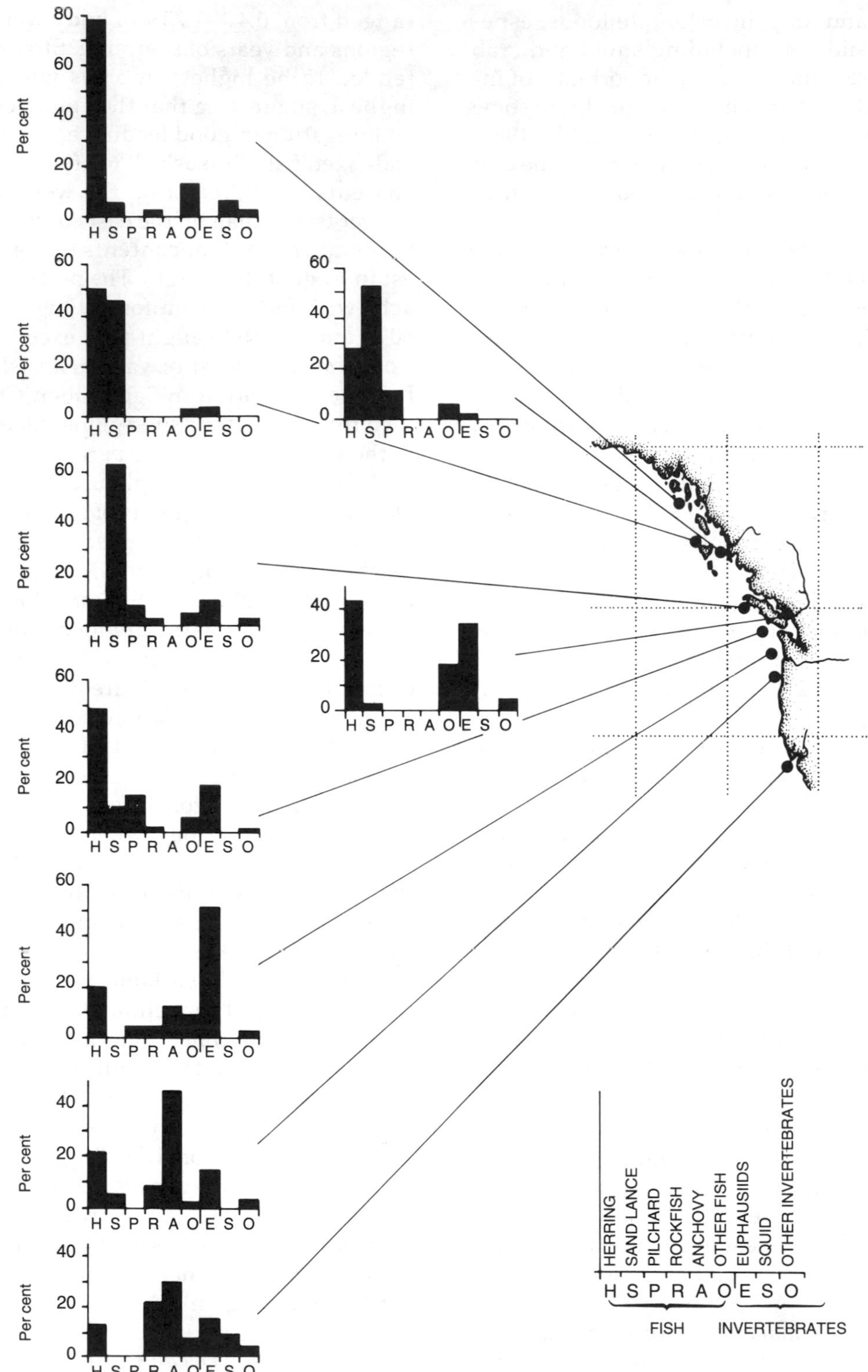

FIGURE 29
Diet composition of chinook from southeastern Alaska to California

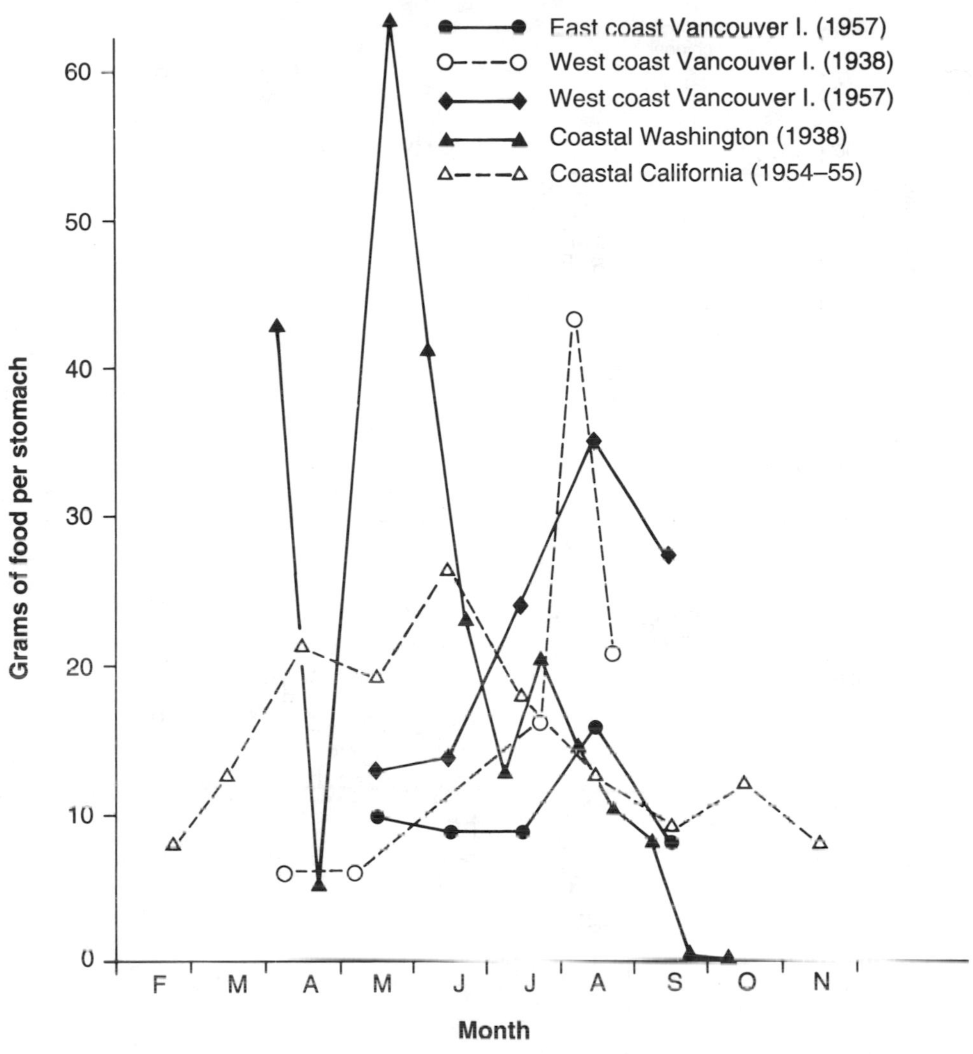

FIGURE 30

Seasonal changes in the weight of stomach contents of chinook captured off Vancouver Island, Washington, and California

first summer at sea. For chinook in the Strait of Georgia in 1976, modal size groups increased in length by 0.75–0.85 mm/d. Chinook captured off the Washington/Oregon coast in May–September 1980 increased in length by 1.74 mm/d. These data must be interpreted cautiously, as there is little likelihood that the same cohort of chinook was sampled throughout the time series of sampling in either study. In particular, the increase in length of chinook off the Washington/Oregon coast seems high. By comparison, chinook sampled by Fisher et al. (1984) in the same area showed little increase in mean length (0.3 mm/d).

Loeffel and Wendler (1969) summarized the known information on growth of chinook at sea based on time series of samples from a single location and on back-calculations from scales. Figure 31, adapted from their report, represents a composite picture of growth of stream- and ocean-type races. The data suggest a definite seasonal pattern of growth in both races, with rapid summer growth and slow winter growth. Faster growing fish mature at a younger age as is clearly shown by the back-calculated lengths of fish that matured at

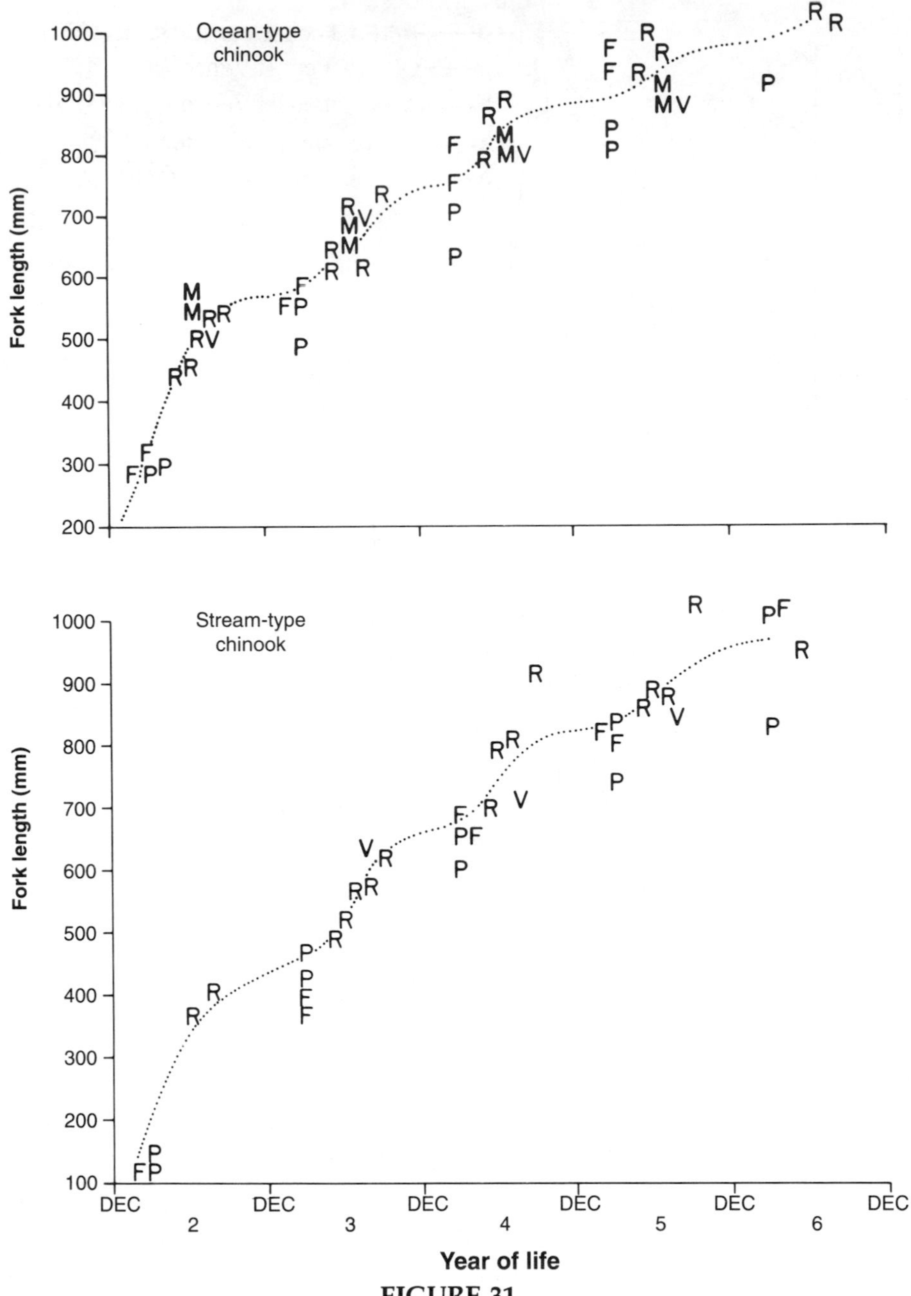

FIGURE 31

Growth curves (fork length at age) for ocean- and stream-type chinook. Letters refer to data from different sources. (Adapted from Loeffel and Wendler 1969)

different ages (Figure 32). Grachev (1967) observed the same pattern in Kamchatka River chinook, and Neilson and Geen (1986) observed that age at maturity was negatively correlated with the size of chinook at the end of their first year in the ocean. The apparent seasonality of growth (Figure 31) may be accentuated, therefore, by the disappearance from the ocean population each summer and autumn of the larger members of each age class that are returning to fresh water to spawn. Ocean-type chinook are larger than stream-type chinook at every calendar age. This is because the stream-

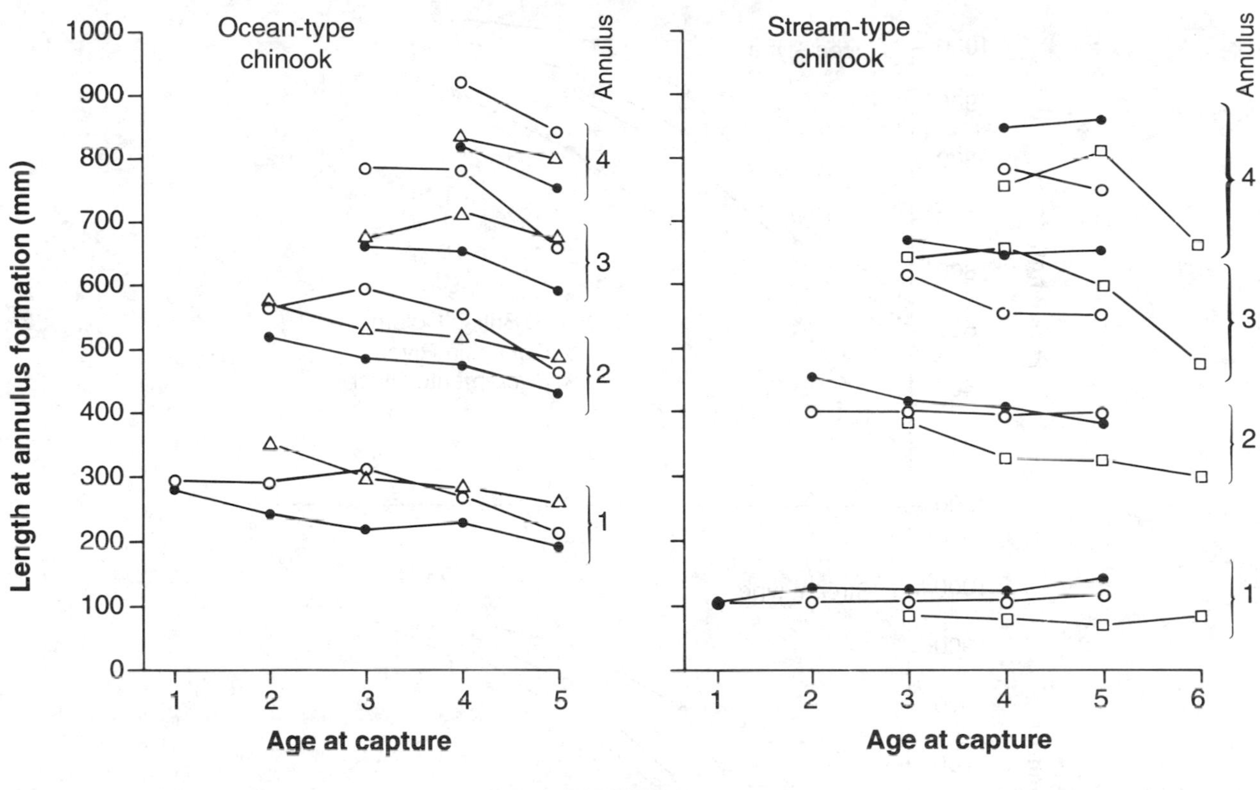

FIGURE 32

Length at annulus formation for ocean- and stream-type chinook which mature at different ages. Data from southeastern Alaska (•), Strait of Georgia (o), west coast of Vancouver Island (Δ), and the Fraser River (□)

type fish grow more slowly than ocean-type during their first year of life. Rates of growth of the two races during ocean life are similar (Figure 31).

Taken at face value, the available data suggest that, for North American populations, length at age is greater in the south than in the north (Figure 33). The data are a little difficult to interpret, however, as the Alaskan samples were from the ocean troll fishery and include many immature fish, of which some would be from Fraser River, Columbia River, and other, more southern stocks (Figures 22-24) (Parker and Kirkness 1956). Size at age of Kamchatka River chinook was similar to chinook caught off southeastern Alaska (Grachev 1967), whereas on the basis of latitude I would have expected them to be more similar to Yukon River chinook. Also, the growth rates of fish aged two to six years are similar for all areas sampled (0.35-0.57 mm/d). In fact, the Columbia and Sacramento River populations of both races had the slowest growth rate over the ages two to six years. The differences in size between regions are entirely due to differences in size at age 0.1 in ocean-type chinook or age 1.1 in stream-type chinook. Conceivably, the differences in size at these ages could be due to differences among investigators in interpretation of annuli. Chinook are notoriously difficult to age from scales (Godfrey et al. 1968). Conversely, it is possible that growth during the first ocean year differs dramatically along the coast. The apparently greater growth of first ocean-year fish off the Columbia River, compared with those in the Strait of Georgia (see earlier), is consistent with this explanation.

Grachev (1967) reported back-calculated growth increments for Kamchatka River chinook spanning 23 brood years, from 1933 to 1955. The most complete samples were of males representing ages 1.2, 1.3, and 1.4. Mean lengths in the year of migration varied from 58.3 to 70.4 cm for fish aged 1.2, and from 72.7 to 88.5 cm for fish aged 1.3. Total length was most closely correlated with the length incre-

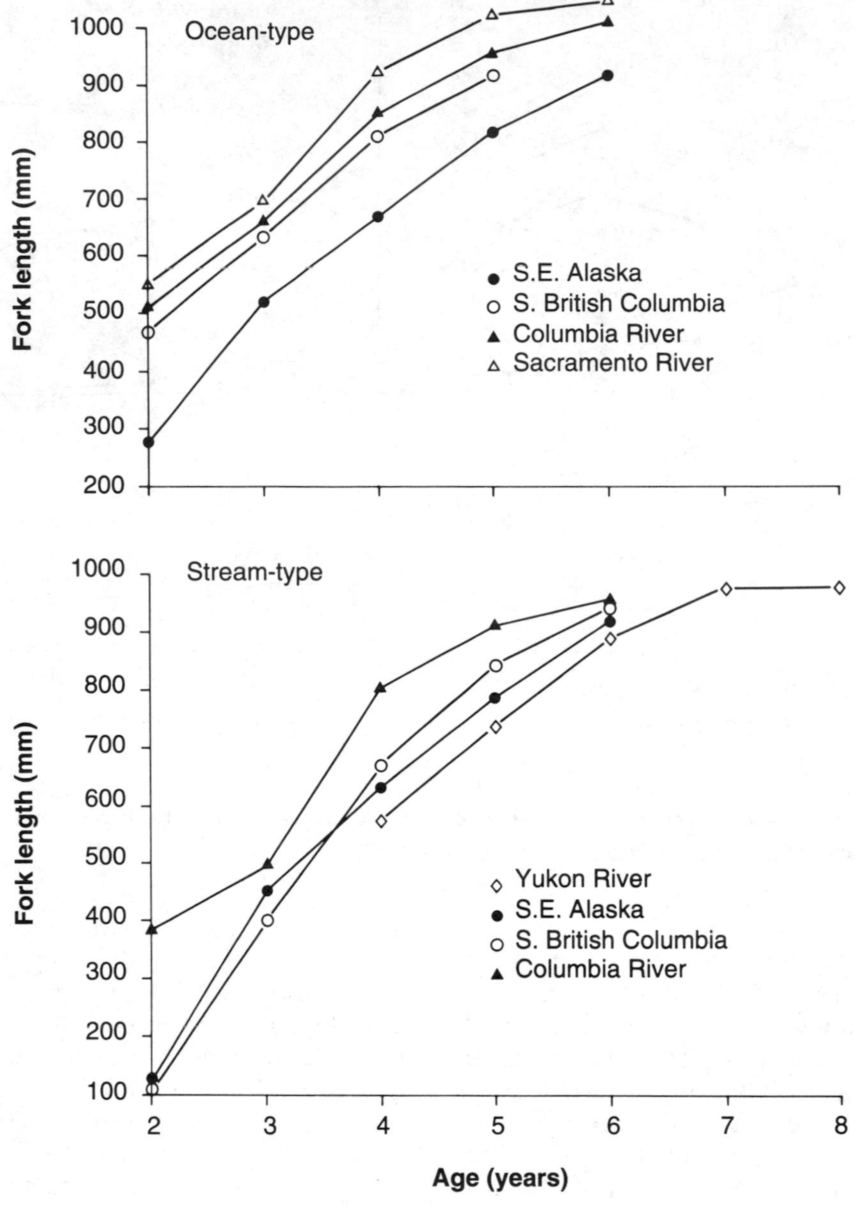

FIGURE 33

Length at age for chinook captured in the southeastern Alaska troll fishery and rivermouth fisheries in southern BC and the Columbia and Sacramento rivers

ment laid down during the final ocean year, suggesting that feeding conditions during this last year at sea were most important in determining final adult size of each age class.

All of the above analyses of growth are incomplete and superficial because they do not take fully into account the complex interrelationships among growth, migration, and maturation in chinook. Samples taken in ocean fisheries consist of representatives of numerous spawning stocks. The younger fish in a sample will be mainly from nearby spawning populations and the older fish, mainly from distant spawning populations. For chinook older than the age of first maturity, those remaining at sea will be the smaller, slower growing component of any spawning stock. Samples taken on the spawning grounds suffer from the opposite problem. The spawning fish in the younger age groups will be the faster growing members of the age class. Any model of growth for chinook must take account of these phenomena. Current data are inadequate for the task. More recent tag recovery data may prove sufficient, but these data are as yet largely unanalysed.

Maturation and Return to Spawn

As was the case with growth rate, estimating age-specific rates of maturity in chinook is a complex problem. Again, current data are inadequate for the task. Three approaches have been taken to date:

1 The proportion of each age class in samples captured at sea that will mature in the year of capture is determined from gonad/body weight ratios or egg size (Rich 1925; Borque and Pitre 1972; Wright and Bernhard 1972; Ito et al. 1974; Baranski 1979).

2 Maturity schedules are estimated from tag returns (Parker and Kirkness 1956; Cleaver 1969).

3 Maturity schedules are inferred from the age composition of spawning runs (various authors, but see Godfrey 1968a; Loeffel and Wendler 1969; Hankin and Healey 1986).

Mature chinook captured on the spawning grounds range from 0.1 to 0.5 in age for ocean-type stocks and from 1.0 to 2.5 for stream-type stocks. Not all populations have all ages, however, and most populations are dominated by a few age classes. Two examples of age-frequency curves will suffice to illustrate the differences between systems, races, and sexes (Figure 34). In the Yukon River (Brady 1983), which has only stream-type chinook, six-year-old fish of both sexes predominate in the spawning run. Males dominate the younger ages (ages 4 and 5), however, and females the older ages (ages 7 and 8). In the Fraser River, which has both stream- and ocean-type fish (Godfrey 1968a), four-year-old fish predominate in the run, and this is the most abundant age class for male and female ocean-type and male stream-type chinook. Female stream-type fish were most abundant at age 5. For both races, males were more common among the younger fish, and females among the older fish.

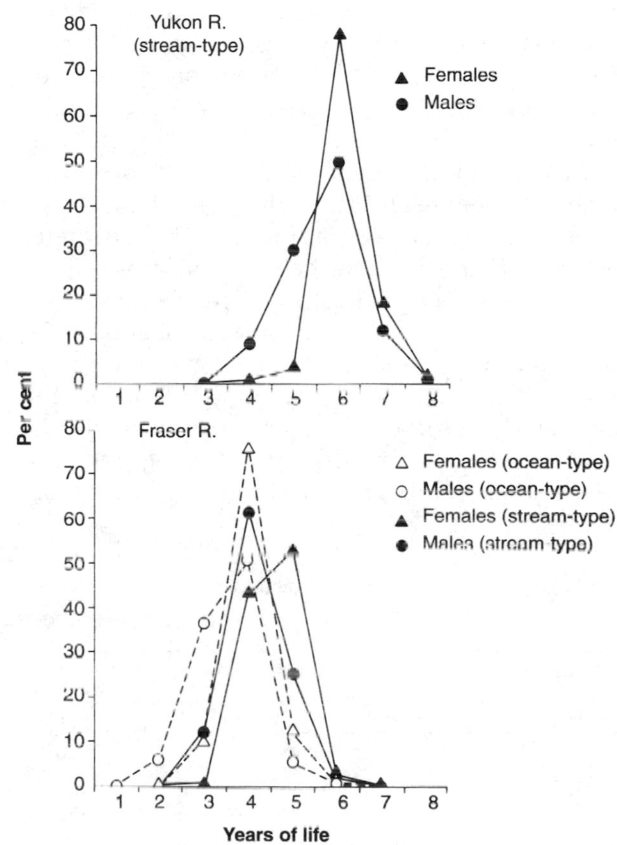

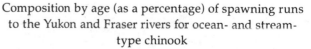

FIGURE 34
Composition by age (as a percentage) of spawning runs to the Yukon and Fraser rivers for ocean- and stream-type chinook

The average age of mature male ocean-type chinook ranged from 3.02 to 3.88 years in river systems for which comparative data exist, and from 3.98 to 4.32 years for mature female ocean-type chinook (Table 7, see table for references). For stream-type chinook, the range was from 3.69 to 5.64 years for males and from 4.39 to 6.12 years for females (Table 7). Thus, males mature at a younger average age than females, regardless of race. The difference in age between males and females was small in some instances (0.1 y) but in other instances it was large (1.3 y) (Table 7). From the Sacramento River (37°45′N) to the Nass River (55°00′N) there was no consistent trend in the age of maturity among sexes and races. There was a weak positive correlation between age and latitude for stream-type females and a weak negative correlation between age and latitude for ocean-type males, but neither correlation was significant. North of the Nass River, however, and in Kamchatka, chinook appear to mature about a year older, on average (Table 7). Thus, the stream-type races, at least on a geographic scale, have a longer generation time than the ocean-type, presumably owing to their longer freshwater residence.

The standard deviation of age at maturity for males ranged from 0.601 to 1.058, with no consistent geographic or racial trend. For females, the standard deviation of age at maturity ranged from 0.206 to 0.698, again with no consistent racial or geographic trend. The variation in male age of return was, therefore, consistently larger than the variance in female age of return. This trend is reflected in the high percentage of females returning to spawn at a single age, compared with males (Figure 34).

The age composition of spawning runs provides a measure of the ages over which chinook mature but gives only a rough indication of age-specific maturation rates. These latter rates depend not only on a knowledge of the number of mature fish in each age group but also on the number of immatures. Such data are provided by sampling populations at sea and assessing the proportion of maturing fish in each age class, and by analysing tag returns to estimate both mortality and maturity rates.

On the basis of tag returns from Columbia River hatchery chinook, Cleaver (1969) estimated maturation rates at various ages for the 1922 brood year, 1954–55 brood years, and the 1961 brood year. Rates of maturation ranged from 1% to 5% for chinook in their second year, from 18% to 35% for chinook in their third year, from 61% to 96% for chinook in their fourth year, and were 100% for chinook in their fifth year (Table 8). There was evidence for a greater proportion of mature fish among the youngest age groups in recent years.

On the basis of tag returns from ocean tagging off southeastern Alaska, Parker and Kirkness (1956)

TABLE 7

Mean age (years) and standard deviation of age for mature stream- and ocean-type chinook from rivers throughout the geographic range of the species

River	N. Latitude	Race	Mean age		Standard deviation		Source
			Male	Female	Male	Female	
Sacramento	37°45′	Ocean	4.18		0.779		Clark (1929)
Klamath	41°30′	Stream	4.40	4.60	0.764	0.592	Snyder (1931)
		Ocean	3.88	3.98	0.618	0.458	
Columbia	46°15′	Stream	3.98	4.39	0.667	0.547	Rich (1925)
		Ocean	3.55	4.29	1.058	0.698	
Fraser	49°15′	Stream	4.13	4.56	0.601	0.563	Godfrey (1968a)
		Ocean	3.59	4.00	0.732	0.206	
Skeena	54°15′	Stream	4.13	4.56	0.677	0.485	Godfrey (1968a)
		Ocean	3.47	4.19	0.827	0.441	
Nass	55°00′	Stream	3.69	4.82	0.876	0.405	Godfrey (1968a)
		Ocean	3.02	4.32	0.745	0.476	
Taku	58°30′	Stream	5.29		0.539		Kissner (1973)
Yukon	62°31′	Stream	5.64	6.12	0.803	0.425	McBride et al. (1983)
Kamchatka	56°00′	Stream	5.39		0.618		Vronskiy (1972)

Note: Sexes not distinguished in Sacramento, Taku, and Kamchatka rivers

TABLE 8

Age-specific maturity schedules (proportion maturing at age) for male and female ocean-type chinook as calculated by various authors for various times and locations

Location	Year(s)	Sex	Age 2	3	4	5	Source
Columbia R.*	1922	male/female	0.009	0.181	0.754	1.000	Cleaver (1969)
	1954–55	male/female	0.050	0.350	0.610	–	
	1961	male/female	0.009	0.338	0.959	1.000	
Wash. coast†	1972	male	0.300	0.700	0.990	1.000	Wright & Bernhard (1972)
		female	0.030	0.490	0.990	1.000	
Puget Sd.†	1975–77	male	0.120	0.480	0.950	–	Baranski (1979)
		female	0.030	0.240	0.950	–	
Vancouver Is.†	1967–70	male	0.360	0.440	0.700	–	Borque & Pitre (1972)
		female	0.020	0.120	0.630	–	
S.E. Alaska*	1950–52	male/female	–	0.320	0.545	0.905	Parker & Kirkness (1956)
Columbia R.‡	1919	female (S)§	0.000	0.000	0.770	0.960	Rich (1925)
		female (O)§	0.000	0.150	0.870	1.000	

Notes: *Results based on tag recaptures
†Results based on ocean sampling and maturity indices
‡Results based on analysis of egg size
§S = stream-type; O = ocean-type

estimated maturity rates for chinook aged three to five years that were similar to, but slightly lower than, Cleaver's (1969) estimates for 1954–55 brood year Columbia River chinook.

Unfortunately, the tag-return data analysed by Parker and Kirkness (1956) and Cleaver (1969) do not permit separate estimates of maturity rate for males and females. Estimates based on maturity indices for fish captured in the ocean do, however, permit separate estimates of maturity by sex. These estimates are consistent with the maturation pattern established from spawning runs in that males show a greater proportion maturing at a younger age than do females (Table 8). Maturity rates of 12%–36% were observed among males in their second year of life, as were maturity rates of 0%–3% for females in their second year and 0%–49% for females in their third year. Although two-year-old jack males are common in some populations, maturity rates of 12%–36% for this age group, as a general proposition, are unlikely, as that would mean two-year-old males would be a dominant age group among mature males in spawning runs to most rivers rather than a minor component (Figure 34). Similarly, two-year-old mature females are virtually unknown, and three-year-old mature females are uncommon in spawning runs except in a few Oregon stocks (Nicholas and Hankin 1988); yet the maturity rate estimates from gonad size indices (Wright and Bernhard 1972; Borque and Pitre 1972; Baranski 1979) suggest that two-year-old mature females should be relatively common and that three-year-old mature females should be a dominant component of spawning runs. Rich's (1925) estimates of female age-specific maturity rates, which are based on egg size rather than gonad size, seem much more appropriate, as they indicate no maturation of two-year-old females, 15% maturation among three-year old ocean-type females, and high rates of maturation among four- and five-year-old females of both ocean- and stream-type races. Also, the patterns of maturity among stream- and ocean-type females described by Rich (1925) are consistent with the pattern observed in spawning runs (Figure 34). Even Rich's (1925) estimates, however, produce run age-composition estimates that include too many young fish in comparison with observed run compositions. Clearly, there is some fundamental problem with the commonly used indices of maturity for fish captured in the ocean, or the samples that have been analysed to date are not representative of the general maturity schedules for chinook. One possible factor contributing to the apparent high proportion of maturing fish in the ocean samples off southwest Vancouver Island, Washington, and in Puget Sound may be the observation by Ito et al. (1974) and Major et al. (1978) that maturing fish constitute a higher proportion of the nearshore catch of chinook. However, this alone is not a sufficient explanation for the high proportion of maturing fish in the younger age groups, particularly the observed numbers of maturing two- and three-year-old females.

Although most chinook do not mature until they have spent at least one summer at sea, a few male chinook may mature without migrating to sea (Rutter 1902; Rich 1920; Burck 1967). Such males may mature at the end of their first summer or, more commonly, after their second summer in fresh water. It is generally the largest males in any sample that show evidence of maturity (Rich 1920). This phenomenon is documented for the Columbia and Sacramento rivers, but probably occurs elsewhere as well. The proportion of stream-dwelling males that become mature is unknown, but Rich (1920) noted that 10%–12% of males in the McCloud River, California, which were stream-resident, matured precociously.

Precocious maturation of males is associated with stream-resident populations in headwater tributaries, suggesting that it is a characteristic of stream-type chinook. It is not known whether precocious males contribute to reproduction, although J.W. Mullan (U.S. Fish and Wildlife Service, Leavenworth, Washington, pers. comm.) suggests that they do. They cannot always do so, however, as some mature outside the normal spawning time for sea-run fish. Nor is it known for sure whether precocious males die after maturing. Rich (1920) claimed that at least some recovered, but Burck (1967) found that, of 259 specimens of mature precocious males that he held in a downstream migrant trap, all died. J.W. Mullan (U.S. Fish and Wildlife Service, Leavenworth, Washington, pers. comm.) suggests that most precocial age 0. males survive whereas most precocial 1. males die.

Mortality in the Ocean

Estimates of natural and fishing mortality of chinook are derived from several sources. Parker and Kirkness (1956) were apparently the first to attempt such an estimate. Their estimate, based on recaptures of fish tagged off southeastern Alaska, was 34.1% annual mortality for all age classes.

Cleaver (1969) estimated probable values of natural mortality and maturity from recaptures of tagged 1961-brood chinook released from several Columbia River hatcheries. Unfortunately, maturity and mortality are confounded in the data so that it was not possible to estimate either uniquely. According to Cleaver (1969), the most realistic solution to the joint estimate of maturity and mortality occurred when instantaneous annual natural mortality was 0.45 (36% annual mortality). This is very close to the value obtained by Parker and Kirkness (1956).

Henry (1978) explored the mortality and maturity schedules for 1961 and 1962 brood year Columbia River hatchery chinook. His analysis differed from Cleaver's (1969) in that he assumed a fixed maturity schedule and calculated age-specific mortality rates. Henry's (1978) estimates suggested some differences in natural mortality between brood years and much higher natural mortality during the first ocean year than during later ocean years. Although these results are intuitively reasonable, the differences in mortality between brood years could be an artifact of the assumption that maturity was fixed. There seems to be no greater reason for assuming a fixed maturity schedule than for assuming a fixed mortality schedule.

Ricker (1976) reviewed all published approaches to estimating the natural mortality rate of Pacific salmon at sea, discussed their strengths and weaknesses, and assessed their biases. The published estimates of natural mortality for chinook, according to Ricker (1976), are likely to be too large, both because of tag loss and mortality of tagged fish, and because they do not take full account of mortality due to catch and release in some fisheries. This latter uncontrolled mortality source is particularly important for chinook, most of which are captured as immature fish in hook-and-line fisheries, and for which there are size limits that forbid landing small fish. At this point in time, it is impossible to state what is the real natural mortality rate of chinook salmon in the ocean. In all probability, the value is considerably less than 35% per year, probably closer to the value of 20% per year estimated to be a reasonable average for sockeye (Ricker 1976). It is also probable that age-specific mortality declines with age in chinook so that most ocean mortality occurs during the first year or two of ocean life.

The available data are too scanty to determine whether ocean mortality schedules differ among populations. In addition, all of these data refer to the ocean-type race so that no interracial comparison may be made. Parker (1962) argued that it was logical to suppose that causes of mortality (especially predators) were more concentrated in the coastal zone, so that one would expect greater mortality among salmon during their residence in coastal waters. Parker (1962) further argued that smaller salmon should be more vulnerable to predators and should, therefore, suffer greater mortality than larger salmon. If these arguments hold, then ocean mortality of ocean-type chinook should be greater than that of stream-type chinook because stream-type chinook enter the ocean at a large size and move offshore quickly, whereas ocean-type chinook enter the ocean at a small size and spend most of their ocean life in coastal waters. Even if their natural mortality rates do not differ substantially, I would expect fishing mortality to be lower for the stream-type race, at least for stream-type populations on the North American coast south of central Alaska. These populations appear not to contribute substantially to the Japanese high-seas fishery and, because they are distributed further offshore than ocean-type chinook, they do not contribute heavily to the intensive coastal troll fishery (Healey 1983).

Homing and Straying

Fish that survive to mature and return to fresh water to spawn must select a spawning stream, ascend it (Plate 15), and then select an appropriate spawning riffle. Upstream migration of mature chinook apparently occurs mainly during daylight hours (Plate 16), at least for the ocean-type race (Neave 1943). A few fish, however, do migrate upstream at night.

Salmon, in general, have well-developed homing behaviour, apparently returning to their natal stream to spawn with considerable fidelity. The choice of spawning river, tributary, and even riffle appears to be guided by long-term memory of specific odours. Groves et al. (1968) demonstrated the apparent importance of olfaction for chinook returning to the Spring Creek Hatchery on the Columbia River. They took chinook that had already returned to the hatchery, occluded their olfactory or visual senses, and released the treated (and untreated control fish) twenty or more kilometres downstream and upstream from the hatchery. The results of the study are summarized in Table 9. About 49% of control fish returned to Spring Creek from downstream releases and 37% from upstream releases. This difference was not statistically significant. About 23% of chinook with their vision destroyed found their way back to Spring Creek, significantly fewer than the control fish ($X^2 = 28.7$; $p < .001$). Only 6 of 193 (3%) chinook with their nasal sacs plugged, however, returned to Spring Creek, significantly less than either the controls ($X^2 = 101.8$; $p < .001$) or chinook with their vision destroyed ($X^2 = 46.1$; $p < .01$). Chinook with both vision and olfaction occluded were hardly recaptured at all, but two still found their way back to Spring Creek. The implication of these results is

TABLE 9

Chinook captured at Spring Creek, tagged, and relased upstream and downstream from Spring Creek, and recaptured at Spring Creek, other Columbia River hatcheries, and in river sport and domestic fisheries

	Treatment			
	Control	Olfactory occlusion	Visual occlusion	Olfactory and visual occlusion
Number released downstream	192	152	192	150
Number recaptured at:				
Spring Creek	94	6	46	2
Other hatcheries	45	27	51	1
Fisheries	1	16	0	15
Number released upstream	49	41	49	41
Number recaptured at:				
Spring Creek	18	0	10	0
Other hatcheries	8	1	4	3
Fisheries	1	2	2	3

Source: From Groves et al. (1968)

that both olfaction and vision are important in selection of home stream, but that olfaction is by far the more important sense. In a related study, Hara et al. (1965) demonstrated an electrophysiological response of the olfactory bulb of chinook to infusion with home-stream water, implying recognition of home-stream odour. The interpretation of electrophysiological results has been cast into doubt, however, by Bodznick (1975) who could find no correlation between the electrophysiological response and the actual migratory behaviour of sockeye salmon.

Quite a large number of the fish released by Groves et al. (1968) were recovered at hatcheries other than Spring Creek, even among the control fish. These cannot be regarded as true strays, however, as most of the hatcheries were operating weirs, and once the fish had entered a hatchery stream, they could not get out. Thus, the fish were unable to correct any mistakes in homing. Similarly, the fish originally selected for the study had been "trapped" by a weir at Spring Creek, and some may not really have been Spring Creek fish.

More representative data on straying are provided by Rich and Holmes (1928) and McIsaac and Quinn (1988), who reported on the return of adults, which had been tagged as fingerlings, to various Columbia River hatcheries and tributaries. None of the fish investigated by Rich and Holmes (1928) was released into a tributary from which its parents had come, whereas McIsaac & Quinn (1988) investigated return rates for transplanted and resident fish. Although all recoveries were made within the Columbia River, Rich and Holmes (1928) found that stream-type chinook showed rather poor fidelity to their particular release tributary. Ocean-type chinook, on the other hand, showed very strong fidelity to their release site. McIsaac and Quinn (1988) found that chinook from an upriver population transplanted to the Bonneville Hatchery returned poorly to the hatchery and that many recaptures came from well upstream. Control fish, by contrast, homed faithfully. Rich and Holmes (1928) suggested that their results implied a contribution of both heredity and learning to homing behaviour, but also that the suitability of the release tributary for chinook was important. It may also be the case that the time of imprinting to home-stream odour of stream-type chinook is different from that of ocean-type. Since stream-type fish undergo several in-stream migrations during their freshwater life, imprinting to the home stream may occur very early, whereas ocean-type fish may delay imprinting until just before ocean migration. Thus, stream-type fish reared for several weeks in a hatchery before being released into a tributary may not imprint to the tributary. McIssac and Quinn's (1988) results, however, suggest much more strongly that heredity is important.

Detailed information on straying of ocean-type chinook is given by Quinn and Fresh (1984) for fish from the Cowlitz River Hatchery on the Columbia River drainage. These authors found that home-stream fidelity among four brood years of chinook averaged 98.6%, and that most strays were from spawning areas close to the Cowlitz River. Ten chinook, however, were recovered from spawning areas in Puget Sound and the Juan de Fuca Strait. Such long-distance strays, although few in number, could have important consequences for the genetic composition of regional spawning populations. In contrast to the results for the Cowlitz River, Uremovich (1977) reported considerable straying of Elk River hatchery chinook into the adjacent Sixes River.

Among the categories of chinook returning to the Cowlitz River, older chinook strayed more than younger chinook, and males that had spent only one summer at sea were the most faithful of all. These males also returned to the river later in the season than older age classes. The proportion of strays varied among brood years, and the amount of straying appeared to be related to brood year success. Higher straying was observed among brood years with the poorest overall returns, suggesting that conditions that were poor for survival were also poor for home-stream fidelity (Quinn and Fresh 1984). Such a behaviour pattern could be coincidental. It could also be a mechanism to provide for wide distribution of spawners during poor survival years. Uremovich (1977) also reported variation in straying between years but offered no explanation.

Concluding Remarks

This completes my summary of life history observations on chinook. Throughout, I have highlighted the degree of variation among chinook populations, and, in particular, the important differences between the stream- and ocean-type races. In my view, there is a sufficient basis for separating the species into these two races. As was outlined in Figure 1, the races are distinguished by fundamental ecological differences in (1) the geographic distribution of their spawning populations; (2) the duration of their freshwater residence as juveniles prior to seaward migration and as adults prior to spawning; (3) their oceanic distribution and dispersal; and (4) timing of their spawning migrations. There is also evidence for genetic segregation between these life history types (Healey 1983; Carl and Healey 1984), for morphological differences between juveniles (Carl and Healey 1984), and for the inheritance of migratory timing (Rich and Holmes 1928).

Assuming that there are two races of chinook salmon, one must then ask how these could arise and how they could persist when the races are sympatric. If it were only the case that stream-type chinook occurred in the headwater tributaries of the larger rivers, then one might presume that stream-dwelling behaviour had arisen independently in each river and was maintained through disruptive selection (Merrell 1981). The environmental heterogeneity producing ocean- and stream-type juvenile life history patterns could be related to a critical time of arrival in the ocean, such as that suggested by Walters et al. (1978) for pink and chum fry from the Fraser River. If such a critical time period existed, it could render seaward migration from the headwaters impractical during the first summer because the length of time required for the young fish to grow to smolt size and traverse the river would bring them to sea at a bad time. The spring and summer redistribution migrations of stream-type chinook within the river system correspond in timing with the spring and summer seaward migrations of ocean-type chinook. These within-river dispersals of the stream-type race could be the remnants of the seaward migratory behaviour of the ocean-type race, and suggest that one type may be derived from the other. Some stream-type populations may, in fact, have arisen through disruptive selection in ocean-type populations. In many instances, the different ocean distributions of stream- and ocean-type chinook, their different migratory behaviour while at sea, and the particular geographic distribution of their spawning populations argue against such an explanation for the origin of stream-type chinook. Recent information on genetic differentiation among populations of chinook, however, does not always reveal patterns consistent with a segregation into stream- and ocean-type races (Gharrett et al. 1987; Utter et al. 1989; Winans 1989).

An alternative explanation is that stream-type chinook developed (or persisted) in the Bering refugium, or on the Asian coast, during the last glaciation, whereas ocean-type chinook developed (or persisted) south of the glaciation on the North American coast. With the retreat of the ice, both races could have expanded to occupy their present distributions. Gharrett et al. (1987) suggested that the genetic composition of chinook populations supports this interpretation. For this explanation to hold, however, selection gradients or reproductive isolating mechanisms must be sufficient to prevent introgression in the natural populations, at least where they are sympatric. It is also necessary to postulate some barrier preventing significant invasion of the ocean-type race north of 56°N. Conditions in the freshwater environment might provide such a barrier if, for example, riverine and estuarine productivity and temperature were too low to permit chinook to reach a critical size for smolting during their first summer. Riverine temperatures, at least during the summer growing season, however, appear adequate for good growth (Healey 1983). Although I am confident that the explanation for the absence of ocean-type chinook north of 56°N lies in the trade-off between survival and growth for small chinook in the river versus in the ocean, the precise mechanism preventing northward invasion of ocean- type chinook is not apparent.

These possible explanations, and others, for the existence of stream- and ocean-type chinook need to be critically tested, and they offer an opportunity for fruitful research into the evolution of Pacific salmon. There is no doubt, however, that Pacific salmon have the capability to adapt quickly to new opportunities. The recent appearance in the Great Lakes of spring-spawning chinook, which must have developed from the fall-spawning introduced race, attests to this (Kwain and Thomas 1984), as does the development of even-year runs among pink salmon from a single odd-year release into the Great Lakes (Kwain and Chappel 1978).

Segregating the species into stream- and ocean-type races accounts for only a small part of the interpopulation variation in chinook. There is also considerable interpopulation (within race) and intrapopulation variation that requires explanation (Healey and Heard 1984). Interpopulation variation includes (1) differences in the proportion of the population migrating seaward at different ages (fry or fingerlings for ocean-type chinook, one- or two-year-old smolts for stream-type); (2) differences in the proportion maturing at each reproductive age; (3) differences in growth; and (4) differences in fecundity. Intrapopulation variation involves the same population attributes and leads to a variety of life history tactics within any population. Variation in length of riverine and estuarine residence has attracted considerable attention and has been interpreted as reflecting life history adaptation (Rich 1925; Reimers 1971; Healey 1980b, 1982b; Levy and Northcote 1981). Differences in the relative length of riverine and estuarine residence by chinook in the Sixes River, Oregon, led Reimers (1971) to postulate five life history types of chinook in that system. From analysis of adult scales, Reimers (1971) concluded that only one of these life history types contributed most to the spawning population in the Sixes River. Healey (1980b, 1982b) recognized two distinct behaviour patterns among ocean-type chinook in Vancouver Island rivers: migration to the river estuary immediately after fry emergence in the spring, and migration to the estuary four to six weeks after emergence.

The existence of this degree of variation indicates considerable plasticity in the species. Whether this plasticity is a consequence of genetic polymorphism, or is largely environmentally induced, remains to be demonstrated. Carl and Healey (1984), however, found genetic differences between fry and fingerling migrants in the Nanaimo River on Vancouver Island, suggesting a genetic basis for the two behaviour patterns of ocean-type chinook recognized by Healey (1980b, 1982b).

Despite the apparent biological and ecological plasticity of the species, the possibility that many populations may be specifically adapted to local conditions cannot be discounted on the basis of present evidence. Such adaptation could have produced unique gene pools in some instances through the fixation of successful mutations. Given the intrapopulation variability of chinook, however, it seems more likely that local adaptations are due to the development of effective gene combinations rather than to fixation of mutations. Furthermore, it suggests that chinook should be able to adapt rapidly to new situations. The success of hatcheries in producing chinook on the Pacific coast, and the success of chinook transplants to New Zealand, Chile, and the Great Lakes testify to the adaptability of the species.

The extent of intrapopulation variation further suggests a degree of "bet hedging" in the life history strategy of the chinook salmon (Stearns 1976). Such a tactic may be very appropriate for a species like chinook, which occurs in many small, possibly locally adapted, spawning populations. Having a variety of life history tactics would serve to spread the risk of mortality across a number of habitats and thus reduce the probability of complete failure of a year class. Indeed, this strategy may partly explain why chinook have been able to persist in the face of continued heavy fishing pressure and, in some systems, significant habitat modification.

References

Alderdice, D.F., and F.P.J. Velsen. 1978. Relation between temperature and incubation time for eggs of chinook salmon (*Oncorhynchus tshawytscha*). J. Fish. Res. Board Can. 35:69–75

Argue, A.W. 1970. A study of factors affecting exploitation of Pacific salmon in the Canadian gauntlet fishery of Juan de Fuca Strait. Fish. Serv. (Can.) Pac. Reg. Tech. Rep. 1970-11:259 p.

Argue, A.W., and D.E. Marshall. 1976. Size and age of chinook and coho salmon for subdivisions of the Strait of Georgia troll fishery, 1966. Fish. Mar. Serv. (Can.) Tech. Rep. PAC/T-76-18:175 p.

Aro, K.V. 1972. Recoveries of salmon tagged offshore in the North Pacific Ocean by Japan and the United States in 1970 and 1971, and additional recoveries from earlier taggings by Canada, Japan, and the United States. Fish. Res. Board Can. MS Rep. Ser. 1186:31 p.

——. Recoveries of salmon tagged offshore in the North Pacific Ocean by Japan and the United States in 1972 and 1973, and additional recoveries from earlier taggings by Canada, Japan, and the United States. Fish. Res. Board Can. MS Rep. Ser. 1328:20 p.

Aro, K.V., and M.P. Shepard. 1967. Pacific salmon in Canada. p. 225–327 *In*: Salmon of the North Pacific Ocean. Part IV. Spawning populations of North Pacific salmon. Int. North Pac. Fish. Comm. Bull. 23

Aro, K.V., J.A. Thomson, and D.P. Giovando. 1971. Recoveries of salmon tagged offshore in the North Pacific Ocean by Canada, Japan, and the United States, 1956 to 1969. Fish. Res. Board Can. MS Rep. Ser. 1147:493 p.

Atkinson, C.E., J.H. Rose, and T.O. Duncan. 1967. Pacific salmon in the United States, p. 43–223. *In*: Salmon of the North Pacific Ocean. Part IV. Spawning populations of North Pacific salmon. Int. North Pac. Fish. Comm. Bull. 23

Ball, E.A.R., and H. Godfrey. 1968a. Lengths and ages of chinook salmon taken in the British Columbia troll fishery and the Fraser River gillnet fishery in 1965. Fish. Res. Board Can. MS Rep. Ser. 952:130 p.

——. 1968b. Lengths and ages of chinook salmon taken in the British Columbia troll fishery and the Fraser River gill net fishery in 1966. Fish. Res. Board Can. MS Rep. Ser. 954:143 p.

——. 1969. Lengths and ages of chinook salmon taken in the British Columbia troll fishery in 1968. Fish. Res. Board Can. MS Rep. Ser. 1073: 41 p.

——. 1970. Lengths and ages of chinook salmon taken in the British Columbia troll fishery in

1969. Fish. Res. Board Can. MS Rep. Ser. 1121: 101 p.

Baranski, C. 1979. Maturity rates for Puget Sound chinook stocks. Wash. Dep. Fish. Tech. Rep. 43:13 p.

Becker, C.D. 1973. Food and growth parameters of juvenile chinook salmon, *Oncorhynchus tshawytscha*, in central Columbia River. Fish. Bull. (U.S.) 71:387-400

Becker, C.D., D.A. Neitzel, and C.S. Abernethy. 1983. Effects of dewatering on chinook salmon redds: tolerance of four development phases to one-time dewatering. N. Am. J. Fish. Manage. 3:373-382

Becker, C.D., D.A. Neitzel, and D.H. Fickeisen. 1982. Effects of dewatering on chinook salmon redds: tolerance of four development phases to daily dewaterings. Trans. Am. Fish. Soc. 111: 624-637

Bell, R. 1958. Time, size, and estimated numbers of seaward migrants of chinook salmon and steelhead trout in the Brownlee-Oxbow section of the middle Snake River. State of Idaho Department of Fish and Game, Boise, ID. 36 p.

Birtwell, I.K. 1978. Studies on the relationship between juvenile chinook salmon and water quality in the industrialized estuary of the Somass River, p. 58-78. *In*: B.G. Shepherd and R.M.J. Ginetz (rapps.). Proceedings of the 1977 Northeast Pacific Chinook and Coho Salmon Workshop. Fish. Mar. Serv. (Can.) Tech. Rep. 759: 164 p.

Bjornn, T.C. 1971. Trout and salmon movements in two Idaho streams as related to temperature, food, stream flow, cover and population density. Trans. Am. Fish. Soc. 100:423-438

Bodznick, D. 1975. The relationship of the olfactory EEG evoked by naturally-occurring stream waters to the homing behaviour of sockeye salmon (*Oncorhynchus nerka* Walbaum). Comp. Biochem. Physiol. 52 (3A):487-495

Borque, S.C., and K.R. Pitre. 1972. Size and maturity of troll chinook salmon (*Oncorhynchus tshawytscha*) caught off the west coast of Vancouver Island in 1969 and 1970. Fish. Mar. Serv. (Can.) Pac. Reg. Tech. Rep. 1972-7:44 p.

Bottom, D. 1984. Food habits of juvenile chinook in Sixes estuary, p. 15-21. *In*: Research and development of Oregon's coastal chinook stocks: annual progress report. Oregon Department of Fish and Wildlife, Portland, OR

Bovee, K.D. 1978. Probability-of-use criteria for the family Salmonidae. U.S. Fish Wildl. Serv. FWS/OBS-78/07; Instream Flow Inf. Paper 4: 80 p.

Brady, J.A. 1983. Lower Yukon River salmon test and commercial fisheries, 1981. Alaska Dep. Fish Game ADF&G Tech. Data Rep. 89:91 p.

Briggs, J.C. 1953. The behaviour and reproduction of salmonid fishes in a small coastal stream. Calif. Dep. Fish Game Fish. Bull. 94:62 p.

Brown, R.L. 1986. 1984 annual report of the Interagency Ecological Studies Program for the Sacramento-San Joaquin estuary. U.S. Fish Wildlife Service, Stockton, CA. 133 p.

Burck, W.A. 1967. Mature stream-reared spring chinook salmon. Res. Briefs Fish Comm. Oreg. 13(1):128 p.

Burck, W.A., and P.E. Reimers. 1978. Temporal and spatial distribution of fall chinook spawning in Elk River. Oreg. Dep. Fish Wildl. Info. Rep. Ser. Fish. 78-3:14 p.

Burner, C.J. 1951. Characteristics of spawning nests of Columbia River salmon. Fish. Bull. Fish Wildl. Serv. 61:97-110

Carl, L.M. 1982. Natural reproduction of coho salmon and chinook salmon in some Michigan streams. N. Am. J. Fish. Manage. 2:375-380

Carl, L.M., and M.C. Healey. 1984. Differences in enzyme frequency and body morphology among three juvenile life history types of chinook salmon (*Oncorhynchus tshawytscha*) in the Nanaimo River, British Columbia. Can. J. Fish. Aquat. Sci. 41:1070-1077

Chapman, D.W., and T.C. Bjornn. 1969. Distribution of salmonids in streams, with special reference to food and feeding, p. 153-176. *In*: T.G. Northcote (ed.). Symposium on Salmon and Trout in Streams. H.R. MacMillan Lectures in Fisheries. Institute of Fisheries, University of British Columbia, Vancouver, BC. 388 p.

Chapman, D.W., D.E. Weitcamp, T.L. Welsh, M.B. Dell, and T.H. Schadt. 1986. Effects of river flow on the distribution of chinook salmon redds. Trans. Am. Fish. Soc. 115:537-547

Chapman, W.M. 1943. The spawning of chinook salmon in the main Columbia River. Copeia 1943:168-170

Chapman, W.M., and E. Quistdorff. 1938. The food of certain fishes of north central Columbia River drainage, in particular, young chinook salmon and steelhead trout. Wash. Dep. Fish. Biol. Rep. 37-A:1-14

Clark, G.H. 1929. Sacramento-San Joaquin salmon (*Oncorhynchus tschawytscha*) [sic] fishery of California. Calif. Dep. Fish Game Fish. Bull. 17:73 p.

Clarke, W.C., and J.E. Shelbourn. 1985. Growth and development of seawater adaptability by juvenile fall chinook salmon (*Oncorhynchus tshawytscha*) in relation to temperature. Aquaculture 45:21-31

Clarke, W.C., J.E. Shelbourn, T. Ogasawara, and T. Hirano. 1989. Effect of initial day length on growth, sea water adaptability and plasma growth hormone levels in underyearling coho, chinook, and chum salmon. Aquaculture 82:51-62

Cleaver, F.C. 1969. Effects of ocean fishing on the 1961-brood fall chinook salmon from Columbia River hatcheries. Res. Rep. Fish Comm. Oreg. 1:76 p.

Clemens, W.A. 1934. The food of young spring salmon in Shuswap Lake, B.C. Can. Field-Nat. 48:142 p.

Clemens, W.A., and J.A. Munro. 1934. The food of the squawfish. Biol. Board Can. Prog. Rep. Pac. Biol. Stn. Nanaimo Fish. Exp. Stn. (Pac.) 19:3-4

Collings, M.R., R.W. Smith, and G.T. Higgins. 1972. Hydrology of four streams in western Washington as related to several Pacific salmon species: Humptulips, Elchoman, Green and Wynoochee rivers: open file report. United States Geological Survey, Tacoma, WA. 128 p.

Congleton, J.L., S.K. Davis, and S.R. Foley. 1981. Distribution, abundance, and outmigration timing of chum and chinook salmon fry in the Skagit salt marsh, p. 153-163. *In*: E.L. Brannon and E.O. Salo (eds.). Proceedings of the Salmon and Trout Migratory Behaviour Symposium. School of Fisheries, University of Washington, Seattle, WA

Conrad, J.F. 1965. Observations on "bad eggs" in Columbia River fall chinook salmon. Prog. Fish-Cult. 27:42-44

Coombs, B.D., and R.E. Burrows. 1957. Threshold temperatures for the normal development of chinook salmon eggs. Prog. Fish-Cult. 19:3-6

Coots, M. 1957. The spawning efficiency of king salmon (*Oncorhynchus tshawytscha*) in Fall Creek, Siskiyou County: 1954-55 investigations. Calif. Dep. Fish Game Inland Fish. Admin. Rep. 57-1:15 p.

Craddock, D.R., T.H. Blahm, and W.D. Parente. 1976. Occurrence and utilization of zooplankton by juvenile chinook salmon in the lower Columbia River. Trans. Am. Fish. Soc. 105:72-76

Cramer, S.P., and J.A. Lichatowich. 1978. Factors influencing the rate of downstream migration of juvenile chinook salmon in the Rogue River, p. 43-48. *In*: B.C. Shepherd and R.M.J. Ginetz (rapps.). Proceedings of the 1977 Northeast Pacific Chinook and Coho Salmon Workshop. Fish. Mar. Serv. (Can.) Tech. Rep. 759:164 p.

Dahlberg, M.L. 1982. Report of incidence of coded-wire tagged salmonids in catches of foreign commercial and research vessels operating in the North Pacific Ocean and Bering Sea during 1980-1982. (Submitted to annual meeting International North Pacific Fisheries Commission, Tokyo, Japan, November 1982.) Auke Bay Laboratory, Northwest and Alaska Fisheries Center, National Marine Fisheries Service, Auke Bay, AK. 11 p.

Dahlberg, M.L., and S. Fowler. 1985. Report of incidence of coded-wire-tagged salmonids in catches of foreign commercial and research vessels operating in the North Pacific Ocean and Bering Sea during 1984-85. (Submitted to annual meeting International North Pacific Fisheries Commission, Tokyo, Japan, November 1985.) Auke Bay Laboratory, Northwest and Alaska Fisheries Center, National Marine Fisheries Service, Auke Bay, AK. 16 p.

Dahlberg, M.L., F.P. Thrower, and S. Fowler. 1986. Incidence of coded-wire-tagged salmonids in catches of foreign commercial and research vessels operating in the North Pacific Ocean and Bering Sea in 1985-1986. (Submitted to annual meeting International North Pacific Fisheries Commission, Anchorage, Alaska, November 1986.) Auke Bay Laboratory, Northwest and Alaska Fisheries Center, National Marine Fisheries Service, Auke Bay, AK. 26 p.

Don Chapman Consultants. 1989. Summer and winter ecology of juvenile chinook salmon and steelhead trout in the Wenatchee River, Washington. Chelan County Public Utility, Wenatchee, WA. 301 p.

Dunford, W.E. 1975. Space and food utilization by salmonids in marsh habitats of the Fraser River estuary. M.Sc. thesis. University of British Columbia, Vancouver, BC. 81 p.

Edmundson, E., F.E. Everest, and D.W. Chapman. 1968. Permanence of station in juvenile chinook salmon and steelhead trout. J. Fish. Res. Board Can. 25:1453-1464

Eisler, R., and R.C. Simon. 1961. Destruction of salmon larvae by *Hydra oligactis*. Trans. Am. Fish. Soc. 90:329-332

Everest, F.H., and D.W. Chapman. 1972. Habitat selection and spatial interaction by juvenile chinook salmon and steelhead trout in two Idaho streams. J. Fish. Res. Board Can. 29:91-100

Fedorenko, A.Y., F.J. Fraser, and D.T. Lightly. 1979. A limnological and salmonid resource study of Nitinat Lake: 1975-1977. Fish. Mar. Serv. (Can.) Tech. Rep. 839:86 p.

Fisher, J.P., W.G. Pearcy, and A.W. Chung. 1983. Studies of juvenile salmonids off the Oregon and Washington coast, 1982. Oreg. State Univ. Coll. Oceanog. Cruise Rep. 83-2:41 p.

——. 1984. Studies of juvenile salmonids off the Oregon and Washington coast, 1983. Oreg. State Univ. Coll. Oceanog. Cruise Rep. 83-2; Oreg. State Univ. Sea Grant Coll. Program ORESU-T-85-004:29 p.

Foerster, R.E., and W.E. Ricker. 1941. The effect of reduction of predaceous fish on survival of young sockeye salmon at Cultus Lake. J. Fish. Res. Board Can. 5:315-336

Fowler, L. 1972. Growth and mortality of fingerling chinook salmon as affected by egg size. Prog. Fish-Cult. 34:66-69

Fraser, C. McL. 1921. Further studies on the growth rate in Pacific salmon. Contrib. Can. Biol. 1918-1920:7-27

Fraser, F.J., P.J. Starr, and A.Y. Fedorenko. 1982. A review of the chinook and coho salmon of the Fraser River. Can. Tech. Rep. Fish. Aquat. Sci. 1126:130 p.

Fry, D.H., Jr., and E.P. Hughes. 1951. The California salmon troll fishery. Pac. Mar. Fish. Comm. Bull. 2:7-42

Galbreath, J.L., and R.L. Ridenhour. 1964. Fecundity of Columbia River chinook salmon. Res. Briefs Fish Comm. Oreg. 10(1):16-27

Gangmark, H.A., and R.G. Bakkala. 1960. A comparative study of unstable and stable (artifical channel) spawning streams for incubating king salmon at Mill Creek. Calif. Fish Game 46:151-164

Gangmark, H.A., and R.D. Broad. 1955. Experimental hatching of king salmon in Mill Creek, a tributary of the Sacramento River. Calif. Fish Game 41:233-242

Gebhards, S.V. 1961. Emergence and mortality of chinook salmon fry in a natural redd. Prog. Fish-Cult. 23:91

Gharrett, A.J., S.M. Shirley, and G.R. Tromble. 1987. Genetic relationships among populations of Alaskan chinook salmon (*Oncorhynchus tshawytscha*). Can. J. Fish. Aquat. Sci. 44:765-774

Gilbert, C.H. 1913. Age at maturity of the Pacific coast salmon of the genus *Oncorhynchus*. Bull. Bur. Fish. (U.S.) 32:1-22

Ginetz, R.M.J. 1976. Chinook salmon in the north coastal division. Fish. Mar. Serv. (Can.) Pac. Reg. Tech. Rep. PAC/T-76-12:96 p.

Godfrey, H. 1968a. Ages and physical characteristics of maturing chinook salmon of the Nass, Skeena, and Fraser Rivers in 1964, 1965, and 1966. Fish. Res. Board Can. MS Rep. Ser. 967:38 p.

——. 1968b. Review of information obtained from the tagging and marking of chinook and coho salmon in coastal waters of Canada and the United States. Fish. Res. Board Can. MS Rep. Ser. 953:172 p.

Godfrey, H.D., D. Worlund, and H.T. Bilton. 1968. Tests on the accuracy of ageing chinook salmon (*Oncorhynchus tshawytscha*) from their scales. J. Fish. Res. Board Can. 25:1971-1982

Goodman, D. 1975. A synthesis of the impacts of proposed expansion of the Vancouver International Airport and other developments on the fisheries resources of the Fraser River estuary. Vol. I and II, Section II. *In*: Fisheries resources and food web components of the Fraser River estuary and an assessment of the impacts of proposed expansion of the Vancouver International Airport and other developments on these resources. Prepared by Department of Environment, Fisheries and Marine Service. Environment Canada, Vancouver, BC

Gordon, D.K., and C.D. Levings. 1984. Seasonal changes of inshore fish populations on Sturgeon and Roberts Bank, Fraser River estuary, British Columbia. Can. Tech. Rep. Fish. Aquat. Sci. 1240:81 p.

Grachev, L.E. 1967. Growth rate of Kamchatka chinook salmon. Izv. Tikhookean. Nauchno-

Issled. Inst. Rybn. Khoz. Okeanogr. 57:89-97. (Transl. from Russian; National Marine Fisheries Service, Northwest Fisheries Center, Seattle, WA)

Groves, A.B., G.B. Collins, and P.S. Trefethen. 1968. Roles of olfaction and vision in choice of spawning site by homing adult chinook salmon (*Oncorhynchus tshawytscha*). J. Fish. Res. Board Can. 25:867-876

Hallock, R.J., and D.H. Fry, Jr. 1967. The five species of salmon, *Oncorhynchus*, in the Sacramento River, California. Calif. Fish Game 53:5-22

Hallock, R.J., D.H. Fry, Jr., and D.A. LaFaunce. 1957. The use of wire fyke traps to estimate the runs of adult salmon and steelhead in the Sacramento River. Calif. Fish Game 43:271-298

Hankin, D.G., and M.C. Healey. 1986. Dependence of exploitation rates for maximum yield and stock collapse on age and sex structure of chinook salmon (*Oncorhynchus tshawytscha*) stocks. Can. J. Fish. Aquat. Sci. 43:1746-1759

Hara, T.J., K. Ueda, and A. Gorbman. 1965. Electroencephalographic studies of homing salmon. Science 149:884-885

Hart, J.L. 1973. Pacific fishes of Canada. Bull. Fish. Res. Board Can. 180:740 p.

Hartt, A.C. 1966. Migrations of salmon in the North Pacific Ocean and Bering Sea as determined by seining and tagging, 1959-1960. Int. North Pac. Fish. Comm. Bull. 19:141 p.

——. 1980. Juvenile salmonids in the oceanic ecosystem; the critical first summer, p. 25-57. *In*: W.J. McNeil and D.C. Himsworth (eds.). Salmonid ecosystems of the North Pacific. Oregon State University Press, Corvallis, OR

Hartt, A.C., and M.B. Dell. 1986. Early oceanic migrations and growth of juvenile Pacific salmon and steelhead trout. Int. North Pac. Fish. Comm. Bull. 46:105 p.

Healey, M.C. 1976. Herring in the diets of Pacific salmon in Georgia Strait. Fish. Res. Board Can. MS Rep. Ser. 1382:38 p.

——. 1980a. The ecology of juvenile salmon in Georgia Strait, British Columbia, p. 203-229. *In*: W.J. McNeil and D.C. Himsworth (eds.). Salmonid ecosystems of the North Pacific. Oregon State University Press, Corvallis, OR

——. 1980b. Utilization of the Nanaimo River estuary by juvenile chinook salmon, *Oncorhynchus tshawytscha*. Fish. Bull. (U.S.) 77:653-668

——. 1982a. Catch, escapement and stock-recruitment for British Columbia chinook salmon since 1951. Can. Tech. Rep. Fish. Aquat. Sci. 1107:77 p.

——. 1982b. Juvenile Pacific salmon in estuaries: the life support system, p. 315-341. *In*: V.S. Kennedy (ed.). Estuarine comparisons. Academic Press, New York, NY

——. 1982c. The distribution and residency of juvenile Pacific salmon in the Strait of Georgia, British Columbia, in relation to foraging success, p. 61-69. *In*: B.R. Melteff and R.A. Nevé (eds.). Proceedings of the North Pacific Aquaculture Symposium. Alaska Sea Grant Rep. 82-2

——. 1983. Coastwide distribution and ocean migration patterns of stream- and ocean-type chinook salmon, *Oncorhynchus tshawytscha*. Can. Field-Nat. 97:427-433

——. 1986. Regional and seasonal attributes of catch in the British Columbia troll fishery. Can. Tech. Rep. Fish. Aquat. Sci. 1494:65 p.

Healey, M.C., and C. Groot. 1987. Marine migration and orientation of ocean-type chinook and sockeye salmon, p. 298-312. *In*: M.J. Dadswell, R.J. Klanda, C.M. Moffitt, R.L. Saunders, R.A. Rulifson, and J.E. Cooper (eds.). Common strategies of anadromous and catadromous fishes. Am. Fish. Soc. Symp. 1

Healey, M.C., and W.R. Heard. 1984. Inter- and intra-population variation in the fecundity of chinook salmon (*Oncorhynchus tshawytscha*) and its relevance to life history theory. Can. J. Fish. Aquat. Sci. 41:476-483

Healey, M.C., and F.P. Jordan. 1982. Observations on juvenile chum and chinook and spawning chinook in the Nanaimo River, British Columbia, during 1975-1981. Can. MS Rep. Fish. Aquat. Sci. 1659:31 p.

Healey, M.C., R.V. Schmidt, F.P. Jordan, and R.M. Hungar. 1977. Juvenile salmon in the Nanaimo area 1975. 2: length, weight, and growth. Fish. Mar. Serv. (Can.) MS Rep. 1438:147 p.

Heg, R., and J. Van Hyning. 1951. Food of the chinook and silver salmon taken off the Oregon coast. Fish Comm. (Oreg.) Res. Briefs 3(2):32-40

Henry, K.A. 1978. Estimating natural and fishing mortalities of chinook salmon, *Oncorhynchus tshawytscha*, in the ocean, based on recoveries of marked fish. Fish. Bull. (U.S.) 76:45-57

Herrmann, R.B. 1970. Food of juvenile chinook and chum salmon in the lower Chehalis River and upper Grays Harbor, p. 59–82. *In*: Grays Harbor cooperative water quality study 1964–1966. Wash. Dep. Fish. Tech. Rep. 7

Hoover, E.E. 1936. Contributions to the life history of the chinook and landlocked salmon in New Hampshire. Copeia 1936:193–198

Hunter, J.G. 1959. Survival and production of pink and chum salmon in a coastal stream. J. Fish. Res. Board Can. 16:835–886

International North Pacific Fisheries Commission (INPFC). 1972–1982. Statistical yearbooks 1970–1979. International North Pacific Fisheries Commission, Vancouver, BC

Irvine, J.R. 1986. Effects of varying discharge on the downstream movement of salmon fry, (*Oncorhynchus tshawytscha*) Walbaum. J. Fish Biol. 28:17–28

Ito, J., Y. Ishida, and S. Ito. 1985. Stock identification of chinook salmon in offshore waters in 1974 based on scale pattern analysis. (Submitted to the annual meeting International North Pacific Fisheries Commission, Tokyo, Japan, November 1985.) Fisheries Agency of Japan, Tokyo, Japan. 14 p.

——. 1986. Further analysis of stock identification of chinook salmon in offshore waters in 1974. (Submitted to the annual meeting International North Pacific Fisheries Commission, Anchorage, Alaska, November 1986.) Fisheries Agency of Japan, Tokyo, Japan. 9 p.

Ito, J., K. Takagi, and S. Ito. 1974. The identification of maturing and immature chinook salmon, *Oncorhynchus tshawytscha* (Walbaum), in the offshore stage and some related information. Bull. Far Seas Fish. Res. Lab. 11:67–75

Jefferts, K.B., P.K. Bergman, and H.F. Fiscus. 1963. A coded wire identification system for macro-organisms. Nature 198:460–462

Kissner, P.D., Jr. 1973. Annual progress report for a study of chinook salmon in southeast Alaska. Study AFS-41-1 Alaska Department of Fish and Game, Division of Sport Fish, Juneau, AK. 24 p.

Kjelson, M.A., P.F. Raquel, and F.W. Fisher. 1981. Influences of freshwater inflow on chinook salmon (*Oncorhynchus tshawytscha*) in the Sacramento-San Joaquin estuary, p. 88–102. *In*: R.D. Cross and D.L. Williams (eds.). Proceedings of the National Symposium on Freshwater Inflow to Estuaries. U.S. Fish Wildl. Serv. Biol. Serv. Prog. FWS/OBS-81/04(2)

——. 1982. Life history of fall-run juvenile chinook salmon, *Oncorhynchus tshawytscha*, in the Sacramento-San Joaquin estuary, California, p. 393–411. *In*: V.S. Kennedy (ed.). Estuarine comparisons. Academic Press, New York, NY

Knudsen, C.M., C.K. Harris, and N.D. Davis. 1983. Origins of chinook salmon in the area of the Japanese mothership and landbased driftnet salmon fisheries in 1980. Univ. Wash. Fish. Res. Inst. FRI-UW-8315:71 p.

Kondo, H., Y. Hirano, N. Nakayama, and M. Miyaki. 1965. Offshore distribution and migration of Pacific salmon (genus *Oncorhynchus*) based on tagging studies (1958–1961). Int. North Pac. Fish. Comm. Bull. 17:213 p.

Kwain, W., and J.A. Chappel. 1978. First evidence for even-year spawning pink salmon, *Oncorhynchus gorbuscha*, in Lake Superior. J. Fish. Res. Board Can. 35:1373–1376

Kwain, W., and E. Thomas. 1984. The first evidence of spring spawning by chinook salmon in Lake Superior. N. Am. J. Fish. Manage. 4:227–228

Levings, C.D. 1982. Short term use of a low tide refuge in a sandflat by juvenile chinook, *Oncorhynchus tshawytscha*, Fraser River estuary. Can. Tech. Rep. Fish. Aquat. Sci. 1111:33 p.

Levings, C.D., C.D. McAllister, and B.D. Chang. 1986. Differential use of the Campbell River estuary, British Columbia, by wild and hatchery-reared juvenile chinook salmon (*Oncorhynchus tshawytscha*). Can. J. Fish. Aquat. Sci. 43:1386–1397

Levy, D.A., and T.G. Northcote. 1981. The distribution and abundance of juvenile salmon in marsh habitats of the Fraser River estuary. Westwater Res. Cent. Univ. Br. Col. Tech. Rep. 25:117 p.

——. 1982. Juvenile salmon residency in a marsh area of the Fraser River estuary. Can. J. Fish. Aquat. Sci. 39:270–276

Levy, D.A., T.G. Northcote, and G.J. Birch. 1979. Juvenile salmon utilization of tidal channels in the Fraser River estuary, British Columbia. Westwater Res. Cent. Univ. Br. Col. Tech. Rep. 23: 70 p.

Lindbergh, J.M. 1982. A successful transplant of Pacific salmon to Chile. Proc. Gulf Caribb. Fish. Inst. 34:81–87

Lister, D.B., and H.S. Genoe. 1970. Stream habitat utilization by cohabiting underyearlings of chinook (*Oncorhynchus tshawytscha*) and coho (*O.*

kisutch) salmon in the Big Qualicum River, British Columbia. J. Fish. Res. Board Can. 27: 1215–1224

Lister, D.B., and C.E. Walker. 1966. The effect of flow control on freshwater survival of chum, coho, and chinook salmon in the Big Qualicum River. Can. Fish Cult. 37:3–25

Lister, D.B., C.E. Walker, and M.A. Giles. 1971. Cowichan River chinook salmon escapements and juvenile production 1965–1967. Fish. Serv. (Can.) Pac. Reg. Tech. Rep. 1971-3: 8 p.

Loeffel, R.E., and H.O. Wendler. 1969. Review of the Pacific coast chinook and coho salmon resources with special emphasis on the troll fishery, p. 1–107. *In*: Informal Committee on Chinook and Coho. Reports by the United States and Canada on the status, ocean migrations, and exploitation of northeast Pacific stocks of chinook and coho salmon, to 1964. Vol. 1: report by the United States Section.

Loftus, W.F., and H.L. Lenon. 1977. Food habits of the salmon smolts, *Oncorhynchus tshawytscha* and *O. keta*, from the Salcha River, Alaska. Trans. Am. Fish. Soc. 106:235–240

Mains, E.M., and J.M. Smith. 1964. The distribution, size, time, and current preferences of seaward migrant chinook salmon in the Columbia and Snake Rivers. Wash. Dep. Fish. Fish. Res. Pap. 2(3):5–43

Major, R.L., J. Ito, S. Ito, and H. Godfrey. 1978. Distribution and origin of chinook salmon (*Oncorhynchus tshawytscha*) in offshore waters of the North Pacific Ocean. Int. North Pac. Fish. Comm. Bull. 38:54 p.

Major, R.L., and J.L. Mighell. 1969. Egg-to-migrant survival of spring chinook salmon (*Oncorhynchus tshawytscha*) in the Yakima River, Washington. Fish. Bull. (U.S.) 67:347–359

Major, R.L., S. Murai, and J. Lyons. 1977. Scale studies to identify Asian and western Alaskan chinook salmon. Int. North Pac. Fish. Comm. Annu. Rep. 1975:68–71

Manzer, J.I., T. Ishida, A.E. Peterson, and M.G. Hanavan. 1965. Salmon of the North Pacific Ocean. Part V. Offshore distribution of salmon. Int. North Pac. Fish. Comm. Bull. 15:452 p.

McBride, D.N., H.H. Hamner, and L.S. Buklis. 1983. Age, sex, and size of Yukon River salmon catch and escapement, 1982. Alaska Dep. Fish Game ADF&G Tech. Data Rep. 90:141 p.

McGregor, E.A. 1922. Observations on the egg yield of Klamath River king salmon. Calif. Fish Game 8:160–164

———. 1923. A possible separation of the river races of king salmon in ocean-caught fish by means of anatomical characters. Calif. Fish Game 9:138–150

McIsaac, D.O., and T.P. Quinn. 1988. Evidence for a hereditary component in homing behavior of chinook salmon (*Oncorhynchus tshawytscha*). Can. J. Fish. Aquat. Sci. 45:2201–2205

McLeod, C.L., and J.P. O'Neil. 1983. Major range extensions of anadromous salmonids and first record of chinook salmon in the Mackenzie River drainage. Can. J. Zool. 61:2183–2184

McPhail, J.D., and C.C. Lindsey. 1970. Freshwater fishes of northwestern Canada and Alaska. Bull. Fish. Res. Board Can. 173:381 p.

Meehan, W.R., and D.B. Siniff. 1962. A study of the downstream migrations of anadromous fishes in the Taku River, Alaska. Trans. Am. Fish. Soc. 91:399–407

Merkel, T.J. 1957. Food habits of the king salmon, *Oncorhynchus tshawytscha* (Walbaum), in the vicinity of San Francisco, California. Calif. Fish Game 43:249–270

Merrell, D.J. 1981. Ecological genetics. University of Minnesota Press, Minneapolis, MN. 500 p.

Miller, D.R., J.G. Williams, and C.W. Sims. 1983. Distribution, abundance, and growth of juvenile salmonids off the coast of Oregon and Washington, summer 1980. Fish. Res. 2:1–17

Milne, D.J. 1955. Selectivity of trolling lures. Fish. Res. Board Can. Prog. Rep. Pac. Coast Stn. 103:3–5

———. 1964. Sizes and ages of chinook, *Oncorhynchus tshawytscha*, and coho, *O. kisutch*, salmon in the British Columbia troll fisheries (1952–1959) and the Fraser River gill-net fishery (1956–1959). Fish. Res. Board Can. MS Rep. Ser. 776:42 p.

Mundie, J.H. 1969. Ecological implications of the diet of juvenile coho in streams, p. 135–152. *In*: T.G. Northcote (ed.). Symposium on Salmon and Trout in streams. H.R. MacMillan Lectures in Fisheries. Institute of Fisheries, University of British Columbia, Vancouver, BC

Murphy, M.L., J. Heifetz, J.F. Thedinga, S.W. Johnson, and K.V. Koski. 1989. Habitat utilization by juvenile Pacific salmon (*Oncorhynchus*) in the glacial Taku River, southeast Alaska. Can. J. Fish.

Aquat. Sci. 46:1677-1685

Myers, K.W. 1986. The effect of altering proportions of Asian chinook stocks on regional scale pattern analysis. (Submitted to the annual meeting International North Pacific Fisheries Commission, Anchorage, Alaska, November 1986.) Univ. Wash. Fish. Res. Inst. FRI-UW-8605:44 p.

Myers, K.W., C.K. Harris, C.M. Knudsen, R.V. Walker, N.D. Davis, and D.E. Rogers. 1987. Stock origins of chinook salmon in the area of the Japanese mothership salmon fishery. N. Am. J. Fish. Manage. 7:459-474

Myers, K.W., and H.F. Horton. 1982. Temporal use of an Oregon estuary by hatchery and wild juvenile salmon, p. 377-392. *In*: V.S. Kennedy (ed.). Estuarine comparisons. Academic Press, New York, NY

Myers, K.W., D.E. Rogers, C.K. Harris, C.M. Knudsen, R.V. Walker, and N.D. Davis. 1984. Origins of chinook salmon in the area of the Japanese mothership and landbased driftnet salmon fisheries in 1975- 1981. (Submitted to annual meeting International North Pacific Fisheries Commission, Vancouver, British Columbia, November 1984.) Fisheries Research Institute, University of Washington, Seattle, WA. 204 p.

Myers, K.W.W. 1980. An investigation of the utilization of four study areas in Yaquina Bay, Oregon, by hatchery and wild juvenile salmonids. M.Sc. thesis. Oregon State University, Corvallis, OR. 233 p.

Neave, F. 1943. Diurnal fluctuations in the upstream migration of coho and spring salmon. J. Fish. Res. Board Can. 6:158-163

——. 1958. The origin and speciation of *Oncorhynchus*. Proc. Trans. R. Soc. Can. Ser. 3, 52(5):25-39

Neilson, J.D., and C.E. Banford. 1983. Chinook salmon (*Oncorhynchus tshawytscha*) spawner characteristics in relation to redd physical features. Can. J. Zool. 61:1524-1531

Neilson, J.D., and G.H. Geen. 1981. Enumeration of spawning salmon from spawner residence time and aerial counts. Trans. Am. Fish. Soc. 110:554-556

——. 1986. First-year growth rate of Sixes River chinook salmon as inferred from otoliths: effects on mortality and age at maturity. Trans. Amer. Fish. Soc. 115:28-33

Neilson, J.D., G.H. Geen, and D. Bottom. 1985. Estuarine growth of juvenile chinook salmon (*Oncorhynchus tshawytscha*) as inferred from otolith microstructure. Can. J. Fish. Aquat. Sci. 42:899-908

Nicholas, J.W., and D.G. Hankin. 1988. Chinook salmon populations in Oregon coastal river basins: description of life histories and assessment of recent trends in run strength. Oreg. Dep. Fish. Wildl. Info. Rep. 88-1:1-359

Northcote, T.G. 1976. Biology of the lower Fraser and ecological effects of pollution, p. 85-119. *In*: A.H.J. Dorcey, I.K. Fox, K.J. Hall, T.G. Northcote, K.G. Peterson, W.H. Sproule-Jones, and J.H. Weins (eds.). The uncertain future of the lower Fraser. Westwater Research Centre, University of British Columbia, Vancouver, BC

Northcote, T.G., N.T. Johnston, and K. Tsumura. 1979. Feeding relationships and food web structure of lower Fraser River fishes. Westwater Res. Cent. Univ. Br. Col. Tech. Rep. 16:73 p.

Novotny, A.J., and C.V.W. Mahnken. 1971. Predation on juvenile Pacific salmon by a marine isopod, *Rocinella belliceps pugettensis* (Crustacea, Isopoda). Fish. Bull. (U.S.) 69:699-701

Paine, J.R., F.K. Sandercock, and B.A. Minaker. 1975. Big Qualicum River project 1972-1973. Fish. Mar. Serv. (Can.) Tech. Rep. PAC/T-75-15: 126 p.

Park, D.L. 1969. Seasonal changes in downstream migration of age-group 0 chinook salmon in the upper Columbia River. Trans. Am. Fish. Soc. 98:315-317

Parker, R.R. 1962. Estimations of ocean mortality rates for Pacific salmon (*Oncorhynchus*). J. Fish. Res. Board Can. 19:561-589

——. 1965. Estimation of sea mortality rates for the 1961 brood-year pink salmon of the Bella Coola area, British Columbia. J. Fish. Res. Board Can. 22:1523-1554

Parker, R.R., and W. Kirkness. 1956. King salmon and the ocean troll fishery of southeastern Alaska. Alaska Dep. Fish. Res. Rep. 1:64 p.

Patten, B.G. 1971. Predation by sculpins on fall chinook salmon, *Oncorhynchus tshawytscha*, fry of hatchery origin. U.S. Nat. Mar. Fish. Serv. Spec. Sci. Rep. Fish. 621:14 p.

Pauley, G.B. 1967. Prespawning adult salmon mortality associated with a fungus of the genus *Dermocystidium*. J. Fish. Res. Board Can. 24: 843-848

Prakash, A. 1962. Seasonal changes in feeding of

coho and chinook (spring) salmon in southern British Columbia waters. J. Fish. Res. Board Can. 19:851–866

Pritchard, A.L., and A.L. Tester. 1944. Food of spring and coho salmon in British Columbia. Bull. Fish. Res. Board Can. 65:23 p.

Quinn, T.P., and K. Fresh. 1984. Homing and straying in chinook salmon (*Oncorhynchus tshawytscha*) from Cowlitz River hatchery, Washington. Can. J. Fish. Aquat. Sci. 41:1078–1082

Raymond, H.L. 1968. Migration rates of yearling chinook salmon in relation to flows and impoundments in the Columbia and Snake rivers. Trans. Am. Fish. Soc. 97:356–359

Real, L.A. 1980. Fitness, uncertainty, and the role of diversification in evolution and behavior. Amer. Nat. 155:623–638

Reid, G. 1961. Stomach content analysis of troll-caught king and coho salmon, southeastern Alaska, 1957–1958. U.S. Fish Wildl. Serv. Spec. Sci. Rep. Fish. 379:8 p.

Reimers, P.E. 1968. Social behaviour among juvenile fall chinook salmon. J. Fish. Res. Board Can. 25:2005–2008

——. 1971. The length of residence of juvenile fall chinook salmon in Sixes River, Oregon. Ph.D. thesis. Oregon State University, Corvallis, OR. 99 p.

Reimers, P.E., and R.E. Loeffel. 1967. The length of residence of juvenile fall chinook salmon in selected Columbia River tributaries. Res. Briefs Fish Comm. Oreg. 13:5–19

Reimers, P.E., J.W. Nicholas, D.L. Bottom, T.W. Downey, K.M. Maciolek, J.D. Rodgers, and B.A. Miller. 1979. Coastal salmon ecology project. Annual progress report, October 1, 1978 to September 30, 1979; completion report, July 1, 1976 to September 30, 1979. Oregon Department of Fish and Wildlife, Portland, OR. 44 p.

Reimers, P.E., J.W. Nicholas, T.W. Downey, R.E. Halliburton, and J.D. Rodgers. 1978. Fall chinook ecology project. Federal aid progress report fisheries AFC-76-2. Oregon Department of Fish and Wildlife, Portland, OR. 52 p.

Rich, W.H. 1920. Early history and seaward migration of chinook salmon in the Columbia and Sacramento Rivers. Bull. Bur. Fish. (U.S.) 37:74 p.

——. 1925. Growth and degree of maturity of chinook salmon in the ocean. Bull. Bur. Fish. (U.S.) 41:15–90

——. 1942. The salmon runs of the Columbia River in 1938. Fish. Bull. Fish Wildl. Serv. 50:101–147

Rich, W.H., and H.B. Holmes. 1928. Experiments in marking young chinook salmon on the Columbia River 1916 to 1927. Bull. Bur. Fish. (U.S.) 44:215–264

Ricker, W.E. 1976. Review of the rate of growth and mortality of Pacific salmon in salt water and noncatch mortality caused by fishing. J. Fish. Res. Board Can. 33:1483–1524

Ricker, W.E., and K.H. Loftus. 1968. Pacific salmon move east. Fish. Counc. Can. Annu. Rev. 43:37–43.

Robinson, C.K., L.A. Lapi, and E.W. Carter. 1982. Stomach contents of spiny dogfish (*Squalus acanthias*) caught near the Qualicum and Fraser Rivers, April-May, 1980–1981. Can. MS Rep. Fish. Aquat. Sci. 1656:21 p.

Rombough, P.J. 1985. Initial egg weight, time to maximum alevin weight, and optimal ponding times for chinook salmon, *Oncorhynchus tshawytscha*. Can. J. Fish. Aquat. Sci. 42:287–291

Rounsefell, G.A. 1957. Fecundity of North American Salmonidae. Fish. Bull. Fish Wildl. Serv. 57:451–468

Russell, L.R., K.R. Conlin, O.K. Johansen, and U. Orr. 1983. Chinook salmon studies in the Nechako River: 1980, 1981, 1982. Can. MS Rep. Fish. Aquat. Sci. 1728:185 p.

Rutter, C. 1902. Natural history of the quinnat salmon: a report of investigations in the Sacramento River 1896–1901. Bull. U.S. Fish. Comm. 22:65–142

Sasaki, S. 1966. Distribution and food habits of king salmon, *Oncorhynchus tshawytscha*, and steelhead rainbow trout, *Salmo gairdnerii*, in the Sacramento-San Joaquin delta, p. 108–114. *In*: J.L. Turner and D.W. Kelley (comp.). Ecological studies of the Sacramento-San Joaquin Delta. Calif. Dep. Fish Game Fish. Bull. 136

Schluchter, M.D., and J.A. Lichatowich. 1977. Juvenile life histories of Rogue River spring chinook salmon, *Oncorhynchus tshawytscha* (Walbaum), as determined by scale analysis. Oreg. Dep. Fish. Wildl. Info. Rep. Ser. Fish. 77-5:24 p.

Shaw, P.A., and J.A. Maga. 1943. The effect of mining silt on yield of fry from salmon spawning beds. Calif. Fish Game 29:29–41

Shelton, J.M. 1955. The hatching of chinook salmon

eggs under simulated stream conditions. Prog. Fish-Cult. 17:20-35

Shelton, J.M., and R.D. Pollock. 1966. Siltation and egg survival in incubation channels. Trans. Am. Fish. Soc. 95:183-187

Shepherd, B. 1975. Upper Skeena chinook stocks: evaluation of the Bear-Sustut, Morice, and lower Babine stocks. Department of Environment, Fisheries and Marine Service, Pacific Region, North Operations Branch, Vancouver, BC. 41 p.

Sibert, J., and B. Kask. 1978. Do fish have diets? p. 48-57. *In*: B.G. Shepherd and R.M. Ginetz (rapps.). Proceedings of the 1977 Northeast Pacific Chinook and Coho Salmon Workshop. Fish. Mar. Serv. (Can.) Tech. Rep. 759

Silliman, R.P. 1941. Fluctuations in the diet of the chinook and silver salmons (*Oncorhynchus tschawytscsha* [sic] and *O. kisutch*) off Washington, as related to troll catch of salmon. Copeia 1941 (2):80-87

——. 1950. Fluctuations in abundance of Columbia River chinook salmon (*Oncorhynchus tschawytscha*) [sic] 1935-45. Fish. Bull. Fish Wildl. Serv. 51:364-383

Silver, S.J., C.E. Warren, and P. Doudoroff. 1963. Dissolved oxygen requirements of developing steelhead trout and chinook salmon embryos at different water velocities. Trans. Am. Fish. Soc. 92:327-343

Slater, D.W. 1963. Winter-run chinook salmon in the Sacramento River, California, with notes on water temperature requirements at spawning. U.S. Fish Wildl. Serv. Spec. Sci. Rep. Fish. 461:9 p.

Smith, A.K. 1973. Development and application of spawning velocity and depth criteria for Oregon salmonids. Trans. Am. Fish. Soc. 102:312-316

Smith, G.R. 1975. Fishes of the Pliocene Glenns Ferry formation, southwest Idaho. Mus. Paleontol. Pap. Paleontol. 14:1-68

Snyder, J.O. 1931. Salmon of the Klamath River, California. Calif. Fish Game Fish. Bull. 34:130 p.

Stearns, S.C. 1976. Life history tactics: a review of the ideas. Q. Rev. Biol. 51:3-47

Stein, R.A., P.E. Reimers, and J.D. Hall. 1972. Social interaction between juvenile coho (*Oncorhynchus kisutch*) and fall chinook salmon (*O. tshawytscha*) in Sixes River, Oregon. J. Fish. Res. Board Can. 29:1737-1748

Taylor, E.B. 1988. Adaptive variation in rheotactic and agonistic behavior in newly emerged fry of chinook salmon, *Oncorhynchus tshawytscha*, from ocean- and stream-type populations. Can. J. Fish. Aquat. Sci. 45:237-243

Taylor, E.B., and P.A. Larkin. 1986. Current response and agonistic behavior in newly emerged fry of chinook salmon, *Oncorhynchus tshawytscha*, from ocean- and stream-type populations. Can. J. Fish. Aquat. Sci. 43:565-573

Taylor, F.H.C. 1969. The British Columbia offshore herring survey, 1968-69. Fish. Res. Board Can. Tech. Rep. 140:54 p.

Thomas, A.E., J.L. Banks, and D.C. Greenland. 1969. Effect of yolk sac absorption on the swimming ability of fall chinook salmon. Trans. Am. Fish. Soc. 98:406-410

Thomas, W.K., R.E. Withler, and A.T. Beckenbach. 1986. Mitochondrial DNA analysis of Pacific salmonid evolution. Can. J. Zool. 64:1058-1064

Thompson, R.B. 1959a. Food of the squawfish, *Ptychocheilus oregonensis* (Richardson), of the lower Columbia River. Fish. Bull. Fish Wildl. Serv. 60:43-58

——. 1959b. A study of localized predation on marked chinook salmon fingerlings released at McNary dam. Wash. Dep. Fish. Fish. Res. Pap. 2:82-83

Tsuyuki, H., and E. Roberts. 1966. Inter-species relationships within the genus *Oncorhynchus* based on biochemical systematics. J. Fish. Res. Board Can. 23:101-107

Tsuyuki, H., E. Roberts, and W.E. Vanstone. 1965. Comparative zone electropherograms of muscle myogens and blood hemoglobins of marine and freshwater vertebrates and their application to biochemical systematics. J. Fish. Res. Board Can. 22:203-213

Uremovich, B. 1977. Straying of fall chinook salmon from Elk River hatchery into Sixes River, 1970-1976. Oreg. Dep. Fish Wildl. Info. Rep. Ser. Fish. 77-6:7 p.

Utter, F., G. Milner, G. Stahl, and D. Teel. 1989. Genetic population structure of chinook salmon (*Oncorhynchus tshawytscha*), in the Pacific northwest. Fish. Bull (U.S.) 87:239-264

Van Hyning, J.M. 1951. The ocean salmon troll fishery of Oregon. Pac. Mar. Fish. Comm. Bull. 2:43-76

Vronskiy, B.B. 1972. Reproductive biology of the Kamchatka River chinook salmon (*Oncorhynchus tshawytscha* (Walbaum)). J. Ichthyol. 12:259-273

Wagner, H.H., F.P. Conte, and J.L. Fessler. 1969. Development of osmotic and ionic regulation in two races of chinook salmon, *Oncorhynchus tshawytscha*. Comp. Biochem. Physiol. 29:325–341

Wahle, R.J., E. Chaney, and R.E. Pearson. 1981. Areal distribution of marked Columbia River basin spring chinook salmon recovered in fisheries and at parent hatcheries. Mar. Fish. Rev. 43(12):1–9

Wahle, R.J., and R.R. Vreeland. 1977. Bioeconomic contribution of Columbia River hatchery fall chinook salmon, 1961 through 1964 broods, to the Pacific salmon fisheries. Fish. Bull. (U.S.) 76:179–208

Waite, D.C. 1979. Chinook enhancement on the Kenai peninsula. Preliminary report. Study No. AFS-46. Alaska Department of Fish and Game, Juneau, AK. 51 p.

Wales, J.H., and M. Coots. 1954. Efficiency of chinook salmon spawning in Fall Creek, California. Trans. Am. Fish. Soc. 84:137–149

Walters, C.J., R. Hilborn, R.M. Peterman, and M.J. Staley. 1978. Model for examining early ocean limitation of Pacific salmon production. J. Fish. Res. Board Can. 35:1303–1315

Waugh, G.D. 1980. Salmon in New Zealand, p. 277–303. *In*: J.E. Thorpe (ed.). Salmon ranching. Academic Press, New York, NY

Weisbart, M. 1968. Osmotic and ionic regulation in embryos, alevins, and fry of the five species of Pacific salmon. Can. J. Zool. 46:385–397

Wertheimer, A.C., and M.L. Dahlberg. 1983. Report of the incidence of coded-wire tagged salmonids in catches of foreign commercial and research vessels operating in the North Pacific Ocean and Bering Sea during 1982–1983. (Submitted to annual meeting International North Pacific Fisheries Commission, Anchorage, Alaska, November, 1983.) Auke Bay Laboratory, Northwest and Alaska Fisheries Center, National Marine Fisheries Service, Auke Bay, AK. 14 p.

——. 1984. Report of the incidence of coded-wire tagged salmonids in catches of foreign commercial and research vessels operating in the North Pacific Ocean and Bering Sea during 1983–1984. (Submitted to annual meeting International North Pacific Fisheries Commission, Vancouver, British Columbia, November, 1984.) Auke Bay Laboratory, Northwest and Alaska Fisheries Center, National Marine Fisheries Service, Auke Bay, AK. 14 p.

Wickett, W.P. 1954. The oxygen supply to salmon eggs in spawning beds. J. Fish. Res. Board Can. 11:933–953

Winans, G.A. 1989. Genetic variability in chinook salmon stocks from the Columbia River basin. N. Am. J. Fish Manage. 9:47–52

Wright, S., and J. Bernhard. 1972. Maturity rates of ocean-caught chinook salmon. Pac. Mar. Fish Comm. Bull. 8:49–59

Wright, S., R. Kolb, and R. Brix. 1972. Size and age characteristics of chinook salmon taken by Washington's commercial troll and ocean sport fisheries 1963–1969. Pac. Mar. Fish. Comm. Bull. 8:37–47

Yancey, R.M., and F.V. Thorsteinson. 1963. The king salmon of Cook Inlet Alaska. U.S. Fish Wildl. Serv. Spec. Sci. Rep. Fish. 440:18 p.

Life History of Coho Salmon

CONTENTS

HEINE

18

PLATE 18. Coho fry in full colour, Rosewall Creek, British Columbia. *Photograph by Marj Trim*

19

PLATE 19. Early coho fry in gravel bed prior to emerging, Rosewall Creek, British Columbia. *Photograph by Marj Trim*

PLATE 17 (*previous page*). Coho salmon life history stages. *Painting by H. Heine*

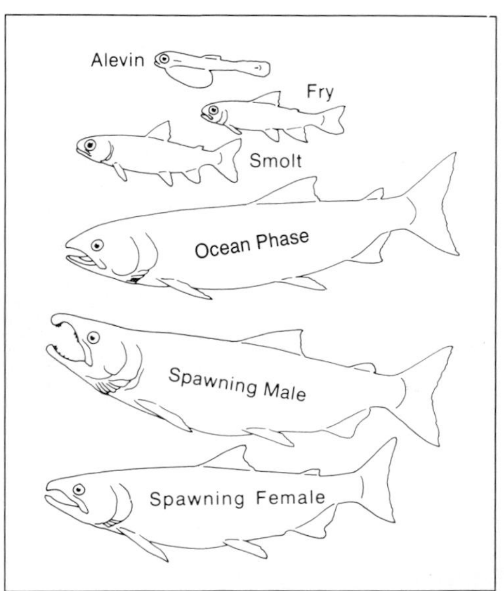

LIFE HISTORY OF COHO SALMON (*Oncorhynchus kisutch*)

F.K. Sandercock*

Introduction

COHO SALMON were first described by Walbaum in 1792 as *Salmo kisutch*, the specific name being the vernacular for coho in Kamchatka, USSR (McPhail and Lindsey 1970). Coho (*Oncorhynchus kisutch*), one of the seven recognized species of Pacific salmon belonging to the genus *Oncorhynchus*, are widely distributed in commercially harvestable quantities throughout their natural range, from the Soviet Far East around the Bering Sea, to Alaska, and south along the North American coast to California (Hart 1973). During the 1970s the world catch of all Pacific salmon averaged just over 400 million kg annually and of this total, approximately 9% consisted of coho, with the North American catch about double the Asian catch (Fredin 1980).

The basic life history pattern for this species begins as adult salmon migrate from the sea into streams to deposit their eggs in gravel. Each female produces several thousand eggs, which are reduced in number by a high mortality during the coho's early life history (Salo and Bayliff 1958). After spawning, the adults die. The eggs incubate during winter in the gravel, and in the spring free-swimming fry emerge. The fry take up residency in the stream for a year or more, migrate to sea as smolts, and then begin their rapid growth phase (Davidson and Hutchinson 1938). After eighteen months or more at sea, the now maturing adults travel hundreds of kilometres across ocean waters, up streams, and through lakes to return to their place of origin (Hoar 1958). Within this basic pattern there are a great many variations that have evolved in response to opportunity and selective pressures.

Relative Abundance

In most areas of the North Pacific, coho occur in small numbers compared to other species of Pacific salmon and represent less than 10% of the total catch (INPFC 1979). About 1.43 million coho of Asian origin were caught per year in the Japanese high-seas fisheries and the Soviet coastal fishery from 1925 to 1951 (Fredin 1980). Pravdin (1940) noted that catches in the early years were not a good indicator of abundance because many of the Kamchatkan fisheries were shut down before the peak of the late-running coho migration. Increased exploitation rates beginning about 1952 expanded the average annual catch in the 1952–76 period to

*Department of Fisheries and Oceans, Salmonid Enhancement Program, 555 West Hastings Street, Vancouver, British Columbia V6B 5G3

about 4.64 million fish (INPFC 1979).

In North America, fishing for coho in western Alaska is considered relatively unimportant, with an annual catch of about 100,000 fish (McPhail and Lindsey 1970). Elsewhere in North America, the commercial catch of coho by American and Canadian fishermen is about 50% greater than the combined USSR–Japan catch (INPFC 1979). From 1952 to 1976 the annual catch averaged 7.46 million coho with a peak in 1986 of 10.6 million (INPFC 1979). The average annual commercial catch of coho by region (from 1920 to 1976) was: central Alaska, 0.7 million; southeastern Alaska, 1.5 million; British Columbia, 2.7 million; Washington, 1.0 million; Oregon, 0.7 million; and California (1963–76), about 330,000 (INPFC 1979).

SPAWNING POPULATIONS

Distribution

Endemic populations of coho are found throughout the North Pacific basin (Figure 1) and they are distributed widely in other cold temperate areas as a result of introductions. The genus *Oncorhynchus* may have evolved from an ancestral *Salmo* in the Sea of Japan during the early Pleistocene (Neave 1958). Geological evidence suggests that the Sea of Japan, the Bering Sea, and the Sea of Okhotsk may have become separated during a later glacial period, thus providing for geographic isolation that

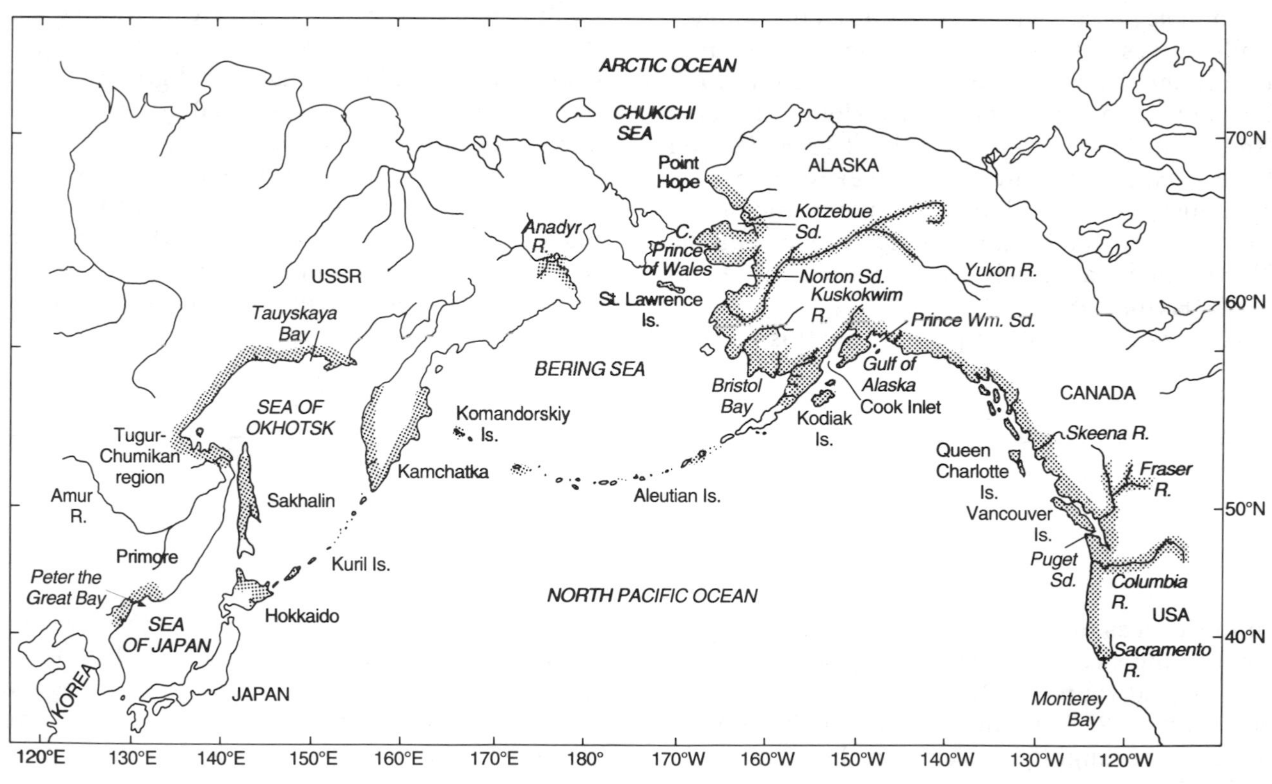

FIGURE 1
Coastal and spawning distribution of coho salmon

led to speciation within the genus *Oncorhynchus*.

Godfrey (1965) noted that coho, like many other species, are less common in the northern and southern fringes of their distribution, and they are most abundant in the central portion of their range. On the Asian side of the Pacific Ocean, coho have been reported as far south as Chongjin on the east coast of North Korea (Matsubara 1955, as cited by Lindberg and Legeza 1965) (Figure 1).

Jordan and Snyder (1902) reported that coho were distributed in the area around Japan, from Otaru on the west coast of Hokkaido, and from Osatsubo and the Ura River. Hikita (1956) indicated that coho were rarely found in Hokkaido, but that some specimens had been examined from the Yurappu and Shokotsu rivers; however, other reports have suggested that coho were absent from both Hokkaido and the waters adjacent to Japan (FRBC 1955).

Coho occur in small numbers along the Kuril Islands chain (Iturup and Etorofu islands) (Okada 1960) and in the Naiba and Tym river systems on Sakhalin Island (Smirnov 1960; Godfrey 1965). Dvinin (1952) reported that the catch of coho from south Sakhalin was negligible. On the mainland side of the Sea of Japan, Lindberg and Legeza (1965) noted that coho are rarely found as far south as Peter the Great Bay. Berg (1948) reported that the most southerly record for Asian coho was from the Suchan River, which enters Peter the Great Bay. Coho are also uncommon on the mainland side of the Sea of Okhotsk, though they are found in the Iski River, just to the north of the mouth of the Amur River (Berg 1948). Coho are rare in the Shantarski Islands and in the Tugur-Chumikan region but they do occur in the most northern part of the Sea of Okhotsk, in the Okhota and Kukhtuy rivers (Shmidt 1950). Moreover, coho runs into Tauyskaya Bay and the Ola River were sufficient to support a commercial fishery (Shmidt 1950).

On the southwest coast of Kamchatka, the most important salmon-producing river is the Bolshaya, which yields 25%–30% of the regional catch. However, more than 90% of the production is pink salmon (*Oncorhynchus gorbuscha*), with coho representing only 1.4% of the catch (Semko 1954). Pravdin (1940) noted that much of the coho run occurred after fishing terminated in early September. Other coho-producing streams on the west coast of Kamchatka include the Icha (Gribanov 1948) and Kikhchik (Pravdin 1940) rivers.

On the east coast of Kamchatka, the major coho producer is the Kamchatka River (Gribanov 1948). Other coho systems are the Paratunka (Popov 1933), and the Kalyger, Kyrganik, and Ozernaya rivers (Gribanov 1948). Further to the east, Berg (1948) reported that coho were uncommon in the Komandorskiy Islands. The most northerly occurring Asian coho were reported in the vicinity of the Anadyr River estuary (Andriashev 1955).

Although Asian coho have not been reported north of 65°N, North American coho have been found above latitude 68°N. In Alaska, the most northern coho population is found in the Kukpuk River near Point Hope on the Chukchi Sea (Wahle and Pearson 1987). Coho have also been found in the Singoolik, Kivalina, Wulik, Noatak, Buckland, and Inmachuk rivers, all of which enter Kotzebue Sound, Alaska. Information on coho from Point Hope to Cape Prince of Wales is sparse because they occur there in relatively small numbers and arrive in the rivers too late to be captured by the native fishermen. South of Cape Prince of Wales on the Bering Strait, coho have been identified from the Agiapuk and Kuzitrin rivers near Port Clarence and from rivers tributary to Norton Sound, such as the Snake, Nome, Fish, Swiniuk, Shaktoolik, Unalakleet, and tributaries of the Yukon River. Wahle and Pearson (1987) also reported coho from the Koozata River on St. Lawrence Island in the northern Bering Sea.

Coho are caught all along the Alaskan coast from Norton Sound to the mouth of the Kuskokwim River (INPFC 1962b). From the Yukon River south to Bristol Bay, coho commonly migrate in coastal streams and well up into the Yukon and Kuskokwim rivers (McPhail and Lindsey 1970). However, Hartt and Dell (1986) concluded that the small numbers of coho found in the eastern Bering Sea were indicative of the relatively small populations of coho that originated in the streams of Bristol Bay and vicinity. Coho are also found in the streams of Attu, Kiska, Adak, Kagalaska, Atka, and Unimak islands of the Aleutian Islands chain (Wahle and Pearson 1987) and have been stocked into five lakes along the Kenai Peninsula (Engel 1972). Throughout the northern Gulf of Alaska and southeastern Alaska, coho spawn in most coastal streams (Atkinson et al. 1967). Coho spawners have been reported in the Klukshu River (Wynne-Edwards

1947) and Village Creek (Hancock and Marshall 1984a), both tributaries of the Tatshenshini River in the Yukon Territory.

In British Columbia, coho are found in most coastal streams. In the larger rivers, including the Fraser, Skeena, Bella-Coola, Nass, and Taku rivers, they migrate some distance inland and spawn in the smaller tributaries. It was estimated by Aro and Shepard (1967) that coho spawned in 970 of the 1,500 known salmon-bearing streams in British Columbia. Coho is the most widespread of the five species of Pacific salmon and no one area of the province is the dominant producer (Milne 1964).

On the east coast of the Queen Charlotte Islands, the Tlell River (Graham Island) and Copper Creek (Moresby Island) are important coho spawning streams. In the northern mainland of British Columbia the most important coho-producing streams are the Lakelse River, a tributary of the Skeena River, and the Bella Coola/Atnarko River system.

The major individual spawning streams in southern British Columbia are the Kingcome, Kakweiken, Nimpkish, Oyster, Toba, Cowichan, San Juan, Squamish, and Chilliwack rivers (Aro and Shepard 1967). Along the west coast of Vancouver Island there are numerous small producers of coho salmon. The Somass River system, tributary to Barkley Sound, is recorded as being the most important of these rivers.

The streams of coastal Washington and Puget Sound are abundant producers of coho (Atkinson et al. 1967). Coho are found in many tributaries along the Washington side of the lower Columbia River and are known to spawn as far northeast as the Wenatchee River in the upper Columbia River basin (Wahle and Pearson 1987).

In Oregon, coho are found in many of the tributaries of the lower Columbia and Willamette rivers, as well as in most coastal streams south to the Rogue River (Atkinson et al. 1967). Wahle and Pearson (1987) observed that coho migrate south in the Willamette River as far as the McKenzie River tributary and east in the Columbia River to the Grande Ronde River, passing through southeastern Washington, southwestern Idaho, and into northeastern Oregon via the Snake River.

The distribution of coho in California has been well documented (Atkinson et al. 1967). They occur in most of the coastal streams from the California/Oregon border south to the San Lorenzo River on Monterey Bay (Wahle and Pearson 1987). Coho are rare in the Sacramento/San Joaquin River system (Hallock and Fry 1967) and occur only in small numbers in the Klamath River (Snyder 1931) and elsewhere. High summer temperatures probably limit their freshwater distribution (Fry 1977). At sea, coho are caught consistently as far south as Monterey Bay, and two coho have been documented from Port Chamalu Bay, Baja California; one was caught 4 August 1942, and a second fish weighing 6.35 kg was caught 20 August 1963 (Messersmith 1965).

In summary, the normal distribution of coho extends from northern Japan through Kamchatka, across the Bering Sea to Alaska, and south through all coastal areas to California (Figure 1). There are, however, a number of locations to which coho eggs or fry have been transplanted in an attempt to establish a landlocked or an anadromous population outside of the North Pacific basin.

Transplants

Coho have been introduced into many areas of North America, Asia, Europe, and South America, with great success in some cases and no success in others. In North America, as early as 1873, attempts were made to introduce coho into the Great Lakes (Scott and Crossman 1973). Thousands of fry were released into Lake Erie between 1873 and 1878, and again in 1933, but although some coho up to 2.3 kg were caught in 1935, the transplant was considered unsuccessful (Scott and Crossman 1973). More recently, coho that were released into two Montana lakes (Anonymous 1951) survived to maturity but were small and had relatively few eggs (about 700 eggs per female); average weight for males was 0.43 kg and for females 0.50 kg (Beal 1955). Coho were released into Parvin Lake and the Granby Reservoir, Colorado, in 1963 (Klein and Finnell 1969). Coho fry were also planted into Lake Berryessa, Lake Almanor, Oroville Lake, and the Merle Collins Reservoir in California in the early 1970s (Wigglesworth and Rawstron 1974). In the Colorado and California releases, no natural reproduction occurred. In 1971, the province of Alberta introduced coho into Cold Lake, some of which were later recovered downstream in Pierce Lake, Saskatchewan (Scott and Crossman 1973). These and other small-scale introduction experiments can be best regarded as unsuccessful or inconclusive.[1]

It was not until the state of Michigan began releasing coho smolts in 1966 into the Great Lakes (Lake Michigan and Lake Superior) that the potential to establish exotic or non-native coho populations was realized. Soon after the unplanned, but successful, introduction of pink salmon into Lake Superior (Kwain and Lawrie 1981), the state of Michigan released 660,000 coho smolts into Lake Michigan (Wells and McLain 1972) and 192,000 into Lake Superior (GLFC 1970). Early jack returns were very promising, so the program was escalated. In Lake Michigan the releases increased to 1.7, 1.2, and 3.3 million smolts in 1967, 1968, and 1969, respectively, with the states of Wisconsin and Illinois contributing to the smolt production in Lake Michigan. In Lake Superior the coho releases were increased from the original 192,000 (from Michigan) to 467,000, 382,000, and 656,000 during the same years, with contributions from the state of Minnesota and the province of Ontario. As the catches and escapements increased there was considerable public pressure to expand the program to the other Great Lakes. Michigan released 402,000 and 667,000 coho smolts into Lake Huron in 1968 and 1969, respectively. The states of Ohio, Pennsylvania, and New York released 111,000 and 236,000 smolts into Lake Erie in 1968 and 1969, and New York and Ontario released 41,000 and 239,000 coho into Lake Ontario in the same two years. By the end of 1977, the following number of coho smolts or fry had been released in each lake: Lake Superior, 5.23 million; Lake Michigan, 30.06 million; Lake Huron, 5.20 million; Lake Erie, 6.80 million; and Lake Ontario 4.25 million; for a total of more than 51 million coho (GLFC 1980).

The success of these transplants was phenomenal. Survival rates of fry to adults (catch plus escapement) for the early releases into Lake Michigan ranged from 19% to 32%, with the fish averaging 74 cm and 4.3 kg (GLFC 1970). In Lake Superior the survival rate for the first three plantings ranged from 2% to 24%. The lower productivity of this lake resulted in slower growth and spawners that averaged only 1.3 kg in weight (Lawrie and Rahrer 1972). In Lake Huron the catch was equally divided between sport and commercial fisheries and totalled about 17% of the juveniles released; average weight for a mature spawner was 4.1 kg (GLFC 1970). In Lake Erie the survival was about 25% and the average weight was about 2.7 kg (Hartman 1972). Most of these fish were caught in the colder, deeper Canadian waters near Long Point, Ontario. Lake Ontario coho reached an average size of 63 cm and 2.3 kg at maturity (Scott and Crossman 1973), but the overall survival rate was low due to lamprey attacks (GLFC 1970).

Following the successful introduction of coho into the Great Lakes, other smaller lakes in the region were planted with coho. These included the Stormy and Pallette lakes in Wisconsin; Hare Lake, Minnesota; and Hemlock Lake in Michigan (Engel and Magnuson 1976; McKnight and Serns 1977).

Long before the introduction of coho into the Great Lakes area, attempts had been made to establish Pacific salmon on the east coast of North America. Most of this effort took place between 1901 and 1910, and a few transplants were carried out as late as 1930. Davidson and Hutchinson (1938), in their review of the geographic and environmental limits of Pacific salmon, reported the following coho egg or fry transplants: Maine, 1.4 million; Maryland, 12,000; New Hampshire, 315,000; New York, 13,500; and to the interior states of Vermont and Pennsylvania, 47,000 and 355,000, respectively. There is no record that any of these releases produced adult salmon.

Ricker (1954) reported that juvenile coho had been introduced into the Ducktrap River in Maine between 1943 and 1948 and that, subsequently, a run of about one hundred adults had returned to spawn in 1952. Ricker was uncertain whether this run would be self-sustaining. The dream of establishing a self-sustaining, east-coast run of coho persists. Symons and Martin (1978) indicated that coho juveniles have been released into a New Hampshire stream since 1969 and into a Massachusetts stream since 1971. In addition, two aquaculture operations have been rearing coho in Maine. One of these sites is thought to have released, intentionally or unintentionally, some coho juveniles that strayed into Frost Fish Creek in New Brunswick in the fall of 1976. No adult coho are known to have subsequently returned to this

1 Hasler and Farner (1942) reported on growth, age, and feeding habits of silver (silverside) salmon (*O. kisutch*) in Crater Lake, Oregon 1935–40. The present author believes this to be an error, and that, in fact, the salmon were kokanee (*O. nerka*).

creek or any other New Brunswick stream.

In the 1970s an attempt was made to introduce Washington coho into South Korean streams (R.J. Wahle, Pacific Marine Fisheries Commission, Seattle, Washington, pers. comm.). The probability of such a transplant succeeding was very low considering that the natural occurrence of coho in this area is extremely rare. Recently, there were a number of attempts to establish self-sustaining runs of coho in the Yurappu River (Ishida et al. 1975), Shibetsu River (Ishida et al. 1976; Nara et al. 1979), and the Ichani River (Umeda et al. 1981); these streams are all located on the east or southeast coast of Hokkaido. Coho eggs were supplied from the 1973–78 (excluding 1975) broods from Washington (University of Washington) and Oregon (Eagle Creek) hatcheries. These transplants appear to have been unsuccessful.

Coho transplant attempts to Europe have included shipments of eggs from British Columbia to Scotland, France, West Germany, and Cyprus, primarily for experimental use or for pen rearing. None of these transplants resulted in the establishment of self-sustaining populations.

In South America, considerable effort has been made to establish coho runs for ocean ranching in Chile. Davidson and Hutchinson (1938) noted that 255,000 coho eggs had been shipped to Chile and 377,000 to Argentina, with the additional comment that "sockeye and coho (were) successfully introduced to Chile." The success of this transplant was not substantiated and, like many others, it probably failed after one or two cycles. In the last decade, large releases of coho smolts have been made in southern Chile (R.E. Noble, Union Carbide, Olympia, Washington, pers. comm.). These smolts were derived from surplus production at Washington hatcheries. Although the adult returns were small, there were sufficient numbers to justify continuation of the experiment in the hope that the transplant might succeed here as well as it did in the Great Lakes.

Spawning Migration

Seasonal Timing

Coho begin to mature during the summer after one winter at sea and arrive at their rivers of origin during late summer and autumn. In some cases the journey is a short one along coastal routes, but in many other instances the spawning run may take one to two months and cover many hundreds of kilometres of open ocean. Successive generations of each stock appear in the estuary and ascend the spawning stream about the same time each year (Royce et al. 1968). In general, the higher the latitude, the earlier the timing (Briggs 1953). In northern Alaska and Kamchatka the migration begins in July and August (Pravdin 1940; Godfrey 1965), in British Columbia the normal timing is September/October (Fraser et al. 1983), whereas in California spawning migrations may be delayed until November/December (Shapovalov and Taft 1954).

Throughout the range of coho there are also many exceptions to the normal timing patterns. They have been observed to leave the marine environment and enter freshwater streams as early as April, e.g., the Capilano River, British Columbia (F.K. Sandercock, unpublished data), and as late as March, e.g., the Kamchatka River (Smirnov 1960) and Waddell Creek, California (Shapovalov and Taft 1954). In one year, Foerster and Ricker (1953) observed coho in early April in Sweltzer Creek, British Columbia. For most stocks, the duration of the spawning migration appears to be three months or more. Fraser et al. (1983) reported a duration of 106 ± 21 days for Big Qualicum River coho. However, Pritchard (1943) reported that the 1942 spawning run in the Cowichan River, British Columbia, took 20–30 days, and the fish were spawned out in 30–60 days.

Coho rarely exhibit seasonal runs to single tributaries (Ricker 1972). Summer and autumn runs into the Paratunka River (USSR) have been noted, as have summer and winter runs into some Kamchatkan rivers (Gribanov 1948). In those cases where coho migrate at unusual times or over a short period, such behaviour appears to have evolved in response to particular flow conditions. For example, obstructions that may be passable under high discharge conditions may be insurmountable during low flows (Neave and Wickett 1953). Conversely, many early-timing runs are thought to have developed because early-entry coho could surmount obstacles during low or moderate flows but not during high flows. It can be concluded that, in such cases, these obstacles might become velocity barriers once the autumn rains begin. In California, many of the smaller coastal streams do not have sufficient flow during the summer and early autumn to breach the sand bars that are thrown up across the mouths of the streams by wave action. Fresh water entering the ocean is by seepage only, and coho cannot enter these streams until the autumn rains produce enough discharge to breach the sand barriers. This usually takes place in October/November (Briggs 1953). Holtby et al. (1984) observed that coho returning to Carnation Creek on the west coast of Vancouver Island moved into the stream on a continual basis, provided that autumn freshets were sustained. In years when freshets were infrequent, the migration was pulsed. When the initial freshet was delayed until late October, 70% of the escapement entered the stream over several days. Returning adults apparently gather at the mouths of shallow coastal streams and then move upstream on high water.

There is also a tendency for fish that migrate early to move further upstream than those that migrate later (Briggs 1953). In the Kamchatka River the early run migrates 25–30 km, whereas the late run migrates only 2–3 km up tributaries (Gribanov 1948). Mid- to late-migrating fish generally return to their natal streams in a more advanced state of maturity and closer to the onset of spawning. However, entry of coho into streams is not necessarily dependent on their state of maturity, as both ripe (mature) and green (sexually immature) fish may occur in all parts of the run (Shapovalov and Taft 1954).

If conditions (flow, temperature, etc.) in the stream are unsuitable, the fish will often mill about in the vicinity of the stream mouth, sometimes waiting weeks or even, in the case of early-timing fish, months for conditions to change. As temperatures decrease and rainfall and flow increase (Gribanov 1948), the coho will make short excursions into the stream and then return to salt water. Coho generally begin their upstream migration when there is a large increase in flow, particularly when combined with a high tide. This was observed by Neave (1943) in the Cowichan River, British Columbia; Sumner (1953) in Sand Creek, Oregon; Shapovalov and Taft (1954) in Waddell and Scott creeks, California; and Fraser et al. (1983) in the Big Qualicum River, British Columbia. The latter authors noted that in some years there were secondary peaks in coho migration during stable or decreasing discharge periods and that migration may be related to factors other than flow. Shapovalov and Taft (1954) also observed that migration occurred on both rising and falling stream flows but not during peak floods.

Reiser and Bjornn (1979) noted that coho normally migrate when water temperature is in the range of 7.2°–15.6°C, the minimum depth is 18 cm, and the water velocity does not exceed 2.44 m/s. This pattern of migration allows coho to reach very small headwater tributaries where good spawning and rearing conditions may be found

Diel Timing

Most coho stocks actively migrate upstream during daylight hours rather than at night. Brett and MacKinnon (1954) observed that 90% of fish moving up a fishway in the Stamp River (Vancouver Island) did so between 0800 and 1830 hours, with a peak between 1400 and 1600 hours. Neave (1943) reported that when there were large numbers of coho moving, the peak migration occurred at midday; when the numbers were smaller there were two peaks, 0700–1000 and 1500–1700 hours. Artificial light had no effect on migration. Other factors, such as water turbidity, degree of sexual maturity, and size of run may influence migration fluctuations (Shapovalov and Taft 1954).

Ellis (1962) observed that, early in the season, coho migrated actively right after dawn, whereas, late in the season (September/October), they began their daily migration with low numbers moving at first light, and with progressively greater numbers moving up to 8 hours later. There was also a tendency for small groups of fish to move earlier in the day than large groups. In the Big Qualicum River, British Columbia, coho moved upstream during all daylight hours, but peak activities occurred at dawn and sunset (Fraser et al. 1983).

Migration Behaviour

During their migration upstream, coho can frequently be seen breaking the surface or jumping clear of the water, whether there is an obstruction nearby or not. Fish holding downstream of an obstruction appear to make a number of feeble attempts to jump, as if to gauge the height or degree of difficulty in overcoming the obstacle. When they repeatedly fail to clear the obstacle, they drop back and spawn downstream in whatever sites are available. In passage through a fishway, Brett and MacKinnon (1965) noted that the fish spent a brief period of reconnoitring in each pool before moving to the next pool. Only infrequently did the fish return downstream. Vertical leaps of more than 2 m are possible.

Shapovalov (1947) noted that coho crossed over low obstructions with a characteristic rolling motion. Briggs (1953) observed coho moving across riffles where the volume of water was as low as 3.4 m^3/min and the water depth only 5 cm.

Since coho are vulnerable to predation while they are migrating through shallow riffle areas, they move through these areas as quickly as possible and seek the deeper, quieter pools. They then rest in these pools before migrating further upstream. Ellis (1962) provided a detailed account of the behaviour of coho migrants in the Somass River, British Columbia. The fish were seen to move quickly through a set of rapids and into a holding pool. On entering the quiet water they then swam steadily along the deepest channel, close to the bottom. The fish held in pools for some duration, periodically jumping or wandering in schools up and down the pool. As they approached the shallow upper end of the pool, a few fish darted over the shallows and up through the next set of rapids.

Migrant Types

Milne (1950) suggested that there are probably two distinct types of coho in British Columbia. The "ocean" type primarily occupy outer coastal or offshore waters, whereas the "inshore" type remain within inside waters during their saltwater life history phase. Taylor and McPhail (1985) also recognized two forms of the fish in their study of the body morphology of coho from the upper Columbia River to Alaska: a "coastal" form, characterized by large median fins and a deep robust body; and an "interior" form with small median fins and a more streamlined body. The characteristics of the latter type were thought to be an adaptation for the long and often arduous migrations in fresh water. In swimming tests, this type outperformed the coastal fish. However, the coastal types exhibited greater morphological variation within and between river systems, which suggests that the straying rate, and hence the gene flow, may be greater among coastal-type coho.

Rate of River Migration

Ellis (1962) reported that at river velocities of 1.0 m/s or less, coho maintained steady swimming speeds of 1.2–1.7 m/s without stopping. Migration speed is equal to swimming speed minus water velocity. Reiser and Bjornn (1979) found that the cruising speed for coho was up to 1.04 m/s, that sustained swimming ranged from 1.04 to 3.23 m/s, and that darting speeds were 3.23–6.55 m/s. Ellis (1962) observed that when stream velocities reach 1.5 m/s or more, the tendency for fish to school broke down, and the steady swimming mode was replaced by resting and darting.

Maximum non-sustainable swimming speeds up to 11 m/s have been recorded (Ellis 1962), and it has been calculated that, under a steady flow condition of 0.4 m/s, coho would be expected to migrate about 2.7 km/h (Ellis 1966). A lower rate of migration was observed by Neave (1949), who noted that coho moved 32 km upstream in two days in the Cowichan River, British Columbia. Assuming that the fish actually migrated 12 hours in each day, this represents a rate of 1.3 km/h.

Upstream Migration

Throughout their range, coho spawn in streams along the coast and in small tributaries of larger rivers (Rounsefell and Kelez 1940). Coho migrate further upstream than pink and chum (*Oncorhynchus keta*) salmon but usually not as far as sockeye (*O. nerka*) and chinook (*O. tshawytscha*). Godfrey (1965) reported that, in general, coho seldom migrate more than 240 km up large rivers to spawn. However, there are some notable exceptions. In the Kamchatka River, coho migrate over 550 km to reach some upper tributaries (Berg 1948). McPhail and Lindsey (1970) reported that coho ascend the main stem of the Yukon River almost to the Alaska/Yukon border, a distance of 1,830 km. More recent work has indicated that coho travelled about 200 km further up the Yukon River to Dawson and beyond, but the location of the spawning grounds was unknown (Ennis et al. 1982). Bryan (1973), in a biological survey of the Yukon Territory, discovered coho at Old Crow on the Porcupine River, and 350 km further upstream on the Fishing Branch River, a total migratory distance from salt water of over 2,200 km. Hancock and Marshall (1984b) reported that the heaviest concentration of spawners on the Fishing Branch River was near the Bear Cave Mountain area. In British Columbia, the long migration stocks travel about 510 km in the Skeena River; and, in the Fraser River system, they migrate 550 km in the South Thompson, 570 km in the North Thompson, and 680 km to the tributaries of the upper Fraser River. Further south, coho have been reported in the Grande Ronde River in northeastern Oregon, a distance of approximately 800 km from the Columbia River estuary (Wahle and Pearson 1987)

Age at Time of Return

Throughout their normal range, the majority of coho mature in their third year of life, having spent about four to six months in incubation and up to fifteen months rearing in fresh water, followed by a sixteen-month growing period in sea water. Based on their scale patterns, these fish are generally designated as age 1.1 (i.e., one winter in fresh water and one winter in salt water). There are, however, many variations to this normal pattern. Some of the males mature precociously and return to spawn after only four to six months in sea water and are referred to as "jacks" (age 1.0), and others may stay in fresh water for two winters and return as age 2.1 fish.

Godfrey (1965) reported that during the period 1926–37, coho from the east coast of Kamchatka were predominantly age 2.1, whereas those from west Kamchatka matured at age 1.1. Since 1959 there has been a shift towards age 2.1 fish.

Many coho from central and southeastern Alaska also mature at age 2.1: up to 60% in the Yukon River (Gilbert 1922), and up to 93% in the Swanson River, Alaska (Engel 1968). However, the proportion may vary from year to year. For example, in Sashin Creek in southeastern Alaska, age 2.1 fish comprised 78%, 59%, 64%, and 62% of the stock in the years 1965 through 1967 and in 1969 (Crone and Bond 1976). In western Alaska, the predominant age at maturity is 1.1 (70.8%), and in the Aleutian Islands age 1.1 fish make up 60% of the stock (INPFC 1962a). Drucker (1972) observed an unusual age distribution for coho in the Karluk River on Kodiak Island, where 56.9% were age 2.1, 41.7% were age 3.1, and 1.4% were age 4.1. Late-maturing age 3.1 fish have been reported from a few other systems in Alaska but the proportion is generally less than 5%. It has been suggested that juveniles that live in Alaskan lakes during their period of freshwater residency may go to sea at an older age than those residing in rivers. However, in Oregon the growth of coho juveniles in lakes is faster than in streams. The lake rearing smolts migrate at the same age as the stream dwellers but are larger in size (A. McGie, Department of Fish and Wildlife, Corvallis, Oregon, pers. comm.).

Pritchard (1940) presented a full spectrum of ages at maturity for coho from British Columbia. He observed 0.1, 0.2, 0.3, 1.0, 1.1, 1.2, 2.0, and 2.1 age groups from scale readings but acknowledged that 97.9% of the coho examined from 6,312 fishery samples were age 1.1. Pritchard did not have the benefit of examining marked fish of known ages to validate age readings from scales, so one might question the accuracy of those observations, particularly with respect to the rarer groups. In general, there is a decrease in the number of age 2.1 fish from north to south (Gilbert 1913), with age 1.1 fish comprising 95% of the stock in British Columbia (Foerster 1955), and virtually no coho of age 2.1 occurring further south (Fry and Hughes 1954). Neave and Pritchard (1942) observed some four-year-old returns (1.2 or 2.1) from their marking experiments on coho from the Cowichan River, British Columbia, but it is possible that the delayed maturity was a result of the trauma associated with the marking.

Precocious males or "jacks," which mature mostly one year earlier than the majority of coho, are a highly variable component of the escapement population. Fraser (1920) reported 28 age 1.0 jacks in a sample of 2,000 coho taken from the Strait of Georgia, the remainder being age 1.1. Marr (1944), in a study of Columbia River coho, observed that about 6% of the jacks were age 1.0, 84% were age 1.1, just under 10% were age 2.1, and a few were age 2.0 (the number of jacks was probably underestimated because of gillnets selectively harvesting larger fish). Neave (1949) observed a jack return to the Cowichan River of 1%–13% over several years. Wickett (1951) reported a return of only 8 jacks among 1,883 coho counted through a fence on Nile Creek, British Columbia, in a six-year period. Foerster and Ricker (1953) observed that the jack count was always greater than the return of age 1.1 fish of the same year class in Sweltzer Creek, British Columbia. In California, Murphy (1952) summarized the counts of coho jacks passing the Benbow Dam on the south fork of the Eel River over the period 1939–51. The number of jacks ranged from 6.9% to 33.8% (average 18%) for a given return year. Morgan and Henry (1959) reported that the jack return to the Ten Mile Lakes (Oregon) in 1955 represented 46% of the total return; this high percentage may be attributed to the larger average size of the smolts migrating out of the lakes (A. McGie, Department of Fish and Wildlife, Corvallis, Oregon, pers. comm.). Salo and Bayliff (1958), in their study of wild coho returns to Minter Creek (Washington), reported jack returns of 21% and 27% in two consecutive years. Andersen and Narver (1975) also observed a high rate (32%) for jack returns in a wild coho population in Carnation Creek, British Columbia.

From these and other studies of both hatchery-produced and wild coho it is obvious that the number of jacks returning to a given system is highly variable between years and between systems. It has been well demonstrated by Bilton et al. (1984) that coho that migrate earlier than average, and at a size larger than average, tend to produce a high rate of jack returns. These larger smolts represent the fast growing component of a specific brood. The jacks are known to contribute to the fertilization of naturally spawned eggs by darting in beside a full-sized male and female during the spawning process. How their genes influence the wild population is not known. During routine hatchery operations, jacks are excluded from fertilization because it is believed that the prevalence of jacks in the subsequent return would increase, and because they are considered less productive due to their small body size.

Berg (1948) suggested that in some coho populations the jacks may not migrate to sea but may mature in fresh water. This possibility is also suggested for those stocks which undergo long freshwater migrations, such as in the Yukon, Fraser, and Columbia rivers. It is speculated that the time involved in the downstream and upstream migration would preclude jacks from travelling much further than down to the estuary and then back to the spawning grounds.

Size at Time of Return

Size at the time of return is variable and may be influenced by sex, age, time position in the run, and perhaps other factors. Marr (1943) observed that males were generally larger than females, older fish larger than younger fish, fish in late runs larger than those in early runs, and fish in southern stocks (on the average) larger than those in northern stocks. In addition, Salo and Bayliff (1958) noted that, for a given stock (Minter Creek, Washington), the fish sampled at the peak of migration tended to be larger than both the early-returning and the late-returning fish.

Fraser et al. (1983), in a study of 2,513 coho from the Big Qualicum River, found that the average length of the three-year-olds (age 1.1) was 52.7 ± 3.2 cm (4 four-year-olds (age 2.1) had a mean length of 56.6 ± 4.7 cm) and that there was no significant difference between males and females. Andersen and Narver (1975) observed that the average length of male coho returning to Carnation Creek (British Columbia) was 58.1 cm compared to 66.9 cm for females. Engel (1968) found that the mean length of coho from Swanson River (Alaska) was 60.4 cm for males and 62.6 cm for females. Gribanov (1948) reported that Kamchatkan coho (males and females combined) ranged from 40 to 88 cm fork length but more commonly averaged 55–69 cm. He also found that males tended to be larger than females.

Over their normal distribution range, there does not appear to be any clear pattern for size. For Kamchatkan coho the average weight is 3.0–3.5 kg with a range of 1.2–6.8 kg (Gribanov 1948); for the

Resurrection Bay area (Alaska) the average is 3.34 kg (McHenry 1981); and for southeastern Alaska, 4.8 kg (females only) (Marriott 1968). In British Columbia the average weight for coho is 4.0 kg in the Cowichan River (Neave 1949), 4.0 kg in the Big Qualicum River (F.K. Sandercock, unpublished data), 3.0 kg in the Capilano River (F.K. Sandercock, unpublished data); and 3.22 kg for all British Columbia commercial fisheries between 1952 and 1961 (Godfrey 1965). For the Columbia River the average weight was 4.5 kg (Cleaver 1951), whereas in California coho commonly weigh 3.2–5.5 kg (Fry and Hughes 1954).

Coho weighing 6.0 kg are not unusual, but any over 9.0 kg are rare. The largest coho caught to date in the Great Lakes was landed in Lake Ontario near Pulaski, New York, on 7 September 1984, and weighed 11.16 kg (G. Radonski, United States Sport Fisheries Institute, Washington, DC, pers. comm.). The largest coho on record was caught off Victoria, British Columbia, in 1947 and weighed 14.0 kg (Hart 1973).

Jacks are substantially smaller than normal adults. Foerster and Ricker (1953) reported that jacks from Sweltzer Creek (British Columbia) averaged 30 cm in length (range 27–34 cm); in Carnation Creek (British Columbia) they averaged 35.6 cm (Andersen and Narver 1975); and in the Big Qualicum River they averaged 34.2 ± 5.1 cm (Fraser et al. 1983). A total of 356 jacks sampled in Waddell Creek (California) averaged 40.9 cm in length over a nine-year period (Shapovalov and Taft 1954).

Sex Ratio

In theory, the sex ratio should be 1:1 males to females. It is assumed that, to at least the migrant stage, there is no differential mortality associated with one or the other sex. Once the fish are at sea, males or females may be subject to different rates of predation, but this seems unlikely. The greatest differential mortality is associated with the commercial and recreational fishery, which is highly selective. Because of gear restrictions and other regulations, there is a strong selection for large fish, especially females, which means more jacks can escape to the spawning grounds. Evidence for selection of females is not strong for coho salmon, but a higher rate of exploitation of large female chinook has been demonstrated (D.E. Marshall, Department of Fisheries and Oceans, Vancouver, British Columbia, pers. comm.).

Foerster and Ricker (1953) noted that it is surprising that there was no constant excess of age 1.1 females over males, because the jacks (age 1.0 males) often outnumbered the age 1.1 males of the same year class. The removal of the jacks from the population must upset the original sex ratio of 1:1 found at the smolt stage.

Marr (1943) thought that the sex ratio of age 1.1 fish in the Columbia River was about 1:1 but noted that there were more males than females in the early part of the run. Sumner (1953) observed a preponderance of males in both the early and late part of the run in Sand Creek, Oregon, but as Godfrey (1965) noted, the higher proportion of males on the spawning grounds at the end of the run may simply reflect the fact that the males live longer than the females.

Some published accounts of sex ratios, especially those in the Cook Inlet area of Alaska, have indicated a greater abundance of males throughout the run. Logan (1967) observed male to female ratios ranging from 0.9:1 to 1.2:1, Engel (1968) found a ratio of 1.2:1 in Swanson River, and McHenry (1981) reported a very high ratio of 2.1:1 for Bear Creek coho. Further south, on the west coast of Vancouver Island, Andersen and Narver (1975) observed a ratio of 1.26:1 in Carnation Creek. Hunter (1949) reported a ratio of 2.07:1 for coho from Port John, British Columbia.

The results of a study by Shapovalov and Taft (1954) on California coho are more typical. They noted that there is characteristically an excess of females in the age 1.1 group, but if the 1.0 and 1.1 males are combined (for a given year class) they will outnumber the females. As an example, Salo and Bayliff (1958) showed that for age 1.1 wild coho returning to Minter Creek in two consecutive years, the females constituted 55.6% and 57.6% of the runs. If, however, the jacks were added to the 1.1 male count, then the males comprised 59.7% and 54.7% of the runs. Based on twelve years of observations, Fraser et al. (1983) noted an average of 55.7% females (range 44.6%–72.7%) among age 1.1 fish. The small sample of age 2.1 fish were 51.9% female. Similar results were reported by Berg (1948) for the Kamchatka River, USSR (55.5% females) and by Godfrey et al. (1954) for the Babine

River, British Columbia (54% females).

Overall, more coho males than females survive, because those that return to spawn as jacks are not subjected to as high a mortality rate as the adults that spend two summers at sea. Presumably, if all coho returned to the home stream at the same age, as do pink salmon, the observed average sex ratio would be 1:1.

Sexual Dimorphism

During the early freshwater and marine stages of their life history there is no apparent external phenotypic difference between male and female coho. However, with the onset of maturity the fish develop markedly different secondary sexual characteristics (Plate 17).

In male coho, the upper jaw forms an elongated hooked snout and the teeth become greatly enlarged. The hook, which turns downward, may be of sufficient size to prevent the mouth from closing. The lower jaw also elongates somewhat and may become hooked (upward) or knobbed. The dorsal area between the head and dorsal fin is projected slightly upward, thereby increasing the body depth (Briggs 1953). The colour of the spawning male is generally brighter than that of the female (Plate 17).

In females, the jaws also elongate, but the development is much less extreme than that observed in the males (McPhail and Lindsey 1970). The dorsal projection seen on the males is absent and the colour of the females is much more subdued (Plate 17). Marr (1943) and Shapovalov and Taft (1954) found that there was a small but consistent tendency for the males of a given year class to be larger than the females.

Adult Colour

Coho captured at sea or shortly after entry into fresh water are mostly silver-coloured on their sides and ventral surfaces. The dorsal surface is a dark metallic blue and there are irregular black spots on the back and the upper lobe of the caudal fin (Hart 1973) (Plate 17).

As spawning time approaches, the males become darker, and the dorsal surface, head, and ventral surface turn bluish green. The sides of the males develop a broad red streak, which, in some populations is very bright (Carl et al. 1959). The females and jacks are not nearly as brightly coloured but appear more brassy green (Shapovalov and Taft 1954). Coho that remain in fresh water until maturity, such as those found in the Great Lakes and the "residuals" observed by Foerster and Ricker (1953), are generally a duller colour at maturity than their anadromous counterparts.

Fecundity

The number of eggs carried by ripe coho females varies with the region and with the size of fish. Rounsefell (1957) provided one of the first reviews of the fecundity of North American salmonids. More recently, Crone and Bond (1976) summarized the available data on fecundity of coho salmon and acknowledged that the numbers given were not strictly comparable because of the various methods used to determine egg number and because of the large variations in sample size (Table 1). There is a definite tendency for fecundity to increase from California to Alaska, and North American stocks generally have a higher number of eggs than Asian stocks.

Apart from the correlation between egg number and latitude there is also a positive correlation between fecundity and length (Drucker 1972). Salo and Bayliff (1958) produced a regression curve for one year's data from Minter Creek of

$$y = -2596 + 84.53x$$

[where x = standard length (cm) and y = number of eggs per female]

Allen (1958) did not observe any relationship between fecundity and the time of entry of coho into fresh water. Females with high egg counts did

TABLE 1
Some fecundities of coho salmon reported in the literature

Stock	Mean no. eggs per female	Source
Asian coho		
Kamchatka	5000	Pravdin (1940)
Kamchatka	4900	Berg (1948)
Kamchatka R.	4883 (range 2881-5974)	Gribanov (1948)
Paratunka R.	4350 (range 2800-7600)	Gribanov (1948)
N. Sakhalin	4570 (range 2995-7110)	Smirnov (1960)
North American coho		
Karluk L., AK	4706 (range 1724-6906)	Drucker (1972)
Swanson R., AK	3149-4023	Engel (1967)
Resurrection Bay, AK	3967 and 3846*	McHenry (1981)
S.E. Alaska	4510	Marriott (1968)
British Columbia		
several stocks	2699	Neave (1948)
Cultus L., BC	2300	Foerster & Ricker (1953)
Nile Cr., BC	2310	Wickett (1951)
Big Qualicum R., BC	2574 ± 549†	Fraser et al. (1983)
Minter Cr., WA	2500‡ (range 1900-3286)	Salo & Bayliff (1958)
Fall Cr., OR	1983 (N = 92)	Koski (1966)

Notes: *Mean no. for two consecutive years
†Average for 14 years
‡Average for 18 years

not have significantly smaller eggs, nor did small females have small eggs. For most stocks the average egg diameter was 4.5-6 mm (McPhail and Lindsey 1970), which is smaller than for most other Pacific salmon. The average egg diameter given for the high fecund Kamchatkan stocks is 4.5 mm (Gribanov 1948). The highly fecund Karluk Lake (Alaska) stock has eggs ranging in size from 4.91 to 6.87 mm with an average of 6.11 mm (Drucker 1972). Scott and Crossman (1973) reported that nonanadromous coho females (of west coast origin), collected in Lake Ontario, produced eggs that were 6.6–7.1 mm in diameter. Drucker (1972), in his study of Karluk Lake (Alaska) coho, found no correlation between egg size and the length of the fish or between egg size and the number of eggs in the ovary.

The weight of the gonadal material as a percentage of total body weight is given as 11% for females (range 5%-32%) and 8% for males (range 5%-12%) by Gribanov (1948). Semko (1954), in his study of Asian coho, estimated that ovaries comprised 22.6% of the body weight. It is assumed that this percentage was determined just prior to spawning.

Spawning

Seasonal Timing

The spawning season for most coho populations is between November and January. However spawning timing, like that for migration, is highly variable. Pravdin (1940) reported that in the Kamchatka River, 90% of the coho had deposited their eggs by December, but that spawning occurred over the period 1 September to 16 March, and in the Bolshaya River spawning occurred from

15 September to January, with a few fish spawning in February. Smirnov (1960) discussed the early and late runs of coho in the Soviet Far East and indicated that spawning occurred from 20 August to mid-March in northern Sakhalin.

In North America, coho also spawn over an extended period from October to March, with the very late spawning often occurring in the smaller, shorter streams (Rounsefell and Kelez 1940). In southeastern Alaska, spawning takes place from early October to mid-November (Crone and Bond 1976). In the Cowichan River, British Columbia, spawning occurs in the November–December period (Neave 1949), and for Oregon coastal streams the timing is generally November to February (Chapman 1965). Severe winter droughts may delay spawning until early March in Oregon coastal streams (A. McGie, Department of Fish and Wildlife, Corvallis, Oregon, pers. comm.).

For the Big Qualicum River, Fraser et al. (1983) observed that the time between the peak of migration and the peak of spawning in this short coastal stream was about 32 days. However, for populations of coho in many rivers there appears to be little correlation between the time of entry to a spawning stream and the spawning date. Early-run fish may spawn early, but many will hold for weeks or even months before spawning. Conversely, late-run fish tend to spawn soon after arrival on the grounds or following a short holding period.

A marking study was conducted at the Capilano Hatchery to determine the relationship between time of arrival and date of spawning of wild, unenhanced coho. Commencing 25 August 1971, female coho arriving at the hatchery were given a distinctive colour-coded tag to identify week of arrival. The majority of the earliest group of fish, those arriving in late August, was ready to spawn on 20 November 1971, when most of the total run was maturing. However, one female marked on 25 August was not ready to spawn until 14 February 1972. Those females that did not arrive at the hatchery until after the period when most of the population had matured were all ready for spawning by 15 January 1972 (F.K. Sandercock, unpublished data).

There appears to be a significant advantage in late spawning with regard to both interspecific and intraspecific competition. Where coho share the same spawning grounds with early migrants, especially sockeye and chum salmon or early-timing coho, the late-running fish can often dig up eggs previously deposited in the gravel, thereby exposing these eggs to almost instant predation (Pravdin 1940). Semko (1954) observed that spawning efficiency is decreased when there is a high density of spawners on the grounds, probably not as a result of congested deposition of eggs but because previously deposited eggs were dug up and lost.

For those eggs that are not dug up, there are other hazards. Low winter flows can result in drying of the redds or may expose the eggs to freezing temperatures. Flooding may cause gravel movement and result in eggs being dislodged and swept downstream. Winter storms often cause excessive siltation that may smother eggs and inhibit intergravel movement of alevins (Neave and Wickett 1953).

Spawning Behaviour

When the fish reach the spawning grounds the female selects a nest site. This first site may not, however, be the only one the female uses. Once the nest site has been selected, she will defend it against other females. One or more males may attend any spawning female but they may initially be chased away from the nest by the female (Briggs 1953).

The female begins digging the nest by rolling onto either side at about a 45°-angle to the current with her head upstream and her body arched. She then commences a series of five to six violent flexes of the body and tail over the gravel on the selected site (Burner 1951). After each digging bout, the female will rest for a few minutes before digging again in the same spot. A depression is created in the stream bottom by the hydraulic suction effect of the tail, which lifts some of the gravel, silt, and sand upward from the bottom. This material is displaced downstream by the current (Briggs 1953). This digging activity may last as long as five days, during which time the female will dig several nests in succession. The males do not participate in the digging.

Although the females may be attended by several males, usually one becomes dominant, stays close to the female, and attempts to drive off other males by assuming a threatening posture and by nipping or biting. As already stated, the teeth become enlarged at breeding time. Damage inflicted by males on each other may result in some pre-spawning mortality. The dominant male is generally the largest of the males in attendance (Shapovalov and Taft 1954). At spawning time the female becomes aggressive towards other females and the extra males. While the female is digging the nest, the dominant male assumes a position to one side and slightly downstream of her each time she resumes her normal position over the nest. He may then move in close alongside her and make quivering movements with his head from left to right, followed by moving over her tail to the other side.

Once the nest is completed the female swims over the depression in the gravel and, while arching her body downward, pushes the anal fin down in the spaces between the stones. At this point the dominant male swims closely alongside the female, and with mouths agape, both bodies quiver, and the sperm and eggs are deposited simultaneously (McPhail and Lindsey 1970). They hold this position for two to three seconds. Accessory males, and, in many cases, jacks, rush in alongside the female or the dominant male and deposit sperm as well. The fact that the nest comprises a depression in the gravel with water flowing over it results in a back eddy of current in the nest. The eggs are nonadhesive and loose, and since they have a specific gravity greater than water they sink to the bottom (Davidson and Hutchinson 1938).

Close to the bottom of the nest the water flow is slow. This permits the sperm to make adequate contact with the eggs, ensuring a high fertilization rate. The eggs are not swept out of the nest and downstream by the current. Shapovalov and Taft (1954) estimated that at least 97% of the eggs deposited in the redd remained there and indicated that the percentage of eggs fertilized was consistently high. Semko (1954), in his study of coho spawning in the Karymaiskiy Spring, USSR, found that, of the total potential egg deposition, 72.2% were actually deposited in nests. Pravdin (1940) noted that if an average female was carrying 5,000 eggs, 1,527–3,600 eggs would be successfully deposited in the nests. By digging up a series of redds, Gribanov (1948) determined that the average number of eggs per nest ranged from 300 to 1,200, the most frequent number being 800–900.

When the spawning act is completed, the female immediately moves about 15 cm upstream of the nest and performs different digging movements. She lays her tail on the gravel and then lifts it up quickly two or three times. After each dig the female circles back into the nest before moving to the front edge of the nest again as she performs another dig. The eggs are buried in about one minute, which minimizes predation. At the same time, a depression is created for the next spawning (Briggs 1953). Successive spawnings take place in a series of nests, each slightly upstream of the earlier one. The female may spawn with other males if the previously dominant male becomes displaced (McPhail and Lindsey 1970).

Post-Spawning Behaviour

The length of time that coho remain in the vicinity of the redd after spawning was reported by Crone and Bond (1976) for several sites. During the two years of study, in Sashin Creek, Alaska, the range of survival time for females was 3–24 days, with an average of 11 and 13 days. Males survived an average of 9 days in Sashin Creek. In Drift Creek, Oregon, the males and females survived just over 13 days. In a study of coho spawners in Spring Creek, Oregon, Willis (1954) found that the average time to death after spawning was 11 days (range 4–15 d) for females, 12 days (range 4–32 d) for three-year-old males, and 15 days (range 3–57 d) for jacks. For the females and the three-year-old males, he found that a regression of the number of days in the stream from date of arrival to death indicated that early arriving fish lived about 5 days longer than late arriving fish. Because males were more abundant than females on the spawning grounds, Gribanov (1948) assumed that males must survive longer than females.

Once the female has deposited her eggs, the attendant male leaves. The spawned-out female may continue to go through digging motions for 10 days until she dies, but nests dug after spawning is complete are shallow and nonfunctional (Burner 1951). She also may continue to guard the redd site until too weak to do so (Briggs 1953). Briggs noted that males continue their courting action until they become too weak to maintain position in the current, and then drift downstream to die.

When spawning is complete, both males and

females exhibit definite external physical deterioration in the form of frayed fins, skin loss, fungus infections, and, at times, blindness. Internally, there is a degeneration of the cardiovascular system, pituitary gland, adrenal gland, stomach, liver, and kidney (Robertson et al. 1961). Following these changes, coho, like other species of Pacific salmon, die. As Shapovalov and Taft (1954) and Sandercock (1969) have pointed out, the physical deterioration is not related to the rigours of a long journey because short and long migrant stocks, as well as non-migrant stocks, undergo the same changes and then die.

Egg Retention

During the spawning phase, coho females in most populations deposit almost all of their eggs. Published reports have indicated varying levels of egg retention, e.g., 4 eggs/female in Prairie Creek, California (Briggs 1953); 4 eggs/female (range 0–38, N = 30) in Fall Creek, Oregon (Koski 1966); 7–16 eggs/female in Kamchatka (Semko 1954); and an average of 60 eggs/female in Waddell Creek, California (Shapovalov and Taft 1954). Estimates of total egg deposition have largely ignored the occurrence of residual eggs, because they usually represent a small percentage of the total. However, Fraser et al. (1983) examined 401 female coho from the Big Qualicum River over a period of nine years and found the average percentage of residual eggs to be 22.0% ± 12.2%, compared to 3.5% ± 2.9% for 290 females sampled from the adjacent Hunts Creek over a four-year period. These extremely high percentages translate into 253–941 eggs retained per Big Qualicum River female. Fraser et al. (1983) suggested that environmental factors in this river may have seriously interfered with spawning success.

Redd Characteristics

Redd sites are characterized by gravel size, water depth, and water velocity. Burner (1951) described in considerable detail the characteristics of coho salmon redds in the upper Toutle and Green River systems in Washington. The coho mostly selected small streams where the flow was 5.0–6.8 m^3/min and the stream width did not exceed 1 m. About 85% of the redds occurred in areas where the substrate was comprised of gravel of 15 cm diameter or smaller. In some situations there was mud or fine sand in the nest site. This material was removed during the digging process. About 10% of the redds occurred in sites where the gravel size exceeded 15 cm, and 5% were located in areas having a high proportion of mud, sand, or silt.

Coho were described by Chamberlain (1907) as being the least particular of all Pacific salmon in their choice of spawning area. They can be found in almost all coastal streams, large rivers, and remote tributaries. The redds may be located on gravel bars of smooth flowing rivers or on whitewater riffles of turbulent mountain streams (Foerster 1935). On the spawning grounds, they appear to seek out sites of groundwater seepage and favour areas where the stream flow is 0.30–0.55 m/s (Gribanov 1948). In Kamchatka, water temperatures at spawning vary from 0.8° to 7.7°C, dissolved oxygen varies from 9.9 to 15.0 mg/l, and pH ranges from 6.3 to 8.6 (Gribanov 1948). California coho may spawn in water ranging from 5.6° to 13.3°C (Briggs 1953), but Davidson and Hutchinson (1938) characterized the optimum temperature for coho egg incubation as 4° to 11°C. The water may be clean or heavily silted and the substrate may vary from fine gravel to coarse rubble (Pritchard 1940). The female generally selects a redd site at the head of a riffle area where there is good circulation of oxygenated water through the gravel (Shapovalov and Taft 1954). By definition, the whole area disturbed by a female is described as a redd, whereas the sites of separate egg depositions within the redd are called nests.

For all salmon, the size of the redd is directly proportional to the size of the female, and is inversely related to the size of the gravel and the degree to which it is compacted. Where the flow rate is high (0.9–1.5 m/s), the redds are broad and oval-shaped. In a study of California coho, Briggs (1953) noted that the eggs were buried to a depth of 25 cm (range 17.8–39.1 cm) in gravel that averaged 9.4 cm in diameter (range 3.9–13.7 cm). Water velocity on the spawning ground averaged 0.58 m/s (range 0.30–0.75 m/s) and the depth of water over the redd was 15.7 cm (range 10.2–20.3 cm). In Kamchatka, Gribanov (1948) observed that coho redds averaged 134 cm in length (range 115–195 cm), 112 cm in width (range 100–135 cm), and 22 cm in depth (range 15–27 cm). The water column over the redd averaged 18 cm (range 4–33 cm).

From the above figures, the average redd size would be 1.5 m^2. Crone and Bond (1976) indicated that the average area of gravel disturbed was 2.6 m^2/redd, which is similar to the average redd size of 2.8 m^2 observed by Burner (1951). During spawning, the redd progressively elongates with successive egg deposits as the female shifts gravel downstream to cover the eggs. The redd increases considerably in length and depth and appears to move upstream as the upper end is displaced into the "tail spill" (Burner 1951). Briggs (1953) reported that coho females often dig false nests both before and after spawning occurs. In Prairie Creek, California, only 46% of nests that were subsequently excavated contained eggs. On this basis, it is assumed that most females may dig at least three to four nests and deposit eggs in each (Godfrey 1965). Once the nest is excavated the water tends to eddy in the depression, causing it to flow slightly upstream and thus ensuring a safe deposition of eggs and good exposure to milt to promote a high rate of fertilization (Figure 2).

Territorial behaviour on the part of the females results in fairly regular spacing of the redds in the stream. Where coho use all parts of a spawning stream the redds will be arranged in diagonal rows across the stream. The reason for this is thought to be related to the behavioural pattern of the female; she will tolerate another female upstream or downstream of her territory but not immediately adjacent (Burner 1951).

For coho, the inter-redd space is usually about three times the size of a redd. However, the size of this space also depends on the number of spawners present, the stream bottom composition, the stream gradient, and the water velocity. A pair of spawning coho requires about 11.7 m^2 for redd and inter-redd space (Burner 1951).

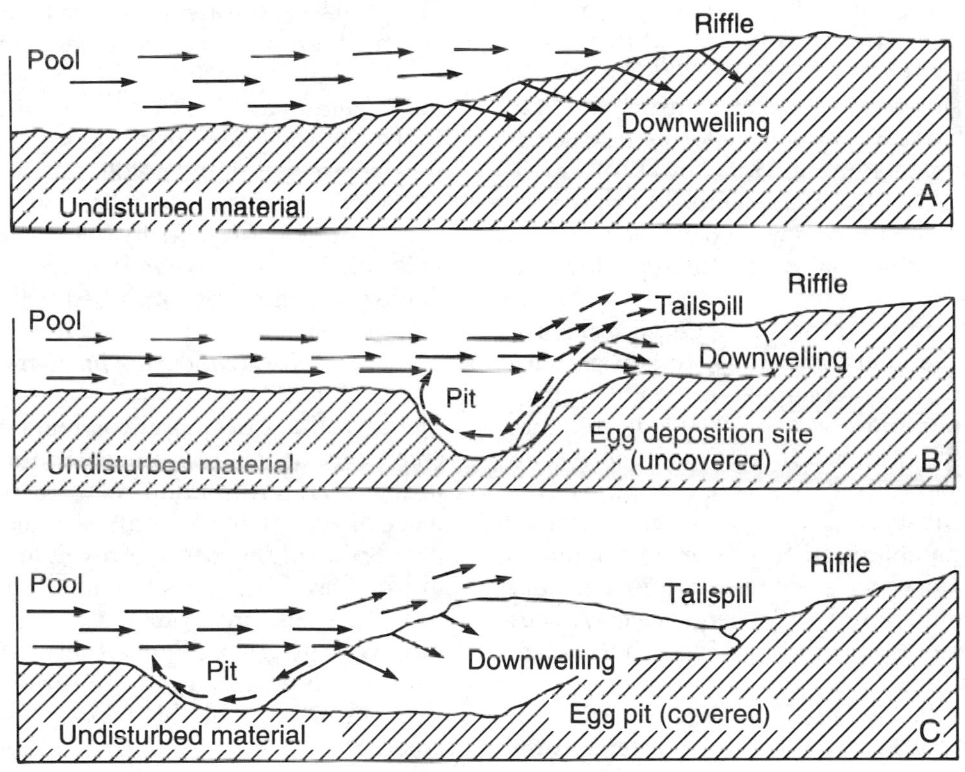

FIGURE 2

Longitudinal sections of spawning area. *A*: pool/riffle relationship results in percolation of water through gravel; *B*: excavation of nest increases flow rate through gravel and creates back eddy in pit; *C*: when the eggs are covered with gravel a second pit is created and water flow to the eggs is enhanced. (From Reiser and Wesche 1977)

Incubation, Survival, and Emergence

Incubation and Hatching

The length of time required for eggs to incubate in the gravel is largely dependent on temperature. The colder the temperature (down to almost freezing), the slower the developmental rate of the embryo and the longer the time to hatching. However, for a given temperature, there may be some variation in hatching time between eggs from different fish sampled on the same day or even between different eggs from the same fish (Shapovalov and Taft 1954). Semko (1954) observed that in Kamchatka, at an average temperature of 2.2°C (range 0.8°–3.5°C), the eggs took 137 days to hatch. Earlier work by Pravdin (1940) and Berg (1948) indicated an average time of 100–115 days for coho eggs to hatch. Gribanov (1948) found that Kamchatkan coho eggs incubating at 4.5°C hatched in 86–101 days. For North American stocks, the time to hatch is shorter than that of Asian stocks, even in the far north. McPhail and Lindsey (1970) gave a range of 42–56 days to hatching for Alaskan stocks, which is similar to the 48 days at 8.9°C and 38 days at 10.7°C for California stocks (Shapovalov and Taft 1954).

The time from hatching, through yolk absorption, to fry emergence is also dependent on temperature and, to a lesser extent, on dissolved oxygen concentration. Semko (1954) indicated that 21 days elapsed from hatching to emergence at an average temperature of 2.2°C, although as Gribanov (1948) observed, 40 days is more typical. In Kamchatka, hatching can occur from mid-January to mid-June, depending on the spawning date and incubation temperature, and emergence may occur from as early as the beginning of March to as late as the end of July.

For coho of the Big Qualicum River, Fraser et al. (1983) found that the total heat requirement for incubation in gravel (spawning to emergence) was 1036 ± 138 degree (°C) days accumulated temperature units, which is the sum of the number of degrees (°C) over zero (of the incubation water) accumulated on a daily basis. The Big Qualicum River coho eggs and alevins were in the gravel from December until May for an average of 167 days (range 149–188 d). Further south, in three Oregon coastal streams, Koski (1966) found that the average time from egg deposition to fry emergence was 110 days (range 104–115 d).

Winter flooding and the associated silt load may reduce water circulation in the gravel to the point where oxygen levels become critical or lethal. In a study of intergravel movement of alevins, Dill (1969) found that, after hatching, the alevins moved a varying distance downward in the gravel, depending on gravel size. Where the gravel was 3.2 cm or smaller in diameter, the alevins moved down 5–10 cm. Where the gravel ranged from 3.2 to 6.3 cm in diameter, and hence there were larger intergravel spaces, the alevins moved down more than 20 cm. This downward movement would appear to be an adaptive mechanism to prevent premature emergence of alevins that are located close to the surface of the gravel bed. At this stage the larval fish have a well-defined yolk sac that is gradually absorbed. There is some indication that at the later stages of development in the gravel the alevins may initiate some exogenous feeding before the yolk sac is completely absorbed (Dill 1969).

Survival during Incubation

The percentage of eggs and alevins that survive to emergence depends on stream and streambed conditions. Winter flooding, with the disruptive effects of gravel movement, accounts for a high proportion of the loss. However, low flows, freezing of gravel, heavy silt loads, bird and insect predators, and infections, such as those by the fungus *Saprolegnia*, all take their toll. Under very harsh conditions, no eggs will survive; under average conditions, probably 15%–27% will survive to emergence (Neave 1949; Crone and Bond 1976); and under very favourable conditions 65%–85% will survive (Shapovalov and Taft 1954). Briggs (1953) examined 22 California coho redds and found that the average egg-to-fry survival was 74.3%; Koski (1966) sampled 21 coho redds in three Oregon coastal streams and found that survival ranged from 0% to 78% with a mean of 27.1%; and Tagart (1984) reported 0.9%–77.3% survival for 19 redds. Neave (1949) observed that the high egg-to-fry survivals achieved by coho in comparison with other salmonids was due to the selection of better spawning sites in areas of good flow stability and to less crowding. However, if the gravel bed had a high concentration (up to 50%) of fine sediment and sand (particle size <0.85 mm), survival was lower (Tagart 1984). Survival to emergence was positively correlated with gravel sizes >3.35 mm and <26.9 mm (Tagart 1984).

Emergence from Gravel

Towards the end of the incubation period, the alevins reverse their downward movement in the gravel and begin making excursions upward (Dill 1969). The alevins do not move directly upward but orient at an angle towards the current flowing over the gravel. The deeper the alevins are found in the gravel, the longer the time for emergence. At this stage of emergence, yolk sac absorption is virtually complete (Plate 19). If the gravel is heavily compacted or loaded with fine sediment and sand, the fry may not be able to get out of the gravel (Koski 1966). Where the gravel/sand mixture was 70% sand (particle size <3.3 mm), survival to emergence was only 8% (Phillips et al. 1975). Tagart (1984) suggested that the high proportion of fines in the gravel effectively reduced the dissolved oxygen levels available and resulted in smaller emergent fry. At 30% fines, the average fry fork length was 37.5 mm; at 10% fines, the average fork length was 39.8 mm (Tagart 1984). Phillips et al. (1975) also observed that a high concentration of fines resulted in early emergence of fry that were smaller and exhibited more yolk.

Shapovalov and Taft (1954) indicated that, in California, coho started emerging from the gravel two to three weeks after hatching, but that the late developers took two to seven weeks longer. Shallow burial in clean, loose gravel, and relatively warm water temperatures all contributed to early emergence. They also found that the fry emerged primarily at night. Koski (1966) observed that the peak of emergence was about 8–10 days after the first fry emerged. Emergence from the redds he sampled continued over a period of 10–47 days with a mean of 35 days. Emergence time was prolonged in those sites having a high proportion of fine sediments.

Freshwater Residence

Fry Behaviour in Streams

When fry emerge from the gravel, they initially congregate in what appear to be schools (Shapovalov and Taft 1954). However, as Hoar (1951) pointed out, this is not true schooling behaviour but a milling about in an aggregation. At this stage, the fry are about 30 mm in length (Gribanov 1948). Fry that emerge first are, on average, larger than those that emerge later (Mason and Chapman 1965). Because of an early growth advantage related to their larger size and better feeding opportunities, they tend to remain larger and ultimately make up a greater proportion of the fingerling population. Small differences in size have a major effect on the outcome of an aggressive encounter and so the larger fish maintain their dominance. Gribanov (1948) observed 30-mm fry emerging as late as 25 July in areas occupied by fry as large as 52 mm in length that had been feeding for a month or more. In cold water systems, fry emerge later in the spring and, in so doing, may avoid spring freshets and the risk of being swept downstream; but they also have a shorter growing period (Scrivener and Andersen 1982). In warm water systems, the freshets may sweep out large numbers of fry, but, for those that remain, growth is fast in the low-density environment and they reach a larger size before the winter (Scrivener and Andersen 1982).

After emergence, the fry continue to hide in gravel and under large stones during daylight hours, but within a few days they progress to swimming close to the banks, taking advantage of any cover that is available. They congregate in quiet backwaters, side channels, and small creeks, especially in shady areas with overhanging branches (Gribanov 1948). Hoar (1958) observed that older coho fry rarely formed schools, except in quiet water, which he thought might be an adaptation to avoid stranding in pools as water levels fell. As the fry become older, they occupy areas along open shorelines and progressively move into areas of higher velocity in midstream and on the stream margins (Lister and Genoe 1970). Coho fry from small tributaries may move upstream or downstream to rear and may occupy areas not accessible to adult coho (Neave 1949). At a length of 38–45 mm (Plate 18), the fry may migrate upstream a considerable distance to reach lakes or other rearing areas (Godfrey 1965). Where they move into lakes to rear they occupy the nearshore littoral zone (Mason 1974). The majority, however, rear in streams where they set up territories. Although coho are found in both pool and riffle areas of the stream, they are best adapted to holding in pools (Hartman 1965). They do not compete well with trout for rearing space on riffles. Stein et al. (1972) observed that coho juveniles at the head of the riffles were able to defend the area against chinook fingerlings.

Territorial behaviour, coupled with the habit of settling to the bottom during darkness, provide a means by which fish may remain in one part of the stream (Allen 1969). Coho fry distribute themselves throughout the stream and, once territories are established, remain in the same locality for relatively long periods (Hoar 1958). If the area is small, an individual occupies a preferred location and defends it by repelling others. However, they do not always display classic territorial behavior (Mundie 1969). Some coho fry form groups in pools, with the large fish at the front and the smaller ones at the back. The large individuals may defend territories (i.e., space in the pool) but the smaller ones may not.

Coho tend to be more aggressive in defense of their territories where the current is fast and where most of the available food is coming from upstream. Where the current is slow or slack, the food can appear from any direction, and the fish tend to move in loose aggregates and scramble for food (Mundie 1969). In the pool areas there is extensive cruising for food, and aggressive behaviour occurs mostly between fish near the faster water (Ruggles 1966). How aggressive the fry become depends on how many are present, their relative size, the amount of food available, and the light intensity. It would appear that when food is scarce, less time is spent on aggression. Conversely, when they are less busy feeding, they have more time for aggression (Mason and Chapman 1965; Chapman and Bjornn 1969). Coho fry are active during daylight hours and seem to tolerate a wide range of light intensities. This adapts them well to the small, shallow streams they normally occupy, where conditions of light and shade are highly variable (Hoar 1958).

After territories are established, the fry do not rest on the bottom during daylight but orient themselves to a particular rock or log in the stream so as to occupy a small space of slow moving water. Failure to rest on the bottom at night may lead to a progressive displacement downstream (Chapman 1962). From their reference points they will make quick excursions of up to 30 cm to grab food or chase intruders but then return to take up their positions. Should they be chased out of their own territory by a larger or more aggressive individual, they will quickly settle into a new space, provided one is available. Small coho juveniles tend to be harassed, chased, and nipped by larger juveniles unless they stay near the bottom or hide under rocks or logs. By avoiding the more aggressive coho, there is much less opportunity to feed, and, consequently, the smaller individuals grow more slowly (Chapman 1962).

Habitat Utilization in Fresh Water

The abundance of coho in a stream is limited by the number of suitable territories that are available (Larkin 1977). More structurally complex streams that contain stones, logs, and bushes in the water support larger numbers of fry (Scrivener and Andersen 1982). Dill et al. (1981) found that the size of a territory was inversely related to the density of benthic food in the area and that territories were smaller where intruder pressure was high. As the fish grow, the size of their defended territory increases; for fry 49 mm in length the average territory is 0.34 m^2, at four months of age it is 0.79 m^2, and at yearling size (110 mm) it is 3.7–5.5 m^2 (Allen 1969). McMahon (1983) reported that pools of 10–80 m^3 or 50–250 m^2 in size were optimum for coho production, provided that there was enough streamside vegetation for shading. However, if the canopy is very dense, then the coho biomass will be reduced (Chapman and Knudsen 1980). Where coho occupy riffle areas, they tend to be uniformly distributed close to the gravel bottom and highly antagonistic. As the amount of aggression increases, so does the emigration rate (Mason and Chapman 1965).

The productive capacity of the freshwater environment for coho has been estimated by a number of investigators. Lister and Walker (1966) determined that in the Big Qualicum River, 19.1 smolts were produced per 100 m^2 of wetted stream area measured at low flow. For three small Oregon streams, Chapman (1965) reported a production of 18–67 smolts per 100 m^2 over a four-year period. Tripp and McCart (1983), in their study on outplants of hatchery coho fry into headwater streams, found that the average production was 8.4–8.5 smolts per 100 m^2, which is low in comparison to other estimates but may be explained by the fact that high-gradient headwater streams are not usually productive areas. In contrast, Armstrong and Argue (1977), in their assessment of some low-gradient side channels of the Cowichan River that appeared to be rich in insect fauna, found that there were 125–141 smolts produced per 100 m^2. Production from average coho streams probably falls between these two extremes. Foerster and Ricker (1953) reported that coho smolt production, expressed as the number of smolts produced per adult female, was seven to ten times greater from streams than from lakes.

Most coho fry move out of river systems with freshets. However, even during periods of stable flow, fry continue to migrate. The numbers of fry moving do not correlate well with the water discharge rate because the first freshet may move most fish, whereas the second freshet, a few days later, may move only the few that are still left in the stream (Hartman et al. 1982). Fish that are unable to find or to defend a territory are generally displaced downstream. If the downstream area is unoccupied, the displaced fry may take up residence. However, if fry are already occupying the space, then the new arrivals will continue to be displaced downstream (Ruggles 1966). Displacement also occurs among larger fish due to demands for territorial space resulting from increased fish size. This displacement probably continues throughout the active growth period unless offset by mortalities (Fraser et al. 1983). Lister and Genoe (1970) found that coho progress through a series of preferred habitats: back eddies, log jams, undercuts or open bank areas, and, finally, fast water. From an evolutionary standpoint, the displacement mechanism may drive fish to explore areas that are some distance from the spawning ground, and if suitable space can be found, they will then make more effective use of the environment by being widely distributed (Allen 1969). However, in most cases, the displacement of the surplus fish is to less favourable sites, where they become vulnerable to predators or may be driven to the estuary.

Mason and Chapman (1965) noted that some coho that were larger than average, as well as smaller fish, were displaced. They suggested that rearing areas suitable for these larger fish were limited. Spring floods will also cause displacement. The preference by coho for shallow water may leave them vulnerable to sudden torrents that sweep them out of their established territory and move them downstream. Some of the coho that are displaced downstream may move back upstream, or they may migrate along the shore in low salinity water and enter other streams (Otto and McInerney 1970).

Chapman (1962), in his work on coho in Oregon, described juveniles that moved downstream between the time of emergence and October of the same year as nomads. Coho migrating downstream from November onward were defined as smolts. In addition to factors such as size, level of aggressive behaviour, and food availability, there may be an innate tendency on the part of nomads to migrate. The fact that some coho fry migrate downstream early in the spring, even when rearing space is available, would support this suggestion.

From the study of scales, it has been concluded that coho that enter the sea in the first spring or summer of life do not generally survive to the adult stage (Crone and Bond 1976). However, Crone and Bond (1976) found that, under experimental conditions, fry could survive salinities as high as 29 ppt. provided that they had been acclimated to lower salinities for 35 days. These authors also noted that the scale patterns developed by fry that reared in salt water were indistinguishable from those of smolts that had spent a year in fresh water. This is contrary to the general observation that fry displaced into salt water do not contribute to production. Underyearling coho are poorly equipped physiologically to survive early emigration into salt water, however, the type of estuary may have a substantial bearing on their ability to survive (Otto 1971). Kennedy et al. (1976) observed that coho fry put into salt water died prematurely and that non-smolted fingerlings either died immediately or grew slowly and then subsequently reverted to their parr condition and died. Weisbart (1968) showed that salinity tolerance was not a function of age but of size. Coho fry up to 5-6 cm in length and five months of age do not survive in sea water. The threshold size for survival seems to be about 7-8 cm.

Although underyearling coho smolts are extremely rare in nature, they are readily produced under hatchery conditions of abundant food, favourable growing temperatures, and proper photoperiod exposure (Brannon et al. 1982; Clarke 1988).

Feeding in Fresh Water

Coho juveniles are highly dependent on visual cues for locating and capturing food (Hoar 1958). Coho rarely feed on non-moving food or off the bottom in streams, preferring to pick off food in suspension or on the surface. At times they dart quickly to the surface and snap at floating particles, some of which are food, but others may be bits of wood, conifer needles, etc. Following these darting movements, coho quickly return to their original position (Shapovalov and Taft 1954). During daylight hours and the morning and evening twilight periods, coho can often be seen jumping clear of the water surface to capture insects flying nearby. During darkness, feeding activity ceases.

Because coho normally occupy the slower moving sections of a stream, this allows the capture of food with the minimum expenditure of energy (Mundie 1969). The most productive coho areas are small streams rather than large rivers, because small streams have the highest proportion of marginal slack water to midstream area. Insect material found in the midstream drift of large streams is generally unavailable to juvenile coho and is lost from production. The wider the stream, the greater the loss of food. Mundie (1969) found that juvenile coho in British Columbia streams were highly dependent on drifting organic material consisting primarily of stream and terrestrial insects. He observed that 38 fish, 3.0-3.5 cm in length and collected through April-May, had consumed 337 different food items. A sample of 30 fish collected in July contained over 900 items. Almost half of the food consisted of chironomids in various stages of development.

Chapman (1965) demonstrated that, in Oregon streams, there was a positive correlation between the amount of terrestrial insect material found in coho stomachs and the extent to which the stream was overgrown with vegetation. In the smallest, most densely shaded stream, terrestrial insects comprised 40% (dry weight) of the food consumed. Where the streams were more open, the percentage was reduced to 21%-29%. The most productive streams are those with alternating pools and riffles about equal in area. A pool to riffle ratio of 1:1 provides optimum food and cover conditions for juvenile coho (Ruggles 1966). Invertebrate food production is maximized in the riffle area and the pool is the optimum environment for coho holding and feeding (Mundie 1969). In Kamchatka, Gribanov (1948) found that the principal food of river-dwelling juvenile coho was adult insects and, only secondarily, insect larvae. In the Paratunka River, coho consumed chironomids, stoneflies, and, occasionally, crustaceans (Pravdin 1940).

Mason (1974) found that, in British Columbia, juvenile coho food could be divided into 21 categories. The most important category was adult winged dipterans, which comprised 80% of the food items eaten. Less than 11% of the lake-dwelling coho population had zooplankton in their stomachs, and this food item represented only 5% of the volume. Zorbidi (1977) observed that coho of Lake Azabache, USSR, at a size of 6.3-7.6 cm, fed primarily on terrestrial insects and chironomids and did not appear to eat zooplankton. In Cultus Lake, British Columbia, young sockeye fry were the principal food item for coho juveniles. Other fish and insects were less consistently a part of the diet of the Cultus Lake coho (Ricker 1941; Foerster and Ricker 1953).

At the yearling stage, coho may become predatory, supplementing their insect diet with the fry of their own or other species (Gribanov 1948). Pritchard (1936) found that yearling coho in British Columbia ate large numbers of pink salmon fry and small numbers of chum and coho fry. Hunter (1959) estimated that coho smolts consumed 1.5–2.0 pink or chum fry per day. Shapovalov and Taft (1954) reported that, in California, steelhead and coho fry were not subject to yearling coho predation, because they emerged from the gravel after the coho smolts had migrated to sea. However, large numbers of chinook fry were taken by coho outmigrants. Chamberlain (1907) and Zorbidi (1977) both reported that larger coho (12.3–13.7 cm) fed on threespine (*Gasterosteus aculeatus*) and ninespine (*Pungitius pungitius*) sticklebacks, in addition to terrestrial insects and aquatic insect larvae.

Fingerling Behaviour in Streams

By late summer and early autumn, as water temperatures begin to decline, juvenile coho feeding activity decreases, and the fish move into the deeper pools of the stream, especially those with overhanging logs (Hartman 1965; Scott and Crossman 1973; Bustard and Narver 1975a). At this time, the number of coho in the stream may be reduced substantially, but those systems with good winter habitat lose fewer juvenile coho (Tschaplinski and Hartman 1983). With the onset of fall freshets, the resident coho undergo high rates of redistribution (Scarlett and Cederholm 1984). The fish spend more time hiding under the cover of logs, exposed tree roots, and undercut banks. Skeesick (1970) reported that coho moved upstream into side creeks that remained clear and stable during the winter. These fish were larger than residents of the side creeks, and this difference in size was maintained over the winter (the average size of the mainstream and side creek coho in the spring was 101 cm and 86 cm, respectively). Tripp and McCart (1983) also observed that juvenile coho would move 200–400 m upstream to enter small tributary streams to overwinter. The distance moved appeared to be governed by the flow rate of the tributary stream. When the discharge was high, the flow could be detected further downstream. Groundwater seepage into these smaller tributaries is thought to be the main attractant. By seeking cover and entering side channels, the fish avoid being swept out of the stream during winter freshets, and they also avoid some predators at a time when their swimming ability is reduced because of a lowered metabolic rate (Hartman 1965; Bustard and Narver 1975a). Brett et al. (1958) demonstrated that the cruising speed of underyearling coho was reduced from 30 cm/s at 20°C to 6 cm/s at just above 0°C.

In some rivers, the fish moved a considerable distance downstream before entering tributaries. In the Clearwater River, Washington, marked coho moved as far as 38 km downstream from summer rearing areas before entering tributaries (Scarlett and Cederholm 1984). The timing of the movement was always in response to freshets and the associated high water velocities, turbidity, and gravel movement. Coho streams with the best over-wintering habitat were those with spring-fed ponds adjacent to the mainstream (Peterson 1980) or protected, slow flowing side channels that may only be wetted in winter (Narver 1978). In unstable coastal systems, coho production may be limited by the lack of side channels and small tributaries to provide protection against winter freshets (Cederholm and Scarlett 1981). Beaver ponds create additional habitat used by coho, both in winter to avoid freshets, and in summer to avoid the stranding caused by low flows (Bryant 1984). However, there may be disadvantages for small coho in these pond-like environments for they become more susceptible to cutthroat trout (*Oncorhynchus clarki*) predation (Peterson 1980). For larger coho, migrating out of the mainstream into tributaries and sloughs may result in a high survival rate, e.g., over 67% in Carnation Creek, British Columbia (Tschaplinski and Hartman 1983).

In December, coho are no longer found on the riffles (Ruggles 1966). Narver (1978) observed that juvenile coho moved into areas with water depths over 45 cm and lower velocities (15 cm/s) when temperatures declined below 7°C. He also found that as temperatures in the stream approached 2°C, the coho moved closer to cover provided by logs, tree roots, undercut banks, etc. Coho also preferred side pools with cover to pools without cover, and clean rubble to silted rubble (Bustard and Narver 1975b). Coho that congregate in the deep pools of the stream form dense groups, and the level of aggression is low. No displacement occurs as subordinate fish are driven back but not out of the group (Hartman 1965). Coho that occupy lakes during the summer migrate out of the lake into inlet streams to overwinter (Gribanov 1948). However, in the Tenmile Lakes, Oregon, coho juveniles moved into the lakes following the fall freshets and reared there until the following spring (A. McGie, Department of Fish and Wildlife, Corvallis, Oregon, pers. comm.). In the spring, there is a strong movement of juvenile coho back to the main stream (Tschaplinski and Hartman 1983).

Growth of Fry and Fingerlings

With moderate water temperatures and an abundant food supply, coho fry will grow from 30 mm at emergence in March to 60–70 mm in September, to 80–95 mm by March of their second year, and to 100–130 mm by May (Rounsefell and Kelez 1940). Mason (1974) described two growth phases for coho of Great Central Lake, British Columbia. From April to mid-June, coho increased in length from 37 mm to 62 mm; in summer the growth slowed; and by October the coho averaged 72 mm in length. By the following April the coho were 90–130 mm in length, which reflects a second spurt of growth in the early spring following the period of no growth in midwinter.

During the winter months, feeding virtually ceases and growth stops. Low winter temperatures are a major cause of growth reduction, but winter floods and turbid water conditions also restrict feeding opportunities. Noggle (1977) observed that coho terminated feeding when sediment concentrations exceeded 300 mg/l (with some variation depending on the type of sediment), but that they did not abandon their territory even when sediment loads approached 4,000 mg/l. Where side channels are fed by groundwater, temperatures may be such that coho continue to feed and grow during the winter (G.F. Hartman, Department of Fisheries and Oceans, Nanaimo, British Columbia, pers. comm.). By March, when temperatures are on the rise, the fish again commence a period of rapid growth. Increasing temperatures and an abundance of insect food stimulate the resumption of feeding. The presmolts complete their final growth phase before starting on their seaward migration (Shapovalov and Taft 1954).

Fry and Fingerling Survival

During their life history stage in freshwater streams, two physical factors play a large role in coho survival: water discharge rate and temperature. Work by Neave (1948, 1949), Smoker (1953), and others has clearly demonstrated a correlation between summer flows and the catch of adult coho salmon two years later. Low summer flows reduce potential rearing areas (less wetted area), cause stranding in isolated pools, and increase vulnerability to predators (Cederholm and Scarlett 1981). High winter flows in typical coastal streams can be particularly hostile to fish 45–70 mm in size (Narver 1978). Coho fry production has been shown to be a function of the stability of winter flows (Lister and Walker 1966). McKernan et al. (1950) stated that winter flooding only had a significant impact when the flow was over 50% greater than the average flood. Extreme floods are almost invariably detrimental. When a flood commences, there is a greater abundance of food available as stream insects are dislodged from the gravel, but this disruption results in a loss of food production in the longer term, as the food sources are destroyed (Mundie 1969).

With low summer flows and high ambient air temperatures, the water temperature can approach or exceed the upper lethal temperature of 25°C for juvenile coho. Brett (1952) found that exposure to temperatures in excess of 25°C or a quick rise in temperature from less than 20°C to 25°C resulted in a high mortality rate. Prolonged exposure to water temperatures close to 0°C was tolerated by coho, but a sharp drop in temperature from 5°C to almost 0°C resulted in mortality. Brett (1952) also observed that juvenile coho preferred a temperature range of 12°–14°C, which is close to optimum for maximum growth efficiency.

Godfrey (1965) summarized the fry-to-smolt survival for two British Columbia streams, one Washington stream, and one California stream. He found that the published values for survival ranged from 0.70% to 9.65% with the average in the range of 1.27%–1.71%. Neave and Wickett (1953) estimated survival from egg to smolt for British Columbia coho to be 1%–2%. Most of the mortality takes place in the first summer. Based on fry outplants, Tripp and McCart (1983) concluded that summer mortality of coho fry was density-independent. In the following spring, the mortality rate was higher than during the winter period, but the mortality was still less than one-third that of the previous summer (Crone and Bond 1976). Survival for the fry-to-smolt stage was estimated by Fraser et al. (1983) at 7.3% for the Big Qualicum River. Drucker (1972) noted that the long period of freshwater residency probably resulted in a higher freshwater mortality but contributed to a lower marine mortality because smolts were larger when they went to sea. Mace (1983) estimated a 2%–4% loss to avian predators after the smolts reached the Big Qualicum River estuary. Because of the relatively low survival rates from fry to smolt, it is obvious that the freshwater environment plays a major role in the fluctuation of coho abundance.

Freshwater Predators

Predation is a major component of the mortality suffered by juvenile coho, but predator species and effect varies with stream system and geographical area. Fry and smolts are subject to predation by a wide variety of predators, especially when coho are aggregated in pools and side channels, or in years when the egg-to-fry survival is high and the fry are very abundant. Larkin (1977) indicated that rainbow trout (*Oncorhynchus mykiss*), cutthroat trout, Dolly Varden charr (*Salvelinus malma*), squawfish (*Ptychocheilus oregonensis*), and Rocky Mountain whitefish (*Prosopium williamsoni*) are all important predators of juvenile coho. Godfrey (1965) suggested that cutthroat trout were the main predators of coho fry in British Columbia, but Chapman (1965), in his studies of Oregon coho populations, found that cutthroat trout were not significant in coho fry mortality because only occasional fry were taken, even when they were abundant. Patten (1977) reported that torrent sculpins (*Cottus rhotheus*) were important predators of coho from the time of emergence at a size of 30 mm until the coho were 45 mm; fry larger than this were rarely taken by sculpins. Logan (1968) found that 31% of the Dolly Varden charr stomachs examined from an Alaskan coastal stream contained coho juveniles. Shapovalov and Taft (1954) observed that predatory fish were responsible for most of the coho loss in California, but that garter snakes (*Thamnophis sirtalis*) were also able to capture coho fry, especially in pools that were drying up.

Dippers (*Cinclus mexicanus*), robins (*Turdus migratorius*), crows (*Corvus brachyrhynchos*), herons (*Ardea herodias*), and fish-eating ducks (e.g., *Mergus merganser*) all consume significant numbers of coho. Wood (1984), in his study of the foraging behaviour and dispersion of common mergansers (*M. merganser*), found that 40-g coho smolts were selected over 2-g coho fry and suggested that the difference in capture frequency could be explained by the difference in conspicuousness due to size. He further observed that, as density of smolts increased, or the amount of cover decreased, the rate of capture by mergansers increased. However, coho smolts, having once been exposed to merganser attacks, were less likely to be captured in subsequent attacks. During the winter months, the avian predation rate is much lower, partly because the migratory species may have departed to southern wintering areas, and also because the coho are hiding. In many streams the presence of an ice cover over the stream makes them less vulnerable (Crone and Bond 1976). Mammals such as mink (*Mustela vison*) and otter (*Lutra canadensis*) prey heavily on over-wintering juveniles and migrating smolts. Predators tend to take a fixed number of prey so that the proportion of prey taken increases as the number of prey decreases. In those situations where salmon fry are reduced to small numbers, the predators can eliminate them entirely (Larkin 1977).

Juvenile Colour

In the alevin stage, young coho have silver- or gold-coloured bodies and large vertically oval blobs of dark brown pigment (parr marks) in a row along the lateral line (Plates 17 and 19). The lateral line bisects most of the parr marks, and the pale area between the parr marks is greater than the width of a parr mark (Scott and Crossman 1973). The back and sides are often cinnamon-yellow and the fins are tinged with orange. Once the fish reach a size of 10–14 cm, the long, narrow, dark brown parr marks along the side (usually 11 per side) are distinctive, the rest of the body is a dull gold colour, and the fins are varying tones of orange (Gribanov 1948) (Plates 17 and 18). The anal fin has a white leading edge, which is followed by a parallel, narrow black stripe (Hart 1973). Stein et al. (1972) described the same white-black striping on the dorsal fin of juvenile coho from the Sixes River, Oregon. As the juveniles approach the migratory smolt stage, the parr marks become less evident and the overall colour of the fish is lighter and more silvery (Plate 17). The fin colours fade from orange to pale yellow, although the tail may retain some of the orange coloration. By the time the smolts reach the estuary, the change to a silvery coloration is almost complete on the sides and bottom, and the dorsals are blue-black.

Residuals

Foerster and Ricker (1953) reported on the occurrence in Cultus Lake, British Columbia, of what they termed "residuals." These are coho that spend their entire life in fresh water. Foerster and Ricker demonstrated that these residuals were derived from anadromous parents and were not produced by a self-sustaining resident population. They thought that the number of residuals was at least equal to, or possibly several times greater than, the number of outmigrant smolts. Because of an unexplained high rate of mortality in their third year, the reproductive potential of the residual population was low, and few survived to maturity. The largest residuals that were sampled ranged from 46.5 to 59.5 cm and had a maximum weight of 2.4 kg. The residuals matured at the same age as the anadromous component and included jacks (age 1.0) and a 1:1 sex ratio among age 2.0 males and females. The colour at maturation of the residuals was more subdued than that of the anadromous fish. Shmidt (1950) reported that lacustrine forms of coho had been collected in the lakes of the middle Okhota River in the Soviet Far East and that they matured at a size of 30–35 cm. These, like the residuals studied by Foerster and Ricker (1953), were not landlocked. Foerster and Ricker indicated that, up until 1953 at least, there were no documented self-sustaining populations of "landlocked" coho. Later, Rounsefell (1958) reported that a dwarfed landlocked coho population had been found in Becharof Lake in the Egegik River system of Bristol Bay, Alaska. Subsequently, it has been confirmed that the Great Lakes coho, though not strictly landlocked, do spend their entire lives in fresh water and have established self-sustaining populations. There is no evidence of residual coho in streams.

Smolt Migration

Seasonal Timing

The migration of coho downstream towards the sea begins in spring. Factors that tend to affect the time of migration include: the size of the fish, flow conditions, water temperature, dissolved oxygen levels, day length, and the availability of food (Shapovalov and Taft 1954).

In the more southerly part of the distribution range for coho, the outmigration begins early. In California, the outmigration of smolts over 10 cm in length occurred as early as mid-March, increased substantially through April, and peaked about mid-May (Shapovalov and Taft 1954). Chapman (1965) reported that smolts migrated in Oregon between early February and May. In Minter Creek, Washington, the smolt migration occurred between 15 April and 1 June, with the peak in May (Salo and Bayliff 1958). In southern British Columbia, Fraser et al. (1983) reported that, over a fifteen-year period in the Big Qualicum River, the migration duration averaged 119 ± 28 days and that the midpoint of the migration occurred on 26 May ± 5 days. Andersen and Narver (1975) found that the midpoint for two-year-old smolts migrating out of Carnation Creek on the west coast of Vancouver Island was reached on 9 May. In northern British Columbia, at Lakelse Lake, the coho outmigration in 1952 took place between 13 May and 14 June (Foerster 1952). In 1981, in the Resurrection Bay area of Alaska, the smolt migration began 28 May, reached its midpoint (50% migration) on 19 June, and ended on 3 September (McHenry 1981).

The smolt migration out of Karluk Lake on Kodiak Island began in mid-May and ended in early July; these are smolts that had resided in the lake one to four years before migrating (Drucker 1972). Drucker observed that, in general, the higher the latitude the later the migration, with about a one-month separation between outmigration peaks of coho from California compared to those from the Gulf of Alaska. In the Soviet Union, Pravdin (1940) found that smolts from the Paratunka River on Kamchatka migrated from June to October, but the main migration occurred between 1 August and 25 September. Churikov (1975), working in northeastern Sakhalin, reported that a net was set in the Bogataya River on 24 May and fished continuously. The first smolt was caught 26 June and catches continued until 5 August. The peak of the migration was between 13–22 July. Scott and Crossman (1973) observed a springtime migration pattern for coho smolts moving downstream to enter the Great Lakes. The fish migrated between late March and June and arrived at the stream mouths between April and August, with the peak in late May.

Tripp and McCart (1983) reported that the main peak of migration for coho coincided with a time of maximum stream discharge and then declined. However, a second peak of migration occurred at a time of decreasing flows but increasing temperature. In some cases, the smolt migration occurred after the spring flood (Churikov 1975). For a single river system there are year-to-year variations in the timing of smolt migration that are related to environmental factors. In years with low flows and higher temperatures, the outmigration is earlier (Shapovalov and Taft 1954). Coho migration in Alaska usually occurs when temperatures are in the range of 5.0°–13.3°C (Drucker 1972) and, in the USSR, at 9.0°–12.0°C (Churikov 1975). In Bear Creek, Alaska, McHenry (1981) found that the migration started when water temperatures were 2.5°C, peaked at 8.1°C, and ended at 9.7°C. Few coho migrated before the temperature reached 3.9°C. In another Alaskan study, Logan (1967) found that most coho migrated when temperatures were 6.7°–11.1°C, with the peak on 11 June at a temperature of 8.0°C.

Most smolts migrate to sea after spending just over one year rearing in fresh water. In the more southerly part of their range, where temperatures and food availability are better, coho reach migrant size in about fifteen months. However, in areas where conditions are less conducive to growth, they may require an additional one or two years to reach this size.

Diel Timing

As indicated earlier, the bulk of the seaward migration takes place at night. Meehan and Siniff (1962) found that on the Taku River (British Columbia) the peak migration occurred daily between 2300 h and 0300 h. Mace (1983) reported that coho smolts were rarely seen descending the Big Qualicum River in daylight, and that they appeared in the transition zone between the river and estuary only in the late afternoon and evening.

Outmigrant fry exhibit quite the opposite behaviour from that of fish that take up residence in the stream. McDonald (1960) showed that with the onset of darkness, the fry moved up to the surface and were distributed across the stream and swam or drifted with the current. Downstream fry movement started at about 2100 hours and concluded at 0400 hours, with a peak migration between midnight and 0100 hours.

Migration Behaviour

Fish that have spent a year or more in the stream, and are about to undergo the physiological changes associated with smoltification, begin by defending their territories less vigorously and by forming aggregations. They rise to the surface at night and begin moving downstream (Hoar 1951). The migrating fish move downstream in schools of 10–50 smolts, and fish of a similar size seem to school together (Shapovalov and Taft 1954). The fish approach falls or rapids, where there is a steep gradient, tail-first, with their heads into the current. They dart forward a few times, each time coming closer to the edge. When they finally go over falls, some turn and head downstream as they are falling. After one is swept over the falls, several others in the school generally follow.

Age at Time of Migration

Throughout a large part of their range in North America coho typically spend one winter in fresh water after emergence from the gravel and migrate downstream as yearling (age 1.) smolts. In some river systems, coho may stay two, three, or even four winters in fresh water before migrating to sea as two- (age 2.), three- (age 3.), or four-year-old (age 4.) smolts, respectively. However, the proportion of these older coho smolts in most freshwater systems is generally low.

In California, Shapovalov and Taft (1954) found no two-year-old coho smolts in their sampling in Waddell and Scott creeks. In British Columbia, Pritchard (1940) examined the scales of 6,312 commercially caught coho and found that the incidence of fish that had migrated to sea as two-year-old smolts was 0.9%; Foerster (1955) estimated a 1% occurrence of two-year-olds; Armstrong and Argue (1977) observed a 2.4% incidence of two-year-old coho smolts in the Cowichan River; and Fraser et al. (1983) reported an average occurrence of 1.6% two-year-olds in the Big Qualicum River for five years sampled during the 1964–73 period, and 7.1% (for 1973) in the Little Qualicum River. An atypical example for British Columbia is the coho stock of Carnation Creek on the west coast of Vancouver Island; in one year (1974) the incidence of two-year-old smolts was 58.4%, and the average of the three previous years was still high at 46.0% (Andersen and Narver 1975). The occurrence of a high proportion of two-year-olds is much more common in the north. For example, the Taku River had a 54% incidence of two-year-olds (Meehan and Siniff 1962); in Hood Bay Creek (Alaska) in 1968 and 1969, respectively, 50% and 45% were two-year-olds, 7% and 5% were three-year-olds, and the balance were one-year-olds (Armstrong 1970); and among Karluk Lake coho, 44%–51% were two-year-olds, 42%–49% were three-year olds, and 1.5%-6% were four-year-olds in the years 1956, 1965, and 1968 (Drucker 1972). However, McHenry (1981) found that the coho smolts coming out of Bear Creek in the northern Gulf of Alaska area were only 27.1% two-year-olds and 0.1% three-year-olds.

In Kamchatka the majority of the coho go to sea as two-year-olds and some as three-year-olds (Gribanov 1948).

Smolt Size

The size of coho smolts is fairly consistent over the geographical range of the species. Gribanov (1948) observed that a fork length of 10 cm seemed to be a threshold for smoltification. Shapovalov and Taft (1954) found that coho from Waddell Creek, California, migrated at an average length of 11.5 cm (range 7.5–16.5 cm). Coho smolts from Sand Creek, Oregon, averaged 10.6 cm (range 3.3–19.3 cm) over a three-year period (Sumner 1953). In Washington, Minter Creek smolts were 9.5–10.6 cm (Salo and Bayliff 1958), and in British Columbia, Foerster and Ricker (1953) observed that Cultus Lake smolts were 11–12 cm, with a few large two-year-old smolts at 26.1 cm. Yearling smolts from Carnation Creek were only 7.4–7.9 cm, whereas the two-year-olds were 9.9–10.3 cm (Andersen and Narver 1975); and Cowichan River yearling smolts were 8.8–9.8 cm, and the two-year-olds were 9.8–10.5 cm (Armstong and Argue 1977). Fraser et al. (1983), in summarizing fifteen years of Big Qualicum data, found that the long-term average smolt size was 9.85 ± 5.8 cm and that these fish weighed 11.18 ± 2.44 g (11 years of data). Alaskan smolts sampled by Logan (1967) were 10.4–15.4 cm, the smaller fish being yearlings, and the larger fish two-year-olds. McHenry (1981) sampled larger yearling smolts from the Resurrection Bay area that averaged 12.2 cm (18 g) and two-year-olds that were 13.5 cm (24 g).

Coho smolts in the eastern USSR also tend to be larger. Pravdin (1940) examined smolts from the Paratunka River that averaged 11 cm as yearlings and 11–15 cm as two-year-olds. Berg (1948) described Kamchatkan coho smolts averaging 13–14 cm in length, and Churikov (1975), working on the Bogataya River, Sakhalin Island, captured smolts averaging 12.3 cm in length, with an average weight of 25.1 g (range 15–34 g).

Growth is obviously very rapid once the smolts reach the estuary, because fish sampled in near-shore areas ranged in size from 14 to 22 cm (Rounsefell and Kelez 1940; Berg 1948; Fisher et al. 1984).

Estuarine Predators

After the smolts reach the estuary they are vulnerable to predation by many of the same predators (cutthroat trout, Dolly Varden charr, herons (*Ardea herodias*), mergansers (*Mergus merganser*), and mink (*Mustela vison*)) that they faced during their freshwater rearing and migration stages. They are also eaten by a variety of new predators. The new ones include dogfish (*Squalus acanthias*), lamprey (*Lampetra* sp.), and sharks (e.g., *Lamna ditropis*) (Larkin 1977); avian predators such as Bonaparte's gulls (*Larus philadelphia*), glaucous-winged gulls (*L. glaucescens*), arctic loons (*Gavia arctica*), and mergansers (Mace 1983); and at least fifteen species of marine mammals (mainly seals (e.g., *Phoca vitulina*), sea lions (*Eumetopias jubatus* and *Zalophus californianus*), and killer whales (*Orcinus orca*)) that consume coho in the lower reaches of rivers, in estuaries, and in nearshore waters (Fiscus 1980). However, for most mammals, salmon constitute only a small proportion of the diet. Other fish, such as the daggertooth (*Anotopterus pharao*), may be important predators in offshore waters (Hartt 1980).

Ocean Life

Fish in their First and Second Ocean Year

One of the first reports of coho smolt behaviour on entering sea water was given by Chamberlain (1907), who concluded that the smolts stayed in the nearshore areas close to their home streams for several months before migrating further. He thought that coho from a particular stream continued to school together. Gribanov (1948) found that migrant Kamchatkan coho first occupied the quiet marine inshore areas, away from the surf zone, and swam in the top layers in discrete schools of 20–30 fish. He concluded that if the smolts remained in areas adjacent to shore during their first summer and winter, it was unlikely that they would undertake a long migration in the second summer before returning to their home stream. Churikov (1975) reported that Asian coho smolts spent only a brief period in estuaries before moving along the shoreline.

Shapovalov and Taft (1954) suggested that California coho remained close to the shoreline after migrating to sea from Waddell Creek, and that they probably stayed there for a few months before beginning to disperse. Recoveries of marked fish at that time indicated that they remained within 150 km of shore, which, in California, is out to the edge of the continental shelf. Milne (1964) noted that the location of the main coho fishing grounds off the coast of British Columbia indicated that the feeding areas were over the continental shelf within sight of land and at depths of less than 90 m. Most coho salmon that were caught for tagging in their first ocean year in the Gulf of Alaska were close to shore, but some were also taken up to 150 km offshore (Godfrey 1965).

Smolts entering the sea from California to British Columbia tend to move northward along the coast. Some reach the coastal waters of central Alaska by late summer. During seining for juvenile salmonids in nearshore and offshore waters, Fisher et al. (1984) found that almost all of the coded-wire tagged coho that had been released from coastal Oregon were recovered further north than Oregon. However, tagged fish from the Columbia River were found both north and south of the Columbia River estuary.

After about twelve months at sea, coho gradually migrate southward along the coast, but some appear to follow a counter-clockwise circuit in the Gulf of Alaska. Royce et al. (1968) reported that coho do not drift, but actively migrate in a circular pattern with the currents. In the open ocean, coho were thought to occupy the area from the surface to a depth of 30 m. There was no evidence of particular stocks schooling together and, in fact, single net sets in some cases contained several salmonid species of different ages (Royce et al. 1968).

Some of the Washington and British Columbia stocks migrate only short distances to good feeding areas and remain there until they approach maturity (Godfrey et al. 1975). Healey (1980), in his study of juvenile salmon in the Strait of Georgia, observed that after coho smolts had entered salt water, they dispersed quickly throughout the strait. The number that remained in the strait to rear to maturity varied from year to year and probably was dependent on smolt density and feeding conditions. Healey suggested that fish that found themselves in poor feeding areas moved to outside waters, whereas those in good feeding areas remained.

Hartt (1980) reported on the results of coho tagging studies conducted along the west coast of North America. Samples were taken from the Juan de Fuca Strait, Washington, to Bristol Bay, Alaska. Coho were found in nearshore and offshore areas. The coho migration out of Puget Sound and the Strait of Georgia took place later in the summer than for other salmon species. Once in outside waters the coho moved rapidly northwest and south along the coast during their first summer. Juvenile coho from California, coastal Oregon, and the Columbia River make up a significant proportion of the coho stocks that follow the coastal belt northward during the summer months, as far as the northeastern section of the Gulf of Alaska (Hartt and Dell 1986). Hartt (1980) also found that some coho followed a counter-clockwise circular route across the open Gulf of Alaska during the first fall and winter. Dahlberg (1982) reported that a coded-wire tagged coho smolt released at Toledo Harbour in southeastern Alaska was later recaptured as a four-year-old at 55°N, 143°W, about 1,600 km west of Baranof Island, Alaska.

Many coho, however, do not take a long migratory route around the Gulf of Alaska but spend their entire marine life in inshore waters (Hartt and Dell 1986). The distribution of juveniles in their first summer in the Gulf of Alaska does not overlap with that of immature salmon that are a year or more older, which may serve to reduce feeding competition. By remaining in inshore areas, coho avoid pelagic predators in the open ocean. Coho that were tagged in the northern Gulf of Alaska, in the vicinity of Kodiak Island, were recovered in almost all areas from Alaska to Oregon. However, coho that were tagged further south, from southeastern Alaska to Cape Flattery, resulted in few Alaskan recoveries (Hartt 1980). As early as 1929 and 1930, Pritchard (1934) found that maturing coho tagged along the north and central coasts of British Columbia were recovered south or east of their release site. He found that the distances travelled between release and recapture sites were much less for coho than for chinook, and he concluded that coho stocks were mostly local.

Clemens (1930) reported that coho tagged in Queen Charlotte Strait were recovered to the south. Fish that were caught and tagged in one area comprised a variety of stocks. There was no evidence from marking studies that fish captured from one school were all headed for a specific area or spawning stream. Coho that were captured and marked along the north coast, and that were subsequently recovered in the Strait of Georgia, used both the Juan de Fuca and Johnstone Strait entrances.

Foerster (1955) summarized coho tagging experiments between 1925 and 1951 and found that coho moved in all directions towards small streams along the coast. He also noted that coho migrated shorter distances and at a slower rate than chinook and concluded that these differences were due to the coho's greater wandering during feeding activity. All tagged coho were recovered in the same year as they were marked and released, whereas chinook tags were recovered over two or more years. Coho tagged in both Alaska and Oregon were recovered in the Fraser River, British Columbia.

Movement and Distribution at Sea

Godfrey (1965) was one of the first to summarize all known data on ocean distribution of coho salmon over the broad expanse of the North Pacific. He noted that, until the 1950s, it was generally believed that coho did not undertake extensive ocean migrations. Not until after the mid-1950s was there any systematic sampling in the offshore areas. One coho tagged in the northern Gulf of Alaska area in September was later recovered in Depoe Bay, Oregon, a distance of 2,200 km (Godfrey 1965). Godfrey (1965) also noted that North American coho have not been found in large numbers in offshore waters and that they probably wintered in areas well to the south of the Gulf of Alaska.

Asian coho are generally found in the southern part of the western Pacific Ocean in spring and early summer, and north of latitude 45°N by mid to late summer (Godfrey 1965). Semko (1958) indicated that most Kamchatkan coho were feeding 1,600-1,800 km southeast of Kamchatka between 165°-173°E and 45°-50°N. He also found that the number of coho in the Bering Sea was insignificant. Godfrey et al. (1975) reported that Asian coho, in their first ocean year, migrated progressively southward from the Kamchatkan peninsula to at least 40°N as water temperatures declined. Of the five species of Pacific salmon common to both North America and Asia, coho have exhibited a preference for the highest minimum ocean temperatures (at 5°-5.9°C) and are not generally found in waters cooler than 7°C (Manzer et al. 1965). As temperatures increase during the summer months, coho move progressively northward throughout the North Pacific Ocean and into the Bering Sea (Figure 3). The following spring, they again move north over a broad expanse from Hokkaido and the Kuril Islands on the west and from as far east as 175°W (Godfrey et al. 1975). With the onset of maturity in mid-July and August, coho that are in waters north of 52°N, near Kamchatka, head for east Kamchatka, while those south of 52°N move towards west Kamchatka, the northern coast of the Sea of Okhotsk, and the east coast of Sakhalin (Kondo et al. 1965).

North American salmon, of all five species, and steelhead intermingle with Asian salmon south of 46°N in the North Pacific Ocean (INPFC 1985). Initial evidence for intermingling of Asian and North American coho stocks came from a 1959 tagging program conducted by Japan. At that time, 225 coho were tagged directly south of Adak Island in the Aleutian Islands (about 176°W); later, eight were recaptured in western Alaska and two in eastern Kamchatka, each having travelled about 1,800 km (Godfrey 1965). More recent studies have shown that coho from western Kamchatka and areas in the northern Sea of Okhotsk occur south of 46°N and between 175°E and 175°W (INPFC 1983); western and central Alaskan coho are also present in this same area (INPFC 1985). In 1983, observers on Japanese research vessels noted that coho were distributed south of 45°N, between 170°E and 180° in May; by June they had moved north to 46°N; and by July they had moved to 51°N between 162°E and 176°W (FAJ 1984). Two coho that were tagged and released in the vicinity of 51°30′N, 177°00′W were subsequently recovered (one each) in southeastern Kamchatka and Bristol Bay (NWAFC 1984). The eastern known limit for Asian coho is about 177°W (at 45°N) for coho recovered in both southeastern and southwestern Kamchatka (NWAFC 1984); the western known limit

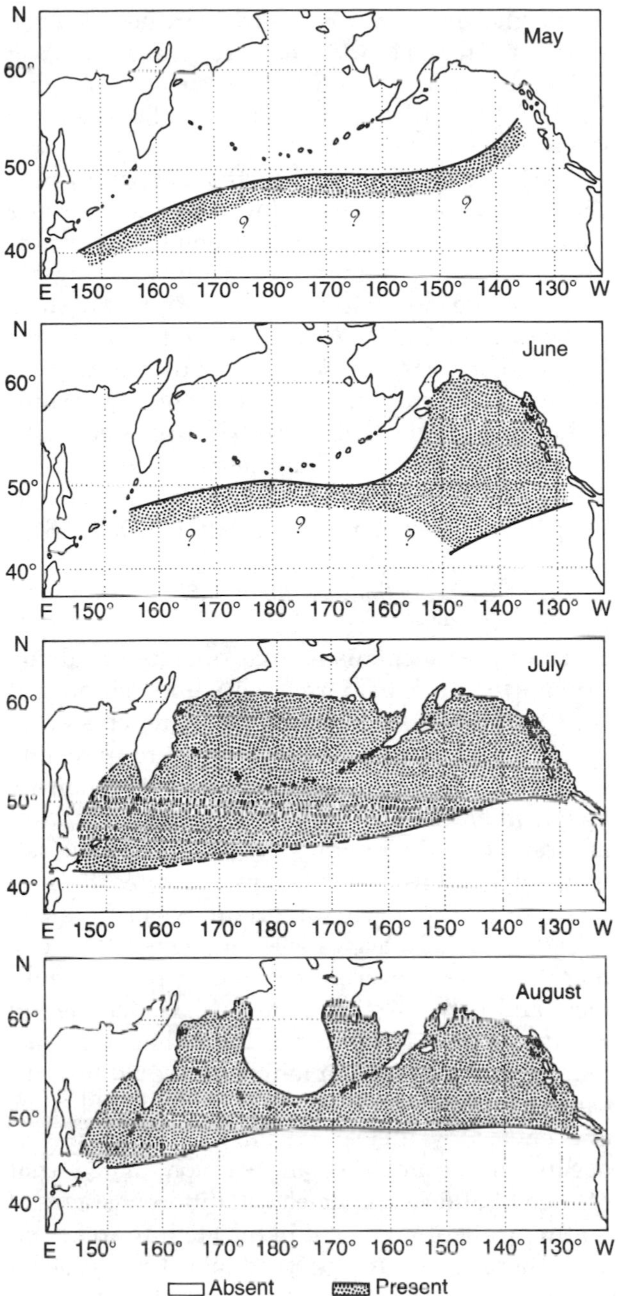

FIGURE 3

Occurrence of coho salmon in the North Pacific Ocean, May through August. (From Manzer et al. 1965)

for North American coho was demarcated by a fish that was tagged and released (June 1983) at 177°33′E (44°30′N) and later recovered (August 1983) in the Mulchatna River, Bristol Bay, Alaska (INPFC 1985).

Alaskan coho from streams that are tributary to the eastern Bering Sea follow a migration route that is similar to Asian coho. When temperatures decline in late summer, immature coho migrate south to the Aleutian Islands (40°N), and some move into the Gulf of Alaska. When temperatures begin to increase in spring, coho commence a northward migration back to their home streams. Coho stocks in the Gulf of Alaska are characteristically highly mixed and are derived from a wide variety of streams from Alaska to as far south as Oregon (Godfrey et al. 1975; Hartt and Dell 1986).

Direction-Finding during Migration

Some of the mechanisms that operate in salmon orientation during migration are understood, but the comprehensive details of the migration mechanisms that direct the fish from smolt to spawning adult are not known. It is not clear how salmon are able to migrate over the great distances they cover and return to their natal streams. During the salmon's ocean life "some awareness of position in relation to the place of origin is maintained" (Neave 1964). Neave (1964) suggested that fish might have an internal clock that records both local and home stream time so that a change in latitude would be detected by a change in day length, and a change in longitude would be indicated by the shift in time at which daylight begins and ends. Royce et al. (1968), in their review of ocean migration of Pacific salmon, noted that salmon of different species and different stages of maturity are often mixed in the feeding areas and that there is no strong tendency to school by group. In fact, the different stocks of coho show as much difference in their distribution as do different species. Royce et al. (1968) thought that the navigation system of salmonids was an inherited series of responses to stimuli because an individual makes the migration circuit only once and must find its way back to the home stream at the right time. They concluded that, because the circuit may be made only once, it was not simply a matter of following the cues in the reverse order. The routes followed are often in the open ocean away from the shoreline, and, in many cases, even beyond the continental shelf.

Ocean temperature gradients are thought to be unlikely as a major guiding mechanism because seasonal variations would potentially disrupt the timing of the return to spawn by two or more weeks (Royce et al. 1968). Celestial navigation has been considered and rejected because during much of the migration period the skies of the North Pacific Ocean are overcast (Royce et al. 1968). Olfactory cues in the open ocean are highly unlikely (Royce et al. 1968). Ocean currents would not seem to be important as the fish are known to migrate actively against, with, and across current patterns, and they do not simply drift with the flow (Royce et al. 1968). It is known that ocean currents generate small electrical potentials across the current (0.1–0.5 microvolts per cm). If fish can detect these potentials, it may provide them with directional cues. It is thought that a sensor associated with the lateral line of the fish may make it possible to use these electromagnetic phenomena for navigation (Royce et al. 1968).

Burgner (1980) suggested that salmon migrating over long distances in the ocean must be relying on an inherited response to guidance stimuli. Presumably, the migration pattern exhibited by a particular stock has evolved to optimize fish growth and survival. Other cues cited by Burgner (1980) are polarized light, photoperiod, pheromones, and electrical and magnetic fields. Although there are a variety of possible open-ocean migration cues, the fish may use a combination of mechanisms, or they may migrate by some means as yet unrecognized. However, after the fish reach the vicinity of their home stream, the guidance mechanism is clearly olfaction (Wisby and Hasler 1954).

Rate of Travel

The speed at which fish travel in the marine environment has only been measured indirectly. Clemens (1930) reported the recovery of two coho that had been tagged at Sooke on the southwest coast of Vancouver Island. One fish was recovered the next day at a distance of 55 km, and a second fish was recovered 11 days later in the Fraser River, having covered a minimum distance of 150 km.

Godfrey (1965) stated that the rate of migration was difficult to judge, but it was known that a few individuals had averaged at least 48 km/d and that a slow rate may be the result of feeding diversions. Jensen (1953) and Allen (1966) estimated a rate of 9.2–13.0 km/d for fish moving from the central Washington coast to the Seattle area. An even slower rate of 6 km/d was reported by Parker and Kirkness (1951), who found that coho marked and released in Alaska had travelled an average of 215 km over a 36-day period. Van Hyning (1951) marked and recaptured coho that were feeding off the Oregon coast and found that they had moved only 3 km/d (range 0–17 km/d). Using some selected recoveries and assuming a fairly direct path between tagging and recovery locations, Godfrey et al. (1975) suggested an average migration rate of just under 30 km/d. Royce et al. (1968) stated that salmon could maintain a rate of 55 km/d over long distances.

Ocean Food Habits

On first entry to salt water, juvenile coho feed mostly on marine invertebrates, but as they grow larger they become more piscivorous (Shapovalov and Taft 1954). A number of studies have shown that coho, during their estuarine and early marine life stages, are important predators on chum and pink salmon fry (Parker 1971; Slaney et al. 1985). In her study of stomach contents of juvenile coho of both wild and hatchery origin from Yaquina Bay, Oregon, Myers (1978) found that coho captured in beach areas had eaten primarily anchovy (*Engraulis mordax*), surf smelt (*Hypomesus pretiosus*), and sand lance (*Ammodytes hexapterus*). In the channel areas, hatchery coho fed primarily on crangonid shrimp and megalopa larvae of Dungeness crab (*Cancer magister*), whereas wild coho concentrated on juvenile surf smelt. Levy and Levings (1978) found that coho smolts in the estuary of the Squamish River, British Columbia, were feeding on unidentified fish, as well as *Anisogammarus* and *Neomysis*. In the Strait of Georgia, Healey (1978) sampled coho that had a mean monthly (May-October 1976) fork length of 11.6–28.1 cm. Examination of stomach contents of these coho and of coho of a similar size caught in 1975 revealed that herring (*Clupea harengus*), sand lance, and unidentified fish remains accounted for 34.6% and 29.0% of the contents (by volume) over the two years and that amphipods accounted for 26.7% and 40.5%. Crab megalops were important in 1975 and made up 26.2% of the diet. Overall, the stomach contents as a percentage of body weight ranged from 0.40% to 1.51%. More recently, Healey (1980) noted that the amount and type of food in juvenile coho stomachs was a function of its availability and that there was a positive correlation between the abundance of juveniles and the amount of food in their stomachs. The fish are obviously attracted to good feeding areas and will remain there as long as the food is in sufficient supply.

Chamberlain (1907), sampling adult coho taken from the northern Gulf of Alaska, found that they had eaten sand lance, sticklebacks (*Gasterosteus aculeatus*), small herring, and the occasional flatfish, cottid, and salmonid. Marine invertebrates, including amphipods, isopods, and crab larvae, were also included in the diet. Pritchard and Tester (1943, 1944) found that coho consumed a wide variety of food items, but that herring and sand lance were the most important components of the diet. Pritchard and Tester (1943, 1944) also found that coho ate sardines (*Sardinops sagax*), anchovies, capelin (*Mallotus villosus*), rockfish (scorpaenids), sable fish (*Anoplopoma fimbria*), lanternfish (myctophids), Pacific saury (*Cololabis saira*), hake (*Merluccius productus*), walleye pollock (*Theragra chalcogrammus*), and other coho salmon. Among the invertebrates eaten by coho, they found euphausiids, squid (*Loligo opalescens*), goose barnacles (*Pollicipes polymerus*), and jellyfish, although the last three items were observed in one year only. The diet of adult coho salmon is very similar to that of chinook, except that invertebrates make up about one-fifth of the diet in coho, and less than 3% in chinook. In some situations, coho may feed more heavily on fish than do other salmonids (FRBC 1955).

On the Oregon coast, 97% of the stomachs of the troll-caught coho contained larvae of the Pacific crab (*C. magister*) (Anonymous 1949). Heg and Van Hyning (1951) found that maturing coho in their second summer at sea consumed herring, anchovies, smelt, euphausiids, and crab larvae (especially *C. magister*), and in some areas, squid. The diet of Washington coastal coho was similar to that of Oregon fish but included sardines and rockfish, with anchovies and smelts occurring rarely (Silliman 1941).

On the British Columbia coast a number of studies have been undertaken to determine the nature of adult coho diets. Herring was found to be the dominant food item for coho off the southwest coast of Vancouver Island (FRBC 1955). Foerster (1955) observed sand lance, sardines, and herring in coho stomachs. Chatham Sound coho stomachs contained mostly herring larvae and sand lance (Manzer 1969).

It was noted earlier that "outside" stocks of coho from the west coast of Vancouver Island were generally larger than "inside" stocks from the east coast of the island. Prakash and Milne (1958) found that the food of coho from the west coast of Vancouver Island consisted primarily of fish, with crustaceans forming a minor part of the diet. On the east coast of the island, fish were less important, and amphipods made up the bulk of the diet. They also noted that, on average, an "inside" fish contained only half the volume of food as an "outside" west coast fish, which was probably relative to the differences in growth. The diet differences reflect different feeding conditions and food availability. In British Columbia waters, coho are opportunistic feeders and show considerable plasticity in what they eat (Prakash 1962).

Senter (1940) noted that the coho of southeastern Alaska fed on a mixture of herring, smelts, and candlefish (*Thaleichthys pacificus*), and that maturing females seemed to have more food in their stomachs than males. Additional items found in the stomachs of coho from the Gulf of Alaska included copepods and chaetognaths (Manzer and Neave 1958, 1959).

Churikov (1975) reported that juvenile coho stomachs sampled in Kamchatkan estuaries contained 69.4% gammarids and 27% winged insects, the balance being made up of miscellaneous food items. Andrievskaya (1968) found that maturing coho in the Sea of Okhotsk consumed young walleye pollock, sand lance, and other fish; and that coho over 25 cm in length on the Bering Sea side of Kamchatka ate young greenlings (hexagrammids), whereas coho sampled more than 150 km offshore were found to contain 90.8% amphipods (primarily *Parathemisto japonica*), with the balance made up of fish. Gribanov (1948) observed that Asian coho continued to feed on fish and invertebrates up to the time they entered their native rivers.

It has been determined by sonar observations and the position of fish in gillnets that salmon feed as individuals. The schools disperse for feeding and then later regroup (Burgner 1980). Before feeding on prey fish, a solitary salmon circles a concentration of prey. Suddenly, it attacks the prey, which surface and churn the area into a froth of bubbles, then quickly disperse. The feeding coho will circle and attack once again when the forage fish have regrouped (Grinols and Gill 1968).

Various estimates have been made of the quantity of food taken by adult coho. LeBrasseur (1966) observed that when coho consumed euphausiids, squid, and fish, the stomach contents were equivalent to about 1% of body weight. He also noted, as did Chapman (1936), that a high proportion of coho sampled had empty stomachs. How the fish are captured during sampling, e.g., hook and line, or gillnet set overnight, will have a strong bearing on what will be found in the stomach. The fish may be caught before they have begun feeding for the day and may have digested all the food captured previously. Coho also frequently egest food from their stomachs when caught.

As a further comment on coho being opportunistic feeders, it is interesting to note that, in the Great Lakes, adult coho feed on rainbow smelt (*Osmerus mordax*) and alewife (*Alosa pseudoharengus*), both of which are abundant in this environment. This may be one of the keys to the success of this introduced exotic (Scott and Crossman 1973).

Ocean Growth

Crone and Bond (1976) found that coho smolts that entered the sea as yearlings ranged in length from 79 mm to 120 mm; those that migrated as three-year-olds were 91 mm to 139 mm, and the four- and five-year-olds were 151 mm and 175 mm, respectively. The latter two unusual age groups were from fish sampled from the Karluk system in Alaska (Drucker 1972).

Coho grow very rapidly after they reach the marine environment. Hartt (1980) reported that the average fork length of smolts that had moved beyond the estuary was 150–270 mm. During their first year at sea, growth was estimated at 1.23 to 1.50 mm per day. For example, fish tagged off southeastern Alaska and the Queen Charlotte Islands grew from an average size of 253 mm in July to 290 mm in August and to 311 mm in September

(Godfrey et al. 1975). In April and May these fish would have been 60–70 mm. Mathews and Buckley (1976) and Healey (1980) reported growth rates of 1.1 mm/d for coho during their first six months in salt water (for a size range of 100–280 mm), with a daily increase in weight of about 2% per day. Healey (1980) calculated that the length-weight relationship for Strait of Georgia juvenile coho was: $W = 1.62 \times 10^{-6}L^{3.42}$, where W = weight in g and L = body length in mm. Phillips and Barraclough (1978) developed a length-weight relationship for Strait of Georgia coho sampled from Saanich Inlet (1966–1975) as: $Log_{10}W = 3.309Log_{10}L - 5.576$, or $W = 2.655 \times 10^{-6}L^{3.309}$.

Clemens (1930) sampled coho from the Strait of Georgia that were entering their second year in salt water. For fish caught April through June, he noted the following progression in average size: 1–15 April, 0.91–1.14 kg; 16–30 April, 1.14–1.35 kg; 1–15 May, 1.35–1.60 kg; 16–31 May, 1.60–1.92 kg; and 1–15 June, 1.92–2.14 kg. By 15 September the fish had reached an average length of 61 cm and a weight of 2.95 kg. He concluded that the growth during the spring months was about 0.45 kg per month and by summer had decreased to half this rate. Milne (1950) and Prakash and Milne (1958) compared the growth rate of coho found in the Strait of Georgia with that of coho from the west coast of Vancouver Island. They observed that fish of the same age from west coast waters were 0.9–1.9 kg heavier than fish from the Strait of Georgia and that both groups doubled their weight between June and September. They attributed this difference to better feeding conditions in west coast waters that may have been a result of higher nutrient levels and more favourable ocean temperatures. Godfrey (1965) reported that coho caught off the west coast of Vancouver Island in June had a fork length of 54.6–58.8 cm, and 64.4–69.8 cm by September. The average weight for commercially caught coho was 3.22 kg. Ricker (1976) concluded that the growth rate of coho in their final year of life was from 0.6 to 2.5 kg in inside waters (Puget Sound, Strait of Georgia) and 0.8 to 4.0 kg in outside areas. These weights are lower than the average of 4.30 kg (2.33–6.76 kg) reported earlier by Rounsefell and Kelez (1940). Shapovalov and Taft (1954) found that coho from Waddell Creek, California, entered the sea at 10–15 cm fork length and returned 16 months later at 62 cm.

Gribanov (1948), in a study of Asian coho, observed that this species grew faster than all other species of salmon except pink salmon. Most coho measured were 55–69 cm, with a range of 40–88 cm in fork length. The usual weight for Kamchatkan coho was 3.0–3.5 kg, with extremes of 1.2–6.8 kg.

Smolt-to-Adult Survival

Various estimates of survival from the smolt to the adult stage have been made. Shapovalov and Taft (1954) observed survivals ranging from 0.98% to 7.72%, with a mean of 4.95%, and suggested that the bulk of the mortality occurred during the first year at sea. Part of the difficulty in making these and other estimates was the necessity of clipping off different combinations of fins. Later it was shown that fin marking resulted in additional mortality, the severity of which depended on which fins were removed.

Foerster (1955) summarized some earlier estimates that had been made on survival rates. For coho from Nile Creek on the east coast of Vancouver Island, the estimates for three different brood years were 3.2%, 4.9%, and 9.9%; for coho from Port John (British Columbia) the rates for a series of four brood years were 7.0%, 7.2%, 3.8%, and 19.1%; for Cultus Lake coho the return rate was 8.1% for marked coho. In a detailed study of Puget Sound hatchery coho, Mathews and Buckley (1976) estimated that, after the first six months at sea, 13% of the smolts survived; after twelve months, survival was down to 9%. Of fish that survived to catchable size, approximately half would be taken in the commercial and recreational fisheries. The numbers that returned to the home stream were about 4% of the smolts that had migrated to sea. This compares to an average survival rate of 3.8% (range 0.9%–9.4%) observed for ten brood returns to Minter Creek (Washington)(Salo and Bayliff 1958), and 5% (range 1.0%–7.7%) for four broods of Waddell Creek (California) coho (Shapovalov and Taft 1954). The smolt-to-adult survival rate for Big Qualicum (British Columbia) coho was much higher at 10.8% (range 5.4%–15.5%) over a 15-year period (Fraser et al. 1983). Ricker (1976) estimated that the mean monthly instantaneous rate of oceanic mortality for coho was 0.013 during the final year of life.

In the early 1970s it was thought that the maximum smolt-to-adult survival (catch plus escapement) was about 20%–25%. However, Bilton et al. (1982) showed clearly that, under optimum conditions of size and time of release, survivals in excess of 40% can be anticipated. From this experiment, maximum adult production (43.5%) was achieved by releasing 25.1-g coho smolts on 22 June (1975). Smolts released at earlier or later dates, and at larger and smaller sizes, did not survive as well. Bilton et al. (1982) also observed that smolts released in April at a size of >20 g produced the largest number of jacks.

Homing and Straying

All anadromous salmonids deposit their eggs in freshwater gravel beds for incubation. The resultant fry, which live in fresh water for varying lengths of time, move downstream to the marine environment as smolts. After a growth period in the sea, fish nearing maturity return or "home back" to their parental stream to spawn. Fish that do not return to their home stream or release site and that spawn in other streams and tributaries are considered to be strays. A return to the parental spawning ground provides a mechanism for enhancing survival by the repeat usage of good sites. Ricker (1972) noted that homing has the further advantage of getting the approximate number of spawners back to a spawning ground or rearing area that can accommodate them. Homing can potentially be a disadvantage if fish return to areas that have marginal spawning areas or poor rearing conditions.

Straying can also be a survival mechanism in that it may protect against the loss of an entire stock due to some environmental catastrophe in the home stream (e.g., the volcanic eruption of Mt. St. Helens, Washington, in 1983). If there is no straying, areas that lack spawners due to poor conditions or restricted access will not become recolonized if conditions become more suitable. Fish are also able to extend their normal ranges through straying.

The cues used by salmon to move from offshore feeding areas to the vicinity of their home stream are not fully understood, but once they reach the point of leaving the estuary and entering fresh water they appear to rely primarily on olfaction.

Harden Jones (1968) developed a hypothesis of sequential imprinting for home-stream detection which states that "young salmon may undergo a series of imprinting processes corresponding to each major change of environment made in fresh water; gravel bed, lake outlet, tributary river, main river. It could be important that the sequence in which these imprints are made should correspond exactly to the reverse sequence of stimuli that the upstream migrant receives on the way home." Thus, distinctive home-stream odours could enable fish to migrate back to their incubation site.

Wisby and Hasler (1954) were the first to demonstrate that fish with their olfactory pits plugged were unable to identify their home stream. Fish with olfactory occlusion continued to migrate upstream, but their choice of direction at critical junctions appeared to be random. Brett and MacKinnon (1954) demonstrated that coho and chinook salmon detect odours at low concentrations. Fish migrating upstream are known to exhibit an avoidance reaction to the presence of mammals by moving back downstream. Odours from the skin of (predatory) mammals or human hands at dilution rates of greater than 1 ppm were sufficient to displace coho downstream. Brett and Groot (1963) reported that, although salmon were highly sensitive to olfactory cues, other cues, such as vision and the sensitivity of the acoustico-lateralis system, are also probably used for successful homing. Groves et al. (1967) demonstrated that in chinook, olfaction was far more important for homing than was vision. Some fish had their olfactory systems experimentally blocked and others were blinded. Of the latter group, at least half of the fish were able to home correctly using olfaction (or other senses). However, when the olfactory system was blocked, fewer than one-tenth of the fish were able to find their home stream.

If the Harden Jones (1968) hypothesis is correct

and olfactory stimuli are imprinted by fish during development, then presumably, by following olfactory cues, the adult fish will get back to the spawning area and the gravel in which it incubated as an egg. In many cases, however, egg-to-fry-to-smolt development is not a simple sequence. In Chilliwack Lake, British Columbia, wild coho juveniles (under-yearlings) were tagged and released in the lake in the fall. In the following spring these marked smolts were found in small tributaries of the lower river below the lake, presumably having over-wintered there. As returning adults, all marked fish were recovered in the main river above the lake and none in the lower river tributaries (B.C. Pearce, Department of Fisheries and Oceans, Vancouver, British Columbia, pers. comm.). In Alaska, coho that were marked while rearing in a small tributary of the Berners River were found spawning the next year in another tributary where conditions were better (Gray et al. 1978). It can be concluded that coho juveniles may rear a long distance from where they emerge as fry but are obviously imprinted with the memory of their natal spawning ground. Imprinting must occur initially in the earliest life history stages. If juveniles are captured or incubated and reared in a hatchery, and then transported for release elsewhere, the adults will return to the point of release. In cases where the point of release is upstream in the same river system in which they were incubated and/or reared, many adults will not bypass the lower site to reach the release point (Lister et al. 1981).

The question of the time required for imprinting has been explored using a large number of hatchery coho. In some cases, the exposure of coho smolts to a specific water source, like spring water, for 36–48 hours has been sufficient to assure a high degree of homing success. Jensen and Duncan (1971) reported on a study that involved transporting marked coho smolts 250 km downstream from the Wenatchee River, Washington, to below Ice Harbour Dam on the Snake River. Both of these rivers are major tributaries to the Columbia River. Several groups of smolts were held up to 48 hours in a spring water source and then released between March and May. By September, jacks were returning to the release site, and the following year jacks (from a second release group) and three-year-old fish returned to this specific water source. This source was flowing at less than 0.76 m^3/min into the Snake River, which was running at many thousands of m^3/min. This, and other examples of very specific homing can be cited, but there are also numerous studies of marked coho that indicate persistent low levels of straying.

Taft and Shapovalov (1938) reported that wild coho were marked (multiple fin clip) and released in Waddell Creek (California). When they returned as adults, approximately 20% of the recoveries were made in Scott Creek, which enters the Pacific Ocean about 8 km south of Waddell Creek. Lister et al. (1981) discussed two experiments on marked wild coho in which coded-wire tags were used rather than the less reliable multiple fin marks. These fish were tagged as smolts and later recovered as adults on the spawning grounds. In two tributaries (13 km apart) of the Squamish River (British Columbia), one of 27 tagged adult coho (3.7%) had strayed between tributaries. In the Cowichan and Koksilah rivers, which enter Cowichan Bay (British Columbia) about 2 km apart, two of 150 tagged adults (1.3%) were from fish that had strayed between systems. Shapovalov and Taft (1954) observed that the rate of straying from a given stream is fairly constant for a given year class but may vary considerably from year class to year class. They hypothesized that the "conditions existing at the time of the migration to the ocean determine the amount of straying that will take place one and two seasons later." Just what these conditions are was not stated, but it was noted that there was a tendency towards (1) a positive correlation between the number of outmigrants and the amount of straying, and (2) a negative correlation between the average size and the number of strays.

A recent study of recoveries of coded-wire tagged coho on the east coast of Vancouver Island indicated that the straying rate for hatchery-released fish (0%–5%) was less than that for wild coho stocks (0.7%), but most straying rates were in the range of 0.1% (M. Labelle, Resource Ecology/Resource Management Science, University of British Columbia, Vancouver, British Columbia, pers. comm.). Fish that were recovered in other than their home streams were found to have strayed, on average, about 12 km. M. Labelle also observed that those stocks that had been manipulated in some way – for example, eggs collected from a particular brood stock, transported to a different incubation and rearing site, and then returned to their native stream for release as fry or smolts – had the highest subsequent straying rates. Deterioration of stream conditions, such as low flows, was also seen to increase the straying rate up to 50%. The furthest stray recorded by M. Labelle was a Quinsam River coho that was recovered in the Quatse River about 190 km to the northwest.

One interesting aspect of the straying question is whether a fish, once having strayed, can sort out the error and find its way to its home stream. In California, Taft and Shapovalov (1938) observed that a steelhead released from Scott Creek returned as an adult to Waddell Creek and was captured at a fence site 2.5 km from salt water, where it was again marked and released. It was subsequently recaptured in its home stream. Similarly, coho marked as juveniles before release from the Capilano Hatchery near Vancouver, British Columbia, were recovered in the Seymour River as adults, about 8 km to the east and 20 km from salt water. These fish (which had previously been marked by an adipose fin clip and a coded-wire tag) were tagged with an external (Petersen disc) tag, and about a week later several of them entered the Capilano Hatchery, a distance of at least 33 km. They had backed out of the Seymour River, reentered salt water, proceeded west to the mouth of the Capilano River, and then migrated up river to the hatchery. There, the coded-wire tags were removed and read to confirm that they were, in fact, fish of Capilano Hatchery origin. Had those fish been killed and examined in the Seymour River, it would have been assumed that they were strays that were about to spawn there. Thus, many records of strays may only indicate that, at the time of capture, the fish was in the wrong place and that, given the opportunity, it may have retraced its route and returned to its home stream (E.T. Stone, Department of Fisheries and Oceans, Vancouver, British Columbia, pers. comm.).

The occurrence of long-distance straying became more evident with the use of coded-wire tags, which proved more reliable than multiple fin marks. Between 1974 and 1980, a total of 70 jacks and 64 adult coho bearing coded-wire tags were recovered at four enhancement sites from which they were not released. On the east coast of Vancouver Island, coho from five release sites (two south and three north) strayed into the Big Qualicum Hatchery. Coho from the Big Qualicum and Puntledge hatcheries entered the Quinsam Hatchery to the north. In the lower mainland of British Columbia, strays occurred in both directions between the Seymour and Capilano rivers, with the majority (by virtue of the numbers released) straying from the Capilano River to the Seymour River. For longer distance strays, a coho from each of the following release sites strayed north into the Big Qualicum Hatchery: Chilliwack Hatchery in the Fraser River valley, British Columbia; Lummi Bay near Blaine, Washington; and the Salmon River, a tributary of the Queets River in northern Washington. Another fish from Lummi Bay entered the Capilano Hatchery (F.K. Sandercock, unpubl. data).

The extensive use of coded-wire tags in recent years has demonstrated beyond doubt that the majority of coho that are native to a particular stream return to that same stream at maturity. In situations in which the survival rate is high or the spawner capacity of the stream is approached, straying may occur into adjacent streams.

Concluding Remarks

In summary, coho are widely distributed over much of the North Pacific basin and have been successfully transplanted to more non-endemic locations than any of the other species of Pacific salmon. Coho have often been described as "opportunistic," a term which is especially appropriate in describing their choice of spawning sites. Their success as a species may be partly attributed to their utilization of a myriad of small coastal streams and to their aggressiveness and apparent determination to reach the small headwater creeks and tributaries of larger rivers to spawn.

In many cases, they overcome difficult obstructions to reach areas inaccessible to other salmon and then share these locations with only migrant steelhead or perhaps resident cutthroat trout. These small headwater streams generally provide cool, clear, well-oxygenated water, with stable flows that are ideal for incubation and subsequent rearing.

Groundwater seepage in these small streams moderates the high temperatures in summer as well as the near-freezing temperatures in winter, thus sustaining a much more stable environment. Aquatic insect production, a prime source of food for juvenile coho, is often rich in small streams and is further supplemented by terrestrial insects that fall into the water from streamside vegetation. Almost anything of an appropriate size that is moving in or on the water column can be considered as food for coho.

Most coho mortality occurs during the rearing stage in fresh water, where the juveniles may be exposed to winter and spring freshets, summer droughts, or simply lack of rearing space. However, by remaining in these streams for a year or more before migrating to sea, juvenile coho avoid the high mortality rate associated with entry, as fry, into sea water, such as is experienced by chum and pink salmon. Once the coho smolts migrate to sea, the survival rate is high.

To take advantage of this high survival rate in the marine environment, coho have been incubated, reared, and released from hatcheries for almost a hundred years. Coho can be readily adapted to the hatchery environment and are potentially the easiest of all the Pacific salmon to domesticate. In fact, the burgeoning fish-farm industry in western North America initially concentrated on coho salmon, partly because of availability of a surplus of coho eggs from hatcheries.

Coho smolts, produced either in the wild or in a hatchery environment, often survive to adults at three or four times (sometimes 10–20 times) the rate of other salmon species. Of all salmon caught commercially in the North Pacific basin, coho make up only a small percentage, which probably reflects on the total amount of freshwater rearing space available.

Coho are taken commercially, as incidental catch in seine and gillnet fisheries (river traps in Asia), and in the directed troll fishery. For trollers, the prized species is chinook, but whatever the coho lack in size, compared to chinook, they make up for in quantity.

The recreational fishery on the Pacific side of North America is highly oriented towards chinook and coho salmon. Although most sportsmen would rather catch a chinook because of the larger size, it is often the coho that fills the bag because of their abundance, their availability in nearshore waters, and their willingness to take the angler's lure. When the coho return to the estuaries and rivers they may be further harvested in the Native food fisheries.

Because spawning stocks of coho are so widespread, it becomes virtually impossible to determine escapement populations and, hence, total stock size. The problem of estimating how many there are is made even more difficult by the fact that adult coho can be found returning to their natal stream in almost every month of the year.

The diversity in life history strategies exhibited by the large number of coho stocks in the North Pacific is reflected in the broad range of migration and spawning timing, the multitude of suitable freshwater habitats, the variety of foods consumed in both fresh water and salt water, and the various strategies followed in ocean rearing. It is this adaptability that would seem to assure the continued survival of this valuable species.

References

Allen, G.H. 1958. Notes on the fecundity of silver salmon (*Oncorhynchus kisutch*). Prog. Fish-Cult. 20:163–169

——. 1966. Ocean migration and distribution of

fin-marked coho salmon. J. Fish. Res. Board Can. 23:1043-1061

Allen, K.R. 1969. Limitations on production in salmonid populations in streams, p. 3-18. *In*: T.G. Northcote (ed.). Symposium on Salmon and Trout in Streams. H.R. MacMillan Lectures in Fisheries. Institute of Fisheries, University of British Columbia, Vancouver, BC

Andersen, B.C., and D.W. Narver. 1975. Fish populations of Carnation Creek and other Barkley Sound streams - 1974: data record and progress report. Fish. Res. Board Can. MS Rep. Ser. 1351:73 p.

Andriashev, A.P. 1955. A contribution to the knowledge of the fishes from the Bering and Chukchi Seas. U.S. Fish Wildl. Serv. Spec. Sci. Rep. Fish. 145:81 p.

Andrievskaya, L.D. 1968. Feeding of Pacific salmon fry in the sea. Izv. Tikhookean. Nauchno-Issled. Inst. Rybn. Khoz. Okeanogr. 64:73-80. (Transl. from Russian; Fish. Res. Board Can. Transl. Ser. 1423)

Anonymous. 1949. Crab larvae as food for silver salmon at sea. Fish. Comm. (Oreg.) Res. Briefs 2(1):17

Anonymous. 1951. Landlocked silver salmon for Montana waters. Prog. Fish-Cult. 13:192

Armstrong, R.H. 1970. Age, food and migration of Dolly Varden smolts in southeastern Alaska. J. Fish. Res. Board Can. 27:991-1004

Armstrong, R.W., and A.W. Argue. 1977. Trapping and coded-wire tagging of wild coho and chinook juveniles from the Cowichan River system, 1975. Fish. Mar. Serv. (Can.) Pac. Reg. Tech. Rep. Ser. PAC/T-77-14:58 p.

Aro, K.V., and M.P. Shepard. 1967. Pacific salmon in Canada, p. 225-327. *In*: Salmon of the North Pacific Ocean. Part IV: spawning populations of North Pacific salmon. Int. North Pac. Fish. Comm. Bull. 23

Atkinson, C.E., J.H. Rose, and T.O. Duncan. 1967. Pacific salmon in the United States, p. 43-223. *In*: Salmon of the North Pacific Ocean. Part IV: spawning populations of North Pacific salmon. Int. North Pac. Fish. Comm. Bull. 23

Beal, F.R. 1955. Silver salmon (*O. kisutch*) reproduction in Montana. Prog. Fish-Cult. 17:79-81

Berg, L.S. 1948. Freshwater fishes of the USSR and adjacent countries, vol. 1. Opred. Faune SSSR 27:466 p. (Transl. from Russian; Israel Program for Scientific Translations, Jerusalem, 1963)

Bilton, H.T., D.F. Alderdice, and J.T. Schnute. 1982. Influence of time and size at release of juvenile coho salmon (*Oncorhynchus kisutch*) on returns at maturity. Can. J. Fish. Aquat. Sci. 39:426-447

Bilton, H.T., R.B. Morley, A.S. Coburn, and J. Van Tine. 1984. The influence of time and size at release of juvenile coho salmon (*Oncorhynchus kisutch*) on returns at maturity: results of releases from Quinsam River Hatchery, B.C. in 1980. Can. Tech. Rep. Fish. Aquat. Sci. 1306:98 p.

Brannon, E., C. Feldmann, and L. Donaldson. 1982. University of Washington zero-age coho salmon smolt production. Aquaculture 28:195-200

Brett, J.R. 1952. Temperature tolerance in young Pacific salmon, genus *Oncorhynchus*. J. Fish. Res. Board Can. 9:265-321

Brett, J.R., and C. Groot. 1963. Some aspects of olfaction and visual responses in Pacific salmon. J. Fish. Res. Board Can. 20:287-303

Brett, J.R., M. Hollands, and D.F. Alderdice. 1958. The effect of temperature on the cruising speed of young sockeye and coho salmon. J. Fish. Res. Board Can. 15:587-605

Brett, J.R., and D. MacKinnon. 1954. Some aspects of olfactory perception in migrating adult coho and spring salmon. J. Fish. Res. Board Can. 11:310-318

Briggs, J.C. 1953. The behavior and reproduction of salmonid fishes in a small coastal stream. Calif. Dep. Fish Game Fish. Bull. 94:62 p.

Bryan, J.E. 1973. The influence of pipeline development on freshwater fishery resources of northern Yukon Territory: aspects of research conducted in 1971 and 1972. Environ. Social Comm. North Pipelines Task Force North Oil Develop. Rep. 73-6:63 p.

Bryant, M.D. 1984. The role of beaver dams as coho salmon habitat in southeast Alaska streams, p. 183-192. *In*: J.M. Walton and D.B. Houston (eds.). Proceedings of the Olympic Wild Fish Conference, March 23-25, 1983. Fisheries Technology Program, Peninsula College, Port Angeles, WA

Burgner, R.L. 1980. Some features of ocean migrations and timing of Pacific salmon, p. 153-164. *In*: W.J. McNeil and D.C. Himsworth (eds.). Salmonid ecosystems of the North Pacific. Oregon State University Press, Corvallis, OR

Burner, C.J. 1951. Characteristics of spawning nests of Columbia River salmon. Fish. Bull. Fish

Wildl. Serv. 61:97–110

Bustard, D.R., and D.W. Narver. 1975a. Aspects of the winter ecology of juvenile coho salmon (*Oncorhynchus kisutch*) and steelhead trout (*Salmo gairdneri*). J. Fish. Res. Board. Can. 32:667–680

——. 1975b. Preferences of juvenile coho salmon (*Oncorhynchus kisutch*) and cutthroat trout (*Salmo clarki*) relative to simulated alteration of winter habitat. J. Fish. Res. Board Can. 32:681–687

Carl, G.C., W.A. Clemens, and C.C. Lindsey. 1959. The fresh-water fishes of British Columbia. 3rd ed. Br. Col. Prov. Mus. Handbk. 5:192 p.

Cederholm, C.J., and W.J. Scarlett. 1981. Seasonal immigrations of juvenile salmonids into four small tributaries of the Clearwater River, Washington, 1977–1981, p. 98–110. *In*: E.L. Brannon and E.O. Salo (eds.) Proceedings of the Salmon and Trout Migratory Behavior Symposium. School of Fisheries, University of Washington, Seattle, WA

Chamberlain, F.M. 1907. Some observations on salmon and trout in Alaska. Rep. U.S. Comm. Fish. 1906 Spec. Pap., U.S. Bur. Fish. Doc. 627:112 p.

Chapman, D.W. 1962. Aggressive behaviour in juvenile coho salmon as a cause of emigration. J. Fish. Res. Board Can. 19:1047–1080

——. 1965. Net production of juvenile coho salmon in three Oregon streams. Trans. Am. Fish. Soc. 94:40–52

Chapman, D.W., and T.C. Bjornn. 1969. Distribution of salmonids in streams with special reference to food and feeding, p. 153–176. *In*: T.G. Northcote (ed.). Symposium on Salmon and Trout in Streams. H.R. MacMillan Lectures in Fisheries. Institute of Fisheries, University of British Columbia, Vancouver, BC

Chapman, D.W., and E. Knudsen. 1980. Channelization and livestock impacts on salmonid habitat biomass in western Washington. Trans. Am. Fish. Soc. 109:357–363

Chapman, W.M. 1936. The pilchard fishery of the state of Washington in 1936 with notes on the food of the silver and chinook salmon off the Washington coast. Wash. Dep. Fish. Biol. Rep. 36C:30 p.

Churikov, A.A. 1975. Features of the downstream migration of young salmon of the genus *Oncorhynchus* from the rivers of the northeast coast of Sakhalin. J. Ichthyol. 15:963–970

Clarke, W.C. 1988. Rearing strategies for zero-age coho salmon *Oncorhynchus kisutch*, p. 387–390. *In*: Proceedings of the Aquaculture International Congress Sept. 6–9, 1988. BC Pavilion Corp., Vancouver, BC

Cleaver, F.C. 1951. Fisheries statistics of Oregon. Contrib. Fish. Comm. Oreg. 16:176 p.

Clemens, W.A. 1930. Pacific salmon migration: the tagging of coho salmon off the east coast of Vancouver Island in 1927 and 1928. Bull. Biol. Board Can. 15:1–19

Crone, R.A., and C.E. Bond. 1976. Life history of coho salmon *Oncorhynchus kisutch*, in Sashin Creek, southeastern Alaska. Fish. Bull. (U.S.) 74:897–923

Dahlberg, M.L. 1982. Report of incidence of coded-wire tagged salmonids in catches of foreign commercial and research vessels operating in the North Pacific Ocean and Bering Sea during 1980–1982. (Submitted to annual meeting International North Pacific Fisheries Commission, Tokyo, Japan, November 1982.) Auke Bay Laboratory, Northwest and Alaska Fisheries Center, National Marine Fisheries Service, Auke Bay, AK. 11 p.

Davidson, F.A., and S.J. Hutchinson. 1938. The geographic and environmental limitations of the Pacific salmon (genus *Oncorhynchus*). Bull. Bur. Fish. (U.S.) 48:667–692

Dill, L.M. 1969. The sub-gravel behaviour of Pacific salmon larvae, p. 89–99. *In*: T.G. Northcote (ed.). Symposium on Salmon and Trout in Streams. H.R. MacMillan Lectures in Fisheries. Institute of Fisheries, University of British Columbia, Vancouver, BC

Dill, L.M., R.C. Ydenberg, and A.H.G. Fraser. 1981. Food abundance and territory size in juvenile coho salmon (*Oncorhynchus kisutch*). Can. J. Zool. 59:1801–1809

Drucker, B. 1972. Some life history characteristics of coho salmon of Karluk River system, Kodiak Island, Alaska. Fish. Bull. (U.S.) 70:79–94

Dvinin, P.A. 1952. The salmon of south Sakhalin. Izv. Tikhookean. Nauchno-Issled. Inst. Rybn. Khoz. Okeanogr. 37:69–108. (Transl. from Russian; Fish. Res. Board Can. Transl. Ser. 120)

Ellis, D.V. 1962. Preliminary studies on the visible migrations of adult salmon. J. Fish. Res. Board Can. 19:137–148

——. 1966. Swimming speeds of sockeye and coho salmon on spawning migration. J. Fish. Res. Board Can. 23:181-187

Engel, L.J. 1967. Egg take investigation in Cook Inlet drainage and Prince William Sound. Prog. Rep. Alaska Dep. Fish Game Sport Fish. Div. 8(1966-67):111-116

——. 1968. Inventory and cataloguing of the sport fish and waters in the Kenai-Cook Inlet-Prince William Sound areas. Prog. Rep. Alaska Dep. Fish Game Sport Fish. Div. 9(1967-68):95-116

——. 1972. Evaluation of sport fish stocking on the Kenai Peninsula-Cook Inlet areas. Annu. Prog. Rep. Alaska Dep. Fish Game Sport Fish Stud. 13(1971-72):67-79

Engel, S., and J.J. Magnuson. 1976. Vertical and horizontal distributions of coho salmon (*Oncorhynchus kisutch*), yellow perch (*Perca flavescens*) and cisco (*Coregonus artedii*) in Pallette Lake, Wisconsin. J. Fish. Res. Board Can. 33:2710-2715

Ennis, G.L., A. Cinader, S. McIndoe, and T. Munsen. 1982. An annotated bibliography and information summary on the fisheries resources of the Yukon River basin in Canada. Can. MS Rep. Fish. Aquat. Sci. 1657:278 p.

Fiscus, C.H. 1980. Marine mammal-salmonid interactions: a review, p. 121-132. *In*: W.J. McNeil and D.C. Himsworth (eds.). Salmonid ecosystems of the North Pacific. Oregon State University Press, Corvallis, OR

Fisher, J.P., W.G. Pearcy, and A.W. Chung. 1984. Studies of juvenile salmonids off the Oregon and Washington coast, 1983. Oreg. State Univ. Coll. Oceanogr. Cruise Rep. 84-2; Oreg. State Univ. Sea Grant Coll. Program ORESU-T-85-004:29 p.

Fisheries Research Board of Canada (FRBC). 1955. Pacific Biological Station, Nanaimo. Annu. Rep. Fish. Res. Board Can. 1954:75-106

Fishery Agency of Japan (FAJ). 1955. On the salmon in waters adjacent to Japan, a biological review. Int. North Pac. Fish. Comm. Bull. 1:57-92

——. 1984. Report on research by Japan for the International North Pacific Fisheries Commission in 1983. Int. North Pac. Fish. Comm. Annu. Rep. 1983:37-77

Foerster, R.E. 1935. Inter-specific cross-breeding of Pacific salmon. Proc. Trans. R. Soc. Can. Ser. 3 29(5):21-33

——. 1952. The seaward-migrating sockeye and coho salmon from Lakelse Lake 1952. Fish. Res. Board Can. Prog. Rep. Pac. Coast Stn. 93:30-32

——. 1955. The Pacific salmon (genus *Oncorhynchus*) of the Canadian Pacific Coast with particular reference to their occurrence in or near fresh water. Int. North Pac. Fish. Comm. Bull. 1:1-56

Foerster, R.E., and W.E. Ricker. 1953. The coho salmon of Cultus Lake and Sweltzer Creek. J. Fish. Res. Board Can. 10:293-319

Fraser, C.M. 1920. Growth rate in the Pacific salmon. Proc. Trans. R. Soc. Can. Ser. 3 13(5):163-226

Fraser, F.J., E.A. Perry, and D.T. Lightly. 1983. Big Qualicum River salmon development project. Volume 1: a biological assessment 1959-1972. Can. Tech. Rep. Fish. Aquat. Sci. 1189:198 p.

Fredin, R.A. 1980. Trends in North Pacific salmon fisheries, p. 59-119. *In*: W.J. McNeil and D.C. Himsworth (eds.). Salmonid ecosystems of the North Pacific. Oregon State University Press, Corvallis, OR

Fry, D.H., Jr. 1977. Information on California salmon fisheries and stocks, p. 15-24. *In*: Additional information on the exploitation, scientific investigation, and management of salmon stocks on the Pacific coast of the United States in relation to the abstention provisions of the North Pacific Fisheries Convention. Int. North Pac. Fish. Comm. Bull. 36

Fry, D.H., Jr., and E.P. Hughes. 1954. Proportion of king and silver salmon in California's 1952 landings. Calif. Dep. Fish Game Fish. Bull. 95:7-16

Gilbert, C.H. 1913. Age at maturity of the Pacific coast salmon of the genus *Oncorhynchus*. Bull. Bur. Fish. (U.S.) 32: 1-22

——. 1922. The salmon of the Yukon River. Bull. Bur. Fish. (U.S.) 38:317-332

Godfrey, H. 1965. Coho salmon in offshore waters, p. 1-39. *In*: Salmon of the North Pacific Ocean. Part IX. Coho, chinook and masu salmon in offshore waters. Int. North Pac. Fish. Comm. Bull. 16

Godfrey, H., K.A. Henry, and S. Machidori. 1975. Distribution and abundance of coho salmon in offshore waters of the North Pacific Ocean. Int. North Pac. Fish. Comm. Bull. 31. 80 p.

Godfrey, H., W.R. Hourston, J.W. Stokes, and F.C. Withler. 1954. Effects of a rock slide on Babine River salmon. Bull. Fish. Res. Board Can. 101:100 p.

Gray, P.L., K.R. Florey, J.F. Koerner, and R.A. Marriott. 1978. Coho salmon fluorescent pigment mark-recovery program for the Taku, Berners and Chilkat rivers in southeastern Alaska (1972–1974). Alaska Dep. Fish Game Inf. Leafl. 176:74 p.

Great Lakes Fishery Commission (GLFC). 1970. 1969 annual report. Great Lakes Fishery Commission, Ann Arbor, MI. 58 p.

——. 1980. 1977 annual report. Great Lakes Fishery Commission, Ann Arbor, MI. 121 p.

Gribanov, V.I. 1948. The coho salmon (*Oncorhynchus kisutch* Walb.)–a biological sketch. Izv. Tikhookean. Nauchno-Issled. Inst. Rybn. Khoz. Okeanogr. 28:43–101. (Transl. from Russian; Fish. Res. Board Can. Transl. Ser. 370)

Grinols, R.B., and C.D. Gill. 1968. Feeding behaviour of three oceanic fishes (*Oncorhynchus kisutch*, *Trachurus symmetricus* and *Anoplopoma fimbria*) from the northeast Pacific. J. Fish. Res. Board Can. 25:825–827

Groves, A.B., G.B. Collins, and P.S. Trefethen. 1967. Roles of olfaction and vision in choice of spawning site by homing adult chinook salmon (*Oncorhynchus tshawytscha*). J. Fish. Res. Board Can. 25:867–876

Hallock, R.J., and D.H. Fry Jr. 1967. Five species of salmon, *Oncorhynchus*, in the Sacramento River, California. Calif. Fish Game 53:5–22

Hancock, M.J., and D.E. Marshall. 1984a. Catalogue of salmon streams and spawning escapements of sub-districts 120 and 130 (Alsek-Stikine-Taku watersheds). Can. Data Rep. Fish Aquat. Sci. 456:233 p.

——. 1984b. Catalogue of salmon streams and spawning escapements of sub-districts 110 and 120 (Yukon-Arctic). Can. Data Rep. Fish. Aquat. Sci. 474:233 p.

Harden Jones, F.R. 1968. Fish migration. St. Martin's Press, New York, NY. 325 p.

Hart, J.L. 1973. Pacific fishes of Canada. Bull. Fish. Res. Board Can. 180:740 p.

Hartman, G.F. 1965. The role of behaviour in the ecology and interaction of underyearling coho salmon (*Oncorhynchus kisutch*) and steelhead trout (*Salmo gairdneri*). J. Fish. Res. Board Can. 22:1035–1081

Hartman, G.F., B.C. Anderson, and J.C. Scrivener. 1982. Seaward movement of coho salmon (*Oncorhynchus kisutch*) fry in Carnation Creek, an unstable coastal stream in British Columbia. Can. J. Fish. Aquat. Sci. 39:588–597

Hartman, W.L. 1972. Lake Erie: effects of exploration, environmental changes and new species on the fishery resources. J. Fish. Res. Board Can. 29:899–912

Hartt, A.C. 1980. Juvenile salmonids in the oceanic ecosystem – the critical first summer, p. 25–57. *In*: W.J. McNeil and D.C. Himsworth (eds.). Salmonid ecosystems of the North Pacific. Oregon State University Press, Corvallis, OR

Hartt, A.C. and M.B. Dell. 1986. Early oceanic migrations and growth of juvenile Pacific salmon and steelhead trout. Int. North Pac. Fish. Comm. Bull. 46:105 p.

Hasler, A.D. and D.S. Farner. 1942. Fisheries investigations in Crater Lake, Oregon, 1937–1940. J. Wildl. Manage. 6:319–327

Healey, M.C. 1978. The distribution, abundance, and feeding habits of juvenile Pacific salmon in Georgia Strait, British Columbia. Fish. Mar. Serv. (Can.) Tech. Rep. 788:49 p.

——. 1980. The ecology of juvenile salmon in Georgia Strait, British Columbia, p. 203–229. *In*: W.J. McNeil and D.C. Himsworth (eds.). Salmonid ecosystems of the North Pacific. Oregon State University Press, Corvallis, OR

Heg, R., and J. Van Hyning. 1951. Food of the chinook and silver salmon taken off the Oregon coast. Fish. Comm. (Oreg.) Res. Briefs 3(2):32–40

Hikita, H. 1956. Pacific salmon (genus: *Oncorhynchus*) known to occur in coasts and rivers within Hokkaido. Sci. Rep. Hokkaido Salmon Hatchery 11:25–44

Hoar, W.S. 1951. The behaviour of chum, pink and coho salmon in relation to seaward migration. J. Fish. Res. Board Can. 8:241–263

——. 1958. The evolution of migratory behaviour among juvenile salmon of the genus *Oncorhynchus*. J. Fish. Res. Board Can. 15:391–428

Holtby, L.B., G.F. Hartman, and J.C. Scrivener. 1984. Stream indexing from the perspective of the Carnation Creek experience, p. 87–111. *In*: P.E.K. Symons and M. Waldichuk (eds.). Proceedings of the Workshop on Stream Indexing for Salmon Escapement Estimation, West Vancouver, British Columbia, 2–3 February, 1984. Can. Tech. Rep. Fish. Aquat. Sci. 1326

Hunter, J.G. 1949. Natural propagation of salmon in the central coastal area of British Columbia. II.

The 1948 run. Fish. Res. Board Can. Prog. Rep. Pac. Coast Stn. 79:33-34

——. 1959. Survival and production of pink and chum salmon in a coastal stream. J. Fish. Res. Board Can. 16:835-886

International North Pacific Fisheries Commission. (INPFC). 1962a. 4. Supplementary information on salmon stocks of the United States: age composition of Pacific salmon, 1934-1955, p. 53-60. *In*: The exploitation, scientific investigation and management of salmon (genus *Oncorhynchus*) stocks on the Pacific coast of the United States in relation to the abstention provisions of the North Pacific fisheries convention. Int. North Pac. Fish. Comm. Bull. 10:53-60

——. 1962b. 5. Supplementary information on salmon stocks of the United States: salmon stocks by species in the area north of Bristol Bay, including catch and effort statistics with regard to the United States fishery, p. 61-66. *In*: The exploitation, scientific investigation and management of salmon (genus *Oncorhynchus*) stocks on the Pacific coast of the United States in relation to the abstention provisions of the North Pacific fisheries convention. Int. North Pac. Fish. Comm. Bull. 10:61-66

——. 1979. Historical catch statistics for salmon of the North Pacific Ocean. Int. North Pac. Fish. Comm. Bull. 39:166 p.

——. 1983. Activities of the Commission concerning salmonids. Int. North Pac. Fish. Comm. Annu. Rep. 1982:6-10

——. 1985. Report of the 31st annual meeting - 1984. Int. North Pac. Fish. Comm. Annu. Rep. 1984:59 p.

Ishida, T., T. Tanaka, S. Kameyama, K. Sasaki, and Y. Nemoto. 1975. On the coho salmon transplanted from USA into Yurappu River, Hokkaido. Sci. Rep. Hokkaido Salmon Hatchery 29:11-15. (In Japanese)

Ishida, T., H. Tsuji, T. Hosokawa, and K. Nara. 1976. On the coho salmon transplanted from North America into the Shibetsu River, Hokkaido. Sci. Rep. Hokkaido Salmon Hatchery 30:47-53. (In Japanese, English abstract)

Jensen, A., and R. Duncan. 1971. Homing of transplanted coho salmon. Prog. Fish-Cult. 33:216-218

Jensen, H.M. 1953. Migrations of silver salmon in Puget Sound. Fish. Res. Pap. Wash. Dep. Fish. 1:13-21

Jordan, D.S., and J.O. Snyder. 1902. A review of the salmonid fishes of Japan. Proc. U.S. Natl. Mus. 24:567-593

Kennedy, W.A., C.T. Shoop, W. Griffioen, and A. Solmie. 1976. The 1975 crop of salmon reared on the Pacific Biological Station experimental fish farm. Fish. Mar. Serv. (Can.) Tech. Rep. 665:20 p.

Klein, W.D., and L.M. Finnell. 1969. Comparative study of coho salmon introductions in Parvin Lake and Granby Reservoir. Prog. Fish-Cult. 31:99-108

Kondo, H., Y. Hirano, N. Nakayama, and M. Miyake. 1965. Offshore distribution and migration of Pacific salmon (genus *Oncorhynchus*) based on tagging studies (1958-1961). Int. North Pac. Fish. Comm. Bull. 17:213 p.

Koski, KV. 1966. The survival of coho salmon (*Oncorhynchus kisutch*) from egg deposition to emergence in three Oregon coastal streams. M.Sc. thesis. Oregon State University, Corvallis, OR. 84 p.

Kwain, W., and A.H. Lawrie. 1981. Pink salmon in the Great Lakes. Fisheries 6(2):2-6

Larkin, P.A. 1977. Pacific salmon, p. 156-186. *In*: J.A. Gulland (ed.). Fish population dynamics. J. Wiley & Sons, New York, NY

Lawrie, A.H., and J.F. Rahrer. 1972. Lake Superior: effects of exploitation and introductions on the salmonid community. J. Fish. Res. Board Can. 29:765-776

LeBrasseur, R.J. 1966. Stomach contents of salmon and steelhead trout in the northeastern Pacific Ocean. J. Fish. Res. Board Can. 23:85-100

Levy, D.A., and C.D. Levings. 1978. A description of the fish community of the Squamish River estuary, British Columbia: relative abundance, seasonal changes, and feeding habits of salmonids. Fish. Mar. Serv. (Can.) MS Rep. 1475:63 p.

Lindberg, G.U., and M.I. Legeza. 1965. Fishes of the Sea of Japan and the adjacent areas of the Sea of Okhotsk and the Yellow Sea. Volume 2. Opred. Faune SSSR 84:389 p. (Transl. from Russian; Israel Program for Scientific Translations, Jerusalem, 1969)

Lister, D.B., and H.S. Genoe. 1970. Stream habitat utilization by cohabiting underyearlings of chinook (*Oncorhynchus tshawytscha*) and coho (*Oncorhynchus kisutch*) salmon in the Big Qualicum River, British Columbia. J. Fish. Res. Board Can.

27:1215-1224

Lister, D.B., D.G. Hickey, and I. Wallace. 1981. Review of the effects of enhancement strategies on the homing, straying and survival of Pacific salmonids. Prepared for Department of Fisheries and Oceans, Salmonid Enhancement Program. D.B. Lister and Associates, West Vancouver, BC. 51 p.

Lister, D.B., and C.E. Walker. 1966. The effect of flow control on freshwater survival of chum, coho and chinook salmon in the Big Qualicum River. Can. Fish Cult. 37:3-25

Logan, S.M. 1967. Silver salmon studies in the Resurrection Bay area. Prog. Rep. Alaska Dep. Fish Game Sport Fish Div. 8(1966-67):83-102

——. 1968. Silver salmon studies in the Resurrection Bay area. Prog. Rep. Alaska Dep. Fish Game Sport Fish Div. 9(1967-68):117- 134

Mace, P.M. 1983. Bird predation on juvenile salmonids in the Big Qualicum estuary, Vancouver Island. Can. Tech. Rep. Fish. Aquat. Sci. 1176:79 p.

Manzer, J.I. 1969. Stomach contents of juvenile Pacific salmon in Chatham Sound and adjacent waters. J. Fish. Res. Board Can. 26:2219- 2223

Manzer, J.I., T. Ishida, A.E. Peterson, and M.G. Hanavan. 1965. Salmon of the North Pacific Ocean. Part V. Offshore distribution of salmon. Int. North Pac. Fish. Comm. Bull. 15:452 p.

Manzer, J.I., and F. Neave. 1958. Data record of Canadian exploratory fishing for salmon in the Northeast Pacific in 1957. Fisheries Research Board of Canada, Pacific Biological Station, Nanaimo, BC. 303 p.

——. 1959. Data record of Canadian exploratory fishing for salmon in the Northeast Pacific in 1958. Fisheries Research Board of Canada, Pacific Biological Station, Nanaimo, BC. 223 p.

Marr, J.C. 1943. Age, length and weight studies of three species of Columbia River salmon (*Oncorhynchus keta, O. gorbuscha*, and *O. kisutch*). Stanford Ichthyol. Bull. 2:157-197

Marriott, R.A. 1968. Inventory and cataloguing of the sport fish and sport fish waters in southwest Alaska. Prog. Rep. Alaska Dep. Fish Game Sport Fish Div. 9(1967-68):81-93

Mason, J.C. 1974. Aspects of the ecology of juvenile coho salmon (*Oncorhynchus kisutch*) in Great Central Lake, B.C. Fish. Res. Board Can. Tech. Rep. 438:7 p.

Mason, J.C., and D.W. Chapman. 1965. Significance of early emergence, environmental rearing capacity and behavioural ecology of juvenile coho salmon in stream channels. J. Fish. Res. Board Can. 22:173-190

Mathews, S.B., and R. Buckley. 1976. Marine mortality of Puget Sound coho salmon (*Oncorhynchus kisutch*). J. Fish. Res. Board Can. 33:1677-1684

McDonald, J. 1960. The behaviour of Pacific salmon fry during their downstream migration to freshwater and saltwater nursery areas. J. Fish. Res. Board Can. 17:655-676

McHenry, E.T. 1981. Coho salmon studies in the Resurrection Bay area. Annu. Prog. Rep. Alaska Dep. Fish Game Fed. Aid Fish Restoration 1980-81:1-52

McKernan, D.L., D.R. Johnson, and J.I. Hodges. 1950. Some factors influencing the trends of salmon populations in Oregon. Trans. N. Am. Wildl. Conf. 15:427-449

McKnight T.C., and S.L. Serns. 1977. Growth and harvest of coho salmon in Stormy Lake, Wisconsin. Prog. Fish-Cult. 39:79-85

McMahon, T.E. 1983. Habitat suitability models: coho salmon. U.S. Fish Wildl. Serv. FWS/OBS 82/10.49:29 p.

McPhail, J.D., and C.C. Lindsey. 1970. Freshwater fishes of northwestern Canada and Alaska. Bull. Fish. Res. Board Can. 173:381 p.

Meehan, W.R., and D.B. Siniff. 1962. A study of the downstream migration of anadromous fishes in the Taku River, Alaska. Trans. Am. Fish. Soc. 91:399-467

Messersmith, J.D. 1965. Southern range extensions for chum and silver salmon. Calif. Fish Game 51:220

Milne, D.J. 1950. The differences in the growth of coho salmon on the east and west coasts of Vancouver Island in 1950. Fish. Res. Board Can. Prog. Rep. Pac. Coast Stn. 85:80-82

——. 1964. The chinook and coho salmon fisheries of British Columbia; with appendix by H. Godfrey. Bull. Fish. Res. Board Can. 142:46 p.

Morgan, A.R., and K.A. Henry. 1959. The 1955-56 silver salmon run into the Tenmile Lakes system. Res. Briefs Fish. Comm. Oreg. 7(1):57-77

Mundie, J.H. 1969. Ecological implications of the diet of juvenile coho in streams, p. 135-152. *In*: T.G. Northcote (ed.). Symposium on Salmon and

Trout in Streams. H.R. MacMillan Lectures in Fisheries. Institute of Fisheries, University of British Columbia, Vancouver, BC

Murphy, G.I. 1952. An analysis of silver salmon counts at Benbow Dam, South Fork of Eel River. Calif. Fish Game 38:105-112

Myers, K.W. 1979. Comparative analysis of stomach contents of cultured and wild juvenile salmonids in Yaquina Bay, Oregon, p. 155-162. *In*: S.J. Lipovsky and C.A. Simenstad (eds.). Gutshop '78: fish food habits studies: proceedings of the Second Pacific Northwest Technical Workshop. Wash. Sea Grant Publ. WSG-WO-79-1

Nara, K., M. Shimizu, G. Okukawa, K. Matsumura, and K. Umeda. 1979. On the coho salmon transplanted from North America into the Shibetsu River (II). Sci. Rep. Hokkaido Salmon Hatchery 33:7-16. (In Japanese)

Narver, D.W. 1978. Ecology of juvenile coho salmon - Can we use present knowledge for stream enhancement? p. 38-43. *In*: B.G. Shepherd and R.M.J. Ginetz (rapps.). Proceedings of the 1977 Northeast Pacific Chinook and Coho Salmon Workshop. Fish. Mar. Serv. (Can.) Tech. Rep. 759:164 p.

Neave, F. 1943. Diurnal fluctuations in the upstream migration of coho and spring salmon. J. Fish. Res. Board Can. 6:158-163

———. 1948. Fecundity and mortality in Pacific salmon. Proc. Trans. R. Soc. Can. Ser. 3 42(5):99-105

———. 1949. Game fish populations of the Cowichan River. Bull. Fish. Res. Board Can. 84:1-32

———. 1958. The origin and speciation of *Oncorhynchus*. Proc. Trans. R. Soc. Can. Ser. 3 52(5):25-39

———. 1964. Ocean migrations of Pacific salmon. J. Fish. Res. Board Can. 21:1227-1244

Neave, F., and A.L. Pritchard. 1942. Recoveries of Cowichan River coho salmon from the 1953 brood year emphasize the value of marking experiments. Fish. Res. Board Can. Prog. Rep. Pac. Coast Stn. 51:3-7

Neave, F., and W.P. Wickett. 1953. Factors affecting the freshwater development of Pacific salmon in British Columbia. Proc. 7th Pac. Sci. Congr. 1949(4):548-556

Noggle, C.C. 1977. Behavioral, physiological and lethal effects of suspended sediment on juvenile salmonids. M.Sc. thesis. University of Washington, Seattle, WA. 87 p.

Northwest and Alaska Fisheries Center, National Marine Fisheries Service (NWAFC). 1984. Investigations by the United States for the International North Pacific Fisheries Commission in 1983. Int. North Pac. Fish. Comm. Annu. Rep. 1983:88-170

Okada, Y. 1960. Studies on the freshwater fishes of Japan. Prefectural University of Mie, Tsu, Japan. 860 p.

Otto, R.G. 1971. Effects of salinity on the survival and growth of pre-smolt salmon (*Oncorhynchus kisutch*). J. Fish. Res. Board Can. 28:343-349

Otto, R.G., and J.E. McInerney. 1970. Development of salinity preference in pre-smolt coho salmon, *Oncorhynchus kisutch*. J. Fish. Res. Board Can. 27:793-800

Parker, R.R. 1971. Size selective predation among juvenile salmonid fishes in a British Columbia inlet. J. Fish. Res. Board Can. 28:1503-1510

Parker, R.R., and W. Kirkness. 1951. Biological investigations. Annu. Rep. Alaska Fish. Board Alaska Dep. Fish. 2(1950):25-41

Patten, B.G. 1977. Body size and learned avoidance as factors affecting predation on coho salmon, *Oncorhynchus kisutch* fry by torrent sculpin, *Cottus rhotheus*. Fish. Bull. (U.S.) 75:457-459

Peterson, N.P. 1980. The role of spring ponds in the winter ecology and natural production of coho salmon (*Oncorhynchus kisutch*) on the Olympic Peninsula, Washington. M.Sc. thesis. University of Washington, Seattle, WA. 96 p.

Phillips, A.C., and W.E. Barraclough. 1978. Early marine growth of juvenile Pacific salmon in the Strait of Georgia and Saanich Inlet, British Columbia. Fish. Mar. Serv. (Can.) Tech. Rep. 830:19 p.

Phillips, R.W., R.L. Lantz, E.W. Claire, and J.R. Moring. 1975. Some effects of gravel mixtures on emergence of coho salmon and steelhead trout fry. Trans Am. Fish. Soc. 104:461-466

Popov, A.M. 1933. Fishes of Avacha Bay on the southern coast of Kamchatka. Copeia 1933:59-67

Prakash, A. 1962. Seasonal changes in feeding of coho and chinook (spring) salmon in southern British Columbia waters. J. Fish. Res. Board Can. 19:851-866

Prakash, A., and D.J. Milne. 1958. Food as a factor affecting the growth of coho salmon off the east and west coasts of Vancouver Island, B.C. Fish. Res. Board Can. Prog. Rep. Pac. Coast Stn. 112:7-9

Pravdin, I.F. 1940. A review of investigations on the far-eastern salmon. Izv. Tikhookean. Nauchno-Issled. Inst. Rybn. Khoz. Okeangr. 18:5–105. (Transl. from Russian; Fish. Res. Board Can. Transl. Ser. 371)

Pritchard, A.L. 1934. Pacific salmon migration: the tagging of the coho salmon in British Columbia in 1929 and 1930. Bull. Biol. Board Can. 40:24 p.

——. 1936. Stomach content analyses of fishes preying upon the young of Pacific salmon during the fry migration at McClinton Creek, Masset Inlet, British Columbia. Can. Field-Nat. 50:104–105

——. 1940. Studies on the age of the coho salmon (*Oncorhynchus kisutch*) and the spring salmon (*Oncorhynchus tschawytscha* [sic]) in British Columbia. Proc. Trans. R. Soc. Can. Ser. 3 34(5):99–120

——. 1943. The behaviour and distribution of the coho salmon above Skutz Falls in the Cowichan River during the spawning run of 1942, p. 10. *In*: Investigators' summaries. Annu. Rep. Pac. Biol. Stn. 1943

Pritchard, A.L., and A.L. Tester. 1943. Notes on the food of coho salmon in British Columbia. Fish. Res. Board Can. Prog. Rep. Pac. Coast Stn. 55:10–11

——. 1944. Food of spring and coho salmon in British Columbia. Bull. Fish. Res. Board Can. 65:23 p.

Reiser, D.W., and T.C. Bjornn. 1979. Influence of forest and rangeland management on anadromous fish habitat in the western United States and Canada. 1. Habitat requirements of anadromous salmonids. U.S. Forest Serv. Gen. Tech. Rep. PNW-96:54 p.

Reiser, D.W., and T.A. Wesche. 1977. Determination of physical and hydraulic preferences of brown and brook trout in the selection of spawning locations. Water Resources Research Institute, University of Wyoming, Laramie, WY. 112 p.

Ricker, W.E. 1941. The consumption of young sockeye salmon by predaceous fish. J. Fish. Res. Board Can. 5:293–313

——. 1954. Pacific salmon for Atlantic waters? Can. Fish Cult. 16:6–11

——. 1972. Hereditary and environmental factors affecting certain salmonid populations, p. 19–160. *In*: R.C. Simon and P.A. Larkin (eds.). The stock concept in Pacific salmon. H.R. MacMillan Lectures in Fisheries. University of British Columbia, Institute of Fisheries, Vancouver, BC

——. 1976. Review of the rate of growth and mortality of Pacific salmon in salt water, and noncatch mortality caused by fishing. J. Fish Res. Board Can. 33:1483–1524

Robertson, O.M., M.A. Krupp, S.F. Thomas, C.B. Favour, S. Hare, and B.C. Wexler. 1961. Hyperadrenocorticism in spawning migratory and non migratory rainbow trout (*Salmo gairdnerii*): comparison with Pacific salmon (genus *Oncorhynchus*). Gen. Comp. Endocrinol. 1:473–484

Rounsefell, G.A. 1957. Fecundity of North American Salmonidae. Fish. Bull. Fish Wildl. Serv. 57:449–468

——. 1958. Anadromy in North American Salmonidae. Fish. Bull. Fish Wildl. Serv. 58:171–185

Rounsefell, G.A., and G.B. Kelez. 1940. The salmon and salmon fisheries of Swiftsure Bank, Puget Sound and the Fraser River. Bull. Bur. Fish. (U.S.) 48:693–823

Royce, W.F., L.S. Smith, and A.C. Hartt. 1968. Models of oceanic migrations of Pacific salmon and comments on guidance mechanisms. Fish. Bull. (U.S.) 66:441–462

Ruggles, C.P. 1966. Depth and velocity as a factor in stream rearing and production of juvenile coho salmon. Can. Fish Cult. 38:37–53

Salo, E.O., and W.H. Bayliff. 1958. Artificial and natural production of silver salmon (*Oncorhynchus kisutch*) at Minter Creek, Washington. Res. Bull. Wash. Dep. Fish. 4:76 p.

Sandercock, F.K. 1969. Bioenergetics of the rainbow trout (*Salmo gairdneri*) and the kokanee (*Oncorhynchus nerka*) populations of Marion Lake, British Columbia. Ph.D. thesis. University of British Columbia, Vancouver, BC. 165 p.

Scarlett, W.J., and C.J. Cederholm. 1984. Juvenile coho salmon fall-winter utilization of two small tributaries of the Clearwater River, Jefferson County, Washington, p. 227–242. *In*: J.M. Walton and D.B. Houston (eds.). Proceedings of the Olympic Wild Fish Conference, March 23–25, 1983. Fisheries Technology Program, Peninsula College, Port Angeles, WA

Scott, W.B., and E.J. Crossman. 1973. Freshwater fishes of Canada. Bull. Fish. Res. Board Can. 184:966 p.

Scrivener, J.C., and B.C. Andersen. 1982. Logging impacts and some mechanisms which determine the size of spring and summer populations of coho salmon fry in Carnation Creek, p. 257–272. *In*: G.F. Hartman (ed.). Proceedings of the Carnation Creek Workshop: a ten year review. Pacific Biological Station, Nanaimo, BC

Semko, R.S. 1954. The stocks of West Kamchatka salmon and their commercial utilization. Izv. Tikhookean. Nauchno- Issled. Inst. Rybn. Khoz. Okeangr. 41:3–109. (Transl. from Russian; Fish. Res. Board Can. Transl. Ser. 288)

——. 1958. Some data on the exploitation, distribution and migration of far eastern salmon in the open ocean, p. 8–30. *In*: Materialy po biologii morskovo perioda zhizni dalnevstochnykh lososei. Vsesoiuznyi Nauchno-Issledovatel'skii institut Morskogo Rybnogo Khoziastva i Okeanografii, Moscow. (Transl. from Russian; Fish. Res. Board Can. Transl. Ser. 179)

Senter, V.E. 1940. Observations on the food of the Pacific salmon. Pac. Fisherman 38:26 p.

Shapovalov, L. 1947. Distinctive characters of the species of anadromous trout and salmon found in California. Calif. Fish Game 33:185–190

Shapovalov, L., and A.C. Taft. 1954. The life histories of the steelhead rainbow trout (*Salmo gairdneri*) and silver salmon (*Oncorhynchus kisutch*) with special reference to Waddell Creek, California, and recommendations regarding their management. Calif. Dep. Fish Game Fish. Bull. 98:375 p.

Shmidt, P.Yu. 1950. Fishes of the Sea of Okhotsk. Tr. Tikhookean. Kom. Akad. Nauk SSSR 6:390 p. (Transl. from Russian; Israel Program for Scientific Translations, Jerusalem, 1965)

Silliman, R.P. 1941. Fluctuations in the diet of chinook and silver salmon (*Oncorhynchus tschawytscha* [sic] and *O. kisutch*) off Washington, as related to the troll catch of salmon. Copeia 1941(2):80–87

Skeesick, D.B. 1970. The fall immigration of juvenile coho salmon into a small tributary. Res. Rep. Fish Comm. Oreg. 2:90–95

Slaney, T.L., J.D. McPhail, D. Radford, and G.J. Birch. 1985. Review of the effects of enhancement strategies on interactions among juvenile salmonids. Can. MS Rep. Fish. Aquat. Sci. 1852:72 p.

Smirnov, A.I. 1960. The characteristics of the biology of reproduction and development of the coho, *Oncorhynchus kisutch* Walbaum. Vestn. Mosk. Univ. Ser. 6 1960(1):9–19. (Transl. from Russian; Fish. Res. Board Can. Transl. Ser. 287)

Smoker, W.A. 1953. Stream flow and silver salmon production in Western Washington. Wash. Dep. Fish. Res. Pap. 1:5–12

Snyder, J.O. 1931. Salmon of the Klamath River, California. I. The salmon and the fishery of the Klamath River. Calif. Dep. Fish Game Fish. Bull. 34:122 p.

Stein, R.A., P.E. Reimers, and J.D. Hall. 1972. Social interaction between juvenile coho (*Oncorhynchus kisutch*) and fall chinook salmon (*O. tshawytscha*) in Sixes River, Oregon. J. Fish. Res. Board Can. 29:1737–1748

Sumner, F.H. 1953. Migrations of salmonids in Sand Creek, Oregon. Trans. Am. Fish. Soc. 82:139–150

Symons, P.E.K., and J.D Martin. 1978. Discovery of juvenile Pacific salmon (coho) in a small coastal stream of new Brunswick. Fish. Bull. (U.S.) 76:487–489

Taft, A.C., and L. Shapovalov. 1938. Homing instinct and straying among steelhead trout (*Salmo gairdneri*) and silver salmon (*Oncorhynchus kisutch*). Calif. Fish Game 24:118–125

Tagart, J.V. 1984. Coho salmon survival from egg deposition to fry emergence, p. 173–181. *In*: J.M. Walton and D.B. Houston (eds.). Proceedings of the Olympic Wild Fish Conference, March 23–25, 1983. Fisheries Technology Program, Peninsula College, Port Angeles, WA

Taylor, E.B., and J.D. McPhail. 1985. Variation in body morphology among British Columbia populations of coho salmon, *Oncorhynchus kisutch*. Can. J. Fish. Aquat. Sci. 42:2020–2028

Tripp, D., and P. McCart. 1983. Effects of different coho stocking strategies on coho and cutthroat trout production in isolated headwater streams. Can. Tech. Rep. Fish. Aquat. Sci. 1212:176 p.

Tschaplinski, P.J., and G.F. Hartman. 1983. Winter distribution of juvenile coho salmon (*Oncorhynchus kisutch*) before and after logging in Carnation Creek, British Columbia, and some implications for overwinter survival. Can. J. Fish. Aquat. Sci. 40:452–461

Umeda, K., K. Matsumura, G. Okukawa, R. Sazawa, H. Honma, M. Arauchi, K. Kasahara, and K. Nara. 1981. On the coho salmon transplanted from North America into the Ichani River. Sci. Rep. Hokkaido Salmon Hatchery 35:9–23. (In Japanese, English abstract)

Van Hyning, J.M. 1951. The ocean salmon troll fishery of Oregon. Pac. Mar. Fish. Comm. Bull. 2:43–76

Wahle, R.J., and R.E. Pearson. 1987. A listing of Pacific coast spawning streams and hatcheries producing chinook and coho salmon (with estimates on numbers of spawners and data on hatchery releases). U.S. Dep. Commerce, NOAA Tech. Memo., NMFS F/NWC-122. 109 p.

Weisbart, M. 1968. Osmotic and ionic regulation in embryos, alevins and fry of the five species of Pacific salmon. Can. J. Zool. 46:385–397

Wells, L., and A.L. McLain. 1972. Lake Michigan: effects of exploitation, introductions and eutrophication on the salmonid community. J. Fish. Res. Board Can. 29:889–898

Wickett, W.P. 1951. The coho salmon population of Nile Creek. Fish. Res. Board Can. Prog. Rep. Pac. Coast Stn. 89:88–89

Wigglesworth, K.A., and R.R. Rawstron. 1974. Exploitation, survival, growth and cost of stocked silver salmon in Lake Berryessa, California. Calif. Fish Game 60:36–43

Willis, R.A. 1954. The length of time that silver salmon spent before death on the spawning grounds at Spring Creek, Wilson R., in 1951–52. Fish Comm. (Oreg.) Res. Briefs 5:27–31

Wisby, W.J., and A.D. Hasler. 1954. Effect of olfactory occlusion on migrating silver salmon (*O. kisutch*). J. Fish. Res. Board Can. 11:472–487

Wood, C.C. 1984. Foraging behavior of common mergansers (*Mergus merganser*) and their dispersion in relation to the availability of juvenile Pacific salmon. Ph.D. thesis. University of British Columbia, Vancouver, BC. 307 p.

Wynne-Edwards, V.C. 1947. The Yukon Territory, p. 6–20. *In*: North West Canadian fisheries surveys in 1944–45. Bull. Fish. Res. Board Can. 72

Zorbidi, Zh. Kh. 1977. Diurnal feeding rhythm of the coho salmon *Oncorhynchus kisutch* of Lake Azabach'ye. J. Ichthyol. 17:166–168

Life Histories of Masu and Amago Salmon

CONTENTS

PLATE 20. Masu salmon life history stages: (*left to right and top to bottom*) alevin, fry, fingerling, stream-type, smolt (*photograph by H. Ida*), ocean phase (*photograph by H. Masuda*), spawning male and spawning female (*photographs by H. Mayamu*)

PLATE 21. Amago salmon life history stages: (*left to right and top to bottom*) alevin, fry, fingerling, mature stream-type (*courtesy Japan Marine Products Photo Materials Association*), immature stream-type, lake-type, and upstream migrating female (*photographs, Freshwater Fish Protection Association*), spawning male (*photograph by H. Matsubara*)

LIFE HISTORIES OF MASU AND AMAGO SALMON
(*Oncorhynchus masou* and *Oncorhynchus rhodurus*)

Fumihiko Kato*

INTRODUCTION

OF THE SEVEN SPECIES of Pacific salmon (genus *Oncorhynchus*), two, *O. masou* (Plate 20) and *O. rhodurus* (Plate 21), occur only in Asia. *Oncorhynchus masou* can be separated into two forms, the anadromous masu salmon, *O. masou* var. *masou* (Brevoort), and the non-anadromous freshwater type, yamame or yamabe, *O. masou* var. *ishikawae* (Jordan and McGregor). *Oncorhynchus rhodurus* also occurs in two forms, the lake-dwelling type, biwamasu, *O. rhodurus* var. *rhodurus* (Jordan and McGregor), and the stream type, amago salmon, *O. rhodurus* var. *macrostomus* (Gunther) (Matsubara 1934; Oshima 1955).

Yamame can be differentiated from biwamasu and amago by the absence of vermillion spots on the sides of the body. Because feeding experiments indicated that these spots are a genotypic character, *O. masou* and *O. rhodurus* are tentatively treated as separate species. However, when masu and amago were crossbred and the offspring mated, the second generation showed good growth (Oshima 1955). This suggests that the two species in question are closely related. There is no known difference in the type of habitat preferred by the two species and, where their distribution overlaps, they may be found in the same stream system, although they are never captured together (Oshima 1955). Imanishi (1951) considered that, from an ecological standpoint, masu and amago represent a single species.

*Seikai Regional Fisheries Research Laboratory, Fisheries Agency of Japan, Kokubumachi 49, Nagasaki City, Japan

Masu has a more southerly geographical distribution than the other species of the genus *Oncorhynchus*. Most of the rivers with spawning populations of masu are in the region of the Sea of Japan and the Sea of Okhotsk. On the Pacific Ocean side masu occur in some rivers of Hokkaido and Honshu. Yamame, the freshwater-type masu, extend even farther south than the anadromous type and their distribution covers all of Japan except for the southern part of Shikoku.

Masu salmon is called sakuramasu in Japan, which means "cherry trout" in Japanese. This refers to the coincidence of the return of the adults with the appearance of the cherry blossoms from March to May. Masu spend the summer in rivers and move to the headwaters for spawning in September or October. They feed actively in the river, which is where gonad development takes place. Most sea-run masu mature at three or four years of age,[1] after spending one or more years in the river and one winter in the ocean. Anadromous masu and non-anadromous yamame have been observed spawning together. All fish that return from the sea die after spawning; however, the freshwater yamame males are known to spawn several times.

Although most *O. rhodurus* (amago) remain in rivers during their life cycle, some migrate to sea. However, the ocean distribution of amago is lim-

1 A fish in its third year of life, including the egg stage, is called a "three-year old." As in the previous chapters, freshwater and ocean ages will be designated by two numbers, separated by a period, e.g., age 1.1. The first number indicates the number of winters spent in fresh water (excluding the egg stage) and the second the winters spent in the sea.

ited to bays and inlets and only a few migrate to the open ocean (Machidori and Kato 1984). Timing of seaward migration of the juveniles takes place in autumn through winter, and upstream migration occurs the following spring. Their period of ocean life is less than six months, which is about half that of masu.

Masu are fished mainly in coastal waters of the Sea of Japan. Some are captured in the offshore drift gillnet and longline fisheries in the Sea of Japan, and a small-scale masu fishery operates on the Pacific Ocean coast of Honshu and Hokkaido. Amago are also fished commercially in rivers.

In this chapter masu will refer to both forms of *O. masou* (masu and yamame), and amago will refer to both forms of *O. rhodurus*, unless otherwise indicated.

Geographical Distribution of Spawning Stocks

Masu. Masu are native to far east Asia only. Rivers in which masu run are located mainly in the coastal areas of the Sea of Japan and the Sea of Okhotsk (Figure 1). The southern limit of masu extends to western Honshu and southeastern Korea, and the northern limit extends to the Amur

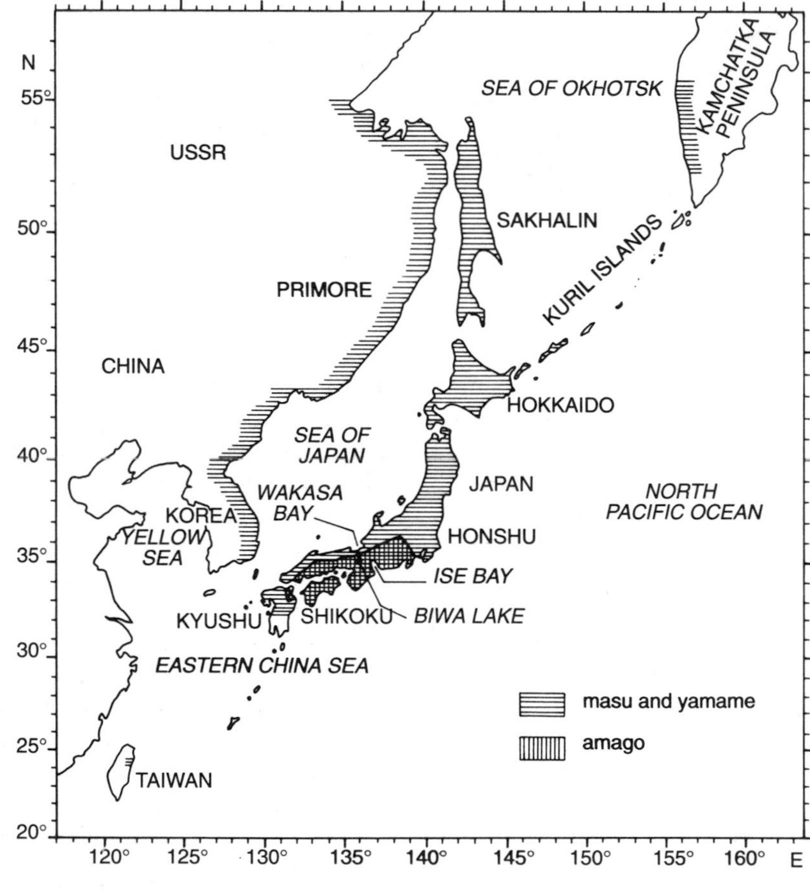

FIGURE 1

Spawning distribution of masu, yamame, and amago salmon in the Far East

River area and western Kamchatka. Fairly large runs of masu occur in Hokkaido, southern Sakhalin, and the Primore areas.

In Japan masu spawn in small and large rivers in Hokkaido and, in Honsu, north of Chiba Prefecture on the Pacific Ocean side and Shimane Prefecture on the Sea of Japan side. A few masu are occasionally observed in Yamaguchi Prefecture in rivers along the Sea of Japan (Katayama and Fujioka 1965), but these fish generally do not migrate into rivers of Kyushu.

On the Korean peninsula, masu have been reported from the Naktong River, which flows into Korea Strait (Mori 1935), and from rivers north of the Hyungsan River (Kyongsang-pukto) on the Sea of Japan side (Atkinson et al. 1967) (Figure 2). Masu occur in many rivers of the northern Korean peninsula, with the largest run in the Tumen River (Yoshida 1942).

In the Soviet Union, masu are found from the Tumen River in the south to the Amur River in the north, with large runs occurring in the rivers of northern Primore (Pravdin 1949) (Figure 2). On Sakhalin, masu migrate more or less into all rivers and are most abundant in the southern part of the island (Dvinin 1956). On the west side of Kamchatka, masu are known to occur from the Bolshaya River to the Icha River (Semko 1956). Only a few masu have been observed in rivers of southeastern and northwestern Kamchatka and in rivers of the northern coastal area of the Sea of Okhotsk (Krykhtin 1962). In the Kuril Islands, masu are found on Kunashir and Iturup islands (Takayasu et al. 1954; Hokkaido Salmon Hatchery 1969), but not on Urup and Simushir islands (Okada and Kawamura 1938; Mihara 1952).

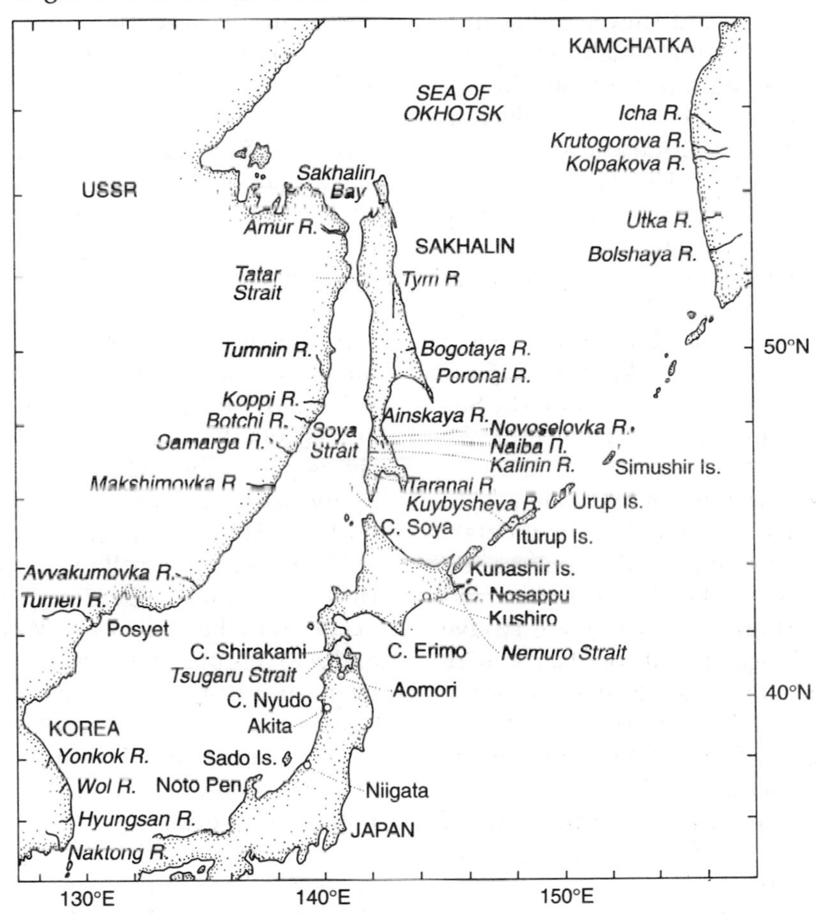

FIGURE 2

Far eastern Asian rivers having anadromous masu salmon in the Korean Peninsula and USSR. (From Machidori and Kato 1984)

The non-anadromous or stream-type masu, the yamame or yamabe, has a similar distribution to the anadromous masu in the north, but in the south these fish extend to Kyushu and to Taiwan in Dai Ko-Kei of the Formosan highland area (Oshima 1955) (Figure 1). On the Korean peninsula, yamame occur in rivers of Cholla-Namdo that enter the western Korea Strait (Jeon et al. 1978).

Amago. The distribution of amago in Japan ranges from the Tokyo Bay area southwestward along the Pacific Ocean side of Honshu to the northern tip of Kyushu, including Shikoku (Oshima 1955) (Figure 1). The climate throughout the amago range is generally somewhat warmer than that throughout the masu range. Amago are able to tolerate warmer water than masu, which is reflected in their more southern distribution.

Abundance of Spawners

Masu. Hokkaido is the main area occupied by anadromous masu in Japan and they migrate into almost all rivers. The rivers into which 500 or more masu were estimated to migrate are shown in Figure 3. According to records of the Hokkaido Salmon Hatchery (1969), the total annual catch of masu spawners in the rivers of Hokkaido peaked at about 6.3 million in 1948 and decreased rapidly during the following years (Figure 4A). Estimates of average annual escapement of masu to areas north of Honshu during 1963–65 were about 130,000, with a breakdown as follows: 67,000 fish in rivers flowing into the Sea of Okhotsk-Nemuro Strait region, 38,000 in rivers flowing into the Sea of Japan, and 26,000 in rivers flowing into the Pacific Ocean-Tsugaru Strait region.

In Hokkaido, large numbers of masu migrate to the districts on the Sea of Japan, Sea of Okhotsk, and Nemuro Strait side, and small numbers migrate to the districts on the Pacific Ocean and Tsugaru Strait side (Figure 2). There are twelve rivers with estimated escapements of 3,000 or more masu. Of these, six (Sarufutsu, Shokotsu, Shari, Churui, Shibetsu, and Nishibetsu) are located in the districts facing the Sea of Okhotsk-Nemuro Strait region. Four (Assabu, Toshibetsu, Shiribetsu, and Teshio) are located in districts facing the Sea of Japan, and two (Tokachi and Shizunai rivers) are located in districts facing the Pacific Ocean (Figure 3). Large runs generally occur in the larger rivers. The Shokotsu River has the largest escapement, estimated at about 10,000 fish.

Masu escapements have not been well investigated in the rivers of Honshu. Records of the Department of Statistics and Information of the Economic Bureau of the Ministry of Agriculture and Forestry indicate that the average annual catch in the main rivers of Honshu was 62.8 t from 1965 to 1975 (Figure 4B). Assuming that the average weight of masu in Honshu is 2.5 kg, this means that about 25,000 masu were caught. If the numbers of masu that escaped and the numbers migrating upstream in rivers that were not investigated are included, the total escapement of masu into the rivers of Honshu was probably at least more than twice that of the catch. On the average, 86% of the catch of masu in Honshu rivers occurred on the Sea of Japan side, because large rivers generally yield large catches of masu and many of these rivers are located in central and northern Honshu facing the Sea of Japan.

Masu runs are small in southern Korea. In 1967 the largest runs, amounting to only 100–200 masu, occurred in the Yonkok and Wol rivers (Atkinson et al. 1967) (Figure 2). There are probably some runs of masu into rivers of northern Korea, but their magnitudes are unknown (Yoshida 1942).

The numbers of migrating masu in rivers of the Soviet Union are also unknown. In the Tumnin, Botchi, and Koppi rivers in northern Primore, there were catches of 100,000, 30,000, and 30,000 masu, respectively, in the early twentieth century (Krykhtin 1962). This could indicate that there were considerable runs of masu in northern Primore at that time.

Gritsenko (1973) estimated that there was an

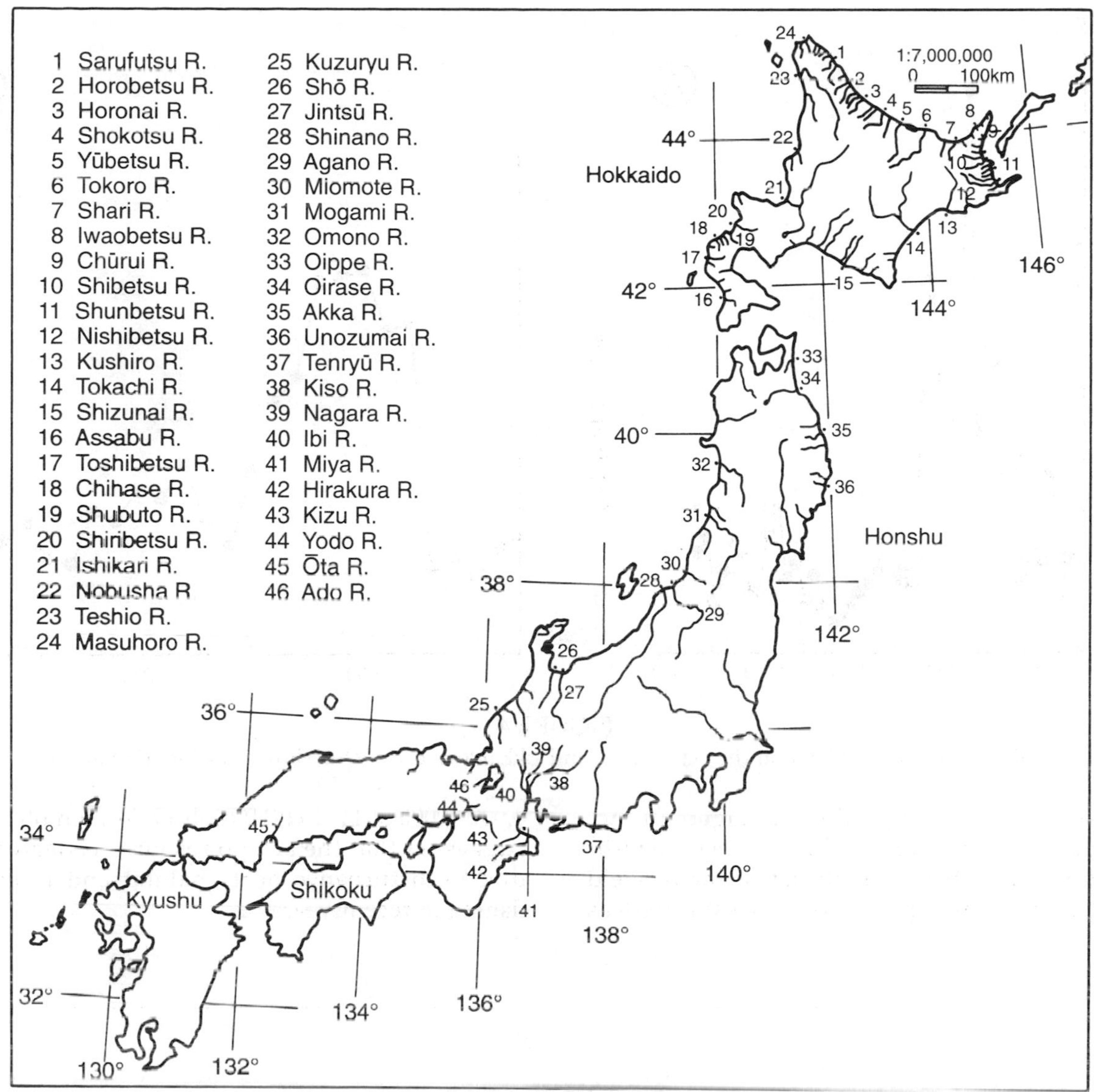

FIGURE 3
Main rivers for anadromous masu and amago salmon in Japan

average annual escapement of 8,000 masu into upstream areas of the Tym River in eastern Sakhalin from 1962 to 1966. On Iturup Island in the Kuril Islands, large runs of masu occur in the Kuybysheva River, and about 6,000 salmon were caught in 1935 (Yagisawa 1970). It is also estimated that considerable runs of masu migrate to rivers in southern Sakhalin and southwestern Kamchatka, but little information about abundance exists.

Amago. Yoshida (1967) reported that there were 30 rivers in which a fishery for amago is carried out and that the total yield of amago in these rivers reached 120 t in 1937. The Yodo River in Osaka Prefecture had the largest harvest, which was estimated to be about 73.7 t. The annual catches in some other rivers were as follows: 15.9 t in the Kiso River, 7.8 t in the Ōta River (Hiroshima Prefecture), 6.9 t in the Tenryū River, 4.8 t in the Nagara River, and 4.0 t in the Ibi River (Figure 3). In recent years there have been hardly any amago captured in the Yodo and many other rivers; only the Kiso, Nagara,

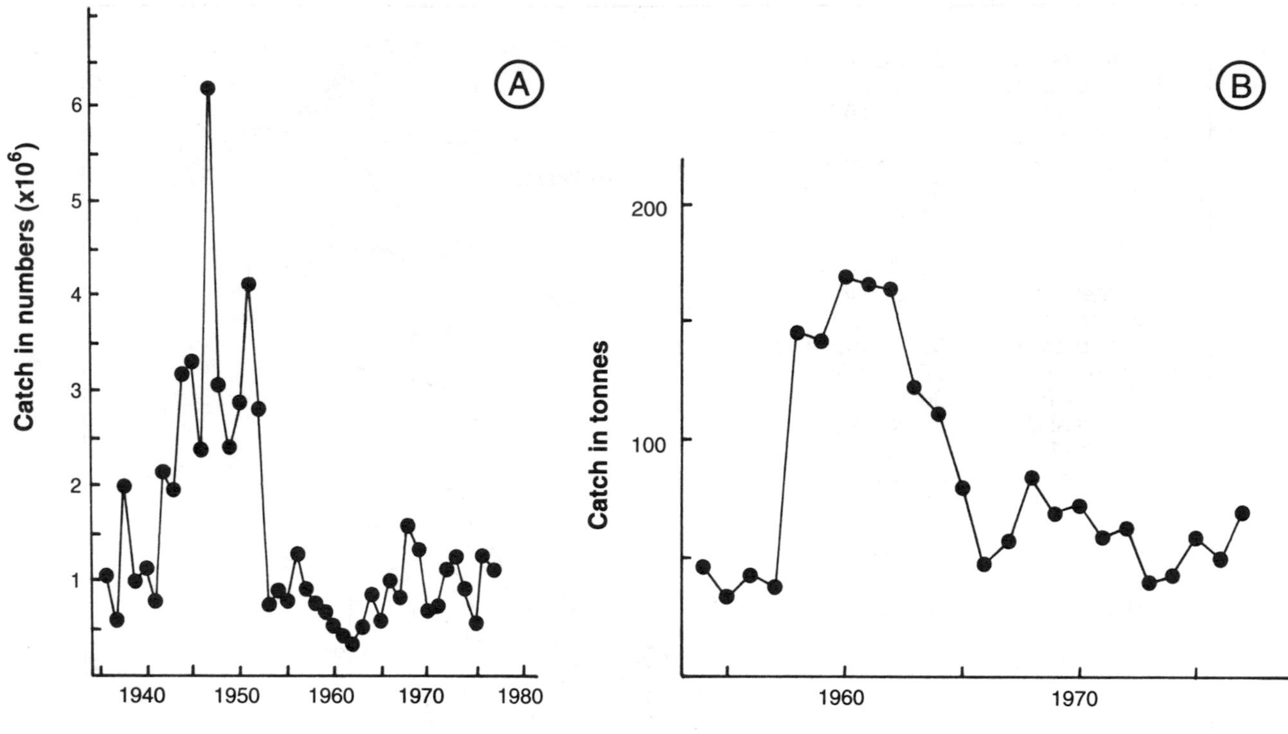

FIGURE 4

Annual catch of masu salmon in the major rivers in Hokkaido and Honshu. (From Machidori et al. 1980)

and Ibi rivers are still producing (Figure 3). According to investigations by Gifu Prefectural Fisheries Experimental Station, the latest annual yield of amago in these three rivers is estimated as 4,000–9,000 or 11–15 t (Honjoh 1985). Honjoh (1977) suggested that the amago resource is decreasing due to disturbance of its habitat and intensive fishing in recent years.

Transplants

Masu. Masu eggs have frequently been transported long distances to be introduced to various areas. In 1965, masu eggs collected in Hokkaido were transported to Canada, from which 5,500 juveniles were ultimately released in Westward Lake, Ontario, in July 1966 (Christie 1970). In 1967 and 1968, 13 masu, having a fork length of 16–42 cm, were caught there. In 1972, masu eggs from Hokkaido were transported to Chile in South America from which 85,000 juvenile fish were ultimately released in 1973 in the Claro River, a tributary of the Simpson River in Aisen province (Nagasawa and Aguilera 1974). No recoveries of these fish as adults have been reported.

Amago. In recent years, eggs of amago have been introduced to river systems without native amago in Kyoto, Fukui, Ishikawa, Toyama, Niigata, and Yamagata prefectures on the west coast of Japan. Kato (1982) collected 33 sea-run amago in the coastal waters of Toyama, Niigata, and Yamagata prefectures and presumed that these specimens originated from the previously transplanted fish.

Upstream Migration of Adults

Season of Ascent

Masu. Masu are the first of the Pacific salmon species to migrate to coastal waters, with the earliest individuals starting their upstream migration as soon as the snow melts. The runs into the rivers generally last from spring to summer. The duration of migration is comparatively long, lasting two to three months, and varies considerably according to area and river. Table 1 presents a summary of the timing of upstream migration to 35 rivers.

TABLE 1

Spawning season of masu salmon in the Far East

Location	Spawning season	Peak spawning
Ishikawa Pref.	Sep–Oct	–
Aomori Pref.	Mid-Sep–late Oct	Early Oct
Hokkaido	Late Aug–early Oct	Early half Sep
Southern Korean Pen.	Sep	–
Posyet District	Sep–early Oct	–
Northern Primore	Jul end of Aug	–
s.w. Sakhalin	Late Jul–early Sep	Latter half of Aug
Central-eastern Sakhalin	Jun–Sep	–
Iturup Island	Early Jul	–
s.w. Kamchatka	Aug	–

Source: From Machidori and Kato (1984)

In general, migration of masu takes place early (April-May) in southern areas and late (June-July) in northern areas. The peaks of the runs are also progressively later northward, between 36°N and 43°N. In higher latitudinal zones, this relationship is weaker and the peak of the run occurs simultaneously through most of June and the first half of July. The timing of the upstream migration of masu bears a close relationship to the advance of the seasons in the Far East.

Variations in timing of runs are large in the vicinity of 41°N to 45°N and the duration of the runs is somewhat longer than at more southern latitudes. At these latitudes, most masu runs start immediately after the snow melts, reach a peak from late spring to early summer, and generally decline slowly thereafter. At these latitudes there also exist masu whose peak run is in late summer to early autumn, as in the Oippe River (Figure 3) and the Posyet District (Table 1, Figure 2) (Machidori and Kato 1984).

After entering rivers in Hokkaido, adult masu usually hide in deep pools in the main streams or in large tributaries and move to spawning areas upstream as spawning time approaches in autumn (Sano 1964). In the Chihase River of Hokkaido (a small river, 22 km long, in a mountainous area, Figure 3), masu have already reached the uppermost part of the stream by early July, whereas fish in other streams in the region are just at the peak of migration. Osanai and Otsuka (1967) reported that the early group in this river rapidly migrates to the uppermost reaches of streams, and the late group, which spawns in the middle and lower reaches of streams and has only a short distance to travel, enters the river just prior to spawning. Upstream migration of adults takes place throughout the day. It is generally more intense in the morning and at dusk and is influenced by changes in water discharge rates (Krykhtin 1962).

Upstream migration occurs over a fairly wide range of water temperatures. In the rivers of Hokkaido, the average monthly water temperatures during migration range from 6° to 17°C (Table 2). The optimum water temperature for migration in

TABLE 2

Average monthly river temperatures during the migratory and spawning season of masu salmon in some rivers in Hokkaido (1964-66)

River	Water temperature (°C)						
	May	Jun	Jul	Aug	Sep	Oct	Nov
Kenichi	8.2	11.5	14.1	17.0	15.7	11.7	–
Toshibetsu	6.8	9.5	12.5	15.3	12.6	9.9	–
Chihase	5.6	8.6	12.6	15.8	13.0	8.5	6.7
Horonai	8.8	14.0	17.1	17.2	13.0	11.2	–
Onnebetsu	–	10.2	11.2	13.9	11.6	8.5	4.4

Source: From Osanai and Otsuka (1969)

these rivers is considered to be from 9.5° to 15.4°C during June and July (Osanai and Otsuka 1967). River temperatures during the peak period of upstream migration of masu along the east coast of Sakhalin are approximately the same. Water temperatures at the time of upstream migration in August in the Tym River are 6°-11°C and in the Poronai River, 10°C in June and 15°C in August (Gritsenko 1973). In the Samarga River in northern Primore, water temperatures at the start of the run of masu salmon in early June are about 6°-8°C (Semenchenko 1978).

Amago and Biwamasu. The timing of upstream migration of the river-dwelling amago or of the amago that occasionally migrate to sea is not well known. Kato (1968) reported that in the Nagara River in Japan (Figure 3), the ascent started in early April, reached a peak in May, and terminated in June (Kato 1968). Water temperatures of the river at the start of upstream migration were about 10°-12°C (Honjoh 1977).

Biwamasu inhabit the hypolimnion in summer in Biwa Lake in Japan (Figure 1) and are found at all depths at other times of the year (Miura 1966). The adults do not leave the lake until the spawning migration starts. Migration from the lake to the river occurs from September to November.

Maturation

Masu. Secondary sexual characters of masu are seldom observed in the early stages of a run and body colour remains silver white. However, maturity is attained soon after entering the river. Body colour darkens in about July, and the head of the male lengthens (Sano 1964). The snout of the males becomes hooked about mid-August and a pattern of pink and olive colours appears on the sides of the body (see Plate 20). At approximately that time, masu actively move upstream to the spawning grounds.

In the rivers of southern Sakhalin, the maturity index (percentage of gonad weight relative to body weight) in early migrants was only 1%-2% for males and 3%-4% for females (Krykhtin 1962). The maturity index of adults caught in the Nobusha River on the west coast of Hokkaido (Figure 3) in June was 0.5% for males and 1.7% for females (Sano 1947). In this river the run of masu occurs from mid-June to the end of July.

The maturity index of masu in the Chihase River on the west coast of Hokkaido was 1% for males and 4%-7% for females in June, 2% for males and 8%-11% for females in July, 13%-19% for females in August, and 1%-6% for males and 24%-31% for females in September (Osanai and Otsuka 1969). Gonad weight in May to June was generally under 40 g for males and 200 g for females, but around August gonads of males and females weighed 30-95 g and more than 200 g, respectively. Gonads of some individual females weighed more than 500 g (Table 3).

TABLE 3

Monthly gonad weight of masu salmon migrating into rivers in Hokkaido

		Gonad weight (g)				
Sex	River	May	Jun	Jul	Aug	Sep
Female	Kenichi	20-30	-	65-325	152-480	-
	Toshibetsu	56-97	26-125	65-232	425-565	310
	Chihase	-	90-261	178-320	210-420	430-905
	Onnebetsu	-	24-59	56-169	210	-
Male	Kenichi	-	-	35-70	30-95	30
	Toshibetsu	-	9	-	-	-
	Chihase	-	35	80	70	14-56
	Onnebetsu	-	-	39-54	-	-

Source: From Osanai and Otsuka (1969)

Amago. The maturity index of female amago that ascended rivers from Ise and Mikawa bays was 1.7% in mid-June, 3.2% in late June, 4.5% in early July, and 4.8 % in late July. The maturity index of female spawners reached 19.6% in October (Honjoh 1985).

Some of the age 1.0 male amago salmon sampled in the Bano River, a tributary of the Kizu River in Honshu (Figure 3), matured in October. The gonads of these fish developed rapidly after August and the gonad indexes (gonad weight × 10/body weight) in August, September, and October were 30.30, 35.80, and 36.10, respectively. The gonad index of age 2.0 males, which had not matured at age 1.0, was between 0.5 and 1.0 in the spring and increased rapidly to 32.0 and 52.8 in September and October, respectively. Female amago do not mature at age 1.0. Their gonads develop very slowly until the spring of age 2.0, and after May the gonad indexes increase gradually from 13.1 in June to 28.4 in August. In autumn, the females mature rapidly and have a gonad index as high as 335.8 in October (Suzuki et al. 1957).

Feeding Habits

Masu. Early migrating (about May) masu to the Nishibetsu and Shibetsu rivers in eastern Hokkaido (Figure 3) are known to feed occasionally in the rivers (Sano 1947). However, it is believed that adult masu do not feed in the river after the summer.

Amago and Biwamasu. The major foods of adult amago in October are adult and larval terrestrial and aquatic insects (Chironomidae, Ephemeroptera, etc.). In August, ayu (*Plecoglossus altivelis*) was the most important food for young biwamasu in Biwa Lake, but macrura and amphipods of the family Gammaridae were also consumed at times (Kato 1978b). Of amago collected in the Nagara and Miya rivers in July, the beginning of the season of upstream migration, 85% had empty stomachs (Honjoh 1985). Loaches, crucian carp, freshwater shrimp, and aquatic insects were found in the stomachs containing food.

Spawning

Timing

Masu. According to Smirnov (1962), masu move actively from deeper water in the river to the spawning areas when water temperature exceeds 12°C; activity reaches a peak in midsummer during periods of high temperature and terminates with the arrival of autumn. Spawning grounds are generally located in upper streams or tributaries of the rivers.

The spawning period for masu reaches its peak from August to September and begins as early as the end of July in the north and as late as mid-September in the south (Table 1). Spawning is completed by the end of August to the beginning of September in the north and by the end of October in the south. Early spawning takes place in high latitude areas such as Sakhalin and northern Primore. Spawning periods for Kamchatka and the Kuril Islands are unknown, but mature masu have been observed in the Bolshaya River in mid- to late August (Semko 1956).

Masu spawn in the upstream areas of the main stem or tributaries of the Kuybysheva River on Iturup Island from early July (Yagisawa 1970). However, the time when spawning ends is unknown. In rivers of southwestern Sakhalin masu spawn from midsummer (the end of July) to the beginning of September, with the peak in the latter half of August (Krykhtin 1962). In tributaries of the Amur River and rivers in northern Primore, the spawning period also extends from the end of July to the end of August (Table 1).

In the rivers of Hokkaido, masu spawning appears to begin around late August and reaches a peak from early to mid-September (Osanai and Otsuka 1967) (Table 1). The duration of the spawning period in each river ranges from 30 to 40 days and terminates around late September in most rivers and in early October in others. Water temperature during the spawning period ranges from 7° or 8° to 20°C, but generally averages from 11° to 15°C (Hokkaido Salmon Hatchery 1969).

Spawning periods in Aomori and Ishikawa prefectures of Honshu are in September to October (Table 1). The peak of spawning in the Oippe River of Aomori Prefecture was observed to be in early October (Aomori Prefectural Fisheries Experimental Station 1976). Masu enter this river late in the season and spawning starts soon after their entry. Water temperatures at spawning time in this river generally range from 13° to 17°C.

Amago. Amago generally spawn later in the season than masu. In the Bano River in Japan, spawning of amago takes place from mid-October to early November (Shiraishi et al. 1957). In the Ado and Nagara rivers (Figure 3), most amago spawn in October. Biwamasu also spawn from October to November (Kato 1978b).

Age at Maturity

Masu. Age determination of masu is based on scales. However, the variety in scale patterns has resulted in confusion in the establishment of criteria for age determination. In the ocean zone of an adult fish scale, one rest zone (a band of closely spaced circuli) is usually observed but two rest zones are occasionally present. In cases where two such zones are observed, they are often close together, and the relationship varies. Variation ranges from one wide rest zone to two rest zones that are completely separated. Japanese scientists generally regard the fainter rest zone as a false annual ring but reasons for formation of the false rings are not known. In contrast, USSR scientists generally tend to regard all rest zones as annuli. In addition, Birman (1972) assumed that even in June there were individuals that had not yet formed the final annulus, and he warned that there is a risk of underestimating ocean age.

Okada and Sakurai (1937) reported that there are large masu (common name, "Taiko masu" or "Ita masu") migrating into the Ishikari River in Hokkaido in which the ocean annulus could not be distinguished. Individuals in which the ocean annulus was difficult to recognize constituted 57% in 1955, 15% in 1965, and 20% in 1966 of those in offshore waters of the Sea of Japan (Hokkaido Fisheries Experimental Station and Hokkaido Regional Fisheries Research Laboratory 1956; Kato 1970). The annulus formed in the marine stage varies from a vestigial one to a clear one. The proportion of fish that show no annulus varies greatly, depending on the criteria used for determination of the annulus. Kato (1970) reported that fish with no ocean annulus had, for some reason, not formed the rings. This conclusion was based on the fact that the proportion of non-ocean fish with an annulus did not decline with any regularity during March to May, even though annulus formation was estimated to take place in January and February.

The scale pattern in the freshwater zone is even more complicated than in the ocean zone and is therefore more troublesome to scientists. The age at downstream migration of masu in Hokkaido and southwestern Sakhalin is estimated as mainly 2 years (age 1.). For masu in Kamchatka, the downstream migration appears to occur at age 2. for 20% or more of the fish. In the Tumnin River of Primore, no masu migrate to sea at age 1. but all migrate at age 2. and older. Semenchenko (1978) also reported that the age composition of downstream migrants in the Primore area, as estimated by scale patterns of adult fish, was 62% at age 2., with lesser percentages of age 1. and age 3. fish.

The age composition of adult masu caught offshore in the Sea of Japan also reflects discrepancies in the criteria for age determination. Although the age of juveniles during downstream migration varies geographically, and the proportion of older fish increases as the latitude increases, it is thought that most masu spend one year in the ocean. Immature masu of ocean age .1, which would remain one more year in the ocean, have not been observed, despite the extensive offshore sampling activities from March to September. Also, tagged ocean age .1 masu were recovered only within the same year, and no later recoveries were made. These facts suggest that most masu reach maturity in their second year of ocean life.

In conclusion, there are two opinions concerning age at maturity of masu, depending on the scale reading criteria used. Some scientists maintain that maturity of masu occurs mainly at three to four years of age (from fertilization of egg to maturity of adult fish) and others suggest that masu reach maturity mainly at four to five or five to six years of age.

Sano (1951) estimated that masu runs in Hokkaido were comprised of 90% or more age 1.1 fish, and 10% or less age 2.1 fish (Table 4). He concluded that most masu attain maturity after spending one winter in fresh water in the juvenile stage (two

TABLE 4
Age composition of masu salmon in Hokkaido rivers in 1946

		Age composition (%)	
River	Number of fish	1.1	2.1
Shubuto	52	91.1	8.9
Tokoro	190	94.6	5.4
Nobusha	196	91.0	9.0
Chitose*	52	100.0	0
Shari	96	92.4	7.6
Average		93.8	6.2

Source: From Sano (1951)
Note: *A tributary of the Ishikari River

winters if the egg and alevin stage are included) and one winter in the sea. However, for masu in the Soviet Union, there are different opinions. Semko (1956) reported that mature masu returning to the west coast of Kamchatka were mainly age 1.2 and 2.2 (Table 5), with most masu spending two winters in the ocean. Birman (1972) estimated that masu in the Tumnin River of Primore were older and included a high proportion of 2.2 and 2.3 fish. Krykhtin (1962) reported on age 1.2 fish on Sakhalin that migrated as smolts to the ocean in the autumn. However, he concluded that the main age of maturity of masu was three or four years. Gritsenko (1973) also estimated that masu in northeastern Sakhalin (Tym and Poronai rivers) consisted of about 80% age 2.1 fish that reach maturity at three to four years of age.

TABLE 5
Age composition of masu salmon in USSR rivers

Region	River	Number of fish	Age composition (%)						
			1.1	2.1	1.2	2.2	3.2	2.3	3.3
Kamchatka	Kolpakova	20	20.0	–	60.0	20.0	–	–	–
	Utka	24	–	8.0	67.0	25.0	–	–	–
Primore	Tumnin	31	–	–	3.2	48.4	6.5	32.3	9.7
Sakhalin	Southwestern rivers	–	69	–	31	–	–	–	–
	Tym	497	17.5	82.5	–	–	–	–	–
	Poronai	221	17.6	76.9	5.4	–	–	–	–

Source: From Machidori and Kato (1984)

The freshwater-type masu (yamame) do not migrate downstream in the spring after one winter in fresh water. They turn a gold colour during the following April and May at age 1.0 or 2.0 and become a darker colour during the summer and autumn. After these colour changes, the fish attain maturity. For yamame males, this generally occurs one year after hatching, at age 1.0. Females mature in the second year after hatching at age 2.0, or, in some cases, even later (Ono 1933). Thus, yamame usually mature one year earlier than sea-run masu.

Amago and Biwamasu. Males of both wild and cultured amago spawn at age 1.0 or 2.0, and females at age 2.0 (Suzuki et al. 1957; Kato 1973a; Honjoh 1977; Kato 1978a). Cultured biwamasu males mature at age 2.0, and the females at age 3.0 (Oshima 1955). Natural mature biwamasu in Biwa Lake range from age 2.0 to 4.0 for males, and from age 3.0 to 4.0 for females. Amago usually mature earlier than biwamasu.

Sex Ratio of Spawners

Masu. Some masu reach maturity as young fish without migrating seaward. Both males and females of such early-maturing adults are found in southern areas of Japan, such as Kyushu, where sea-run types do not occur at all (Kimura 1972). In more northern areas, such as Hokkaido, only males of this non-migratory type are found (Ono 1933). Males tend to be more of the stream type than females because the sex ratio of masu returning from the ocean predominantly favours females in almost all areas. On the Pacific Ocean side of Honshu, such as in Iwate and Aomori prefectures, females account for about 90% of the total run. On the Sea of Japan side of Honshu, the proportion of females in Toyama Prefecture, located more to the south, reaches 80%, and in Ishikawa and Fukui prefectures, even further south, only females occur (Table 6).

In areas west of Cape Erimo and in the Sea of Japan, the proportion of females was 70% or more of the total catch. These high female percentages are associated with the relatively warm weather in the southern and western coastal areas off Hokkaido due to the influence of the warm Tsushima Current. In the Sea of Okhotsk and the Nemuro Strait areas, where winters are long and cold, the proportion of females is lower and amounts to about 65%. Masu caught in the rivers of Hokkaido

TABLE 6
Sex composition of anadromous masu salmon caught in Japanese rivers

Region*	Number of fish	Sex composition (%)	
		Female	Male
Pacific Ocean side of Honshu			
Aomori Prefecture	1 210	88.9	11.1
Iwate Prefecture	688	96.5	3.5
Subtotal	1 898	91.8	8.2
Sea of Japan side of Honshu			
Akita Prefecture	230	74.3	25.7
Niigata Prefecture	1 302	69.8	30.2
Toyama Prefecture	603	80.6	19.4
Ishikawa Prefecture	–	98.3	1.7
Fukui Prefecture	–	99.0	1.0
Sub-total	2 135	73.3	26.7
Hokkaido			
Sea of Japan areas	33 789	71.0	29.0
Areas west of C. Erimo	852	88.6	11.4
Areas east of C. Erimo	4 436	69.2	30.8
Nemuro Strait areas	96 251	65.9	34.1
Sea of Okhotsk areas	102 947	63.9	36.1
Sub-total	238 275	65.9	34.1
All areas†	242 308	66.2	33.8

Source: Adapted from Machidori and Kato (1984)
Notes: **Pacific Ocean side of Honshu*: areas east of Cape Tappi, Aomori Prefecture. *Sea of Japan side of Honshu*: areas west of Cape Tappi, Aomori Prefecture. *Hokkaido* – Sea of Japan areas: Cape Soya-Cape Shirakami; areas west of Cape Erimo: Cape Shirakami-Cape Erimo; areas east of Cape Erimo: Cape Erimo-Cape Nosappu; Nemuro Strait areas: Cape Nosappu-Cape Shiretoko; Sea of Okhotsk areas: Cape Shiretoko-Cape Soya
†Excluding Ishikawa Prefecture and Fukui Prefecture

for use as brood stock in hatcheries are composed of 64%–80% females.

The sex ratios of masu that migrate into rivers in Sakhalin are similar to those in the rivers of eastern and northern Hokkaido. About 70% of anadromous adults in some rivers in southern Sakhalin (Bolotnaya and Taranai rivers) are female. The proportion of males is sometimes higher in the early period of the run, but after the mid-period, the proportion of females increases (Krykhtin 1962). For anadromous adult masu in two rivers of northeastern Sakhalin (Tym and Poronai rivers), the proportion of females is about 66% (Table 7).

For fish migrating into certain rivers in northern Primore, the proportion of females is 65% (Semenchenko 1978). For adult masu caught off the west coast of Kamchatka, females constitute 60%–70% according to Semko (1956). Japanese research conducted offshore of Kamchatka found female proportions of about 60%, which was somewhat lower than in other areas. Thus, in summary, the sex ratio of anadromous adults ranges from almost all females in the southern limits of distribution to almost a 1:1 ratio of female/male in the north.

TABLE 7
Sex composition of adult masu salmon caught in USSR rivers and in offshore areas of the west coast of Kamchatka

Region	Number of fish	Sex composition (%)	
		Female	Male
Sakhalin			
Southern area	6 511	69.5	30.5
Northeastern area	716	65.5	34.5
Kamchatka			
Off the west coast	1 722*	59.8	40.2

Source: From Machidori and Kato (1984)
Note: *The catch by gillnets of 111–121 mm mesh sizes in 1969–72

Amago and Biwamasu. Sex ratios of adult amago vary among rivers. Of 16 amago collected in the Ado River, Shiga Prefecture in October, 11 were females (69%) and 5 males (31%) (Kato 1978b), whereas in the Bano River, Mie Prefecture, the proportion of females in the samples of mature fish was 44% and 32% in October and November, respectively (Suzuki et al. 1957). In the Ado River, the dominance of males (65%) was also observed in mature biwamasu in November (Kato 1978b).

Size of Spawners

Masu. The size of masu migrating upstream varies greatly. The smallest are only about 35 cm in fork length and about 0.3–0.4 kg in weight, and the larger individuals reach 70 cm or more in fork length and 5 kg or more in weight. Masu grow quickly in the ocean and the ultimate size of mature fish is largely affected by the length of their ocean life. Body length varies with age. In general, older fish have a larger body (Table 8), however, there are no standard criteria for age determina-

TABLE 8
Average fork length by age of masu salmon returning to USSR rivers

Region	River	Fork length (cm) by age 1.1	2.1	1.2	2.2	2.3
Sakhalin						
Southwestern	Many rivers	47.2	–	53.4	–	–
Northeastern	Tym	52.9	54.1	–	–	–
	Poronai	52.4	52.8	52.9	–	–
Pimore	Tumnin	–	–	–	61.2	61.9
Kamchatka	Kolpakova	38.5	–	44.9	46.1	–
	Utka	–	38.0	47.1	47.0	–

Source: From Machidori and Kato (1984)

tion of masu and thus the relationship between age and body length is not clear.

Masu in the rivers of Primore, Honshu, and the east coast of the northern part of the Korean peninsula have a tendency to have greater average fork lengths (Tables 9 and 10). The average fork length and weight of masu in the central to northern Primore areas are almost 57–58 cm and about 3.0 kg, respectively. In the Tumnin River and Posyet areas, the average fork length reaches 60 cm or more. In some rivers in Honshu the average fork length exceeds 55 cm.

In contrast, for masu from the west coast of Kamchatka and the Sea of Okhotsk side of Hokkaido, average fork lengths are rather small, about 46–47 cm, or 10 cm less than masu in the Primore area. The average weight ranges from about 1.0 to 1.5 kg and is one-half or less of that of masu in the Primore area. The reason for the differences in fish size by area is unknown.

Masu in rivers of central and southern Hokkaido and Sakhalin are of medium size and weight with average fork lengths ranging from about 47 to 48 cm to 52 to 53 cm, whereas those from rivers in southern Hokkaido and northeastern Sakhalin are somewhat larger with an average fork length exceeding 50 cm. Masu that migrate into rivers of south and west Sakhalin are relatively small; their

TABLE 9
Fork length and weight of masu salmon from USSR rivers

Region and river	Year	Age	Sex	Fork length (cm) Average	Fork length (cm) Range	Weight (kg) Average	Weight (kg) Range
Primore							
Posyet area	–	–	F	61	52–68	–	–
Tumnin	–	–	M	63	max. 71	4.0	max. 6.0
Amur	–	–	M	56.8	46–67	2.3	1.8–3.2
Amur	–	–	F	54.4	47–62	2.3	1.6–3.1
Samarga	–	–	M	66	52–78	4.8	2.2–8.5
Samarga	–	–	F	58	54–67	3.7	2.9–4.4
Makshimovka	–	–	M	56.8	–	3.1	–
Makshimovka	–	–	F	57.2	–	2.8	–
Zheltaya	–	–	M	58.8	–	3.2	–
Zheltaya	–	–	F	57.5	–	3.0	–
Sakhalin							
Southern area	1953–54	1.1	F & M	47.2	37–57	1.9	0.5–2.9
Southern area	1953–54	2.1	F & M	53.4	47–60	2.1	1.3–3.0
Tym	1962–66	–	M	54.1	40.0–71.0	2.2	0.8–3.9
Tym	1962–66	–	F	53.8	45.0–62.0	2.0	1.2–3.6
Poronai	1965–68	–	M	52.4	41.0–64.2	2.0	1.0–3.3
Poronai	1965–68	–	F	52.5	45.6–62.5	1.9	1.2–2.9
Kamchatka							
Kolpakova	–	–	F & M	44.0	–	1.3	–
Utka	–	–	F & M	46.2	–	1.5	–
Southern Kuril Islands	1962–64	–	F	47.2	–	–	–

Source: From Machidori and Kato (1984)

TABLE 10
Fork length and body weight of masu salmon from Japanese rivers

Region	River	Year	Number of fish	Fork length (cm)*		Weight (kg)	
				Average	Range	Average	Range
Honshu							
Toyama Pref.	Shō	1933-36	10	58.3	50.6-58.3	2.56	1.6-3.5
Aomori Pref.	Oippe	1963-67	56	55.7	44.4-69.0	–	–
Hokkaido							
Sea of Japan side	Masuhoro	1966	9	40.2	33.6-43.4	–	–
	Teshio	1952	129	50.9	39.1-60.8	1.57	0.7-3.0
	Nobusha	1946	190	49.7	–	–	–
	Atsuta	1966	9	47.6	42.3-51.8	–	–
	Ishikari	1946	52	50.9	–	–	–
	Shubuto	1946	96	48.7	–	–	–
	Chihase	1964-66	126	51.7	37.6-62.8	–	–
	Toshibetsu	1964-66	49	52.8	42.3-62.3	–	–
	Kenichi	1964-66	130	50.4	35.9-66.6	–	–
Pacific Ocean areas	Horobetsu	1966	10	49.8	39.1-58.2	–	–
	Tokachi	1950	109	51.3	35.4-64.5	1.94	0.3-4.1
Nemuro Strait areas	Shunbetsu	1950	150	47.9	41.3-61.3	1.28	0.6-2.5
	Shibetsu	1952	145	48.2	37.0-65.6	1.31	0.5-2.8
Sea of Okhotsk areas	Iwaobetsu	1950-52	187	46.0	33.3-60.3	1.55	0.3-3.2
	Onnebetsu	1965-66	37	45.7	40.4-62.4	–	–
	Shari	1946	52	47.9	–	–	–
	Mokoto	1952	75	42.8	34.1-49.7	0.94	0.5-1.8
	Tokoro	1946	186	42.3	30.0-50.2	1.24	0.5-2.6
	Yūbetsu	1950	120	47.1	33.8-55.0	1.56	0.5-2.8
	Horonai	1966	6	46.9	33.0-55.0	–	–

Source: From Machidori and Kato (1984)
Note: *Standard length was converted to fork length by the following formula: fork length = 0.0582 + 1.0576 standard length.

average fork length is 50 cm or less and their average body weight about 2 kg or less. Masu caught in rivers on the Sea of Japan side of northern Hokkaido and Nemuro Strait area are also small, similar to those in southern Sakhalin.

The relationship between sex and size varies by area. There is little difference in average fork length by sex for masu from rivers in Sakhalin and Primore (Table 11). However, for masu from rivers in Japan, the average fork length of females is frequently larger, sometimes as much as 5-6 cm. In some rivers, the differences in average fork length by sex are slight, and in others the males are larger than the females.

The variance in fork length for masu males is usually larger than that for females. For fish caught in rivers of Hokkaido, the coefficient of variation for fork length of males is about 1% greater than that for females (Machidori and Kato 1984). The variance of fork length of male masu in USSR rivers is probably also larger than that for females because the ranges of fork lengths of males are larger than those for females (Table 9).

Yamame show very poor growth compared with masu. By the end of their second year after hatching, yamame are about 20 cm in length and after three years they are only about 25 cm.

In conclusion, there are differences in fork length and body weight according to age and sex of anadromous masu caught in rivers, but frequently only small samples were taken in a particular year. Yearly changes and many other elements associated with sampling undoubtedly affected the estimates. There is also a large individual and geographical variation in length. In general, the anadromous form of masu grows at least twice as large as its freshwater type, yamame.

Amago and Biwamasu. Mature lake-type *O. rhodurus* (biwamasu) are larger than the stream-type

TABLE 11

Average fork length by sex of masu salmon returning to Far East rivers

Region	River	Year	Female		Male	
			Number	Average length (cm)	Number	Average length (cm)
Primore	Samarga	–	–	58.0	–	66.0
	Makshimovka	–	–	57.2	–	56.8
	Zheltaya	–	–	57.5	–	58.8
	Amur	–	–	54.4	–	56.8
Sakhalin, west coast		1947–53	–	45.4	–	46.6
Northeastern Sakhalin	Tym	1962–66	334	53.8	166	54.1
	Poronai	1965–68	135	52.5	81	52.4
Hokkaido*	Kenichi	1964–66	82	50.1	48	43.4
	Chihase	1964–66	71	51.2	55	45.7
	Teshio	1952	108	48.2	22	47.3
	Tokachi	1950	51	48.0	58	47.9
	Shunbetsu	1952	92	46.2	58	44.0
	Shibetsu	1952	70	47.3	75	43.9
	Iwaobetsu	1950–52	134	42.2	53	44.9
	Mokoto	1952	26	37.9	75	40.4
	Tokoro	1950	115	41.5	71	37.5
	Yūbetsu	1950	75	44.6	45	44.3

Source: From Machidori and Kato (1984)
Note: *For Hokkaido rivers, standard length was converted to fork length by the following formula: fork length = 0.0582 + 1.0576 standard length.

(amago). For example, in the Ado River, body length ranges from 22 to 24 cm for biwamasu, and from 12 to 18 cm for amago (Kato 1978b). Thus, the lake-type biwamasu is much larger than the river-type amago at maturity.

Fecundity

Masu. Mature eggs of masu are orange with dark red pigment. The diameter of eggs of mature masu in southern Sakhalin are, on the average, 6.0 mm, and range from 4.5 to 7.7 mm (Krykhtin 1962). Larger females generally have larger eggs. Egg diameters for mature masu in rivers in western Hokkaido are larger than in southwestern Sakhalin. Ripe eggs measured in September (in the spawning period) were 6.15–7.07 mm in the Chihase River and 6.00–6.77 mm in the Toshibetsu River (Osanai and Otsuka 1969). Such differences reflect the fact that the body sizes of masu in western Hokkaido are larger than in southern Sakhalin.

Fecundity also varies by area, and the larger the female, the higher the fecundity. For masu in rivers in Japan, the average fecundity varies among rivers and districts from approximately 1,900 to 3,850 (Table 12). In the Honshu area, where larger fish occur, average fecundity is about 3,000 to 3,850 with an average of about 3,400. However, average fecundity on the Sea of Okhotsk side of Hokkaido is only 1,900. In Hokkaido, fecundity is higher in areas of the Sea of Japan (average 3,000) and west of Cape Erimo (average 2,600), and lower in the eastern (average 2,100) and northern (average 1,900) areas.

The average fecundity of masu in the USSR also varies by area, which is consistent with the geographic variation in fish size. Fish size is large in the Primore area and the average fecundity is 3,200 (Table 13). Fish size in the northeastern Sakhalin area (Tym River) is moderate and the average fecundity is also moderate, about 2,800. In southern Sakhalin and the Kuril Islands, masu are small and the average fecundity is low, about 1,600 to 1,700.

The fecundity of female yamame is lower than for masu, averaging only 200 to 300 eggs. The average diameter of the mature ova is approximately 5 mm (Ono 1933).

Amago and Biwamasu. Mature eggs of amago are

TABLE 12

Average number of eggs collected and the estimated fecundity* of masu salmon in Japanese salmon hatcheries

Region	River	Number of fish	Average eggs collected	Estimated fecundity
Honshu†				
Pacific Ocean side	Unozumai	48	2 667	3 040
	Akka	10	2 700	3 078
	Oirase	290	2 869	3 271
	Oippe	450	2 718	3 099
Sea of Japan side	Shinano	510	3 376	3 849
	Agano	80	2 600	2 964
	Miomote	82	2 646	3 016
	Omono	56	2 946	3 358
All of Honshu		1 526	2 963	3 378
Hokkaido†				
Sea of Japan areas	Some rivers	19 880	2 597	2 961
Areas west of C. Erimo	Some rivers	638	2 251	2 566
Areas east of C. Erimo	Some rivers	203	1 803	2 055
Nemuro Strait areas	Some rivers	44 168	1 912	2 180
Sea of Okhotsk areas	Some rivers	41 499	1 662	1 895
All of Hokkaido		106 388	1 944	2 216
All of Japan		107 914	1 959	2 233

Source: Honshu, Japan Salmon Enhancement Association (1973–77); Hokkaido, Hokkaido Salmon Hatchery (1953–72)
Notes: *According to Sano (1959), the number of eggs collected constituted about 88% of fecundity. Therefore, the fecundity was estimated as 1.14 times the number collected.
†For eggs collected in Honshu in 1972–76 and in Hokkaido in 1952–71

TABLE 13

Fecundity of masu salmon in USSR rivers

Region	River	Number of fish	Fecundity	
			Average	Range
Primore	Tumnin	–	3200	1381–3261
	Amur	81	3200	–
Sakhalin	Southern rivers	325	1662	520–3059
	Taranai	431	1600	–
	Poronai	96	2672	–
	Tym	284	2850	–
Southern Kuril Is.	Some rivers	107	1629	–
Kamchatka	Kolpakova	–	2025	–
	Utka	–	3700	–

Source: From Machidori and Kato (1984)

light yellow. The mean diameter of eggs from mature amago sampled in the Bano River (Mie Prefecture) in October was 4.25 mm. Adult females, from 11.8–17.4 cm long, had 95–209 eggs with an average of 148.8. It was estimated that only 13% of the eggs in the ovary would ripen because many unripe eggs were observed in the gonads of mature females (Suzuki et al. 1957).

The diameter of eggs of mature amago in the Ado River (Honshu) is from 4 to 5 mm, and the fecundity of five fish, 12–18 cm in length, ranged from 78 to 205 eggs (Kato 1978b). Mature eggs of biwamasu are yellowish orange and are from 6 to 8 mm in diameter. The fecundity of five biwamasu of the Ado River, with lengths from 22 to 24 cm, ranged from 428 to 815 eggs (Kato 1978b). Thus, in the same river, mature biwamasu have larger and more eggs than amago (Kato 1978b).

Spawning Act

Masu. Spawning grounds of masu are generally located in upstream areas of the main stream or in tributaries, specifically in shallow waters with swift currents at the lower end of deep pools or in shallow waters with swift currents along the river banks.

The digging of the nest is conducted only by females. Several test diggings are made before she settles on a final site and digging is halted if clay, rock, or large pebbles are encountered. Each female is accompanied by two or three ocean-type and several stream-type males (Osanai and Otsuka 1967) that swim around the nest location to keep other males away. The females do not react when a male is replaced. In the case of approach by another female, the female with the nest actively challenges and turns away the invader.

Females dig nests by turning on their side and beating the stream bed with their caudal fin while moving 1–2 m ahead. They repeat this type of action many times, at intervals of 1–5 minutes, to make a depression in the gravel. When the nest is finished, egg deposition and fertilization takes place with the female and male coming together in the nest and releasing ova and sperm simultaneously. Nest-digging activity starts between 0400 hours and 1000 hours in the morning, and some fish complete spawning activities as early as 1800 hours. Most fish, however, continue on the following day until 1000-1500 hours. Time spent from start of digging to covering of the redd is 10-23 hours (Osanai and Otsuka 1967). After spawning, the female protects the redd for two to eight days to prevent other females from dislocating the eggs by spawning in the same area. Gradually she loses swimming power, drifts downstream, and dies.

Spawned-out masu generally have only a few eggs remaining in the abdominal cavity, indicating that egg retention is low (Table 14). For example, of 40 individuals examined in some Hokkaido rivers, 22 contained no eggs, whereas the remaining 18 had an average of 49.8 unreleased eggs.

TABLE 14

Number of eggs remaining in the abdominal cavities of masu salmon spawners in some Hokkaido rivers

River	Number of fish sampled	Eggs remaining	Average number remaining
Chihase	20	1434	71.7
Kenichi	13	44	3.3
Toshibetsu	7	516	73.7
Total	40	1994	49.8

Source: From Osanai and Otsuka (1967)

Males of yamame also mate with masu females that have migrated from the ocean to the spawning grounds.

It is generally accepted that, under natural conditions, all Pacific salmon die after spawning. However, the scales of some yamame have what is considered a spawning check (Ono 1933). In some rearing experiments, yamame have spawned successfully two or more times.

Amago. From observations in the Bano River (Honshu), Shiraishi et al. (1957) described the spawning act of amago as follows: In preparation for spawning, pairs of mature amago swam together in deep pools of the river. Subsequently, the female began to construct a nest under the protection of the male. The biggest male was the dominant fish and courted the female. He maintained a lateral and posterior position with respect to the female and, when another male entered the nest territory, he attacked and drove off the invader. The nest territory covered an area of about one metre. The female dug the gravel of the streambed every 30 seconds by turning on her side and rapidly flexing her body. The dug-up gravel was shifted downstream by the water flow and a pocket resulted in the streambed. When the nest was completed, male and female spawned at the centre of the excavation. After spawning, the female covered the eggs with gravel.

Description of a Redd

Masu. Osanai and Otsuka (1967) reported that the water depth where redds of masu were built was 12-45 cm on average, with a minimum of 3.5 cm and a maximum of about 64 cm. Current speeds at locations of redds averaged 46-56 cm/s with minimum and maximum speeds ranging from 12 to 101 cm/s. Redds were generally oval in shape, particularly in slow currents. Those in faster currents were longer, being slender upstream and fan-shaped downstream. The average length was about 124-210 cm with widths of about 74-84 cm. The longest redd observed was 410 cm and the shortest about 95 cm. The maximum and minimum widths were 165 cm and about 45 cm, respectively. Completed redds were dome shaped.

Redds are generally built in the shallows in the central part of the stream where gravel size varies greatly. An analysis of three redds in the Kenichi and Chihase rivers showed gravel composition by weight to be 9.3% with a diameter less than 6 mm, 30.0% from 7 to 25 mm, 14.1% from 26 to 45 mm, 15.8% from 46 to 60 mm, and 30.8% over 61 mm (Osanai and Otsuka 1967).

Masu generally spawn upstream in clear and neutral water. Water on the spawning grounds in southwest Sakhalin has a dissolved oxygen content of 8-10 mg/l and a pH of 6.8-7.2 (Krykhtin 1962).

Redds in some rivers in Hokkaido and Sakhalin were excavated after masu had spawned, in order to count the number of eggs. The average number of eggs per redd ranged from 1,062 to 1,520 (Table 15). Osanai and Otsuka (1967) assumed that the actual number of eggs was probably substantially higher than shown in Table 15 because some eggs were broken and disappeared during excavation.

Amago. Shiraishi et al. (1957) reported that the water depth where amago built redds ranged be-

TABLE 15
Number of eggs found in spawning beds of masu salmon in rivers in Hokkaido and Sakhalin

Hokkaido			Sakhalin
Kenichi River	Toshibetsu River	Chihase River	Tym River
1722	1030	1005	746
1855	–	1425	1066
1557	–	1645	981
1665	–	1054	1227
1082	–	–	1024
1240	–	–	835
–	–	–	1133

Source: Hokkaido, Osanai and Otsuka (1967); Sakhalin, Gritsenko (1973)

tween 10 and 30 cm, averaging 22 cm. The average current speed at the locations of the redds was 13.9 cm/s, with a minimum and maximum ranging from 0 to 30 cm/s. Redds of amago were oval in shape, like those of masu, and had an average length of 52.6 cm, an average width of 43.1 cm, and covered an approximate area of 1,400 cm^2. The average number of eggs found in the excavated redds was 56. This suggested that, if the total number of ripe eggs per female was 100–140, she laid her eggs in two or three batches during spawning. Shiraishi et al. (1957) estimated that about 2% of the eggs remained in the abdominal cavity of the females and that 3% of the deposited eggs died in the redds.

Incubation and Emergence

Hatching and Emergence

Masu. In the rivers of southern Hokkaido spawning of masu generally takes place from late August to early October. Eggs develop in the redds, and the alevins hatch from November to December. Fry emerge from the gravel from late March to early May. In this region, stream water temperatures drop rapidly after December, and cold temperatures continue until March. In the Chihase River, for example, water temperatures from December to February are generally below 1°C and even in March reach only about 2°C (Table 16).

TABLE 16
Water temperature in the Chihase River, Hokkaido, during the period of development of masu salmon eggs

	Temperature (°C)		
Month	Average	Maximum	Minimum
December	0.3	1.0	0.2
January	0.2	0.2	0.2
February	0.5	1.0	0.1
March	2.0	–	–

Source: From Osanai and Otsuka (1967)

Development of masu eggs and emergence of fry are closely related to water temperature, and there is, therefore, a great difference in rate of incubation and development between natural and hatchery conditions. Under constant water temperature of 8°C (well water), as used in many hatcheries, masu eggs reach the eyed stage at the end of the fourth week after fertilization and hatch out after 55–60 days. For eggs laid in natural streams with lower water temperatures than in hatcheries, development may take as long as three months (Ono 1933). In the Chihase River, for example, eggs that were fertilized artificially on 29 September and reared separately in river water and warmer well water showed a 68-day difference in the emergence date of the fry (Table 17) (Osanai and Otsuka 1967). The incubation of eggs held in river water was slower and hatching took place 59 days after fertilization, and emergence occurred 91 days following hatching.

The incubation period for masu requires 430–480 C degree-days (above 0°C) to time of hatching. Japanese hatchery managers generally allow for 55–60 days to hatch masu at 8°C. Masu in USSR hatcheries require about 475 degree-days to hatch out, which is 33–35 days after fertilization under the warm hatchery water temperature conditions (Smirnov 1962).

TABLE 17
Period of development of masu salmon eggs collected in the Chihase River, Hokkaido

Classification	Spring water group	River water group
Fertilization	29 Sep	29 Sep
Eyed stage	29 Oct	30 Oct
Hatching	26 Nov	25 Dec
Emerging	25 Feb	4 May
Days	150	218
Temperature	8°C	-

Source: From Osanai and Otsuka (1967)

Development of yamame eggs from fertilization to hatching takes 85-90 days at approximately 6°C, about 50 days at approximately 10°C, and about 35 days at about 14°C.

Amago. Amago eggs require about 450 degree-days until hatching and 800 degree-days until emergence of the fry from the gravel (Honjoh 1977).

Factors Affecting Incubation

Masu. The rate of development and loss of masu eggs within a redd depends primarily on the temperature of the water bathing the eggs. In one series of experiments, eggs collected from yamame captured in a stream tributary to Kizaki Lake (Nagano Prefecture) were hatched under various water temperatures, and it was found that for water temperatures between 7°C and 11°C there was a high rate of survival to hatching accompanied by a very low abnormality rate (Kawajiri 1927). When the water temperature was kept below 3°C, 100% mortality occurred.

Dissolved oxygen, light, fluctuation in stream flow, predation, shock, and superimposition of redds may also contribute to the rate of development and mortality of eggs, but the precise effect of these factors is not known.

Artificial Reproduction

Masu. To facilitate the natural reproduction of masu, restrictions have been established on the catch of these fish in rivers of Japan and the USSR. In Japan the catch of masu has been completely prohibited in specific rivers, and in other rivers only catch of spawners for artificial reproduction has been permitted. Catch restrictions also apply to juveniles during their river stage.

Production of masu is maintained mainly by natural spawning but some hatchery production also occurs. Hatcheries for masu in Japan have operated since 1878 and are as old as those for chum salmon (Akiba 1980). Artificial reproduction of masu did not progress as well as that for chum because of technical and economical difficulties in breeding the spawners and producing the juvenile fish. These problems have not yet been solved and hatchery activities for masu are still on a small scale.

The number of juvenile masu released from hatcheries per year on Hokkaido ranged from 1.3 to 13.6 million between 1936 and 1982 and on Honshu from 0.1 to 2.5 million between 1959 and 1973 (Hokkaido Salmon Hatchery 1956; Hokkaido Salmon Hatchery 1957–75; Honshu Salmon Enhancement Association 1962–75; Hiroi 1984). For the USSR, information is only available for the years 1955–58, during which period between 2.6 and 7.3 million young masu were released (Sakiura et al. 1964). Releases are probably no longer being made.

Initially, the juvenile fish were released into rivers at the time of yolk absorption. Recently, releases have been made after feeding commences. However, retention during the feeding period is generally no more than a few months and is shorter than the stream life of juvenile masu. Most fish are released into the rivers during spring and early summer. Migration to sea generally occurs the following spring. The effectiveness of artificial releases of masu has not yet been adequately evaluated.

Stream Life of Juveniles

Migration

Masu. After hatching, the alevins of masu in the southern rivers of Hokkaido remain in the gravel until some time between the end of April and the end of May. Fry emerge from the gravel beds from late March to about early May. Emerged fry spend some time in relatively dense schools in shallow backwaters near the spawning grounds (Kubo 1980). As the fry grow they disperse and take up residence at the downstream end of deep pools, either near the bottom or in mid-water layers. They feed actively and form a size-dominance hierarchy during daylight hours. At night they are found at the edge of pools or behind rocks where the current is slower (Kubo 1980). This pattern is established by mid-June and early July and persists until the end of September (Figure 5). Fry leaving the redds are about 3 cm in fork length. They have grown to 7–8 cm by July and 9–11 cm by September.

In the southern rivers of Hokkaido, water temperature begins to drop in late September or Octo-

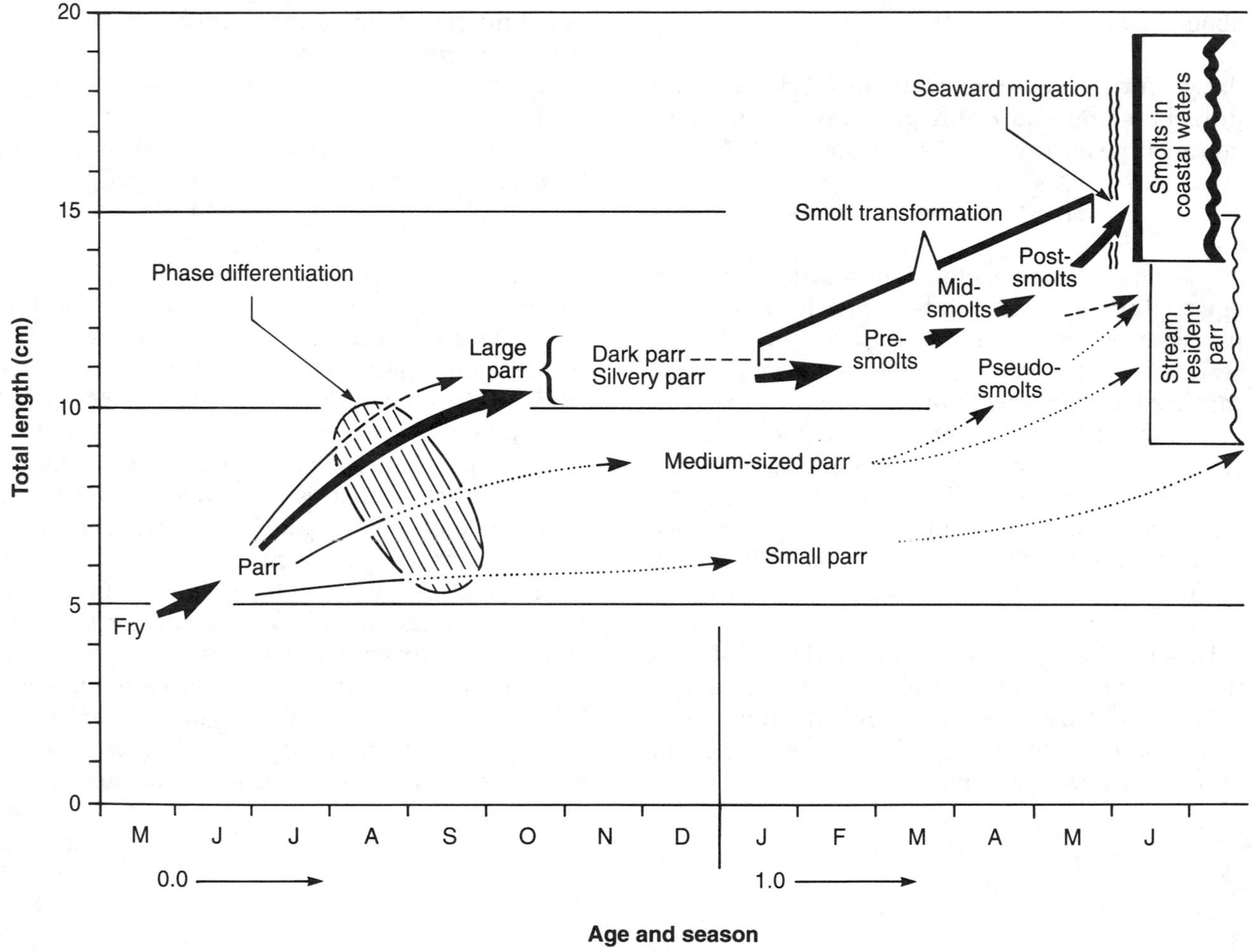

FIGURE 5

Semi-diagrammatic representation of body growth, phase differentiation, and smolt transformation of masu salmon. (From Kubo 1980)

ber and precipitation increases. At about that time, juveniles tend to move from their pool habitat to warmer water areas downstream or to tributaries fed by warmer spring water, and hide under banks that may be covered with bamboo or dead grass during the winter (Inoue and Ishigaki 1968). In this period, physical differences, which distinguish the sea-run type from the stream type, start to develop. The sea-run type becomes slender and the body grows longer (fork length, 10–13 cm) and begins to turn slightly silver or white. They are active during daytime, even in a severe winter, and will migrate to sea the following spring (Kubo 1976). The stream type is shorter (fork length, 7.5–11 cm) and increases in body depth, with some male fish showing signs of maturing. The immatures retain their parr characteristics, whereas the maturing individuals become dark with red coloration along the lateral line (Plate 20). Mature individuals of the stream type have a greater body length than immatures and can be observed until early November in southern Hokkaido (Kubo 1980).

The stream life of juveniles in Sakhalin is the same. In the Bolotnaya and Novoselovka rivers in southwestern Sakhalin, for example, most fry (3–3.5 cm in fork length) leave the redds in April or the first half of May and soon after occur throughout upstream areas (Krykhtin 1962). The juveniles live in schools in shallow areas (5–10 cm deep) with slow currents and where the gravel is covered by mud. After they reach 5–6 cm in fork length (about July) they move to deeper water locations, where the current is stronger. Growth in the May to August period is rapid, and the juveniles reach 8.2 cm in average fork length and 6.4 g in average weight by the end of the summer season.

Amago. The fry of amago emerge from the redds in spring and grow rapidly after they start to free-swim and feed (Shiraishi and Suzuki 1957). During the summer, juvenile amago tend to gather in the tributaries in which water temperatures are comparatively lower (below 20°C). In winter, juveniles move to upper or midstream areas of the main stream (Kato 1978b).

Growth

Masu. The average fork length and body weight of fry and juveniles of masu collected from the upper part of the Yurappu River in Hokkaido were examined by the JIBP-PF Research Group of the Yurappu River (1975). Almost all the juveniles sampled were hatchery-reared fish that had been released in mid- or late May. At the time of release the fry were about 4 cm in length and weighed 1 g.

The increase in body weight during the summer and fall showed interannual differences (Figure 6). A linear increase in growth was observed in 1968, but there was little change in body weight after October for the three years from 1970 to 1972. It is assumed that differences in growth patterns arose from differences in population density and availability of food organisms.

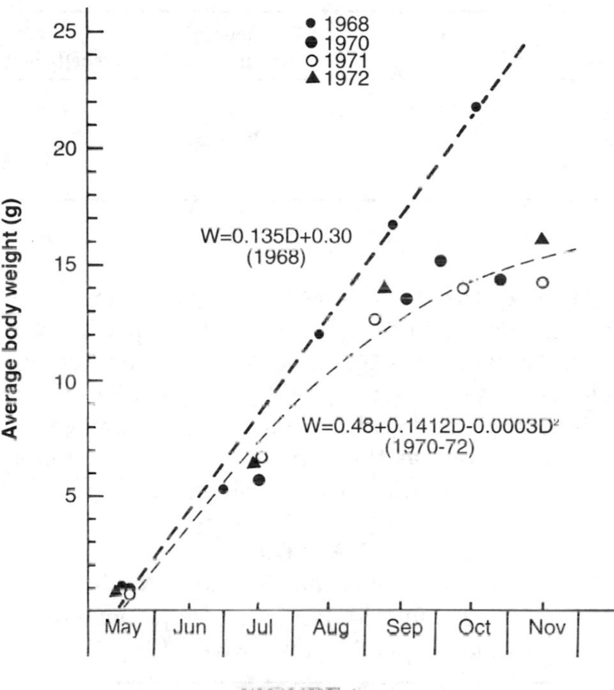

FIGURE 6

Seasonal growth of juvenile masu and its annual variations. (From JIBP-PF Research Group of Yurappu River 1975)

The relation between days after release and increase in body weight in 1968 is represented by the linear equation $W = 0.135D + 0.30$, where W = body weight (g) and D = days after release. For the years 1970–72, the relation between growth and time has the form of the quadratic equation $W = 0.48 + 0.1412D - 0.0003D^2$.

The coefficient of growth for juvenile masu in the Yurappu River (for the years 1970–72), which was calculated by dividing the total quantity of food consumed in July, September, and October by the total increase in weight, varied from 4.04 to 4.93 with an average of 4.5 (Table 18). This suggests that 67.5 g of food were required from late May to early November by a juvenile masu to reach a weight of 16 g during the growing season. If the above coefficient is applied to the data for 1968, then about 110 g of food would have been necessary to produce the 26 g of body weight in that year.

TABLE 18
Monthly growth coefficient of juvenile masu salmon in the Yurappu River, Hokkaido

Month	Growth (g)	Total feeding quantity (g)	Growth coefficient
Jul	3.20	14.47	4.52
Sep	1.88	7.59	4.04
Oct	1.32	6.51	4.93

Source: From JIBP-PF Research Group of Yurappu River (1975)

The relation between fork length (*L*) and body weight (*W*) can be expressed by the equation $W = aL^b$. By using this formula, the calculated coefficients and exponents for the years 1968–72, as given in Table 19, show that body weight varies approximately as the cube of the fork length and that the distribution is parabolic in form.

TABLE 19
Coefficients of equation $W = aL^b$ (W = body weight (g), *L* = fork length (cm)) for juvenile masu salmon in the Yurappu River, 1968–72

Year	Coefficients	
	a	*b*
1968	1.130×10^{-2}	3.110
1969	1.161×10^{-2}	3.091
1970	1.119×10^{-2}	3.031
1971	1.102×10^{-2}	3.076
1972	1.118×10^{-2}	3.073

Source: From JIBP-PF Research Group of Yurappu River (1975)

Feeding Habits

Masu. Until July newly emerged fry of masu feed actively on small-sized benthos, such as larval Chironomidae (Tendipedidae), and terrestrial insects which have fallen into the water. In the Yurappu River, Hokkaido, they eat many chironomid insects of the genus *Baetiella* in this period (JIBP-PF Research Group of the Yurrapu River 1975). In September, they actively feed on terrestrial insects of the order Hymenoptera or on adult *Baetiella*. Aquatic insects and insects dropping from the air are important food sources for juveniles throughout the seasons.

Juvenile masu generally feed by holding position in flowing water and by darting to approaching small-sized items. For example, stomach contents of juveniles collected during early winter from the Shiodomari River in the suburbs of Hakodate were not appreciably different from those of summer and autumn. The main food items were small-sized insects of the order Ephemeroptera and the family Chironomidae (Tendipedidae) (Kubo 1974).

Masu fry in Sakhalin also primarily eat aquatic insects such as Chironomidae (Tendipedidae) and Ephemeroptera soon after emergence (Table 20). They feed actively in areas where the current is slow and the water shallow (Volovik 1963). Juveniles at the so-called parr stage eat mainly insects that drop on the water during warm seasons and these constitute 60%–90% of the total weight of stomach contents. Juvenile masu also gather with other species on spawning grounds to eat drifting eggs of spawning salmon (Krykhtin 1962; Gritsenko 1973). During the autumn and winter seasons they feed only on aquatic insects (Krykhtin 1962).

In general, juvenile masu show a positive reaction to any small object that appears to be a food organism drifting downstream. They follow it and eat it (if it is food), while keeping themselves orientated rheotactically in the current. They also feed on insects that fly close to the surface of the water by jumping towards them.

Juvenile masu in fresh water show diel differences in their stream behaviour. During the day they feed actively and occupy the middle and lower depths of the pools. After sunset, however, they remain close to the edge of the pools or under rocks, where the current is slow. Since feeding activity of salmonids is generally controlled by visual rather than olfactory cues (Uchihashi 1953), it is expected that differences exist in the amount of food eaten at different times of day.

Diurnal fluctuations in feeding activity for juve-

TABLE 20
Weight composition of stomach contents of juvenile masu salmon in Sakhalin rivers

	Stomach content weight composition (%)							
Growth stage	Fry	Fingerling	Parr			Silvery parr	Smolt	Dwarf male
Age	0.	0.	0.	1.	1.	–	–	–
Period of research	End of May	End of May–early Jun	Early Jul–end Oct	May–Jun	Jul–Sep	May–Sep	May–Sep	Aug
Number of samples	10	106	250	21	26	23	28	47
Chironomidae	24.4	9.7	1.5	17.8	1.3	–	3.8	0.1
Emphemeroptera	61.0	23.6	18.9	33.0	3.4	4.4	28.9	10.1
Trichoptera		5.0	2.0	7.7	1.4	9.3	7.5	3.4
Plecoptera	–	1.3	5.7	2.5	1.4	6.4	0.4	0.3
Other benthos	–	41.8	5.7	19.0	1.7	59.2	30.1	22.5
Geoinvertebrata	14.6	18.6	66.2	20.0	90.7	20.7	29.3	63.6
Number of empty stomachs	1	–	2	–	–	2	8	–

Source: From Volovik (1963)

nile masu were studied by the JIBP-PF Research Group of the Yurappu River (1975). The relationship between average wet weight of the stomach contents by species groups and amount of drifting organisms in the river at the same time that the stomach content index samples were collected is shown in Figure 7. The drifting organisms were sampled every three hours by a No. 52 mesh net with an opening of 25 cm^2 and a length of 1 m. The period of sampling was from 1200 hours on 15 July to 1500 hours on 16 July 1970. The sample taken at 0300 hours contained the lowest amount of discernible species (45%), which indicated that the fish were not eating at that time but merely digesting food they had eaten some time before.

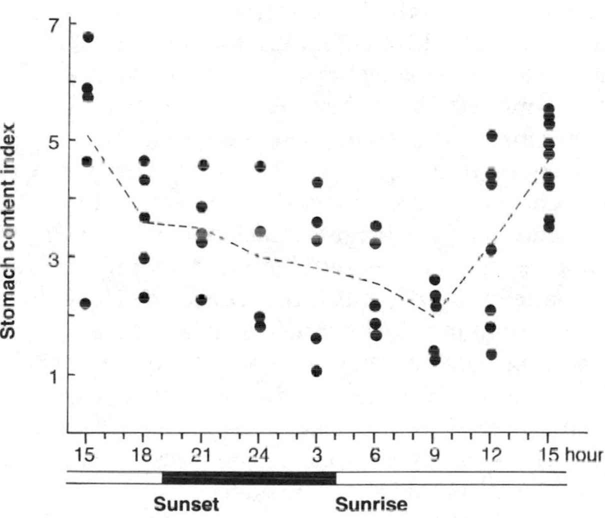

FIGURE 7
Diurnal change of stomach content index (weight of food/body weight × 100) for juvenile masu, 15–16 July 1970. (From JIBP-PF Research Group of Yurappu River 1975)

A large portion of food in the stomach during the day is typically composed of Ephemeroptera and "fallen" insects, without any particularly marked difference in composition related to time. The weight of these two groups together was about 70%–80% of the total. The order of importance in the stomach contents is Trichoptera, Diptera, and Plecoptera. Drifting organisms are most abundant at night (2100–0300 h), and the stomachs of the fish contain approximately 80% of such organisms during this period. In contrast, only 20% or less of drifting organisms are present in the stomachs during daytime. There is general agreement between species composition of drifting organisms and stomach contents during the daytime hours, but not at night (JIPB-PF Research Group of the Yurappu River 1975).

The rate of digestion for juvenile masu is estimated by the decrease of weight of stomach contents over time (JIPB-PF Research Group of the Yurappu River 1975). The weight of the stomach contents decreases almost linearly for 22 hours after the beginning of fasting (Figure 8).

The relation between stomach contents and rate of digestion can be expressed by the formula $F = 4.65 - 0.1812t$, where F = weight of food/body weight × 100 and t = hours of fasting. Using this formula it is estimated that it takes 25 hours for food eaten by a fish to pass through the stomach completely. This means that food eaten one morning will not be completely out of the stomach until about the same time the next day.

The amount of food taken daily by juvenile masu in May can be expressed by the quadratic equation $(SCI/CFI) \times 100 = 126.85 - 6.99t + 0.15t^2$, where t = total time of feeding activity, SCI = stomach content index, and CFI = cumulative feeding index.

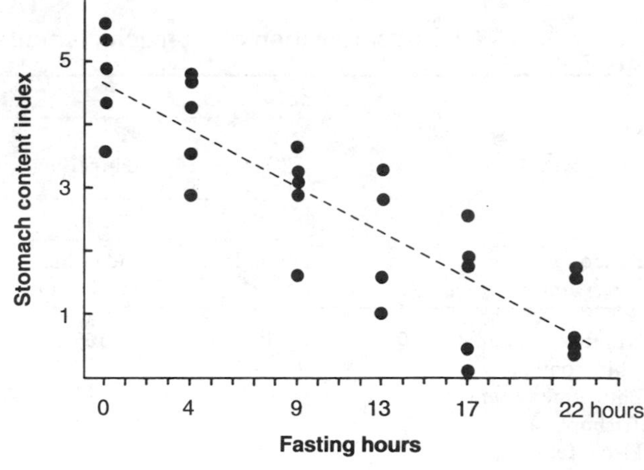

FIGURE 8

Decrease of the stomach content index (weight of food/body weight × 100) for starved juvenile masu, 15-16 July 1970. (From JIBP-PF Research Group of Yurappu River 1975)

Freshwater Residence

Growth and Maturation

Masu. During the first summer after emergence, masu fry feed in the streams and grow rapidly. After spending one winter in fresh water, most migrate downriver to sea the following spring. A few remain in the rivers for another winter. Many males and females mature and complete their life cycle in fresh water. Residual masu occur in some man-made lakes in Japan, i.e., Syumarinai and Nukabira lakes (Osanai 1962, 1982) and Okutadami Lake (Honda et al. 1980).

Female masu generally mature after two years of age whereas males can reach the spawning stage after one or two years (Utoh 1976). Immature and mature age 1.0 males collected in July from rivers of southern Hokkaido were separated by means of the gonado-somatic index (Figure 9) (Utoh 1977). In maturing males, spermatozoa were observed in the testes at the end of July, and in September the testicular lobules were filled with spermatozoa. The testes of immature males were filled only with spermatogonia during the spawning season. The maturing age 1.0 males were longer than the immatures. The smallest maturing males were almost the same size (7.0-8.0 cm in fork length in July) in the upper, middle, and lower reaches of the rivers. The occurrence of age 1.0 maturing males was dependent on growth in relation to length up to the end of July. All age 2.0 male masu were larger than 9.0 cm in fork length and had a high gonado-somatic index by June. Immature age 2.0 male fish were not found during the spawning period.

Young masu that have migrated from river to lake have shiny silver sides and a blue back. The parr marks are still visible when the scales are removed and these do not disappear when the fish mature. Young masu that take up residence in a lake do not disperse but stay near the entrance of the river into the lake. They move to the main body of the lake and disperse when the water temperature rises. Schools of young masu can be observed near the surface of lakes in spring and autumn. In summer they occur at the lower strata (Osanai 1982).

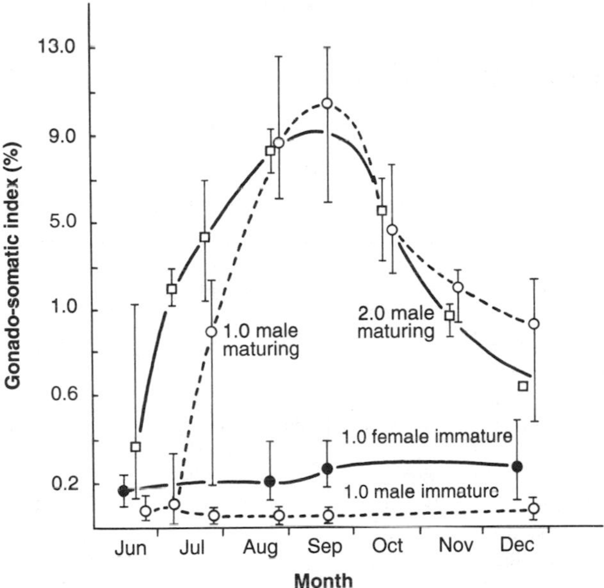

FIGURE 9

Seasonal changes in the gonado-somatic index (gonad weight x 100/body weight) of juvenile masu. Vertical bars indicate the range of values. (From Utoh 1977)

Masu that do not migrate seaward but stay in the lake grow well and are of a relatively large size. For example, juveniles, 5.1–9.2 cm long, released in Okutadami Lake (Niigata Prefecture), grew to average sizes of 18.7 cm (105 g), 31.3 cm (498 g), and 43.0 cm in May after one, two, and three years, respectively. Almost all resident female masu in lakes mature at age 2.0 (Honda et al. 1983).

Amago. There are sea-run populations of amago but the species occurs more commonly as river- and lake-resident stocks (Shiraishi et al. 1957; Kato 1973a, 1973b, 1975, 1977, 1978a, 1982; Kato et al. 1982).

Growth of amago fry in the Bano River reaches its maximum in May and June, decreases in autumn, and is negligible during December to February. Body lengths of fingerlings increase from 7.5 cm in July to 7.9 cm in August, and to 9.5 cm in November (Shiraishi and Suzuki 1957). Amago collected in the Nagara River (Honshu) by Kato (1982) in October were immature and ranged from 6.7 to 13.0 cm (average 9.5 cm) in length. In February, body lengths ranged from 12.1 to 14.8 cm (average 13.6 cm), and in May from 9.5 to 22.0 cm (average 17.3 cm).

Feeding Habits

Masu. Osanai (1962) analysed the stomach contents of 175 residual masu from Uryu Lake (Hokkaido) and found that 32 were empty, 142 contained only pond smelt (*Hypomesus transpacificus*), and one fish had eaten terrestrial insects. The major food items of residual masu in Okutadami Lake (Niigata Prefecture) are pond smelt, Japanese dace (*Tribolodon hakonensis*), and, rarely, terrestrial insects (Honda et al. 1980).

Amago. For residual amago, aquatic and terrestrial insects are the most important food items. Stomachs of residual young amago caught in the Nagara River in December contained 86.3% Trichoptera and Diptera, 13.3% Ephemeroptera and Plecoptera, and no terrestrial insects (Kato 1982). In March the Trichoptera and Diptera decreased to 59.3% of stomach contents, whereas Ephemeroptera and Plecoptera increased to 40.0%. Terrestrial insects were rarely found, averaging 0.4%. In May, 54.8% of the stomach contents were terrestrial insects and 30.5% Ephemeroptera and Plecoptera.

In the Hirakura River (Mie Prefecture) the smaller amago feed mainly on smaller aquatic insects, such as Baetidae and Ecdyonuridae, whereas larger amago have a strong preference for larger aquatic and terrestrial insects. During the summer, smaller fish have little chance of obtaining terrestrial insects because the larger fish eat them as soon as they fall on the water surface (Nagoshi and Sakai 1980).

Seaward Migration

Season of Migration

Masu. In southern Hokkaido, snow begins to melt in midstream areas of the rivers generally by early March. By mid-March, juvenile masu develop the characteristic silver and white coloration of smolts (Kubo 1976). When the water temperature reaches about 10°C in mid- to late April the parr marks are almost gone, the pink colour on the lateral line and elsewhere has disappeared, the tip of the dorsal fin has become dark black, and most masu have turned into smolts. In the early smolt stages, masu hold position in the current. As smoltification progresses, schools become more concentrated and seaward migration commences.

In Hokkaido, the migration season of smolts varies by area. In southern areas, it occurs in April to May, in central areas in mid- to late May, and in eastern areas mainly during the early half of June. Although information on the season of seaward migration for smolts is incomplete, it can be concluded that, in Hokkaido, smoltification generally occurs from March to May and that smolts migrate seaward during the three months from April to June (Sano 1964).

In rivers in the suburbs of Hakodate, late April to late May is the main season for seaward migration. This migration peaks from early to mid-May. Water temperature at this time is generally above 13° or 14°C. The downstream movement of smolts is most intense for several hours after sunset. Some fish may migrate downstream during daylight hours on sunny days (Kubo 1976).

Downstream migration of smolts occurs earlier in the season in southern areas than in northern areas. For example, smolts migrate one month later in Hokkaido than in Fukui and Toyama prefectures of central Honshu and about one month earlier than in southwestern Sakhalin (Table 21). In the Oippe River of Aomori Prefecture, the migration of smolts peaked principally in early May from 1963 to 1967 (Aomori Prefectural Fisheries Experimental Station 1964-68). In 1964, downstream migration of smolts in the Oippe River started when the water temperature reached 7°C. However, observations generally suggest that there is no direct relationship between the start of migration and water temperature (Krykhtin 1962).

TABLE 21
Seaward migration season of juvenile masu salmon

Region	River	Season of seaward migration	Peak
Fukui Pref.	Kuzuryu	Apr	-
Toyama Pref.	-	Apr-May	-
Aomori Pref.	Oippe	12 Apr-10 May	Early May
S. Hokkaido	Shiodomari	Late April to late May	Early half of May
	Mitsuishi	20 May-10 Jun	Late May
Central Hokkaido	Atsuta	-	Mid-May
E. Hokkaido	Nemuro area	-	Early half of June
Korean Pen.	Tumen	Late May	-
s.w. Sakhalin	Kalinin	June	-
	Porotonaya	Spring to early half of summer	-
E. Sakhalin	Tym	July	-
	Bogataya	8 Jul-5 Aug	Late July
s.w. Kamchatka	Bolshaya	July	-

Source: From Machidori and Kato (1984)

In the Bolotnaya and Teremok rivers on Sakhalin, smolts migrate mainly in spring and the early half of summer with a few fish moving to sea as late as August (Krykhtin 1962). In northeastern Sakhalin the downstream migration is later still. For example, in the upper Tym River, many smolts were observed from the end of May, and seaward migration took place much later in July (Gritsenko 1973). Observations in the Bogataya River in 1971 showed that smolts migrated from 8 July to 5 August, after the rise in river level in spring (Churikov 1975.)

The season of seaward migration of smolts in Kamchatka is still unknown; however, some smolts have been caught in the Bolshaya River on 21 July (Semko 1956).

Amago. Smolt transformation of amago in the wild, e.g., the Nagara River, takes place in November and December in one-year-old (age 1.0) fish (Kato 1973a). In winter, mainly December, the sea-run type of young amago descend to the sea.

Juveniles of amago cultured in ponds begin to turn into smolts in autumn (Honjoh et al. 1975). After release in the river, these amago smolts migrate downstream within a week.

Size of Smolts

Masu. Juvenile masu are identified as smolts when they have grown beyond a certain body length at a certain time. In rivers in southwestern Sakhalin, smolts migrate at fork lengths of 10-11 cm and greater (Krykhtin 1962). Downstream migration at 8-9 cm in fork length seldom occurs. The fork length of smolts in Hokkaido ranges from 9 to 19 cm (Sano 1964). Geographical variations in fork length are generally small (Table 22). For example, the fork length of smolts in the Oippe River of northern Honshu, a comparatively southern area of distribution of masu, ranges from about 11 to 14 cm, whereas in main streams and tributaries in high latitude areas of Sakhalin and Kamchatka, it ranges from about 11 to 13 cm. Sugiwaka and Kojima (1979) and Volovik (1963) noted that the fork length of smolts also showed variation with age at migration.

TABLE 22
Fork length and body weight of masu salmon smolts

Region	River	Age	Fork length (cm)			Body weight (g)
			Average	Main range	Range	Average or range
N.E. Honshu	Oipppe R (mainstream)*	All	–	13.1–14.1	9.3–14.4	–
	Oippe R (tributary)*	All	–	11.2 13.1		
Pacific coast of Hokkaido	Mitsuishi*	All	17.0	–	–	–
Hokkaido	Atsuta	1.	13.0	–	12.0–14.0	–
	Atsuta	2.	14.5	–	12.8–16.2	–
	–*	All	–	11.2–14.1	9.3–18.9	
Korean Pen.	Tumen*	All	15.4	–	–	
Primore	–	All	–	11.0–12.0	–	
s.w. Sakhalin	–	All	–	11.0–12.0	8.0–12.0	12.0–18.0
S. Sakhalin	Naiba	0.	7.1	–	6.7–7.4	4.7
	Naiba	1.	10.6	–	9.2–11.4	13.4
	Naiba	2.	12.6	–	11.6–16.0	25.9
N.E. Sakhalin	Bogataya	All	13.7	–	12.5–16.0	30.1
s.w. Kamchatka	Bolshaya	All	11.1		–	–

Source: From Machidori and Kato (1984)
Note: *Total length converted to fork length by the formula: fork length = –3.3154 + 0.9611 total length

Amago. The body lengths of amago smolts sampled in the Nagara River in December ranged from 11.2 to 19.1 cm and averaged 15.5 cm (Figure 10). The smolts were sampled just before downstream migration, as was evident from their slender body shape, the blackened dorsal fin tip, the silvery body surface, and the absence of parr marks (Kato 1982).

Age Composition of Smolts

Masu. The results of determining the age of downstream migrating juveniles (smolts) using scale pattern analysis vary. The smolts in the Bogataya River of eastern Sakhalin were all age 2.0 fish, according to Churikov (1975). It is known that there are two age types of smolts in rivers and coastal waters in Hokkaido. Most juveniles are age 1.0 fish, and some mixed groups contain age 2.0 fish with a fairly high rate of age 1.0 fish (Table 23). Such differences in age at downstream migration appear to reflect geographical differences. However, Kato (1973) pointed out that the diversity in

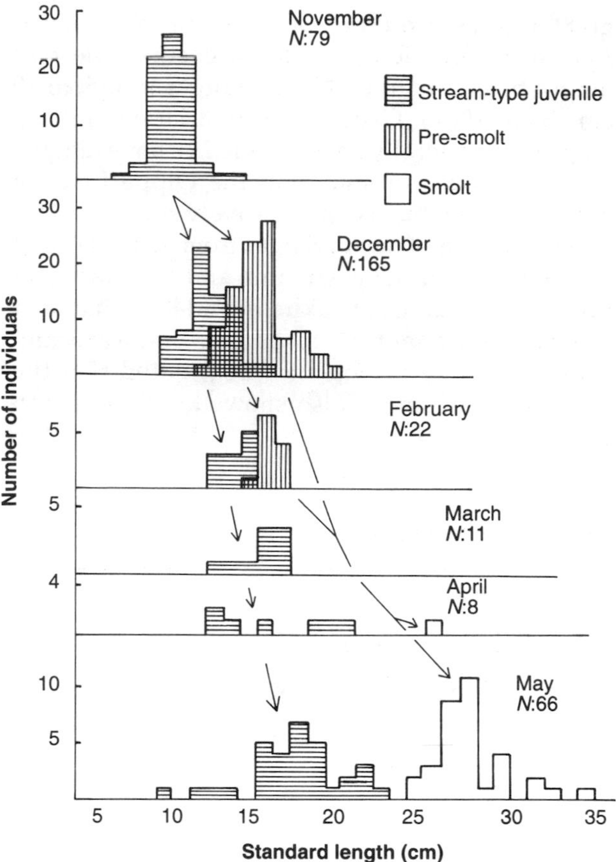

FIGURE 10

Standard-length distributions of residual and smolt-type amago salmon caught in the Nagara River. (From Kato 1982)

TABLE 23

Age composition of masu salmon juveniles caught in coastal waters or rivers of Hokkaido

Region	Number of fish	Age composition %	
		1.0	2.0
Coastal areas of Shizunai	129	97	3
Kushiro River mouth	116	100	0
Coastal area of Konbumori	61	100	0
Atsuta River	75	59	41

Source: From Machidori and Kato (1984)

ages might have resulted from scientists using different criteria for age determination. The apparent differences may also be caused by sampling error. Sugiwaka and Kojima (1979), for example, expressed concern about selective sampling of older fish by angling.

Sex Ratio of Smolts

Masu. Smolts generally consist of more females than males because some males remain in the streams and become adult fish while retaining the appearance of juveniles. Sano and Ozaki (1969) found that 43.8% of juvenile masu originating from the Shiribetsu River of western Hokkaido, and incubated and reared in the hatchery at the Yurrapu River of southern Hokkaido, became smolts one year after hatching; of these, 70.7% were females and 29.3% males. Of the females, the mid- and large-sized individuals became smolts, and small-sized fish with poor growth became smolts in the following year. Of the males, only the mid-sized fish became smolts and the large and small individuals did not. Rearing experiments by Kubo (1968) showed that the difference in sex ratio of the smolts changed from 17% to 70% according to rearing conditions. This varying sex ratio appeared to be associated with many elements, such as water temperature, light, structure of ponds, density during rearing, quality and quantity of feed, and growth.

Observations on the sex ratio of smolts in streams are limited. Smolts in Hokkaido and southwestern Sakhalin, where masu are abundant, consisted of 70%–80% females in Hokkaido and about 55% females in southwestern Sakhalin (Table 24). There is a geographical cline in sex ratio of smolts with the proportion of females increasing from north to south. This indicates that in southern areas most males become mature in the

TABLE 24

Sex ratio of migrating masu salmon smolts

Region or river	Sex ratio (%)	
	Female	Male
Honshu-Oippe River	85–90	10–15
Hokkaido-Mitsuishi River	72	28
Hokkaido-Atsuta River	61–77	23–39
Hokkaido	70–80	20–30
S.W. Sakhalin	54.4	45.6
N.E. Sakhalin	55.0	45.0

Source: From Machidori and Kato (1984)

streams and that in northern areas about half of the males migrate to sea.

Amago. Sex ratio of amago smolts sampled at the Nagara River was slightly more than 2:1 in favour of females in December and almost 3.5:1 in favour of females from January to April (Kato 1973a). As with masu, the females of amago outnumber the males in the smolt stage.

Early Sea Life

Coastal Distribution of Young

Masu. With the arrival of spring, temperatures rise and water levels in rivers increase. In waters north of central Honshu, the downstream migration of juvenile masu begins at this time. Juveniles are often caught incidentally from April to May in coastal waters of the Sea of Japan from Niigata Prefecture to Aomori Prefecture (Figure 11) by the small-sized setnets fixed close to shore for the purpose of catching other small coastal fishes. On the Pacific Ocean side many young masu are caught in May in coastal areas of Aomori Prefecture. Sea surface temperatures in coastal waters of Honshu at this time are 8°–12°C. When the surface water temperature rises to 14° or 15°C in June, almost no young masu are caught.

The colour pattern of young fish that have migrated to sea consists of shiny silver on the sides of the body, white on the abdomen, and blue on the back with black spots on the tips of the dorsal and caudal fins. The body of the fish is comparatively slender at this stage, the head is somewhat round, and scales are easily lost. Specimens preserved in formalin lose their silver colour in about ten days, at which time the parr marks appear (Kawakami and Machida 1933).

Farther south, in Kyushu, only some fish appear to migrate downstream early in the season. In Ariake Bay, one masu caught in December and one caught in March measured 17.2 cm and 22.5 cm in total length, respectively (Kimura and Tsukahara 1969). Because of their small size it is believed that these fish had only recently entered the marine environment. In Shiranuhi Bay, in the coastal area of Kumamoto Prefecture of Kyushu, some masu have been caught in early April (Matsubara 1934). Of these, one fish was 31 cm in total length and weighed 420 g. Several other masu were caught in Tateyama Bay in Chiba Prefecture on the Pacific Ocean coast of Honshu in March (Ebina 1935). One measured 27.6 cm in total length, and the ovary was immature and contained very small eggs. The body size was too small for a masu that had spent one year at sea. Also, specimens of these fish kept in formalin revealed parr marks, which suggests that only a short time had elapsed between their seaward migration and their capture at sea. These masu probably migrated to sea in winter or fall several months before they were caught. Juvenile masu with smolt characteristics have been caught in the lower reaches of the Edo River in Tokyo in December (Nakamura 1958) and in the Chikugo River in autumn (Kimura and Tsukahara 1969). Thus, in southern areas of Japan, where anadromous masu usually do not occur, seaward migration appears to take place in autumn and winter when water temperatures drop in the rivers and coastal waters.

In coastal waters of Hokkaido young masu are caught from May to July, which is about one month later than in the coastal waters of Honshu. Catches generally occur in May in coastal waters of Tsugaru Strait and southwestern Hokkaido, which are affected early in the season by the warm Tsushima Current, and in June in northern Hokkaido waters. A few young masu are caught in July in coastal areas of eastern Hokkaido, where spring comes latest.

Sea surface temperatures at the time young masu are caught off northwestern Hokkaido and in the Sea of Okhotsk range from 7° to 15°C (Far Seas Fisheries Research Laboratory 1979–80). The greatest number are captured in water temperatures of 11°–12°C. It is generally inferred that the seaward migration of young masu is related to water tem-

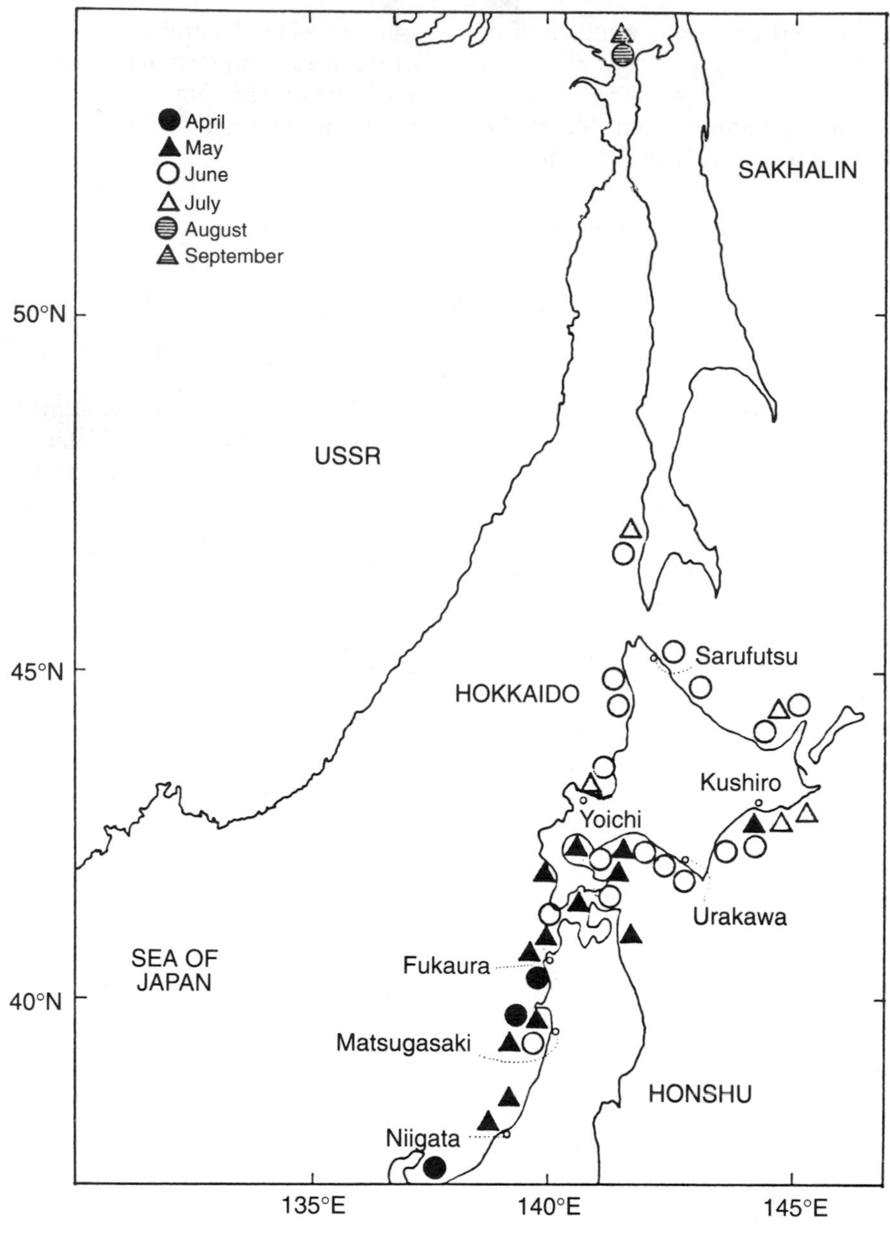

FIGURE 11

Seasonal distribution of young masu in coastal areas. (From Machidori and Kato 1984)

perature or some associated elements (Sano and Abe 1967). For example, Sano and Abe (1967) reported that in coastal areas near the Kushiro River young masu were first caught to the west of the river mouth in late May to mid-June because of higher temperatures there. The reverse occurred by about June when more young masu were captured east of the river mouth where water temperatures had been colder than on the west side (Figure 12). The coastal area on the west side is strongly affected by river water and temperatures on that side are generally 2°–3°C higher than those in the coastal areas on the east side.

The timing of the appearance of young masu in coastal waters in Sakhalin is not well known. Juvenile masu have been caught by setnets in coastal

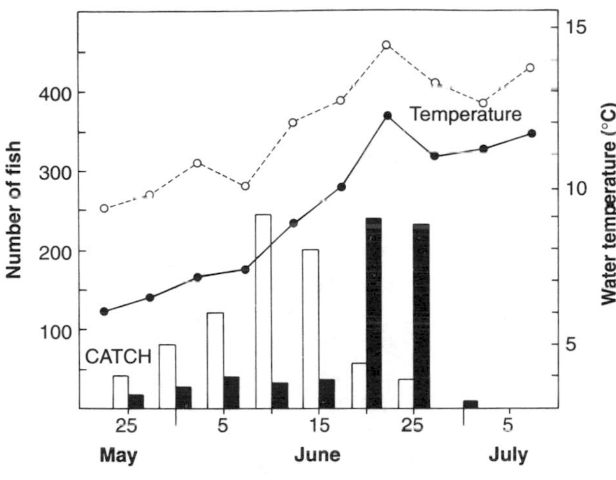

FIGURE 12

Changes in numbers of young masu caught by small-sized setnets in coastal areas east (black histograms) and west (open histograms) of the Kushiro River mouth, and changes in surface water temperature in the coastal area east (black circles) and west of the river mouth (open circles). (From Sano and Abe 1967)

waters of southwestern Sakhalin around June and July (Dvinin 1956) and in Sakhalin Bay of northern Sakhalin in August and September (Semko 1956).

The seasons when young masu appear in coastal waters of Primore, the Kuril Islands, and Kamchatka are unknown, but the timing of occurrence is probably similar to that in the coastal waters of Japan and Sakhalin.

Amago. In Mikawa, Ise, and Wakasa bays of Honshu, some young amago are caught in December in the estuary of rivers in which artificially reared amago smolts are released. After remaining in the river mouth for about one month, the young amago disperse all over the bay. Amago do not migrate as widely as other Pacific salmon species (Honjoh 1985).

Size and Age of Coastal Young

Masu. Because masu smolts are predominantly females, the young fish caught in coastal waters are also mainly females. For example, in coastal waters of Hokkaido, from 70% to 90% (mean 81%) of fish caught are females (Table 25). Small individuals average 12–13 cm in fork length and 17–18 g in body weight, and large individuals are 25–26 cm in fork length and average about 200 g in body weight (Table 26). The small-sized fish seem to be those taken soon after their downstream migration.

TABLE 25

Sex ratio of young masu salmon caught by small setnets in the coastal waters of Hokkaido

Location	Number of fish	Sex ratio (%) Female	Male
Various coastal waters in the Pacific	210	79.5	20.5
Shizunai coast	129	90.7	9.3
Kushiro coast	110	78.1	21.8
Konbumori coast	61	70.5	29.5

Source: From Machidori and Kato (1984)

Young masu caught on the Pacific Ocean coast of Aomori Prefecture and off Sarufutsu-Hamatonbetsu and Konbumori in Hokkaido are generally large. Even in May they average about 20 cm. The average fork length and body weight of young masu taken throughout the coast of Japan are 17.5 cm and 55.6 g, respectively. Along the coast of southwestern Sakhalin young masu average 17 cm in fork length and 74 g in body weight.

Although problems remain with scale reading, it is generally considered that young masu caught in coastal waters vary in age at the time of their downstream migration. Semko (1956) reported that juveniles with an average fork length of 22.2 cm caught in Sakhalin Bay were age 1.0, and those with an average fork length of 26.4 cm were age 2.0. Three fish caught in the Kushiro River estuary in June, with an average fork length of 17.0 cm and an average body weight of 47.9 g, were age 2.0, and others caught during the same period, with an average length and weight of 15.6 cm and 44.3 g, respectively, were age 1.0 (Table 27). The smallest individuals of the age 1.0 fish, with fork lengths of 12–13 cm, did not grow during the whole season. However, the largest individuals showed increases in length of about 6 cm between late May to early June and mid- to late June, reaching sizes of 21–22 cm in fork length by mid- to late June.

Because early migrants are joined by newly migrated groups, it is not particularly useful to estimate growth of young masu using seasonal change of average fork length. The following formula was obtained by using the greatest fork length for each

TABLE 26
Fork length and body weight of young masu salmon caught in coastal waters

Location	Month	Number of Fish	Fork length (cm)		Body weight (g)	
			Average	Range	Average	Range
Sea of Japan coast of Honshu						
Toyama Pref. coast	Apr	3	16.4*	6.3-16.6*	39	38-41
Niigata Pref. coast	May	8	–	7.4-19.4	–	–
Akita Pref. coast	Apr	2	14.0	13.9-14.0	33	32-33
Akita Prefecture coast	May	1	19.1	–	91	–
Aomori Pref. coast	May	5	16.7	13.0-18.7	56	21-77
Pacific coast of Honshu						
Aomori Pref. coast	May	73	19.7	16.5-26.4	90	50-278
Hokkaido coast						
Ishikari-Teshio coast	Jun	2	12.7	12.7-12.7	27	24-30
Sarufutsu-Hamatonbetsu coast	Jun	6	17.3	15.9-19.7	52	38-69
Monbetsu-Shiretoko Peninsula coast	Jun	6	12.9	11.6-15.2	22	16-31
Monbetsu-Shiretoko Peninsula coast	Jul	1	14.1	–	30	–
Kushiro coast	May	52	15.1	12.6-17.4	35	20-59
Kushiro coast	Jun	58	15.6	12.0-22.9	44	20-144
Konbumori coast	Jul	61	18.8	15.6-22.6	49	38-56
Shizunai coast	Jun	125	18.1*	13.9-26.6*	–	–
Sakhalin coast						
Southwestern coast	Jun-Jul	–	17.0	13.5-26.5	74	30-184
Sakhalin Bay	Aug-Sep	–	22.2†	–	–	–
Sakhalin Bay	Aug-Sep	–	26.4†	–	–	–

Source: From Machidori and Kato (1984)
Notes: *The relationship between fork length and total length of young masu in the 15.0-25.0 cm fork length range, caught in coastal areas, was: fork length = 0.9611 × total length - 3.3154. Lengths of masu caught on the Toyama and Shizunai coasts were originally given as total length and have been converted to fork length using this formula.
†The upper row for Sakhalin Bay are age 1.0, the lower row age 2.0. Almost all fish caught in coastal waters in Japan are age 1.0.

TABLE 27
Fork length and body weight by season of young masu salmon caught in small-sized setnets in coastal waters adjacent to the Kushiro River mouth, eastern Hokkaido

Season	Age	Number of fish	Fork length (cm)		Body weight (g)	
			Average	Range	Average	Range
Late May	1.0	52	15.12	12.6-17.4	35.2	20.0-59.0
Early June	1.0	7	14.83	13.7-15.2	33.3	26.0-54.0
Mid-June	1.0	31	15.93	12.0-22.9	47.6	20.0-144.0
Late June	1.0	20	15.77	13.1-21.0	43.9	26.3-118.5
June	2.0	3	16.97	16.1-17.4	49.4	37.8-56.2

Source: From Sano and Abe (1967)

ten-day period from late May to late June: $L = 154.25 + 1.85\,D$, where L = fork length in mm and D is the days elapsed since 25 May. This gives an increase in fork length of 1.85 mm/d. Because the greatest fork length reflects growth of an individual which has good growth, it probably overestimates the average growth rate for the group as a whole. It is also possible that individuals that migrated earlier may have left the estuary, and, thus, the value of 1.85 mm/d may underestimate the growth for the largest individual.

Amago. As with sea-run masu, young amago caught in the coastal waters are mainly females. From December to February, young amago in coastal waters feed little and, as a result, there is little or no growth during winter. After March, young amago begin to grow rapidly, reaching a maximum body weight of 1,200 g in May when they ascend the river again. There are some variations in sex ratio between wild and artifically reared amago. For example, of young wild and artificially reared amago caught at Mikawa Bay and Ise Bay, 81.6%–88.9% and 76.0%–98.1%, respectively, were females (Honjoh 1985). The amago from these two bays weighed from about 36 to 114 g when they entered estuaries.

There are variations in size by area among young amago caught in various coastal waters. For example, the body weight of fish caught in early and mid-May 1978 were, on average, 711 g in Ise Bay, 607 g in Mikawa Bay, and 820 g in Wakasa Bay. These differences in fish size are attributed to varying temperatures of the sea and abundance of food (Honjoh 1985).

Food of Coastal Young

Masu. The main food items of young masu in coastal waters are generally crustaceans and fish, with sand lance (*Ammodytes personatus*) and sand eel (*Hypoptychus dybowskii*) as the most common food items. Stomach contents of young masu off Honshu in April and May are mainly small sand lance, with the remainder being amphipods (Table 28). In Sakhalin Bay in August and September, the main food item was sand lance (4–6 cm in body length), with herring (*Clupea harengus pallasi*) and smelt (Osmeridae) also having been consumed (Semko 1956). It is probable that young masu encounter juvenile chum and pink salmon (*O. keta* and *O. gorbuscha*) in the coastal waters, but these salmon species have never been found in their stomachs. Scales of young masu do occur in stomach contents but Kawakami and Machida (1933) did not find fish remains. As these young fish were caught in setnets, it is possible that scales were swallowed while in the nets.

In some areas, crustaceans were at times the main food for young masu. Sano and Abe (1967) observed that in the Kushiro River estuary, many individuals were satiated with euphausiids. According to Dvinin (1949), young masu caught in the coastal waters of southwestern Sakhalin in June had also eaten large quantities of crustaceans.

Many other species of crustaceans and insects, both freshwater and terrestrial forms, have been found in the stomachs of young masu. However, these food items constituted only a small part of the total quantity of food.

Amago. Over twenty items of food organisms are

TABLE 28
Stomach contents of young masu salmon in coastal waters

			Stomach Contents*						
Area	Month	Number of fish	Fish	Euphausiids	Crustaceans†	Amphipods	Shrimp	Terr. insects	Aquatic insects
Honshu									
Shinano R. mouth	May	8	••	–	–	–	–	–	–
Miomote R. mouth	May	–	••	–	–	–	–	–	–
Matsugasaki coast	Apr	2	••	–	–	–	–	–	–
Kisagata coast	May	1	••	–	–	–	–	–	–
Kisagata coast	Jun	2	••	–	–	–	–	–	–
Fukaura coast	May	–	••	–	–	•	–	–	–
Shiranuka coast	May	73	••	–	–	•	•	–	–
Hokkaido									
Shizunai coast	May-Jun	129	••	–	–	•		•	•
Kushiro coast	May-Jun	–	•	••	–	–	–	–	–
Sakhalin									
s.w. coast	Jun-Jul	–	•	–	••	–	–	–	–
Central and eastern coast	Jul-Aug	–	–	–	–	••	–	–	–
Sakhalin	Aug-Sep	33	••	–	•	–	–	•	–

Source: From Machidori and Kato (1984)
Notes: *Double circles show the species eaten more frequently.
†In the case of crustaceans, details are not clear.

found in stomachs of young amago taken in the coastal waters of Ise and Mikawa bays. During late December to early March, crustaceans such as pea crabs (*Pinnixa rathbuni*) and larvae of Brachyura, Macrura, and Mysidacea, are the dominant items in Ise Bay. Gobies (family Gobiidae), ghost shrimp (*Callianassa japonica*), polychaetes, and algae are also present in the stomachs. From mid-March to mid-May, sand lance, anchovy (*Engraulis japonica*), and sardines (*Sardinops melanostictus*) are the most abundant food items, whereas in late April, larvae of chub mackerel (*Scomber japonicus*), and in mid- and late May, greenling (*Hexagrammos otakii*) were the main food of young amago. In Wakasa Bay, young amago primarily eat anchovies or other small fish. The length of consumed anchovies increases as the young amago grow.

Indices of the weight of stomach contents (stomach contents weight/body weight × 100) of young amago collected in Ise Bay varied from 0.1% to 0.8% during December to February, 0.1% to 2.3% in March, 0.6% to 2.5% in April, and 0.3% to 1.5% in May (Honjoh 1985). The seasonal increase of the stomach contents index corresponds to the growth curve of young amago.

Offshore Migration

Northward Migration of Young Masu in Spring

Masu. The migratory behaviour of young masu after entering the sea varies considerably. Because young masu were only fished near coastal waters with small-sized setnets, information on movements in spring is limited, and the extent to which the young fish migrate to offshore waters is not known.

Some masu smolts released in the rivers of southern Hokkaido were recovered in coastal waters in the vicinity of the release river over fairly long periods of time, and others were recovered long distances from the river. Tag and recovery studies in Tsugaru Strait and along the Pacific Ocean side of Hokkaido showed that masu juveniles migrated east on the south side of Hokkaido, and it is presumed that all young fish that migrate east on the Pacific Ocean side of Hokkaido move north past Nemuro Strait (Figure 2). There is no information about their migration beyond that point. In the Pacific Ocean east of 145°E, sampling with small-meshed gillnets has been conducted for many years, but no young masu have been caught. The migration may continue either through the passes of the southern Kuril Islands into the Sea of Okhotsk or north along the Kuril Islands. Young masu tagged along the west coast of Hokkaido migrated north in the Sea of Japan. However, recoveries showed that some fish migrated along the coast and then returned to the river mouth area from which they had initially migrated. Smolts released in the Assabu River (Figure 3) were recovered in coastal waters of the Sea of Okhotsk side of Hokkaido in autumn of the same year. This suggests that the young fish migrate to the Sea of Okhotsk through Soya Strait. However, some young masu migrating northward along the west coast of Hokkaido may also move to Tatar Strait. Young masu that migrate from areas off the west coast of Honshu probably migrate east through Tsugaru Strait or north along the west coast of Hokkaido, but no firm evidence is available.

The average fork length of young masu caught in the various coastal waters of Japan is relatively small (12–16 cm), suggesting that they are captured soon after downstream migration to sea. Large individuals (more than 20 cm in fork length) are relatively rare, indicating that young masu do not stay long in shallow coastal waters where the small setnets are placed. Young fish in the coastal waters of Honshu are 16–19 cm in fork length in May. An average fork length of 18 cm in mid-May suggests an increase in length of 1.8 mm/d and that the fork length will be 23–24 cm by mid-June. Such large-sized fish have rarely been caught in coastal waters of Hokkaido north of Honshu, even in late June. If young masu originating in Honshu migrate to coastal waters of Hokkaido, then they probably do so some distance from shore.

A summary of the fragmentary information on migration routes of young masu in spring in coastal waters is presented in Figure 13. The direction of migration of young fish along Honshu and around Hokkaido corresponds to the seasonal movement of surface temperature isotherms and to the general flow of the current. Around Japan, where cold and warm currents mix, there are many counter-currents and eddies, so the migration of young fish is probably not simple.

In the coastal waters of the USSR, young fish appear in June and July, which coincides with the rising temperature and increasing abundance of food organisms. The northward movement of the 10°–12°C isotherm coincides with the time of occurrence and the northward migration of young fish. It is presumed that young fish that move downstream from Primore and from the west coast of Sakhalin migrate towards the north in the Sea of Japan and in Tatar Strait, and that young fish that move downstream from areas off the east coast of Sakhalin and the west coast of Kamchatka migrate slowly towards the north.

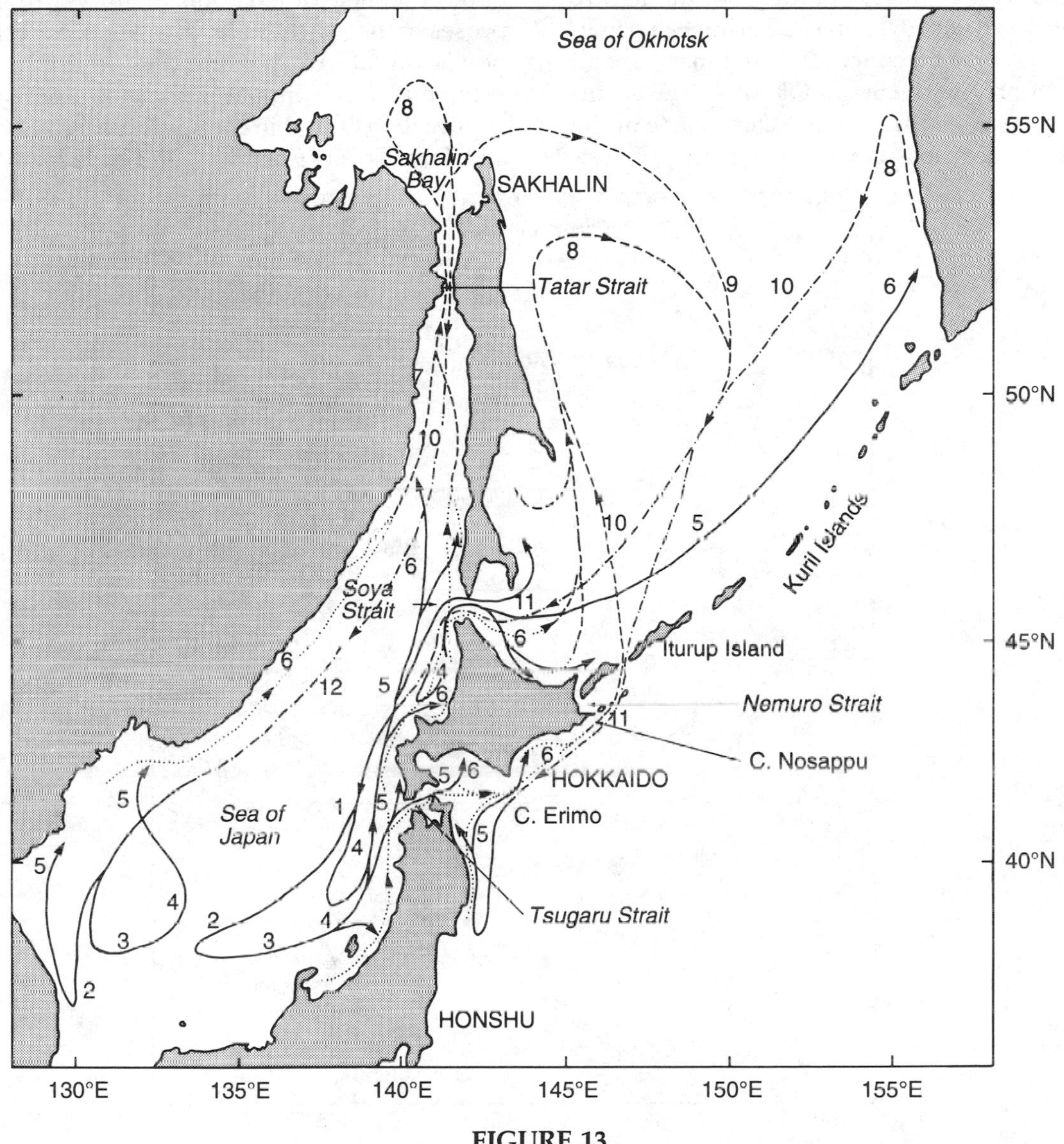

FIGURE 13

Presumed migration routes of masu salmon. Numbers represent month of the year. Dotted lines: spring (April to June), broken lines: summer (July to September), chain lines: autumn (October to December), solid lines: winter and spring (January to June). (From Machidori and Kato 1984)

HIGH-SEAS PHASE

Immature Masu

Summer. The Sea of Okhotsk is the only known region where immature masu commonly occur in summer. In the Pacific Ocean, only a few fish are found in the vicinity of Iturup Island in October (Okazaki 1977). Neither mature nor immature masu have been captured in the Bering Sea. However, it cannot yet be concluded that masu spend the summer only in the Sea of Okhotsk. The northern Sea of Japan and the Pacific Ocean side of the Kuril Islands are possible summer areas of residence for immature masu, though no clear evidence is yet available.

The immature masu that originate in Japan and move north in the Sea of Japan may migrate to the Sea of Okhotsk mainly through Soya Strait, and the fish that move north in the Pacific Ocean may enter the Sea of Okhotsk mainly through the passes of the southern Kuril Islands, especially the pass south of Iturup Island (Figure 13).

In mid-July, immature masu are probably mainly distributed in the southern Sea of Okhotsk and Tatar Strait (Figure 14) and have likely not yet

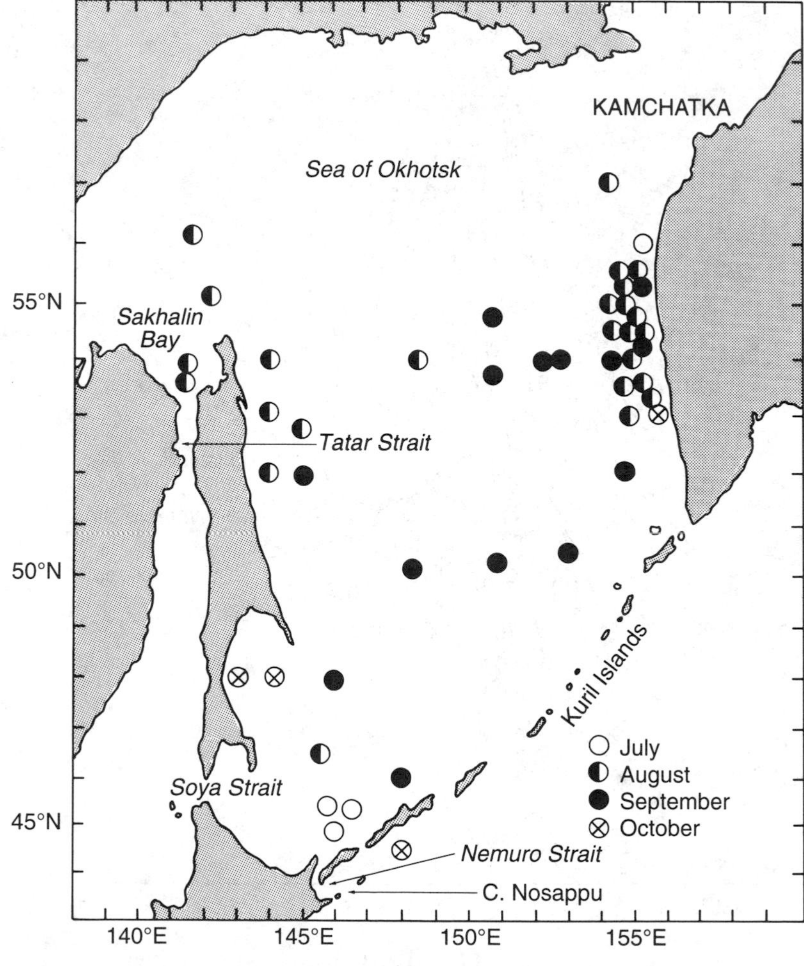

FIGURE 14

Distribution of catches of immature masu in offshore waters during summer and autumn. (From Machidori and Kato 1984)

reached the offshore areas that extend from the central to the northern Sea of Okhotsk.

In August, almost all masu appear to be distributed in the Sea of Okhotsk close to the east and west coasts. The catch in the central Sea of Okhotsk at this time is generally low and it is possible that some immature masu still occur in Tatar Strait in August.

In September, the distribution of fish changes. Beginning in mid-September, the surface water temperature gradually decreases and immature masu become widely distributed throughout all offshore areas in the Sea of Okhotsk.

Masu that originate in Primore and the west coast of Sakhalin may either migrate towards the north in Tatar Strait and to offshore waters in the Sea of Okhotsk through the Strait of Nevelskoy, or spend the summer in Tatar Strait and Sakhalin Bay. The former group may migrate south in the Sea of Okhotsk with cooling of the water during September and October, whereas the latter may migrate south in Tatar Strait (Machidori and Kato 1984).

When immature masu were caught in offshore waters of the Sea of Okhotsk from July to September, the surface water temperature ranged from 9.7° to 19.4°C. More specifically, the temperature at time of catch was 13°-19°C in the southern Sea of Okhotsk, 11°-15°C in western waters, and 9°-14°C in eastern waters. The surface water temperatures where the catch was abundant were 13°-16°C in southern waters, 13°-15°C in western waters, and 10°-14°C in eastern waters. A marked thermocline develops in the Sea of Okhotsk by summer and, because the depth distribution of masu in these waters is unknown, it is difficult to determine their water temperature habitat in summer. In general, masu are seldom found in waters in which the surface temperature is 16°C and over.

Biological information on immature masu distributed in offshore waters is limited. The sex ratio in the Sea of Okhotsk in summer is close to 1:1 but the numbers examined are small (Table 29). The scales of the immature fish show no ocean winter zone, which indicates that all have migrated to sea in the year of capture. Most fish were age 1.0 and some were age 2.0 (Table 30) (Machidori and Kato 1984).

The fork length of immature masu caught in offshore waters of the Sea of Okhotsk from July to September ranged from 18.6 to 42.4 cm and averaged 22.6 cm (Table 31). The average body weight was 255 g with the smallest fish weighing 75 g and the largest 800 g. Although only a few months had passed since their downstream migration, some fish had grown to sizes recorded for small adults taken the following spring. Four immature masu caught on the Pacific Ocean side of Iturup Island in mid- to late October had an average fork length of 42.1 cm and an average body weight of 968 g. Three were females and all were age 1.0 (Machidori and Kato 1984).

Fish length varies by area. Immature fish in the eastern Sea of Okhotsk are more than 10 cm shorter in average fork length and weigh about one-third less than those in other waters at the same time. The average fork length of immature masu in eastern waters in August and September

TABLE 29

Sex composition of immature masu salmon caught in offshore waters of the Sea of Okhotsk from July to September

Area	Number of fish	Sex composition (%)	
		Female	Male
Southern areas*	10	70	30
Western areas†	5	20	80
Eastern areas‡	19	47	53
All areas	34	50	50

Source: From Machidori and Kato (1984)
Notes: *South of 50°N
†North of 50°N and west of 150°E
‡North of 50°N and east of 150°E

TABLE 30

Age composition of immature masu salmon caught in offshore waters of the Sea of Okhotsk from July to September

Area	Number of fish	Age composition (%)	
		1.0	2.0
Southern areas*	14	93	7
Western areas†	4	75	25
Eastern areas‡	13	54	46
All areas	31	74	26

Source: From Machidori and Kato (1984)
Notes: *South of 50°N
†North of 50°N and west of 150°E
‡North of 50°N and east of 150°E

TABLE 31
Fork length and body weight of immature masu salmon caught in offshore waters of the Sea of Okhotsk from July to September

	Southern areas		Western areas*	Eastern areas†	Entire area‡
	Late July	Late Aug/early Sep	Early Sep	Late Aug/early Sep	Jul-Sep
Fork Length (cm)					
Number	16	4	5	19	44
Average	28.0	34.6	34.5	21.6	26.6
Range	20.5-35.8	29.4-42.4	31.2-39.6	18.6-26.3	18.6-42.4
Body weight (g)					
Number	16	3§	5	19	43
Average	284	425	555	125	255
Range	100-700	310-500	440-800	75-230	75-800

Source: From Machidori and Kato (1984)
Notes: *South of 50°N
†North of 50°N and west of 150°E
‡North of 50° and east of 150°E
§Body weight of the largest individual is unknown.

is about 6 cm less than for those in the southern Sea of Okhotsk in late July (Machidori and Kato 1984).

The small body size of the immature masu in the eastern part of the Sea of Okhotsk in summer suggests that they are from the west coast of Kamchatka, and that they migrate downstream late, thus having only a short growing period in the ocean. Masu from more southern areas migrate downstream earlier in the season, and their growing period to midsummer is longer, resulting in larger sizes.

Growth of immature masu in offshore waters is rapid because immature fish in the southern Sea of Okhotsk in late July average 28 cm in fork length and are about 6.6 cm longer one month later. The average fork lengths observed at intervals over a 40-day period for immature masu caught off the west coast of Kamchatka in August and September 1973 are shown in Table 32. Data in the table yield the formula $L = 183.5 + 1.80D$, where L = fork length (mm) and D = number of days from 1 August. It is estimated that the increase in fork length of immature masu off the west coast of Kamchatka in summer approaches 1.8 mm/d.

The relationship between fork length and body weight (Figure 15) of immature masu caught in the Sea of Okhotsk (calculated from data presented in Table 31) is described by the formula $W = 0.0066\ L^{3.1876}$, where W = body weight (g) and L = fork length (cm). No variations by area in the relationship between fork length and body weight were observed.

TABLE 32
Average fork length by day of immature masu salmon caught in offshore waters of the west coast of Kamchatka during August and September, 1973

Date	Latitude	Longitude	Number of fish	Average fork length (cm)
Aug 9	53°00′N	154°55′E	3	19.53
10	53°30′N	154°44′E	7	20.39
11	54°00′N	154°44′E	39	20.77
12	54°30′N	154°52′E	7	20.30
13	55°00′N	154°21′E	1	19.10
14	55°31′N	154°40′E	4	17.50
25	55°12′N	142°12′E	1	21.70
27	54°00′N	144°00′E	2	24.30
30	54°00′N	148°24′E	3	23.17
Sep 1	54°00′N	152°48′E	2	22.15
2	54°00′N	154°45′E	2	23.00
13	52°00′N	154°50′E	16	27.14
14	50°30′N	153°00′E	2	22.10
15	50°20′N	150°40′E	1	28.10
16	50°10′N	148°20′E	1	26.10

Source: From Shimazaki (1975)

Autumn and Winter. In October, immature masu occur in the Sea of Okhotsk off southeastern Sak-

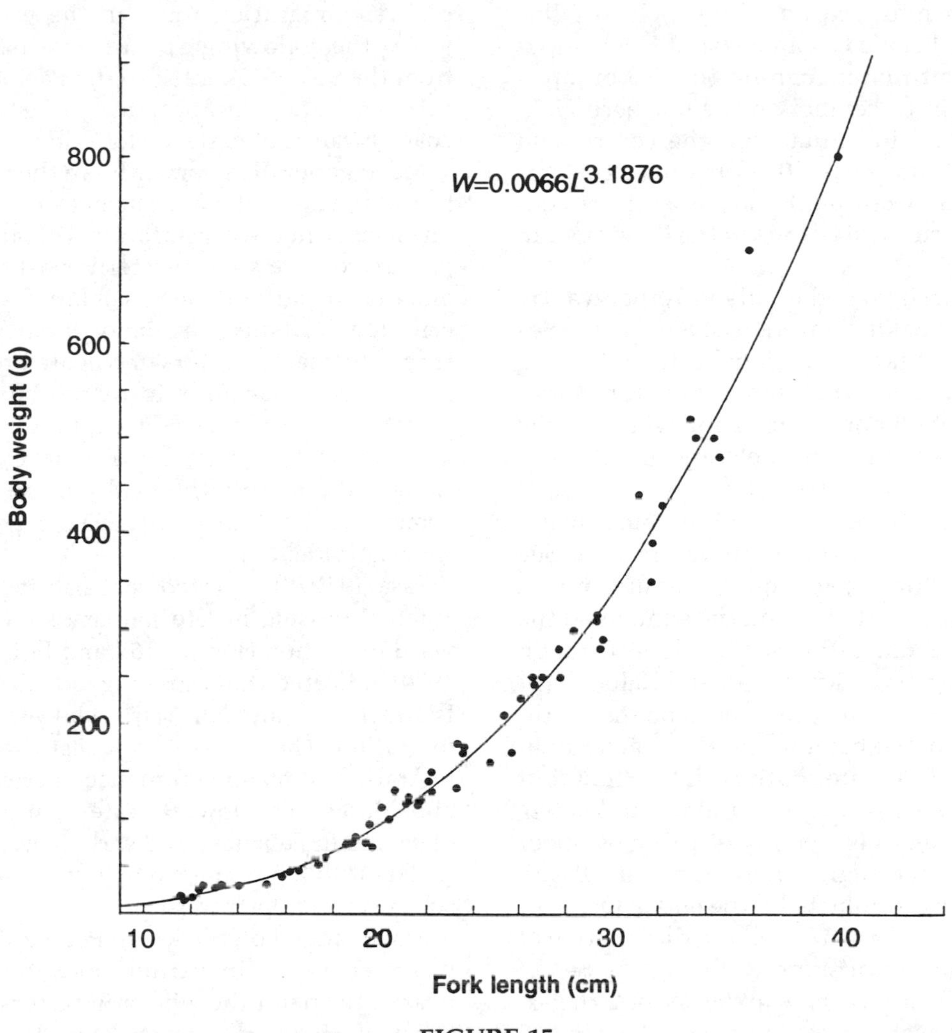

FIGURE 15

Relationship between fork length and body weight of .0 age masu caught in coastal waters of Japan and in offshore waters of the Sea of Okhotsk from June to September. (From Machidori and Kato 1984)

halin and southwestern Kamchatka and on the Pacific Ocean side of Iturup Island in the Kuril Islands. Scattered catches in other areas suggest that they are distributed over a wide area in the Sea of Okhotsk and in waters of the southern Kuril Islands in autumn.

By mid- to late October, immature masu begin to be caught incidentally in setnets in the coastal area on the Sea of Okhotsk side of Hokkaido (Sasaki 1978). From about late October, immature masu occur in Nemuro Strait, and, by November, they appear on the Sea of Japan side of Hokkaido and Honshu.

The pole-and-line fishery for young masu on the Sea of Japan side of Hokkaido begins in early November in the north, early December in the central region, and late December in the south. The start of the fishing season in the coastal waters of Aomori Prefecture in Honshu is in mid- to late January. From these observations, it is estimated that the first southward migrating groups of immature masu reach the vicinity of the Sea of Okhotsk coastal waters of Hokkaido by about the latter half of October, pass through Soya Strait in November, and gradually migrate south along the west coast of Hokkaido or offshore (Figure 13).

The peak fishing season of the pole-and-line fishery on the Pacific Ocean coast of Hokkaido is about one month later than on the Sea of Japan side. It starts in December on the northern Hokkaido coast, and in January on the central and southern Hokkaido coast. It is presumed that immature masu move through the passes in the central and southern Kuril Islands to the Pacific Ocean in October.

Immature masu tagged in July in Terpeniya Bay in southeastern Sakhalin were recovered in the Sea of Japan west of northern Honshu the following April. This suggests that some immature masu from the Sea of Okhotsk spend the winter in the Sea of Japan. Because the pole-and-line fishery becomes more productive on the west coast of Hokkaido after November, and because many masu are distributed in offshore waters of the Sea of Japan the following spring, it is assumed that most immature fish that spend the summer in the Sea of Okhotsk enter the Sea of Japan through Soya Strait in autumn (Machidori and Kato 1984).

The occurrence of immature masu on the southwestern coast of Sakhalin in mid-November indicates that masu migrate south in Tatar Strait, but the course of southward migration is unknown. Because the Strait of Nevelskoy is frozen by about mid-November, the migration from Sakhalin Bay to Tatar Strait must terminate by the end of October. There is no firm information on the distribution of masu in autumn in offshore waters of the Sea of Japan, in the Pacific Ocean, or in the Sea of Okhotsk.

The surface water temperatures where masu have been found in coastal areas of Hokkaido range from about 5° to 14°C, and they seem to be most abundant in offshore areas with surface water temperature of about 8°C. This suggests that masu migrate south through Tatar Strait in October and are distributed throughout the northern Sea of Japan by about November. The masu that migrate south along the Kuril Islands are probably distributed in the vicinity of the eastern coast of Hokkaido by about November. By mid-December, surface water temperatures in the southern Sea of Okhotsk have dropped to 3°-5°C or less, and in mid-January to 1°-2°C and less. By the end of December most masu have left the Sea of Okhotsk because of such low temperatures.

Some masu may migrate south in Tatar Strait close to Primore, as there are fish that show a different migration route in the western Sea of Japan the following spring. Masu that migrate from the Sea of Okhotsk to the Pacific Ocean and turn at Cape Nosappu may migrate south in coastal waters of eastern Hokkaido.

Masu generally overwinter in the southern waters of the Sea of Japan. The peak of the pole-and-line fishery in the coastal waters of Japan occurs in January on the southwestern coast of Hokkaido, and in February on the Sea of Japan side of northern Honshu (such as Aomori and Akita prefectures). In coastal waters of Niigata Prefecture, the peak season is from mid-March onward. In the eastern part of Tsugaru Strait, between Hokkaido and Honshu, and on the Pacific Ocean side of Aomori Prefecture, the peak season is in March, somewhat later than on the Sea of Japan side at the same latitude.

Sasaki (1978) reported that fish tagged off Shakotan Peninsula in late January showed a southward migration (Figure 16), and Fukataki (1976b, 1970) indicated that those tagged adjacent to Sado Island in mid- and late March showed a northward migration. These results suggest that northward migration of masu commences between January and March. The lowest water temperatures are observed in February and early March, which corresponds to the change in migration direction from southward to northward.

Masu tagged off Shakotan Peninsula in January were recovered in various localities. Some appeared to spend the winter in waters on the west coast of Hokkaido and to start their spawning migration when spring arrived. Others were recovered in waters as far south as 40°N in the central Sea of Japan or in the vicinity of Sado Island. Because these recoveries were made in April, it is possible that these fish had migrated even farther south or west during February and March and had then started to move northward in April.

The distribution of masu in winter, as inferred from the distribution of surface water temperature, suggests that a migratory corridor running from the west coast of Hokkaido obliquely south to the central and southern Sea of Japan exists from November to February. Some migrants move south along the coast of Hokkaido to the west coast of Honshu (Figure 13), whereas others enter Tsugaru Strait and then migrate south along the east coast of Honshu (Machidori and Kato 1984).

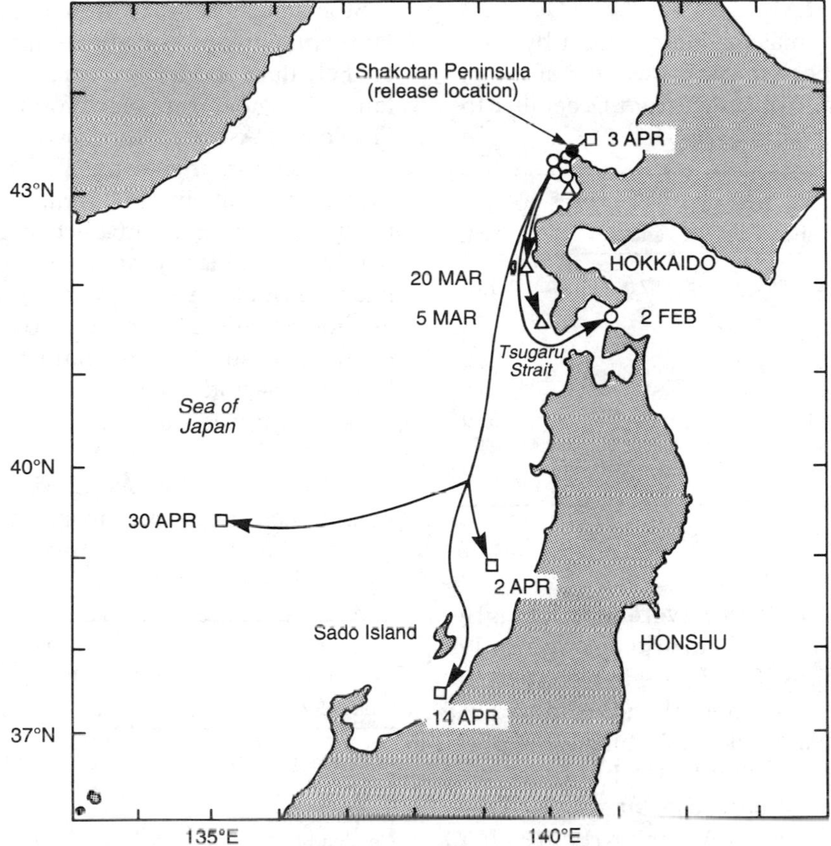

FIGURE 16

Locations of recovery of masu released in January 1977 and recovered in February (o), March (Δ), and April (□) of the same year. (From Sasaki 1978)

There is no direct evidence that masu migrate from the Sea of Okhotsk through Nemuro Strait and then turn south at Cape Nosappu to continue on the Pacific Ocean side of Hokkaido. One large masu (43 cm fork length and 1.2 kg in weight) was incidentally caught by a trawl net in the coastal waters of Tokachi (in the vicinity of 42°N, 144°E) in mid-January. Masu larger than those on the Sea of Japan side in the same season are sometimes caught in waters adjacent to Cape Esan (on the eastern tip of Hokkaido in Tsugaru Strait) after December. These large masu may be passing from the Pacific Ocean side through Tsugaru Strait to the Sea of Japan. Since there is also a group of relatively large fish that migrate north along the Pacific Ocean coast of Honshu in May, it is assumed that the fish migrating southward along the Pacific Ocean side of Hokkaido may spend a winter in the Pacific Ocean south of Miyagi Prefecture (Machidori and Kato 1984).

On the east coast of the Korean peninsula, masu are caught by setnets from February to about June. The relation between the coastal southward migrants from February to March and the migrants that showed a northward migration in the western Sea of Japan during April to May is not known. However, it is possible that some of the southward moving masu migrating early in the year overwinter in the southwestern Sea of Japan and then move northward later in the season along the western side of the Sea of Japan.

The sex composition of 395 masu caught in the coastal waters of the Shakotan Peninsula off the west coast of Hokkaido from December to February in 1976–78 was 65% females and 35% males (Table 33), and remained relatively stable through-

TABLE 33

Sex composition of masu salmon caught by the pole-and-line fishery in coastal waters of the Shakotan Peninsula, Hokkaido, from December to February, 1976–78

Date	Number of fish	Sex composition (%)	
		Female	Male
Late Dec	23	87.0	13.0
Early Jan	–	–	–
Mid-Jan	29	62.1	37.9
Late Jan	80	67.5	32.5
Early Feb	98	61.2	38.8
Mid-Feb	103	62.1	37.9
Late Feb	62	66.0	34.0
Total	395	65.0	35.0

Source: From Sasaki (1978)

out this period (Sasaki 1978). Average fork length of the above masu was 34.3 cm and average body weight was 553 g (Table 34). These fish were 8 cm longer and weighed 300 g more than the immature fish caught in the Sea of Okhotsk during the previous July to September (Table 31), but their size was similar to that of immature masu caught in the southern Sea of Okhotsk in August to September and off northeastern Sakhalin in September. As mentioned above, the average fork length and average body weight of immature masu do not show much change from December to February, indicating a cessation of growth during this period.

TABLE 34

Fork length and body weight of immature masu salmon caught in coastal waters of the Shakotan Peninsula, Hokkaido, from December to February, 1976–78

	Number of fish	Average fork length (cm)	Average weight (g)
Late Dec	23	33.7	540.3
Early Jan	–	–	–
Mid-Jan	29	33.0	468.4
Late Jan	80	34.1	516.5
Early Feb	98	34.3	580.7
Mid-Feb	103	34.8	582.2
Late Feb	62	34.6	552.0
Total	395	34.3	553.0

Source: From Sasaki (1978)

Stomach contents of masu caught near Shakotan Peninsula during the winter months of December to February consisted predominantly of sand lance, the most important food item in this region (Table 35). As in spring, two or three sand lance were found in the stomachs of individual masu, indicating that, in this area, masu were actively feeding in winter. Sand lance migrate for spawning from mid- and late January along the coastal areas near the Shakotan Peninsula (Sasaki 1978). When anchovies migrate to the coastal areas in December, masu also eat quantities of this species, but the period in which anchovies remain in coastal waters is short.

TABLE 35

Stomach contents of masu salmon caught in coastal waters of the Shakotan Peninsula, Hokkaido, from December to February, 1976–78

Stomach contents	Composition (%)					
	Late Dec	Mid-Jan	Late Jan	Early Feb	Mid-Feb	Late Feb
Plankton	–	32.8	8.1	2.2	1.9	4.8
Euphausiids	–	27.6	6.9	0.8	0.7	2.4
Amphipods	–	5.2	–	1.4	1.2	2.4
Copepods	–	–	1.2	–	–	–
Fishes (total)	100.0	15.5	59.4	85.5	59.7	75.0
Ammodytes personatus	56.5	13.8	59.4	84.7	59.5	75.0
Engraulis japonica	43.5	–	–	–	–	
Hexagrammos sp.	–	1.7	–	–	0.2	–
Hemilepidotus sp.	–	–	–	0.8	–	–
Squid	–	–	–	–	0.5	7.3
Fish eggs	–	3.4	–	–	–	–
Miscellaneous (digested)	–	48.3	32.5	12.3	37.9	12.9
Number sampled	23	29	80	98	103	62

Source: From Sasaki (1978)

The percentage of crustaceans in the stomach contents of masu is low and they are generally a secondary food item. Sometimes a fairly high percentage of euphausiids occurs in masu stomachs in mid-January.

Adult Masu

Horizontal Distribution and Oceanographic Conditions. Seasonal changes in catch in the pole-and-line fishery and oceanographic conditions in the Sea of Japan indicate that masu migrate from north to south in the Sea of Japan from December to about February, and reach the southern or central part of the Sea of Japan by March. February and March are the months of lowest water temperature in the Sea of Japan. The 0°C isotherm extends into Tatar Strait and the southwestern coast of Sakhalin, and the Sea of Okhotsk is covered by ice.

A review of the distribution of masu relative to surface water temperatures based on gillnet catches and oceanographic observations in the Sea of Japan in 1970–72 suggests that there is a close relationship between the two. Cold water in the north and warm water in the south form a distinctive polar front with marked temperature gradients (Figure 17). The centre of distribution of masu in March is an area sandwiched between the offshore frontal zone on the south and the polar front on the north. Isotherms in the polar front area generally run in a northeasterly direction but move in a zigzag direction in the main area of concentration of masu.

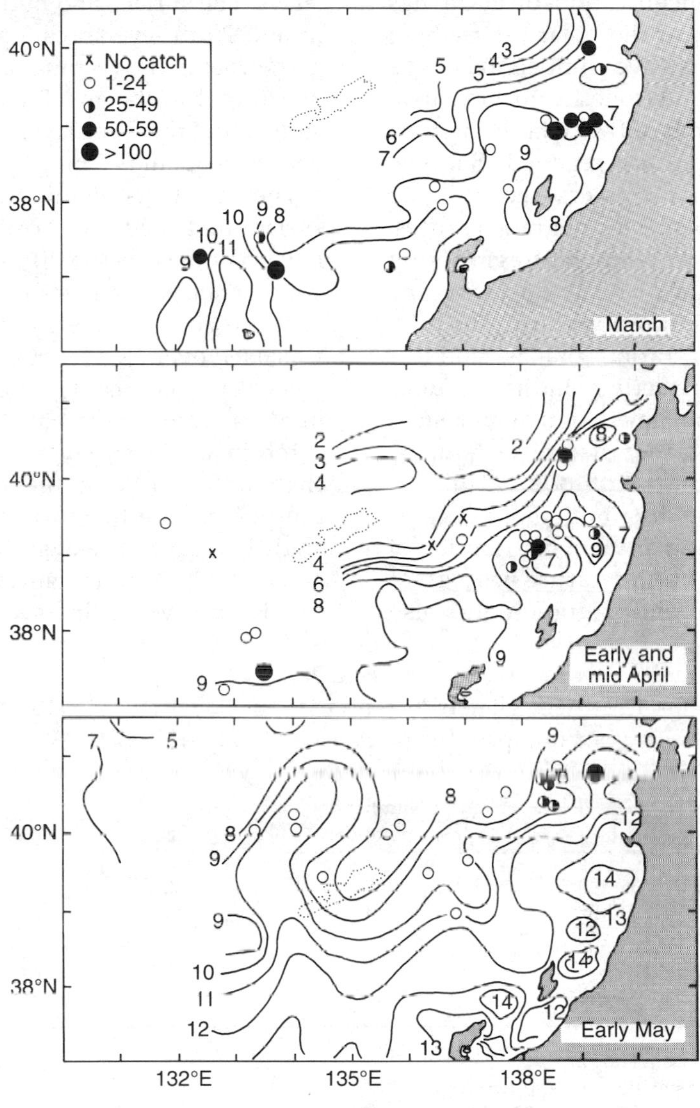

FIGURE 17

Distributions of catch of masu per 100 tans of gillnet, and surface water temperatures (°C) in the Sea of Japan in the spring of 1970. (From Machidori and Kato 1984)

In waters north of Sado Island where masu are concentrated in March, the polar front is close to the coast of Aomori Prefecture or near Oga Peninsula of Akita Prefecture. To the south of this front there is an oblong-shaped area of mixed water. In waters of 3°-6°C, where the horizontal temperature gradient is large, masu are seldom observed. This situation still exists by early April, with a high density of masu in the warm southern waters along the polar front where the temperature is about 8°-11°C.

In May, the main concentration of masu has moved to the north side of the polar front where sea surface temperatures have climbed to 7°-8°C or even higher. It is inferred from this that masu do not migrate to areas north of the polar front zone until these temperatures are reached. Because masu return to their home rivers according to certain schedules and by specific migration routes, it is possible that low water temperatures in spring may constitute temporary obstacles, delaying the fish at places along the migration route. The relationship between concentrations of masu and water temperature, either directly or indirectly, does not always hold. It is assumed that under such circumstances, elements other than water temperature, such as the abundance of food organisms, may have an effect on the distribution of fish.

In general, the limits and areas of distribution of masu suggest that water temperature has a great influence on distribution and migration. It is also believed that water temperature influences regional changes in fish density, but this aspect needs further clarification.

Depth Distribution. Observations on the swimming depths of masu at night were obtained with surface and mid-layer gillnets by the Niigata Prefectural Fisheries Experimental Station in offshore waters of the Sea of Japan in April and May of 1972-74. Two gillnets of 5.6-5.8 m depth were fixed at target depths with hang ropes at 0, 10, and 20 m depth levels. The surface and mid-layer gillnets (at the 10 m and 20 m levels, respectively) were linked together, set in the evening, and hauled the following morning (Karube and Takahashi 1975a). It was assumed that the three layers of gillnets caught fish at depths of about 0-6 m, 10-16 m, and 20-26 m.

The first tests, conducted with gillnets at the surface and at 20 m in April 1972, resulted in much lower catches in the 20 m net (12%) than in the surface net (88%). In subsequent sampling only a combination of gillnets at the surface and at 10 m was used over a wide area in the Sea of Japan. The tests were conducted in the same season each year in about the same areas. In late April, the masu catch in the 10 m layer gillnet was higher (68%) than in the surface gillnet (32%). In other seasons, the percentage taken by the surface gillnet was slightly higher. The overall average percentage of catch was 48% for the surface gillnets and 54% for the 10 m layer gillnets (Table 36) (Karube and

TABLE 36

Catch of masu salmon in surface gillnets (0-6 m) and mid-layer gillnets (10-16 m) fished in the Sea of Japan during April and May, 1972-74

Period	Area	Number of operations	CPUE	CPUE proportions by depth 0-6 m*	10-16 m†
Mid-Apr	38-41°N, 136-140°E	8	30.2	0.528	0.472
Late Apr	39-40°N, 133-137°E	4	43.3	0.319	0.681
Mid-May	39-41°N, 135-136°E	9	2.4	0.545	0.455
Late May	40-43°N, 132-135°E	7	8.4	0.556	0.444
Throughout	38-43°N, 132-140°E	28	17.8	0.458	0.542

Source: From Karube and Takahashi (1975a)

Notes:

$$\text{CPUE} = \frac{\text{*Numbers caught per 100 tans at 0-6 m depth}}{\text{Number of research operations}}$$

$$\frac{\text{†Numbers caught per 100 tans at 10-16 m depth}}{\text{Number of research operations}}$$

Takahashi 1975a), and the ratios of numbers of masu caught by gillnets at the surface, 10 m, and 20 m were 1.0:1.18:0.14.

The gillnets were designed to have neutral buoyancy at the depths fished, but it was difficult to maintain proper distance between the hanging ropes, and there was a tendency for the net to bag because of variations in current strength at different depth levels. It is believed that the results may underestimate the relative abundance of fish in deeper layers (Karube and Takahashi 1975a).

Images of masu that appeared on the fish finder in offshore waters in the Sea of Japan extended to 40 m during the day but were found in shallow layers from evening to night, with many images appearing at depths shallower than 20 m after sunset (Karube and Takahashi 1975b).

Biological Information. The fork length frequencies of masu caught in offshore waters of the Sea of Japan in March, April, and May from 1962 to 1972 are shown in Figure 18. The frequencies are not

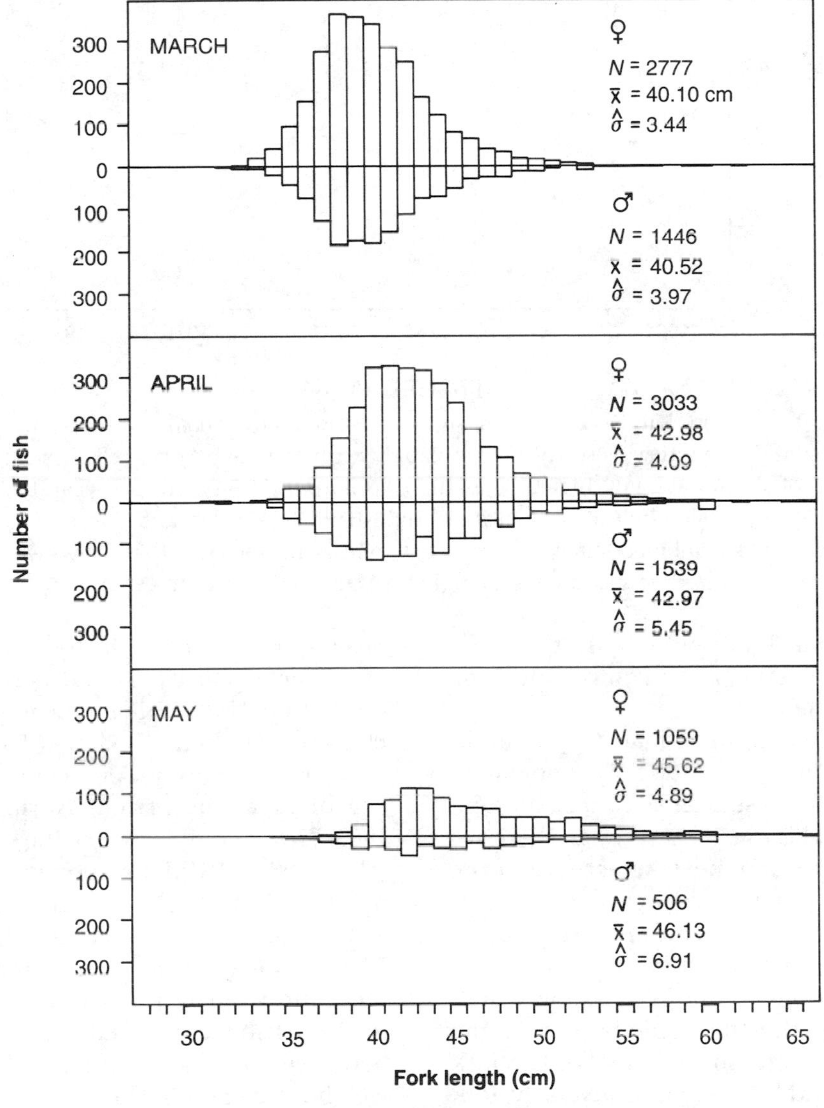

FIGURE 18

Monthly fork length distributions for adult masu caught with gillnets by research vessels in offshore waters in the Sea of Japan, 1962–72. (From Machidori and Kato 1984)

symmetrical but skewed to larger sizes. The difference between maximum and minimum sizes of individuals was fairly large, ranging from 31 to 53 cm in March, 30 to 60 cm in April, and 34 to 60 cm in May. The average of males and females over the eleven years of sampling was similar.

The average fork lengths and their 95% confidence intervals for samples obtained in April between 1962 and 1972 show that there are interannual variations in body length. The maximum average fork length of masu in April, for sexes combined, was 45.1 cm in 1968, and the minimum was 40.2 cm in 1965, a difference of 4.9 cm. Minimum average lengths were found in 1965 and 1971 (Figure 19). The lowest average sea surface temperatures in the Sea of Japan occurred in 1963 and 1969 (Naganuma 1977), which are the years when most of the masu salmon sampled in 1965 and 1971 were hatched.

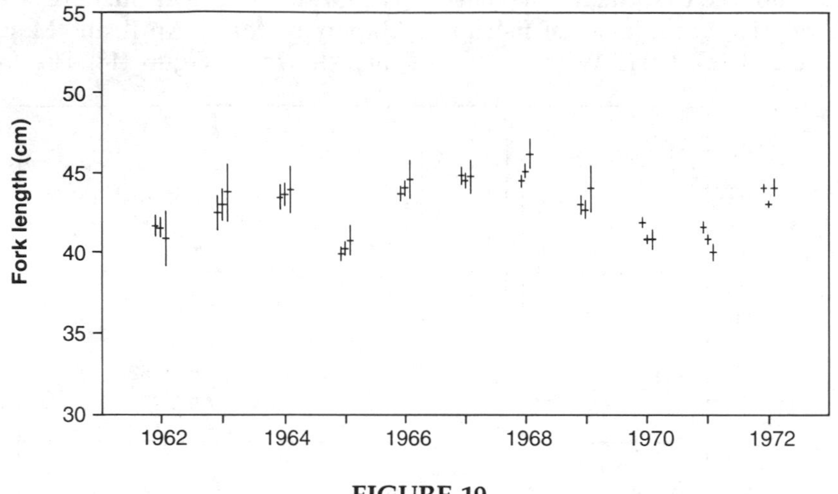

FIGURE 19

Average fork lengths of adult masu by year (horizontal lines) and their 95% confidence intervals (vertical lines) caught with gillnets by research vessels in offshore waters of the Sea of Japan in April from 1962–72. In each year, the left-hand side indicates females, the right-hand side males, and the centre indicates combined female and male fish, which includes fish that were measured only and not sexed. (From Machidori and Kato 1984)

The average body weight of masu captured in spring for the eleven years (1962–72) pooled increased regularly as the season advanced by 1 cm per ten-day period (Figure 20). Although there are problems with respect to mesh selectivity of the gillnets used, such increases in average body weight reflect the growth of fish during the season.

The geographical distribution of average fork lengths of masu in the Sea of Japan shows that, from March to May, the closer the fish are to the northwest coast of Honshu, the smaller the fork length (Machidori and Kato 1984). Conversely, fork lengths are greater in waters west of the central Sea of Japan. However, the average fork lengths of masu are small in the western ridge areas of the Sea of Japan. Although the distribution of different-sized masu by period is complicated, it appears that the smaller fish migrate north along Honshu Island and the larger fish along Primore.

The range in body weight of masu collected in the Sea of Japan in March, April, and May during 1962–72 was similar. Size of the smallest individuals (above 500 g) did not change during the three months, but unusually large individuals with weights up to 4,000 g in March, 5,000 g in April, and 7,100 g in May, which were greatly different from the mode, were frequently observed. The values for the large fish were 4.3, 4.0, and 5.0 times the average body weight in each month. The range in body weights was similar for each sex, but the average weight of males in all three months of sampling was larger than that of females. Differences between males and females ranged from 29 to 144 g.

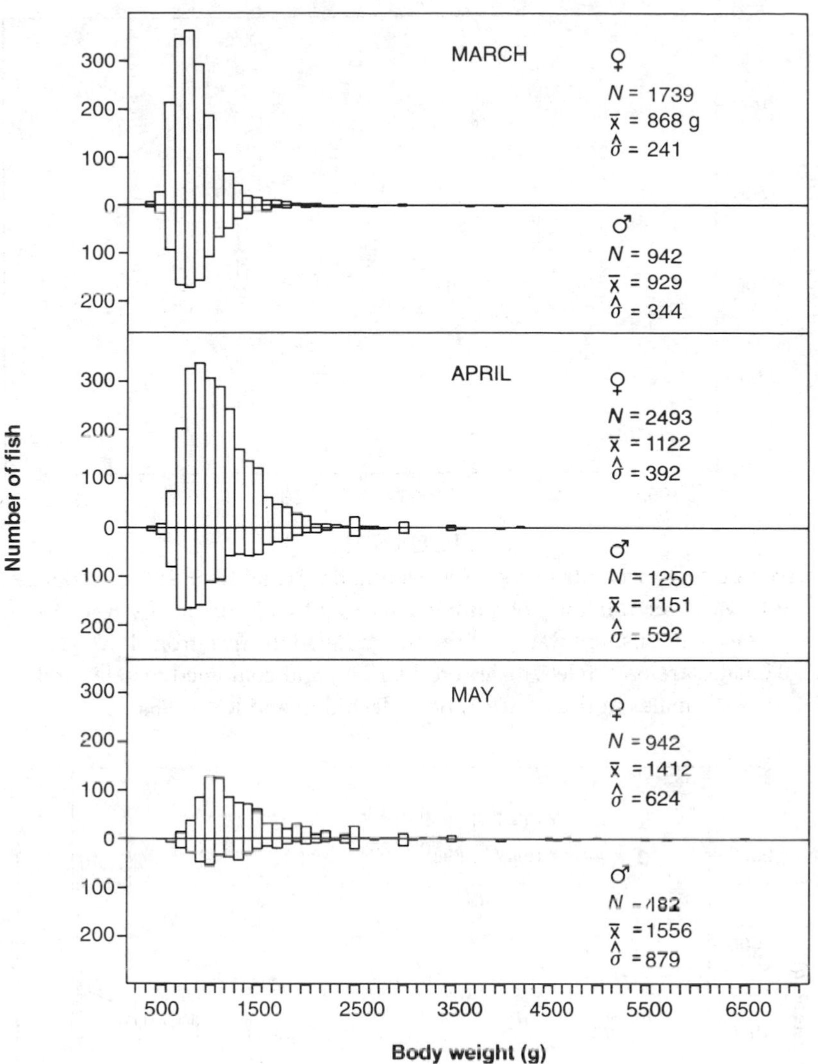

FIGURE 20

Body weight composition by month of adult masu caught with gillnets by research vessels in offshore waters of the Sea of Japan, 1962-72. (From Machidori and Kato 1984)

The average weights and 95% confidence intervals for samples obtained in April of each year from 1962 to 1972 for males and females combined are shown in Figure 21. The weights ranged from a low of 867 g in 1965 to a high of 1,412 g in 1967. The minimum average weights were recorded in 1965 and 1971, which were also years of minimum average fork length. Similar to the case for fork lengths, the average weight of masu in each month from March to June showed a positive correlation with the average weight in April, and it is apparent that the average weight of the fish early in the fishing season is related to the weight later in the season.

The seasonal change in average weight was exponential (Figure 22). The average weight per ten-day period increased by 67-80 g for females and 80-100 g for males in March, 87-102 g for females and 111-134 g for males in April, and 111-131 g for females and 149-181 g for males in May. The weight of males increased more rapidly than that of females.

The geographical distribution of average

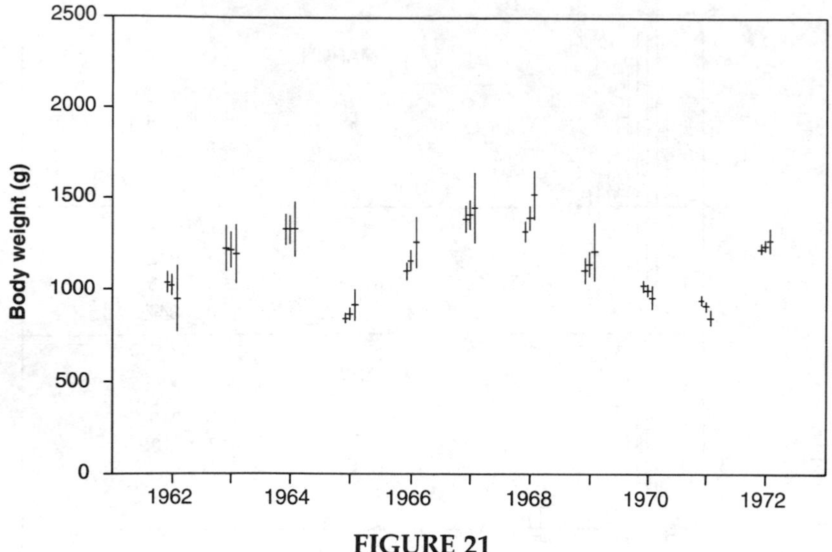

FIGURE 21

Average body weights by year (horizontal lines) and their 95% confidence intervals (vertical lines) of adult masu caught with gillnets by research vessels in offshore waters of the Sea of Japan in April from 1962–72. Females are on the left, males on the right, and combined females and males in the centre. (From Machidori and Kato 1984)

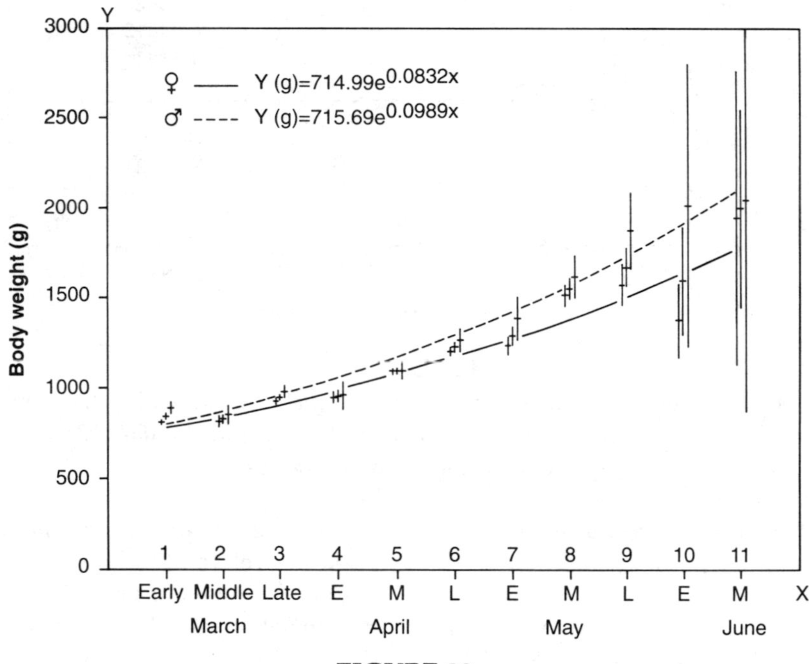

FIGURE 22

Average body weights by ten-day period (horizontal lines) and their 95% confidence intervals (vertical lines) of adult masu caught with gillnets by research vessels in offshore waters of the Sea of Japan, 1962–72. Females on the left, males on the right, and combined females and males in the centre. (From Machidori and Kato 1984)

weights of masu in the Sea of Japan shows that the closer the fish are distributed to the northwest coast of Honshu, the lower the average weight. Conversely, in areas west of the central Sea of Japan, the average weight is greater (Machidori and Kato 1984).

The average gonad weights and their 95% confidence intervals for masu caught in April of each year from 1962 to 1972 varied considerably during the eleven years (Figure 23). The average gonad weight of females in late April was smallest at 16.4 g in 1965 and largest at 37.4 g in 1968, and for males, the smallest was 2.3 g in 1965 and the largest was 13.0 g in 1963 (Machidori and Kato 1984, Table 60). Females had low average gonad weights in 1965 and 1970 and yearly variations which were similar to those of average fork length. Ovaries were relatively large at the northern limit of distribution in early May, in the western areas in mid-May, and in the areas north of 42°N in late May (Machidori and Kato 1984).

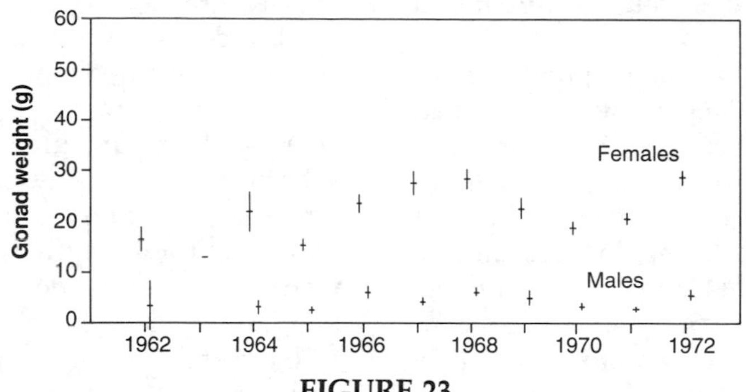

FIGURE 23

Average gonad weights by year (horizontal lines) and their 95% confidence intervals (vertical lines) for adult masu caught by gillnets in offshore waters of the Sea of Japan in April from 1962-72. (From Machidori and Kato 1984)

The average gonad weight of masu in the Sea of Japan from 1962 to 1972 during March to May increased exponentially with time for both sexes (Figure 24). The average gonad weights per ten-day period for females increased by 2.1-3.6 g in

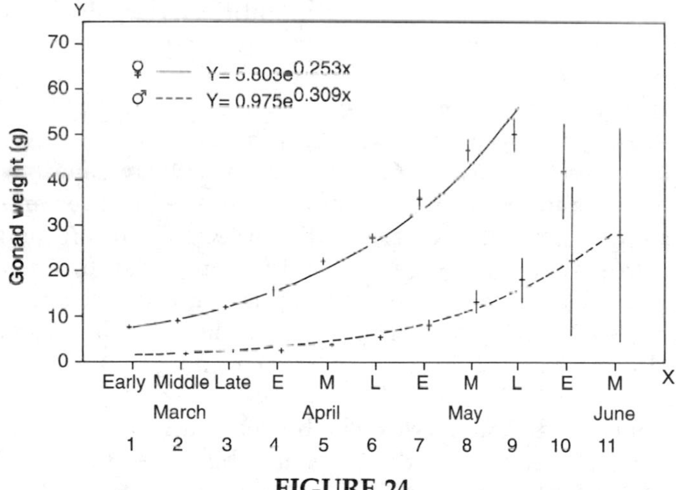

FIGURE 24

Average gonad weights by ten-day period (horizontal lines) and their 95% confidence intervals (vertical lines) for adult masu caught with gillnets by research vessels in offshore waters of the Sea of Japan, 1962-72. (From Machidori and Kato 1984)

March, 4.6–7.6 g in April, and 9.8–12.6 g in May. For males, the increase was 0.5–0.9 g in March, 1.2–2.3 g in April, and 3.1–5.7 g in May.

No particular variations by area were observed in average gonad weights of males, although in the northern limit of distribution in early and mid-May, masu with large testes weights were found (Machidori and Kato 1984).

Kato (1971) reported that the fecundity of 147 female masu caught in 1969 in the Sea of Japan ranged from 1,100 to 5,600 eggs, with an average of 2,600 eggs and a mode around 2,200–2,400 eggs (Table 37). These eggs ranged in diameter from 1.1 to 2.4 mm in March, 1.3 to 4.2 mm in April, and 2.1 to 5.2 mm in May. Seasonal trends in egg numbers were not observed. Thirteen masu females captured by Ogata (1960) in the Sea of Japan in 1959 carried an average of 3,244 eggs.

TABLE 37

Average fecundity of adult masu salmon caught in the Sea of Japan

Area	Number of fish	Fecundity	
		Average	Range
Polar front area	147	2610	1100-5600
Adjacent to Okushiri Island	21	2054	1161-2726
West coast of Sakhalin	-	1900	739-2683

Source: From Machidori and Kato (1984)

The sex composition of adult masu sampled in the Sea of Japan from 1962 to 1972 was, on the average, 33.7% males and 66.3% females for all years (Table 38). The average percentage of males was highest in 1971 (36.0%) and lowest in 1962 (31.4%). No obvious seasonal changes were observed in sex ratios by ten-day period. The sex composition in offshore waters in the Sea of Japan was 33.5% males from data collected in 1957–61 (Watanabe and Ouchi 1962) and 34.8% males for data collected in 1965–69 (Kato 1971). This is very similar to the results from the 1962–72 data presented above. It can be concluded that the sex composition of adult masu in offshore waters of the Sea of Japan shows little annual fluctuation.

Food. Masu eat mainly small-sized fish and squid, and large-sized plankton, including amphipods and euphausiids, during their northward migration in offshore waters of the Sea of Japan from the wintering grounds to the spawning regions (Table 39). Fish found in the stomachs were Japanese pearlside (*Maurolicus japonicus*), saury (*Cololabis saira*), sand lance, Atka mackerel (*Pleurogrammus azonus*),[2] sailfin sandfish (*Arctoscopus japonicus*), anchovies, greenlings (*Hexagrammos otakii* or *H. agrammus*), and sculpins (*Hemilepidotus sp.*). The first five species constituted high percentages in the weight of stomach contents (Table 40). Japanese pearlsides, sculpins, and greenlings found in the stomachs of masu were generally 5 cm in length or less. Japanese pearlsides are small fish, ranging from 1.5 to 5 cm in total length, that inhabit mid-layers in waters south of the polar front. Sand lance, sandfish, and anchovies were mainly 10–20 cm, and saury were 20–30 cm in length. Various sizes of Atka mackerel were eaten by masu. Lengths ranged from about 2 to 23 cm with peaks at 4–5 cm and 12–16 cm (Fukataki et al. 1961).

Crustaceans (amphipods and euphausiids) were also important food for masu in offshore areas of the Sea of Japan and constituted 44.1% of the weight of stomach contents of samples in 1965–66. Of the amphipods, both *Parathemisto japonica* and *Primno macropa* were eaten by masu, but the former was predominant. Of the euphausiids, *Thysanoessa longipes* was more abundant than *Euphausia pacifica*. Other crustaceans, such as larval decapods and copepods, were also observed but in negligible quantities. The planktonic crustaceans that appeared frequently in stomachs were relatively large in size.

The average weight of stomach contents of mature masu in offshore waters of the Sea of Japan was 5.5 g. On average, 68% and 57% of the individuals sampled in 1965 and 1966, respectively, had 5 g or less of food in their stomachs. Such a high rate of empty stomachs may be the result of the fact that the samples were obtained at night, and the masu may have been tangled in the gillnets from evening to the following morning.

The type of food in stomachs of masu changed by season. In March, amphipods were predominant, but after mid-April, fish were often more

2 *Editors' Note*: Atka mackerel is the recognized common name of the closely related *Pleurogrammus monopterygius*. Apparently no English common name exists for *P. azonus*, hence, the author used Atka mackerel. The Japanese name for this fish is "hokke," and for *P. monopterygius* it is "kitano-hokke."

TABLE 38
Percentage of males among adult masu salmon caught in gillnets by research vessels in offshore waters of the Sea of Japan, 1962–72

	Male masu (% of catch)											
	March			April			May			June		
Year	Early	Mid	Late	Early	Mid	Late	Early	Mid	Late	Early	Mid	Average
1962	–	–	35.3	24.2	0.0	20.0	47.1	31.0	–	–	–	31.4
1963	–	–	–	–	40.5	22.6	32.6	–	–	–	–	34.6
1964	38.2	13.3	44.5	100.0	34.7	13.3	33.3	100.0	–	–	–	35.9
1965	50.0	33.3	42.9	24.2	31.2	33.3	47.4	35.0	22.2	–	–	34.8
1966	37.4	10.5	34.3	42.9	40.7	19.7	28.6	0.0	0.0	16.7	–	35.8
1967	33.3	37.1	30.6	–	29.6	38.6	34.6	29.2	25.0	42.9	–	34.1
1968	51.4	27.2	32.1	28.6	35.7	36.4	40.9	44.4	–	66.7	–	34.8
1969	32.2	30.4	36.6	26.7	30.8	28.3	26.9	43.9	34.3	–	–	31.8
1970	20.7	31.0	33.0	31.3	38.3	34.9	53.8	21.3	33.8	31.6	40.0	32.7
1971	23.3	29.5	45.3	44.9	37.0	32.1	40.0	35.8	28.8	50.0	–	36.0
1972	29.6	34.3	34.0	28.6	29.7	34.8	53.8	21.3	33.8	0.0	66.6	32.1
Total	36.2	29.6	34.8	31.6	34.2	33.3	34.6	30.8	30.5	34.0	50.0	33.7

Source: From Machidori and Kato (1984)

TABLE 39
Weight composition of stomach contents of adult masu salmon caught in the Sea of Japan

			Stomach content weight composition (%)				
Area and period	Number of fish	scw* (g)	Fish	Squid	Amphipods	Euphausiids	Other†
Sea of Japan offshore area, March to June, 1965–66	493	5.5	41.3	7.5	27.4	16.7	7.1
Off southwest coast of Hokkaido, April to May, 1955	213	7.9	36.9	12.8	26.7	23.5	0.1

Source: Sea of Japan, Fukutaki (1967a, 1969); Hokkaido, Hokkaido Fisheries Experimental Station and Hokkaido Regional Fisheries Laboratory (1956)
Notes: *Average stomach content weight per fish
†Copepods, decapod larvae, pteropods, and digested items

TABLE 40
Seasonal change in weight composition of stomach contents of masu salmon caught by gillnets in offshore waters of the Sea of Japan, March to June, 1965–66

	Weight composition (%)*									
	March			April			May			June
Stomach contents	Early	Mid	Late	Early	Mid	Late	Early	Mid	Late	Early
Fish	28.5	41.0	21.0	20.7	54.1	35.4	22.8	31.1	51.3	68.5
Engraulis japonica	–	–	–	–	1.1	–	–	–	–	–
Maurolicus japonicus	5.1	5.5	0.6	–	–	0.1	–	–	–	–
Cololabis saira	19.6	–	–	–	21.5	–	–	11.6	–	46.7
Arctososcopus japonicus	–	30.6	–	–	12.1	15.1	–	–	–	–

(continued on next page)

TABLE 40 (continued)

Stomach contents	Weight composition (%)*									
	March			April			May			June
	Early	Mid	Late	Early	Mid	Late	Early	Mid	Late	Early
Ammodytes personatus	–	4.9	0.2	–	11.5	16.0	–	–	–	–
Pleurogrammus azonus	1.6	–	10.8	11.5	6.0	3.1	15.8	16.9	49.9	21.8
Hexagrammos sp.	–	–	–	–	0.0	–	–	–	–	–
Hemilepidotus sp.	–	–	–	–	0.0	–	–	–	–	–
Unidentified (digested)	2.2	–	9.4	9.2	1.9	1.1	7.0	2.6	1.4	–
Squid	–	0.8	0.1	13.4	7.9	10.2	56.0	7.6	23.1	–
Amphipods	61.9	50.2	53.9	47.3	8.2	18.0	3.8	9.7	8.9	–
Parathemisto japonica	61.4	49.9	53.8	46.5	6.9	17.6	3.5	9.4	6.9	–
Primno macropa	0.5	0.3	0.1	0.8	1.3	0.4	0.3	0.3	2.0	–
Euphausiids	1.4	–	15.9	–	25.6	30.1	9.4	28.5	14.4	30.6
Thysanoessa longipes	–	–	15.8	–	25.6	30.1	9.4	28.5	14.4	30.6
Euphausia pacifica	1.4	–	0.1	–	–	0.0	–	–	–	–
Decapods	–	–	–	–	–	0.0	–	–	–	–
Chionoecetes zoeae	–	–	–	–	–	0.0	–	–	–	–
Chionoecetes megalops	–	–	–	–	–	0.0	–	–	–	–
Copepods	–	–	0.1	–	0.0	0.0	–	–	–	–
Calanus cristatus	–	–	0.0	–	0.0	0.0	–	–	–	–
Euchaeta elongata	–	–	0.0	–	–	–	–	–	–	–
Terrestrial insects	–	–	–	–	0.0	0.0	–	7.7	0.0	–
Digested items	8.2	8.0	9.0	18.6	4.2	6.3	8.0	15.4	2.3	0.9
Number of fish	93	27	50	30	108	116	15	32	15	7
Average stomach content weight composition per fish (g)	6.3	5.3	5.2	2.5	9.4	9.5	5.8	4.9	12.0	16.6

Source: From Fukataki (1969)
Note: *0.0% implies less than 0.05%

important. Of the crustaceans, the amphipod *P. japonica* was a major food item until early April. The relative importance of crustaceans declined after mid-April and shifted to the euphausiid *T. longipes*. Squid became relatively important after mid-April.

Fish were important as food organisms from March through June but the dominant species varied by season. Japanese pearlsides were only eaten by masu in early and mid-March, and sandfish and sand lance were eaten mainly from mid-March to late April. The relative importance of Atka mackerel in stomachs increased after May. Saury were eaten from early March to early June but not consistently. Terrestrial insects were eaten only in April and May.

The Japanese pearlsides are one of the most important food items for masu in their wintering period but do not appear to be utilized during the fastest growing season in spring (about April and May). A decrease in the amount of Japanese pearlsides in the stomachs takes place when masu migrate northward, possibly because they are leaving the region of concentration of this food species.

The availability of crustaceans as food items changes by area, which is reflected in changes in composition of plankton in stomachs. The predominant food by weight in masu stomachs for a number of sampling stations is shown in Figure 25. Amphipods, rather than euphausiids, were the predominant item in waters farther south. *Parathemisto japonica*, the major species of amphipod, favours cold water and mid-layers and is generally ubiquitous in waters affected by the Oyashio Current, which runs down the eastern side of the southern Kuril Islands and Hokkaido. *Thysanoessa longipes* is distributed in the northern Sea of Japan and in the Pacific Ocean north of 40°N, in the cold water areas farther north than *P. japonica*. The change of the major crustaceans eaten by masu from *P. japonica* to *T. longipes* as the season advances suggests that these fish move from the mixed water areas south of the polar front to the polar front or further north in early mid-April.

The areas where squid were predominant in stomachs were limited to the polar front zone.

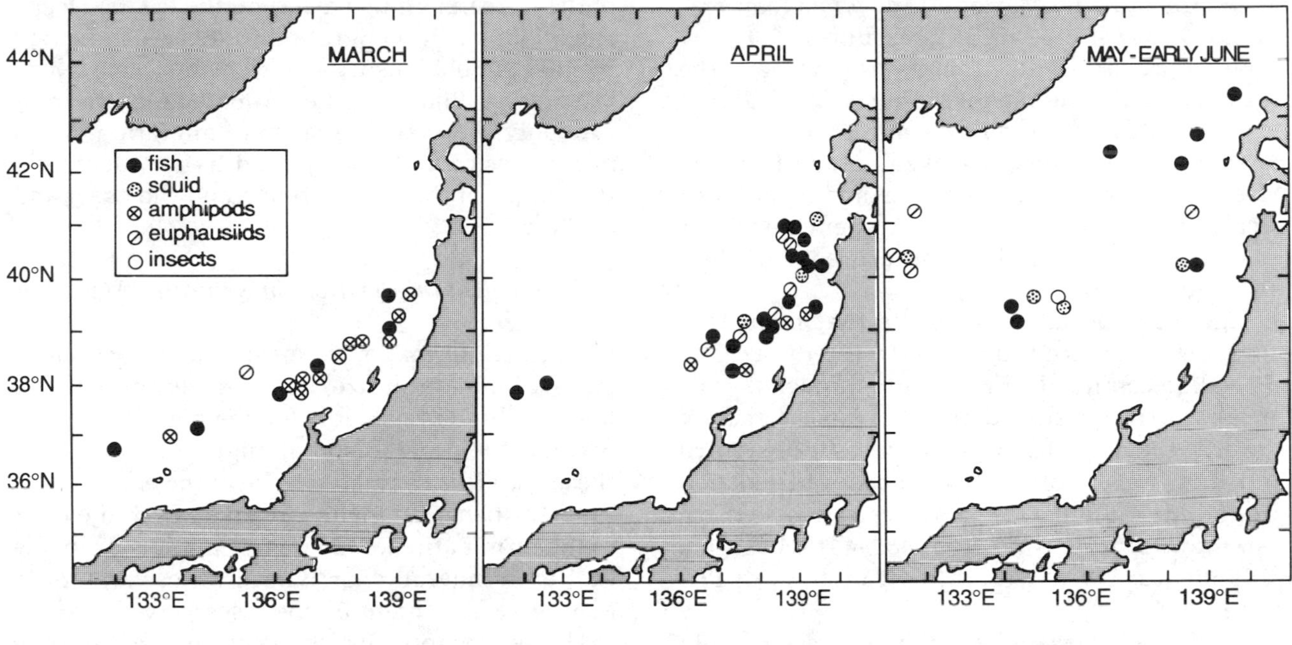

FIGURE 25
Distribution of food organisms of first rank in the stomach contents (weight composition) of masu caught by gillnet during 1965 and 1966. (From Machidori and Kato 1984)

Maturation and Return

Return Migration of Masu

The northward migration of masu in the Sea of Japan begins in the latter half of May when water temperatures are rising. The migratory routes are related to oceanographic conditions, which are governed by a cold water area intruding from the north on the Primore side and by the warm Tsushima Current flowing northwestward into the southern end of the Sea of Japan and extending northward along the west coast of Honshu and Hokkaido. This sets up a SW-NE running frontal zone in the southern Sea of Japan, with high surface water temperatures and high salinity on the Japan side and low sea surface temperatures and low salinity on the continental side.

Tagging and recovery operations show that masu actively migrate in April south of the frontal zone towards the east and then gradually turn north near Noto Peninsula and Sado Island and continue northward along the Honshu coast to the west coast of Hokkaido. Thus, masu from the central and eastern Sea of Japan take a counterclockwise route around the cold water area and more or less follow the Tsushima Current. It is assumed that the masu that winter in the northern colder water area off the west coast of Hokkaido migrate farther north than those that winter in the southern warmer water area. Some masu that overwintered in the Sea of Japan migrated as far as the west coast of Kamchatka, because fish off western Kamchatka were recovered with hooks from the

Sea of Japan longline fishery embedded in them. These masu must have migrated through the Sea of Okhotsk during May and June when temperatures were still 2°-3°C. It is assumed that they stayed in relatively low temperature areas of the Sea of Japan during the preceding winter and that they probably left earlier than masu that returned to spawning streams of Hokkaido and Primore. Masu that originated from streams in southern areas, such as Honshu, may have spent the winter in warm water near the southern edge of their distribution range.

Tagging information suggests that masu that migrate to northern Primore and western Sakhalin also take a counterclockwise route through the Sea of Japan. It is assumed that they move mainly towards the north from waters near 40°N in April and from waters off the west coast of Hokkaido around May (Figure 13). It is uncertain how these fish move in waters north of 43°-44°N but they probably turn west with the current towards Primore.

Masu that migrate to Primore are generally large. Changes in spatial and temporal distribution of such large fish, and their catch per unit effort offshore from the Korean peninsula in early May, indicate that they return from the southern part of the Sea of Japan to southern Primore spawning streams via a clockwise route (Figure 13). There is also an indication that some large masu migrate northwest from the southwestern Sea of Japan and approach Peter the Great Bay.

Masu that migrate from the Sea of Japan to the Pacific Ocean side of Tsugaru Strait pass through the strait during April and May. The release location of tagged fish recovered east of Cape Shirakami (the western extent of the Strait) indicated that they spent a winter in an area south of the polar front, from the west coast of Honshu to the central and southern Sea of Japan. On the Pacific Ocean side, recoveries of these fish have been limited to areas west of Cape Erimo, with no recoveries from the east coast of Honshu and off Hokkaido east of Cape Erimo. This suggests that these fish mainly spawn in streams of southern Hokkaido.

The spawning migration routes of masu shown in a very generalized way in Figure 13 should be viewed with caution. The movements were deduced from tagging results and the time and space distribution of catches. Information is poor in waters from the western Sea of Japan to the east coast of the Korean peninsula and on the Pacific Ocean side of Japan. The migration routes are indicated by lines but obviously the movements occur over a broad front. Similarly, the estimated months indicate only the average season of migration. Masu migrate over a fairly long period, as is evident from the length of the fishing season on the coast and offshore.

Biological Information on Returning Masu

Biological information on masu caught by gillnets (111-121 mm mesh size) in the eastern Sea of Okhotsk in June is available for the period 1969-72. Because the peak season of migration of masu to these waters is estimated to be in the early half of June, the fish examined were taken from the peak to the termination of the migration. It is probable that early migrants, arriving in these waters in May, were not caught by the research vessels.

The sex composition of masu captured in June was 60.4% females and 39.6% males. The percentage of males in the eastern Sea of Okhotsk was higher than that for masu caught in the Sea of Japan. The average fork length for sexes combined was 45.9 cm and the average weight was 1,452 g (Table 41). Although bodies were relatively small, the gonads were well developed, with an average ovary weight of 92 g and an average maturity index of 6.5. Testes of males were small, with an average weight of 24 g and an average maturity index of 1.6. The maturity index for adult masu in the Sea of Japan, by comparison, was 1.72 for females and 0.18 for males in April, and 3.01 for females and 0.44 for males in May (Tanaka 1965). Thus, the masu caught in the eastern Sea of Okhotsk in June were obviously more advanced in maturity than those in the Sea of Japan.

Sockeye and chum salmon that have reached 15-20 g or more in ovary weight and 2-3 g and greater in testes weight around June are regarded as adult fish that will reach full maturity that season (Takagi 1961; Ishida et al. 1961). Because masu caught in the eastern Sea of Okhotsk in June had ovary weights of 35 g or more and testes weights of 5 g or more, they appeared to be adults that would reach maturity in the year in which they were caught.

TABLE 41

Fork length, body weight, gonad weight, and maturity index of adult masu salmon caught by gillnets of 111-121 mm mesh size in the eastern Sea of Okhotsk in June, 1969-72

	Number of fish	Average	Standard deviation	Range
Fork length (cm)				
Female	1054	45.79	2.32	39.2-53.6
Male	690	46.12	3.52	36.2-56.3
All	1744	45.92	2.86	36.2-56.3
Body weight (g)				
Female	1016	1441	240	880-2410
Male	696	1474	362	640-2700
All	1712	1452	296	640-2700
Gonad weight (g)				
Female	1031	91.7	25.9	35-255
Male	704	23.7	13.7	5-100
Maturity Index*				
Female	1031	6.53	1.46	2.2-12.9
Male	681	1.61	0.77	0.3-5.0

Source: From Machidori and Kato (1984)
Note: *Maturity index = (gonad weight/body weight) × 100

Although the average fork length of males was somewhat greater than that of females, the difference was extremely small. However, the ranges in fork length and body weight were much greater for males than for females. The range in size of females was relatively narrow, but for males, extremely small and relatively large individuals were observed. The smallest male was 36.2 cm long and weighed 880 g and the largest was 56.3 cm and 2,700 g, differences of 1.6 times in fork length and 3.1 times in body weight.

The seasonal changes in these biological data showed interesting features. The proportion of male fish decreased slightly between early and mid-June and then rapidly after 20 June, when most of the migration was completed. The average fork lengths and average weights also decreased as the season advanced (Machidori and Kato 1984). This suggests that groups of fish approached the coastal waters and that the groups sampled changed within the period of sampling. It also suggests that large-sized fish migrated first, and small-sized fish, including many females, constituted the tail end of the migration. A difference of about 200 g was observed in the average weight between early June and about 20 June, indicating that there were substantial differences in body size between the early and late migrants. Although the late arriving fish were small in size, maturity was advanced. The ovaries of early and late fish showed almost the same average weight, but ovary weight increased as a percentage of body weight, as did testes weight of males (Machidori and Kato 1984).

Adult masu off the southwest coast of Sakhalin eat Atka mackerel, Japanese pearlsides, and herring (Table 42). Sampling of masu by Japanese research vessels off the southwest coast of Kamchatka in June showed that in this area they mainly eat young pollock (*Theragra chalcogramma*). Semko (1956) reported that masu also eat various coldwater fishes such as capelin (*Mallotus catervarius*), saffron cod (*Eleginus gracilis*), and Dolly Varden charr (*Salvelinus malma*) in coastal areas of Kamchatka.

TABLE 42

Species eaten by adult masu salmon caught in coastal waters of the USSR

Southwest coast of Sakhalin	Southwest coast of Kamchatka
Fish	
Pleurogrammus azonus (Atka mackerel)*	*Mallotus catervarius* (capelin)
Clupea harengus pallasi (herring)	*Eleginus gracilis* (saffron cod)
Cololabis saira (saury)	*Salvelinus malma* (Dolly Varden charr)
Engraulis japonica (anchovy)	
Ammodytes personatus (sand lance)	
Crustacea	
Euphausiids	Mysidacea
Calanus cristatus	
Calanus tonsus	
Hyperiidae	

Source: Sakhalin, Dvinin (1956); Kamchatka, Semko (1956)
Note: *See Editors' Note, p. 498

Coastal and Upstream Migration of Adults

Masu. Masu generally appear in coastal areas in spring but the timing varies considerably according to area.

On the west coast of Kamchatka, from the Bolshaya to the Icha river, migration into the estuarine area takes place between mid-June and mid-July and peaks in early July (Semko 1956). Catches of masu by coastal trap nets are made earlier than those of other species of salmon in the area.

Masu appear on the west coast of southern Sakhalin in mid- and late May. They are most abundant in June and are usually present until early July. Almost as soon as they appear in the coastal waters, they start their upstream migration into the rivers. In general, the timing of the spawning migration of masu is earlier than that of other species of salmon in the area (Dvinin 1952).

There is a tendency for migration to take place somewhat earlier in the south than in the north. In Hokkaido, good coastal catches are made in mid-May and continue until late June. On the Sea of Japan coast of Ishikawa Prefecture (Honshu) the fishing season for the coastal longline fishery is from late February to mid-April. The number of fish taken by coastal trap-nets reaches a peak in March and April. On the Pacific Ocean coast of Iwate Prefecture (Honshu) some catches of masu are, with the exception of the winter months, made throughout the year. The largest catches are made in May, June, and July. A second, smaller peak occurs in September. Along the northern Hamkyong coast of North Korea, masu are caught by trap nets from February to June.

In summary, the coastal migration of adult masu for the Sea of Japan coast of Honshu and North Korea is complete by the end of June; peaks in June and July for Hokkaido, Sakhalin, the Amur River, and the west coast of Kamchatka; and occurs mainly in August and September for the Pacific Ocean coast of Honshu.

The water temperatures observed in coastal waters during the season of spawning runs vary from approximately 5° to 12°C in Hokkaido (Ono 1933) and from 7° to 12°C, with an average of 9°–10°C, on the west coast of Kamchatka (Semko 1956). These temperature ranges in coastal waters where masu are present are similar to those observed in offshore areas where masu are distributed.

The size of masu migrating to coastal waters varies from a minimum fork length of 40 cm (weighing less than 1,000 g) to a maximum of about 70 cm (about 9,000 g). This significant difference in size can probably be attributed to the great variations in the marine environments that masu occupy, from the southern Sea of Japan to the Sea of Okhotsk and the Pacific Ocean side of Honshu and Hokkaido.

Masu generally migrate upstream from spring to summer. The duration of the run is comparatively long, lasting two to three months or more and varying considerably according to area and river.

In the rivers of Honshu, the runs are early with a peak in April to May. In Ishikawa Prefecture (Honshu) upstream migration occurs between February and June (Tabata et al. 1959). In the Tumen River, which has the largest masu run in the Korean peninsula, the run starts in early April immediately after the ice melts, peaks in May, and continues until June (Yoshida 1942). In some rivers on the Sea of Japan side of southern Korea, masu have been found in the rivers from early June to late July (Atkinson et al. 1967) but the timing of the run is not clear.

A late run is found in the Oippe River (Aomori Prefecture) of northeastern Honshu, with the peak occurring about early September. In some rivers on the Pacific Ocean side of Aomori Prefecture, such as the Oirase and Mabechi rivers, masu migrate from late summer to early fall. In the rivers in Iwate Prefecture, masu runs start in May, with a peak in July and another smaller peak in September shortly before the run is completed (Tanaka 1965). Late season runs of masu occur in areas of southern Hokkaido facing Tsugaru Strait. Such late-run (September) stocks are also known to occur in southern Primore and the Posyet District (Pravdin 1949).

The timing of the runs in Hokkaido, northern Primore, and southern Sakhalin, the main reproductive areas for masu, are generally from June to

July. In many rivers in Hokkaido the runs start in April–May, reach a peak in June–July, and terminate in August to September. The seasonal catch patterns for masu in two rivers on the Sea of Okhotsk side of Hokkaido are shown in Figure 26. In these rivers, masu runs occur in spring. River temperatures at the peak of runs are about 9°–16°C (Osanai and Otsuka 1967). In the Samarga River in northern Primore, the masu run starts in early June, after the spring tides, and peaks in late June (Semenchenko 1978). Water temperatures at the start of the run are about 6°–8°C. The upstream migration of masu into the Amur River starts in May and is complete about mid-July (Kryklitin 1962).

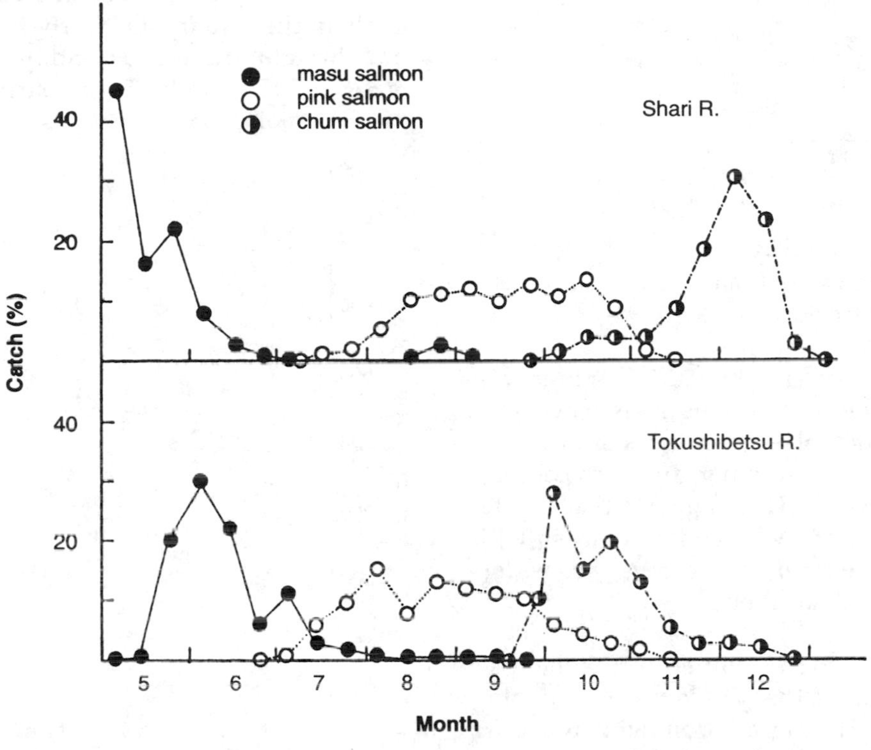

FIGURE 26

Seasonal changes in salmon caught in two rivers on the Sea of Okhotsk side of Hokkaido (average for 1971–74). (From Japan Salmon Resources Conservation Association 1976)

In summary, upstream migration of masu generally takes place earlier in southern than in northern areas. The peaks of runs are also progressively later, moving northward between 36°N and 43°N. In higher latitudinal zones (above 43°N) this relationship is weaker and the peak of the runs occurs through most of June and the first half of July (Figure 27). In many cases, the timing of migration is not well known and only a general relationship can be determined. Taking a broad view, however, it is clear that the timing of runs of masu bears a close relationship to seasonal changes in the Far East.

Variations in timing of runs are large in the vicinity of 41°N–45°N and the duration of runs is somewhat longer than in other areas. At these latitudes, masu runs start immediately after the snow melts, reach a peak from late spring to early summer, and generally decline slowly thereafter. At these latitudes, there also exist masu that show a peak in the run in late summer to early fall (as in the Oippe River and the Posyet District).

Temperatures during summer in the coastal regions of these latitudes reach about 20°C and higher and exceed the water temperatures of zones where masu generally live in offshore waters. River

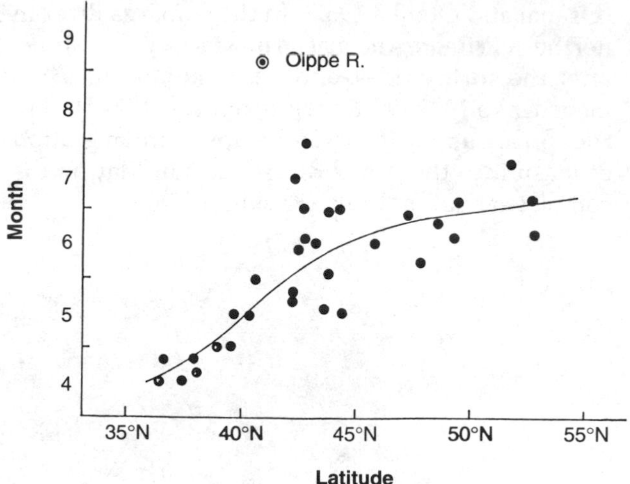

FIGURE 27

Relationship between latitude and peak of run of masu. (From Machidori and Kato 1984)

temperatures also rise to about 20°C or higher in summer. Under these conditions, runs of masu are known to occur at cool temperatures in spring, at warm temperatures in summer, and at cooler temperatures in early fall. This suggests that coastal and upstream migratory behaviour of masu with respect to water temperature can vary considerably in these latitudinal zones.

Amago. In April, when the temperature of the river water rises as high as that of sea water, maturing amago (see Plate 21) migrate from winter feeding areas, such as Ise and Mikawa bays, to estuarine waters. In May, they move to the mouth of the river they descended half a year earlier. Widely migrating or large amago show an early homing migration. Little time is spent in the estuary, and upstream migration starts in mid-April and peaks between mid- and late May (Honjoh 1985). In Wakasa Bay, homing migration of maturing amago starts in mid- and late March with the rise in temperature of sea water.

The mean body weight of amago in Ise and Mikawa bays increases rapidly and reaches 290 g, 367–411 g, and 540–557 g in early, mid-, and late April, respectively (Figure 28). Although the average body weight attains a maximum of 711 g in early May, migrating fish grow very slowly, or not at all, in the estuary and in the stream. The mean body weight of amago ascending the Nagara River is 678 g, 715 g, and 705 g in early, mid-, and late May (Honjoh 1985).

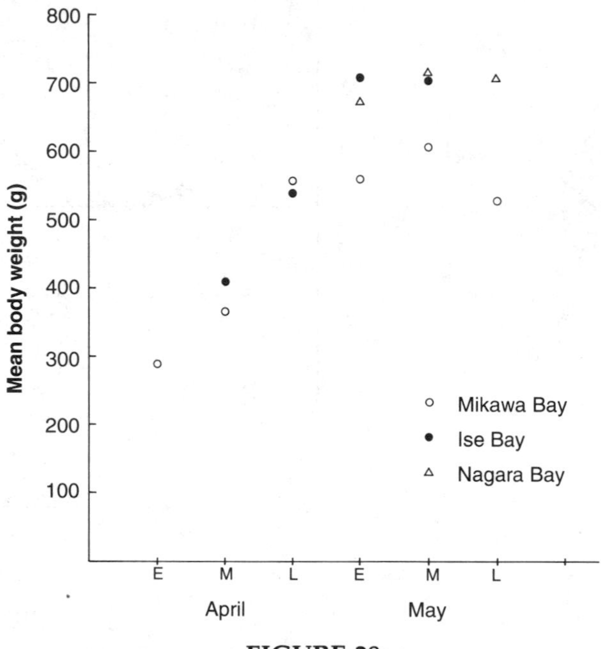

FIGURE 28

Seasonal change in body weight of amago migrating in Ise and Mikawa bays and ascending the Nagara River. (From Honjoh 1985)

Homing and Straying

As with the other species of Pacific salmon, it is believed that the great majority of adult masu and amago return to the streams in which they originated. Some straying, however, does occur, as is clear from marking experiments.

Masu. Mayama et al. (1985) marked 72,700 hatchery-produced masu smolts (averaging 13.8

cm in fork length and 28.0 g in body weight) by removing the adipose fin. They released them into the Mena River, a tributary of the Shiribetsu River system in Hokkaido, during late April to early May in 1981. It was estimated that about 61,000 fish, except resident-type parr, migrated to the sea as smolts. Some returned the following year as three-year-old adults after spending one winter in the ocean. Four hundred and eighty-one marked fish were recovered on the Suttsu coast around the home river during mid-February to late June, and 361 were caught in the Mena River during late August to mid-October in 1982. In autumn of 1983, only one marked female, which had spent two years in the river before seaward migration, returned to the Mena river at age 2.1. Although straying to rivers other than parental streams was not positively demonstrated, it appears that this experiment supports the homing hypothesis.

Amago. Honjoh (1985) presented data on return rates of amago smolts that were marked and released in the Kiso, Nagara, and Miya rivers during 1976–79. Of all fish that returned, more than 95% were recovered as adults in their parental streams, indicating a strong homing tendency (Table 43).

TABLE 43

Recovery composition by river for marked and released smolts of amago

River	Released		Recovered		Recovery composition (%)			
	Year	Number	Number	Per cent	Kiso R.	Nagara R.	Ibi R.*	Miya R.
Kiso	1978	29 488	1 555	5.3	95.82	3.99	0.19	0
Nagara	1976	29 517	1 313	4.4	0.30	99.47	0.23	0
	1977	23 255	2 098	9.0	0.29	95.33	4.39	0
	1979	7 623	965	12.7	0.62	98.55	0.21	0.62
Miya	1977	6 198	1 814	29.3	0	0.39	0	99.61
	1979	8 803	1 530	17.4	0	1.31	0	98.69

Source: From Honjoh (1985)
Note: *A tributary of the Nagara River

Summary of Life History Patterns

Masu. The distribution of anadromous masu is limited to Far East areas, with the centre of abundance in the Sea of Japan and the Sea of Okhotsk (Figure 29). Their summer distribution is in the Sea of Okhotsk, the Sea of Japan is the wintering area, and they migrate between the two areas in spring and autumn. On the Pacific Ocean side of Japan, masu are distributed from Hokkaido to the coastal areas of northern Honshu and rarely occur offshore. No masu are found in the Bering Sea.

The geographical distribution of yamame, the freshwater-type masu, extends farther south than the anadromous form and covers all of Japan, except for some southern parts.

Masu migrate into both large and small rivers. The statistics on catch and escapement are not adequate for all areas. In the early 1960s, about 130,000 masu returned to Hokkaido and some tens of thousands to the Honshu areas annually. Runs of masu to the Korean peninsula are sparse in southern rivers and greater in northern rivers. It is estimated that there are large runs in the northern areas of Primore, but no detailed information exists. In Sakhalin, anadromous masu runs occur in most areas. In the 1960s, about 8,000 masu migrated annually into the Tym River of northeastern Sakhalin. Although it is known that there are rather good runs to the west coast of Kamchatka

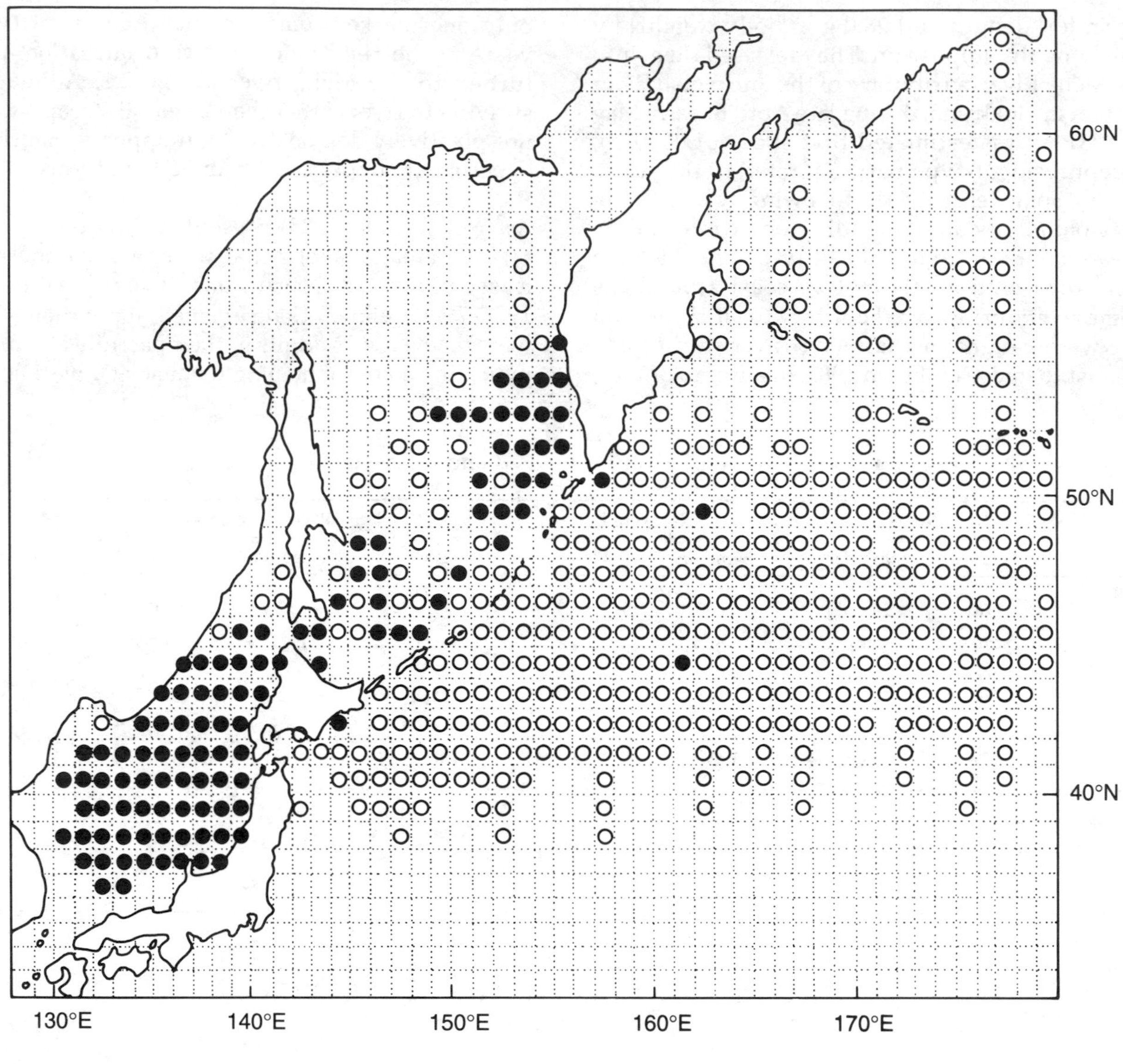

FIGURE 29

Distribution of Japanese research vessel operations and locations where adult masu were found from February to June (black circles). (From Machidori and Kato 1984)

and the southern Kuril Islands, no quantitative information exists.

Masu return to rivers for spawning mainly from spring to summer, returning early in southern and late in northern areas. Runs occur in Honshu about May, and in Hokkaido, Sakhalin, and Primore in June and July (Table 44). On the west coast of Kamchatka, the peak of the runs occurs in the early half of July. The migration period in Japan and Sakhalin continues for two or three months or more, with runs continuing up to the spawning season in autumn. There are peculiar groups that show a peak in migration activity just prior to their spawning season on the Pacific Ocean side of Aomori Prefecture (northern Honshu), the area fronting the eastern Tsugaru Strait of Hokkaido, and the Posyet area of southern Primore.

River water temperatures during upstream migration range from 9° to 16°C in Hokkaido. Migrating masu hold in deep pools in the main stream or

TABLE 44
Variation and range of life history patterns of masu and amago salmon

Life stage	Item	Masu salmon			Amago salmon
		Honshu	Hokkaido	USSR	
Coastal adult	Migration season	Feb–Apr	Mid-May–late Jun	Mid-May–early Jul (Sakhalin)	Apr–May
	Water temperature		5°–12°C	7°–12°C (Kamchatka)	
Upstream adult	Upstream season	April–May (peak) Early Sep (Oippe R.) May–Sep (Iwate Pref.)	Apr–Sep with peak in Jun–July	May–mid-Jul (Amur R.) Early Jun–late Jun (peak, Samarga R.)	Apr–Jun
	Water temperature		6°–17°C	6°–15°C	10°–23°C
	Maturity index				
	Female		4%–7% (Jun) 8%–11% (Jul) 13%–19% (Aug) 24%–31% (Sep)	3%–4% (Sakhalin, early migrants)	1.7% (mid-Jun) 3.2% (late Jun) 4.5% (early Jul) 4.8% (late Jul) 19.6% (Oct)
	Male		1% (Jun) 2% (Jul) 1%–6% (Sep)	1%–2% (Sakhalin, early migrants)	
	Feeding		Occasional feeding (May) No feeding (after summer)		No feeding (Miya R., Nagara R.)
Spawner	Spawning season	Sep–Oct	Late Aug–early Oct	Jul–end Aug (Primore, Sakhalin) Aug (Kamchatka)	Oct–Nov
	Age at maturity	Yamame 2.0 (female) 1.0 or 2.0 (male)	1.1 (90%), 2.1 (10%)	Mainly 2.2. & 2.3 (Primore) 2.1 (80%, Sakhalin) Mainly 1.2 & 2.2 (Kamchatka)	2.0 (female) 1.0 or 2.0 (male)
	Female ratio	80%	65%–70%	65% (Primore) 66%–70% (Sakhalin) 60% (Kamchatka)	69%
	Fork length and body weight	57–53 cm 3 kg	47–53 cm 2 kg	57–58 cm, 3 kg (Primore) 47–53 cm, 2 kg (Sakhalin) 46–47 cm, 1.0–1.5 kg (Kamchatka)	12–18 cm
	Fecundity (eggs)	3000–4000 200–300 (Yamame)	1900	3200 (Primore) 2800 (N. Sakhalin) 1600–1700 (S. Sakhalin)	95–209

(continued on next page)

TABLE 44 (continued)

Life stage	Item	Masu salmon			Amago salmon
		Honshu	Hokkaido	USSR	
Embryo, alevin	Incubation period		430°–480°C-days	475°C-days	450°C-days
	Hatch-out		Nov-Dec	Apr-May (Sakhalin)	Apr
	Emergence		Late Mar–early May		
Juvenile	Growth		7–8 cm (Jul) 9–11 cm (Sep)	5–6 cm (Jul) 8.2 cm (Aug) (Sakhalin)	
	Food		Small benthos (emergence) Aquatic organisms Terrestrial insects (Jul) Terrestrial insects (Sep)	Aquatic insects (emergence) Terrestrial insects (warm seasons)	
Freshwater resident	Growth (fork length)	5.1–9.2 cm (1.0 fish) 18.7 cm, 105 g (2.0 fish) 31.3 cm, 498 g (3.0 fish) 43.0 cm (4.0 fish) (lake residual, in May)			7.5 cm (Jul) 7.9 cm (Aug) 9.5 cm (Nov) 13.6 cm (Feb) 17.3 cm (May)
	Food	Pond smelt Japanese dace (lake residual)	Pond smelt (lake residual)		Aquatic and terrestrial insects
Seaward migrant	Season	Apr-May	May	Jun-Jul (Sakhalin) July (Kamchatka)	Dec
	Fork length	11–14 cm	11–14 cm	11–13 cm (Kamchatka)	11.2–19.1 cm (Body L.)
	Female ratio		70%–80%	55% (Sakhalin)	67%–79%
Coastal young	Appearance season	Apr-May	May-Jul	Jun-Jul Aug-Sep (Sakhalin)	Dec
	Surface water temperature	8°–10°C	11°–12°C		
	Female ratio		70%–90%		81.6%–88.9% (wild fish) 76.0%–98.1% (hatchery fish)
	Fork length	6.3–19.4 cm	11.6–26.6 cm	13.5–26.5 cm	
	Body weight	21–77 g	16–144 g	30–184 g	36–114 g
	Food	Sand lance	Sand lance Sand eel Euphausiids	Sand lance Herring Crustaceans	Crustaceans (Dec) Small fish (Mar-May)

TABLE 44 (continued)

Life stage	Item	Masu salmon			Amago salmon
		Honshu	Hokkaido	USSR	
Immature in summer (for all masu stocks)	Distribution	Sea of Okhotsk, Pacific Ocean vicinity of Iturup Is.			
	Season	Jul-Sep			
	Water temperature	9.7°-19.4°C			
	Sex ratio	1:1			
	Age composition	1.0 (most), 2.0 (some)			
	Fork length	18.6-42.4 cm (range), 22.6 cm (average)			
	Body weight	75-800 g (range), 255 g (average)			
Southward migrant (for all masu stocks)	Season	Oct-Dec			
	Water temperature	5°-14°C			
	Female ratio	65%			
	Fork length	34.3 cm (average)			
	Body weight	553 g (average)			
Immature in winter (for all masu stocks)	Distribution	South of the polar front zone, east of the central Sea of Japan; Pacific Ocean south of Miyagi Prefecture			
	Season	Jan-Feb			
Adult in the Sea of Japan (for all masu stocks)	Season	Mar-May			
	Water temperature	8°-11°C			
	Depth distribution	Mainly 20-40 m during daytime			
	Fork length-female	40.1 cm (Mar), 43.0 cm (Apr), 45.6 cm (May) on average			
	-male	40.5 cm (Mar), 43.0 cm (Apr), 46.1 cm (May) on average			
	Body weight-female	868 g (Mar), 1122 g (Apr), 1412, g (May) on average			
	-male	929 g (Mar), 1151 g (Apr), 1556 g (May) on average			
	Gonad weight	9.1 g (Mar), 22.1 g (Apr), 46.6 g (May) on average			
	Testes weight	1.9 g (Mar), 4.0 g (Apr), 13.3 g (May) on average			
	Fecundity	1100-2400 eggs (range), 2600 eggs (average)			
	Female ratio	66.3% (years 1962-72)			
	Food	Atka mackerel,* saury, sailfish, sandfish, sand lance (fish) *Parathemisto japonica, Thysanoessa longipes* (crustaceans)			
Returning migrant (for all masu stocks)	Timing	Begin latter half of March			
	Fork length	45.9 cm (average, caught in the eastern Sea of Okhotsk in June)			
	Body weight	1452 g			
	Female ratio	60.4%			
	Gonad weight	92 g (average)			
	Testes weight	24 g (average)			

Note: *See Editors' Note, p. 498

large tributaries and move to upstream areas as spawning time approaches.

Body size of anadromous masu varies according to area. Fish are large in Primore and Honshu with average fork lengths of about 55–66 cm, and average body weights of about 2,000–4,000 g. They are small in Hokkaido, Sakhalin, and Kamchatka, with average fork lengths of about 42–54 cm and average weights of about 1,000–2,000 g. The range in the sizes of males is greater than that of females, but the average sizes are not appreciably different between males and females.

Most adult masu returning to Japan are considered to be age 1.1; however, for masu from USSR rivers, ages of 1.1 to 3.3 are reported. Such discrepancies in estimated age result from differences of criteria used in age determination brought about by variation in scale patterns.

Male masu that remain in the rivers without migrating downstream are found in all areas. As a result, the sex ratio in the ocean favours female fish. Masu running into Honshu rivers consist of 70%–100% females, with the percentage of females higher in areas to the south. In Hokkaido, the sex ratio varies by area. In the Sea of Okhotsk and Nemuro Strait areas, females constitute about 65%, and in southern areas fronting Tsugaru Strait and vicinity they make up about 80% of the masu. Masu running into the rivers of Sakhalin are about 65%–70% females, and in offshore waters of southwest Kamchatka about 60% are females.

The spawning season of masu is earlier in the northern than in the southern areas. The peak of spawning takes place in August in Sakhalin, in the first half of September in Hokkaido, and in September and October in Honshu. During the early period of upstream migration, gonads are not well developed and no secondary sex characteristics are observed. Gonads grow rapidly from about July and August, and the distinctive body colours appear in August. During this time the migrating fish rarely feed.

Mature eggs of masu have an average diameter of 6.0–7.1 mm. Fecundity is proportional to body size, so in areas where the anadromous fish are large, average fecundity is high. In Honshu and Primore, where fish are usually large, average fecundity ranges from 3,000 to 4,000 eggs, and on the Sea of Okhotsk side of Hokkaido and southwestern Sakhalin, where fish are usually small, average fecundity ranges from 1,600 to 1,900 eggs.

The fecundity of yamame, the freshwater-type masu, is lower than that of anadromous masu, averaging only 200–300 eggs. The average diameter of the mature ova is approximately 5 mm.

The redds of masu are constructed in shallow upstream areas. In Hokkaido and Sakhalin, redds contain about 700–1,900 eggs, which is much less than absolute fecundity.

Artificial rearing of masu fry is conducted at some hatcheries in Japan. The average number released from hatcheries in the five-year period 1969–73 was 9.3 million fish. In Sakhalin, annual releases during 1955–58 averaged 4.8 million fish, and there seem to have been no releases in recent years.

In southern Hokkaido, spawning of masu takes place in September, eggs hatch during November and December, and fry leave the redds during the following March to May. Newly emerged fry live in the shallow backwaters in schools, and from about June they settle at the edge of pools and in rapids. In their early stream life, they feed on small-sized benthos, such as chironomid larvae, and grow to 7–8 cm in fork length by July. Some fast growing males mature in September. About October, juvenile masu move either to downstream areas that have relatively warm temperatures or to spring-fed tributaries. In winter they hide in deep waters along river banks that have some cover. Juveniles that will migrate seaward in the following spring are active in the pools when the water temperature rises during the day, even during winter. They feed on Ephemeroptera or Chironomidae but do not grow during the winter.

In southern Hokkaido, the silvery coloration of juvenile masu becomes pronounced from about mid- to late April, and the peak of their seaward migration occurs in early and mid-May. The seaward migration in central and northern Honshu and in northern Hokkaido takes place in the early half of June, and in northeastern Sakhalin it occurs in the latter half of July. The geographical and individual variations in size of smolts that migrate seaward is small, and their fork lengths and body weights are relatively uniform at about 11–14 cm and 20 g, respectively.

Young masu are distributed in the coastal waters of Honshu in April and May, and in the coastal waters of Hokkaido in May and June. The surface temperatures in the coastal waters at this season are about 10°-12°C. Young fish in coastal waters average about 17 cm in fork length and about 60 g in weight. However, there is a wide range in size for fish in these waters, and young masu can be 10-25 cm in fork length. The main food for these young fish is small fish, such as sand lance, and crustaceans, such as euphausiids.

The Sea of Okhotsk is the only area known to be occupied by immature masu in summer. Although Tatar Strait, the Sea of Japan, and the Pacific Ocean side of the Kuril Islands are considered possible summer areas for immature fish, none have been captured in these waters in midsummer. Immature masu are present in the southern Sea of Okhotsk in July and in the areas close to the east and west coasts of the Sea of Okhotsk in August. Those present in the eastern sea of Okhotsk in August are assumed to be of Kamchatkan origin. In September they occur throughout this sea, except for the northernmost waters. Immature fish caught in the Sea of Okhotsk during July and September vary considerably in size, averaging 26.6 cm (range 18.6-42.4 cm) in fork length and 255 g in weight. They grow rapidly during the summer and increase in fork length about 1.8 mm/d.

It is assumed that masu are widely distributed in the Sea of Okhotsk in October, and begin their southward migration as water temperature declines in autumn. At this time of year they begin to be caught by setnets in coastal waters on the Sea of Okhotsk side of Hokkaido.

In coastal areas on the Sea of Japan side of Hokkaido and Honshu, the southward migration of masu lasts from December to the following February. Their average fork length and average weight is 34.3 cm and 553 g, respectively. Masu in this area eat mainly sand lance and, secondarily, anchovies and euphausiids.

Young and adult masu are caught by setnets in the coastal waters of northern Japan, the Korean peninsula, and the USSR in spring and summer, and a pole-and-line fishery targets masu in the coastal waters of Hokkaido and Honshu. Gillnet and longline fisheries for masu operate in offshore waters of the Sea of Japan during March and June.

The average annual catch of masu by Japan was 3,478 t from 1973 to 1976, with 3,671 t taken at sea and 77 t in fresh water (FAO 1977). Forty to eighty per cent of the catch at sea was taken in coastal waters of Honshu and Hokkaido. Catches of masu of up to 100 t were made off the southern coast of the Korean peninsula in the 1960s. The catch in the USSR and the northern Korean peninsula is unknown. Annual catches of about 1,000 t were obtained in Primore in the 1930s, but the fishery for masu in this area is now prohibited. Fishing for masu in rivers is strictly regulated or prohibited by both Japan and the USSR to protect the resource.

In March, masu are generally distributed in the Sea of Japan in the area of intermixing of warm and cold currents south of the polar front. Concentrations extend somewhat obliquely across the Sea of Japan near 38°N and curve north on the north side of Noto Peninsula and Sado Island, and along western Honshu. The area from the north side of Noto Peninsula to the west of Cape Nyudo is an especially important wintering region for masu. The distribution of masu in April is not appreciably different from that in March. The main concentration occurs south of 39°N and has generally moved somewhat to the north, particularly in waters close to Japan. From about late April, abundance decreases in the central Sea of Japan. By mid-May, substantial concentrations are observed only in waters close to both the east and west coasts in the Sea of Japan north of 40°N. In June, masu are not found in the Sea of Japan south of 45°N. The distribution north of 45°N has not been determined because of lack of research.

In the southern Sea of Okhotsk, the ice melts in mid- to late April and sometimes later. It is assumed that masu migrate from the Sea of Japan to the Sea of Okhotsk through Soya Strait. In May, they are present in the Sea of Okhotsk on the southeastern side of a line from Cape Terpeniya on the east coast of Sakhalin to near the Kolpakova River on the west coast of Kamchatka. In offshore areas of the west coast of Kamchatka, concentrations are observed about early and mid-June, but in late June the abundance declines markedly, and almost no fish are caught in July. The water temperatures in areas where adult masu occur in the Sea of Okhotsk range from 2° to 6°C, which is 4°-5°C, or more, lower than for the areas of distribution in offshore waters in the Sea of Japan.

The depths at which adult masu are found in offshore waters of Japan during winter and spring are mainly 20-40 m during the day, but vary from about 10 to 100 m. At night, in offshore waters in the Sea of Japan during April and May, fish density declines rapidly below 20 m and the majority are found from the surface to 20 m.

Masu caught in offshore waters of the Sea of Japan during March and June feed on fish, squid, amphipods, and euphausiids with fish and crustaceans the most important food items. The principal fish consumed are Atka mackerel, saury, sailfin sandfish, and sand lance, and the principal crustaceans are *Parathemisto japonica* and *Thysanoessa longipes*. In comparison, the main fish species eaten by masu in the Sea of Okhotsk during June and July are pollock, capelin, saffron cod, and Dolly Varden charr. The stomach contents of adult masu vary by season and area, indicating changes in the available food organisms. Japanese pearlside and the crustacean *P. japonica* are mostly taken by masu in relatively southern waters in March, whereas squid are eaten at the polar front zone during April and May. The crustacean *T. longipes* is eaten by masu in waters north of the polar front after April.

The minimum and maximum fork lengths and weights of adult masu caught with research gillnets in offshore waters in the Sea of Japan during March to May in 1962–72 showed fairly large variations. The average fork lengths observed during March, April, and May were 40.1 cm, 43.0 cm, and 45.6 cm for females, and 40.5 cm, 43.0 cm, and 46.1 cm for males, respectively. The average weights of females were 868 g, 1,122 g, and 1,412 g, and of males were 929 g, 1,151 g, and 1,556 g, respectively. Males were slightly larger than females. The average fork length in April of each year for males and females combined ranged from 40.2 to 45.1 cm, and the average weight ranges from 867 to 1,412 g. The average gonad weights for male and female fish ranged from 2.3 to 13.0 g, and 16.4 to 37.4 g, respectively.

The annual changes in magnitude of these three characteristics showed common features with minimum values observed in 1965 and 1971. The increase in average fork length from March to May was about 1 cm per ten days. The increase in average weight per ten days was 67–80 g for females and 80–100 g for males in March, and 87–131 g for females and 149–181 g for males in May. The average gonad weight increment per ten days for females was 2.1–3.6 g in March, 4.6–7.6 in April, and 9.8–12.6 g in May, and for males it was 0.51–0.9 g in March, 1.2–2.3 g in April, and 3.1–5.7 g in May. For masu occurring in the Sea of Japan in March to May, the average fork length in waters offshore of the northwest coast of Honshu was smaller throughout each season than for those occurring in waters west of the central Sea of Japan. Average weights and average gonad weights showed a similar geographic trend.

The average fecundity of adult female masu in the Sea of Japan was 3,244 eggs in a 1959 sample and 2,600 eggs in a 1969 sample. The sex composition of adult masu calculated for the eleven years from 1962 to 1972 was 66.3% females and 33.7% males, with small annual fluctuations.

Adult masu caught in the eastern Sea of Okhotsk in June are relatively small. Average fork length is 45.9 cm, average weight 1,452 g, and the average maturity index 6.5% for females and 1.6% for males. The percentage of female fish (60.4%) caught in the Sea of Okhotsk is lower than that caught in the Sea of Japan. In this sea, the average lengths decline as the season advances and the percentage of females increases.

In anadromous masu, female fish are predominant in the Honshu area, but the sex ratio approaches 1:1 farther northward. The distribution of sex ratios in offshore waters is quite complicated, and it is difficult to define subpopulations from their spatio-temporal distributions. The average fork length of anadromous fish is large in Primore and Honshu and small in northeastern Hokkaido, southern Sakhalin, and Kamchatka. In the latter areas anadromous fish are mainly 40 cm and less in fork length. In offshore waters of the Sea of Japan, the average fork lengths are small (less than 40 cm) in the area close to Japan, indicating that fish originating in regions that produce small-sized fish are in fairly high abundance in this area. Very small masu (less than 40 cm in fork length) appear mainly in this region. Very large fish appear in broad areas in the central Sea of Japan and are not limited to any specific area.

Masu that migrate from the Sea of Japan to Tsugaru Strait and east are widely distributed south of the polar front zone in the central and eastern Sea of Japan during March and April, and they enter Tsugaru Strait during April and May. There are two groups of fish in the western Sea of Japan west of 134°E. One group migrates in a northeasterly direction and the other group migrates north and northwest. It is believed that the former are mainly middle-sized and small fish that migrate to areas such as northern Primore, Sakhalin, Hokkaido, and Honshu, and the latter migrate to southern Primore.

The migration range of masu in the ocean is extremely limited compared to that of other Pacific salmon. The limits of their distribution are close to the coastal waters throughout their marine life. Many live in the coastal areas not only immediately after their seaward migration and before entering rivers but also during other periods in their life. It is not known whether their affinity for coastal waters is by preference, or whether it results from the complicated topography and oceanographic conditions within their migration range. Other young Pacific salmon that spend a summer in the Sea of Okhotsk migrate mostly to the Pacific Ocean with the cooling in fall and spend a winter in offshore waters. In contrast, masu migrate south to the Sea of Japan and never migrate to offshore waters of the Pacific Ocean.

Amago. The distribution of amago ranges from Shikoku and the Pacific Ocean side of Honshu from the Tokyo Bay area southwestward to and including the northern tip of Kyushu. The climate throughout the range of amago is somewhat warmer than throughout most of the range of masu.

There are thirty rivers in which a fishery for amago operates. Although the total yield in these rivers reached 120 t in the year 1937, in recent years the yield in many rivers has been small. The Kiso, Nagara, and Ibi rivers are exceptions, with recent annual catches ranging from 4,000 to 9,000 fish or 11 to 15 t.

In the rivers of Honshu, the ascent of amago starts in early April, reaches a peak in May, and terminates in June. Water temperatures of the river at the start of the ascent are about 10°-12°C.

For both feral and cultured amago, the males mature at age 1.0 or 2.0 and the females at age 2.0. Mature eggs of amago are light yellow. The diameters of eggs of mature amago in the Ado River, Honshu, are from 4 to 5 mm. The fecundity of amago, 12-18 cm long, ranged from 78 to 205 eggs.

Redds of amago are oval, like those of masu. The average egg content per redd is about 56.

In Mie Prefecture, the spawning season of amago is October to December, and body length of alevins reaches only 19.6-22.0 mm in February and March. Fry emerge in April. The stream residual amago feed on terrestrial insects and grow to 9.5 cm by October and to 17.3 cm in average fork length by May. In the Nagara River, the smolt transformation of amago takes place from the end of October in age 1.0 fish. Although most amago stay in the rivers, the sea-run type descend to salt water mainly in December.

In Mikawa, Ise, and Wakasa bays, young amago are distributed in the estuaries of rivers in December. After remaining there for about one month, they disperse all over the bays. Amago do not migrate as widely as other Pacific salmon species.

After the feeding migration in a bay, maturing amago gather in estuarine waters of their parental river during April to May. Upstream migration starts in mid-April and reaches a maximum between mid- and late May. Mean body weight of amago increased from 290 g to approximately 550 g from early to late April. Although body weight of amago can attain a maximum of 711 g in early May, estuary and upstream migrating fish grow very slowly or not at all.

As with the other species of Pacific salmon, it is believed that the great majority of adult masu and amago return to the parental streams. Marking experiments of smolts have shown that over 95% of recovered amago adults return to their parental streams. The rate of straying in both species is, thus, fairly small.

References

Akiba, T. 1980. Chitose Hokkaido salmon coterie. Chitose, Japan. 196 p. (In Japanese)

Akita Prefectural Fisheries Experimental Station. 1963-73. Annual reports, 1962-72. Fisheries

Agency of Japan, Akita Prefectural Fisheries Experimental Station, Ogashi, Akitaken, Japan. (In Japanese)

——. 1973-76. Reports of observation program on salmon fry passing the river mouths, 1972-75. Fisheries Agency of Japan, Akita Prefectural Fisheries Experimental Station, Ogashi, Akitaken, Japan. (In Japanese)

Aomori Prefectural Fisheries Experimental Station. 1964-68. Reports of research on the conservation program on masu salmon, 1963-67. Aomori Prefectural Fisheries Experimental Station, Akashi, Amoriken, Japan. (In Japanese)

——. 1972-76. Reports of observation program on salmon fry passing the river mouth. Coastal waters research report, 1971-75. Aomori Prefectural Fisheries Experimental Station, Akashi, Amoriken, Japan. (In Japanese)

——. 1976. Report of research on the conservation program on masu salmon, 1973-75. Aomori Prefectural Fisheries Experimental Station, Akashi, Amoriken, Japan. 40 p. (In Japanese)

Atkinson, C.E., S. Chun, E.R. Jefferies, J. Kim, K. Kim, and R.T. Pressey. 1967. A survey of the salmon and trout resources of the Republic of Korea: a feasibility study. U.S. Department of State, Agency for International Development; Office of Fisheries, Republic of Korea, Seoul, Korea. 89 p.

Birman, I.B. 1972. Some questions on the biology of masu salmon (*Oncorhynchus masu* Brevoort). Izv. Tikhookean. Nauchno-Issled. Inst. Rybn. Khoz. Okeanogr. 82:235-247. (In Russian, English summary)

Christic, W.J. 1970. Introduction of the cherry salmon (*Oncorhynchus masou*) in Algonquin Park, Ontario. Copeia 70:378-379

Churikov, A.A. 1975. Features of the downstream migration of young salmon of the genus *Oncorhynchus* from the rivers of the northeast coast of Sakhalin. J. Ichthyol. 15:963-970

Dvinin, P.A. 1949. Mass aggregations of juvenile salmonids on the shores of Sakhalin. Rybn. Khoz. 25(7):39-41. (In Russian)

——. 1952. The salmon of southern Sakhalin. Izv. Tikhookean. Nauchno-Issled. Inst. Rybn. Khoz. Okeanogr. 37:69-108. (Transl. from Russian; Fish. Res. Board Can. Transl. Ser. 120)

——. 1956. Distinctive features of the biology of masu salmon (*Oncorhynchus masou* Brevoort) of Sakhalin. Vopr. Ikhtiol. 7:33-35. (Transl. from Russian; Fish. Res. Board Can. Transl. Ser. 350)

Ebina, K. 1935. Salmon caught in Tateyama Bay. J. Fish Cult. Assoc. (Yoshoku-kai-shi) 5 (9/10):166-167. (In Japanese)

Food and Agriculture Organizaton of the United Nations (FAO). 1977. Catches and landings, 1976. FAO Yearb. Fish. Stat. 42:320 p.

Far Seas Fisheries Research Laboratory. 1979-80. Reports of research on young salmon conducted by the Hokusei Maru, Fisheries Department, Hokkaido University, 1978-80. Far Seas Fisheries Research Laboratory, Fisheries Agency of Japan, Shimuzu, Shizuoka, Japan. (In Japanese)

Fukataki, H. 1967. Notes on migration of the masu salmon, *Oncorhynchus masou* (Brevoort), in the Japan Sea as determined by tagging. Bull. Jpn. Sea Reg. Fish. Res. Lab. 18:1-11. (In Japanese)

——. 1969. Stomach contents of the masu salmon, *Oncorhynchus masou* (Brevoort), in the offshore regions of the Japan Sea. Bull. Jpn. Sea Reg. Fish. Res. Lab. 21:17-34. (In Japanese)

——. 1970. Further notes on migration of masu salmon, *Oncorhynchus masou* (Brevoort), in the Japan Sea as determined by tagging. Bull. Jpn. Sea Reg. Fish. Res. Lab. 22:1-14. (In Japanese)

Fukataki, H., T. Ogata, A. Ouchi, and S. Machida. 1961. Fisheries biological study on salmon in the Sea of Japan, p. 120-134. *In*: Study on the Japan Sea Polar Front fishing grounds, 2nd Year (1960). Japan Sea Regional Fisheries Research Laboratory, Niigata, Niigata, Japan. (In Japanese)

Gritsenko, O.F. 1973. Biology of masu and coho salmon in northern Sakhalin. Nauch. Otchet Teme Vses. Nauchno-Issled. Inst. Morsk. Rybn. Khoz. Okeanogr. 10:40 p. (Transl. into Japanese from Russian; Japan Fisheries Association, Translation Series of USSR Reports on North Pacific Fisheries:10)

Hiroi, O. 1984. Technical approaches in production of the seedlings for the artificial propagation of masu salmon, p. 120-128. *In*: Marine Ranching Program, Progress Reports on Masu Salmon Production 4. (In Japanese)

Hokkaido Fisheries Experimental Station and Hokkaido Regional Fisheries Research Laboratory. 1956. Study on exploitation of the Tsushima current system, No. 3, p. 165-232. *In*: Summary Report of Population Studies of the Hokkaido Area 13. (In Japanese)

Hokkaido Salmon Hatchery (Hokkaido Sakemasu Fakajo). 1953-76. Annual reports, 1952-75. Hokkaido Salmon Hatchery, Sapporo, Hokkaido, Japan. (In Japanese)

——. 1956. Reports on the numbers of salmon eggs collected and of fry released. Hokkaido Salmon Hatchery, Sapporo, Hokkaido, Japan. 91 p. (In Japanese)

——. 1969. Reports on research of the pink and masu salmon returning to rivers in Hokkaido (pink and masu salmon). Sci. Rep. Hokkaido Salmon Hatchery 23:29-44. (In Japanese)

Honda N., T. Kataoka, M. Hoshino, and Y. Seki. 1983. Studies on the reproduction of landlocked masu salmon, *Oncorhynchus masou*, in Okutadami reservoir. v. Studies on the growth and age of maturity of the landlocked masu salmon in Okutadami reservoir. Rep. Niigata Pref. Inland Water Fish. Exp. Stn. 10:13-20. (In Japanese)

Honda N., A. Suzuki, K. Amita, T. Kataoka, and K. Emura. 1980. Studies on the reproduction of landlocked masu salmon, *Oncorhynchus masou* (Brevoort), in Okutadami reservoir. I. Vertical distribution and feeding habits of fish (landlocked masu salmon, etc.). Rep. Niigata Pref. Inland Water Fish. Exp. Stn. 8:5-15. (In Japanese)

Honjoh, T. 1977. Studies on the culture and transplantation of amago salmon, *Oncorhynchus rhodurus*. Rep. Gifu Pref. Fish. Exp. Stn. 22:104 p. (In Japanese)

——, ed. 1985. Artificial propagation of sea-run amago. Jpn. Fish. Resour. Prot. Assoc. Ser. Fish Cult. Propag. 32:101 p. (In Japanese)

Honjoh, T., M. Okazaki, and S. Mori. 1975. Studies on the effective stocking of Japanese native salmonoid fishes. III. On the sea-run and homing migration of amago salmon, *Oncorhynchus rhodurus*. Rep. Gifu Pref. Fish. Exp. Stn. 20:12 p. (In Japanese)

Honshu Salmon Enhancement Association (Honshu Keison Zoshoku Shinko Kyokai). 1953-77. Annual reports on the numbers of salmon eggs collected and of fry released, 1952-76. (In Japanese)

Imanishi, K. 1951. Iwana and yamame. Tech. Assoc. Jpn. For. Guidebk. For. Ser. 35:36 p. (In Japanese)

Inoue, S., and K. Ishigaki. 1968. Notes on the biology of juvenile masu salmon (*Oncorhynchus masou*) during winter in the Chihase River, Hokkaido. Jpn. J. Limnol. 29:27-36. (In Japanese)

Ishida, R., K. Takagi, and S. Arita. 1961. Criteria for the discrimination of mature and immature forms of chum and sockeye salmon in northern areas. Int. North Pac. Fish. Comm. Bull. 5:23-40

Japan Salmon Resources Conservation Association (Nippon Keison Shigen Hogo Kyokai). 1976. Analysis of the condition on reproduction and utilization for Japanese salmon resources. Japan Salmon Resources Conservation Association. 186 p. (In Japanese)

Jeon, S., H. Yang, I. Kim, E. Choi, G. Chang, G. Li, T. Gweon, Y. Kim, and J. Bag. 1978. The atlas of Korean fresh-water fishes. Korean Institute of Fresh-water Biology. 35 p.

JIBP-PF Research Group of Yurappu River. 1975. Productivity of biotic communities in the Yurappu River. *In*: S. Mori and G. Yamamoto (eds.). Productivity of communities in Japanese inland waters. Jpn. Comm. Int. Biol. Program JIBP Synthesis 10:287-338. (In Japanese)

Karube, S., and O. Takahashi. 1975a. On salmon gillnets in the Sea of Japan (efficiency of the sunken gillnets), p. 99-111. *In*: Material for the 7th All Japan Council of Promotion of Fishing Gear and Fishing Method Study. Fisheries Agency of Japan, Tokyo, Japan. (In Japanese)

——. 1975b. On salmon gillnet fishery in the Sea of Japan: on swimming layers for pink salmon. Bull. Niigata Pref. Fish. Exp. Stn. 4:6-9. (In Japanese)

Katayama, M., and Y. Fujioka. 1965. Salmonid species and distribution in Yamaguchi Prefecture. Yamaguchi Univ. Dep. Educ. Res. Ser. 15:65-76. (In Japanese)

Kato, F. 1968. On the salmon *Oncorhynchus rhodurus* of the Nagara River, p. 895-904. *In*: Report of a Survey of the Fishery Resources in Estuary Waters of the Kiso, Nagara and Ibi Rivers 5. (In Japanese)

——. 1973a. Ecological study on the sea-run form of *Oncorhynchus rhodurus* found in Ise Bay. Jpn. J. Ichthyol. 20:225-234. (In Japanese)

——. 1973b. On the sea-run form of *Oncorhynchus rhodurus* obtained in Ise Bay. Jpn. J. Ichthyol. 20:107-112. (In Japanese)

——. 1975. On the distribution of a sea-run form of the salmonid fish *Oncorhynchus rhodurus* found in southwestern Japan. Jpn. J. Ichthyol. 21:191-197. (In Japanese)

——. 1977. On a sea-run specimen of *Oncorhyn-*

chus rhodurus collected on the Echizen coast, facing the Japan Sea. Jpn. J. Ichthyol. 25:71–72. (In Japanese)

——. 1978a. Lepidological study on sea-run specimens of *Oncorhynchus rhodurus*. Jpn. J. Ichthyol. 25:51–57. (In Japanese)

——. 1978b. Morphological and ecological studies on two forms of *Oncorhynchus rhodurus* found in Lake Biwa and adjoining inlets. Jpn. J. Ichthyol. 25:197–204. (In Japanese)

——. 1982. Ecological study on the amago, *Oncorhynchus rhodurus*, going down and resident in the mid-stream of Nagara River, p. 104–111. *In*: Special Report on Freshwater Fish: *Oncorhynchus masou* and *O. rhodurus*. (In Japanese)

Kato, F., Y. Hida, E. Noda, and Y. Tsuno. 1982. Specimens of sea-run type *Oncorhynchus rhodurus* collected in the coastal waters of the Hokuriku and Tohoku regions, in the Japan Sea. Bull. Jpn. Sea Reg. Fish. Res. Lab. 33:56–62. (In Japanese)

Kato, M. 1970. Review of some problems of age determination of masu salmon, *Oncorhynchus masou* (Brevoort), by scale reading. Bull. Jpn. Sea Reg. Fish. Res. Lab. 22:15–29. (In Japanese)

——. 1971. Sex ratio, fecundity and maturation of masu salmon, *Oncorhynchus masou* (Brevoort), during their marine life. Bull. Jpn. Sea Reg. Fish. Res. Lab. 23:55–67. (In Japanese)

——. 1973. Age determination of masu salmon, *Oncorhynchus masou* (Brevoort), by scale reading, with special reference to a criterion to classify the age during their freshwater life. Bull. Jpn. Sea Reg. Fish. Res. Lab. 24:53–66. (In Japanese)

Kawajiri, M. 1927. On the preservation of the eggs and sperm of *Oncorhynchus masou* (Brevoort). J. Imper. Fish. Inst. 23:11–13

Kawakami, S., and S. Machida. 1933. On number of circuli on scales of masu salmon formed immediately after their downstream migration. Tenday Rep. Hokkaido Fish. Exp. Stn. 229:272–276. (In Japanese)

Kimura, S. 1972. On the spawning behavior of the fluvial dwarf form of masu salmon, *Oncorhynchus masou*. Jpn. J. Ichthyol. 19:111–119. (In Japanese)

Kimura, S., and H. Tsukahara. 1969. On the smolt of the salmon, *Oncorhynchus masou* (Brevoort), obtained in Ariake Sound in Kyushu. Jpn. J. Ichthyol. 16:131–134. (In Japanese)

Krykhtin, M.L. 1962. Data on the stream life of masu salmon. Izv. Tikhookean. Nauchno-Issled. Inst. Rybn. Khoz. Okeanogr. 48:84–132. (In Russian)

Kubo, T. 1968. On enhancement of masu salmon. Salmon and Trout (Sake to Masu) 15:11–18. (In Japanese)

——. 1974. Notes on the phase differentiation and smolt transformation of juvenile masu salmon (*Oncorhynchus masou*). Sci. Rep. Hokkaido Salmon Hatchery 28:9–26. (In Japanese)

——. 1976. Behavior and movements of the juvenile "masu" salmon (*Oncorhynchus masou*) during stream life in Hokkaido. Physiol. Ecol. Jpn. 17:411–417. (In Japanese)

——. 1980. Studies on the life history of "masu" salmon (*Oncorhynchus masou*) in Hokkaido. Sci. Rep. Hokkaido Salmon Hatchery 34:1–95. (In Japanese)

Machidori S., and F. Kato. 1984. Spawning populations and marine life of masu salmon (*Oncorhynchus masou*). Int. North Pac. Fish. Comm. Bull. 43:138 p.

Machidori, S., S. Shirahata, T. Kobayashi, T. Ishida, F. Kato, T. Tokui, T. Kato, M. Tanaka, R. Ishida, and M. Kato. 1980. Resource management of river spawning type pelagic species, p. 1–18. *In*: Marine Ranching Program: the 1979 analytical report. Agriculture, Forestry, and Fishery Techological Council Secretariat. (In Japanese)

Matsubara, K. 1934. On masu salmon caught off Kumamoto Prefecture. J. Fish Cult. Assoc. (Yoshoku-kai-shi) 4:114–117. (In Japanese)

Mayama H., K. Ohkuma, T. Nomura, and K. Matsumura. 1985. Experimental release of masu salmon (*Oncorhynchus masou*) smolts into the Shiribetsu River: adult returns of marked fish released in the spring of 1981. Sci. Rep. Hokkaido Salmon Hatchery 39:1–16. (In Japanese)

Mihara, T. 1952. Tokotan Lake on Urup Island (the central Kuril Islands): report on lakes and rivers research. Sci. Rep. Hokkaido Salmon Hatchery 7:11–91. (In Japanese)

Miura, T. 1966. Ecological notes of the fishes and interspecific relations among them in Lake Biwa. Jpn. J. Limnol. 27:1–24. (In Japanese)

Mori, T. 1935. On the geographical distribution of Korean salmonid fishes. Bull. Biogeogr. Soc. Jpn. 6:1–9

Naganuma, K. 1977. Oceanographic fluctuations in the Sea of Japan. Ocean Sci. 9:65-69. (In Japanese)

Nagasawa, A., and P. Aguilera. 1974. Introduction into Aysen Chile, of the Pacific salmon. 1: transportation and rearing trials with Japanese cherry salmon (*Oncorhynchus masou*) 1972-73. Servicio Nacional de Pesca, Republica de Chile; Japan International Cooperation Agency, Santiago, Chile. 21 p.

Nagoshi, M., and T. Sakai. 1980. Relation between body size and food of the amago (*Oncorhynchus rhodurus*) from Hirakura Stream, Mie Prefecture, Japan. Jpn. J. Ichthyol. 26:342-350. (In Japanese)

Nakamura, M. 1958. A juvenile masu salmon caught downstream in the Edo River. Angler (Tsuri-bito) 13:40-41. (In Japanese)

Ogata, T. 1960. Biological research on salmon in the Sea of Japan, p. 117-134. *In*: Study on Polar Front fishing grounds in the Sea of Japan, 1st Year (1959). Japan Sea Regional Fisheries Research Laboratory, Niigata, Japan. (In Japanese)

Okada, S., and T. Kawamura. 1938. Salmonid fishes in the waters of Shimshir Island. Plant Anim. 6:851-857. (In Japanese)

Okada, S., and M. Sakurai. 1937. Several instances of an incomplete second winter zone on the scales of masu salmon. Salmon J. 9:19-25. (In Japanese)

Okazaki, T. 1977. Research report on the adult salmon approaching coastal waters in the Sea of Okhotsk by the Hokko Maru in 1977. Far Seas Fisheries Research Laboratory, Shimizu, Japan. 14 p. (In Japanese)

Ono, I. 1933. The life history of masu salmon of Hokkaido. Salmon J. 5(2):15-26, 5(3):13-25. (In Japanese)

Osanai, M. 1962. Ecological studies on the land-locked masu salmon, *Oncorhynchus masou* (Brevoort). 1. Ecological succession of the limnological conditions and feeding habit of the lake-locked form at Uryu reservoir. Sci. Rep. Hokkaido Fish Hatchery 17:21-29. (In Japanese)

——. 1982. Masu salmon in lakes, p. 92-96. *In*: Special Report on Freshwater Fish; *Oncorhynchus masou* and *O. rhodurus*. (In Japanese)

Osanai, M., and M. Otsuka. 1967. Ecological studies on the masu salmon, *Oncorhynchus masou* (Brevoort), of Hokkaido. 1. Morphology and spawning habit of masu salmon which ascend the river. Sci. Rep. Hokkaido Fish Hatchery 22:17-32. (In Japanese)

——. 1969. Ecological studies on the masu salmon, *Oncorhynchus masou* (Brevoort), of Hokkaido. 2. On the effect of movement upstream and spawning behavior on the sexual maturity of masu salmon. Sci. Rep. Hokkaido Fish Hatchery 24:45-54. (In Japanese)

Oshima, M. 1955. Masu salmon, *Oncorhynchus masou* (Brevoort), and Biwa salmon, *Oncorhynchus rhodurus* (Jordan and McGregor). Nireshobo, Tokyo, Japan. 79 p. (In Japanese)

Pravdin, I.F. 1949. Masu salmon (*Oncorhynchus masu* Brevoort), p. 168-174. *In*: L.S. Berg, A.S. Bogdanov, N.I. Kozhin, and T.S. Rass (eds.). Commercial fishes of the USSR. Pishchepromizdat, Moscow, USSR. (In Russian)

Sakiura, H., S. Yubashi, and Y. Koyama. 1964. On the reproduction of salmon stocks of the USSR. Japan Fish. Resour. Conserv. Assoc. (Nihon Suisan Shigen Hogo Kyokai) Overseas Fish. Ser. 1:67 p. (In Japanese)

Sano, S. 1947. Changes in masu salmon during the non-feeding season. Salmon J. 44:9-14. (In Japanese)

——. 1951. Scale patterns of masu salmon. Salmon J. 52:8-12. (In Japanese)

——. 1959. The ecology and propagation of the genus *Oncorhynchus* found in northern Japan. Sci. Rep. Hokkaido Salmon Hatchery 14:21-90. (In Japanese)

——. 1964. The ecology and protection of reproduction. Fish and Egg (Uo to Tamago) 104:1-7. (In Japanese)

Sano, S., and S. Abe. 1967. Ecological studies of masu salmon (*Oncorhynchus masou*) (Brevoort): observations on smolts in coastal waters. Sci. Rep. Hokkaido Salmon Hatchery 21:1-10. (In Japanese)

Sano, S., and Y. Ozaki. 1969. An ecological study of the masu salmon (*Oncorhynchus masou*) (Brevoort): artificial rearing and marking of masu smolts. Sci. Rep. Hokkaido Salmon Hatchery 23:1-8. (In Japanese)

Sasaki, F. 1978. The results of recovery of tagged fish and several biological informations of masu salmon, *Oncorhynchus masou* (Brevoort), migrating to the Shakotan waters. *In*: Materials of the

Research Council for Masu Salmon in the Sea of Japan. (In Japanese)

Semenchenko, A.U. 1978. Biological characteristics of the spawning part of the cherry salmon population of North Primorski Territory, p. 48-49. *In*: Biology of salmons: abstracts. Pacific Research Institute of Fisheries and Oceanography, Vladivostok, USSR

Semko, R. S. 1956. New data on masu salmon from western Kamchatka. Zool. Zh. 35:1017-1022. (In Russian, English summary)

Shimazaki, K. 1971. Notes on biological characteristics and migration of the masu salmon (*Oncorhynchus masou*) (Brevoort), in the offshore areas of the west coast of the Kamchatka Peninsula. Bull. Fac. Fish. Hokkaido Univ. 22:37-46. (In Japanese)

——. 1975. Notes on the distribution and biological characteristics of the masu salmon in the Sea of Okhotsk. Bull. Jpn. Soc. Fish. Oceanogr. 27:104-110. (In Japanese)

Shiraishi Y., and N. Suzuki. 1957. Fishery biology of amago salmon in the Bano River of Mie Prefecture. I. Morphology. Rep. Freshwater Fish. Res. Lab. 9:1-18. (In Japanese)

Shiraishi Y., K. Suzuki, and G. Tamada. 1957. Fishery biology of amago salmon in the Bano River of Mie Prefecture. II. Spawning. Rep. Freshwater Fish. Res. Lab. 14:1-17. (In Japanese)

Smirnov. A. I. 1962. Reproductive ecology of masu salmon, *Oncorhynchus masu* (Brevoort). Dokl. Akad. Nauk. SSSR. 143:1449-1452. (In Russian)

Sugiwaka, K., and H. Kojima. 1979. Studies on the smolts of juvenile masu salmon (*Oncorhynchus masou*) in the Atsuta River: age and ecology of smolts in 1978. Sci. Rep. Hokkaido Fish Hatchery 34:25-39. (In Japanese)

Suzuki N., Y. Shiraishi, and J. Yoshihara. 1957. Fishery biology of amago salmon in the Bano River of Mie Prefecture. III. Gonad. Rep. Freshwater Fish. Res. Lab. 15:1-17. (In Japanese)

Tabata, Y., S. Hashida, and S. Michitaka. 1959. A survey of the salmon runs to the coast of the Sea of Japan. (Mimeo). Ishikawa Prefectural Fisheries Experiment Station, Notomachi, Ishakawa, Japan. (In Japanese)

Takagi, K. 1961. The seasonal change of gonad weight of sockeye and chum salmon in the North Pacific Ocean, especially with reference to mature and immature fish. Bull. Hokkaido Reg. Fish. Res. Lab. 23:17-34. (In Japanese)

Takayasu, M., K. Kondo, S. Ohigashi, and S. Watari. 1954. Limnological studies on the lake of Iturup Island. Sci. Rep. Hokkaido Fish Hatchery 9:1-85. (In Japanese)

Tanaka, S. 1965. A review of the biological information on masu salmon (*Oncorhynchus masou*), p. 75-135. *In*: Salmon of the North Pacific Ocean. Part IX. Coho, chinook and masu salmon in offshore waters. Int. North Pac. Fish. Comm. Bull. 16

Uchihashi, K. 1953. Ecological studies of Japanese teleosts in relation to brain morphology. Bull. Jpn. Sea Reg. Fish. Res. Lab. 2:1-166. (In Japanese)

Utoh, H. 1976. A study of the mechanism of differentiation between the stream resident form and the seaward migration form of masu salmon, *Oncorhynchus masou* (Brevoort). I. Growth and sexual maturity of precocious masu salmon parr. Bull. Fac. Fish. Hokkaido Univ. 26:321-326. (In Japanese)

——. 1977. A study of the mechanism of differentiation between the stream resident form and the seaward migration form of masu salmon, *Oncorhynchus masou* (Brevoort). II. Growth and sexual maturity of precocious masu salmon parr. Bull. Fac. Fish. Hokkaido Univ. 28:66-73. (In Japanese)

Volovik, S.P. 1963. Data on the biology of juvenile masu salmon, *Oncorhynchus masou* (Brevoort), in some Sakhalin rivers. Vopr. Ikhtiol. 3:506-512. (In Russian)

Watanabe, K., and A. Ouchi. 1962. Biological information of masu salmon during their marine life, p. 60-67. *In*: Study on Polar Front fishing grounds in the Sea of Japan, 3rd Year (1961). Japan Sea Regional Fisheries Research Laboratory, Niigata, Japan. (In Japanese)

Yagisawa, Y. 1970. The hatchery program of salmon on Iturup Island. Fish and Egg (Uo to Tamago) 133:35-79. (In Japanese)

Yoshida, H. 1942. The masu salmon in the sea off North Korea. Fish Res. (Suisan Kenkyu-shi) 37:152-156. (In Japanese)

——. 1967. Masu salmon, p. 1397-1406. *In*: Report of a Survey of the Fishery Resources in Estuary Waters of the Kiso, Nagara and Ibi Rivers 4. (In Japanese)

Indexes

GEOGRAPHICAL INDEX

When looking up place names, note that this index is arranged according to the first word of the place name as it is commonly used. For example, the entry "Cultus Lake" would be found under "Cultus" but "Lake Erie" would be found under "Lake Erie," not "Erie, Lake."

Co-ordinates with asterisks indicate approximations. Dashes indicate that precise co-ordinates for localities were not available.

The following references were used in determining place names, general locations, and co-ordinates:

The Columbia Lippincott Gazetteer of the World. 1962 with 1961 supplement. L.E. Seltzer (ed.). New York: Columbia University Press

Gazetteer of Canada: British Columbia. 1985. 3rd ed. Canadian Permanent Committee on Geographical Names. Energy, Mines, and Resources, Canada

International North Pacific Fisheries Commission. 1967. Salmon of the north Pacific Ocean. Part IV. *Int. North Pac. Fish Comm. Bull*. 23:327 p.

Morskoi Atlas (Marine Atlas) and Index. 1950, 1952. Vol. 1. Izdanie Morskogo Generalnogo Shtaba, Moscow, USSR

The Times Atlas of the World. 1957, 1959. Mid-century ed. Vol. 1, *World, Australasia and East Asia*; Vol. 2, *South-west Asia and Russia*; Vol. 5, *The Americas*. Edinburgh, Scotland: John Bartholomew and Sons

The Times Atlas of the World. 1981. Comprehensive ed. 6th ed. Edinburgh, Scotland: John Bartholomew and Sons

U.S. Geological Survey, Branch of Geographic Names, for Alaska, Washington, and Oregon place names and co-ordinates

Place names	**Location**	**Co-ordinates**
Abashiri Bay	Japan/northern Hokkaido	44°02′N 144°17′E
Abernathy River	USA/Washington/Columbia River system	46°11′N 123°09′W
Adak Bay	USA/Alaska/Aleutian Islands/Adak Island	51°46′N 176°40′W
Adak Island	USA/Alaska/Aleutian Islands	51°50′N 176°40′W
Adams Lake	Canada/British Columbia/Fraser River system	51°00′N 119°45′W
Adams River	Canada/British Columbia/Fraser River system	50°54′N 119°33′W
Admiralty Inlet	USA/Washington	48°10′N 122°40′W
Admiralty Island	USA/southeastern Alaska/Alexander Archipelago	57°40′N 134°20′W
Ado River	Japan/Honshu/Shiga Prefecture	35°10′N 136°00′E
Afognak Island	USA/Alaska/Gulf of Alaska	58°10′N 152°50′W
Agano River	Japan/Honshu/Sea of Japan coast	37°58′N 139°05′E
Agiapuk River	USA/Alaska/Port Clarence	65°15′N 166°40′W*
Agulowak River	USA/Alaska/Bristol Bay	59°24′N 158°54′W
Agulukpak River	USA/Alaska/Bristol Bay/Wood River system	59°38′N 158°32′W
Ainskaya River	USSR/eastern Sakhalin	48°22′N 142°09′E
Aisen Province	Chile	46°00′N 73°00′W
Akita	Japan/Honshu/Sea of Japan coast	39°44′N 140°05′E
Akita Prefecture	Japan/northern Honshu	39°50′N 140°50′E

Place names	Location	Co-ordinates
Akive River	USA/Alaska/Malaspina Glacier	59°50′N 140°40′W
Akka River	Japan/northeastern Honshu/Iwate Prefecture	40°03′N 141°52′E
Alagnak River	USA/southwestern Alaska/Bristol Bay	59°00′N 156°53′W
Alaska Peninsula	USA/southwestern Alaska	57°00′N 158°00′W
Alastair Lake	Canada/British Columbia/Skeena River system	54°08′N 129°14′W
Albion	Canada/British Columbia/lower Fraser River system	49°11′N 122°33′W
Aleutian Islands	USA/Alaska (also known as Aleutian Chain)	52°00′N 176°00′W
Alsek River	Canada/British Columbia/Yukon-USA/southeastern Alaska	59°26′N 137°58′W
Amanka Lake	USA/southwestern Alaska/Bristol Bay	59°03′N 159°16′W
Amgun River	USSR/Amur River system	52°56′N 139°38′E
Amur District	USSR/Far East	52°00′N 137°00′E
Amur River	USSR/Far East	53°10′N 140°44′E
Anadyr Gulf	USSR/northwestern Bering Sea	64°00′N 177°05′E
Anadyr River	USSR/Far East/Bering Sea drainage	64°41′N 177°32′E
Anana Lagoon	USSR/Cape Olyutorskiy	59°58′N 170°10′E
Anapka River	USSR/eastern Kamchatka	60°01′N 163°55′E
Anebetsu River	Japan/Hokkaido/Furen River system	43°20′N 145°25′E
Aniva Bay	USSR/southern Sakhalin Island	46°30′N 143°30′E
Aomori	Japan/Honshu	40°50′N 140°43′E
Aomori Prefecture	Japan/Honshu	41°00′N 140°30′E
Ariake Bay	Japan/Kyushu	33°00′N 130°20′E
Assabu River	Japan/Hokkaido/Sea of Japan coast	41°55′N 140°09′E
Atka Island	USA/Alaska/Aleutian Islands	52°05′N 174°40′W
Atnarko River	Canada/British Columbia/Bella Coola River system	52°24′N 126°06′W
Atsuta River	Japan/Hokkaido/Sea of Japan coast	43°25′N 141°24′E
Attu Island	USA/Alaska/Aleutian Islands	52°55′N 173°00′E
Auke Bay	USA/southeastern Alaska	58°24′N 134°40′W
Auke Creek	USA/southeastern Alaska	58°22′N 134°38′W
Auke Lake	USA/southeastern Alaska	58°24′N 134°40′W
Avacha Bay	USSR/southeastern Kamchatka	53°00′N 158°30′E
Avvaakumouka River	USSR/Primore	43°21′N 134°46′E
Ayakulik	USA/Alaska/Kodiak Island	57°10′N 154°35′W
Babine Lake	Canada/British Columbia/Skeena River system	54°40′N 126°00′W
Babine River	Canada/British Columbia/Skeena River system	55°41′N 127°42′W
Baja California	Mexico	30°00′N 115°00′W
Baker Lake	Canada/Northwest Territories	64°10′N 95°30′W
Bakhura River	USSR/southeastern Sakhalin	47°13′N 143°01′E
Baltic Sea	Northern Europe	56°00′N 18°00′E
Bano River	Japan/Honshu/Mie Prefecture/Kizu River system	34°44′N 136°14′E
Baranof Island	USA/southeastern Alaska	57°00′N 135°00′W
Bare Lake	Canada/British Columbia/Fraser River system	51°11′N 120°32′W
Barents Sea	USSR/Europe/Arctic Ocean	74°00′N 36°00′E
Barkley Sound	Canada/British Columbia/Vancouver Island	48°40′N 125°10′W
Bathurst Inlet	Canada/Northwest Territories/Arctic	66°49′N 108°00′W
Bear Cave Mountain	Canada/Yukon	67°30′N 138°00′W
Bear Creek	USA/Alaska/Cook Inlet	60°15′N 150°50′W
Bear Harbor	USA/Alaska/Kuiu Island	56°06′N 134°13′W

Place names	Location	Co-ordinates
Bear Lake	USA/Alaska/Kodiak Island	57°20′N 153°20′W*
Bear River	Canada/British Columbia/Vancouver Island/north of Campbell River	50°01′N 125°15′W*
Bear River	Canada/British Columbia/Skeena River system	56°17′N 126°58′W
Beaufort Sea	USA-Canada/Arctic Ocean	72°00′N 140°00′W
Becharof Lake	USA/southwestern Alaska/Bristol Bay	58°00′N 156°30′W
Bella Bella	Canada/central coast of British Columbia	52°07′N 128°05′W
Bella Coola	Canada/central coast of British Columbia	52°23′N 126°45′W
Bella Coola River	Canada/central coast of British Columbia	52°22′N 126°43′W
Bellingham	USA/Washington	48°45′N 122°29′W
Benbow Dam	USA/California/south fork of Eel River	40°00′N 123°50′W*
Bering River	USA/Alaska/central coast of Alaska	60°10′N 144°19′W
Bering Sea	USA-USSR	60°00′N 175°00′W
Bering Strait	USSR-USA	65°30′N 169°00′W
Berners River	USA/Alaska	58°51′N 135°05′W
Beshenaya River	USSR/Amur River system	51°20′N 139°12′E
Big Beef Creek	USA/Washington/Hood Canal	47°39′N 122°46′W
Big Creek	USA/Oregon/lower Columbia River system	46°10′N 123°39′W
Big Lake	USA/Alaska/Cook Inlet	61°25′N 149°55′W
Big Qualicum River	Canada/British Columbia/Vancouver Island	49°24′N 124°37′W
Bira River	USSR/Amur River system	48°05′N 133°00′E
Biwa Lake	Japan/southern Honshu	35°15′N 136°05′E
Black Lake	USA/Alaska/Alaska Peninsula	56°29′N 159°00′W
Black Sea	Southeast Europe	43°00′N 35°00′E
Blaine	USA/Washington	49°00′N 122°44′W
Bogataya River	USSR/eastern Sakhalin	50°20′N 143°45′E
Bolotnaya River	USSR/southern Sakhalin	48°30′N 142°06′E
Bolshaya River	USSR/southwestern Kamchatka	52°40′N 156°10′E
Bonneville Hatchery	USA/northern Oregon/Columbia River	45°38′N 121°56′W
Botchi River	USSR/northern Primore	48°00′N 139°30′E
Bowron Lake	Canada/British Columbia/Fraser River system	53°14′N 121°23′W
Brinnon	USA/Washington/Hood Canal	47°41′N 122°54′W
Bristol Bay	USA/southwestern Alaska	58°00′N 159°00′W
Brooks Lake	USA/southwestern Alaska/Bristol Bay/Naknek River system	58°30′N 156°00′W
Buckland River	USA/Alaska/Kotzebue Sound	66°10′N 161°20′W
Burke Channel	Canada/central coast of British Columbia	51°55′N 127°53′W
Campbell River	Canada/British Columbia/Vancouver Island	50°01′N 125°15′W
Cannery Creek Hatchery	USA/Alaska/northwestern Prince William Sound/Port Wells	60°46′N 149°04′W*
Canyon Creek	USA/Alaska/Kodiak Island	57°16′N 153°59′W
Cape Chaplin	USSR/Chukchi Peninsula	64°24′N 172°10′E
Cape Erimo	Japan/Hokkaido	41°55′N 143°13′E
Cape Esan	Japan/southern Hokkaido	41°49′N 141°12′E
Cape Flattery	USA/Washington	48°24′N 124°43′W
Cape Navarin	USSR/Kamchatka	62°17′N 179°13′E
Cape Nosappu	Japan/Hokkaido	43°29′N 146°00′E
Cape Nyudo	Japan/northwestern Honshu	40°00′N 139°42′E
Cape Olyutorskiy	USSR/Far East	59°58′N 170°25′E
Cape Prince of Wales	USA/northwestern Alaska	65°35′N 168°05′W

Place names	Location	Co-ordinates
Cape Shipunskiy	USSR/east coast of Kamchatka	53°07′N 160°05′E
Cape Shirakami	Japan/Hokkaido	41°25′N 140°03′E
Cape Shiretoko	Japan/Hokkaido	44°24′N 145°20′E
Cape Soya	Japan/Hokkaido	45°33′N 141°58′E
Cape Tappi	Japan/northern Honshu	41°14′N 140°21′E
Cape Terpeniya	USSR/east coast of Sakhalin Island	48°38′N 144°43′E
Capilano Hatchery	*See* Capilano River	
Capilano River	Canada/south coast of British Columbia	49°19′N 123°08′W
Carnation Creek	Canada/British Columbia/Vancouver Island	48°55′N 125°00′W
Cedar River	USA/Washington/Lake Washington	47°28′N 122°15′W
Central Ferry	USA/Washington/Snake River/Columbia River system	46°37′N 117°49′W
Chatham Sound	Canada/British Columbia/Dixon Entrance	54°22′N 130°35′W
Chatham Strait	USA/southeastern Alaska	57°00′N 134°40′W
Chauekuktuli Lake	USA/Alaska/Bristol Bay/Nushagak River system	60°05′N 158°30′W
Cheakamus River	Canada/south coast of British Columbia	49°47′N 123°10′W
Chehalis River	USA/Washington/Olympic Peninsula	46°58′N 123°49′W
Chehalis River	Canada/British Columbia/lower Fraser River system	49°16′N 121°56′W
Chemainus River	Canada/British Columbia/Vancouver Island	48°55′N 123°43′W
Chiba Prefecture	Japan/Honshu	35°45′N 140°07′E
Chichagof Island	USA/Alaska	57°40′N 136°00′W
Chignik Lake	USA/Alaska/Alaska Peninsula	56°18′N 158°45′W
Chignik River	USA/Alaska/Alaska Peninsula	56°17′N 158°38′W
Chihase River	Japan/west coast of Hokkaido	42°31′N 139°30′E
Chikugo River	Japan/Kyushu	33°10′N 130°20′E
Chilkat Lake	USA/southeastern Alaska/Lynn Canal	59°21′N 135°56′W
Chilkat River	USA/southeastern Alaska/Haines	59°11′N 135°23′W
Chilko Lake	Canada/British Columbia/Fraser River system	51°16′N 124°04′W
Chilko River	Canada/British Columbia/Fraser River system	52°06′N 123°27′W
Chilkoot Lake	USA/southeastern Alaska/Lynn Canal	59°20′N 135°33′W
Chilliwack Lake	Canada/British Columbia/Fraser River system	49°10′N 122°01′W
Chilliwack River	Canada/southern British Columbia-USA/Washington	49°09′N 121°54′W
Chitose River	Japan/Hokkaido	42°50′N 141°39′E
Chollanamdo	Korean Peninsula/western Korea Strait	– –
Chongjin	East coast of North Korea	41°05′N 129°55′E
Chukchi Sea	USSR-USA/Arctic Ocean	69°00′N 171°00′W
Chumikan	USSR/west coast of Sea of Okhotsk	54°40′N 135°15′E
Churui River	Japan/Hokkaido	43°43′N 145°06′E
Clark Lake	USA/Alaska/Bristol Bay/Kvichak River system	60°15′N 154°20′W
Claro River	Chile/Aisen Province/Simpson River system	45°25′S 72°23′W*
Clearwater River	USA/Washington	47°35′N 124°17′W
Cold Bay	USA/Alaska/Aleutian Islands/Adak Island	55°20′N 162°40′W
Cold Lake	Canada/Alberta	54°28′N 110°15′W
Columbia River	USA/Washington/Oregon-Canada/British Columbia	46°15′N 124°05′W
Colville River	USA/Alaska/Beaufort Sea	70°20′N 150°10′W
Cook Inlet	USA/central coast of Alaska	60°30′N 152°00′W
Coos Bay	USA/Oregon	43°23′N 124°12′W
Copper Creek	Canada/British Columbia/Queen Charlotte Islands/Moresby Island	53°10′N 131°48′W
Copper River	USA/Alaska/central coast of Alaska	60°20′N 145°00′W

Place names	Location	Co-ordinates
Coppermine River	Canada/Northwest Territories	67°50′N 115°12′W
Courtenay River	Canada/British Columbia/Vancouver Island	49°41′N 125°12′W
Cowichan Bay	Canada/British Columbia/Vancouver Island	48°42′N 123°32′W
Cowichan River	Canada/British Columbia/Vancouver Island	48°46′N 123°38′W
Cowlitz River	USA/Washington/Columbia River system	46°08′N 122°56′W
Cowlitz River Hatchery	USA/Washington/Columbia River system	46°05′N 122°53′W*
Crater Lake	USA/Oregon	42°50′N 122°10′W
Crescent Lake	USA/southwestern Alaska/Cook Inlet	– –
Crooked Creek	USA/Alaska/Prince William Sound	61°07′N 146°20′W
Cross Sound	USA/Alaska	58°08′N 136°35′W
Cultus Lake	Canada/British Columbia/lower Fraser River system	49°03′N 121°59′W
Current River	Canada/Ontario/Lake Superior	48°27′N 89°11′W
Dai Ko-Kei	Taiwan/Formosan Highland	24°19′N 120°33′E
Datlamen Creek	Canada/British Columbia/Queen Charlotte Islands/Graham Island	53°34′N 132°29′W
Datta Bay	USSR/Primore	49°18′N 140°23′E
Dawson	Canada/Yukon	64°04′N 139°24′W
Delta River	USA/Alaska/Yukon River system	64°50′N 148°00′W
Dennys River	USA/Maine	44°54′N 67°14′W
Departure Bay	Canada/British Columbia/Vancouver Island	49°12′N 123°57′W
Depoe Bay	USA/Oregon	44°00′N 124°04′W
Deschutes River	USA/Oregon/Columbia River system	45°40′N 120°57′W
Disappearance Creek	USA/Alaska/southeastern Alaska	55°05′N 132°25′W
Discovery Bay	USA/Washington/Puget Sound	– –
Dixon Entrance	Canada/British Columbia/north of Queen Charlotte Islands	54°25′N 132°00′W
Docee River	Canada/central coast of British Columbia	51°10′N 127°30′W
Dog Salmon River	USA/Alaska/Kodiak Island/Fraser Lake	57°15′N 154°08′W*
Drift Creek	USA/Oregon	44°25′N 124°00′W
Duckabush River	USA/Washington/Puget Sound/Hood Canal	47°38′N 122°55′W
Ducktrap River	USA/Maine	– –
Dungeness River	USA/Washington/Puget Sound/Hood Canal	48°08′N 123°07′W
Eagle Creek	USA/Oregon/Columbia River system	44°45′N 117°10′W
East River	USA/southeastern Alaska/Alsek River system	59°10′N 138°10′W
East Siberian Sea	USSR/Arctic Ocean	72°00′N 170°00′E
Edo River	Japan/Honshu/Tokyo	35°37′N 139°53′E
Eel River	USA/California	40°34′N 124°10′W
Egavik River	USA/Alaska/Norton Sound	64°02′N 160°55′W
Egegik River	USA/southwestern Alaska/Bristol Bay	58°11′N 157°24′W
Egushik River	USA/southwestern Alaska/Bristol Bay	58°50′N 158°55′W
Elk River	USA/Oregon	42°45′N 124°30′W
Elokomin River	USA/Washington/lower Columbia River system	46°10′N 123°40′W
Elrington Passage	USA/central coast of Alaska	60°00′N 148°04′W
Elva Creek	USA/southwestern Alaska/Bristol Bay/Wood River system	59°38′N 159°09′W
Elva Lake	USA/southwestern Alaska/Bristol Bay/Wood River system	59°38′N 159°09′W
Erimo Point	Japan/Hokkaido	41°55′N 143°13′E
Eshamy Lake	USA/central coast of Alaska/Prince William Sound	60°29′N 148°00′W
Etolin Island	USA/southeastern Alaska	56°10′N 132°30′W

Place names	Location	Co-ordinates
Etorofu Island	Same as Iturup Island	45°00′N 148°00′E
Eva Lake	USA/southeastern Alaska/Baranof Island	57°24′N 135°06′W
Evans Island	USA/Alaska/Prince William Sound	60°03′N 148°04′W
Everett	USA/Washington	47°59′N 122°14′W
Excursion Inlet	USA/southeastern Alaska/Icy Strait	58°23′N 135°23′W
Fall Creek	USA/California	– –
Fall Creek	USA/Oregon	– –
Falls Creek	USA/southeastern Alaska	– –
Farallon Islands	USA/California	37°40′N 123°00′W
Finch Creek	USA/Washington/Hood Canal	47°24′N 123°08′W
Finnmark	Northern Norway	70°00′N 22°00′E
Fish River	USA/Alaska/Norton Sound	64°30′N 163°00′W*
Fishing Branch River	Canada/Yukon	67°00′N 138°00′W*
Forfar Creek	Canada/British Columbia/Fraser River system	55°00′N 125°25′W
Formosan Highlands	Taiwan	23°30′N 121°00′E
Francois Lake	Canada/British Columbia/Fraser River system	54°03′N 125°45′W
Fraser Canyon	Canada/British Columbia	49°38′N 121°25′W
Fraser Lake	Canada/British Columbia/Fraser River system	54°03′N 124°31′W
Fraser River	Canada/British Columbia	49°09′N 123°12′W
Frazer Lake	USA/Alaska/Kodiak Island	57°15′N 154°08′W
Frost Fish Creek	Canada/New Brunswick	– –
Fukaura	Japan/Honshu/Sea of Japan coast/Aomori Prefecture	40°40′N 139°52′E
Fukui Prefecture	Japan/central Honshu	36°00′N 137°00′E
Fukuoka Prefecture	Japan/Kyushu	33°40′N 130°40′E
Fulton River	Canada/British Columbia/Skeena River system	54°48′N 126°09′W
Galiano Island	Canada/British Columbia/Gulf Islands	48°55′N 123°30′W
Gifu Prefecture	Japan/Honshu	35°30′N 136°50′E
Gizhiga River	USSR/north coast of Sea of Okhotsk	62°00′N 160°34′W
Glendale River	Canada/central coast of British Columbia	50°39′N 125°44′W
Goat Creek	USA/central coast of Alaska/Copper River system	62°23′N 143°22′W
Goodnews River	USA/western Alaska	59°06′N 161°38′W
Goose Creek	Canada/Hudson Bay	– –
Grace Creek	USA/Alaska	– –
Graham Island	Canada/British Columbia/Queen Charlotte Islands	53°30′N 132°30′W
Granby Reservoir	USA/Colorado	40°09′N 105°50′W
Grande Ronde River	USA/northeastern Oregon	46°06′N 117°02′W
Grays Harbor	USA/Washington/Olympic Peninsula	46°55′N 124°05′W
Grays River	USA/Washington/lower Columbia River system	46°22′N 123°36′W
Great Central Lake	Canada/British Columbia/Vancouver Island	49°21′N 125°15′W
Great Lakes	USA-Canada	45°00′N 83°00′W
Green River	USA/Washington	47°24′N 122°15′W
Grosvenor River	USA/Alaska/Bristol Bay/Naknek River system	58°40′N 155°00′W
Gulf Islands	Canada/British Columbia/Strait of Georgia	48°45′N 123°00′W
Gulf of Alaska	USA/Alaska/eastern North Pacific Ocean	58°00′N 145°00′W
Gulf of Kamchatka	USSR/Kamchatka	55°35′N 162°21′E
Gulf of Korf	USSR/Kamchatka	60°20′N 165°50′E

Place names	Location	Co-ordinates
Gulf of Kronotskiy	USSR/Kamchatka	54°00′N 160°30′E
Gulf of Riga	USSR/Baltic Sea	57°30′N 23°35′E
Gulkana River	USA/central coast of Alaska/Copper River system	63°13′N 145°23′W
Hagan Arm	Canada/British Columbia/Babine Lake/Skeena River system	54°59′N 126°12′W
Hagemeister Island	USA/Alaska/Bristol Bay	58°39′N 160°54′W
Haines	USA/Alaska/Lynn Canal	59°11′N 135°23′W
Hakodate	Japan/Hokkaido	41°46′N 140°44′E
Hamatonbetsu	Japan/northeastern Hokkaido	45°08′N 142°24′E
Hamkyong	North Korea	– –
Hammond Bay	Canada/British Columbia/Vancouver Island	49°14′N 123°58′W
Hanson Creek	USA/southwestern Alaska/Bristol Bay/Wood River system	59°20′N 158°45′W
Hare Lake	USA/Minnesota	– –
Harris River	USA/southeastern Alaska	55°27′N 132°41′W
Harrison Lagoon Creek	USA/Alaska/Prince William Sound	60°59′N 148°12′W
Harrison Lake	Canada/British Columbia/lower Fraser River system	49°33′N 121°50′W
Harrison Rapids	Canada/British Columbia/lower Fraser River system/Harrison River	49°19′N 121°48′W
Harrison River	Canada/British Columbia/lower Fraser River system	49°13′N 121°57′W
Hecate Strait	Canada/British Columbia/between Queen Charlotte Islands and mainland British Columbia	53°00′N 131°00′W
Hell's Gate	Canada/British Columbia/Fraser River system	49°47′N 121°27′W
Hemlock Lake	USA/Michigan	42°43′N 77°36′W
Henderson Lake	Canada/British Columbia/Vancouver Island	49°06′N 125°03′W
Herman Creek	USA/southeastern Alaska/on east Behm canal	– –
Hidaka District	Japan/southern Hokkaido	42°30′N 142°30′E
Hidden Creek	USA/southwestern Alaska/Bristol Bay/Lake Brooks	58°30′N 156°00′W*
Hidden Lake	USA/southwestern Alaska/Cook Inlet/Skilak Lake	60°20′N 150°30′W*
Hirakura River	Japan/Honshu/Mie Prefecture	34°34′N 136°16′E
Hiroshima Prefecture	Japan/Honshu	34°30′N 133°00′E
Hobiton Lake	Canada/British Columbia/Vancouver Island	48°45′N 124°49′W
Hobo Creek	USA/Alaska/Prince William Sound	60°57′N 148°14′W
Hokkaido	Japan	44°00′N 143°00′E
Holman Island	Canada/Arctic	70°42′N 117°41′W
Honshu	Japan	36°00′N 138°00′E
Hood Bay Creek	USA/southeastern Alaska/Admiralty Island	57°23′N 134°24′W
Hood Canal	USA/Washington/Puget Sound	47°35′N 123°00′W
Hood Canal Hatchery	USA/Washington/Hood Canal/Finch Creek	47°33′N 122°59′W
Hoodsport Hatchery	USA/Washington/Hood Canal	47°24′N 123°10′W
Hooknose Creek	Canada/central coast of British Columbia	52°07′N 127°50′W
Hope	Canada/British Columbia	49°21′N 121°28′W
Horobetsu River	Japan/Hokkaido/Pacific Ocean coast	42°12′N 141°15′E
Horonai River	Japan/Hokkaido/Sea of Okhotsk coast	42°14′N 143°08′E
Horsefly River	Canada/British Columbia/Fraser River system	52°28′N 121°23′W
Hudson Bay	Canada/Ontario/Manitoba/Northwest Territories	60°00′N 86°00′W
Humpback Creek	USA/Alaska/Prince William Sound	60°37′N 145°41′W
Humptulips River	USA/Washington/Grays Harbor	47°04′N 124°02′W
Hunts Creek	Canada/British Columbia/Vancouver Island/Big Qualicum River system	49°23′N 124°39′W

Place names	Location	Co-ordinates
Hyunghsan River	South Korea/Sea of Japan coast	36°00′N 129°26′E
Ibi River	Japan/Honshu	35°03′N 136°42′E
Ice Harbour Dam	USA/Washington/Snake River/Columbia River system	46°16′N 118°51′W
Icha River	USSR/western Kamchatka	55°40′N 156°00′E
Ichani River	Japan/Hokkaido	43°39′N 145°08′E
Icy Strait	USA/southeastern Alaska	58°18′N 135°30′W
Igushik River	USA/southwestern Alaska/Bristol Bay	58°40′N 158°50′W
Ikpikpuk drainage	USA/northern Alaska	70°49′N 154°23′W
Iliamna Lake	USA/southwestern Alaska/Bristol Bay/Kvichak River system	59°30′N 155°00′W
Iliuk Arm	USA/southwestern Alaska/Bristol Bay/Naknek Lake	58°25′N 155°40′W
Im River	USSR/Amur River system	52°48′N 138°22′E
Inanusi River	USSR/Sakhalin	– –
Inmachuk River	USA/Alaska/Kotzebue Sound	66°04′N 162°42′W
Ise Bay	Japan/Honshu/Pacific Coast	34°40′N 136°40′E
Ishikari River	Japan/Hokkaido	43°00′N 141°50′E
Ishikawa Prefecture	Japan/Honshu	36°45′N 136°45′E
Iski River	USSR/Amur River system	53°26′N 140°56′E
Iskut River	Canada/British Columbia/Stikine River system	56°45′N 131°47′W
Italio River	USA/southeastern Alaska	59°25′N 139°00′W
Iturup Island	USSR/part of Kuril Islands	45°00′N 148°00′E
Iwaobetsu River	Japan/Hokkaido/Sea of Okhotsk coast	44°06′N 145°02′E
Iwate Prefecture	Japan/Honshu	39°37′N 141°22′E
Jintsu River	Japan/Honshu/Sea of Japan coast	36°45′N 137°14′E
Johnstone Strait	Canada/British Columbia/between Vancouver Island and mainland British Columbia	50°50′N 126°00′W
Jonah Creek	USA/Alaska/Prince William Sound	60°50′N 148°10′W
Jones Creek	Canada/British Columbia/lower Fraser River system	49°18′N 118°28′W
Juan de Fuca Strait	USA-Canada/between Vancouver Island and Washington State	48°20′N 124°00′W
Kachemak Bay	USA/Alaska/Cook Inlet	59°40′N 151°30′W
Kagalaska Island	USA/Alaska/Aleutian Islands	51°49′N 176°20′W
Kakweiken River	Canada/British Columbia/Thompson Sound	50°40′N 126°02′W
Kalama River	USA/Washington/Columbia River system	46°01′N 122°40′W
Kalininka River	USSR/west coast of Sakhalin	47°00′N 142°05′E
Kallum River	Canada/British Columbia/Skeena River system	54°31′N 128°39′W
Kalyger River	USSR/Kamchatka/Kamchatka River system	52°35′N 141°16′E
Kamchatka Peninsula	USSR/Far East	55°00′N 160°40′E
Kamchatka River	USSR/east coast of Kamchatka	56°15′N 162°30′E
Kanektok River	USA/western Alaska/Kuskokwim Bay	59°45′N 161°59′W
Kara Sea	USSR/Arctic	75°00′N 65°00′E
Karaginskiy District	USSR/eastern Kamchatka	58°00′N 164°00′E
Karaginskiy Gulf	USSR/eastern Kamchatka	59°02′N 164°00′E
Karluk Lake	USA/Alaska/Kodiak Island	57°20′N 154°05′W
Karluk River	USA/Alaska/Kodiak Island	57°34′N 154°32′W
Karymaiskiy Spring	USSR/Kamchatka/Bolshaya River system	52°30′N 157°00′E
Kasilof Lake	USA/Alaska/Cook Inlet	60°20′N 151°15′W

Place names	Location	Co-ordinates
Kasilof River	USA/Alaska/Kenai Peninsula	60°21′N 151°17′W
Kelly River	USA/Alaska/Kotzebue Sound	67°55′N 162°21′W
Kenai Lake	USA/Alaska/Cook Inlet	60°25′N 149°30′W
Kenai Peninsula	USA/central coast of Alaska	60°10′N 150°00′W
Kenai River	USA/Alaska/Cook Inlet	60°33′N 151°16′W
Kenichi River	Japan/Hokkaido/Sea of Japan coast	42°05′N 139°58′E
Kennedy Lake	Canada/British Columbia/Vancouver Island	49°04′N 125°34′W
Ketchikan	USA/southeastern Alaska	55°25′N 131°40′W
Kettle Falls	USA/Washington	48°38′N 118°03′W
Khabarovsk Region	USSR/Far East	52°00′N 135°00′W
Khor River	USSR/Amur River system	48°00′N 135°00′W
Khungari River	USSR/Amur River system	50°06′N 136°57′E
Khvostovka River	USSR/Sakhalin	46°08′N 142°13′E
Kikhchik River	USSR/southwestern Kamchatka	53°57′N 156°14′E
Kingcome River	Canada/central coast of British Columbia	50°56′N 126°12′W
Kisa Kata	Japan/Honshu	35°13′N 139°54′E
Kiska Island	USA/Alaska/Aleutian Islands	52°00′N 177°30′W
Kiso River	Japan/Honshu	39°30′N 137°00′E
Kispiox River	Canada/British Columbia/Skeena River system	55°21′N 127°42′W
Kitimat River	Canada/north coast of British Columbia	54°00′N 128°40′W
Kitoi Bay Hatchery	USA/Alaska/60 km from Seal Bay Creek	58°11′N 152°21′W
Kitoi Lake	*See* Little Kitoi Lake	
Kitwanga River	Canada/British Columbia/Skeena River system	55°06′N 128°04′W
Kivalina River	USA/Alaska/Kotzebue Sound	67°45′N 164°40′W
Kizaki Lake	Japan/Honshu/Nagano Prefecture	36°33′N 137°51′E
Kizu River	Japan/Honshu/Mie Prefecture	34°40′N 135°50′E
Klamath River	USA/northern California	41°32′N 124°03′W
Kleanza Creek	Canada/British Columbia/Skeena River system	54°36′N 128°25′W
Klickitat River	USA/Washington/Columbia River system	45°42′N 121°15′W
Klukshu River	Canada/Yukon/Tatshenshini River system	60°05′N 137°00′W
Knutson Bay	USA/southwestern Alaska/Iliamna Lake	59°48′N 154°15′W
Kobuk River	USA/Alaska/Kotzebue Sound	66°50′N 161°40′W
Kochi Prefecture	Japan/Shikoku	33°30′N 133°30′E
Kodiak Island	USA/Alaska/Gulf of Alaska	57°20′N 153°20′W
Koksilah River	Canada/British Columbia/Vancouver Island	48°45′N 123°39′W
Kola Peninsula	USSR/Murmansk Region	68°50′N 33°00′E
Kolpakova River	USSR/west coast of Kamchatka	54°30′N 155°30′E
Kolyma River	USSR/Siberia/East Siberian Sea	68°45′N 161°15′E
Komandorskiy Islands	USSR/Bering Sea off east coast of Kamchatka	55°00′N 167°00′E
Konbumori	Japan/Hokkaido	42°57′N 144°32′E
Koozata River	USA/Alaska/St. Lawrence Island	63°20′N 170°30′W
Koppi River	USSR/northern Primore	48°30′N 140°00′E
Korea Strait	Japan-South Korea	35°00′N 129°00′E
Kotzebue Sound	USA/northwestern Alaska	66°45′N 163°00′W
Kronotsk Lake	USSR/east coast of Kamchatka	54°10′N 160°15′E
Kronotsk River	USSR/east coast of Kamchatka	54°30′N 160°55′E
Krutogorovo River	USSR/west coast of Kamchatka	55°05′N 155°40′E
Kuiu Island	USA/southeastern Alaska	56°40′N 134°00′W

Place names	Location	Co-ordinates
Kukaklek Lake	USA/Alaska/Bristol Bay/Alagnak River system	59°14′N 155°20′W
Kukhtuy River	USSR/northern coast of Sea of Okhotsk	59°38′N 143°10′E
Kukpuk River	USA/northwestern Alaska/Chukchi Sea	68°19′N 166°20′W
Kumamoto Prefecture	Japan/Kyushu	32°50′N 130°42′E
Kunashir Island	USSR/part of Kuril Islands	44°00′N 146°00′E
Kur River	USSR/Amur River system	48°49′N 134°15′E
Kuril Islands	USSR/Sea of Okhotsk	46°00′N 150°00′E
Kurilka River	USSR/Kuril Islands/Iturup Island	45°13′N 147°52′E
Kushiro	Japan/southeastern Hokkaido	42°58′N 144°24′E
Kushiro District	Japan/southeastern Hokkaido	43°00′N 144°00′E
Kushiro River	Japan/southeastern Hokkaido	42°58′N 144°24′E
Kuskokwim River	USA/western Alaska	60°41′N 161°59′W
Kutlushnaya River	USSR/northeastern Kamchatka	60°19′N 165°50′E
Kuybysheva River	USSR/Kuril Islands/Iturup Island	45°00′N 147°45′E
Kuzitrin River	USA/northwestern Alaska	65°10′N 165°25′W
Kuzuryu River	Japan/Honshu/Fukui Perfecture	36°14′N 136°09′E
Kvichak River	USA/southwestern Alaska/Bristol Bay	58°52′N 157°03′W
Kyongsan	South Korea	35°48′N 128°43′E
Kyoto Prefecture	Japan/Honshu	35°02′N 135°45′E
Kyrganik River	USSR/southeastern Kamchatka	54°00′N 158°00′E
Kyushu	Japan	33°00′N 131°00′E
La Jolla	USA/California	32°50′N 117°16′W
La Perouse Strait	Japan-USSR/between Hokkaido and Sakhalin	45°50′N 142°00′E
Lake Akan	Japan/Hokkaido	43°26′N 144°05′E
Lake Aleknagik	USA/southwestern Alaska/Bristol Bay/Wood River system	59°17′N 158°37′W
Lake Almanor	USA/California	40°15′N 121°10′W
Lake Azabache	USSR/eastern Kamchatka/Kamchatka River system	56°00′N 161°50′E
Lake Berryessa	USA/California	38°35′N 122°35′W
Lake Beverley	USA/southwestern Alaska/Bristol Bay/Wood River system	59°40′N 158°50′W
Lake Blizhnee	USSR/southeastern Kamchatka/Paratunka River system	52°30′N 157°50′E
Lake Brooks	USA/southwestern Alaska/Bristol Bay/Naknek River system	58°21′N 155°47′W
Lake Coville	USA/southwestern Alaska/Bristol Bay/Naknek River system	58°50′N 155°00′W
Lake Dalnee	USSR/southeastern Kamchatka/Paratunka River system	52°30′N 157°54′E
Lake Erie	Canada-USA	42°00′N 82°00′W
Lake Grosvenor	USA/southwestern Alaska/Bristol Bay/Naknek River system	58°40′N 155°00′W
Lake Huron	Canada-USA	45°00′N 82°00′W
Lake Krasivoye	USSR/Kuril Islands/Iturup Island	45°00′N 148°00′E*
Lake Kulik	USA/southwestern Alaska/Bristol Bay/Wood River system	59°49′N 158°50′W
Lake Kuril	USSR/southern Kamchatka/Ozernaya River system	51°30′N 157°30′E
Lake Michigan	USA-Canada	44°00′N 87°00′W
Lake Nerka	USA/southwestern Alaska/Bristol Bay/Wood River system	59°35′N 159°05′W
Lake Nunavaugaluk	USA/southwestern Alaska/Bristol Bay	59°15′N 158°55′W
Lake Nuyakuk	USA/southwestern Alaska/Bristol Bay/Nushagak River system	59°05′N 158°10′W
Lake Ontario	Canada-USA	43°30′N 78°00′W
Lake Palana	USSR/northwestern Kamchatka	59°05′N 160°00′E*
Lake Pekulheyskoye	USSR/northern Bering Sea	62°35′N 177°26′E
Lake Quinault	USA/Washington/Olympic Peninsula	47°27′N 123°50′W

Place names	Location	Co-ordinates
Lake Sammamish	USA/west-central Washington	47°36′N 122°22′W
Lake Sarannoye	USSR/Komandorskiy Islands	55°17′N 166°13′E
Lake Superior	Canada/Ontario-USA/Minnesota	48°00′N 87°00′W
Lake Ualik	USA/southwestern Alaska/Bristol Bay	59°07′N 159°30′W
Lake Vaamochkino	USSR/northern Bering Sea	62°30′N 176°40′E
Lake Washington	USA/Washington	47°37′N 122°15′W
Lake Whatcom	USA/Washington	48°45′N 122°25′W
Lakelse Lake	Canada/British Columbia/Skeena River system	54°23′N 128°33′W
Lakelse River	Canada/British Columbia/Skeena River system	54°26′N 128°47′W
Laptev Sea	USSR/Arctic Ocean	75°00′N 125°00′E
Larson Lake	USA/Alaska/Cook Inlet	62°30′N 150°30′W
Lemhi River	USA/Idaho/Columbia River system	45°12′W 113°53′W
Lena River	USSR/Siberia/Laptev Sea	73°20′N 126°20′E
Lesnaya Hatchery	USSR/southeastern Sakhalin	46°54′N 143°05′E
Lesnaya River	USSR/Sakhalin	48°34′N 142°46′E
Lewis River	USA/Washington	46°10′N 122°00′W
Liard River	Canada/Northwest Territories/upper Mackenzie River system	60°00′N 123°48′W
Little Goose Dam	USA/southeastern Washington/Snake River/Columbia River system	46°35′N 118°01′W
Little Kitoi Lake	USA/Alaska/Kodiak Island	58°11′N 152°27′W
Little Port Walter	USA/southeastern Alaska	56°23′N 134°38′W
Little Qualicum River	Canada/British Columbia/Vancouver Island	49°22′N 124°30′W
Little River	Canada/British Columbia/Shuswap Lake/Fraser River system	50°52′W 119°37′W
Little Shuswap Lake	Canada/British Columbia/Fraser River system	50°51′N 119°38′W
Little Togiak Lake	USA/southwestern Alaska/Bristol Bay/Wood River system	59°15′N 159°00′W
Little Togiak River	USA/southwestern Alaska/Bristol Bay/Wood River system	59°15′W 159°00′W
Long Lake	Canada/central coast of British Columbia	50°36′N 127°16′W
Long Point	Canada/Ontario/Lake Erie	42°33′N 80°04′W
Lost Creek	USA/southeastern Alaska	55°27′N 139°30′W
Lovers Cove Creek	USA/southeastern Alaska	56°23′N 134°43′W
Lovetskaya River	USSR/Primore	–
Lower Babine River	Canada/British Columbia/Skeena River system	56°05′N 126°35′W
Lumni Bay	USA/Washington	48°45′N 122°45′W
Lumni River	USA/Washington	48°48′N 122°45′W
Lynn Canal	USA/southeastern Alaska	59°00′N 135°00′W
Lynx Lake	USA/southwestern Alaska/Bristol Bay/Wood River system	59°30′N 158°50′W
Lyutoga River	USSR/Sakhalin	– –
Mabechi River	Japan/Honshu/Aomori Prefecture	40°30′N 141°30′E
McClinton Creek	Canada/British Columbia/Queen Charlotte Islands	53°38′N 132°36′W
McCloud River	USA/California	40°46′N 122°18′W
Mackenzie Delta	Canada/Northwest Territories	67°30′N 135°00′W
Mackenzie River	Canada/Northwest Territories	69°15′N 134°00′W
Magadan District	USSR/Far East	65°00′N 160°00′E
Magunkotan River	USSR/southern Sakhalin	– –
Makshimovka River	USSR/Primore	46°04′N 137°48′E
Malaspina Glacier	USA/Alaska	59°50′N 140°40′W
Mamin River	Canada/British Columbia/Queen Charlotte Islands	53°37′N 132°19′W
Mashikova River	USSR/Primore	47°00′N 138°22′E

Place names	Location	Co-ordinates
Masset Inlet	Canada/British Columbia/Queen Charlotte Islands	53°43′N 132°20′W
Masuhoro River	Japan/Hokkaido/Sea of Japan coast	45°30′N 141°55′E
Mathers Creek	Canada/British Columbia/Queen Charlotte Islands	53°01′N 131°46′W
Matsugasaki	Japan/Honshu/Sea of Japan coast/Akita Prefecture	39°40′N 140°04′E
Maybeso Creek	USA/southeastern Alaska	55°29′N 132°39′W
Meadow Creek	USA/Alaska/Kodiak Island/Karluk Lake	57°18′N 154°00′W
Mena River	Japan/Hokkaido/Shiribetsu River system	42°48′N 140°27′E
Merle Collins Reservoir	USA/California/Yuba County	39°16′N 121°17′W*
Methow River	USA/Washington/upper Columbia River system	48°03′N 119°55′W
Meziaden Lake	Canada/British Columbia/Nass River system	56°03′N 129°17′W
Mie Prefecture	Japan/Honshu	34°30′N 136°30′E
Mikawa Bay	Japan/Honshu	34°45′N 137°00′E
Mill Creek	USA/California/Sacramento River system	38°03′N 121°56′W*
Minter Creek	USA/Washington/Puget Sound	47°22′N 122°41′W
Miomote River	Japan/Honshu/Sea of Japan coast/Niigata Prefecture	38°10′N 139°24′E
Mission	Canada/British Columbia	49°14′N 122°20′W
Mitsuishi River	Japan/southern Hokkaido	42°15′N 142°33′E
Miya River	Japan/Honshu/Mie Prefecture	34°30′N 136°30′E
Miyagi Prefecture	Japan/Honshu	38°50′N 141°00′E
Miyako	Japan/Honshu/Pacific Ocean coast	39°38′N 141°59′E
Mogami River	Japan/Honshu/Sea of Japan coast	38°55′N 139°51′E
Mokoto River	Japan/Hokkaido/Sea of Okhotsk coast	43°55′N 144°30′E
Monbetsu	Japan/northern Hokkaido	44°20′N 143°20′E
Monbetsu	Japan/southern Hokkaido	42°28′N 142°10′E
Monterey Bay	USA/California	36°40′N 122°00′W
Moose Bay	USA/southwestern Alaska/Iliamna Lake/Kvichak River system	– –
Moresby Island	Canada/British Columbia/Queen Charlotte Islands	52°25′N 131°30′W
Morice River	Canada/British Columbia/Skeena River tributary	54°24′N 126°45′W
Morris Creek	Canada/British Columbia/Fraser River system/Harrison River tributary	49°18′N 121°53′W
Morris Slough	*See* Morris Creek	– –
Morrison Creek	Canada/British Columbia/Vancouver Island	49°41.N 125°01.W
Morrison Lake	Canada/British Columbia/Skeena River system	55°14′N 126°22′W
Motykleyka River	USSR/northern coast of Sea of Okhotsk	– –
Mt. St. Helens	USA/Washington	46°12′N 122°11′W
Muchka River	USSR/Kola Peninsula	68°50′N 33°00′E*
Mulchatna River	USA/southwestern Alaska/Bristol Bay/Nushagak River system	59°55′N 156°25′W
My River	USSR/Amur River system	– –
Nadina River	Canada/British Columbia/Fraser River system	53°59′N 126°31′W
Nagano Prefecture	Japan/Honshu	36°00′N 138°00′E
Nagara River	Japan/Honshu/Pacific Ocean coast	35°01′N 136°43′E
Nagasaki Prefecture	Japan/Kyushu	33°00′N 129°30′E
Naiba River	USSR/Sakhalin	47°26′N 142°40′E
Naknek Lake	USA/southwestern Alaska/Bristol Bay/Naknek River system	58°40′N 156°25′W
Naknek River	USA/southwestern Alaska/Bristol Bay	58°40′N 157°00′W
Naktong River	South Korea/flows into Korea Strait	35°07′N 129°03′E
Nanaimo River	Canada/British Columbia/Vancouver Island	49°08′N 123°54′W

Place names	Location	Co-ordinates
Nass River	Canada/northern coast of British Columbia	54°59′N 129°52′W
Nechako River	Canada/British Columbia/upper Fraser River system	53°56′N 122°42′W
Neiden River	Norway	69°42′N 29°25′E
Nemuro Peninsula	Japan/Hokkaido	43°20′N 146°00′E
Nemuro Strait	Japan/Hokkaido	44°00′N 145°36′E
Nenana River	USA/Alaska/Yukon River system	64°35′N 149°20′W
Nestucca River	USA/Oregon	45°16′N 123°57′W
Netarts Bay	USA/Oregon/Tillamook Bay	45°24′N 123°56′W
Niigata	Japan/Honshu/Sea of Japan coast	37°58′N 139°02′E
Niigata Prefecture	Japan/Honshu	37°09′N 138°30′E
Nile Creek	Canada/British Columbia/Vancouver Island	49°25′N 124°38′W
Nilkitkwa Lake	Canada/British Columbia/Skeena River system	55°22′N 126°39′W
Nimpkish Lake	Canada/British Columbia/Vancouver Island	50°25′N 126°59′W
Nimpkish River	Canada/British Columbia/Vancouver Island	50°34′N 126°59′W
Nishibetsu River	Japan/eastern Hokkaido	43°30′N 145°30′E
Nisqually River	USA/Washington/Puget Sound	47°05′N 122°45′W
Nitinat Hatchery	Canada/British Columbia/Vancouver Island	48°50′N 124°50′W
Nitinat Lake	Canada/British Columbia/Vancouver Island	48°45′N 124°45′W
Nitinat River	Canada/British Columbia/Vancouver Island	48°49′N 124°37′W
Noatak River	USA/Alaska/Kotzebue Sound	67°00′N 162°40′W
Nobusha River	Japan/west coast of Hokkaido	43°20′N 141°25′E
Nome River	USA/Alaska/Norton Sound	64°30′N 165°30′W
None River	Japan/Shikoku/Kochi Prefecture	33°30′N 134°17′E
Nonvianuk Lake	USA/Alaska/Bristol Bay/Alagnak River system	59°02′N 155°20′W
Nooksack River	USA/Washington	48°46′N 122°35′W
North Harbor River	Canada/Newfoundland/Saint Mary's Bay	47°12′N 53°40′W
North Nandai River	North Korea	– –
North Thompson River	Canada/British Columbia/Fraser River system	52°41′N 120°20′W
Norton Sound	USA/Alaska/Bering Sea	64°00′N 164°00′W
Noto Peninsula	Japan/Honshu	37°20′N 137°00′E
Novoselovka River	USSR/southwestern Sakhalin	47°39′N 141°59′E
Noyes Island	USA/southeastern Alaska	55°30′N 133°40′W
Nukabira Lake	Japan/Hokkaido	43°24′N 143°13′E
Nushagak Bay	USA/southwestern Alaska/Bristol Bay	58°30′N 158°30′W
Nushagak River	USA/southwestern Alaska/Bristol Bay	59°03′N 158°23′W
Nuyakuk River	USA/southwestern Alaska/Bristol Bay	59°55′N 158°15′W
Ochikho River	USSR/Sakhalin	– –
Oga Peninsula	Japan/Honshu/Akita Prefecture	39°55′N 139°45′E
Ogden Channel	Canada/northern coast of British Columbia	53°52′N 130°18′W
Ohkawa River	Japan/Honshu/Miyagi Prefecture	38°52′N 141°36′E
Oippe River	Japan/Honshu/Aomori Prefecture	41°10′N 141°23′E
Oirase River	Japan/Honshu/Aomori Prefecture	40°36′N 141°30′E
Ojika Peninsula	Japan/Honshu/Pacific Ocean coast	38°18′N 141°30′E
Okanagan River	USA-Canada/Columbia River system	49°00′N 119°25′W
Okhota River	USSR/northern coast of Sea of Okhotsk	60°00′N 142°30′E
Okhotsk	USSR/northern coast of Sea of Okhotsk	59°20′N 143°00′E
Okushiri Island	Japan/Sea of Japan	42°10′N 139°28′E

Place names	Location	Co-ordinates
Okutadami Lake	Japan/Honshu/Niigata Prefecture	37°09′N 139°00′E
Ola River	USSR/northern coast of Sea of Okhotsk	59°35′N 151°15′E
Old Crow	Canada/Yukon/Porcupine River	67°34′N 139°43′W
Old Toms Creek	USA/southeastern Alaska	55°24′N 132°23′W
Olsen Creek	USA/Alaska/Prince William Sound	60°51′N 146°05′W
Olya River	USSR/Kuril Islands/Iturup Island	44°54′N 147°30′E
Olympic Peninsula	USA/Washington	47°30′N 123°30′W
Olyutorskiy District	USSR/Kamchatka	60°30′N 169°30′E
Omono River	Japan/Honshu/Sea of Japan coast/Akita Prefecture	39°40′N 140°04′E
Onekotan Island	USSR/part of Kuril Islands	49°20′N 154°40′E
One-Shot Creek	USA/Alaska/Bristol Bay/Naknek River system	58°30′N 156°00′W*
Onnebetsu River	Japan/Hokkaido/Sea of Okhotsk coast	44°00′N 145°00′E
Oroville Lake	USA/California	– –
Osaka Prefecture	Japan/Honshu	34°30′N 135°30′E
Oshatsubo	Japan/Hokkaido/Tokachi District	43°20′N 143°30′E
Osoyoos Lake	USA/Washington-Canada/British Columbia	49°02′N 119°25′W
Ota River	Japan/Honshu/Hiroshima Prefecture	34°22′N 132°25′E
Otaru	Japan/Hokkaido	43°14′N 140°59′E
Otsuchi River	Japan/Honshu/Iwate Prefecture	39°22′N 141°54′E
Owikeno Lake	Canada/central coast of British Columbia	51°40′N 126°55′W
Oxbow River	USA/Washington	45°45′N 121°50′W
Oyster River	Canada/British Columbia/Vancouver Island	49°52′N 125°07′W
Ozernaya River	USSR/eastern Kamchatka	57°17′N 162°40′E
Ozernaya River	USSR/southwestern Kamchatka	51°23′N 156°30′E
Ozerpakh	USSR/Far East/mouth of Amur River	53°04′N 141°41′E
Packers Lake	USA/Alaska/Cook Inlet/Kalgin Island	60°30′N 151°56′W
Pallant Creek	Canada/British Columbia/Queen Charlotte Islands	53°03′N 132°02′W
Pallette Lake	USA/Wisconsin	46°00′N 89°30′W
Paramushir Island	USSR/part of Kuril Islands	50°30′N 155°30′E
Paratunka River	USSR/southeastern Kamchatka	52°58′N 158°14′E
Parotonaya River	USSR/southwestern Sakhalin	– –
Parvin Lake	USA/Colorado	40°35′N 105°05′W
Pedro Bay	USA/southwestern Alaska/Iliamna Lake/Kvichak River system	50°46′N 154°10′W
Pembroke	USA/Maine	44°55′N 67°08′W
Penobscot	USA/Maine	44°25′N 68°42′W
Peril Strait	USA/southeastern Alaska	57°30′N 135°13′W
Peter The Great Bay	USSR/Primore	43°00′N 132°00′E
Pick Creek	USA/Bristol Bay/Alaska/Wood River system/Lake Nerka	59°35′N 159°00′W*
Pierce Lake	Canada/Saskatchewan	54°30′N 109°42′W
Pinkut Creek	Canada/British Columbia/Skeena River system	54°27′N 125°27′W
Pitt River	Canada/British Columbia/Fraser River system	49°14′N 122°45′W
Point Barrow	USA/Alaska/Beaufort Sea	71°10′N 156°40′W
Point Hope	USA/Alaska/Chukchi Sea	68°20′N 166°50′W
Point No Point	USA/Washington/Admiralty Inlet	48°10′N 122°40′W
Porcupine Creek	USA/southeastern Alaska	56°07′N 132°39′W
Porcupine River	Canada/Yukon Territory	67°35′N 135°00′W
Poronya River	USSR/eastern Sakhalin	49°13′N 143°08′E

Place names	Location	Co-ordinates
Port Arthur Hatchery	Canada/Ontario/Lake Superior/Thunder Bay	48°27′N 89°12′W
Port Chamalu Bay	Mexico/Baja California	– –
Port Clarence	USA/northwestern Alaska	65°15′N 166°40′W
Port Fidalgo	USA/Alaska/Prince William Sound	60°47′N 146°45′W
Port Heiden	USA/Alaska/Bristol Bay	56°55′N 158°41′W
Port John	Canada/central coast of British Columbia	52°07′N 127°50′W
Port San Juan Hatchery	USA/Alaska/Prince William Sound	– –
Port Wells	USA/Alaska/Prince William Sound	60°46′W 149°04′W
Portland Canal	USA/southeastern Alaska-Canada/British Columbia	55°20′N 130°00′W
Posyet	USSR/southern Primore	42°40′N 130°50′E
Prairie Creek	USA/California	41°18′N 124°05′W
Pravda	USSR/southern Sakhalin	46°50′N 142°00′E
Priest Rapids	USA/Washington	46°39′N 119°55′W
Primore Region	USSR/Far East	46°00′N 135°00′E
Prince of Wales Island	USA/southeastern Alaska	55°40′N 133°00′W
Prince William Sound	USA/central coast of Alaska	60°45′N 147°00′W
Principe Channel	Canada/northern coast of British Columbia	53°29′N 129°59′W
Pritornaya River	USSR/eastern Sakhalin	49°38′N 144°02′E
Proster Gulf	USSR/Kuril Islands/Iturup Island	45°30′N 148°20′E
Puget Sound	USA/Washington	47°50′N 122°30′W
Pulaski	USA/New York/Lake Ontario	43°34′N 76°06′W
Puntledge Hatchery	Canada/British Columbia/Vancouver Island	49°42′N 125°00′W
Puntledge River	Canada/British Columbia/Vancouver Island	49°42′N 125°00′W
Puyallup River	USA/Washington/Puget Sound	47°16′N 122°30′W
Qualicum River	Canada/British Columbia/Vancouver Island	49°24′N 124°37′W
Queen Charlotte Islands	Canada/northwestern British Columbia	53°00′N 132°00′W
Queen Charlotte Strait	Canada/British Columbia/north of Vancouver Island	50°45′N 127°15′W
Queets River	USA/Washington	47°32′N 124°19′W
Quesnel Lake	Canada/British Columbia/Fraser River system	52°30′N 121°00′W
Quesnel River	Canada/British Columbia/Fraser River system	52°58′N 122°31′W
Quinault River	USA/Washington	47°20′N 124°17′W
Quinsam River	Canada/British Columbia/Vancouver Island	50°02′N 125°18′W
Raft River	Canada/British Columbia/Fraser River system	51°38′N 119°59′W
Rampart River	USA/Alaska/Yukon River system	65°29′N 150°18′W
Red Lake	USA/Alaska/Kodiak Island	57°15′N 154°17′W
Red River	USA/Alaska/Kodiak Island	57°16′N 154°37′W
Redding	USA/California/Sacramento River	40°35′N 122°24′W
Reidovaya River	USSR/southern Kuril Islands	46°10′N 152°00′E
Resurrection Bay	USA/central coast of Alaska	59°50′N 149°30′W
Rio Santa Maria	Southern Chile/Magallanes Province	– –
Rishiri Island	Japan/west of northern Hokkaido/Sea of Japan	45°10′N 141°15′E
Rivers Inlet	Canada/central coast of British Columbia	51°28′N 127°25′W
Roberts Banks	Canada/British Columbia/mouth of the Fraser River	49°05′N 123°11′W
Robertson Creek	Canada/British Columbia/Vancouver Island	49°00′N 125°00′W
Rogue River	USA/Oregon	42°26′N 124°26′W
Rudyard (Rudyerd) Bay	USA/southeastern Alaska	55°35′N 130°52′W

Place names	Location	Co-ordinates
Russian Lake	USA/Alaska/Cook inlet	60°20′N 150°00′W
Russian River	USA/Alaska/Cook Inlet	60°29′N 150°00′W
Ruth Lake	USA/Alaska/Kodiak Island	– –
Rybatskaya River	USSR/Kuril Islands/Iturup Island	44°54′N 147°30′E
Saanich Inlet	Canada/British Columbia/Vancouver Island	48°37′N 123°30′W
Sacramento	USA/California	38°33′N 121°30′W
Sacramento River	USA/California	38°03′N 121°56′W
Sado Island	Japan/Honshu/Niigata Prefecture	38°00′N 138°25′E
Salmon Banks	USA/Washington	– –
St. Lawrence Island	USA/Alaska/Bering Sea	63°30′N 170°30′W
St. Lawrence River	Canada/Ontario-Quebec	47°30′N 70°00′W
St. Mary's Bay	Canada/Newfoundland	46°30′N 53°45′W
Sakhalin Bay	USSR/northern Sakhalin	54°00′N 141°00′E
Sakhalin Island	USSR/Sea of Okhotsk	51°00′N 143°00′E
Salcha River	USA/Alaska/Yukon River system	64°29′N 147°00′W
Salmon River	USA/Washington/Queets River system	47°30′N 124°10′W
Samarga River	USSR/Primore	47°15′N 138°45′E
Samnya River	USSR/Amur River system	– –
San Francisco	USA/California	37°48′N 122°24′W
San Joaquin River	USA/California	38°03′N 121°50′W
San Juan Islands	USA/northwestern Washington	48°30′N 123°00′W
San Juan River	Canada/British Columbia/Vancouver Island	48°34′N 124°24′W
San Lorenzo River	USA/California/Monterey Bay	36°58′N 122°03′W
Sand Creek	USA/Oregon	45°30′N 123°55′W
Sandy River	USA/Oregon/Columbia River system	45°34′N 122°24′W
Sarufutsu	Japan/northeastern Hokkaido	45°17′N 142°17′E
Sarufutsu River	Japan/northeastern Hokkaido	45°17′N 142°17′E
Sashin Creek	USA/southeastern Alaska/Baranof Island	57°23′N 134°39′W
Satsop River	USA/Washington/Grays Harbor	46°58′N 123°28′W
Scott Creek	USA/California	37°02′N 122°13′W
Scully Creek	Canada/British Columbia/Skeena River system/Lakelse Lake	54°23′N 128°33′W
Sea of Japan	Japan-USSR	40°00′N 135°00′E
Sea of Okhotsk	USSR-Japan	55°00′N 150°00′E
Seal Bay Creek	USA/Alaska/Afognak Island	58°10′N 152°50′W*
Seattle	USA/Washington	47°35′N 122°20′W
Semga River	USSR/Sakhalin	46°51′N 142°17′E
Seton Creek	Canada/British Columbia/Fraser River system	50°41′N 121°56′W
Seymour River	Canada/British Columbia/Fraser River system	49°20′N 123°00′W
Shakotan Peninsula	Japan/west coast of Hokkaido	43°24′N 140°19′E
Shantarskie Islands	USSR/Sea of Okhotsk	55°00′N 138°00′E
Shari River	Japan/Hokkaido/Sea of Okhotsk coast	43°56′N 144°41′E
Shaktoolik River	USA/Alaska/Norton Sound	64°20′N 161°10′W
Sheenjek River	USA/Alaska/Yukon River system	66°40′N 144°40′W
Shiashkotan Island	Japan/part of Kuril Islands	48°45′N 154°00′E
Shibetsu River	Japan/eastern Hokkaido/Nemuro Strait coast	43°40′N 145°10′E
Shiga Prefecture	Japan/southern Honshu	35°15′N 136°00′E
Shikoku	Japan	33°50′N 133°30′E

Place names	Location	Co-ordinates
Shimane Prefecture	Japan/Honshu	35°00′N 132°00′E
Shinano River	Japan/Honshu/Sea of Japan coast	37°58′N 139°02′E
Shiodomari River	Japan/southwestern Hokkaido/Hakodate	41°46′N 140°51′E
Shiranuhi Bay	Japan/Kyushu/Kamamoto Prefecture	31°20′N 130°20′E
Shiranuka	Japan/southern Hokkaido	42°58′N 144°02′E
Shiretoko Peninsula	Japan/Hokkaido/Sea of Okhotsk coast	44°34′N 145°20′E
Shiribetsu River	Japan/western Hokkaido	42°43′N 140°22′E
Shizunai	Japan/southern coast of Hokkaido	42°20′N 142°23′E
Shizunai River	Japan/Hokkaido	42°20′N 142°23′E
Sho River	Japan/Honshu/Toyama Prefecture	36°50′N 137°03′E
Shokotsu River	Japan/Hokkaido	44°20′N 143°20′E
Shubuto River	Japan/Hokkaido/Sea of Japan coast	42°50′N 140°20′E
Shumshu Island	USSR/part of Kuril Islands	50°40′N 156°20′E
Shunbetsu River	Japan/Hokkaido/Nemuro Strait coast	43°40′N 145°10′E
Shuswap Lake	Canada/British Columbia/Fraser River system	50°57′N 119°15′W
Simpson River	Chile/Aisen Province	45°25′S 72°32′W
Simushir Island	USSR/part of Kuril Islands	47°00′N 152°00′E
Singoolik River	USA/Alaska/Kotzebue Sound	66°45′N 163°00′W*
Situk River	USA/southeastern Alaska	59°30′N 139°30′W
Siuslaw River	USA/Oregon	44°01′N 124°08′W
Six Mile Creek	Canada/British Columbia/Babine Lake/Skeena River system	55°16′N 127°32′W
Sixes River	USA/Oregon	42°45′N 124°30′W
Skagit Bay	USA/Washington/Puget Sound	48°23′N 122°21′W
Skagit River	USA/Washington/Puget Sound	48°23′N 122°21′W
Skeena River	Canada/British Columbia	54°09′N 130°02′W
Skykomish River	USA/Washington	47°49′N 121°32′W
Smith Inlet	Canada/central coast of British Columbia	51°19′N 127°25′W
Snake Creek	USA/southeastern Alaska/Etolin Island	56°12′N 132°20′W
Snake River	USA/Alaska/Norton Sound	64°00′N 164°00′W*
Snake River	USA/Washington-Oregon-Idaho/Columbia River system	46°13′N 119°02′W
Snake River	USA/southwestern Alaska/Bristol Bay	58°50′N 158°45′W
Snohomish River	USA/Washington	47°46′N 122°13′W
Sokolov Hatchery	USSR/Sakhalin	47°00′N 142°40′E
Somass Estuary	Canada/British Columbia/Vancouver Island	49°14′N 124°49′W
Somass River	Canada/British Columbia/Vancouver Island	49°14′N 124°49′W
Sooke	Canada/British Columbia/Vancouver Island	48°20′N 123°42′W
South Cholla Province	South Korea/Korea Strait	34°45′N 127°00′E
South Sound	USA/Washington/Puget Sound	– –
South Thompson River	Canada/British Columbia/Fraser River system	50°41′N 120°20′W
Soya Strait	Same as La Perouse Strait	45°50′N 142°00′E
Spokane River	USA/Washington/upper Columbia River system	47°54′N 118°19′W
Spring Creek	USA/Oregon/Columbia River system	– –
Spring Creek Hatchery	USA/Oregon/Columbia River system	– –
Sproat Lake	Canada/British Columbia/Vancouver Island	49°16′N 125°00′W
Squamish River	Canada/southern coast of British Columbia	49°41′N 123°11′W
Stamp River	Canada/British Columbia/Vancouver Island	49°18′N 124°53′W
Steel River	Canada/Ontario/Lake Superior	49°00′N 86°58′W
Stellako River	Canada/British Columbia/Fraser River system	54°03′N 124°53′W

Place names	Location	Co-ordinates
Stephens Passage	USA/southeastern Alaska	57°50′N 133°50′W
Stikine River	USA/southeastern Alaska-Canada/northern British Columbia	56°38′N 132°20′W
Stillaguamish River	USA/Washington/Puget Sound	48°14′N 122°22′W
Stormy Lake	USA/Wisconsin	46°00′N 89°30′W
Strait of Georgia	Canada/British Columbia/between Vancouver Island and mainland	49°30′N 124°00′W
Strait of Magellan	Argentina	53°00′S 71°00′W
Strait of Nevelskoy	USSR/Sakhalin	52°00′N 141°30′E
Stuart Lake	Canada/British Columbia/Fraser River system	54°26′N 124°15′W
Stuart River	Canada/British Columbia/Fraser River system	53°59′N 123°32′W
Sturgeon Bank	Canada/British Columbia/mouth of the Fraser River	49°11′N 123°14′W
Suchan River	USSR/Primore/Peter the Great Bay	42°30′N 133°05′E
Summit Lake	USA/central coast of Alaska/Copper River system	63°08′N 145°30′W
Sumner Strait	USA/southeastern Alaska	56°24′N 133°48′W
Sungari River	People's Republic of China	47°44′N 132°32′E
Susitna River	USA/Alaska/Cook Inlet	61°16′N 150°30′W
Suttsu Coast	Japan/Hokkaido	42°49′N 140°04′E
Swanson River	USA/Alaska/Cook Inlet	60°48′N 151°02′W
Sweltzer Creek	Canada/British Columbia/Vancouver Island	49°06′N 121°58′W
Swiniuk River	USA/Alaska/Norton Sound	64°00′N 164°00′W*
Syumarinai Lake	Japan/Hokkaido	44°18′N 142°12′E
Tacoma	USA/Washington	47°16′N 122°30′W
Tacoma Narrows	USA/Washington/Puget Sound	47°16′N 122°32′W
Taku River	Canada/northern British Columbia	58°35′N 133°39′W
Tanana River	USA/central Alaska/Yukon River system	65°11′N 152°10′W
Taranai River	USSR/southern Sakhalin	46°37′N 142°26′E
Tatar Strait	USSR/west of Sakhalin	50°00′N 141°00′E
Tateyama Bay	Japan/Honshu/Chiba Prefecture	34°59′N 139°50′E
Tatshenshini River	Canada/Yukon Territory	59°27′N 137°43′W
Taui River	USSR/northern coast of Sea of Okhotsk	59°37′N 149°04′E
Tauyskaya Bay	USSR/northern coast of Sea of Okhotsk	59°20′N 150°00′E
Tazlina Lake	USA/Alaska/Copper River system	61°53′N 146°30′W
Ten Mile River	USA/New York	41°40′N 73°30′W
Ten Mile Lakes	USA/Oregon	43°32′N 124°14′W
Tenryu River	Japan/Honshu	35°40′N 137°50′E
Teremok River	USSR/Sakhalin	– –
Terpeniya Bay	USSR/southeastern Sakhalin	49°00′N 143°30′E
Teshio River	Japan/Hokkaido	44°53′N 141°44′E
Thompson River	Canada/British Columbia/Fraser River system	50°14′N 121°35′W
Thunder Bay	Canada/Ontario	48°25′N 89°00′W
Tikchik Lake	USA/southwestern Alaska/Bristol Bay/Nushagak River system	59°58′N 158°20′W
Tikchik River	USA/southwestern Alaska/Bristol Bay/Nushagak River system	59°59′N 158°22′W
Tillamook Bay	USA/Oregon	45°28′N 123°50′W
Tlell River	Canada/British Columbia/Queen Charlotte Islands/Graham Island	53°35′N 131°56′W
Tlupana River	Canada/British Columbia/west coast of Vancouver Island	49°45′N 126°23′W
Toba River	Canada/south coast of British Columbia	50°30′N 124°21′W
Togiak Lake	USA/southwestern Alaska/Bristol Bay/Togiak River system	59°40′N 159°38′W
Togiak River	USA/southwestern Alaska/Bristol Bay	59°05′N 160°30′W

Place names	Location	Co-ordinates
Tokachi River	Japan/Hokkaido/Pacific Ocean coast	42°44′N 143°42′E
Toklat River	USA/Alaska/Yukon River system	64°29′N 150°30′W
Tokoro River	Japan/eastern Hokkaido	44°07′N 144°04′E
Tokum Lake	USA/central coast of Alaska/Copper River system	– –
Tokyo	Japan/Honshu	35°40′N 139°50′E
Tokyo Bay	Japan/Honshu	35°40′N 139°55′E
Toledo Harbour	USA/southeastern Alaska	56°22′N 134°38′W
Toshibetsu River	Japan/Hokkaido	42°20′N 140°00′E
Toutle River	USA/Washington	46°20′N 122°52′W
Toyama Prefecture	Japan/Honshu	36°30′N 137°30′E
Traitors Cove	USA/southeastern Alaska	55°42′N 131°39′W
Traitors Creek	USA/southeastern Alaska	55°44′N 131°30′W
Triangle Island	USA/southwestern Alaska/Iliamna Lake/Kvichak River system	59°43′N 154°28′W
Tsolum River	Canada/British Columbia/Vancouver Island	49°42′N 124°59′W
Tsugaru Strait	Japan/southwestern Hokkaido	41°30′N 140°30′E
Tugur	USSR/Amur District	53°44′N 136°45′E
Tumansk River	USSR/Far East/south of Anadyr River estuary	64°03′N 178°10′E
Tumen River	USSR-North Korea	42°18′N 130°41′E
Tumnin River	USSR/Primore	49°18′N 140°24′E
Tunagoruk River	USA northern Alaska/north of 71.00°N	– –
Turomcha River	USSR/northern coast of Sea of Okhotsk	–
Tustumena Lake	USA/Alaska/Cook Inlet	60°16′N 151°12′W
Tym River	USSR/northeastern Sakhalin	51°50′N 143°10′E
Ualik Lake	USA/southwestern Alaska/Bristol Bay	59°07′N 159°30′W
Ugashik Lake (Upper and Lower)	USA/southwestern Alaska/Bristol Bay/Ugasik River system	57°32′N 157°00′W
Ugashik River	USA/southwestern Alaska/Bristol Bay	57°30′N 157°37′W
Ul River	USSR/Amur River system	49°00′N 140°20′E*
Ulika River	USSR/Primore Region/Amur River system	48°48′N 134°11′E
Ulkhan River	USSR/Kamchatka/Karymaiskiy Spring/Bolshaya River system	52°30′N 157°00′E*
Umpqua River	USA/southwestern Oregon	42°56′N 123°24′W
Unalakleet River	USA/Alaska/Norton Sound	63°52′N 160°50′W
Unimak Island	USA/Alaska/Aleutian Islands	55°00′N 164°00′W
Unozumai River	Japan/Honshu/Pacific Ocean coast/Iwate Prefecture	39°20′N 141°50′E
Upper Babine River	Canada/British Columbia/Skeena River system	55°19′N 126°34′W
Upper Pitt River	Canada/British Columbia/Fraser River system	49°30′N 122°45′W
Upper Seton Creek	Canada/British Columbia/Fraser River system	50°41′N 121°56′W
Ura River	Japan/Honshu/Niigata Prefecture	37°41′N 139°57′E
Urakawa	Japan/Hokuido/Pacific Ocean coast	42°10′N 142°46′E
Urup Island	USSR/part of Kuril Islands	46°00′N 150°00′E
Uryu Lake	Japan/Hokkaido	44°20′N 142°05′E
Ussuri River	USSR/Amur River system	48°27′N 135°04′E
Usuktuk River	USA/northern Alaska/Meade River system	70°30′N 157°28′W
Utka River	USSR/western Kamchatka	53°10′N 156°04′E
Utukok drainage	USA/northern Alaska	70°10′N 162°06′W
Valdez Arm	USA/central coast of Alaska/Prince William Sound	61°07′N 146°16′W
Vancouver Island	Canada/southwestern British Columbia	49°30′N 125°30′W

Place names	Location	Co-ordinates
Varangerfjord	Norway	70°00′N 28°45′E
Vedder River	Canada/British Columbia/Fraser River system	49°08′N 122°06′W
Victoria	Canada/British Columbia/Vancouver Island	48°25′N 123°22′W
Vilkitskiy Strait	USSR/Arctic Ocean	77°55′N 103°00′E
Village Creek	Canada/Yukon/Tatshenshini River system	60°00′N 138°15′W
Waddell Creek	USA/California	37°10′N 122°20′W
Wakasa Bay	Japan/Honshu/Sea of Japan coast	35°45′N 135°40′E
Walcott Slough	USA/Washington/Hood Canal	47°40′N 122°50′W
Washougal River	USA/Washington/Columbia River system	45°35′N 122°21′W
Weaver Creek	Canada/British Columbia/Fraser River system/Harrison River	49°19′N 121°53′W
Wenatchee Lake	USA/Washington/upper Columbia River system	47°50′N 120°48′W
Wenatchee River	USA/Washington/upper Columbia River system	47°26′N 120°20′W
Westport	USA/Washington	46°53′N 124°06′W
Westport	USA/Oregon	46°10′N 123°23′W
Westward Lake	Canada/Ontario	45°29′N 78°47′W
Whale Channel	Canada/central coast of British Columbia	53°11′N 129°08′W
Whannock River	Canada/central coast of British Columbia/Owikeno Lake	51°40′N 127°05′W
Whidbey Island	USA/Washington/Puget Sound	48°20′N 122°40′W
White Salmon River	USA/Washington/Columbia River system	45°45′N 121°28′W
White Sea	USSR/Arctic Ocean	65°30′N 38°00′E
Willamette River	USA/Oregon	45°38′N 122°50′W
Willapa Harbor	USA/Washington/Olympic Peninsula	46°40′N 123°45′W
Wol River	South Korea	37°00′N 129°26′E
Wood River	USA/southwestern Alaska/Bristol Bay	59°03′N 158°30′W
Wrangell district	USA/southeastern Alaska	56°20′N 132°10′W
Wulik River	USA/Alaska/Kotzebue Sound	67°50′N 164°00′W
Yakima River	USA/Washington/Columbia River system	46°15′N 119°02′W
Yakoun River	Canada/British Columbia/Queen Charlotte Islands	53°39′N 132°12′W
Yakutat	USA/southeastern Alaska	59°30′N 139°45′W
Yakutat Bay	USA/southeastern Alaska	59°45′N 140°45′W
Yamagata Prefecture	Japan/Honshu	38°15′N 140°15′E
Yamaguchi Prefecture	Japan/Honshu	34°20′N 131°30′E
Yamato Bank	Sea of Japan	39°30′N 134°00′E
Yana River	USSR/Siberia/Arctic Ocean	71°30′N 136°30′E
Yaquina Bay	USA/Oregon	44°37′N 124°04′W
Yellow Sea	China-North and South Korea	35°00′N 124°00′E
Yenisey River	USSR/Siberia/Arctic Ocean	70°08′N 83°13′E
Yes Bay	USA/southeastern Alaska	55°55′N 131°48′W
Yodo River	Japan/Honshu/Osaka Prefecture	34°41′N 135°25′E
Yoichi	Japan/Hokkaido/Sea of Japan coast	43°14′N 140°42′E
Yonkok River	South Korea	38°00′N 128°40′E
Yubetsu River	Japan/Hokkaido/Sea of Okhotsk coast	44°15′N 143°37′E
Yukon River	USA/Alaska-Canada/Yukon Territory	63°00′N 165°00′W
Yurappu River	Japan/southwestern Hokkaido	42°15′N 140°01′E
Yuzhnaya Gavan	USSR/Amur River system	– –
Zheltaya River	USSR/Primore	– –

SUBJECT INDEX